# Materials

# Materials: engineering, science, processing and design

## Third edition

Mike Ashby, Hugh Shercliff, and David Cebon

*Department of Engineering, University of Cambridge, UK*

ELSEVIER

AMSTERDAM • BOSTON • HEIDELBERG • LONDON
NEW YORK • OXFORD • PARIS • SAN DIEGO
SAN FRANCISCO • SINGAPORE • SYDNEY • TOKYO
Butterworth-Heinemann is an imprint of Elsevier

Butterworth-Heinemann is an imprint of Elsevier
The Boulevard, Langford Lane, Kidlington, Oxford, OX5 1GB, UK
225 Wyman Street, Waltham, 02451, USA

First published 2007
Second edition 2010
Third edition 2014

**Notice**

No responsibility is assumed by the publisher for any injury and/or damage to persons or property as a matter
of products liability, negligence or otherwise, or from any use or operation of any methods, products, instructions
or ideas contained in the material herein. Because of rapid advances in the medical sciences, in particular,
independent verification of diagnoses and drug dosages should be made

**British Library Cataloguing in Publication Data**
A catalogue record for this book is available from the British Library

**Library of Congress Cataloging-in-Publication Data**
A catalog record for this book is available from the Library of Congress

ISBN: 978-0-08-097773-7

For information on all Butterworth-Heinemann publications
visit our website at books.elsevier.com

Printed and bound in Italy by Printer Trento S.r.l.

14 15 16 17 18   10 9 8 7 6 5 4 3 2 1

Working together
to grow libraries in
developing countries

www.elsevier.com • www.bookaid.org

# Contents

## Guided Learning Unit 1: Simple ideas of crystallography    **GL1-1**

## Guided Learning Unit 2: Phase diagrams and phase transformations    **GL2-1**

# Preface to 3rd edition

## Science-led or design-led? Two approaches to materials teaching

Most things can be approached in more than one way. In teaching this is especially true. The way to teach a foreign language, for example, depends on the way the student wishes to use it—to read the literature, say, or to find hotel accommodations, order meals and buy beer. So it is with the teaching of this subject, Materials.

The figure shows the enrolment in engineering and materials-related departments in US Universities in 2006. Mechanical, Civil and Chemical Engineering account for two-thirds of the total. Aerospace, Manufacturing and General Engineering account for a further 20%. The more science-related subjects—Materials Science, Engineering Science and Physics—total 3%. All of these courses carry requirements for Materials teaching, but the way the students in some courses will use it differs from those in others.

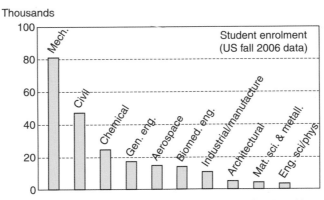

*Enrolment in engineering-related courses in US universities.*

The traditional approach to Materials teaching starts with fundamentals: the electron, the atom, atomic bonding, and packing, crystallography and crystal defects. Onto this is built alloy theory, the kinetics of phase transformations and the development of microstructure on scales made visible by electron and optical microscopes. This sets the stage for the understanding and control of properties at the millimeter or centimeter scale at which they are usually measured. This science-led approach emphasises the physical basis but gives little emphasis to the behaviour of structures and components in service or methods for material selection and design.

The alternative approach is design-led. The starting point is the requirements that materials must meet if they are to perform properly in a given design. To match material to design requires a perspective on the range of properties they offer, how these properties combine to limit performance, the influence of manufacturing processes on properties, and ways of accessing the data needed to evaluate all of these. Once the importance of certain properties is established there is good reason and a clear context from which to 'drill down', so to speak, to examine the science that lies behind them—valuable because an understanding of the fundamentals itself informs material choice, processing and usage.

Each approach has its place. The choice depends on the way the student will wish to use the information. If the intent is pure scientific research, the first is the logical way to go. If it is engineering design and applied industrial research, the second makes better sense. This book follows the second.

## What is different about this book?

There are many books about the science of engineering materials; many more about design. What is different about this one?

First is its *design-led approach*, specifically developed to guide material selection and understanding for a wide spectrum of engineering courses. The approach is systematic, leading from design requirements to a prescription for optimised material choice. The approach is illustrated by numerous case studies. Practice in using it is provided by worked Examples in the text and Exercises at the end of each Chapter.

Second is its emphasis on *visual communication* through a unique graphical presentation of material properties as *material property charts* and numerous *schematics*. These are a central feature of this approach, helpful in utilising visual memory as a learning tool, understanding the origins of properties, their manipulation and their fundamental limits, and providing a tool for selection and for understanding the ways in which materials are used.

Third is its *breadth*. We aim here to present the properties of materials, their origins and the way they enter engineering design. A glance at the contents pages will show sections dealing with:

- Physical properties
- Mechanical characteristics
- Thermal behaviour
- Electrical, magnetic and optical response
- Durability
- Processing, and the way it influences properties
- Environmental issues, and the broader issues of Sustainable Technology

Throughout we aim for a simple, straightforward presentation, developing the materials science as far as it is helpful in guiding engineering design, avoiding detail where this does not contribute to this end.

The fourth feature is that of guided self-leaning. Certain topics lend themselves to self-instruction with embedded exercises to build systematic understanding. This works particularly well for topics that involve a contained set of concepts and tools. Thus Crystallography, as an example, involves ideas of symmetry and three-dimensional geometry that are most easily grasped by drawing and problem-solving. And understanding Phase Diagrams and Phase Transformations relies on interpreting graphical displays of compositional and thermodynamic information. Their use to understand and predict microstructure follows procedures that are best learned by application. Both topics can be packaged into self-contained guided learning units, with each new concept being presented and immediately practiced with exercises, thereby building confidence. Students who have worked through a package can feel that they have mastered the topic and know how to apply the ideas it contains. We have chosen to present Crystallography and Phase Diagrams and Phase Transformations in this way here. Both topics appear briefly in the main text to give a preliminary overview. The full Guided Learning Units follow later in the book, providing for those courses that require a deeper understanding.

## What's new in the 3rd edition

The main features that are new to the 3rd edition of this book are:

- The number of worked examples in the text has been increased by 50%, to provide broader illustration of key concepts and equations.
- The number of standard end-of-chapter exercises has been doubled to 400 (complemented with 150 exercises using the CES EduPack software, see below), with a full solution manual available to instructors.
- The text and the figures have been revised and updated throughout, with particular additions in the following topic areas:
  - *Elastic properties*: expanded coverage of elastic stress analysis and polymer moduli, and a new section on acoustic properties (Chapter 4).
  - *Fracture behaviour*: the compressive and tensile failure of ceramics is covered, including Weibull analysis (Chapter 8).
  - *Material behaviour in processing*: discussion of several topics has been extended to the context of manufacturing processes and control of properties, e.g. hot strength in metal forming, friction processing, transient heat flow and diffusion in bulk and surface treatments (Chapters 7, 11, 12, 13, 19).
  - *Material design at temperature*: diffusion and creep are covered in more depth, introducing constitutive modelling for metals and polymers (Chapter 13).
  - *Functional properties*: Chapter 14 (electrical properties) includes a discussion of superconductors, electron mobility and semiconductors; Chapter 15 (magnetic materials) includes flexible magnets and magnets for use above room temperature; Chapter 16 (optical devices) has a new discussion of colour and optical interference.
  - *Electrochemistry*: Chapter 17 (on durability) now includes the Nernst equation and the Faraday equation.
  - *Materials and the environment*: the coverage has been updated, with a new section on Sustainability and Sustainable Technology (Chapter 20).

## This book and the CES EduPack Materials and Process Information software

Engineering design today takes place in a computer-based environment. Stress analysis (finite element method, or FEM, codes for instance), computer-aided design (CAD), design for manufacture (DFM) and product data management (PDM) tools are part of an engineering education. The CES

Materials and Process Information software[1] for education (the CES EduPack) provides a computer-based environment for optimised materials selection.

This book is self-contained and does not depend on computer support for its use. But at the same time it is designed to interface closely with the CES software, which implements the methods developed in it. This enables realistic design problems and selection studies to be addressed, properly managing the multiple constraints on material and process attributes. The methodology also provides the user with novel ways to explore how properties are manipulated. And with sustainability becoming a core topic in most materials-related teaching, the book and the software introduce students to the ideas of life-cycle assessment. Using the book with CES EduPack enhances the learning experience and provides a solid grounding in many of the domains of expertise specified by the various professional engineering accreditation bodies (analysis of components, problem-solving, design and manufacture, economic, societal and environmental impact, and so on). In addition, the CES Elements database documents the fundamental physical, crystallographic, mechanical, thermal, electrical, magnetic and optical properties of all 111 stable elements of the periodic table. This allows the scientific origins and interrelationships between many properties, developed in the text, to be explored in greater depth.

The design-led approach is developed to a higher level in three further textbooks, the first relating to Mechanical design[2], the second to Design for the environment[3] and the third to Industrial design[4].

---

[1] The CES EduPack, Granta Design Ltd., Rustat House, 62 Clifton Court, Cambridge CB1 7EG, UK. www.grantadesign.com.

[2] Ashby, M. F. (2011), *Materials selection in mechanical design*, 4th edition, Butterworth Heinemann, Oxford, UK. ISBN 978-1-85617-663-7. (An advanced text developing material selection methods in detail.)

[3] Ashby, M. F. (2012), *Materials and the environment: Eco-informed material choice*, 2nd edition, Butterworth Heinemann, Oxford, UK. ISBN 978-0-12-385971-6. (A teaching text introducing students to the concepts and underlying facts about concerns for the environment and the ways in which materials and products based on them can both help and harm it.)

[4] Ashby, M. F., and Johnson, K. (2014), *Materials and design: The art and science of material selection in product design*, 3rd edition, Butterworth Heinemann, Oxford, UK. ISBN 978-0-08-098205-2. (A text that complements those above, dealing with the aesthetics, perceptions and associations of materials and their importance in product design.)

# Resources that accompany this book

*Exercises*   Each chapter ends with exercises of three types: the first rely only on information, diagrams and data contained in the book itself; the second makes use of the CES software in ways that use the methods developed here and the third explores the science more deeply using the CES Elements database that is part of the CES system.

*The CES EduPack*   CES EduPack is the software-based package to accompany this book, developed by Michael Ashby and Granta Design. Used together, *Materials: Engineering, Science, Processing and Design* and CES EduPack provide a complete materials, manufacturing and design course. For further information please visit www.grantadesign.com.

## Resources available to adopting instructors who register on the Elsevier textbook website, http://textbooks.elsevier.com:

*Instructor's manual*   The book itself contains a comprehensive set of exercises. Worked-out solutions to the exercises are freely available online to teachers and lecturers who adopt this book.

*Image Bank*   The Image Bank provides adopting tutors and lecturers with various electronic versions of the figures from the book that may be used in lecture slides and class presentations.

*PowerPoint Lecture Slides*   Use the available set of lecture slides in your own course as provided, or edit and reorganise them to meet your individual course needs.

*Interactive Online Materials Science Tutorials*   To enhance students' learning of materials science, instructors adopting this book may provide their classes with a free link to a set of online interactive tutorials. There are 12 available modules, including Bonding, Tensile Testing, Casting and Recrystallization, Fracture and Fatigue, Corrosion, Phase Equilibria, Wear, Ceramics, Composites, Polymers, Light Alloys and Dislocations. Each module includes a selection of self-test questions. The link is available to instructors who register on the Elsevier textbook website.

# Acknowledgements

No book of this sort is possible without advice, constructive criticism and ideas from others. Numerous colleagues have been generous with their time and thoughts. We would particularly like to recognize the suggestions made by Professors Mick Brown, Archie Campbell, David Cardwell, Ken Wallace, and Ken Johnson, all of Cambridge University, and Professor John Abelson of the University of Illinois.

Equally valuable has been the contribution of the team at Granta Design, Cambridge, responsible for the development of the CES software that has been used to make the material property charts that are a feature of this book.

We gratefully acknowledge the following *Materials, Third Edition* reviewers and advisors, whose valuable insight throughout the development of the third edition helped to shape this text.

## Reviewers

Elizabeth M. Dell, Rochester Institute of Technology (Guided Learning Section 1)
Trevor S. Harding, California Polytechnic State University (Ch. 20)
Susan L. Holl, California State University, Sacramento (Ch. 14)
Ketul C. Popat, Colorado State University (Chs. 2 and 3)
Daniel J. Lewis, Rensselaer Polytechnic Institute (Ch. 3)
Lewis Rabenberg, University of Texas at Austin (Chs. 15 and 16)
Scott Ramsay, University of Toronto (Guided Learning Section 1)
Daniel Samborsky, Montana State University (Chs. 4, 12, and 13)

## Advisors

Pranesh B. Aswath, University of Texas at Arlington
Stacy G. Birmingham, Grove City College
Elizabeth M. Dell, Rochester Institute of Technology
Trevor S. Harding, California Polytechnic State University
Susan L. Holl, California State University, Sacramento
John S. Kallend, Illinois Institute of Technology
Blair London, California Polytechnic State University
Todd A. Palmer, Pennsylvania State University
Ketul C. Popat, Colorado State University
Scott Ramsay, University of Toronto
Daniel Samborsky, Montana State University

# Chapter 1
# Introduction: materials—history and character

Professor James Stuart, the first Professor of Engineering
at Cambridge. Note the cigar.

# **1.1** Materials, processes and choice

Engineers *make* things. They make them out of *materials,* and they shape, join and finish them using *processes.* The materials have to support loads, to insulate or conduct heat and electricity, to accept or reject magnetic flux, to transmit or reflect light, to survive in often-hostile surroundings and to do all this without damage to the environment or costing too much.

There is also a partner in all this. To make something you also need a process. And not just any process—it has to be compatible with the material you plan to use. Sometimes the process is the dominant partner and a compatible material-mate must be found—it is a marriage. Compatibility is not easily found—many marriages fail—and material failure can be catastrophic, with issues of liability and compensation. This sounds like food for lawyers, and sometimes it is: some lawyers make their living as expert witnesses in court cases involving failed materials. But our aim here is not contention; rather, it is to give a vision of the universe of materials (since even on the remotest planets you will find the same elements) and of the universe of processes, and to provide methods and tools for choosing them to ensure a happy, durable union.

But, you may say, engineers have been making things out of materials for centuries, and successfully so—think of Isambard Kingdom Brunel, Thomas Telford, Gustave Eiffel, Henry Ford, Karl Benz and Gottlieb Daimler, the Wright brothers. Why do we need new ways to choose them? A little history helps here. Glance at the portrait with which this chapter starts: it shows James Stuart, the first Professor of Engineering at Cambridge University from 1875 to 1890. In his day the number of materials available to engineers was small, a few hundred at most. There were no synthetic polymers—there are now over 45,000 of them. There were no light alloys (aluminum was first established as an engineering material only in the twentieth century)—now there are thousands. There were no high-performance composites—now there are hundreds of them. The history is developed further in Figure 1.1, the time-axis of which spans 10,000 years. It shows roughly when each of the main classes of materials first evolved. The time-scale is non-linear—almost all the materials we use today were developed in the last 100 years. And this number is enormous: over 160,000 materials are available to today's engineer, presenting us with a problem that Professor Stuart did not have—that of optimally

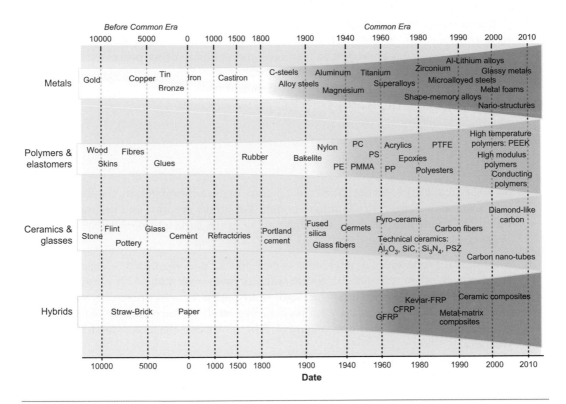

**Figure 1.1** The development of materials over time. The materials of pre-history, on the left, all occur naturally; the challenge for the engineers of that era was one of shaping them. The development of thermo-chemistry, and (later) of polymer chemistry, enabled man-made materials, shown in the colored zones. Three — stone, bronze and iron — were of such importance that the era of their dominance is named after them.

selecting the best one. With the ever-increasing drive for performance, economy and efficiency and the imperative to avoid damage to the environment, making the right choice becomes very important. Innovative design means the imaginative exploitation of the properties offered by materials.

These properties, today, are largely known and documented in handbooks; one such—the ASM Materials Handbook—runs to 22 fat volumes, and it is just one of many. How are we to deal with this vast body of information? Fortunately another thing has changed since Professor Stuart's day: we now have digital information storage and manipulation. Computer-aided design is now a standard part of an engineer's training, and it is backed up by widely-available packages for solid modelling, finite-element analysis, optimisation and material and process selection. Software for the last of these—the selection of materials and processes—draws on databases of the attributes of materials and processes, documents their mutual compatibility and allows them to be searched and displayed in ways that enable selections that best meet the requirements of a design.

If you travel by foot, bicycle or car, you take a map. The materials landscape, like the terrestrial one, can be complex and confusing; maps, here, are also a good idea. This text presents a design-led approach to materials and manufacturing processes that makes use of such maps: novel graphics to display the world of materials and processes in easily accessible ways. They present the properties of materials in ways that give a global view, that reveal relationships between properties and that enable selection.

## **1.2** Material properties

So what are these properties? Some, like density (mass per unit volume) and price (the cost per unit volume or weight) are familiar enough, but others are not, and getting them straight is essential. Think first of those that have to do with carrying load safely—the *mechanical properties*.

*Mechanical properties*   A steel ruler is easy to be bend *elastically*—'elastic' means that it springs back when released. Its elastic stiffness (here, resistance to bending) is set partly by its shape—thin strips are easy to bend—and partly by a property of the steel itself: its *elastic modulus, E*. Materials with high $E$, like steel, are intrinsically stiff; those with low $E$, like polyethylene, are not. Figure 1.2(b) illustrates the consequences of inadequate stiffness.

The steel ruler bends elastically, but if it is a good one, it is hard to give it a permanent bend. Permanent deformation has to do with *strength*, not stiffness. The ease with which a ruler can be permanently bent depends, again, on its shape and also on a different property of the steel—its *yield strength*, $\sigma_y$. Materials with large $\sigma_y$, like titanium alloys, are hard to deform permanently even though their stiffness, coming from $E$, may not be high; those with low $\sigma_y$, like lead, can be deformed with ease. When metals deform, they generally get stronger (this is called 'work hardening'), but there is an ultimate limit, called the *tensile strength*, $\sigma_{ts}$, beyond which the material fails (the amount it stretches before it breaks is called the *ductility*). The hardness, $H$, is closely related to the strength, $\sigma_y$. High hardness gives scratch resistance and resistance to wear. Figure 1.2(c) gives an idea of the consequences of inadequate strength.

So far so good. There is one more property, and it is a tricky one. If the ruler were made not of steel but of glass or of PMMA (Plexiglas, or Perspex), as transparent rulers are, it is not possible to bend it permanently at all. The ruler will fracture suddenly, without warning, before it acquires a permanent bend. We think of materials that break in this way as brittle, and materials that do not as tough. There is no permanent deformation here, so $\sigma_y$ is not the right property. The resistance of materials to cracking and fracture is measured instead by the *fracture toughness*, $K_{1c}$. Steels are tough—well, most are (steels *can* be made brittle)—and they have a high $K_{1c}$. Glass epitomises brittleness; it has a very low $K_{1c}$. Figure 1.2(d) suggests the consequences of inadequate fracture toughness.

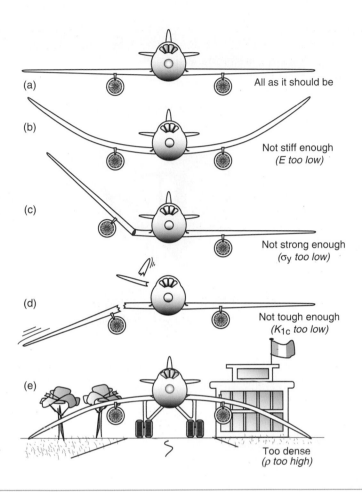

(a)    All as it should be

(b)    Not stiff enough
(*E too low*)

(c)    Not strong enough
($\sigma_y$ *too low*)

(d)    Not tough enough
($K_{1c}$ *too low*)

(e)    Too dense
($\rho$ *too high*)

**Figure 1.2**   Mechanical properties.

We started with the material property *density*, mass per unit volume, symbol $\rho$. Density, in a ruler, is irrelevant. But for almost anything that moves, weight carries a fuel penalty, which is modest for automobiles, greater for trucks and trains, greater still for aircraft and enormous in space vehicles. Minimizing weight has much to do with clever design (we will get to that later) but equally with choice of material. Aluminum has a low density, lead a high one. If our little aircraft were made of lead, it would never get off the ground at all (Figure 1.2(e)).

### Example 1.1 Design requirements (1)

You are asked to select a material for the teeth of the scoop of a digger truck. To do so you need to prioritize the materials properties that matter. What are they?

*Answer.* The teeth will be used in a brutal way to cut earth, scoop stones, crunch rock, often in unpleasant environments (ditches, sewers, fresh and salt water and worse) and their maintenance will be neglected. These translate into a need for high hardness, $H$, to resist wear, and high fracture toughness, $K_{1c}$, so they don't snap off. Does the cost of the material matter? Not much—it is worth paying for good teeth to avoid expensive downtime.

These are not the only mechanical properties, but they are the most important ones. We will meet them, and others, in Chapters 4–11.

*Thermal Properties*  The properties of a material change with temperature, usually for the worse. Its strength falls, it starts to 'creep' (to sag slowly over time) and it may oxidize, degrade or decompose (Figure 1.3(a)). This means that there is a limiting temperature called the *maximum service temperature, $T_{max}$,* above which its use is impractical. Stainless steel has a high $T_{max}$ and can be used up to 800 °C; most polymers have a low $T_{max}$ and are seldom used above 150 °C.

Most materials expand when they are heated, but by differing amounts depending on their *thermal expansion coefficient, $\alpha$.* The expansion is small, but its consequences can be large. If, for instance, a rod is constrained, as in Figure 1.3(b), and then heated, expansion forces the rod against the constraints, causing it to buckle. Railroad track buckles in this way if provision is not made to cope with it. Bridges have expansion joints for the same reason.

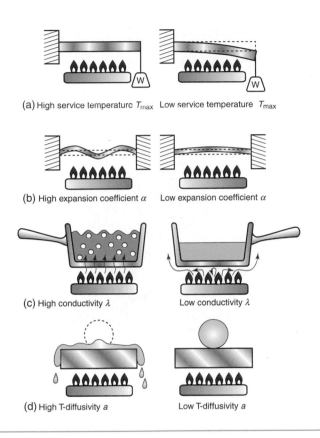

**Figure 1.3** Thermal properties.

Some materials, metals for instance, feel cold; others, like woods, feel warm. This feel has to do with two thermal properties of the material: *thermal conductivity* and *heat capacity*. The first, thermal conductivity, $\lambda$, measures the rate at which heat flows through the material when one side is hot and the other cold. Materials with high $\lambda$ are what you want if you wish to conduct heat from one place to another, as in cooking pans, radiators and heat exchangers; Figure 1.3(c) suggests consequences of high and low $\lambda$ for the cooking vessel. But low $\lambda$ is useful too—low $\lambda$ materials insulate homes, reduce the energy consumption of refrigerators and freezers and enable space vehicles to re-enter the earth's atmosphere.

These applications have to do with long-time, steady heat flow. When time is limited, the other thermal property matters—*heat capacity* , $C_p$. It measures the amount of heat that it takes to make the temperature of material rise by a given amount. High-heat capacity materials—copper, for instance—require a lot of heat to change their temperature; low-heat capacity materials, like polymer foams, take much less. Steady heat flow has, as we have said, to do with thermal conductivity. There is a subtler property that describes what happens when heat is first applied. Think of lighting the gas under a cold slab of material with a ball of ice cream on top (here, lime ice cream) as in Figure 1.3(d). An instant after ignition, the bottom surface is hot but the rest is cold. After a bit, the middle gets hot, then later still, the top begins to warm up, and only then does the ice cream start to melt. How long does this take? For a given thickness of slab, the time is inversely proportional to the *thermal diffusivity, a,* of the material of the slab. It differs from the conductivity because materials differ in their heat capacity, in fact, diffusivity is proportional to $\lambda/C_p$.

---

### Example 1.2 Design requirements (2)

You are asked to select a material for energy-efficient cookware. What material properties are you looking for?

*Answer.* To be energy-efficient the pan must have a high thermal conductivity, $\lambda$, to transmit and spread the heat well, and it must resist corrosion by anything that might be cooked in it, including hot salty water, dilute acids (acetic acid, vinegar) and mild alkalis (baking soda).

---

There are other thermal properties, which we'll meet in Chapters 12, 13, and 17, but these are enough for now. We turn now to matters electrical, magnetic and optical.

***Electrical, magnetic and optical properties***   Start with electrical conduction and insulation (Figure 1.4(a)). Without electrical conduction we would lack the easy access to light, heat, power, control and communication that we take for granted today. Metals conduct well—copper and aluminum are the best of those that are affordable. But conduction is not always a good thing. Fuse boxes, switch casings and the suspensions for transmission lines all require insulators that must also carry some load, tolerate some heat and survive a spark if there is one. Here the property we want is *resistivity*, $\rho_e$, the inverse of electrical conductivity, $\kappa_e$. Most plastics and glass have high resistivity (Figure 1.4(a)); they are used as insulators, although with special treatment they can be made to conduct a little.

Figure 1.4(b) suggests further electrical properties: the ability to allow the passage of microwave radiation, as in the radome, or to reflect it, as in the passive reflector of the boat.

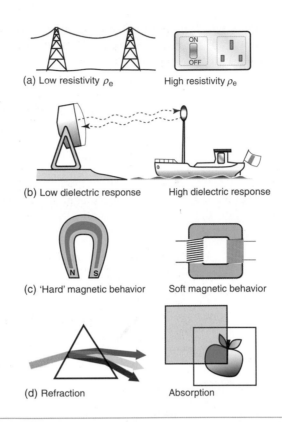

(a) Low resistivity $\rho_e$    High resistivity $\rho_e$

(b) Low dielectric response    High dielectric response

(c) 'Hard' magnetic behavior    Soft magnetic behavior

(d) Refraction    Absorption

**Figure 1.4**  Electrical, magnetic and optical properties.

Both have to do with *dielectric* properties, particularly the *dielectric constant* $\varepsilon_D$. Materials with high $\varepsilon_D$ respond to an electric field by shifting their electrons about, even reorienting their molecules; those with low $\varepsilon_D$ are immune to the field and do not respond. We explore this and other electrical properties in Chapter 14.

Electricity and magnetism are closely linked. Electric currents induce magnetic fields; a moving magnet induces, in any nearby conductor, an electric current. The response of most materials to magnetic fields is too small to be of practical value. But a few— called *ferro-magnets* and *ferri-magnets*—have the capacity to trap a magnetic field permanently. These are called 'hard' magnetic materials because, once magnetized, they are hard to demagnetize. They are used as permanent magnets in headphones, motors and dynamos. Here the key property is the *remanence*, a measure of the intensity of the retained magnetism. A few others—'soft' magnetic materials—are easy to magnetize and demagnetize. They are the materials of transformer cores and the deflection coils of an old TV tube. They have the capacity to conduct a magnetic field, but not to retain it permanently (Figure 1.4(c)). For these a key property is the *saturation magnetization*, which measures how large a field the material can conduct. These we meet again in Chapter 15.

Materials respond to light as well as to electricity and magnetism—hardly surprising, since light itself is an electromagnetic wave. Materials that are opaque *reflect* light; those that are transparent *refract* it; and some have the ability to *absorb* some wavelengths (colors) while allowing others to pass freely (Figure 1.4(d)). These are explored in more depth in Chapter 16.

## Example 1.3 Design requirements (3)

What are the essential and the desirable requirements of materials for eyeglass (spectacle) lenses?

*Answer.* Essentials: The optical qualities are paramount, so a material with optical quality transparency is the first requirement. It is also essential that it can be molded or ground with precision to the required prescription. And it must resist sweat and be sufficiently scratch-resistant to cope with normal handling.

Desirables: A high refractive index and a low density allow thinner, and thus lighter, lenses. Given that, a material that is cheap allows either a lower cost for the consumer or a greater profit-margin for the maker.

*Chemical properties*   Products often have to function in hostile environments, being exposed to corrosive fluids, hot gases or radiation. Damp air is corrosive; so is water; the sweat of your hand is particularly corrosive, and of course there are far more aggressive environments than these. If the product is to survive for its design-life it must be made of materials, or at least coated with materials, that can tolerate the surroundings in which they operate. Figure 1.5 illustrates some of the commonest of these: fresh and salt water, acids and alkalis, organic solvents, oxidizing flames and ultra-violet radiation. We regard the intrinsic resistance of a material to each of these as material

**Figure 1.5**  Chemical properties: resistance to water, acids, alkalis, organic solvents, oxidation and radiation.

properties, measured on a scale of 1 (very poor) to 5 (very good). Chapter 17 deals with the material durability.

*Environmental properties*  Making, shaping, joining and finishing materials consumes nearly one-third of global energy demand. The associated emissions are already a cause for international concern, and demand for material extraction and processing is likely to double in the next 40 years. It's important, therefore, to understand the environmental properties of materials and to seek ways to use them more sustainably than we do now. Chapter 20 introduces the key ideas of material life-cycle analysis, material efficiency and material sustainability, all of which are central to the way we will use materials in the future.

### Example 1.4 Design requirements (4)

What are the essential and the desirable requirements of materials for a single-use (disposable) water bottle that does minimal environmental harm?

*Answer.* Essentials: The health aspects come first: the material of the bottle must be non-toxic and able to be processed in a way that leaves no contaminants. The bottle will meet its intended use only if its material and the process used to shape it are cheap.

Desirables: The material of the bottle should be recyclable and, if possible, biodegradable. It makes handling easier if the material is not brittle, although glass bottles are widely used.

## **1.3** Design-limiting properties

The performance of a component is limited by certain of the properties of the materials of which it is made. This means that, to achieve a desired level of performance, the values of the design-limiting properties must meet certain targets, and those that fail to do so are not suitable. In the cartoon graphic of Figure 1.2, stiffness, strength and toughness are design-limiting—if any one of them are too low, the plane won't fly. In the design of power transmission lines, electrical resistivity is design-limiting; in the design of a camera lens, it is optical quality and refractive index.

Materials are chosen by identifying the design-limiting properties, applying limits to them, and screening out those that do not meet the limits (Chapter 3). Processes, too, have properties, although we have not met them yet. These can be design-limiting as well, leading to a parallel scheme for choosing viable processes (Chapters 18 and 19).

## **1.4** Summary and conclusions

Engineering design depends on *materials* that are shaped, joined and finished by *processes*. Design requirements define the performance required of the materials, expressed as target values for certain *design-limiting properties*. A material is chosen because it has properties that meet these targets and is compatible with the processes required to shape, join and finish it.

This chapter introduced some of the design-limiting properties: *physical properties* like density, *mechanical properties* like modulus and yield strength, and *functional properties*, such as those describing the thermal, electrical, magnetic and optical behavior. We examine all of these in more depth in the chapters that follow, but those just introduced are enough to proceed with. We turn now to the materials themselves: the families, classes and members.

## **1.5** Further reading

### The history and evolution of materials

Delmonte, J. (1985). *Origins of Materials and Processes*. Pennsylvania, USA: Technomic Publishing Company. ISBN 87762-420-8. (A compendium of information about materials in engineering, documenting the history.)

Hummel, R. (2004). *Understanding Materials Science: History, Properties, Applications* (2nd ed.). New York, USA: Springer Verlag. ISBN 0-387-20939-5.

Singer, C., Holmyard, E. J., Hall, A. R., Williams, T. I., & Hollister-Short, G. (Eds.), (1954–2001). *A History of Technology,* 21 volumes. Oxford, UK: Oxford University Press. ISSN 0307–5451. (A compilation of essays on aspects of technology, including materials.)

Tylecoate, R. F. (1992). *A History of Metallurgy* (2nd ed.). London, UK: The Institute of Materials. ISBN 0-904357-066. (A total-immersion course in the history of the extraction and use of metals from 6000 BC to 1976, told by an author with forensic talent and a love of detail.)

## **1.6** Exercises

Exercise E1.1    Use a search engine such as Google to research the history and uses of one of the following materials:

Tin.
Glass.
Cement.
Titanium.
Carbon fibre.

Present the result as a short report of about 100–200 words (roughly half a page).

Exercise E1.2  What is meant by the *design-limiting properties* of a material in a given application?

Exercise E1.3  There have been many attempts to manufacture and market plastic bicycles. All have been too flexible. Which design-limiting property is insufficiently large?

Exercise E1.4  What, in your judgment, are the design-limiting properties for the material of the blade of a knife that will be used to gut fish?

Exercise E1.5  What, in your judgment, are the design-limiting properties for the material of an oven glove?

Exercise E1.6  What, in your judgment, are the design-limiting properties for the material of an electric lamp filament?

Exercise E1.7  A material is needed for a tube to carry fuel from the fuel tank to the carburetor of a motor-powered mower. The design requires that the tube be flexible, and that the fuel be visible. List what you think would be the design-limiting properties.

Exercise E1.8  A material is required as the magnet for a magnetic soap holder. Soap is mildly alkaline. List what you would judge to be the design-limiting properties.

Exercise E1.9  The cases in which most CDs are sold have an irritating way of cracking and breaking. Which design-limiting property has been neglected in selecting the material of which they are made?

Exercise E1.10  List three applications that, in your judgment, need high stiffness and low weight. Think of things that must be light (as they are moved, perhaps rapidly) but must not be too 'bendy'.

Exercise E1.11  List three applications that, in your judgment, need optical-quality glass. Think of products that rely on distortion-free imaging.

Exercise E1.12  List three applications that you think would require high thermal conductivity. Think of things that you have to get heat into or out of.

Exercise E1.13  List three applications that you think would require low thermal expansion. Think of things that will lose accuracy or won't work if they distort.

# Chapter 2
# Family trees: organising materials and processes

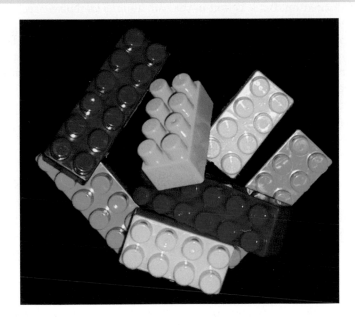

Chapter contents

## **2.1** Introduction and synopsis

A successful product—one that performs well, is good value for the money and gives pleasure to the user—uses the best materials for the job and fully exploits their potential and characteristics.

The families of materials—metals, polymers, ceramics and so forth—are introduced in Section 2.2. What do we need to know about them if we are to design products using them? That is the subject of Section 2.3, in which distinctions are drawn between various types of materials information. But it is not, in the end, a *material* that we seek; it is a certain *profile of properties*—the one that best meets the needs of the design. Each family has its own characteristic profile, or the 'family likeness', which is useful to know when deciding which family to use for a given design. Section 2.2 explains how this provides the starting point for a classification scheme for materials, allowing information about them to be organised and manipulated.

Choosing a material is only half the story. The other half is the choice of a process route to shape, join and finish it. Section 2.3 introduces process families and their attributes. Choice of material and process are tightly coupled: a given material can be processed in some ways but not others, and a given process can be applied to some materials but not to others. On top of that, the act of processing can change, even create, the properties of the material. Process families, too, exhibit family likenesses—commonality in the materials that members of a family can handle, or the shapes they can make. Section 2.3 introduces a classification for processes that parallels that for materials.

Family likenesses are most strikingly seen in *material property charts*, which are a central feature of this book (Section 2.5). These are charts with material properties as axes showing the location of the families and their members. Materials have many properties, which can be thought of as the axes of a 'material-property' space—each chart is a two-dimensional slice through this space. Each material family occupies a discrete part of this space, distinct from the other families. The charts give an overview of materials and their properties, they reveal aspects of the science underlying the properties, and provide a powerful tool for materials selection. Process attributes can be treated in a similar way to create process-attribute charts—which we leave for Chapter 18.

The classification systems of Sections 2.2 and 2.3 provide a structure for computer-based information management, which is introduced in Section 2.6. The chapter ends with a summary, further reading and exercises.

## **2.2** Getting materials organised: the materials tree

*Classifying materials*   It is conventional to classify the materials of engineering into the six broad families shown in Figure 2.1: metals, polymers, elastomers, ceramics, glasses and hybrids—composite materials made by combining two or more of the others. There is sense in this: the members of a family have certain features in common—similar properties, similar processing routes and, often, similar applications. Figure 2.2 shows examples of each family.

Figure 2.3 illustrates how the families are expanded to show classes, sub-classes and members, each of which is characterised by a set of *attributes*: its properties. As an example,

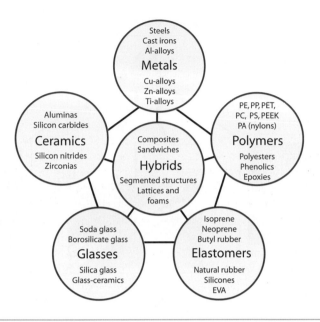

**Figure 2.1**   The menu of engineering materials. The basic families of metals, ceramics, glasses, polymers and elastomers can be combined in various geometries to create hybrids.

the Materials universe contains the family *Metals*, which in turn contains the class *Aluminum alloys*, which contains the sub-class the *6000 series*, within which we find the particular member *Alloy 6061*. It, and every other member of the universe, is characterised by a set of attributes that include not only the properties mentioned in Chapter 1, but also its processing characteristics, the environmental consequences of its use, and its typical applications. We call this its *property profile*. Selection involves seeking the best match between the property profiles of the materials in the universe and those required by the design. As already mentioned, the members of one family have certain characteristics in common. Here, briefly, are some of them.

*Metals* have relatively high stiffness, measured by the modulus $E$. Most, when pure, are soft and easily deformed, meaning that $\sigma_y$ is low. They can be made stronger by alloying and by mechanical and heat treatment, increasing $\sigma_y$, but they remain ductile, allowing them to be formed by deformation processes. And, broadly speaking, they are tough, with a usefully high $K_{1c}$. They are good electrical and thermal conductors. But metals have weaknesses too: they are reactive, and most corrode rapidly if not protected.

*Ceramics* are non-metallic, inorganic solids, like porcelain or alumina—the material of spark-plug insulators. They have many attractive features. They are stiff, hard and abrasion-resistant; they retain their strength in high temperatures; and they resist corrosion well. Most are good electrical insulators. They, too, have their weaknesses: unlike metals, they are brittle, with low $K_{1c}$. This gives ceramics a low tolerance for stress concentrations (like holes or cracks) or for high contact stresses (at clamping points, for instance). For this reason it is more difficult to design with ceramics than with metals.

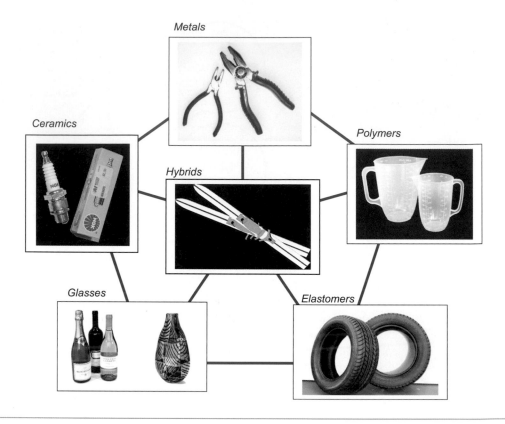

**Figure 2.2** Examples of each material family. The arrangement follows the general pattern of Figure 2.1. The central hybrid here is a sandwich structure made by combining stiff, strong face sheets of aluminum with a low-density core of balsa wood.

*Glasses* are non-crystalline ('amorphous') solids, a term explained more fully in Chapter 4. The commonest are the soda-lime and boro-silicate glasses familiar as bottles and Pyrex ovenware, but there are many more. The lack of crystal structure suppresses plasticity, so, like ceramics, glasses are hard and remarkably corrosion resistant. They are excellent electrical insulators and, of course, they are transparent to light. But like ceramics, they are brittle and vulnerable to stress concentrations.

*Polymers* are organic solids based on long chains of carbon (or, in a few, silicon) atoms. Polymers are light—their densities $\rho$ are less than those of the lightest metals. Compared with other families they are floppy, with moduli $E$ that are roughly 50 times less than those of metals. But they can be strong, and because of their low density, their strength per unit weight is comparable to that of metals. Their properties depend on temperature, so a polymer that is tough and flexible at room temperature may be brittle at the $-4\,^{\circ}\text{C}$ of a household freezer, yet turn rubbery at the $100\,^{\circ}\text{C}$ of boiling water. Few have useful strength above $150\,^{\circ}\text{C}$. If these aspects are allowed for in the design, the advantages of polymers can be exploited. And there

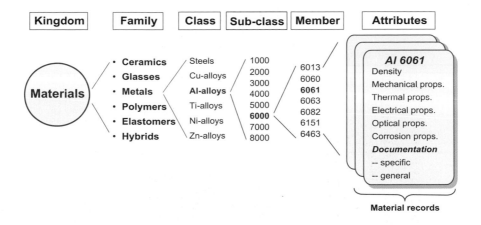

**Figure 2.3** The taxonomy of the kingdom of materials and their attributes. Computer-based selection software stores data in a hierarchical structure like this.

are many. They are easy to shape (which is why they are called 'plastics'): Complicated parts performing several functions can be moulded from a polymer in a single operation. Their properties are well suited for components that snap together, making assembly fast and cheap. And by accurately sizing the mould and precoloring the polymer, no finishing operations are needed. Good design exploits these properties.

*Elastomers*, the material of rubber bands and running shoes, are polymers with the unique property that their stiffness, measured by $E$, is extremely low—500 to 5000 times less than those of metals—and the ability to be stretched to many times their starting length yet recovering their initial shape when released. Despite their low stiffness they can be strong and tough—think of car tires.

*Hybrids* are combinations of two (or more) materials in an attempt to get the best of both. Glass and carbon-fiber reinforced polymers (GFRP and CFRP) are hybrids; so, too, are sandwich structures, foams and laminates. And almost all the materials of nature—wood, bone, skin, leaf—are hybrids. Bone, for instance, is a mix of collagen (a polymer) with hydroxyapatite (a mineral). Hybrid components are expensive, and they are relatively difficult to form and join. So despite their attractive properties, the designer will use them only when the added performance justifies the added cost. Today's growing emphasis on high performance and fuel efficiency provide increasing drivers for their use.

## 2.3 Organising processes: the process tree

A *process* is a method of shaping, joining or finishing a material. *Casting, injection moulding, fusion welding* and *electro-polishing* are all processes; there are hundreds of them (see Figures 2.4 and 2.5). It is important to choose the right process-route at an early stage in the design before the cost-penalty of making changes becomes large. The choice for a given component depends on the material of which it is to be made; on its shape, dimensions and precision; and on how many are to be made—in short, on the *design requirements*.

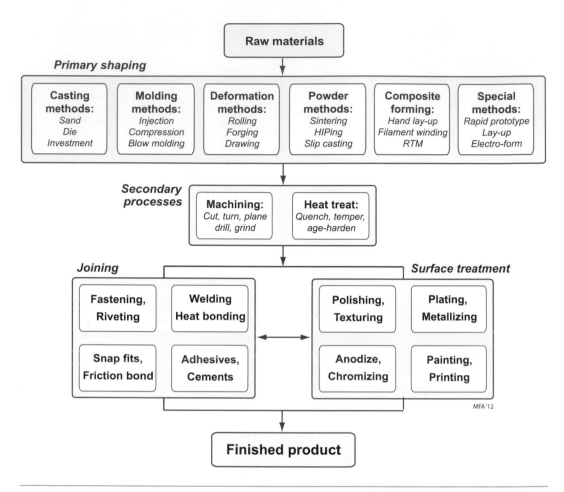

**Figure 2.4**  The classes of process. The first row contains the primary shaping processes; below lie the secondary processes of machining and heat treatment, followed by the families of joining and finishing processes.

The choice of material limits the choice of process. Polymers can be moulded; other materials cannot. Ductile materials can be forged, rolled and drawn, but those that are brittle must be shaped in other ways. Materials that melt at modest temperatures to low-viscosity liquids can be cast; those that do not have to be processed by other routes. Shape, too, influences the choice of process. Slender shapes can be made easily by rolling or drawing but not by casting. Hollow shapes cannot be made by forging, but they can by casting or moulding.

*Classifying processes*  Manufacturing processes are organised under the headings shown in Figure 2.4. *Primary processes* create shapes. The first row lists six primary forming processes: casting, moulding, deformation, powder methods, methods for forming composites and special methods including rapid prototyping. *Secondary processes* modify shapes or properties;

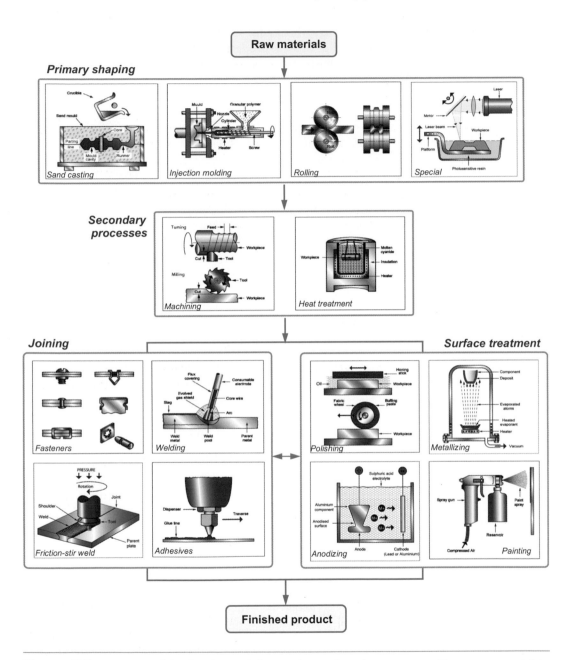

**Figure 2.5** Examples of the families and classes of manufacturing processes. The arrangement follows the general pattern of Figure 2.4.

here they are shown as *machining*, which adds features to an already shaped body, and *heat treatment*, which enhances surface or bulk properties. Below these come *joining*, and, finally, *surface treatment*. Figure 2.5 illustrates some of these; it is organised in the same way as Figure 2.4. The merit of Figure 2.4 is that it is a flow chart: a progression through a manufacturing route. It should not be treated too literally: the order of the steps can be varied to suit the needs of the design.

To organise information about processes, we need a hierarchical classification like that used for materials, giving each process a place. Figure 2.6 shows part of the hierarchy. The Process universe has three families: *shaping, joining* and *surface treatment*. In this figure, the shaping family is expanded to show classes: casting, deformation, moulding and so on. One of these, *moulding*, is again expanded to show its members: rotation moulding, blow moulding, injection moulding and so forth. Each process is characterised by a set of *attributes*: the materials it can handle, the shapes it can make, their size, precision and an economic batch size (the number of units that it can make most economically).

The other two families are partly expanded in Figure 2.7. There are three broad classes of joining process: adhesives, welding and fasteners. In this figure one of them, *welding*, is expanded to show its members. As before each member has attributes. The first is the material or materials that the process can join. After that the attribute list differs from that for shaping. Here the geometry of the joint and the way it will be loaded are important, as are requirements that the joint can or cannot be disassembled, be watertight or be electrically conducting.

The lower part of the figure expands the family of surface treatment processes. Some of the classes it contains are shown; one, *coating*, is expanded to show some of its members. Finishing adds cost: the only justification for applying a finishing process is that it hardens, protects or decorates the surface in ways that add value. As with joining, the material to be coated is an important attribute, but the others again differ from those for shaping.

We will return to process selection in Chapters 18 and 19.

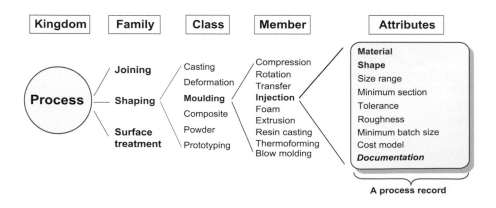

**Figure 2.6** The taxonomy of the kingdom of process with part of the shaping family expanded. Each member is characterised by a set of attributes. Process selection involves matching these to the requirements of the design.

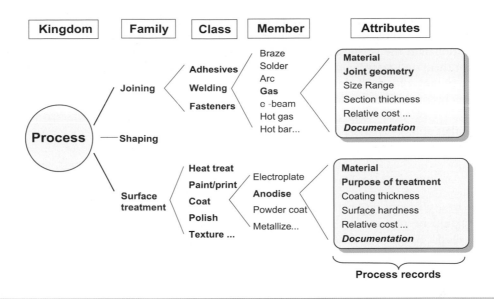

**Figure 2.7** The taxonomy of the process kingdom again, with the families of joining and surface treatment partly expanded.

## 2.4 Process-property interaction

Processing can change properties. If you hammer a metal ('forging') it gets harder; if you then heat it up it gets softer again ('annealing'). If polyethylene—the stuff of plastic bags—is drawn to a fiber, its strength is increased by a factor of 5. Soft, stretchy rubber is made hard and brittle by vulcanising. Heat-treating glass in a particular way can give it enough impact resistance to withstand a projectile ('bullet-proof glass'). And composites like carbon-fiber reinforced epoxy have no useful properties at all until processed—prior to processing they are just a soup of resin and a sheaf of fibers.

Joining, too, changes properties. Welding involves the local melting and resolidifying of the faces of the parts to be joined. As you might expect, the weld zone has properties that differ from those of the material far from the weld—usually worse. Surface treatments, by contrast, are generally chosen to improve properties: electroplating to improve corrosion resistance, or carburising to improve wear.

Process-property interaction appears in a number of chapters. We return to it specifically in Chapter 19.

## 2.5 Material property charts

Data sheets for materials list their properties, but they give no perspective and present no comparisons. The way to achieve these is to plot *material property charts*. They are of two types: bar charts and bubble charts.

A *bar chart* is simply a plot of one property for all the materials of the universe. Figure 2.8 shows an example: it is a bar chart for modulus, $E$. The largest is more than ten million times greater than the smallest—many other properties have similar ranges—so it makes sense to plot them on logarithmic[1] not linear scales, as here. The length of each bar shows the range of the property for each material, here segregated by family. The differences between the families now become apparent. Metals and ceramics have high moduli. Those of polymers are smaller, by a factor of about 50, than those of metals, while those of elastomers are some 500 times smaller still.

---

### Example 2.1 Use of bar charts

In a cost-cutting exercise, one designer suggests that certain die-cast zinc components could be replaced by cheaper moulded polyethylene (PE) components with the same shape. Another member of the team expresses concern that the PE replacement might be too flexible. By what factor do the moduli of the two materials differ?

*Answer.*    The bar chart of Figure 2.8 shows that the modulus of PE is less, by a factor of about 100, than that of zinc alloys. The concern is a real one.

---

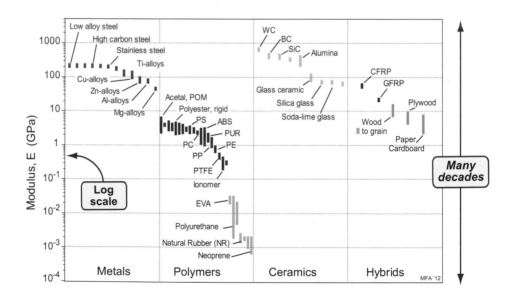

**Figure 2.8**    A bar chart of modulus. It reveals the difference in stiffness between the families.

---

[1] *Logarithmic* means that the scale goes up in constant multiples, usually of ten. We live in a logarithmic world—our senses, for instance, all respond in this way.

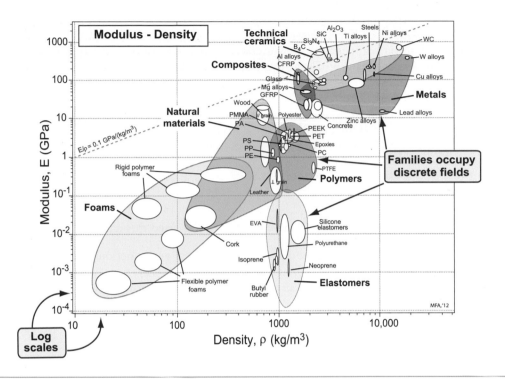

**Figure 2.9**    A bubble chart of modulus and density. Families occupy discrete areas of the chart.

More information is packed into the picture if two properties are plotted to give a *bubble chart*, as in Figure 2.9, here showing modulus $E$ and density $\rho$. As before, the scales are logarithmic. Now families are more distinctly separated: all metals lie in the reddish zone near the top right; all polymers lie in the dark blue envelope in the center, elastomers in the lighter blue envelope below, ceramics in the yellow envelope at the top. Each family occupies a distinct, characteristic field. Within these fields, individual materials appear as smaller ellipses.

### Example 2.2 Use of bubble charts

Steel is stiff (big modulus $E$) but heavy (big density $\rho$). Aluminum alloys are less stiff but also less dense. One criterion for lightweight design is a high value of the ratio $E/\rho$, defining materials that have a high stiffness per unit weight. Does aluminum have a significantly higher value of $E/\rho$ than steel? What about carbon-fiber reinforced plastic (CFRP)?

*Answer.* The $E/\rho$ chart of Figure 2.9 can answer the question in seconds. All three materials appear on it. Materials with equal values of $E/\rho$ lie along a line of slope 1—one such line is shown. Materials with high $E/\rho$ lie towards the top left, those with low values towards the bottom right. If a line with the same slope is drawn through Al-alloys on the chart, it passes almost exactly through steels: the two materials have almost the *same* value of $E/\rho$, a surprise when you think that aircraft are made of aluminum, not steel (the reason for this will become clear in Chapter 5). CFRP, by contrast, has a much higher $E/\rho$ than either aluminum or steel.

Material property charts like these are a core tool used throughout this book:

- They give an overview of the physical, mechanical and functional properties of materials, presenting the information about them in a compact way.
- They reveal aspects of the physical origins of properties, which are helpful in understanding the underlying science.
- They become a tool for optimised selection of materials to meet given design requirements, and they help us understand the use of materials in existing products.

## 2.6 Computer-aided information management for materials and processes

Software is now available to manage information about materials and processes, making it easier to find data and to manipulate it. Figure 2.10 shows part of a typical record for a material; Figure 2.11 shows the same for a process. Each record contains data of two types. *Structured data* are numeric, or Boolean (Yes / No) or discrete (e.g. Low / Medium / High), and can be stored in tables. Later chapters show how structured data are used for selection. *Unstructured data* take the form of text, images, graphs and schematics. Such information cannot so easily be used for selection but it is essential for documentation in making a final choice of material. Such software can be used alone or coupled with finite-element analysis systems, product data-management systems and environmental design and life-cycle analysis systems to provide the data they require in a semi-automatic way. We will encounter one material data-management package later in the book. Here is a selection of others. Some are free, and some require a license.

## Acrylonitrile butadiene styrene (ABS)

### The Material
ABS (Acrylonitrile-butadiene-styrene) is tough, resilient, and easily molded. It is usually opaque, although some grades can now be transparent, and it can be given vivid colors. ABS-PVC alloys are tougher than standard ABS and, in self-extinguishing grades, are used for the casings of power tools.

### General properties
| | | | | |
|---|---|---|---|---|
| Density | 1e3 | - | 1.2e3 | kg/m$^3$ |
| Price | 2 | - | 2.7 | USD/kg |

### Mechanical properties
| | | | | |
|---|---|---|---|---|
| Young's modulus | 1.1 | - | 2.9 | GPa |
| Hardness - Vickers | 5.6 | - | 15 | HV |
| Elastic limit | 19 | - | 51 | MPa |
| Tensile strength | 28 | - | 55 | MPa |
| Compressive strength | 31 | - | 86 | MPa |
| Elongation | 1.5 | - | 1e2 | % |
| Endurance limit | 11 | - | 22 | MPa |
| Fracture toughness | 1.2 | - | 4.3 | MPa.m$^{1/2}$ |

### Thermal properties
| | | | | |
|---|---|---|---|---|
| Thermal conductivity | 0.19 | - | 0.34 | W/m.K |
| Thermal expansion | 85 | - | 230 | µstrain/°C |
| Specific heat | 1400 | - | 1900 | J/kg.K |
| Glass Temperature | 88 | - | 130 | °C |
| Max service temp. | 62 | - | 90 | °C |

### Electrical properties
| | | | | |
|---|---|---|---|---|
| Resistivity | 2.3e21 | - | 3e22 | µohm.cm |
| Dielectric constant | 2.8 | - | 2.2 | |

### Typical uses
Safety helmets; camper tops; automotive instrument panels and other interior components; pipe fittings; home-security devices and housings for small appliances; communications equipment; business machines; plumbing hardware; automobile grilles; wheel covers; mirror housings; refrigerator liners; luggage shells; tote trays; mower shrouds; boat hulls; large components for recreational vehicles; weather seals; glass beading; refrigerator breaker strips; conduit; pipe for drain-waste-vent (DWV) systems.

**Figure 2.10**  Part of a record for a material, ABS. It contains numeric data, text and image-based information.

## Injection molding

### The process
No other process has changed product design more than INJECTION MOULDING. Injection molded products appear in every sector of product design: consumer products, business, industrial, computers, communication, medical and research products, toys, cosmetic packaging and sports equipment. The most common equipment for molding thermoplastics is the reciprocating screw machine, shown schematically in the figure. Polymer granules are fed into a spiral press where they mix and soften to a dough-like consistency that can be forced through one or more channels ('sprues') into the die. The polymer solidifies under pressure and the component is then ejected.

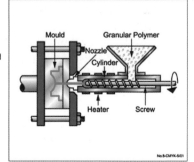

### Physical Attributes
| | | | | |
|---|---|---|---|---|
| Mass range | 1e-3 | - | 25 | kg |
| Range of section thickness | 0.4 | - | 6.3 | mm |
| Surface roughness (A=v. smooth) | A | | | |

### Economic Attributes
| | | | |
|---|---|---|---|
| Economic batch size (units) | 1e4 | - | 1e6 |
| Relative tooling cost | very high | | |
| Relative equipment cost | high | | |
| Labor intensity | low | | |

### Shape
| | |
|---|---|
| Circular Prismatic | True |
| Non-circular Prismatic | True |
| Solid 3-D | True |
| Hollow 3-D | True |

### Typical uses
Extremely varied. Housings, containers, covers, knobs, tool handles, plumbing fittings, lenses, etc.

**Figure 2.11**  Part of a record for a process, injection moulding. The image shows how it works, and the numeric and Boolean data and text document its attributes.

---

### Computer-based resources for materials

*CES Edu* (www.Grantadesign.com, license required). A comprehensive suite of databases for materials and processes with editions for general engineering, aerospace, polymers engineering, environmental design and industrial design. It includes powerful search, selection and eco-auditing tools. (The property charts in this book were made with this software).

*Matbase* (www.matbase.com, to become www.matbase.nl, free). A database of the technical properties of materials, originally from the Technical University of Denmark.

*Matdata* (www.matdata.com, limited access is free, full access requires a license). A well-documented database of the properties of metals.

*Materia* (www.materia.nl, free). A database aimed at industrial design, with high-quality images of some 2000 products.

*Material Connexion* (www.materialconnexion.com, license required). A materials library emphasising industrial design with records for some 7000 materials, each with an image, a description and a supplier.

*MatWeb* (www.matweb.com, limited access is free, full access requires a license). A large database of the engineering properties of materials, drawn from suppliers' data sheets.

*Rematerialise* (www.rematerialise.org, free) A database of 'sustainable' materials chosen because they are derived from renewable (biological) materials or use recycled materials.

---

## 2.7 Summary and conclusions

Classification is the first step in creating an information management system for materials and processes. In it the records for the members of each universe are indexed, so to speak, by their position in the tree-like hierarchies of Figures 2.3, 2.6 and 2.7. Each record has a unique place, making retrieval easy.

There are six broad families of materials for mechanical design: metals, ceramics, glasses, polymers, elastomers and hybrids that combine the properties of two or more of the others. Processes, similarly, can be grouped into families: those that create shape, those that join, and those that modify the surface to enhance its properties or to protect or decorate it. The members of the families can be organised into a hierarchical tree-like catalogue, allowing them to be 'looked up' in much the same way that you would look up a member of a company in a management sheet. A record for a member stores information about it: numeric and other tabular data for its properties, and text, graphs and images to describe its use and applications. This structure forms the basis of computer-based selection systems of which the CES system is an example. It enables a unique way of presenting data for materials and processes as property charts, two of which appear in this chapter. They become one of the central features of the chapters that follow.

## **2.8** Further reading

Askeland, D. R., & Phulé, P. P. (2006). *The Science and Engineering of Materials* (5th ed.). Toronto: Canada. Thomson. ISBN 0-534-55396-6.

Bralla, J. G. (1998). *Design for Manufacturability Handbook* (2nd ed.). New York: USA: McGraw-Hill. ISBN 0-07-007139-X. (Turgid reading, but a rich mine of information about manufacturing processes.)

Budinski, K. G., & Budinski, M. K. (2010). *Engineering Materials, Properties and Selection* (9th ed.). New York, USA: Prentice Hall. ISBN 978-0-13-712842-6. (A well-established materials text that deals well with both material properties and processes.)

Callister, W. D. (2010). *Materials Science and Engineering: An Introduction* (8th ed.). New York, USA: John Wiley & Sons. ISBN 978-0-470-41997-7. (A well-established text taking a science-led approach to the presentation of materials teaching.)

Dieter, G. E. (1991). *Engineering Design: A Materials and Processing Approach* (2nd ed.). New York, USA: McGraw-Hill. ISBN 0-07-100829-2. (A well-balanced and respected text focusing on the place of materials and processing in technical design.)

Farag, M. M. (2008). *Materials and Process Selection for Engineering Design* (2nd ed.). London, UK: CRC Press, Taylor and Francis. ISBN 9-781-420-06308-0. (A materials science approach to the selection of materials.)

Kalpakjian, S., & Schmid, S. R. (2003). *Manufacturing Processes for Engineering Materials* (4th ed.). New Jersey, USA: Prentice Hall, Pearson Education. ISBN 0-13-040871-9. (A comprehensive and widely used text on material processing.)

Shackelford, J. F. (2009). *Introduction to Materials Science for Engineers* (7th ed.). New Jersey, USA: Prentice Hall. ISBN 978-0-13-601260-4. (A well-established materials text with a design slant.)

## **2.9** Exercises

Exercise E2.1   *Material properties from experience.* List the six main classes of engineering materials. Use your own experience to rank them approximately:

(a) by stiffness (modulus, $E$). A sheet of a material that has a high modulus is hard to bend when in the form of a sheet. A sheet of material with a low modulus is floppy.

(b) by thermal conductivity ($\lambda$). Materials with high conductivity feel cold when you pick them up on a cold day. Materials with low conductivity may not feel warm, but they don't freeze your hands.

Exercise E2.2   *Classification (1).* A good classification looks simple—think, for instance, of the periodic table of the elements. Creating it in the first place, however, is another matter. This chapter introduced two classification schemes that work, meaning that every member of the scheme has a unique place in it, and any new member can be inserted into its proper position without disrupting the whole. Try one for yourself. Here are some scenarios. Make sure that each level of the hierarchy properly contains all those below it. There may be more than one way to do this, but one is usually better than the others. Test it by thinking how you would use it to find the information you want.

How many different ways can two sheets of paper be attached to each other, temporarily or permanently? Classify these, using 'Mechanism of joining' as the top level of the classification. Then try to develop the next level, based on your observations of the ways in which sheets of paper are joined.

Exercise E2.3   *Classification (2)*. In how many ways can wood be treated to change its surface appearance? Classify these, using the generic finishing technique as the top level of the classification. Then try to develop the next level, based on your observations of the ways in which wood products are finished.

Exercise E2.4   *Classification (3)*. You run a bike shop that stocks bikes of many types (children's bikes, street bikes, mountain bikes, racing bikes, folding bikes, etc.), prices and sizes. You need a classification system to allow customers to look up your bikes on the Internet. How would you do it?

Exercise E2.5   *Classification (4)*. You are asked to organise the inventory of fasteners in your company. There are several types (snap, screw, bolt, rivet), and within each, a range of materials and sizes. Devise a classification scheme to store information about them.

Exercise E2.6   *Shaping*. What is meant by a shaping process? Look around you and ask yourself how the things you see were shaped.

Exercise E2.7   *Joining*. Almost all products involve several parts that are joined. Examine products immediately around you and list the joining methods used to assemble them.

Exercise E2.8   *Surface treatment*. How many different surface treatment processes can you think of, based on your own experience? List them and annotate the list with the materials to which they are typically applied.

Exercise E2.9   *Use of bar charts*. Examine the material property chart of Figure 2.8. By what factor, on average, are polymers less stiff than metals? By what factor is Neoprene less stiff than Polypropylene (PP)?

Exercise E2.10   *Use of bar and bubble charts (1)*. Wood is a natural polymer, largely made up of cellulose and lignin, and they, like engineering polymers, are almost entirely hydrogen, carbon, oxygen and a little nitrogen. Has nature devised a cocktail of these elements that has a higher modulus than any bulk man-made polymer? Engineering polymers and woods appear on the charts of Figures 2.8 and 2.9. Use either one to answer the question.

Exercise E2.11   *Use of bubble charts (2)*. Windows can be made of glass (houses); poly-carbonate, PC (conservatories); or poly-methyl-methacrylate, PMMA (aircraft). Would a glass window be more flexible than a replacement of the same thickness made of polycarbonate? Would it be heavier? Use the bubble chart of Figure 2.9 to find out.

Exercise E2.12  *Use of bubble charts (3)*. Do zinc alloys have a higher specific stiffness $E/\rho$ than Polypropylene (PP)? Use the bubble chart of Figure 2.9 to find out.

## 2.10 Exploring design using CES

Designers need to be able to find data quickly and reliably. That is where the classifications come in. The CES system uses the classification scheme described in this chapter. Before trying these exercises, open the Materials Universe in CES and explore it. The opening menu offers three or more options—take the first, 'Level 1'.

Exercise E2.13  Use the 'Browse' facility in Level 1 of the CES Software to find the record for *Copper*. What is its thermal conductivity? What is its price?

Exercise E2.14  Use the 'Browse' facility in Level 1 of the CES Software to find the record for the thermosetting polymer *Phenolic*. Is it cheaper or more expensive than *Epoxies*?

Exercise E2.15  Use the 'Browse' facility to find records for the polymer-shaping processes *Rotational moulding*. What, typically, is it used to make?

Exercise E2.16  Use the 'Search' facility to find out what *Plexiglas* is. Do the same for *Pyroceram*.

Exercise E2.17  Use the 'Search' facility to find out about the process *Pultrusion*. Do the same for *TIG welding*. Remember that you need to search the Process Universe, not the Materials Universe. To do this change the Table from 'MaterialUniverse' to 'ProcessUniverse' using the tab at the top of the Browse window.

Exercise E2.18  Compare Young's modulus $E$ (the stiffness property) and thermal conductivity $\lambda$ (the heat-transmission property) of *aluminum alloys* (a non-ferrous metal), *alumina* (a technical ceramic), *polyethylene* (a thermoplastic polymer) and *neoprene* (an elastomer) by retrieving values from CES Level 1. Which has the highest modulus? Which has the lowest thermal conductivity?

## 2.11 Exploring the science with CES Elements

The CES system contains a database for the periodic table called 'Elements'. The records contain fundamental data for each of the elements. We will use this in the book to delve a little deeper into the science that lies behind material properties.

Exercise E2.19    Refresh your memory of the periodic table, perhaps the most significant classification system of all time. Open the Elements database by clicking on Change in the Browse window and selecting *Elements* from the menu that is offered (or go to *File > Change database > Elements*) and double-click on periodic table to see the table. This database, like the others described in this chapter, has a tree-like structure. Use this to find the record for aluminum (Row 3, Atomic number 13) and explore its contents. Many of the properties won't make sense yet. We introduce them gradually throughout the book.

# Chapter 3
# Strategic thinking: matching material to design

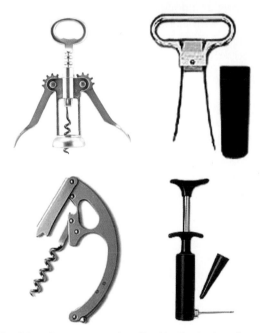

Images embodying the concepts described in the text: pull, geared pull, shear and pressure. (Image courtesy of A-Best Fixture Co., 424 West Exchange Street, Akron, Ohio, 44302, USA.)

## Chapter contents

## **3.1** Introduction and synopsis

Our aim in this chapter is to develop a strategy for selecting materials and processes that is *design-led*; that is, the strategy uses the requirements of the design as inputs. To do so we must first look briefly at design itself. This chapter introduces some of the vocabulary of design, the stages in its implementation and the ways in which materials selection links with these.

Design starts with a *market need*. The need is analysed, expressing it as a set of *design requirements*. Ways to meet these requirements (*Concepts*) are sought, developed (*Embodied*) and refined (*Detailed*) to give a *product specification*. The choice of material and process evolves in parallel with this process, in the way detailed in this chapter.

With this background we can develop the selection strategy. It involves four steps: *translation*, *screening*, *ranking* and *documentation*. These steps are explained, and the first, that of translation, is illustrated with examples.

## **3.2** The design process

*Original design* starts from a new concept and develops the information necessary to implement it. *Evolutionary design* (or *redesign*) starts with an existing product and seeks to change it in ways that increase its performance, reduce its cost, or both.

***Original design*** This starts from scratch, with a new idea or working principle (the vinyl disk, the audio tape, the compact disc, the MP3 player were all, in their day, completely new). Original design can be stimulated by new materials. High-purity silicon enabled the transistor; high-purity glass, the optical fibre; high coercive-force magnets, the miniature earphone and now the electric car; solid-state lasers, the barcode. Sometimes the new material suggests the new product. More often new products or enterprises demand the development of a new material: nuclear technology drove the development of new zirconium alloys and new stainless steels; space technology stimulated the development of beryllium alloys and lightweight composites; turbine technology today drives development of high-temperature alloys and ceramics; concern for the environment drives the development of bio-polymers.

The central column of Figure 3.1 shows the design process. The starting point is a *market need* or a *new idea*; the end point is the full *product specification* for a product that fills the need or embodies the idea. A need must be identified before it can be met. It is essential to define the need precisely, that is, to formulate a *need statement*, often in the form: 'a device is required to perform task X', expressed as a set of *design requirements*. Between the need statement and the product specification lie the set of stages shown in Figure 3.1: *conceptual design*, *embodiment design* and *detailed design*, explained in a moment.

At the conceptual design stage, all options are open: the designer considers alternative concepts and the ways in which these might be separated or combined. The next stage, embodiment, takes the promising concepts and seeks to analyse their operation at an approximate level. This involves sizing the components and a preliminary selection of materials and processes, and examining the implications for performance and cost. The embodiment stage ends with a feasible layout that becomes the input to the detailed design stage. Here, specifications and dimensions for each component are drawn up. Critical components may be subjected to precise

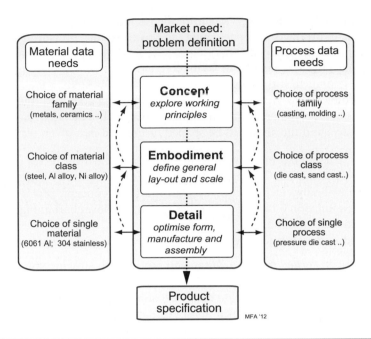

**Figure 3.1** The design flow chart, showing how material and process selection enter. Information about materials is needed at each stage, but at very different levels of breadth and precision. The broken lines suggest the iterative nature of original design and the path followed in redesign.

mechanical or thermal analysis. Optimisation methods are applied to components and groups of components to maximise performance, minimise cost and ensure safety. A final choice of geometry and material is made, and the methods of production are analysed and costed. The stage ends with a detailed production specification.

*Redesign* Most design is not 'original' in the sense of starting from a totally new idea. It starts with an existing product and corrects its shortcomings, enhances its performance or reduces its cost or environmental impact, making it more competitive. Here are some scenarios that call for redesign:

- The 'product-recall' scenario: if a product, once released to the market, fails to meet safety standards, urgent redesign is required. Often the problem is a material failure; then an alternative must be found that retains the desirable features of the original but overcomes its weaknesses.
- The 'poor value for money' scenario: the product performs safely but offers performance that, at its price, is perceived to be mediocre, requiring redesign to enhance performance.
- The 'inadequate profit margin' scenario: the cost of manufacture exceeds the price that the market will bear. Much of the cost of a mass-produced product derives from the materials of which it is made and the processes chosen to make it; the response is to re-examine both with cost-cutting as the objective.

- The 'sustainable technology' scenario: the response of the designer to the profligate use of materials in products and packaging, and to consumer pressure for production that does not harm the environment.
- The 'Mac-effect' scenario: in a market environment in which many almost-identical products compete for the consumer's attention, it is style, image and character that sets some products above others. Much creative thinking goes into this 'industrial design', and in it the choice of material, or of a change of material, is dictated mainly by aesthetics: color, texture, feel and the ability to be shaped or finished in a given way.

Much of redesign has to do with detail—the last of the three boxes in the central window of Figure 3.1—but not all. The necessary changes may require a change of configuration and layout (the embodiment phase) or even of the basic concept, replacing one of the ways of performing a function by another. In other words, the flow chart at Figure 3.1 remains a useful summary to keep in mind.

Described in the abstract, these ideas are not easy to grasp; an example will help.

***Devices to open corked bottles*** When you buy a bottle of wine you find, generally, that it is sealed with a cork. This creates a market need: it is the need to gain access to the wine inside. We might state it thus: 'a device is required to allow access to wine in a corked bottle,' and we might add, 'with convenience, at modest cost, and without contaminating the wine'.

Three concepts for doing this are shown in Figure 3.2. In order, they are to remove the cork by axial traction (= pulling); to remove it by shear tractions; to push it out from below. In the first, a screw is threaded into the cork to which an axial pull is applied; in the second, slender elastic blades inserted down the sides of the cork apply shear tractions when pulled; and in the third, the cork is pierced by a hollow needle through which a gas is pumped to push it out. The opening picture of this chapter has photos of bottle openers based on all three concepts.

Figure 3.3 shows, left, embodiment sketches for devices based on concept (a), that of axial traction. The first is a direct pull; the other three use some sort of mechanical advantage—levered pull, geared pull and spring-assisted pull. The embodiments suggest the layout, the mechanisms and the scale. In the final, detailed, stage of design, the components are dimensioned so that they carry the working loads safely, their precision and surface

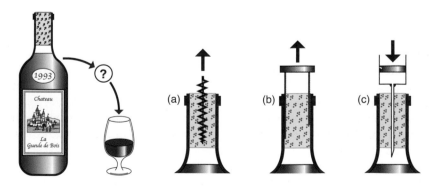

Figure 3.2 A market need—that of gaining access to wine in corked bottles—and three concepts for meeting the need. Devices based on all three of these concepts exist and can be bought.

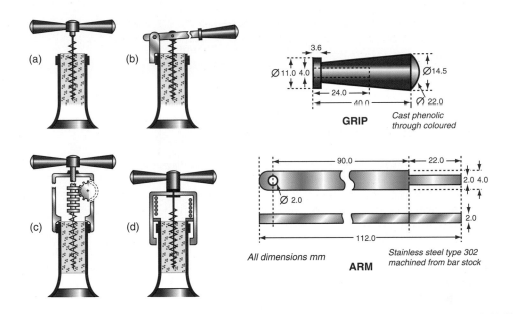

**Figure 3.3** Embodiment sketches for the first concept: direct pull, levered pull, geared pull and spring-assisted pull. Each system is made up of components that perform a sub-function. Detailed design drawings for the lever of embodiment (b) are shown on the right.

finish are defined, and a final choice of material and manufacturing route is made (Figure 3.3, right). Let us examine how this is done.

## 3.3 Material and process information for design

Materials selection enters each stage of the design (Figure 3.1, left). The nature of the data needed in the early stages differs greatly in its level of precision and breadth from that needed later on. At the concept stage, the designer requires only approximate property-values, but for the widest possible range of materials. All options are open: a polymer may be the best choice for one concept, a metal for another. The problem, at this stage, is not precision and detail, it is breadth and speed of access: how can the vast range of data be presented to give the designer the greatest freedom in considering alternatives?

At the embodiment stage a concept has been selected and the materials options narrow. Here we need data for a subset of materials, but at a higher level of precision and detail. These are found in more specialised handbooks and software that deal with a single class or sub-class of materials—metals, or just aluminum alloys, for instance. The risk now is that of losing sight of the bigger spread of materials to which we must return if the details don't work out; it is easy to get trapped in a single line of thinking when others have potential to offer better solutions.

The final stage of detailed design requires a still higher level of precision and detail, but for only one or a very few materials. A given material (polyethylene, for instance) has a range of properties that derive from differences in the ways different producers make it. At the detailed design stage suppliers must be identified and the properties of their materials used in the design

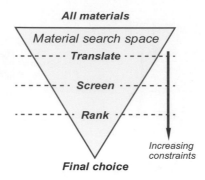

**Figure 3.4** The narrowing of material search space as design constraints are applied.

calculations. And sometimes even this is not good enough. If the component is a critical one (meaning that its failure could, in some sense or another, be disastrous) then it may be prudent to conduct in-house tests to measure the critical properties, using a sample of the material that will be used to make the product itself. The process is one of narrowing the materials search space by screening out materials that cannot meet the design requirements, ranking those that remain for identifying the most promising choice (Figure 3.4).

The material decisions do not end with the establishment of production. Products fail in service, and failures contain information. It is an imprudent manufacturer who does not collect and analyse data on failures. Often this points to the misuse of a material, one that redesign or reselection can eliminate.

The selection of a material cannot be separated from that of process and of shape. To make a shape, a material undergoes a chain of processes that involve shaping, joining and finishing. Figures 2.4 and 2.5 of Chapter 2 introduced these processes. The selection of process follows a route that runs parallel to that of material (Figure 3.1, right). The starting point is a catalogue of all processes, which is then narrowed by screening out those that fail to make the desired shape or are incompatible with the choice of material. Material, shape and process interact (Figure 3.5). Process choice is influenced by the material: by its formability, machinability, weldability, heat-treatability and so on. Process choice is influenced by the requirements for shape: the process determines the shape, the size, the precision and, to a large extent, the cost of a component. The interactions are two-way: specification of shape restricts the choice of material and process; but equally, the specification of process limits the materials you can use and the shapes they can take. And it is often processing that is used to develop the properties on which the material has been selected. The more sophisticated the design, the tighter the specifications and the greater the interactions. The interaction between material, shape and process lies at the heart of the selection process. To tackle it we need a strategy.

## 3.4 The strategy: translation, screening, ranking and documentation

Selection involves seeking the best match between the attribute profiles of the materials and processes—bearing in mind that these must be mutually compatible—and those required by

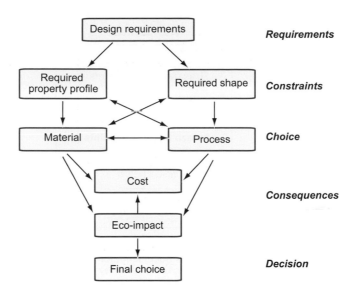

**Figure 3.5** The interaction between design requirements, material, shape and process.

the design. The strategy, applied to materials, is sketched in Figure 3.6. The first task is that of *translation*: converting the design requirements into a prescription for selecting a material. This proceeds by identifying the *constraints* that the material must meet and the *objectives* that the design must fulfil. These become the filters: materials that meet the constraints and rank highly in their ability to fulfil the objectives pass through the filters, those that do not are filtered out. The second task, then, is that of *screening*: eliminating the materials that cannot meet the constraints. This is followed by the *ranking* step: ordering the survivors by their ability to meet a criterion of excellence, such as that of minimising cost. The final task is to explore the most promising candidates in depth, examining how they are used at present, case histories of failures, and how best to design and manufacture with them—a step we call *Documentation*.

Process selection follows a parallel route. In this case translation means identifying the geometric and other constraints that must be met—dimensions, shape, precision and material compatibility—and using these to screen out processes that cannot provide them. We return to process selection in Chapters 18 and 19, but for now we stick to materials.

***Translation*** Any engineering component has one or more *functions*: to support a load, to contain a pressure, to transmit heat, and so forth. This must be achieved subject to *constraints*: that certain dimensions are fixed, that the component must carry the design loads without failure, the need to insulate against or to conduct heat or electricity, that it can function in a certain range of temperature and in a given environment, and many more. In designing the component, the designer has one or more *objectives*: to make it as cheap as possible, perhaps, or as light, or as environmentally benign, or some combination of these. Certain parameters can be adjusted in order to optimise the objective—the designer is free to vary dimensions that

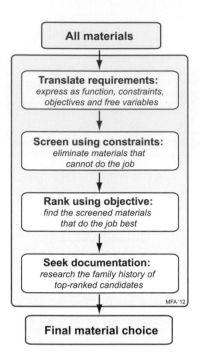

**Figure 3.6** The strategy applied to materials. The same strategy is later adapted to select processes. There are four steps: translation, screening, ranking and supporting information. All can be implemented in software, allowing large populations of materials to be investigated.

are not constrained by design requirements and, most importantly, free to choose the material for the component and the process to shape it. We refer to these as *free variables*.

Constraints, objectives and free variables (Table 3.1) define the boundary conditions for selecting a material and—in the case of load-bearing components—a shape for its cross-section.

It is important to be clear about the distinction between constraints and objectives. A constraint is an essential condition that must be met, usually expressed as a limit on a material or process attribute: the material must not cost more than $4/kg, for example. An objective is a quantity for which an extreme value (a maximum or minimum) is sought, frequently cost, mass or volume, but there are others (Table 3.2). Getting it right can take a little thought. In

**Table 3.1** Function, constraints, objectives and free variables

| | |
|---|---|
| *Function* | • What does the component do? |
| *Constraints* | • What non-negotiable conditions must the material meet? |
| *Objectives* | • What aspects of performance are to be maximised or minimised? |
| *Free variables* | • What parameters of the problem is the designer free to change? |

Table 3.2  Common constraints and objectives

| Common constraints* | Common objectives |
|---|---|
| **Meet a target value of** | **Minimise** |
| Stiffness | Cost |
| Strength | Mass |
| Fracture toughness | Volume |
| Thermal conductivity | Impact on the environment |
| Electrical resistivity | Heat loss |
| Magnetic remanence | |
| Optical transparency | **Maximise** |
| Cost | Energy storage |
| Mass | Heat flow |

* All these properties, and products that rely on them, appear in later chapters.

choosing materials for a super-light sprint bicycle, for example, the objective is to minimise mass, with an upper limit on cost, thus treating cost as a constraint. But in choosing materials for a cheap 'shopping' bike the two are reversed: now the objective is to minimise cost with a (possible) upper limit on mass, thus treating it as a constraint.

The outcome of the translation step is a list of the design-limiting properties and the constraints they must meet. The first step in relating design requirements to material properties is therefore a clear statement of function, constraints, objectives and free variables.

*Screening*   Constraints are gates: meet the constraint and you pass through the gate; fail to meet it and you are out. Screening (Figure 3.6) does just that: it eliminates candidates that cannot do the job at all because one or more of their attributes lies outside the limits set by the constraints. As examples, the requirement that 'the component must function in boiling water', or that 'the component must be transparent', imposes obvious limits on the attributes of *maximum service temperature* and *optical transparency* that successful candidates must meet. We refer to these as *attribute limits*.

*Ranking*   To rank the materials that survive the screening step we need a criterion of excellence. They are found in the *material indices*, developed below and in later chapters, which measure how well a candidate that has passed the screening step can do the job (Figure 3.6 again). Performance is sometimes limited by a single property, sometimes by a combination of them. Thus the best materials for buoyancy are those with the lowest density, $\rho$; those best for thermal insulation are the ones with the smallest values of the thermal conductivity, $\lambda$—provided, of course, that they also meet all other constraints imposed by the design. Here, maximising or minimising a single property maximises performance. Often, though, it is not one but a group of properties that are relevant. Thus the best materials for a light, stiff tie-rod are those with the greatest value of the *specific stiffness*, $E/\rho$, where $E$ is Young's modulus. The best materials for a spring are those with the greatest value of $\sigma_y^2/E$, where $\sigma_y$ is the yield strength. The property or property group that maximises performance for a given design is called its *material index*. There are many such indices, each associated with maximising some aspect of performance—subsequent chapters will explain all about them. They provide criteria of excellence that allow ranking of materials by their ability to perform well in the given application.

To summarise: *screening* isolates candidates that are capable of doing the job; *ranking* identifies those among them that can do the job best.

*Documentation*    The outcome of the steps so far is a ranked short-list of candidates that meet the constraints and that maximise or minimise the criterion of excellence, whichever is required. You could just choose the top-ranked candidate, but what hidden weaknesses might it have? What is its reputation? Has it a good track record? To proceed further we seek a detailed profile of each: its *documentation* (Figure 3.6, bottom).

What form does documentation take? Typically, it is descriptive, graphical or pictorial: case studies of previous use of the material, details of its corrosion behaviour in particular environments, of its availability and pricing, warnings of its environmental impact or toxicity, or sensitivity in some of its properties to the way it is processed. Such information is found in handbooks, suppliers' data sheets, CD-based data sources and high-quality websites. Documentation helps narrow the short-list to a final choice, allowing a definitive match to be made between design requirements and material and process attributes.

Why are all these steps necessary? Without screening and ranking, the candidate pool is enormous and the volume of documentation is overwhelming. Dipping into it, hoping to stumble on a good material, gets you nowhere. But once a small number of potential candidates have been identified by the screening-ranking steps, detailed documentation can be sought for these few alone, and the task becomes viable.

## 3.5 Examples of translation

The following examples illustrate the translation step for a number of problems, starting with the lever for the corkscrew.

---

### Example 3.1 A corkscrew lever

Figure 3.3 shows the lever for one of the corkscrews in the design case study. In use it is loaded in bending. It must carry the bending moment without deflecting to an awkward degree, which means a high modulus, $E$. It must not bend permanently (though some cheap corkscrews do), which means a high yield strength, $\sigma_y$. And it must not snap off altogether, which means it must have adequate fracture toughness, $K_{1c}$. Finally, it must not corrode in wine or water. The length of the lever is specified, but the cross-section is not—we are free to choose a section that is sufficient to bear the use-loads. Given all these, the lever should be as cheap as possible. Formulate the translation.

*Answer.*    Table 3.3 lists the translation.

The *design-limiting properties* are those directly relating to the constraints: modulus $E$, strength $\sigma_y$, fracture toughness $K_{1c}$ and corrosion resistance.

Table 3.3   Translation for the corkscrew lever

| Function | • Lever (beam loaded in bending) | |
|---|---|---|
| Constraints | • Stiff enough<br>• Strong enough<br>• Some toughness<br>• Resist corrosion in wine and water | } Functional constraints |
| | • Length $L$ specified | A geometric constraint |
| Objective | • Minimise cost | |
| Free variable | • Choice of material<br>• Choice of cross-section area | |

Example 3.2 Redesign of a CD case

Music lovers will affirm that CDs—the best of them—are divine. But the cases they come in are the work of the devil (Figure 3.7). They are called 'jewel' cases for reasons of their optical clarity, but in performance they are far from jewels. They are usually made of polystyrene, PS, chosen for its low cost and water-clear transparency; and they are made by injection moulding and that, too, is cheap if you are making millions. Polystyrene can, at least in principle, be recycled. But PS jewel cases crack easily, jam shut, the hinges break, and the corners of the case are hard and sharp enough to inflict terminal damage on a CD. They badly need redesign. Decide on the features you think really matter, and formulate constraints, objective and free variables for the redesign of a CD case.

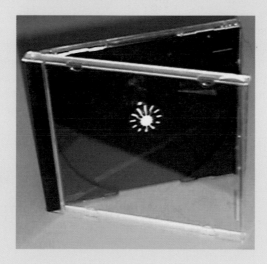

Figure 3.7   A polystyrene CD case. It is cheap, but it is brittle and cracks easily.

*Answer.* The way to tackle redesign is to seek a replacement material that retains the good properties of the old one, but without the bad. Thus we seek a material that is optically transparent to allow the label to be read, is able to be injection moulded because this is the most economic way to make large numbers, and is recyclable. But it must be tougher than polystyrene. Of the materials that meet these constraints, we want the cheapest. Table 3.4 summarises the translation.

Table 3.4 Translation for the redesigned CD case

| Function | • Contain and protect a CD | |
|---|---|---|
| Constraints | • Optically clear<br>• Able to be injection moulded<br>• Recyclable<br>• Tougher than polystyrene | } Functional constraints |
| | • Dimensions identical with PS case | A geometric constraint |
| Objective | • Minimise cost | |
| Free variable | • Choice of material | |

Potential design-limiting properties are optical transparency, fracture toughness $K_{1c}$ (must be better than PS) and the ability to be injection moulded and recycled.

Example 3.3 Heat sinks for microchips

A microchip may only consume milliwatts, but this power is dissipated in a tiny volume, making the *power-density* high. As chips shrink and clock speeds grow, overheating becomes a problem. The chips of today's PCs require forced cooling to limit temperatures to 85 °C. Multiple-chip modules (MCMs) pack as many as 130 chips on to a single substrate, and they get even hotter—up to 180 °C. Heating is kept under control by attaching the chips to a heat sink (Figure 3.8), taking pains to ensure good thermal contact between chip and sink. The heat sink is a critical component, limiting further miniaturisation of the electronics. How can its performance be maximised?

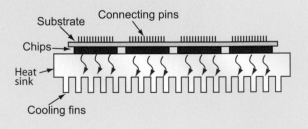

Figure 3.8 A heat sink. It must conduct heat well, but be electrically insulating.

To prevent electrical coupling and stray capacitance between chip and heat sink, the heat sink must be a good electrical insulator. If it is to work with one surface at 180 °C, it must have a maximum service temperature (the temperature at which it can operate continuously without damage) that is at least as great as 180 °C. These define the constraints. To drain heat away from the chip as fast as possible, it must also have the highest possible thermal conductivity, $\lambda$, defining the objective. Formulate the translation.

*Answer.*    The translation step is summarised in Table 3.5, where we assume that all dimensions are constrained by other aspects of the design.

Table 3.5    Translation for the heat sink

| Function | • Heat sink | |
|---|---|---|
| Constraints | • Material must be good electrical insulator | ⎫ |
| | • Maximum operating temperature > 200 °C | ⎬ Functional constraints |
| | • All dimensions are specified | ⎭ Geometric constraints |
| Objective | • Maximise thermal conductivity | |
| Free variable | • Choice of material | |

The design-limiting properties, clearly, are maximum service temperature $T_{max}$, electrical resistivity $\rho_e$ and thermal conductivity $\lambda$.

### Example 3.4 HF transformer cores

An electrical transformer uses electromagnetic induction to convert one AC voltage to another (Figure 3.9). To minimise energy loss the material must be a soft magnet—one that is easy to magnetise and de-magnetise (Chapter 14). And to avoid eddy current losses at high frequencies it must also be an electrical insulator. The constraints of 'soft magnetic

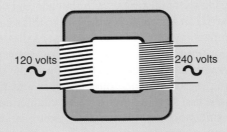

120 volts        240 volts

Figure 3.9    A transformer. The core must be a soft magnetic material, and if this is a high-frequency transformer, it must be an electrical insulator.

material' and 'electrical insulator' are very restrictive—they will screen out all but a small number of candidates. If the transformer is for an everyday product, the objective would be to minimise the cost. Formulate the translation.

*Answer.* Table 3.6 lists the translation.

Table 3.6 Translation for the transformer core

| Function | • HF transformer core | |
|---|---|---|
| Constraints | • Soft magnetic material <br> • Electrical insulator | Functional constraints |
| | • All dimensions are specified | Geometric constraints |
| Objective | • Minimise cost | |
| Free variable | • Choice of material | |

These translations are the first step in selection. In them we have identified the constraints; they will be used for screening. We have also identified the objective; it will be used for ranking. We will return to all four of these examples in later chapters when we know how to screen and rank.

## 3.6 Summary and conclusions

The starting point of a design is a *market need* captured in a set of *design requirements. Concepts* for a product that meet the need are devised. If initial estimates and exploration of alternatives suggest that the concept is viable, the design proceeds to the *embodiment* stage: working principles are selected, size and layout are decided and initial estimates of performance and cost are made. If the outcome is successful, the designer proceeds to the *detailed design* stage: optimisation of performance, full analysis of critical components and preparation of detailed production drawings (usually as a CAD file), showing dimensions, specifying precision and identifying material and manufacturing path. But design is not a linear process, as Figure 3.1 might suggest. Some routes lead to a dead end, requiring reiteration of earlier steps. And, frequently, the task is one of redesign, requiring that constraints be rethought and objectives realigned.

The selection of material and process runs parallel to this set of stages. Initially the search space for both is wide, encompassing all possible candidates. As the design requirements are formulated in increasing detail, constraints emerge that both must meet, and one or more objectives is formulated. The constraints narrow the search space and the objective(s) allow ranking of those that remain. Identifying the constraints, the objectives and the free variables (the process we called 'translation') is the first step in selection. This chapter ended with examples of translation when the task was that of choosing a material; the exercises suggest more. When the task is the choice of process, a similar translation is needed; we return to this in Chapter 18. The other steps—screening, ranking and documentation—are discussed in the chapters that follow.

## **3.7** Further reading

Ashby, M. F. (2011). *Materials Selection in Mechanical Design* (4th ed.). Oxford, UK: Butterworth Heinemann. ISBN 978-1-85617-663-7. (An advanced text developing material selection methods in detail.)

Cross, N. (2000). *Engineering Design Methods* (3rd ed.). Chichester, UK: Wiley. ISBN 0-471-87250-3. (A durable text describing the design process, with emphasis on developing and evaluating alternative solutions.)

Pahl, G., Beitz, W., Feldhusen, J., Grote, & K. H. (2007). *Engineering Design: A Systematic Approach*. Translated by K. Wallace and L. Blessing (3rd ed.). London, UK: The Design Council, and Berlin, Germany: Springer Verlag. ISBN 978-1-84-628318-5. (The Bible—or perhaps more exactly the Old Testament—of the technical design field, developing formal methods in the rigorous German tradition.)

Ullman, D. G. (1992). *The Mechanical Design Process*. New York, USA: McGraw-Hill. ISBN 0-07-065739-4. (An American view of design, developing ways in which an initially ill-defined problem is tackled in a series of steps, much in the way suggested by Figure 3.1.)

Ulrich, K. T. (2011). *Design: Creation of Artefacts in Society*. Philadelphia, USA: University of Pennsylvania Press. ISBN 978-0-9836487-0-3. (An excellent short introduction to the kind of structured reasoning that lies behind good product design. The text is available on http://www.ulrichbook.org/ )

Ulrich, K. T., & Eppinger, S. D. (2008). *Product Design and Development* (4th ed.). New York, USA: McGraw Hill. ISBN 978-007-125947-7. (A readable, comprehensible text on product design, as taught at MIT. Many helpful examples but almost no mention of materials.)

## **3.8** Exercises

| | |
|---|---|
| Exercise E3.1 | What are the steps in developing an original design? |
| Exercise E3.2 | What is meant by an *objective* and what by a *constraint* in the requirements for a design? How do they differ? |
| Exercise E3.3 | Describe and illustrate the 'translation' step of the material selection strategy. |
| Exercise E3.4 | *Translation (1).* The teeth of a scoop for a digger truck, pictured in Example 1.1 of Chapter 1, must cut earth, scoop stones, crunch rock, often in the presence of fresh or salt water and worse. Translate these requirements into a prescription of Function, Constraints, Objectives and Free variables. |
| Exercise E3.5 | *Translation (2).* A material for an energy-efficient saucepan, pictured in Example 1.2 of Chapter 1, must transmit and spread the heat well, resist corrosion by foods, and withstand the mechanical and thermal loads expected in normal use. The product itself must be competitive in a crowded market. Translate these requirements into a prescription of Function, Constraints, Objectives and Free variables. |
| Exercise E3.6 | *Translation (3).* A material for eyeglass lenses, pictured in Example 1.3 of Chapter 1, must have optical-quality transparency. The lens may be moulded |

or ground with precision to the required prescription. It must resist sweat and be sufficiently scratch-resistant to cope with normal handling. The mass-market end of the eyeglass business is very competitive so price is an issue. Translate these requirements into a prescription of Function, Constraints, Objectives and Free variables.

Exercise E3.7 Bikes come in many forms, each aimed at a particular sector of the market:
- Sprint bikes.
- Touring bikes.
- Mountain bikes.
- Shopping bikes.
- Children's bikes.
- Folding bikes.

Use your judgment to identify the primary objective and the constraints that must be met for each of these.

Exercise E3.8 A material is required for the windings of an electric air furnace capable of operating at temperatures up to 1000 °C. Think out what attributes a material must have if it is to be made into windings and function properly in a furnace. List the function and the constraints; set the objective to 'minimise material price' and the free variables to 'choice of material'.

Exercise E3.9 A material is required to manufacture office scissors. Paper is an abrasive material, and scissors sometimes encounter hard obstacles like staples. List function and constraints; set the objective to 'minimise material price' and the free variables to 'choice of material'.

Exercise E3.10 A material is required for a heat exchanger to extract heat from geo-thermally heated saline water at 120 °C (and thus under pressure). List function and constraints; set the objective to 'minimise material price' and the free variables to 'choice of material'.

Exercise E3.11 A material is required for a disposable fork for a fast-food chain. List the objective and the constraints that you would see as important in this application.

Exercise E3.12 Formulate the constraints and objective you would associate with the choice of material to make the forks of a racing bicycle.

Exercise E3.13 Cheap coat hangers used to be made of wood, but now only expensive ones use this material. Most coat hangers are now metal or plastic, and both differ in shape from the wooden ones, and from each other. Examine wood, metal and plastic coat hangers, comparing the designs, and comment on the ways in which the choice of material has influenced them.

Exercise E3.14 Cyclists carry water in bottles that slot into bottle holders on their bicycles. Examine metal and plastic bottle holders, comparing the designs, and comment on the ways in which the choice of material has influenced them.

## **3.9** Exploring design using CES

Exercise E3.15   A company wishes to enhance its image by replacing oil-based plastics in its products with polymers based on natural materials. Use the 'Search' facility in CES to find *biopolymers*. List the materials you find.

Exercise E3.16   A maker of garden furniture is concerned that the competition is stealing part of his market with furniture made by RTM, a term with which he is unfamiliar. Use the 'Search' facility in CES to find out what *RTM* is, and whether it is used to make things like garden furniture.

Exercise E3.17   Use the 'Search' facility in CES Level 2 to find materials for *furnace windings*.

Exercise E3.18   Use the 'Search' facility in CES Level 2 to find materials for *scissors* and *knife blades*.

Exercise E3.19   Use the 'Search' facility in CES Level 2 to find materials for *heat exchangers*.

Exercise E3.20   Use the 'Search' facility in CES Level 2 to find materials for *flooring*.

Exercise E3.21   Eyeglass lenses require a material with optical-quality transparency that can be moulded with precision and will resist sweat. A high refractive index (greater than 1.5, say) allows thinner, lighter lenses. Open CES Edu Materials Level 2, choose the data subset Materials with Durability properties, and open a Limit stage. Open *Optical properties* and select Optical quality and set a lower limit of 1.5 for the refractive index. Open *Processability* and put a lower limit of 5 (the best possible) on Mouldability. *Open Durability: water and aqueous environments* and click on Excellent in Fresh water and in Salt water (sweat). Finally, click *Apply*. What is the selected short list? (It appears in the lower left of the CES user-interface.)

# Chapter 4
# Stiffness and weight: density and elastic moduli

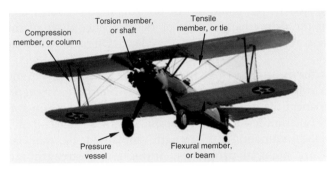

Modes of loading. (Image of Stansted Airport courtesy of Norman Foster and Partners, London, UK)

## **4.1** Introduction and synopsis

Stress causes strain. If you are human, the ability to cope with stress without undue strain is called *resilience*. If you are a material, it is called *elastic modulus*.

Stress is something that is applied to a material by loading it. Strain—a change of shape—is its response; it depends on the magnitude of the stress and the way it is applied—the *mode of loading*. The cover picture illustrates the common ones. Ties carry tension—often, they are cables. Columns carry compression—tubes are more efficient as columns than solid rods because they don't buckle as easily. Beams carry bending moments, like the wing spar of the plane or the horizontal roof beams of the airport. Shafts carry torsion, as in the drive shaft of cars or the propeller shaft of the plane. Pressure vessels contain a pressure, as in the tires of the plane. Often they are shells: curved, thin-walled structures.

*Stiffness* is the resistance to change of shape that is *elastic*, meaning that the material returns to its original shape when the stress is removed. *Strength* (Chapter 6) is its resistance to permanent distortion or total failure. Stress and strain are not material properties; they describe a stimulus and a response. Stiffness (measured by the elastic modulus $E$, defined in a moment) and strength (measured by the elastic limit $\sigma_y$ or tensile strength $\sigma_{ts}$) *are* material properties. Stiffness and strength are central to mechanical design, often in combination with the density, $\rho$. This chapter introduces stress and strain and the elastic moduli that relate them. These properties are neatly summarised in a *material property chart*—the modulus–density chart—the first of many that we shall explore in this book.

Density and elastic moduli reflect the mass of the atoms, the way they are packed in a material, and the stiffness of the bonds that hold them together. There is not much you can do to change any of these, so the density and moduli of pure materials cannot be manipulated at all. If you want to control these properties you can either mix materials together, making composites, or disperse space within them, making foams. Property charts are a good way to show how this works.

## **4.2** Density, stress, strain and moduli

*Density*   Many applications (e.g. sports equipment, transport systems) require low weight and this depends in part on the density of the materials of which they are made. Density is mass per unit volume. It is measured in $kg/m^3$ or sometimes, for convenience, $Mg/m^3$ ($1\ Mg/m^3 = 1000\ kg/m^3$).

The density of samples with regular shapes can be determined using a precision mass balance and accurate measurements of the dimensions (to give the volume), but this is not the best way. Better is the 'double weighing' method: the sample is first weighed in air and then when fully immersed in a liquid of known density. When immersed, the sample feels an upward force

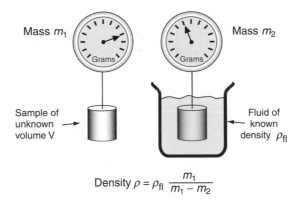

**Figure 4.1**  Measuring density by Archimedes' method

equal to the weight of liquid it displaces (Archimedes' principle[1]). The density is then calculated as shown in Figure 4.1.

*Modes of loading*  Most engineering components carry loads. Their elastic response depends on the way the loads are applied. As explained earlier, the components in both structures shown on the front page of this chapter are designed to withstand different modes of loading: tension, compression, bending, torsion and internal pressure. Usually one mode dominates, and the component can be idealised as one of the simply loaded cases in Figure 4.2—*tie, column, beam, shaft* or *shell*. Ties carry simple axial tension, shown in (a); columns do the same in simple compression, as in (b). Bending of a beam (c) creates simple axial tension in elements on one side the neutral axis (the center-line, for a beam with a symmetric cross-section) and simple compression in those on the other. Shafts carry twisting or torsion (d), which generates shear rather than axial load. Pressure difference applied to a shell, like the cylindrical tube shown in (e), generates bi-axial tension or compression.

*Stress*  Consider a force $F$ applied normal to the face of an element of material, as in Figure 4.3 on the left of row (a). The force is transmitted through the element and balanced by an equal but opposite force on the other side, so that it is in equilibrium (it does not move). Every plane normal to $F$ carries the force. If the area of such a plane is $A$, the *tensile stress* $\sigma$ in the element (neglecting its own weight) is

$$\sigma = \frac{F}{A} \tag{4.1}$$

---

[1]  Archimedes (287–212 BC), Greek mathematician, engineer, astronomer and philosopher, designer of war machines, the Archimedean screw for lifting water, evaluator of $\pi$ (as $3 + 1/7$) and conceiver, while taking a bath, of the principle that bears his name.

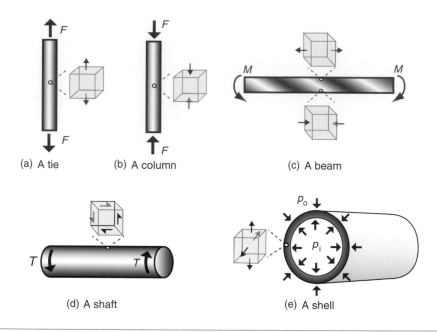

(a) A tie    (b) A column    (c) A beam

(d) A shaft    (e) A shell

Figure 4.2    Modes of loading and states of stress

If the sign of $F$ is reversed, the stress is compressive and given a negative sign. Forces[2] are measured in newtons (N), so stress has the dimensions of N/m$^2$. But a stress of 1 N/m$^2$ is tiny—atmospheric pressure is $10^5$ N/m$^2$—so the usual unit is MN/m$^2$ ($10^6$ N/m$^2$), called *megapascals*, symbol[3] MPa.

---

### Example 4.1

A brick chimney is 50 m tall. The bricks have a density of $\rho = 1800$ kg/m$^3$. What is the axial compressive stress at its base? Does the shape of the cross-section matter?

*Answer.* Let the cross-section area of the chimney be $A$ and its height be $h$. The weight of the chimney is $F = \rho A h g$, where $g$ is the acceleration due to gravity. Using equation (4.1), the axial compressive stress at the base is $\sigma = -F/A = -\rho g h$ where the '−' indicates compression (the area $A$ has cancelled out). This is the same formula as for the static pressure at depth $h$ in a fluid of density $\rho$. The stress is independent of the shape or size of the cross-section. Assuming $g = 10$ m/s$^2$, $\sigma = -1800 \times 10 \times 50 = -9 \times 10^5$ N/m$^2 = -0.9$ MPa.

---

[2] Isaac Newton (1642−1727), scientific genius and alchemist, formulator of the laws of motion, the inverse-square law of gravity (though there is some controversy about this), laws of optics, the differential calculus and much more.

[3] Blaise Pascal (1623−1662), philosopher, mathematician and scientist, who took a certain pleasure in publishing his results without explaining how he reached them. Almost all, however, proved to be correct.

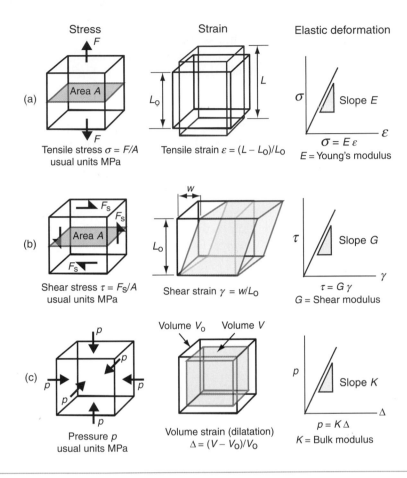

**Figure 4.3** The definitions of stress, strain and elastic moduli

If, instead, the force lies parallel to the face of the element, three other forces are needed to maintain equilibrium (Figure 4.3, row (b)). They create a state of shear in the element. The shaded plane, for instance, carries the *shear stress τ* of

$$\tau = \frac{F_s}{A} \tag{4.2}$$

The units, as before, are MPa.

One further state of multi-axial stress is useful in defining the elastic response of materials: that produced by applying equal tensile or compressive forces to all six faces of a cubic element, as in Figure 4.3, row (c). *Any* plane in the cube now carries the same state of stress—it is equal to the force on a cube face divided by its area. The state of stress is one of *hydrostatic pressure*, symbol *p*, again with the units of MPa. There is an unfortunate convention here. Pressures are positive when they push—the reverse of the convention for simple tension and compression.

Engineering components can have complex shapes and can be loaded in many ways, creating complex distributions of stress. But no matter how complex, the stresses in any small element within the component can always be described by a combination of tension, compression and shear. Commonly the simple cases of Figure 4.3 suffice, using super-position of two cases to capture, for example, bending plus compression.

*Strain*  Strain is the response of materials to stress (second column of Figure 4.3). A tensile stress $\sigma$ applied to an element causes the element to stretch. If the element in Figure 4.3(a), originally of side length $L_o$, stretches by $\delta L = L - L_o$, the nominal *tensile strain* is

$$\varepsilon = \frac{\delta L}{L_o} \tag{4.3}$$

A compressive stress shortens the element; the nominal compressive strain (negative) is defined in the same way. Since strain is the ratio of two lengths, it is dimensionless.

A shear stress causes a *shear strain* $\gamma$ (Figure 4.3(b)). If the element shears by a distance $w$, the shear strain

$$\tan(\gamma) = \frac{w}{L_o} = \gamma \tag{4.4}$$

In practice $\tan \gamma \approx \gamma$ because strains are almost always small. Finally, a hydrostatic pressure $p$ causes an element of volume $V$ to change in volume by $\delta V$. The volumetric strain, or *dilatation* (Figure 4.3(c)), is

$$\Delta = \frac{\delta V}{V} \tag{4.5}$$

*Stress–strain curves and moduli*  Figure 4.4 shows typical tensile stress–strain curves for a ceramic, a metal and a polymer. The initial part, up to the elastic limit $\sigma_{el}$, is approximately linear (Hooke's[4] law), and it is elastic, meaning that the strain is recoverable—the material returns to its original shape when the stress is removed. Stresses above the elastic limit cause permanent deformation (ductile behavior) or brittle fracture.

Within the linear elastic regime, strain is proportional to stress (Figure 4.3, third column). The tensile strain is proportional to the tensile stress:

$$\sigma = E\varepsilon \tag{4.6}$$

and the same is true in compression. The constant of proportionality, $E$, is called *Young's[5] modulus*. Similarly, the shear strain $\gamma$ is proportional to the shear stress $\tau$:

$$\tau = G\gamma \tag{4.7}$$

---

[4] Robert Hooke (1635–1703), able but miserable man, inventor of the microscope, and perhaps, too, of the idea of the inverse-square law of gravity. He didn't get along with Newton.

[5] Thomas Young (1773–1829), English scientist, expert on optics and deciphering ancient Egyptian hiero-glyphs (among them, the Rosetta stone). It seems a little unfair that the modulus carries his name, not that of Hooke.

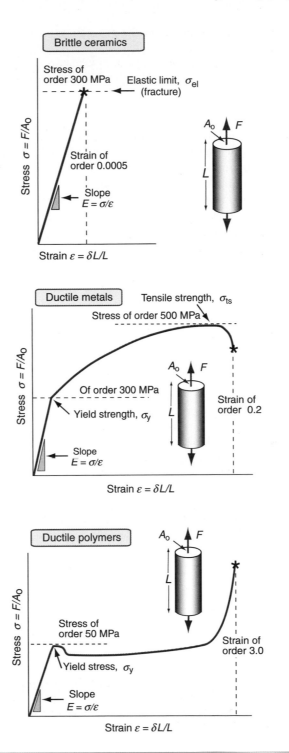

**Figure 4.4**   Tensile stress–strain curves for ceramics, metals and polymers

and the dilatation $\Delta$ is proportional to the pressure $p$:

$$p = k\Delta \tag{4.8}$$

where $G$ is the *shear modulus* and $K$ the *bulk modulus*, as illustrated in the third column of Figure 4.3. All three of these moduli have the same dimensions as stress, that of force per unit area ($N/m^2$ or Pa). As with stress it is convenient to use a larger unit, this time an even bigger one, that of $10^9$ $N/m^2$, *gigapascals*, or GPa.

Young's modulus, the shear modulus and the bulk modulus are related, but to relate them we need one more quantity, *Poisson's[6] ratio*. When stretched in one direction, the element of Figure 4.3(a) generally contracts in the other two directions, as it is shown doing here. Poisson's ratio, $\nu$, is the negative of the ratio of the lateral or transverse strain, $\varepsilon_t$, to the axial strain, $\varepsilon$, in tensile loading:

$$\gamma = -\frac{\varepsilon_t}{\varepsilon} \tag{4.9}$$

Since the transverse strain itself is negative, $\nu$ is positive—it is typically about 1/3.

---

### Example 4.2

The chimney in Example 4.1 has bricks with Young's modulus $E = 25$ GPa, and Poisson's ratio $\nu = 0.2$. What are the axial and transverse strains at the bottom of the chimney?

*Answer.* Using the result of Example 4.1 with equation (4.6), the axial strain is $\varepsilon = \sigma / E$ $= -9 \times 10^5/25 \times 10^9 = -3.6 \times 10^{-5}$ (often written as 36 microstrain, $\mu\varepsilon$). From equation (4.9), the transverse strain is $\varepsilon_t = -\nu\varepsilon = 7.2$ $\mu\varepsilon$. This time the strain is positive, indicating a small expansion at the base of the chimney.

---

In an isotropic material (one for which the moduli do not depend on the direction in which the load is applied) the moduli are related in the following ways:

$$G = \frac{E}{2(1 + v)}; \quad K = \frac{E}{3(1 - 2v)} \tag{4.10}$$

Commonly $v \approx 1/3$ so that

$$G = \frac{3}{8}E \text{ and } K = E \tag{4.11a}$$

---

[6] Siméon Denis Poisson (1781—1840), French mathematician, known both for his constant and his distribution. He was famously uncoordinated, failed geometry at university because he could not draw, and had to abandon experimentation because of the disasters resulting from his clumsiness.

Elastomers are exceptional. For these $\nu \approx 1/2$ when

$$G = \frac{1}{3}E \text{ and } K \gg E \tag{4.11b}$$

This means that rubber (an elastomer) is easy to stretch in tension (low $E$), but if constrained from changing shape, or loaded hydrostatically, it is very stiff (large $K$)—a feature for which designers of shoes have to allow.

Data sources like CES list values for all three moduli. In this book we examine data for $E$; approximate values for the others can be derived from equations (4.11a) and (4.11b) when needed.

*Hooke's Law in three dimensions*   The simple loading states in Figure 4.2 lead to stresses in one or two perpendicular directions. But, as we have now seen, even under uniaxial load the strain is inherently three-dimensional, thanks to Poisson's ratio. So it is helpful to relate stress and strain in a general way, for stresses that could in principle be acting in all three perpendicular directions at once. All of the simpler cases can then be derived from these relationships, by setting suitable stresses to zero.

Figure 4.5 shows our general cubic element of material, now subjected to three unequal stresses $\sigma_1$, $\sigma_2$ and $\sigma_3$, resulting in strains $\varepsilon_1$, $\varepsilon_2$ and $\varepsilon_3$. To relate stress to strain we use the principle of superposition—applying each stress in turn, finding the strains, and then summing these to find the overall strain when all three stresses act together. First applying a stress $\sigma_1$, the resulting strains (from equations (4.6) and (4.9)) are

$$\varepsilon_1 = \frac{\sigma_1}{E}, \quad \varepsilon_2 = \varepsilon_3 = -\nu\varepsilon_1 = -\frac{\nu\sigma_1}{E} \tag{4.12}$$

Repeating for stresses $\sigma_2$ and $\sigma_3$, and summing the strains gives us *Hooke's Law in three dimensions*:

$$\varepsilon_1 = \frac{1}{E}\left(\sigma_1 - \nu\sigma_2 - \nu\sigma_3\right)$$

$$\varepsilon_2 = \frac{1}{E}\left(-\nu\sigma_1 + \sigma_2 - \nu\sigma_3\right) \tag{4.13}$$

$$\varepsilon_3 = \frac{1}{E}\left(-\nu\sigma_1 - \nu\sigma_2 + \sigma_3\right)$$

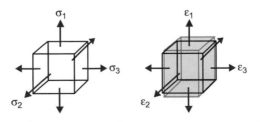

**Figure 4.5**   General 3-dimensional states of stress and strain

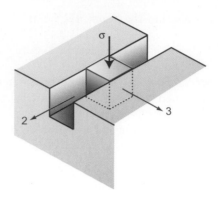

Figure 4.6 Compression of a block constrained from straining in one direction

These results are particularly helpful in design problems in which strain is constrained. Figure 4.6 shows a cube of material located in a slot of exactly the same size as the cube. A vertical downwards stress $\sigma_1 = -\sigma$ is applied to the top face of the cube. At first sight this looks like conventional uniaxial compression. But the cube wishes to expand laterally, due to Poisson's ratio. In the 2-direction it can do so, but perpendicular to the slot there is constraint from the surrounding material—if we assume this is much stiffer, then the strain in the 3-direction is effectively zero. What does Hooke's Law tell us about the stresses and strains?

For the 3-direction, we can write (noting that $\sigma_2 = 0$)

$$\varepsilon_3 = 0 = \frac{1}{E}(+\nu\sigma + \sigma_3), \quad \text{so} \quad \sigma_3 = -\nu\sigma$$

The effect of constraint is therefore to induce a stress $\nu\sigma_1$ in the constrained direction, perpendicular to the applied stress $\sigma_1$. But this stress will itself contribute a Poisson strain in the 1-direction, so the strain vertically is

$$\varepsilon_1 = \frac{1}{E}(\sigma_1 - \nu\sigma_3) = \frac{-\sigma}{E}\left(1 - \nu^2\right)$$

Now look at the ratio of stress to strain in the 1-direction:

$$\frac{\sigma_1}{\varepsilon_1} = \frac{E}{(1-\nu^2)} \tag{4.14}$$

So the constrained cube behaves like a material with an 'effective modulus' which is greater than $E$ (since the factor $(1 - \nu^2)$ is less than unity).

### Example 4.3

By what factor is the modulus effectively increased for the loading geometry of Figure 4.6, for a cube of (a) foam ($\nu \approx 0$); (b) metal ($\nu \approx 0.33$); (c) rubber ($\nu \approx 0.5$)?

*Answer.* From equation (4.14) the factor is $1/(1 - v^2)$. Hence (a) for foam, factor $= 1$ (no Poisson expansion, so no induced stress: the loading is purely uniaxial compression); (b) for metal, 1.12; (c) for rubber, 1.33.

The effect is even more marked when there is constraint in both 2- and 3-directions (see Exercises at the end of the chapter), and the effect is most significant for values of Poisson's ratio near 0.5. As noted earlier, the solid material with the lowest modulus—rubber—displays the most significant stiffening effect when it is loaded in a geometry that imposes constraint.

*Elastic energy*  If you stretch an elastic band, energy is stored in it. The energy can be considerable: catapults can kill people. The super-weapon of the Roman arsenal at one time was a wind-up mechanism that stored enough elastic energy to hurl a 10 kg stone projectile 100 yards or more.

How do you calculate this energy? A force $F$ acting through a displacement d$L$ does work $F$ d$L$. A stress $\sigma = F/A$ acting through a strain increment d$\varepsilon = $ d$L/L$ does work per unit volume

$$dW = \frac{FdL}{AL} = \sigma d\varepsilon \tag{4.15}$$

with units of J/m$^3$. If the stress is acting on an elastic material, this work is stored as elastic energy. The elastic part of all three stress–strain curves of Figure 4.4—the part of the curve before the elastic limit—is linear; in it $\sigma = E\varepsilon$. The work done per unit volume as the stress is raised from zero to a final value $\sigma^*$ is the area under the stress–strain curve:

$$W = \int_0^{\sigma^*} \sigma d\varepsilon = \int_0^{\sigma^*} \frac{\sigma d\sigma}{E} = \frac{1}{2}\frac{(\sigma^*)^2}{E} \tag{4.16}$$

This is the energy that is stored, per unit volume, in an elastically strained material. The energy is released when the stress is relaxed.

## Example 4.4

A steel rod has length $L_o = 10$ m and a diameter $d = 10$ mm. The steel has Young's modulus $E = 200$ GPa, elastic limit (the highest stress at which it is still elastic and has not yielded) $\sigma_y = 500$ MPa and density $\rho = 7800$ kg/m$^3$. Calculate the force in the rod, its extension and the elastic energy per unit volume it stores when it is stretched so that the stress in it just reaches the elastic limit. Compare the elastic energy per unit mass with the chemical energy stored in gasoline, which has a heat of combustion (calorific value) of 43 000 kJ/kg.

*Answer.* The cross-section area of the rod is $A = \pi d^2/4 = 7.85 \times 10^{-5}$ m$^2$.
At the elastic limit, the force in the rod is (4.1): $F = \sigma_y A = 39.2$ kN (3.92 tonnes).
The corresponding strain is (4.6): $\varepsilon = \sigma/E = 500 \times 10^6/200 \times 10^9 = 0.0025$ (i.e. 0.25%).
The extension of the rod is (4.3): $\delta L = \varepsilon L_o = 0.0025 \times 10 = 25$ mm.
The elastic energy per unit volume, from equation (4.13), is $W = \sigma^2/2E = (500 \times 10^6)^2/(2 \times 200 \times 10^9) = 625$ kJ/m$^3$, giving an elastic energy per kg of $W/\rho = 80$ J/kg. This is less, by a factor of more than 500 000 than that of gasoline.

*Measurement of Young's modulus*  You might think that the way to measure the elastic modulus of a material would be to apply a small stress (to be sure to remain in the linear elastic region of the stress–strain curve), measure the strain and divide one by the other. In reality, moduli measured as slopes of stress–strain curves are inaccurate, often by a factor of 2 or more, because of contributions to the strain from material creep or deflection of the test machine. Accurate moduli are measured dynamically: by measuring the frequency of natural vibrations of a beam or wire, or by measuring the velocity of sound waves in the material. Both depend on $\sqrt{E/\rho}$, so if you know the density $\rho$ you can calculate $E$. We return to sound transmission in materials at the end of this chapter.

*Stress-free strain*  Stress is not the only stimulus that causes strain. Certain materials respond to a magnetic field by undergoing strain—an effect known as magneto-striction. Others respond to an electrostatic field in the same way—they are known as piezo-electric materials. In each case a material property relates the magnitude of the strain to the intensity of the stimulus (Figure 4.7). The strains are small but can be controlled with great accuracy and, in

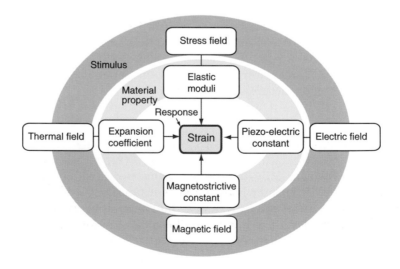

Figure 4.7   Stimuli leading to strain

the case of magneto-striction and piezo-electric strain, can be changed with a very high frequency. This is exploited in precision positioning devices, acoustic generators and sensors—applications we return to in Chapters 14 and 15.

A more familiar effect is that of thermal expansion: strain caused by change of temperature. The thermal strain $\varepsilon_T$ is linearly related to the temperature change $\Delta T$ by the expansion coefficient, $\alpha$:

$$\varepsilon_T = \alpha \Delta T \tag{4.17}$$

where the subscript $T$ is a reminder that the strain is caused by temperature change, not stress.

The term 'stress-free strain' is a little misleading. It correctly conveys the idea that the strain is not *caused* by stress but by something else. But these strains can nonetheless give rise to stresses if the body suffering the strain is constrained. Thermal stress—stress arising from thermal expansion—particularly, can be a problem, causing mechanisms to jam and railway tracks to buckle. We analyse it in Chapter 12.

## 4.3 The big picture: material property charts

We met the idea of material property charts in Section 2.5. Now is the time to use them. If we want materials that are stiff and light, we first need an overview of what's available. What moduli do materials offer? What are their densities? The modulus–density chart shows them.

*The modulus–density chart*    Figure 4.8 shows that the modulus $E$ of engineering materials spans seven decades[7], from 0.0001 to nearly 1000 GPa; the density $\rho$ spans a factor of 2000, from less than 0.01 to 20 Mg/m$^3$. The members of the ceramics and metals families have high moduli and densities; none has a modulus less than 10 GPa or a density less than 1.7 Mg/m$^3$. Polymers, by contrast, all have moduli below 10 GPa and densities that are lower than those of any metal or ceramic—most are close to 1 Mg/m$^3$. Elastomers have roughly the same density as other polymers but their moduli are lower by a further factor of 100 or more. Materials with a lower density than polymers are porous: man-made foams and natural cellular structures like wood and cork.

This property chart gives an overview, showing where families and their members lie in $E-\rho$ space. It helps in the common problem of material selection for stiffness-limited applications in which weight must be minimised. Chapter 5 provides more on this.

*The modulus–relative cost chart*    Often it is minimising cost, not weight, that is the overriding objective of a design. The chart of Figure 4.9 shows, on the $x$-axis, the relative prices per unit volume of materials, normalised to that of the metal used in larger quantities than any other: mild steel. Concrete and wood are among the cheapest; polymers, steels and aluminum alloys come next; special metals like titanium, most technical ceramics and a few polymers like PTFE and PEEK are expensive. The chart allows the selection of materials that are stiff and cheap. Chapter 5 gives examples.

---

[7] Very low density foams and gels (which can be thought of as molecular-scale, fluid-filled, foams) can have lower moduli than this. As an example, gelatine (as in Jell-O) has a modulus of about $10^{-5}$ GPa.

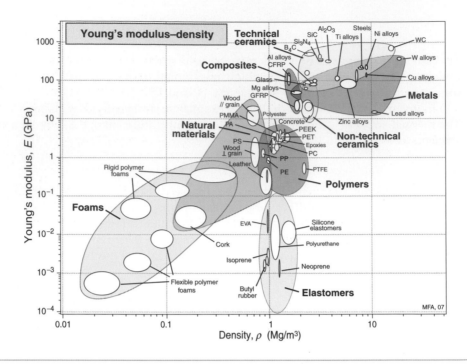

**Figure 4.8** The modulus–density chart

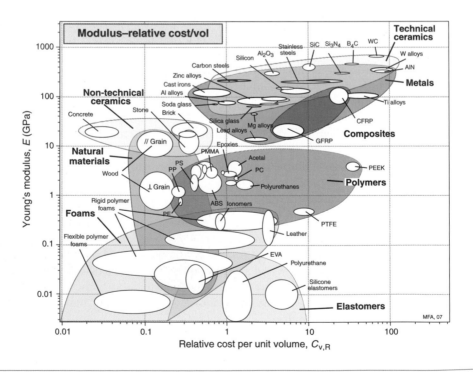

**Figure 4.9** The modulus–relative cost chart. (The CES software contains material prices, regularly updated)

*Anisotropy* Glasses and most polymers have disordered structures with no particular directionality about the way the atoms are arranged. They have properties that are *isotropic*, meaning the same no matter which direction they are measured. Most materials are crystalline—made up of ordered arrays of atoms. Metals and ceramics are usually polycrystalline—made up of many tiny, randomly oriented crystals. This averages out the directionality in properties, so a single value is enough. Occasionally, though, anisotropy is important. Single crystals, drawn polymers and fibres are *anisotropic*; their properties depend on the direction in the material in which they are measured. Woods, for instance, are much stiffer along the grain than across it. Figures 4.8 and 4.9 have separate property bubbles for each of the two loading directions. Fibre composites are yet more extreme: the modulus parallel to the fibres can be larger by a factor of 20 than that perpendicular to them. Anisotropy must therefore be considered when wood and composite materials are selected.

## **4.4** The science: what determines stiffness and density?

*Introduction to Guided Learning Unit 1: Simple ideas of crystallography* Many of the properties of materials depend directly on the way the atoms or molecules within them are packed. In *glasses* the arrangement is a disordered one with no regularity or alignment, and *polymers* are largely made up of tangled long-chain molecules. *Metals* and *ceramics*, however, are *crystalline*, with a regularly repeating pattern of atomic packing in structural units. Crystals are described using the language and methods of *crystallography*, which provides an elegant framework for understanding their three-dimensional geometry. The key ideas of crystallography are introduced briefly in the following pages. Later in the book you will find *Guided Learning Unit 1:* Simple ideas of crystallography, which develops the ideas in greater depth. The rest of this chapter will be intelligible without reference to the unit, but working through it and its exercises will give a more complete understanding and confidence.

*Atom packing in metals and the unit cell* Atoms often behave as if they were hard, spherical balls. The balls on a pool table, when set, are arranged as a close-packed layer, as shown in Figure 4.10(a). The atoms of many metals form extensive layers packed in this way. There is no way to pack the atoms more closely than this, so this particular arrangement is called

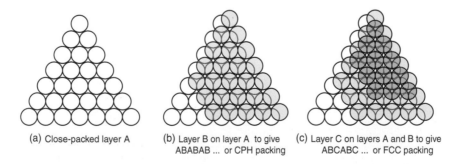

(a) Close-packed layer A     (b) Layer B on layer A to give ABABAB ... or CPH packing     (c) Layer C on layers A and B to give ABCABC ... or FCC packing

Figure 4.10 (a) A close-packed layer of spheres, layer A; atoms often behave as if hard and spherical. (b) A second layer, B, nesting in the first; repeating this sequence gives ABAB ... or CPH stacking. (c) A third layer, C, can be nested so that it does not lie above A or B; if repeated this gives ABCABC ... or FCC stacking

'close-packed'. These two-dimensional layers of atoms can then be close-packed in three dimensions. Surprisingly, there are two ways to do this. Where three atoms meet in the first layer, layer A, there are natural depressions for atoms of a second close-packed layer B. A third layer can be added such that its atoms are exactly above those in the first layer, so that it, too, is in the A orientation, and the sequence is repeated to give a crystal with ABABAB … stacking, as in Figure 4.10(b); it is called *close-packed hexagonal,* or *CPH* (sometimes *HCP*) for short, for reasons explained in a moment. But there is an alternative stacking sequence. In placing the third layer, layer C, there are two choices of position—one aligned with A (as before), but another offset from A. If the third layer, C, is nested onto B so that it lies in this alternative position, the stacking becomes (on repeating) ABCABCABC … as shown in Figure 4.10(c); it is called *face-centred cubic* (FCC for short). Many metals, such as copper, silver, aluminum and nickel, have the FCC structure; many others, such as magnesium, zinc, and titanium have the CPH structure. The two alternative structures have exactly the same packing fraction, 0.74, meaning that the spheres occupy 74% of the available space (as calculated in Guided Learning Unit 1). But the small difference in layout influences properties, particularly those having to do with plastic deformation (Chapter 6).

Not all structures are close-packed. Figure 4.11 shows one of these, made by stacking square-packed layers with a lower packing density than the hexagonal layers of the FCC and CPH structures. An ABABAB … stacking of these layers builds the *body-centred cubic* structure, *BCC* for short, with a packing fraction of 0.68. Iron and most steels have this structure. There are many other crystal structures, but for now these three are enough, covering most important metals.

Any regular packing of atoms that repeats itself is called a *crystal*. It is possible to pack atoms in a non-crystallographic way to give what is called an *amorphous* structure, sketched in Figure 4.12. This is not such an efficient way to fill space with spheres: the packing fraction is 0.64 at best.

The characterising unit of a crystal structure is called its *unit cell*. Figure 4.13 shows three; the red lines define the cell (the atoms have been shrunk to reveal it more clearly; in reality they are touching in close-packed directions). In the first, shown at Figure 4.13(a), the cell is a hexagonal prism. The atoms in the top, bottom and central planes form close-packed layers like that of Figure 4.10(b), with ABAB … stacking—this explains why this structure is called

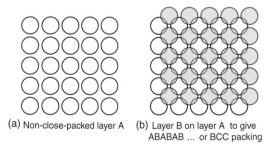

(a) Non-close-packed layer A     (b) Layer B on layer A to give ABABAB … or BCC packing

**Figure 4.11**   (a) A square grid of spheres; it is a less efficient packing than that of the previous figure. (b) A second layer, B, nesting in the first, A; repeating this sequence gives ABAB … packing. If the sphere spacing is adjusted so that the blue spheres lie on the corners of a cube, the result is the non-close-packed BCC structure

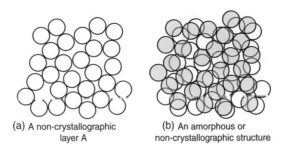

(a) A non-crystallographic
layer A

(b) An amorphous or
non-crystallographic structure

Figure 4.12  (a) An irregular arrangement of spheres. (b) Extending this in three dimensions gives a random or amorphous structure

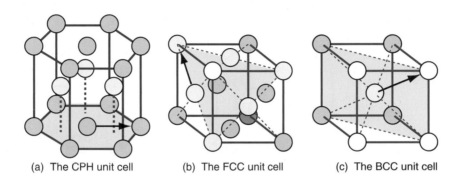

(a) The CPH unit cell          (b) The FCC unit cell          (c) The BCC unit cell

Figure 4.13  Unit cells. All the atoms are of the same type, but are shaded differently to emphasise their positions. (a) The close-packed hexagonal (CPH) structure. (b) The close-packed face-centred cubic (FCC) structure. (c) The non-close-packed body-centred cubic (BCC) structure. Arrows show nearest neighbors

close-packed hexagonal (CPH). The second, shown at Figure 4.13(b), is also made up of close-packed layers, though this is harder to see: the shaded triangular plane is one of them. If we think of this as an A plane, atoms in the plane above it nest in the B position, and those in the plane above that, in the C position, giving ABCABC … stacking, as in Figure 4.10(c). The unit cell itself is a cube with an atom at each corner and one at the center of each face—for this reason it is called face-centred cubic (FCC). The final cell, shown at Figure 4.13(c), is the characterising unit of the square-layer structure of Figure 4.11; it is a cube with an atom at each corner and one in the middle, and it is called, appropriately, body-centred cubic (BCC).

Unit cells pack to fill space as in Figure 4.14: the resulting array is called the *crystal lattice*; the points at which cell edges meet are called *lattice points*. The crystal itself is generated by attaching one or a group of atoms to each lattice point so that they form a regular, three-dimensional, repeating pattern. The cubic and hexagonal cells are among the simplest; there are many others with edges of differing lengths meeting at differing angles. The one thing they have in common is their ability to stack with identical cells to completely fill space.

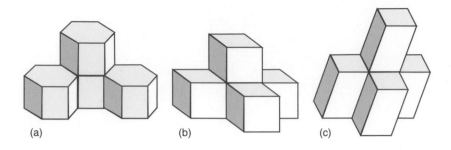

**Figure 4.14** Unit cells stacked to fill space. (a) The hexagonal cell. (b) The cubic cell. (c) A cell with edges of differing length that do not meet at right angles

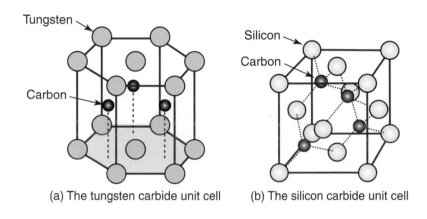

(a) The tungsten carbide unit cell     (b) The silicon carbide unit cell

**Figure 4.15** Unit cells of compounds. (a) Tungsten carbide. (b) One form of silicon carbide

***Atom packing in ceramics*** Most ceramics are compounds, made up of two or more atom types. They, too, have characteristic unit cells. Figure 4.15 shows those of two materials that appear on the property charts: tungsten carbide (WC), and silicon carbide (SiC). The cell of the first is hexagonal, that of the second is cubic, but now a pair of different atoms is associated with each lattice point: a W-C pair in the first structure, and a Si-C pair in the second. More examples are investigated in Guided Learning Unit 1.

***Atom packing in glasses*** The crystalline state is the lowest energy state for elements and compounds. Melting disrupts the crystallinity, scrambling the atoms and destroying the regular order. The atoms in a molten metal look very like the amorphous structure of Figure 4.12. On cooling through the melting point most metals crystallize easily, though by cooling them exceedingly quickly it is sometimes possible to trap the molten structure to give an amorphous metallic 'glass'. With compounds it is easier to do this, and with one in particular, silica ($SiO_2$), crystallization is so sluggish that its usual state is the amorphous one. Figure 4.16 shows, on the left, the atom arrangement in crystalline silica: identical hexagonal Si—O rings, regularly

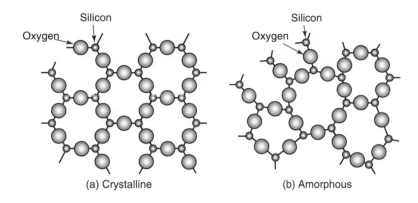

Figure 4.16 Two alternative structures for silica, the basis of most glasses. (a) Crystalline silica. (b) Glassy or amorphous silica

arranged. On the right is the more usual amorphous state. Now some rings have 7 sides, some have 6, some 5, and there is no order—the next ring could be any one of these. Amorphous silica is the basis of almost all glasses; it is mixed with $Na_2O$ to make soda glass (windows, bottles) and with $B_2O_5$ to make borosilicate glasses (Pyrex), but it is the silica that gives the structure. It is for this reason than the structure itself is called 'glassy', a term used interchangeably with 'amorphous'.

*Atom packing in polymers* Polymer structures are quite different. The backbone of a 'high' polymer ('high' means high molecular weight) is a long chain of carbon atoms, to which side groups are attached. Figure 4.17 shows a segment of the simplest: polyethylene (PE) $(-CH_2-)_n$. The chains have ends; the ends of this one are capped with a $CH_3$ group. PE is made by the polymerisation (snapping together) of ethylene molecules, $CH_2 = CH_2$, where the = sign is a double bond, broken by polymerisation to give covalent links to more carbon neighbours to the left and right. Figure 4.18 shows the chain structure of five of the most widely used linear polymers.

Polymer molecules bond together to form solids. The chains of a linear polymer, $10^3$ to $10^6$ $(-CH_2-)$ units in length, are already strongly, covalently bonded along the chain itself.

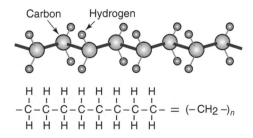

Figure 4.17 Polymer chains have a carbon–carbon backbone with hydrogen or other side groups. The figure shows three alternative representations of the polyethylene molecule

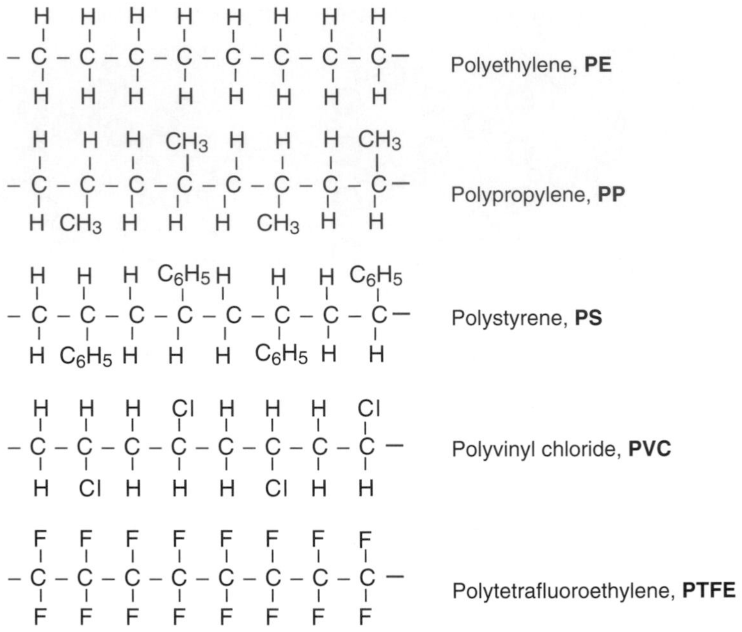

H   H   H   H   H   H   H   H
|   |   |   |   |   |   |   |
– C – C – C – C – C – C – C – C —        Polyethylene, **PE**
|   |   |   |   |   |   |   |
H   H   H   H   H   H   H   H

H   H   H   CH₃  H   H   H   CH₃
|   |   |   |   |   |   |   |
– C – C – C – C – C – C – C – C —        Polypropylene, **PP**
|   |   |   |   |   |   |   |
H   CH₃  H   H   H   CH₃  H   H

H   H   H   C₆H₅ H   H   H   C₆H₅
|   |   |   |   |   |   |   |
– C – C – C – C – C – C – C – C —        Polystyrene, **PS**
|   |   |   |   |   |   |   |
H   C₆H₅ H   H   H   C₆H₅ H   H

H   H   H   Cl  H   H   H   Cl
|   |   |   |   |   |   |   |
– C – C – C – C – C – C – C – C —        Polyvinyl chloride, **PVC**
|   |   |   |   |   |   |   |
H   Cl  H   H   H   Cl  H   H

F   F   F   F   F   F   F   F
|   |   |   |   |   |   |   |
– C – C – C – C – C – C – C – C —        Polytetrafluoroethylene, **PTFE**
|   |   |   |   |   |   |   |
F   F   F   F   F   F   F   F

**Figure 4.18**   Five common polymers, showing the chemical make-up. The strong carbon–carbon bonds are shown in red

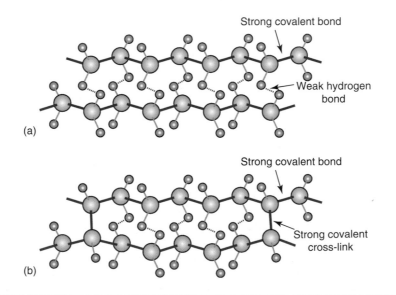

**Figure 4.19**   (a) Polymer chains have strong covalent 'backbones', but bond to each other only with weak hydrogen bonds unless they become cross-linked. (b) Cross-links bond the chains tightly together. The strong carbon–carbon bonds are shown as solid red lines

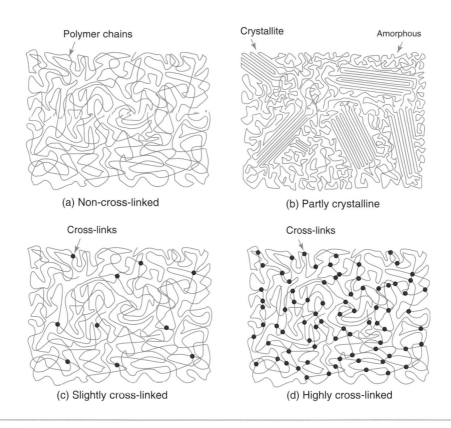

Figure 4.20 (a) Chains in polymers like polypropylene form spaghetti-like tangles with no regular repeating pattern—that structure is amorphous or 'glassy'. (b) Some polymers have the ability to form regions in which the chains line up and register, giving crystalline patches. The sketch shows a partly crystalline polymer structure. (c) Elastomers have occasional cross-links between chains, but these are far apart, allowing the chains between them to stretch. (d) Heavily cross-linked polymers like epoxy inhibit chain sliding

Separated chains attract each other, but only weakly (Figure 4.19(a))—they are sticky, but not very sticky, as the weak 'hydrogen' bonds that make them stick are easily broken or rearranged. The resulting structure is sketched in Figure 4.20(a): a dense spaghetti-like tangle of molecules with no order or crystallinity; it is amorphous. This is the structure of *thermoplastics* like those of Figure 4.18; the weak bonds melt easily, allowing the polymer to be moulded, retaining its new shape on cooling.

The weak bonds of thermoplastics do, however, try to keep the bond lengths short by lining the molecules up, as they are shown in Figure 4.19(a). The molecules are very long and their shape somewhat irregular and bumpy so that total alignment is not possible, but where segments of molecules manage it there are small crystalline regions, as in Figure 4.20(b). These *crystallites*, as they are called, are small—often between 1 and 10 microns across, just the right size to scatter light. So amorphous polymers with no crystallites can be transparent: polycarbonate (PC), polymethyl-methacrylate (PMMA, Plexiglas) and polystyrene (PS) are examples. Those that are *semi-crystalline*, like polyethylene (PE) and nylon (PA), scatter light and are translucent.

The real change comes when chains are cross-linked by replacing some of the weak hydrogen bonds by much more muscular covalent C-C bonds, as in Figure 4.19(b), making the whole array into one huge multiply-connected network. *Elastomers* (rubbery polymers) have relatively few cross-links, as in Figure 4.20(c). *Thermosets* like epoxies and phenolics have many cross-links, as in Figure 4.20(d), making them stiffer and stronger than thermoplastics. The cross-links are not broken by heating, so once the links have formed, elastomers and thermosets cannot be thermally moulded or (for that reason) recycled.

*Cohesive energy and elastic moduli*   Atoms bond together, some weakly, some strongly. The *cohesive energy* measures the strength of this bonding. It is defined as the energy per mol (a *mol* is $6.022 \times 10^{23}$ atoms) required to separate the atoms of a solid completely, giving neutral atoms at infinity. Equally it is the energy released if the neutral, widely spaced atoms are brought together to form the solid.

The greater the cohesive energy, the stronger are the bonds between the atoms and the higher is the modulus. Think of the bonds as little springs (Figure 4.21). The atoms have equilibrium spacing $a_0$; a force $F$ pulls them apart a little, to $a_0 + \delta$, but when it is released they pull back to their original spacing. The same happens in compression, because the potential energy of the bond increases no matter in which direction the force is applied, as the curve in the figure suggests. The bond energy is a minimum at the equilibrium spacing. A spring that stretches by $\delta$ under a force $F$ has a stiffness, $S$, of

$$S = \frac{F}{\delta} \tag{4.18}$$

and this is the same in compression as in tension.

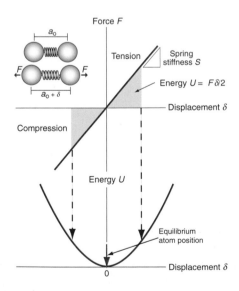

**Figure 4.21**   Atoms in solids are linked by atomic bonds that behave like springs. The bond stiffness is $S = F/\delta$. Stretching or compressing the bond by a displacement $\delta$ stores energy $U = F\delta/2$. The equilibrium (no force) atom separation is at the bottom of the energy well

How is the modulus related to the bond stiffness? When a force $F$ is applied to a pair of atoms, they stretch apart by $\delta$. A force $F$ applied to an atom of diameter $a_o$ corresponds to a stress $\sigma = F/a_o^2$, assuming each atom occupies a cube of side $a_o$. A stretch of $\delta$ between two atoms separated by a distance $a_o$ corresponds to a strain $\varepsilon = \delta/a_o$. Substituting these into equation (4.18) gives

$$\sigma = \frac{S}{a_o}\,\varepsilon \tag{4.19}$$

Comparing this with equation (4.6) reveals that Young's modulus, $E$, is approximately

$$E = \frac{S}{a_o} \tag{4.20}$$

Note that the atom-scale force-displacement response is linear-elastic around the equilibrium atomic spacing (Figure 4.21), and this is directly responsible for the linear elasticity of the bulk material, characterised by $E$.

Table 4.1 lists the stiffnesses of the different bond types, and corresponding ranges for Young's modulus, $E$. The covalent bond is particularly stiff ($S = 20-200$ N/m; diamond has a very high modulus because the carbon atom is small, giving a high bond density) and its atoms are linked by very strong springs ($S = 200$ N/m). For most engineering metals, the metallic bond is somewhat less stiff ($S = 8-60$ N/m), but metal atoms are often close-packed, giving them high moduli, though not as high as that of diamond. Ionic bonds, found in many ceramics, have stiffnesses comparable with those of metals, giving them high moduli too. Polymers contain both strong, diamond-like covalent bonds along the polymer chain and weak hydrogen or Van der Waals bonds ($S = 0.5-5$ N/m) between the chains; it is the weak bonds that stretch when the polymer is deformed, giving them low moduli. Only if the polymer is first drawn out to a very large strain can the molecules be sufficiently aligned for the covalent bonds to be stretched, giving a much higher modulus (see the right-hand end of the polymer stress-strain curve of Figure 4.4). This is exploited in the production of polymer fibres, giving properties that are far superior to a bulk polymer. This will be discussed further in Chapter 19.

Table 4.1   Bond stiffnesses, S

| Bond type | Examples | Bond stiffness S (N/m) | Young's modulus E (GPa) |
|---|---|---|---|
| Covalent | Carbon–carbon bond | 20–200 | 100–1000 |
| Metallic | Engineering metals | 8–60 | 20–210 |
| Ionic | Sodium chloride | 4–100 | 30–400 |
| Hydrogen bond / Van der Waals | Polyethylene | 0.5–5 | 0.5–5 |

Table 4.1 enables us to estimate the lower limit for Young's modulus for a true solid. The largest atoms ($a_o = 4 \times 10^{-10}$ m) bonded with the weakest bonds ($S = 0.5$ N/m) will have a modulus of roughly

$$E = \frac{0.5}{4 \times 10^{-10}} \approx 1 \; GPa \tag{4.21}$$

Many polymers do have moduli of about this value. But as the $E/\rho$ chart (Figure 4.8) shows, materials exist that have moduli that are much lower than this limit. They are either *foams* or *elastomers*. Foams have low moduli because the cell walls bend easily (allowing large displacements) when the material is loaded (Section 4.5). The origin of the moduli of elastomers takes a little more explaining.

***The elastic moduli of elastomers*** An elastomer is a tangle of long-chain molecules with occasional cross-links, as on the left of Figure 4.22. The bonds between the molecules, apart from the cross-links, are weak—so weak that, at room temperature, they have melted (which will be discussed further in the next section). Segments are free to slide over each other, and were it not for the cross-links, the material would have no stiffness at all; it would be a viscous liquid.

The tangled structure of Figure 4.22(a) has high randomness, or as expressed in the language of thermodynamics, its *entropy* is high. Stretching it, as in Figure 4.22(b), aligns the molecules: some parts of it now begin to resemble the crystallites of Figure 4.20(b). Crystals are ordered, the opposite of randomness; their entropy is low. So here there is a resistance to stretching—a stiffness—that has nothing to do with bond-stretching but with strain-induced molecular ordering. The cross-links give the elastomer a 'memory' of the disordered shape it had to start with, so very

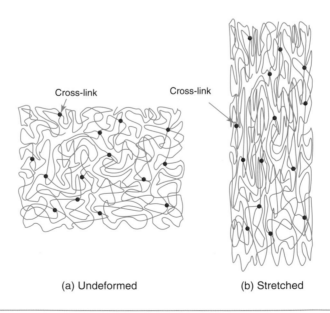

(a) Undeformed        (b) Stretched

**Figure 4.22** The stretching of an elastomer. Here the structure has been stretched to twice its original length. The stretching causes alignment, producing crystal-like regions. Thermal vibration drives the structure back to the one on the left, restoring its shape

large strains are elastic and fully recoverable. A full theory is complicated—it involves the statistical mechanics of long-chain tangles—so it is not easy to calculate the value of the modulus. The main thing to know is that the moduli of elastomers are low because they have this strange origin. There is a further odd effect. Higher temperature favours randomness—it is why crystals melt into disordered fluids at their melting point, and evaporate into even more random gases at the boiling point. As the modulus of elastomers is controlled by the ordering of their random structure, the modulus increases with temperature (because of the increasing tendency to randomness). This is in complete contrast to the behaviour of all other solids, whose moduli decrease on heating (because of thermal expansion), something we will revisit in Chapter 13.

***Temperature-dependence of polymer moduli: the glass transition temperature***   Crystalline solids have well-defined melting points, governed by the thermodynamics of the bonding energy as temperature rises. Polymer bonding is weaker and more diffuse, with the hydrogen bonding spread over a range of atomic spacing because of the tangled molecular structure. The temperature at which the weak inter-chain bonds start to melt is called the *glass transition temperature*, $T_g$. This affects the amorphous regions, but in semi-crystalline thermoplastics the crystallites melt at a higher temperature, typically $1.5 \times T_g$, due to the closer packing of the chains. Elastomers and thermosets have a glass transition, but do not melt on heating—because of the cross-linking, they decompose and burn instead.

How does the glass transition affect the Young's modulus? The answer is dependent on the type of polymer, since even if the weak bonding has gone, there can still be elastic behaviour above $T_g$. Figure 4.23 shows schematics of the variation of modulus with temperature. Below $T_g$, polymers are referred to as glassy, with the modulus controlled by the hydrogen bonding.

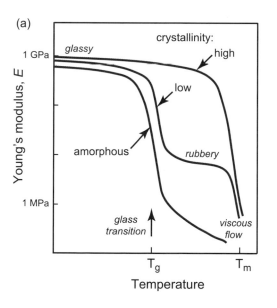
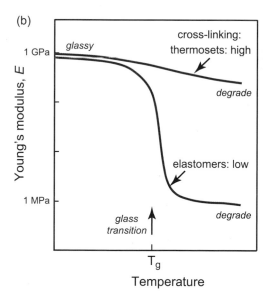

Figure 4.23   Schematic temperature-dependence of Young's modulus for: (a) thermoplastics (amorphous and semi-crystalline); (b) elastomers and thermosets. The glass transition temperature $T_g$ marks the position of the main variation in stiffness

Above $T_g$, amorphous thermoplastics (left-hand figure) show a tiny residual elasticity that stems from entanglements in the molecular structure, blurring into viscous flow. Semi-crystalline thermoplastics show a significant drop in modulus across the glass transition, of the order of 100 to 1000 times, but with a plateau in modulus that is sometimes referred to, somewhat misleadingly, as 'rubbery' behaviour. The degree of crystallinity determines how stiff the material remains above $T_g$, but eventually all thermoplastics melt and flow. Cross-linked behaviour is shown in Figure 4.23(b). Elastomers also display a dramatic drop in modulus by a factor of order 1000: the very low modulus of rubbers reflects the fact that at room temperature the material is already above its glass transition. The effect is well-illustrated by immersing a rubber tube in liquid nitrogen! Thermosets have a much higher degree of cross-linking, giving a range of stiffness above $T_g$ depending on the cross-link density: heavily cross-linked thermosets barely show a glass transition at all.

Polymer deformation shows other unusual characteristics. Since stretching the material can be accompanied by some sliding of the molecules past one another, the stiffness is particularly sensitive to the rate of loading. Rapid loading does not give time for chain sliding; slow loading enables it, giving a quite different response. The glass transition temperature is therefore sensitive to the deformation rate. So at the same temperature, a polymer can switch from being floppy and resistant to fracture, to being much stiffer and glass-like, by pulling on it rapidly. Design to cope with the effect of temperature on polymers is discussed further in Chapter 13.

*Origin of density*    Atoms differ greatly in weight but little in size. Among solids, the heaviest stable atom, uranium (atomic weight 238), is about 35 times heavier than the lightest, lithium (atomic weight 6.9), yet when packed to form solids their diameters are almost exactly the same (0.32 nm). The largest atom, cesium, is only 2.5 times larger than the smallest, beryllium. Thus the density is mainly determined by the atomic weight and only to a lesser degree by the atom size and the way in which they are packed. Metals are dense because they are made of heavy atoms, packed densely together (iron, for instance, has an atomic weight of 56). Ceramics, for the most part, have lower densities than metals because they contain light Si, O, N or C atoms. In Guided Learning Unit 1, we show that for metals and ceramics, a knowledge of the unit cell packing and atomic weights allows the bulk material density to be calculated directly from the atomic scale. Polymers have low densities because they are largely made of light carbon (atomic weight: 12) and hydrogen (atomic weight: 1) in low-density amorphous or semi-crystalline packings. The lightest atoms, packed in the most open way, give solids with a density of around 1 $Mg/m^3$—the same as that of water (see Figure 4.8). Materials with lower densities than this are *foams* (and woods), made up of cells containing a large fraction of pore space.

*Density and modulus of metallic alloys*    Most metallic alloys are not pure but contain two or more different elements. Often they dissolve in each other, like sugar in tea, but as the material is solid we call it a *solid solution*: examples are brass (a solution of Zn in Cu) and stainless steel (a solution of Ni and Cr in Fe).

As we shall see later, some material properties are changed a great deal by making solid solutions; modulus and density are not. As a general rule the density $\tilde{\rho}$ of a solid solution lies between the densities $\rho_A$ and $\rho_B$ of the materials that make it up, following a *rule of mixtures* (an arithmetic mean, weighted by volume fraction) known, in this instance, as *Vegard's law*:

$$\tilde{\rho} = f\rho_A + (1 - f)\,\rho_B \qquad\qquad (4.22)$$

where $f$ is the fraction of A atoms. Modulus is a bit more complicated: atoms in pure materials have only one type of neighbour to which they are bonded, A—A say; mixtures of A and B atoms have three, A—A, B—B and A—B. But mixtures of metallic bonds generally average out to stiffness values between those of the pure elements.

## **4.5** Manipulating the modulus and density

The underlying science of atomic packing and bonding shows that changing the composition offers little scope for manipulating the modulus and density of a given class of metallic alloys (steels, Al alloys and so on). Bulk polymers are more varied in their stiffness response, via changes in the degree of crystallinity or cross-linking. And very substantial changes in modulus can be achieved by aligning the polymer molecules in a fibre (by cold drawing, for example). But to produce new combinations of modulus and density across the whole range of materials, mixtures must be made at a more macroscopic scale, to make a *hybrid* material. For example, we can mix two discrete solids together to make *composites*, or we can mix in some space to make *foams*. The effects on modulus and density are nicely illustrated using property charts.

*Composites*  Composites are made by embedding fibres or particles in a continuous matrix of a polymer (polymer matrix composites, PMCs), a metal (MMCs) or a ceramic (CMCs), as in Figure 4.24. The development of high-performance composites is one of the great material developments of the last 40 years, now reaching maturity. Composites have high stiffness and strength per unit weight, and—in the case of MMCs and CMCs—high-temperature performance. Here, then, we are interested in stiffness and weight.

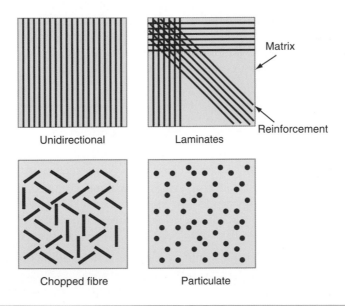

Figure 4.24  Manipulating the modulus by making composites, mixing stiff fibres or particles into a less-stiff matrix

When a volume fraction $f$ of a reinforcement $r$ (density $\rho_r$) is mixed with a volume fraction $(1-f)$ of a matrix $m$ (density $\rho_m$) to form a composite with no residual porosity, the composite density $\tilde{\rho}$ is given exactly by the rule of mixtures:

$$\tilde{\rho} = f\rho_r + (1-f)\rho_m \tag{4.23}$$

The geometry or shape of the reinforcement does not matter except in determining the maximum packing fraction of reinforcement and thus the upper limit for $f$ (typically 50%).

The modulus of a composite is bracketed by two *bounds*—limits between which the modulus must lie. The upper bound, $\tilde{E}_U$, is found by assuming that, on loading, the two components strain by the same amount, like springs in parallel. The stress is then the average of the stresses in the matrix and the stiffer reinforcement, giving, once more, a rule of mixtures:

$$\tilde{E}_U = fE_r + (1-f)E_m \tag{4.24}$$

where $E_r$ is the Young's modulus of the reinforcement and $E_m$ that of the matrix. To calculate the lower bound, $\tilde{E}_L$, we assume that the two components carry the same stress, like springs in series. The strain is the average of the local strains and the composite modulus is

$$\tilde{E}_L = \frac{E_m E_r}{fE_m + (1-f)E_r} \tag{4.25}$$

---

### Example 4.5

A composite material has a matrix of polypropylene and contains 10% glass reinforcement. Estimate the Young's modulus and density of the composite: (i) if the glass is in the form of long parallel fibres (calculate the properties parallel to the fibres); (ii) if the glass is in the form of small particles. Compare the specific moduli ($E/\rho$) of the constituent materials and the two composites and plot the four materials on a copy of Figure 4.25.
    Use the following property data:

| | $E$ (GPa) | $\rho$ (Mg/m$^3$) | $E/\rho$ |
|---|---|---|---|
| Polypropylene (m) | 1 | 0.9 | 1.1 |
| Glass (r) | 70 | 2.2 | 31.8 |

*Answer.* The density of both composites is the same. Using equation (4.23) with $f = 0.1$ gives $\tilde{\rho} - 1.03$ Mg/m$^3$ (it would just sink in water). Using equation (4.24), Young's modulus of the fibre-reinforced composite (parallel to the fibres) is $\tilde{E}_{PRP} = 7.9$ GPa, with a specific modulus of $\tilde{E}_{PRP}/\rho = 7.9/1.03 = 7.7$. Using equation (4.25), the modulus of the particle-reinforced composite would be $\tilde{E}_{PRP} = 1.1$, with a specific modulus $\tilde{E}_{PRP}/\rho = 1.07$. The specific modulus of the fibre-reinforced composite is seven times that of the matrix, which is an enormous benefit if you want to build a low mass component or structure. That of the particle-reinforced composite is marginally lower than that of the matrix alone.

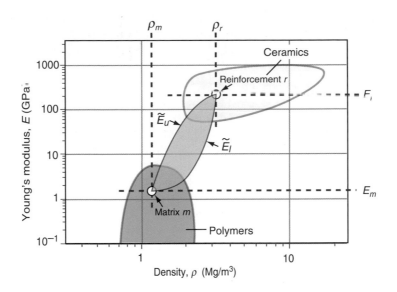

**Figure 4.25** Composites made from a matrix *m* with a reinforcement *r* have moduli and densities, depending on the volume fraction and form of the reinforcement, that lie within the gray shaded lozenge bracketed by equations (4.24) and (4.25). Here the matrix is a polymer and the reinforcement a ceramic, but the same argument holds for any combination

Figure 4.25 shows the range of composite properties that could, in principle, be obtained by mixing two materials together—the boundaries are calculated from equations (4.23)–(4.25). Composites can therefore fill in some of the otherwise empty spaces on the $E$–$\rho$ chart, opening up new possibilities in design. There are practical limits of course—the matrix and reinforcement must be chemically compatible and available in the right form, and processing the mixture must be achievable at a sensible cost. Fibre-reinforced polymers are well-established examples—Figure 4.8 shows GFRP and CFRP, sitting in the composites bubble between the polymers bubble and the ceramics bubble. They are as stiff as metals, but lighter.

*Foams* Foams are made much as you make bread: by mixing a matrix material (the dough) with a foaming agent (the yeast), and controlling what then happens in such a way as to trap the bubbles. Polymer foams are familiar as insulation and flotation and as the filler in cushions and packaging. You can, however, make foams from other materials: metals, ceramics and even glass. They are light for the obvious reason that they are, typically, 90% space, and they have low moduli. This might make them sound as though they are of little use, but that is mistaken: if you want to cushion or to protect a delicate object (such as yourself) what you need is a material with a low, controlled, stiffness and strength. Foams provide it.

Figure 4.23 shows an idealised cell of a low-density foam. It consists of solid cell walls or edges surrounding a void containing a gas. Cellular solids are characterised by their

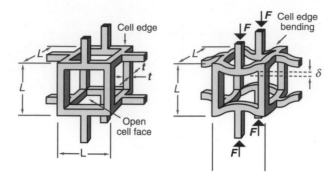

**Figure 4.26** Manipulating the modulus by making a foam—a lattice of material with cell edges that bend when the foam is loaded

*relative density*, the fraction of the foam occupied by the solid. For the structure shown here (with $t << L$) it is

$$\frac{\tilde{\rho}}{\rho_s} - \left(\frac{t}{L}\right)^2 \tag{4.26}$$

where $\tilde{\rho}$ is the density of the foam, $\rho_s$ is the density of the solid of which it is made, $L$ is the cell size and $t$ is the thickness of the cell edges.

When the foam is loaded, the cell walls bend, as shown on the right of the figure. This behavior can be modeled (we will be able to do so by the end of the next chapter), giving the foam modulus $\tilde{E}$

$$\frac{\tilde{E}}{E_s} = \left(\frac{\tilde{\rho}}{\rho_s}\right)^2 \tag{4.27}$$

where $E_s$ is the modulus of the solid from which the foam is made. This gives us a second way of manipulating the modulus, in this case, decreasing it. At a relative density of 0.1 (meaning that 90% of the material is empty space), the modulus of the foam is only 1% of that of the material in the cell wall, as sketched in Figure 4.27. The range of modulus and density for real foams is illustrated in the chart of Figure 4.8—as expected, they fall below and to the left of the polymers of which they are made.

## **4.6** Acoustic properties

Sound is transmitted through materials as an elastic wave. The wavelength $\lambda$ is related to the frequency $f$ by

$$f = \frac{v}{\lambda} \tag{4.28}$$

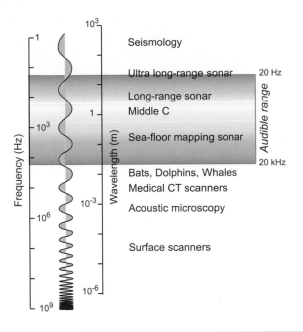

**Figure 4.27** The acoustic spectrum from the extreme ultra-sonic to the seismic, a range of $10^8$. The audible part is a thick slice of this. The figure is based on acoustic waves in air, velocity 343 m/s; for water, shift the wavelength scale up relative to the frequency scale by a factor of 3; for rock, shift it up by a factor of 10

where $v$ is the velocity of sound in the medium in which it is travelling. The human ear responds to frequencies from 20 to about 20,000 Hz, corresponding to wavelengths in air[8] between 17 m and 17 mm. The range of acoustic frequency is far greater than this (Figure 4.27) with wavelengths that extend into the nano range. The vibrations that cause sound produce a change of air pressure in the range $10^{-4}$ Pa (low amplitude sound) to 10 Pa (the threshold of pain). It is usual to measure this on a relative, logarithmic scale, with units of *decibels* (dB). The decibel scale compares two sound intensities using the *threshold of hearing* as the reference level (0 dB).

***Sound velocity and wavelength*** The longitudinal sound velocity in a long rod of a solid material, such that the thickness is small compared with the wavelength, is

$$v_1 = \sqrt{\frac{E}{\rho}} \tag{4.29}$$

---

[8] The velocity of sound in air is 343 m/s at 20 °C and sea level.

where $E$ is Young's modulus and $\rho$ is the density. If the thickness of the rod is large compared with the wavelength, the velocity, instead, is

$$v_B = \sqrt{\frac{E(1-v)}{(1-v-2v^2)\rho}}$$

where $v$ is Poisson's ratio. They don't differ much—a maximum of 25% at most.

Elastically anisotropic solids (fibre composites like CFRP, and all woods, are anisotropic) have sound velocities which depend on direction. If $E_{\parallel}$ is the modulus along the direction of the fibres or grain and $E_{\perp}$ is that perpendicular to it, the ratio of the two velocities, as given by equation (4.29), is

$$\frac{v_{\parallel}}{v_{\perp}} = \sqrt{\frac{E_{\parallel}}{E_{\perp}}} \tag{4.30}$$

*Sound management*   The mechanism for reducing sound intensity within a given enclosed space (a room, for instance) depends where it comes from. If it is generated within the room, one seeks to *absorb* the sound. If it is airborne and comes from outside, one seeks to *insulate* the space to keep the sound out. And if it is transmitted through the frame of the structure itself, one seeks to *isolate* the structure from the source of vibration. Cellular, porous solids are good for absorbing sound; and—in combination with other materials—they can help in isolation. They are very poor at providing insulation.

Soft porous materials absorb incident, airborne sound waves, converting them into heat. Sound power, even for very loud noise, is small, so the temperature rise is negligible. Porous or highly flexible materials such as low-density polymeric foams, plaster and fibreglass absorb well; so do woven polymers in the form of carpets and curtains. Several mechanisms of absorption are at work here. First, there is the viscous loss as air is pumped into and out of the open, porous structures—sealing the surface (with a film of paint, for instance) greatly reduces the absorption. Second, there is the intrinsic damping in the material. It measures the fractional loss of energy of a wave, per cycle, as it propagates within the material itself. Intrinsic damping in most metals and ceramics is low, typically $10^{-6}$ to $10^{-4}$; in polymers and foams made from them it is high, in the range 0.01 to 0.2. The proportion of sound absorbed by a surface is called the *sound-absorption coefficient*. A material with a coefficient of 0.8 absorbs 80% of the sound that is incident on it, and material of coefficient 0.03 absorbs only 3% of the sound, reflecting 97%.

Sound insulation requires materials of a completely different sort. The degree of insulation is proportional to the mass of the wall, floor or roof through which sound has to pass. This is known as the *mass-law*: the heavier the material, the better it insulates. The lightweight walls of modern buildings, designed for good thermal insulation, do not in general provide good sound insulation. Foams, plaster or fibreglass, for obvious reasons, are not a good choice here; instead, the practice is to add mass to the wall or floor with an extra layer of brick or concrete, or a lead cladding.

Impact noise is transmitted directly into the structure or fabric of a building. Elastic materials, and particularly a steel frame if there is one, transmit vibrations throughout the building. Unlike airborne sound, noise of this sort is not attenuated by additional mass. Since it is transmitted by the continuous solid part of the structure, it can be reduced by interrupting the sound path by floating the floor or building foundation on a resilient material. Here, low modulus materials can be useful. The impact sound of footsteps can be suppressed by using cork, porous rubber or plastic tiles. On a larger scale, buildings are isolated by setting the entire structure on resilient pads, such as a composite of rubber filled with cork particles. The low shear modulus of the rubber isolates the building from shear waves, and the compressibility of the cork adds a high impedance to compressive waves.

*Sound wave impedance and radiation of sound energy*   If a sound-transmitting material is interfaced with a second one with different properties, part of the amplitude of the sound wave is transmitted across the interface and part is reflected back into the first material. The transmission and reflection factors are determined by the relative impedances of the two materials. The impedance is defined by

$$Z = \sqrt{\rho E} \tag{4.31}$$

where $E$ is the appropriate modulus, and $\rho$ the density. The reflection coefficient $R$ is the fraction of the acoustic energy that is reflected, given

$$R = \left( \frac{Z_2 - Z_1}{Z_2 + Z_1} \right)^2 \tag{4.32}$$

where $Z_1$ is the impedance in the medium in which the sound originates and $Z_2$ is that of the material into which it is transmitted. The energy that is not reflected is transmitted, so the transmission coefficient $T$ is

$$T = 1 - R = \frac{4 Z_1 Z_2}{(Z_2 + Z_1)^2} \tag{4.33}$$

Thus if the two impedances are about equal most of the sound is transmitted, but if they differ greatly most is reflected. This is the origin of the 'mass-law' cited earlier: a heavy wall gives a large impedance mismatch with air, so that most of the sound is reflected and does not penetrate to the neighbouring room.

Table 4.2 lists typical values of the acoustic impedance. In the design of sound boards (the front plate of a violin, the sound board of a harpsichord, the panel of a loudspeaker), the intensity of sound radiation is an important design parameter. This intensity, $I$, is proportional to the surface velocity, and for a given driving function, this scales with modulus and density as:

$$I \propto \sqrt{\frac{E}{\rho^3}} \tag{4.34}$$

A high value of the combination of properties $\sqrt{E/\rho^3}$, called the 'radiation factor', is used by instrument makers to select materials for sound boards. When the material is elastically anisotropic (as is wood), $E$ is replaced by $\overline{E} = \sqrt{E_{\parallel} E_{\perp}}$, where $E_{\parallel}$ is the modulus parallel to the grain and $E_{\perp}$ is that across the grain. The last column of Table 4.2 gives some representative values for the radiation factor. Spruce, widely used for the front plates of violins, has a

Table 4.2 Acoustic impedances and radiation factors*

| Material | Density, $\rho$ (Mg/m$^3$) | Young's modulus, $E^*$ (GPa) | Acoustic impedance, $\sqrt{\rho E}$ (MN s/m$^3$) | Radiation factor, $\sqrt{E/\rho^3}$ (m$^5$/N·s$^3$) |
|---|---|---|---|---|
| PS foam, low-density | 0.03 | 0.01 | 0.01 | 19 |
| PS foam, high-density | 0.78 | 1.6 | 1.1 | 2.0 |
| Solid PS | 1.04 | 3.6 | 1.9 | 1.8 |
| Nylon 6/6 | 1.14 | 2.4 | 1.7 | 1.3 |
| Spruce, ∥ to grain | 0.51 | 16 | | |
| Spruce, ⊥ to grain | 0.51 | 0.84 $\bar{E}= 3.7$ | 1.4 | 5.3 |
| Maple, ∥ to grain | 0.52 | 8.6 | | |
| Maple, ⊥ to grain | 0.52 | 0.93 $\bar{E}= 2.8$ | 1.2 | 4.4 |
| Aluminum | 2.7 | 69 | 13.6 | 1.87 |
| Steel | 7.9 | 210 | 40.7 | 0.65 |
| Silver | 10.4 | 75 | 27.9 | 0.26 |
| Glass | 2.24 | 46 | 10.1 | 2.0 |

* Data from American Institute of Physics, *Handbook* (1972), Meyer (1995) and CES (2013).

particularly high value—20% greater than that of maple, which is used for the back plate, the function of which is to reflect, not radiate.

***Visualising acoustic properties*** Figure 4.28 is a copy of the Modulus–Density chart. A characteristic of the log scales of these charts is that simple ratios and products of the properties plot as straight lines—the figure shows contours of wave speed $\sqrt{E/\rho}$ and acoustic impedance $\sqrt{\rho E}$. Values for water and for air also appear (by extending the Density axis); for these it is the bulk modulus that is relevant. The chart gives an overview of acoustic behaviour. Very little sound is reflected when a sound wave passes from one material to another if their values of impedance are very similar. By contrast, reflection is strong when impedances differ greatly.

## 4.7 Summary and conclusions

When a solid is loaded, it initially deforms elastically. 'Elastic' means that, when the load is removed, the solid springs back to its original shape. The material property that measures stiffness is the elastic modulus, and because solids can be loaded in different ways, we need three of them:

- *Young's modulus*, $E$, measuring resistance to stretching and compressing.
- *Shear modulus*, $G$, measuring resistance to twisting.
- *Bulk modulus*, $K$, measuring resistance to hydrostatic compression.

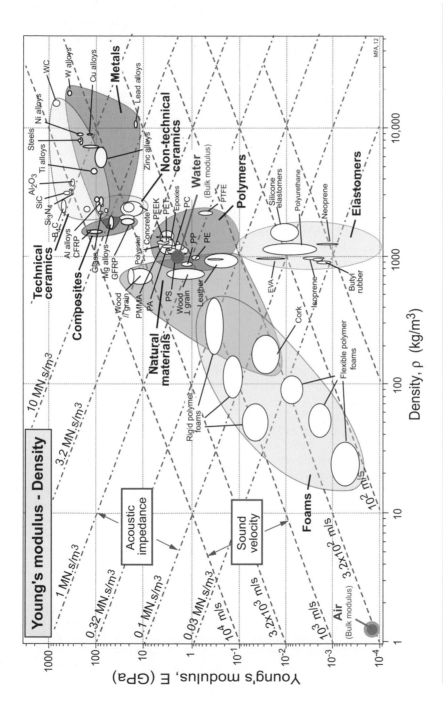

**Figure 4.28** The sound velocity and acoustic impedence of materials superimposed on a Modulus – Density chart. The Density axis has been extended to include air; water is also shown (both in red). For these, it is the bulk modulus that is relevant

Many applications require stiffness at low weight, particularly ground, air and space vehicles, and that means a high modulus and a low density. Property charts were introduced to help visualise and rationalise the relative profiles of the material classes.

The moduli have their origins in the stiffness of the bonds between atoms in a solid and in the packing of the atoms, and thus the number of bonds per unit area. The atomic packing does not vary much from one solid to another, so the moduli mainly reflect the stiffness of the bonds. Bonding can take several forms, depending on how the electrons of the atoms interact. Metallic, covalent and ionic bonds are stiff, hydrogen and Van der Waals bonds are much less so—that is why steel has high moduli and polyethylene has low.

The density of a material is the weight of its atoms divided by the volume they occupy. Atoms do not differ much in size and packing, but they differ a great deal in weight. Thus the density is principally set by the atomic weight; the further down the periodic table we go the greater the density.

There is very little that can be done to change the bond stiffness or atomic weight of a solid, so at first sight we are stuck with the moduli and densities of the materials we already have. But there are two ways to manipulate them: by mixing two materials to make composites, or by mixing a material with space to make foams. Both are powerful ways of creating 'new' materials that occupy regions of the $E-\rho$ map that were previously empty.

Finally, sound transmission and absorption also depend on modulus and density. This chapter concluded with a look at material acoustics, illustrating another way to use the property charts to identify combinations of properties—a technique we will use again for material selection, starting in the next chapter.

## **4.8** Further reading

American Institute of Physics. (1972). In E. E. Gray (Ed.), *Handbook*, 3rd ed. NY USA: McGraw-Hill.

Ashby, M. F., & Jones, D. R. H. (2006). *Engineering Materials 1* (3rd ed.). Oxford, UK: Elsevier Butterworth-Heinemann. ISBN 0-7506-6380-4. (One of a pair of introductory texts dealing with the engineering properties and processing of materials.).

Askeland, D. R., & Phule, P. P. (2006). *The Science of Engineering Materials* (5th ed.). Toronto, Canada: Thompson Publishing. (A mature text dealing with the science of materials, and taking a science-led rather than a design-led approach.).

Beranek, L. L. (Ed.). (1960). *Noise Reduction* (pp. 349–395). New York, USA: McGraw-Hill.

Callister, W. D., Jr. (Ed.). (2007). *Materials Science and Engineering, An Introduction* (7th ed.). New York, USA.: Wiley. ISBN 0-471-73696-1. (A long established and highly respected introduction to materials, taking the science-based approach.).

Gray, E. E. (Ed.). (1972). *American Institute of Physics Handbook* (3rd ed). New York, USA: McGraw-Hill.

Fletcher, N. H., & Rossing, T. D. (1991). *The Physics of Musical Instruments*. Berlin, Germany: Springer-Verlag.

Lauriks, W. (1994). Acoustic Characteristics of Low Density Foams. In N. C. Hilyard, & A. Cunningham (Eds.), *Low Density Cellular Plastics*. London, UK: Chapman and Hall.

Meyer, H. G. (1995). A Practical Approach to the Choice of Tone Wood in the Instruments of the Violin Family. *Catgut Acoust. Soc. J, 2*(No. 7), 9–12.

Parkin, P. H., Humphreys, J. R., & Cowell, J. R. (1979). *Acoustics, Noise and Buildings* (4th ed.). London, UK: Faber. 279–83.

# **4.9** Exercises

| | |
|---|---|
| Exercise E4.1 | Identify which of the five modes of loading (Figure 4.2) is dominant in the following components:<br>• Fizzy drinks container.<br>• Overhead electric cable.<br>• Shoe soles.<br>• Wind turbine blade.<br>• Climbing rope.<br>• Bicycle forks.<br>• Aircraft fuselage.<br>Can you think of another example for each mode of loading? |
| Exercise E4.2 | The cable of a hoist has a cross-section of 80 mm$^2$. The hoist is used to lift a crate weighing 500 kg. What is the stress in the cable? The free length of the cable is 3 m. How much will it extend if it is made of steel (modulus 200 GPa)? How much if it is made of polypropylene, PP (modulus 1.2 GPa)? |
| Exercise E4.3 | Water has a density of 1000 kg/m$^3$. What is the hydrostatic pressure at a depth of 100 m? How many times atmospheric pressure is this (1 atm = 0.1 MPa)? |
| Exercise E4.4 | Figure 4.5 shows a cubic element of material with Young's modulus $E$ and Poisson's ratio $\nu$ which is subjected to normal stresses $\sigma_1$, $\sigma_2$ and $\sigma_3$, resulting in strains $\varepsilon_1$, $\varepsilon_2$ and $\varepsilon_3$ (equation (4.13)).<br>(a) Find an expression for the dilatation $\Delta$ when a unit cube experiences strains of $\varepsilon_1$, $\varepsilon_2$ and $\varepsilon_3$, assuming that the strains are elastic and therefore small ($<< 1$).<br>(b) Find the dilatation $\Delta$ for an element subjected to a uniaxial tensile stress $\sigma_1 = \sigma$ (i.e. $\sigma_2 = \sigma_3 = 0$). For what value of Poisson's ratio is volume conserved?<br>(c) Find the dilatation $\Delta$ for an element subjected to equal pressure $p$ in all directions, that is, $\sigma_1 = \sigma_2 = \sigma_3 = -p$. Hence derive the formula relating the bulk modulus, $K$, to Young's modulus, $E$ and Poisson's ratio, $\nu$ (equation (4.10)). |
| Exercise E4.5 | A mounting block made of butyl rubber of modulus = 0.0015 GPa is designed to cushion a sensitive device against shock loading. It consists of a cube of the rubber of side length $L = 40$ mm located in a slot in a rigid plate, exactly as shown in Figure 4.6. The slot and cube have the same cross-sectional dimensions, and the surrounding material is much stiffer, so the strain in one direction is constrained to be zero. A vertical compressive load $F$ is applied, and the deflection of the top block is $\delta$. The maximum expected value of $F$ is 50 N. |

Use equation (4.14) to find the stiffness of the cube ($F/\delta$) in this constrained condition, assuming $\nu \approx 0.5$. Hence find the maximum deflection $\delta$ of the top face of the block. What would this deflection be without any constraint?

**Exercise E4.6**  A cube of linear elastic material is again subjected to a vertical compressive stress $\sigma_1$ in the 1-direction, but it is now constrained ($\varepsilon = 0$) in both the 2- *and* the 3-directions.

(a) Find expressions for the induced transverse stresses, $\sigma_2$ and $\sigma_3$ in terms of $\sigma_1$ . Hence derive an expression for the 'effective modulus' ($\sigma_1/\varepsilon_1$) in this case.
(b) Sketch the variation of effective modulus with $\nu$ and comment on the limiting values when $\nu = 0$ (foam) and $\nu \approx 0.5$ (rubber).
(c) Explain why the rubber soles of running shoes are designed with some combination of air or gel pockets, partially foamed rubber and a tread.

**Exercise E4.7**  A catapult has two rubber arms, each with a square cross-section with a width 4 mm and length 300 mm. In use its arms are stretched to three times their original length before release. Assume the modulus of rubber is $10^{-3}$ GPa and that it does not change when the rubber is stretched. How much energy is stored in the catapult just before release?

**Exercise E4.8**  Use the modulus-density chart of Figure 4.8 to find, from among the materials that appear on it:

(a) The material with the highest density.
(b) The metal with the lowest modulus.
(c) The polymer with the highest density.
(d) The approximate ratio of the modulus of woods measured parallel to the grain and perpendicular to the grain.
(e) The approximate range of modulus of elastomers.

**Exercise E4.9**  Use the modulus–relative-cost chart of Figure 4.9 to find, from among the materials that appear on it:

(a) The cheapest material with a modulus greater than 1 GPa.
(b) The cheapest metal.
(c) The cheapest polymer.
(d) Whether magnesium alloys are more or less expensive than aluminum alloys.
(e) Whether PEEK (a high-performance engineering polymer) is more or less expensive than PTFE.

**Exercise E4.10**  What is meant by:

- A crystalline solid?
- An amorphous solid?
- A thermoplastic?
- A thermoset?
- An elastomer?

Exercise E4.11    The stiffness $S$ of an atomic bond in a particular material is 50 N/m and its centre-to-centre atom spacing is 0.3 nm. What, approximately, is its elastic modulus?

Exercise E4.12    Derive the upper and lower bounds for the modulus of a composite quoted as equations (4.24) and (4.25) of this chapter. To derive the first, assume that the matrix and reinforcement behave like two springs in parallel (so that each must strain by the same amount), each with a stiffness equal to its modulus $E$ multiplied by its volume fraction, $f$. To derive the second, assume that the matrix and reinforcement behave like two springs in series (so that both are stressed by the same amount), again giving each a stiffness equal to its modulus $E$ multiplied by its volume fraction, $f$.

Exercise E4.13    A volume fraction $f = 0.2$ of silicon carbide (SiC) particles is combined with an aluminum matrix to make a metal matrix composite. The modulus and density of the two materials are listed in the following table. The modulus of particle-reinforced composites lies very close to the lower bound, equation (4.25), discussed in the text. Calculate the density and approximate modulus of the composite. Is the specific modulus, $E/\rho$, of the composite greater than that of unreinforced aluminum? How much larger is the specific modulus if the same volume fraction of SiC in the form of continuous fibres is used instead? For continuous fibres the modulus lies very close to the upper bound, equation (4.24).

|  | Density, Mg/m$^3$ | Modulus, GPa |
|---|---|---|
| Aluminum | 2.70 | 70 |
| Silicon carbide | 3.15 | 420 |

Exercise E4.14    Medical prosthetic implants such as hip replacements have traditionally been made of metals such as stainless steel or titanium. These are not ideal as they are much stiffer than the bone, giving relatively poor transfer of load into the bone. New composite materials are being developed to provide a much closer stiffness match between the implant material and the bone. One possibility uses high-density polyethylene (HDPE) containing particulate hydroxyapatite (HA), the natural mineral in bone, which can be produced artificially. Data for some experimental composites are provided in the table below, and those for the bulk materials are HDPE 0.65 GPa and Hydroxyapatite 80 GPa.

(a) Plot the upper and lower bounds for the Young's modulus of HDPE-HA composites against the volume fraction of reinforcement, $f$ (from 0 to 1), together with the experimental data.

(b) Which of the bounds is closer to the data for the particulate composite? Use this bound as a guide to extrapolate the experimental data to identify the volume fraction of hydroxyapatite particulate that is required to match the modulus of bone, $E = 7$ GPa. Is this practical?

| Volume fraction of HA | Young's modulus (GPa) |
|---|---|
| 0 | 0.65 |
| 0.1 | 0.98 |
| 0.2 | 1.6 |
| 0.3 | 2.73 |
| 0.4 | 4.29 |
| 0.45 | 5.54 |

**Exercise E4.15**  (a) Use the material property data in the table below to find the composite density $\tilde{\rho}$ and upper bound modulus $\tilde{E}_U$ for the following composites: (i) carbon-fibre/epoxy resin ($f=0.5$); (ii) glass-fibre/polyester resin ($f = 0.5$).

| Material | Density (Mg m$^{-3}$) | Young's modulus (GPa) |
|---|---|---|
| Carbon fibre | 1.90 | 390 |
| Glass fibre | 2.55 | 72 |
| Epoxy resin | 1.15 | 3 |
| Polyester resin | 1.15 | 3 |

(b) A magnesium company has developed an experimental metal-matrix composite (MMC), by casting magnesium containing 20% (by volume) of particulate SiC. Use the lower bound estimate for Young's modulus to estimate $\tilde{E}_L$ for this MMC, with the following data for matrix and particles: Mg alloys, $E = 44.5$ GPa, $\rho = 1.85$ Mgm$^{-3}$; SiC, $E = 380$ GPa, $\rho = 3.1$ Mgm$^{-3}$.

(c) Find the *specific stiffness* $\tilde{E}/\tilde{\rho}$ for the three composites in (a, b). Compare the composites with steels, for which $E/\rho \approx 28$ GPa Mg$^{-1}$ m$^3$.

**Exercise E4.16**  Read the modulus $E$ for polypropylene (PP) from the $E-\rho$ chart of Figure 4.8. Estimate the modulus of a PP foam with a relative density $\tilde{\rho}/\rho_s$ of 0.2.

**Exercise E4.17**  Use the relationship $\tilde{E} = E_s \left(\frac{\tilde{\rho}}{\rho_s}\right)^2$ (equation (4.27)) to estimate the ranges of modulus and density that might be achievable by foaming the following metals to relative densities in the range $0.1 - 0.3$: (a) Ni alloys ($E = 205$ GPa, $\rho = 8.89$ Mg/m$^3$); (b) Mg alloys ($E = 44.5$ GPa, $\rho = 1.85$ Mg/m$^3$). Locate approximate 'property bubbles' for these new metallic foams on a copy of the modulus–density chart. With which existing materials would these foams appear to compete, in terms of modulus and density? Can you think of a property for which the metallic foams will differ significantly from the competition?

*Exercise E4.18*   The speed of longitudinal waves in a material is proportional to $\sqrt{E/\rho}$. Plot contours of this quantity onto a copy of a modulus–density chart allowing you to read off approximate values for any material on the chart. Which metals have about the same sound velocity as steel? Does sound move faster in titanium or glass?

*Exercise E4.19*   The density $\rho$ of carbon steel is 7,500 kg/m$^3$ and its Young's $E$ modulus is 210 GPa. What is the velocity of longitudinal sound waves in steel? Watch the units!

*Exercise E4.20*   Given the data in the previous question, what is the acoustic impedance of steel?

*Exercise E4.21*   The acoustic impedance of water is $1.5 \times 10^6$ N.s/m$^3$. If a sound wave of unit power in water impinges on a steel ship's hull with an impedance of $4 \times 10^7$ N.s/m$^3$, what fraction of the power is transmitted into the steel?

*Exercise E4.22*   Ultrasonic testing relies on the reflection of sound waves from cracks or imperfections in castings and forgings. If the resolution (the smallest imperfection it can detect) of the method is the wavelength, will a tester operating at 1 megahertz be able to detect a 1-mm-sized imperfection of an aluminum casting?

*Exercise E4.23*   Sonar depends on the reflection of sound waves in water. Would sonar more easily detect a large object made of polypropylene (PP) or one of similar size made of aluminum? Use Figure 4.28 to find out.

*Exercise E4.24*   Practice visualising radiation factor and acoustic impedance. Make a copy of the modulus–density chart of Figure 4.8 and plot on it a set of contours showing the radiation factor

$$I = \sqrt{\frac{E}{\rho^3}}$$

and a set for acoustic impedance

$$Z = \sqrt{\rho E}$$

An innovative guitar-maker plans to make acoustic guitars with CFRP or GFRP composite front faces instead of the more usual choice of spruce, chosen for its high radiation factor. What can you learn from the chart that might help the guitar-maker choose between CFRP and GFRP?

**Exercise E4.25** High buildings on either side of busy streets reflect the noise of traffic and add to the stress of city life. Buildings can be faced with wood, concrete or glass. Use approximate data from Figure 4.28 to estimate the relative reflectance of these three types of facing.

## 4.10 Exploring design with CES

Use Level 2, Materials, throughout the following exercises.

**Exercise E4.26** Make an $E - \rho$ chart using the CES software. Use a box selection to find three materials with densities between 1000 and 3000 kg/m$^3$ and the highest possible modulus.

**Exercise E4.27** Explore data estimation. The modulus $E$ is approximately proportional to the melting point $T_m$ in Kelvin (because strong inter-atomic bonds give both stiffness and resistance to thermal disruption). Use CES to make an $E - T_m$ chart for metals and estimate a line of slope 1 through the data. Use this line to estimate the modulus of cobalt, given that it has a melting point of 1760 K.

**Exercise E4.28** Using sanity checks for data. A text reports that nickel, with a melting point of 1720 K, has a modulus of 5500 GPa. Use the $E - T_m$ correlation of the previous question to check the sanity of this claim. What would you expect it to be?

**Exercise E4.29** Explore the potential of PP–SiC (polypropylene–silicon carbide) fibre composites in the following way. Make a modulus–density ($E - \rho$) chart and change the axis ranges so that they span the range $1 < E < 1000$ GPa and $500 < \rho < 5000$ kg/m$^3$. Find and label PP and SiC, then print it. Retrieve values for the modulus and density of PP and of SiC from the records for these materials (use the means of the ranges). Calculate the density $\tilde{\rho}$ and upper and lower bounds for the modulus $\tilde{E}$ at a volume fraction $f$ of SiC of 0.5 and plot this information on the chart. Sketch by eye two arcs starting from $(E, \rho)$ for PP, passing through each of the $(\tilde{E}, \tilde{\rho})$ points you have plotted and ending at the $(E, \rho)$ point for SiC. PP–SiC composites can populate the area between the arcs roughly up to $f = 0.5$ because it is not possible to insert more than this.

**Exercise E4.30** Explore the region that can be populated by making PP foams. Expand an $E - \rho$ plot so that it spans the range $10^{-4} < E < 10$ GPa and $10 < \rho < 2000$ kg/m$^3$. Find and label PP, then print the chart. Construct a band starting with the PP bubble by drawing lines corresponding to the scaling law for foam modulus $\tilde{E} \propto \tilde{\rho}^2$ (equation 4.27) touching the top and the bottom of the PP bubble. The zone between these lines can be populated by PP foams.

# **4.11** Exploring the science with CES Elements

**Exercise E4.31**  This chapter cited the following approximate relationships between the elastic constants Young's modulus, $E$, the shear modulus, $G$, the bulk modulus $K$ and Poisson's ratio, $\nu$.

$$G = \frac{E}{2(1+\nu)}; \qquad K = \frac{E}{3(1-2\nu)}$$

Use CES to make plots with the bit on the left-hand side of each equation on one axis and the bit on the right on the other. To do this you will need to use the 'Advanced' facility in the dialog box for choosing the axes to create functions on the right of the two equations. How good an approximation are they?

**Exercise E4.32**  The cohesive energy $H_c$ is the energy that binds atoms together in a solid. Young's modulus $E$ measures the force needed to stretch the atomic bonds, and the melting point, $T_m$, is a measure of the thermal energy needed to disrupt them. Both derive from the cohesion, so you might expect $E$ and $T_m$ to be related. Use CES to plot one against the other to see if this is so. (Use absolute melting point, not centigrade or Fahrenheit.)

**Exercise E4.33**  The force required to stretch an atomic bond is

$$F = \frac{dH}{da}$$

where $dH$ is the change in energy of the bond when it is stretched by $da$. This force corresponds to a stress

$$\sigma = \frac{F}{a_o^2} = \frac{1}{a_o^2}\frac{dH}{da} = \frac{1}{a_o^3}\frac{dH}{d(a/a_o)}$$

The modulus $E$ is

$$E = \frac{d\sigma}{d\varepsilon} = \frac{d\sigma}{d(a/a_o)} = \frac{1}{a_o^3}\frac{d^2H}{d(a/a_o)^2}$$

The binding energy per atom in a crystal, $H_a$ is

$$H_a = \frac{H_c}{N_A}$$

where $H_c$ is the cohesive energy and $N_A$ is Avogadro's number ($6.022 \times 10^{23}$). If we assume that a stretch of 2% is enough to break the bond, we can make the approximation

$$\frac{d^2H}{d(a/a_o)^2} \approx \frac{H_c}{(0.02)^2}$$

giving

$$E \approx \frac{1}{a_o^3}\left(\frac{N_c}{0.0004.N_A}\right)$$

Make a plot of Young's modulus $E$ against the quantity on the right of the equation (using the 'Advanced' facility in the dialog box for choosing the axes) to see how good this is. (You will need to multiply the right by $10^{-9}$ to convert it from Pascals to GPa.)

Exercise E4.34 Make a chart of the sound velocity $(E/\rho)^{1/2}$ for the elements. To do so, construct the quantity $(E/\rho)^{1/2}$ on the y-axis using the 'Advanced' facility in the axis-choice dialog box, and plot it against atomic number $A_n$. Use a linear scale for $A_n$. To do so, change the default log scale to linear by double-clicking on the axis name to reveal the axis-choice dialog box and choose 'Linear'. And multiply $E$ by $10^9$ to give the sound velocity in m/s.)

# Chapter 5

# Flex, sag and wobble: stiffness-limited design

A pole vaulter—the pole stores elastic energy. (Image courtesy of Gill Athletics, 2808 Gemini Court, Champaign, IL 61822-9648, USA)

## **5.1** Introduction and synopsis

A few years back, with the millennium approaching, countries and cites around the world turned their minds to iconic building projects. In Britain there were several. One was—well, is—a new pedestrian bridge spanning the river Thames, linking St Paul's Cathedral to the Museum of Modern Art facing it across the river. The design was—oops, is—daring: a suspension bridge with suspension cables that barely rise above the level of the deck instead of the usual great upward sweep. The result was visually striking: a sleek, slender, span like a 'shaft of light' (the architect's words). Just one problem: it wasn't stiff enough. The bridge opened but when people walked on it, it swayed and wobbled so alarmingly that it was promptly closed. A year and $5 000 000 later it reopened, much modified, and now it is fine.

The first thing you tend to think of with structures, bridges included, is *strength*: they must not fall down. *Stiffness*, often, is taken for granted. But, as the bridge story relates, that can be a mistake—stiffness *is* important. Here we explore stiffness-limited design and the choice of materials to achieve it. This involves the modeling of material response in a given application. The models can be simple because the selection criteria that emerge are insensitive to the details of shape and loading. The key steps are those of identifying the constraints that the material must meet and the objective by which the excellence of choice will be measured.

We saw at the beginning of Chapter 4 that there are certain common modes of loading: *tension*, *compression*, *bending* and *torsion*. The loading on any real component can be decomposed into some combination of these. So it makes sense to have a catalog of solutions for the standard modes, relating the component response to the loading. In this chapter, the response is elastic deflection; in later chapters it will be yielding or fracture. You don't need to know how to derive all of these—they are standard results—but you *do* need to know where to find them. You will find the most useful of them here; the sources listed under 'Further reading' give more. And, most important, you need to know how to use them. That needs practice—you get some of that here too.

The first section of this chapter, then, is about standard solutions to elastic problems. The second is about their use to derive *material limits* and *indices*. The third explains how to plot them onto material property charts. The last illustrates their use via case studies.

## **5.2** Standard solutions to elastic problems

Modeling is a key part of design. In the early, conceptual stage, approximate models establish whether a concept will work at all and identify the combination of material properties that maximise performance. At the embodiment stage, more accurate modeling brackets values for the design parameters: forces, displacements, velocities, heat fluxes and component dimensions. And in the final stage, modeling gives precise values for stresses, strains and failure probabilities in key components allowing optimised material selection and sizing.

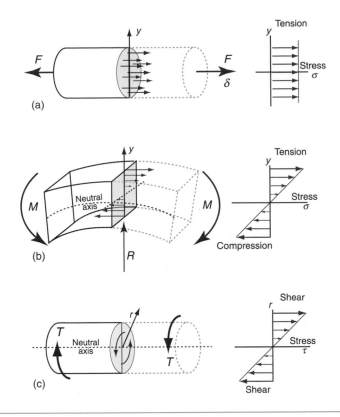

**Figure 5.1**   (a) A tie with a cross-section $A$ loaded in tension. Its stiffness is $S = F/\delta$.
(b) A beam of rectangular cross-section loaded in bending. The stress $\sigma$ varies linearly from
tension to compression, changing sign at the neutral axis, resulting in a bending moment $M$.
(c) A shaft of circular cross-section loaded in torsion.

Many common geometries and load patterns have been modeled already. A component can
often be modeled approximately by idealising it as one of these. There is no need to reanalyse
the beam or the column or the pressure vessel; their behaviour under all common types of
loading has already been analysed. The important thing is to know that the results exist, where
to find them and how to use them.

This section is a bit tedious (as books on the strength of materials tend to be) but the results
are really useful. Here they are listed, defining the quantities that enter and the components to
which they apply.

***Elastic extension or compression***   A tensile or compressive stress $\sigma = F/A$ applied axially to
a tie or strut of length $L_0$ and constant cross-section area $A$ suffers a strain $\varepsilon = \sigma/E$. The strain
$\varepsilon$ is related to the extension $\delta$ by $\varepsilon = \delta/L_0$ (Figure 5.1(a)). Thus, the relation between the load
$F$ and deflection $\delta$ is

$$\delta = \frac{L_0 F}{AE} \tag{5.1}$$

The stiffness $S$ is defined as

$$S = \frac{F}{\delta} = \frac{AE}{L_o} \qquad (5.2)$$

Note that the shape of the cross-section area does not matter because the stress is uniform over the section.

*Elastic bending of beams*  When a beam is loaded by a bending moment $M$, its initially straight axis is deformed to a curvature $\kappa$ (Figure 5.1(b))

$$\kappa = \frac{d^2u}{dx^2} = \frac{1}{R}$$

where $u$ is the displacement parallel to the y-axis. The curvature generates a linear variation of axial strain $\varepsilon$ (and thus stress $\sigma$) across the section, with tension on one side and compression on the other—the position of zero stress being the *neutral axis*. The stress increases linearly with distance $y$ from the neutral axis. Material is more effective at resisting bending the farther it is from that axis, so the shape of the cross-section is important. Elastic beam theory gives the stress $\sigma$ caused by a moment $M$ in a beam made of material of Young's modulus $E$, as

$$\frac{\sigma}{y} = \frac{M}{I} = E\kappa$$
$$= E\frac{d^2u}{dx^2} \qquad (5.3)$$

where $I$ is the second moment of area, defined as

$$I = \int_{\text{section}} y^2 b(y)\mathrm{d}y \qquad (5.4)$$

The distance $y$ is measured vertically from the neutral axis and $b(y)$ is the width of the section at $y$. The moment $I$ characterises the resistance of the section to bending—it includes the effect of both size and shape. Examples for four common sections are listed in Figure 5.2 with expressions for the cross-section area $A$ and the second moment of area, $I$.

---

### Example 5.1

(a) A beam has a rectangular cross-section with height $h$ and width $b$. Show that the second moment of area is $I = bh^3/12$.

(b) A steel ruler is 300 mm long with a width $w = 25$ mm and a thickness $t = 1$ mm. Calculate the second moments of area $I_{XX}$ and $I_{YY}$.

*Answer.*

(a) For bending about the X-X axis, consider a narrow horizontal strip of height $dy$ and width $b$. Inserting this into (5.4) gives:

$$I_{XX} = \int_{-b/2}^{b/2} y^2 b \, dy = \frac{b}{3}[y^3]_{-b/2}^{b/2} = \frac{bb^3}{12}$$

(b) For the steel ruler, bending about the (flexible) X-X axis:

$$I_{XX} = \frac{wt^3}{12} = \frac{25 \times 1^3}{12} = 2.1 \text{ mm}^4$$

whereas for bending about the (stiff) Y-Y axis:

$$I_{YY} = \int_{-b/2}^{b/2} x^2 b \, dy = \frac{tw^3}{12} = \frac{1 \times 25^3}{12} = 1300 \text{ mm}^4$$

$I_{YY}$ is more than 600 times greater than $I_{XX}$, because the material of the ruler is located close to the X-X axis but (on average) farther away from the Y-Y axis.

Moments of inertia for complex sections can be calculated by superimposing the results for simple shapes like the rectangle (though sometimes you may need to use the 'parallel axis theorem'—see books on mechanics of structures).

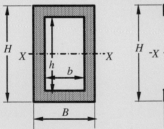 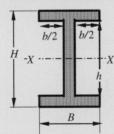

For both the rectangular hollow section and the I-section shown, $I_{XX}$ can be calculated by subtracting $I_{XX}$ for a small rectangle ($b \times h$) from the value of $I_{XX}$ for the large rectangle ($B \times H$), giving:

$$I_{XX} = \frac{BH^3}{12} - \frac{bh^3}{12}$$

Moments of area are tabulated in books for standard cases, and in tables of structural sections for commercially available section shapes.

The ratio of moment to curvature, $M/\kappa$, is called the *flexural rigidity*, $EI$. Figure 5.3 shows three possible distributions of load $F$, each creating a distribution of moment $M(x)$. The maximum deflection, $\delta$, is found by integrating equation (5.3) twice for a particular $M(x)$. For a beam of length $L$ with a transverse load $F$, the stiffness is

$$S = \frac{F}{\delta} = \frac{C_1 EI}{L^3} \tag{5.5}$$

| Section shape | Area $A$ m² | Moment $I$ m⁴ | Moment $K$ m⁴ |
|---|---|---|---|
| 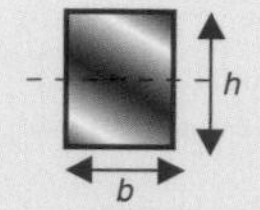 | $bh$ | $\dfrac{bh^3}{12}$ | $\dfrac{bh^3}{3}(1 - 0.58\dfrac{b}{h})$ $(h > b)$ |
| 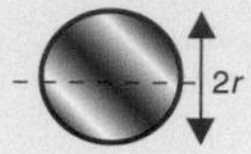 | $\pi r^2$ | $\dfrac{\pi}{4}r^4$ | $\dfrac{\pi}{2}r^4$ |
| 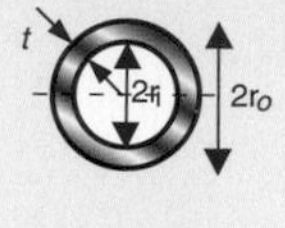 | $\pi(r_o^2 - r_i^2)$ $\approx 2\pi rt$ | $\dfrac{\pi}{4}(r_o^4 - r_i^4)$ $\approx \pi r^3 t$ | $\dfrac{\pi}{2}(r_o^4 - r_i^4)$ $\approx 2\pi r^3 t$ |
| 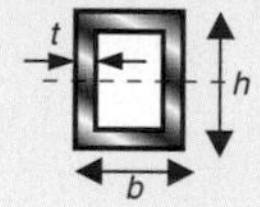 | $2t(h + b)$ $(h, b \gg t)$ | $\dfrac{1}{6}h^3 t(1 + 3\dfrac{b}{h})$ | $\dfrac{2tb^2h^2}{(b + h)}(1 - \dfrac{t}{h})^4$ |

**Figure 5.2** Cross-section area and second moments of sections for four section shapes.

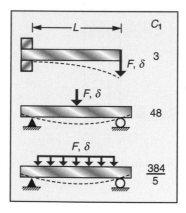

**Figure 5.3** Elastic deflection of beams. The deflection δ of a span $L$ under a force $F$ depends on the flexural stiffness $EI$ of the cross-section and the way the force is distributed. The constant $C_1$ is defined in equation (5.5).

The result is the same for all simple distributions of load; the only thing that depends on the distribution is the value of the constant $C_1$; the figure lists values for the three distributions. We will find in Section 5.3 that the best choice of material is independent of the value of $C_1$ with the happy result that the best choice for a complex distribution of loads is the same as that for a simple one.

---

### Example 5.2

(a) The ruler in Example 5.1 is made of stainless steel with Young's modulus of 200 GPa. A student supports the ruler as a horizontal cantilever, with 250 mm protruding from the edge of a table and hangs a weight of 10 N on the free end. Calculate the vertical deflection of the free end if the ruler is mounted with: (i) the X-X axis horizontal; (ii) the Y-Y axis horizontal. (Ignore the deflection caused by the self-weight of the ruler.)

*Answer.*

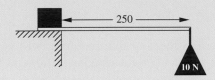

Use equation (5.5) with $C_1 = 3$ (from Figure 5.3), and use $I_{XX}$ and $I_{YY}$ from Example 5.1.

(i) With X-X horizontal: $\delta = \dfrac{FL^3}{C_1 E I_{XX}} = \dfrac{10(0.25^3)}{3(200 \times 10^9)(2.1 \times 10^{-12})} = 124$ mm.

(ii) With Y-Y horizontal: $\delta = \dfrac{FL^3}{C_1 E I_{YY}} = \dfrac{10(0.25^3)}{3(200 \times 10^9)(1300 \times 10^{-12})} = 0.02$ mm.

This shows the dramatic benefit of locating the material of a beam as far as possible from the bending axis, so as to maximise the second moment of area. Consequently 'I-sections' are often used for structural girders.

(b) A new ruler is to be made from polystyrene, with Young's modulus of 2 GPa. Its length and width are to be the same as the stainless steel ruler, but its thickness can be changed by the designer. The polystyrene ruler is to be mounted with its X-X axis horizontal and its deflection must not be any greater than that of the stainless steel ruler under the 10 N end load. Calculate the thickness needed to achieve this. Compare the masses of the two rulers. (The densities of stainless steel and polystyrene are 7800 kg/m$^3$ and 1040 kg/m$^3$.)

*Answer.* Using the same equations,

$$I_{XX} = \frac{FL^3}{C_1 E \delta} = \frac{10(0.25^3)}{3(2 \times 10^9)(0.124)} = 210 \times 10^{-12} \text{ m}^4 = 210 \text{ mm}^4$$

The thickness can be found from

$$I_{XX} = \frac{wt^3}{12} : t \geq \left[\frac{12\,I_{XX}}{w}\right]^{1/3} = \left[\frac{12 \times 210}{25}\right]^{1/3} = 4.6 \text{ mm.}$$

The masses are: $m_{SS} = 7800 \times 0.3 \times 0.025 \times 0.001 = 59$ g and $m_{PS} = 1040 \times 0.3 \times 0.025 \times 0.0046 = 36$ g. Despite its much lower density, the polystyrene ruler is only 40% lighter than the stainless steel one, because of its much greater thickness.

*Torsion of shafts* A torque, $T$, applied to the ends of an isotropic bar of a uniform section generates a shear stress $\tau$ (Figures 5.1(c) and 5.4). For circular sections, the shear stress varies with radial distance $r$ from the axis of symmetry is

$$\frac{\tau}{r} = \frac{T}{K} \tag{5.6}$$

where $K$ measures the resistance of the section to twisting (the torsional equivalent to $I$, for bending). $K$ is easiest to calculate for circular sections, when it is equal to the polar second moment of area:

$$J = \int\limits_{\text{section}} 2\pi r^3 \, \mathrm{d}r \tag{5.7}$$

where $r$ is measured radially from the centre of the circular section. For non-circular sections, $K$ is less than $J$. Figure 5.2 gives expressions for $K$ for four standard shapes.

The shear stress acts in the plane normal to the axis of the bar. It causes the bar, with length $L$, to twist through an angle $\theta$. The twist per unit length, $\theta/L$, is related to the shear stress and the torque by

$$\frac{\tau}{r} = \frac{T}{K} = \frac{G\theta}{L} \tag{5.8}$$

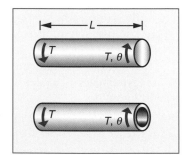

Figure 5.4 Elastic torsion of circular shafts. The stress in the shaft and the twist per unit length depend on the torque $T$ and the torsional rigidity $GK$.

where $G$ is the shear modulus. The ratio of torque to twist, $T/\theta$, per unit length, is equal to $GK$, called the *torsional rigidity*.

---

### Example 5.3

(a) Derive an expression for the polar second moment of area of a tube having a hollow circular section with inner radius $r_i$ and outer radius $r_o$.
(b) A brass rod with shear modulus 40 GPa, length 200 mm, and having a solid circular cross-section with diameter 10 mm, is twisted with a torque of 10 Nm. What is the angle of twist?

*Answer.*

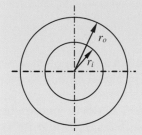

(a) Applying equation (5.7): $J = \int_{r_i}^{r_o} 2\pi r^3 \, dr = \frac{\pi}{2}[r_o^4 - r_i^4]$

(b) For a rod with a solid circular section of radius $R = 5$ mm, $J = \frac{\pi}{2}0.005^4 = 0.98 \times 10^{-9}\,\mathrm{m}^4$. So the angle of twist is given by:

$$\theta = \frac{TL}{GJ} = \frac{10 \times 0.2}{(0.98 \times 10^{-9})(40 \times 10^9)} = 0.051 \text{ rad} = 2.9 \text{ deg}$$

---

*Buckling of columns and plates*   If sufficiently slender, an elastic column or plate, loaded in compression, fails by elastic buckling at a critical load, $F_{crit}$. Beam theory shows that the critical buckling load depends on the length $L$ and flexural rigidity, $EI$:

$$F_{crit} = \frac{n^2 \pi^2 EI}{L^2} \tag{5.9}$$

where $n$ is a constant that depends on the end constraints: clamped, or free to rotate, or free also to translate (Figure 5.5). The value of $n$ is just the number of half-wavelengths of the buckled shape (the half-wavelength is the distance between inflection points).

Some of the great engineering disasters have been caused by buckling. It can occur without warning and slight misalignment is enough to reduce the load at which it happens. When dealing with compressive loads on columns or in-plane compression of plates it is advisable to check that you are well away from the buckling load.

*Vibrating beams and plates*   Any undamped system vibrating at one of its natural frequencies can be reduced to the simple problem of a mass $m$ attached to a spring of stiffness $k$—the restoring force per unit displacement. The lowest natural frequency of such a system is

$$f = \frac{1}{2\pi}\sqrt{\frac{k}{m}} \tag{5.10}$$

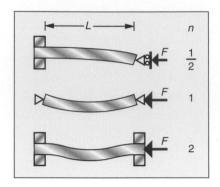

**Figure 5.5** The buckling load of a column of length $L$ depends on the flexural rigidity $EI$ and on the end constraints; three are shown here, together with the value of $n$.

Different geometries require appropriate estimates of the effective $k$ and $m$—often these can be estimated with sufficient accuracy by approximate modeling. The higher natural frequencies of rods and beams are simple multiples of the lowest.

Figure 5.6 shows the lowest natural frequencies of the flexural modes of uniform beams or plates of length $L$ with various end constraints. The spring stiffness, $k$, is that for bending, given by equation (5.5), so the natural frequencies can be written:

$$f = \frac{C_2}{2\pi} \sqrt{\frac{EI}{m_0 L^4}} \qquad (5.11)$$

where $C_2$, listed in the figure, depends on the end constraints and $m_0$ is the mass of the beam per unit length. The mass per unit length is just the area times the density, $A\rho$, *so the natural frequency* becomes

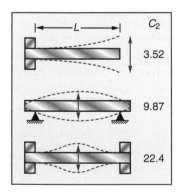

**Figure 5.6** The natural vibration modes of beams clamped in different ways.

$$f = \frac{C_2}{2\pi}\sqrt{\frac{1}{AL^4}}\sqrt{\frac{E}{\rho}} \tag{5.12}$$

Thus, frequencies scale as $\sqrt{E/\rho}$.

## 5.3 Material indices for elastic design

We can now start to implement the steps outlined in Chapter 3 and summarised in Figure 3.6:

- *Translation*.
- *Screening*, based on constraints.
- *Ranking*, based on objectives.
- *Documentation*, to give greater depth.

The first two were fully described in Chapter 3. The third—ranking based on objectives—requires simple modeling to identify the *material index*.

An *objective*, remember, is a criterion of excellence for the design as a whole, something to be minimised (like cost, weight or volume) or maximised (like energy storage). Here we explore those for elastic design.

***Minimising weight: a light, stiff tie-rod*** Think first of choosing a material for a cylindrical tie-rod like one of those in the cover picture of Chapter 4. Its length $L_o$ is specified and it must carry a tensile force $F$ without extending elastically by more than $\delta$. Its stiffness must be at least $S^* = F/\delta$ (Figure 5.7(a)). This is a load-carrying component, so it will need to have some

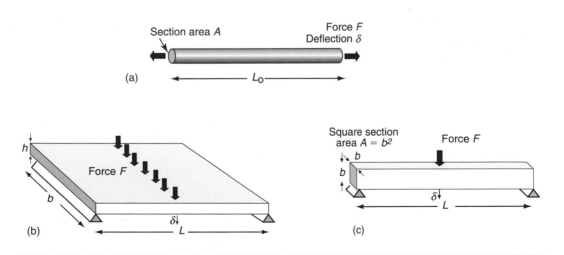

**Figure 5.7** (a) A tie with cross-section area $A$, loaded in tension. Its stiffness is $S = F/\delta$ where $F$ is the load and $\delta$ is the extension. (b) A panel loaded in bending. Its stiffness is $S = F/\delta$, where $F$ is the total load and $\delta$ is the bending deflection. (c) A beam of square section, loaded in bending. Its stiffness is $S = F/\delta$, where $F$ is the load and $\delta$ is the bending deflection.

Table 5.1   Design requirements for the light, stiff tie

| Function | • Tie-rod | |
|---|---|---|
| Constraints | • Stiffness S* specified | Functional constraints |
| | • Some toughness | |
| | • Length $L_o$, specified | Geometric constraint |
| Objective | • Minimise mass | |
| Free vanables | • Choice of material | |
| | • Choice of cross-section area $A$ | |

toughness. The objective is to make it as light as possible. The cross-section area $A$ is free. The design requirements, translated, are listed in Table 5.1.

We first seek an equation that describes the quantity to be maximised or minimised, here the mass $m$ of the tie. This equation, called the *objective function*, is

$$m = AL_o\rho \tag{5.13}$$

where $A$ is the area of the cross-section and $\rho$ is the density of the material of which it is made. We can reduce the mass by reducing the cross-section, but there is a constraint: the section area $A$ must be sufficient to provide a stiffness of $S^*$, which, for a tie, is given by equation (5.2):

$$S^* = \frac{AE}{L_o} \tag{5.14}$$

If the material has a low modulus, a large $A$ is needed to give the necessary stiffness; if $E$ is high, a smaller $A$ is needed. But which gives the lower mass? To find out, we eliminate the free variable $A$ between these two equations, giving

$$m = S^* L_o^2 \left(\frac{\rho}{E}\right) \tag{5.15}$$

Both $S^*$ and $L_o$ are specified. The lightest tie that will provide a stiffness $S^*$ is that made of the material with the smallest value of $\rho/E$. We could define this as the material index of the problem, seeking the material with a minimum value, but it is more usual to express indices in a form for which a maximum is sought. We therefore invert the material properties in equation (5.15) and define the material index $M_t$ (subscript 't' for tie) as:

$$M_t = \frac{E}{\rho} \tag{5.16}$$

It is called the *specific stiffness*. Materials with a high value of $M_t$ are the best choice, provided that they also meet any other constraints of the design, in this case the need for some toughness.

The mode of loading that most commonly dominates in engineering is not tension, but bending—think of floor joists, of wing spars, of golf club shafts. The index for bending differs from that for tension, and this (significantly) changes the optimal choice of material. We start by modeling panels and beams, specifying stiffness and seeking to minimise weight.

**Table 5.2**  Design requirements for the light, stiff panel

| Function | • Panel | |
|---|---|---|
| Constraints | • Stiffness $S^*$ specified | Functional constraint |
| | • Length $L$ and width $b$ specified | Geometric constraint |
| Objective | • Minimise mass | |
| Free variables | • Choice of material | |
| | • Choice of panel thickness $h$ | |

*Minimising weight: a light, stiff panel*   A panel is a flat slab, like a table top. Its length $L$ and width $b$ are specified but its thickness $h$ is free. It is loaded in bending by a central load $F$ (Figure 5.7(b)). The stiffness constraint requires that it must not deflect more than $\delta$ under the load $F$ and the objective is again to make the panel as light as possible. Table 5.2 summarises the design requirements.

The objective function for the mass of the panel is the same as that for the tie:

$$m = AL\rho = bhL\rho \tag{5.17}$$

Its bending stiffness $S$ is given by equation (5.5). It must be at least:

$$S^* = \frac{C_1 EI}{L^3} \tag{5.18}$$

The second moment of area, $I$, for a rectangular section (Table 5.2) is

$$I = \frac{bh^3}{12} \tag{5.19}$$

The stiffness $S^*$, the length $L$ and the width $b$ are specified; only the thickness $h$ is free. We can reduce the mass by reducing $h$, but only so far that the stiffness constraint is still met. Using the last two equations to eliminate $h$ in the objective function for the mass gives

$$m = \left(\frac{12S^*}{C_1 b}\right)^{1/3} (bL^2) \left(\frac{\rho}{E^{1/3}}\right) \tag{5.20}$$

The quantities $S^*$, $L$, $b$ and $C_1$ are all specified; the only freedom of choice left is that of the material. The best materials for a light, stiff panel are those with the smallest values of $\rho/E^{1/3}$ (again, so long as they meet any other constraints). As before, we will invert this, seeking instead large values of the material index $M_p$ for the panel:

$$M_p = \frac{E^{1/3}}{\rho} \tag{5.21}$$

This doesn't look much different from the previous index, $E/\rho$, but it is. It leads to a different choice of material, as we shall see in a moment. For now, note the procedure. The length of the panel was specified but we were free to vary the section area. The objective is to minimise its mass, $m$. Use the stiffness constraint to eliminate the free variable, here $h$. Then read off the combination of material properties that appears in the objective function—the equation for the mass. It is the index for the problem.

It sounds easy, and it is—so long as you are clear from the start what the constraints are, what you are trying to maximise or minimise, and which parameters are specified and which are free.

Now we look at another bending problem, in which the freedom to choose shape is rather greater than for the panel.

*Minimising weight: a light, stiff beam*   Beams come in many shapes: solid rectangles, cylindrical tubes, I-beams and more. Some of these have too many free geometric variables to apply the previous method directly. However, if we constrain the shape to be *self-similar* (such that all dimensions change in proportion as we vary the overall size), the problem becomes tractable again. We therefore consider beams in two stages: first, to identify the optimum materials for a light, stiff beam of a prescribed simple shape (such as a square section); then, second, we explore how much lighter it could be made, for the same stiffness, by using a more efficient shape.

Consider a beam of square section $A = b \times b$ that may vary in size but with the square shape retained. It is loaded in bending over a span of fixed length $L$ with a central load $F$ (Figure 5.7(c)). The stiffness constraint is again that it must not deflect more than $\delta$ under the load $F$, with the objective that the beam should again be as light as possible. Table 5.3 summarises the design requirements.

Proceeding as before, the objective function for the mass is:

$$m = AL\rho = b^2L\rho \tag{5.22}$$

The beam bending stiffness $S$ (equation (5.5)) is

$$S^* = \frac{C_1 EI}{L^3} \tag{5.23}$$

The second moment of area, $I$, for a square section beam is

$$I = \frac{b^4}{12} = \frac{A^2}{12} \tag{5.24}$$

For a given length $L$, the stiffness $S^*$ is achieved by adjusting the size of the square section. Now eliminating $b$ (or $A$) in the objective function for the mass gives

$$m = \left(\frac{12S^*L^3}{C_1}\right)^{1/2} (L) \left(\frac{\rho}{E^{1/2}}\right) \tag{5.25}$$

The quantities $S^*$, $L$ and $C_1$ are all specified—the best materials for a light, stiff beam are those with the smallest values of $\rho/E^{1/2}$. Inverting this, we require large values of the material index $M_b$ for the beam:

$$M_b = \frac{E^{1/2}}{\rho} \tag{5.26}$$

Table 5.3   Design requirements for the light, stiff beam

| Function | • Beam | |
|---|---|---|
| Constraints | • Stiffness $S^*$ specified | Functional constraint |
| | • Length $L$ } | |
| | • Section shape square | Geometric constraints |
| Objective | • Minimise mass | |
| Free variables | • Choice of material | |
| | • Area $A$ of cross-section | |

This analysis was for a square beam, but the result in fact holds for any shape, so long as the shape is held constant. This is a consequence of equation (5.24)—for a given shape, the second moment of area $I$ can always be expressed as a constant times $A^2$, *so changing the shape just changes the constant* $C_1$ in equation (5.25), not the resulting index.

As noted earlier, real beams have section shapes that improve their efficiency in bending, requiring less material to get the same stiffness. By shaping the cross-section it is possible to increase $I$ without changing $A$. This is achieved by locating the material of the beam as far from the neutral axis as possible, as in thin-walled tubes or I-beams. Some materials are more amenable than others to being made into efficient shapes. Comparing materials on the basis of the index in equation (5.26) therefore requires some caution—materials with lower values of the index may 'catch up' by being made into more efficient shapes. So we need to get an idea of the effect of shape on bending performance.

Figure 5.8 shows a solid square beam, of cross-section area $A$. If we turned the same area into a tube, as shown in the left of the figure, the mass of the beam is unchanged (equation (5.22)). The second moment of area, $I$, however, is now much greater—and so is the stiffness (equation (5.23)). We define the ratio of $I$ for the shaped section to that for a solid square section with the same area (and thus mass) as the *shape factor* $\Phi$. The more slender the shape the larger is $\Phi$, but there is a limit—make it too thin and the tube will buckle—so there is a *maximum shape factor* for each material that depends on its properties. Table 5.4 lists some typical values.

Shaping is also used to make structures lighter: it is a way to get the same stiffness with less material (Figure 5.8, right). The mass ratio is given by the reciprocal of the square root of the maximum shape factor, $\Phi^{-1/2}$ (because $C_1$, which is proportional to the shape factor, appears as $(C_1)^{-1/2}$ in equation (5.25)). Table 5.4 lists the factor by which a beam can be made lighter, for the same stiffness, by shaping. Metals and composites can all be improved significantly (though the metals do a little better), but wood has more limited potential because it is more difficult to shape it into efficient, thin-walled shapes. So, when comparing materials for light, stiff beams using the index in equation (5.26), the performance of wood is not as good as it looks because other materials can be made into more efficient shapes. As we will see, composites (particularly CFRP) have very high values of all three indices $M_t$, $M_p$ and $M_b$, but for beams this advantage relative to metals is reduced a little by the effect of shape.

Shaping offers exactly the same benefits under torsional loading—a tube of the same mass has a greater resistance to twisting (or, alternatively, a lighter tube can do the same job as a solid circular rod). The following example illustrates the effect.

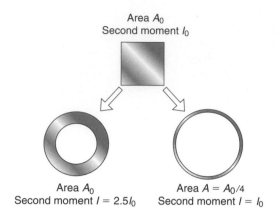

Area $A_0$
Second moment $I_0$

Area $A_0$
Second moment $I = 2.5I_0$

Area $A = A_0/4$
Second moment $I = I_0$

**Figure 5.8** The effect of section shape on bending stiffness $EI$: a square-section beam compared, left, with a tube of the same area (but 2.5 times stiffer) and, right, a tube with the same stiffness (but four times lighter).

**Table 5.4** The effect of shaping on stiffness and mass of beams in different structural materials

| Material | Typical maximum shape factor (stiffness relative to that of a solid square beam) | Typical mass ratio by shaping (relative to that of a solid square beam) |
| --- | --- | --- |
| Steels | 64 | 1/8 |
| Al alloys | 49 | 1/7 |
| Composites (GFRP, CFRP) | 36 | 1/6 |
| Wood | 9 | 1/3 |

### Example 5.4

The brass rod in Example 5.3 is formed into a hollow tube with an outer radius of 25 mm, and the same length, using the same amount of material. It is twisted with the same torque. What is its angle of twist? Compare its torsional stiffness with the solid rod in Example 5.3.

*Answer.* To have the same amount of material, the cross-section area of the tube $\pi(r_o^2 - r_i^2)$ must equal that of the solid bar $\pi R^2$. Therefore the inner radius of the tube must be

$$r_i = \sqrt{r_o^2 - R^2} = \sqrt{12.5^2 - 5^2} = 11.5 \text{ mm}$$

(i.e. a wall thickness of 1 mm). The polar moment of area of the tube is then:

$$\frac{\pi}{2}\left[0.0125^4 - 0.0115^4\right] = 10.9 \times 10^{-9} \text{m}^4$$

and the angle of twist is

$$\theta = \frac{10 \times 0.2}{(10.9 \times 10^{-9})(40 \times 10^9)} = 4.6 \times 10^{-3} \text{ rad} = 0.26 \text{ deg}$$

This compares to 2.9 deg for the solid shaft in Example 5.3. Therefore the hollow shaft, which has the same mass as the solid one, has about 11 times the torsional stiffness ($T/\theta$).

*Minimising material cost* When the objective is to minimise cost rather than weight, the indices change. If the material price is $C_m$ $/kg, the cost of the material to make a component of mass $m$ is just $mC_m$. The objective function for the material cost $C$ of the tie, panel or beam then becomes

$$C = mC_m = ALC_m\rho \tag{5.27}$$

Proceeding as in the three previous examples then leads to indices that are just those of equations (5.16), (5.21) and (5.26), with $\rho$ replaced by $C_m\rho$.

The material cost is only part of the cost of a shaped component; there is also the manufacturing cost—the cost to shape, join and finish it. We leave these to a later chapter.

## 5.4 Plotting limits and indices on charts

*Screening: attribute limits on charts* Any design imposes certain non-negotiable demands (constraints) on the material of which it is made. These limits can be plotted as horizontal or vertical lines on material property charts, as illustrated in Figure 5.9, which shows a schematic of the $E$−*Relative cost* chart of Figure 4.9. We suppose that the design imposes limits on these of $E > 10$ GPa and *Relative cost* $< 3$, shown in the figure. All materials in the window defined by the limits, labeled 'Search region', meet both constraints.

Later chapters of this book show charts for many other properties. They allow limits to be imposed on other properties.

*Ranking: indices on charts* The next step is to seek, from the subset of materials that meet the property limits, those that maximise the performance of the component. We will use the design of light, stiff components as examples; the other material indices are used in a similar way.

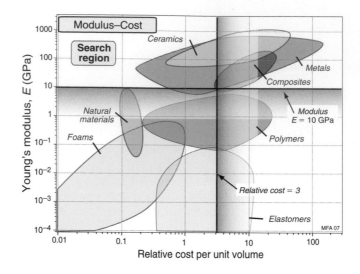

**Figure 5.9**   A schematic *E–Relative cost* chart showing a lower limit for *E* and an upper one for *Relative cost.*

Figure 5.10 shows a schematic of the $E-\rho$ chart of Figure 4.8. The logarithmic scales allow all three of the indices $E/\rho$, $E^{1/3}/\rho$ and $E^{1/2}/\rho$, derived in the last section, to be plotted onto it. Consider the condition

$$M = \frac{E}{\rho} = \text{constant}, C$$

that is, a particular value of the specific stiffness. Taking logs,

$$\log(E) = \log(\rho) + \log(C) \tag{5.28}$$

This is the equation of a straight line of slope 1 on a plot of $\log(E)$ against $\log(\rho)$, as shown in the figure. Similarly, the condition

$$M = \frac{E^{1/3}}{\rho} = \text{constant}, C$$

becomes, on taking logs,

$$\log(E) = 3\log(\rho) + \log(C) \tag{5.29}$$

This is another straight line, this time with a slope of 3, also shown. And by inspection, the third index $E^{1/2}/\rho$ will plot as a line of slope 2. We refer to these lines as *selection guidelines*. They give the slope of the family of parallel lines belonging to that index.

It is now easy to read off the subset of materials that maximise performance for each loading geometry. For example, all the materials that lie on a line of constant $M = E^{1/3}/\rho$ perform equally well as a light, stiff panel; those above the line perform better, those below less

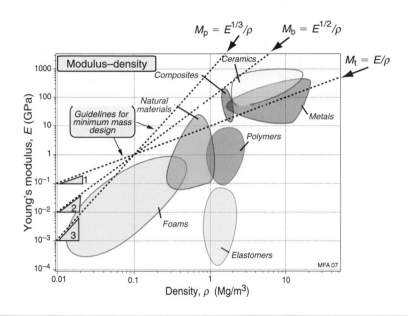

**Figure 5.10**   A schematic $E$–$\rho$ chart showing guidelines for three material indices for stiff, lightweight structures.

well. Figure 5.11 shows a grid of lines corresponding to values of $M = E^{1/3}/\rho$ from $M = 0.22$ to $M = 4.6$ in units of $GPa^{1/3}/(Mg/m^3)$. A material with $M = 3$ in these units gives a panel that has one-tenth the weight of one with $M = 0.3$. The case studies in the next section give practical examples.

*Computer-aided selection*    The charts give an overview, but the number of materials that can be shown on any one of them is obviously limited. Selection using them is practical when there are very few constraints, but when there are many—as there usually are—checking that a given material meets them all is cumbersome. Both problems are overcome by a computer implementation of the method.

The *CES* material and process selection software[1] is an example of such an implementation. Its database contains records for materials, organised in the hierarchical manner shown in Figure 2.3 in Chapter 2. Each record contains property data for a material, each property stored as a range spanning its typical (or, often, permitted) values. It also contains limited documentation in the form of text, images and references to sources of information about the material. The data are interrogated by a search engine that offers the search interfaces shown schematically in Figure 5.12. On the left is a simple query interface for screening on single attributes. The desired upper or lower limits for constrained properties are entered; the search engine rejects all materials with attributes that lie outside the limits. In the centre is shown a second way of interrogating the data: a bar chart, constructed by

---

[1] Granta Design Ltd, Cambridge, UK (www.grantadesign.com).

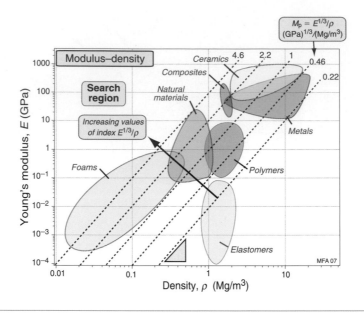

**Figure 5.11** A schematic $E$–$\rho$ chart showing a grid of lines for the index $E^{1/3}/\rho$. The units are $(\text{GPa})^{1/3}/(\text{Mg/m}^3)$.

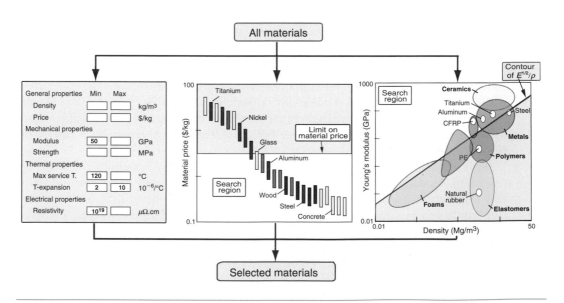

**Figure 5.12** Computer-aided selection using the CES software. The schematic shows the three types of selection window. They can be used in any order and any combination. The selection engine isolates the subset of materials that pass all the selection stages.

the software, for any numeric property in the database. It, and the bubble chart shown on the right, are ways both of applying constraints and of ranking. For screening, a selection line or box is superimposed on the charts with edges that lie at the constrained values of the property (bar chart) or properties (bubble chart). This eliminates the materials in the shaded areas and retains the materials that meet the constraints. If, instead, ranking is sought (having already applied all necessary constraints) an index-line like that shown in Figure 5.11 is positioned so that a small number—say, 10—of materials are left in the selected area; these are the top-ranked candidates. The software delivers a list of the top-ranked materials that meet all the constraints.

## 5.5 Case studies

Here we have case studies using the two charts of Chapter 4. They are deliberately simplified to avoid obscuring the method under layers of detail. In most cases little is lost by this: the best choice of material for the simple example is the same as that for the more complex.

*Light levers for corkscrews*   The lever of the corkscrew (Figure 5.13) is loaded in bending: the force $F$ creates a bending moment $M = FL$. The lever needs to be stiff enough that the bending displacement, $\delta$, when extracting a cork, is acceptably small. If the corkscrew is intended for travelers, it should also be light. The section is rectangular. We make the assumption of self-similarity, meaning that we are free to change the scale of the section but not its shape. The material index we want was derived earlier as equation (5.26). It is that for a light, stiff beam:

$$M = \frac{E^{1/2}}{\rho} \tag{5.30}$$

where $E$ is Young's modulus and $\rho$ is the density. There are other obvious constraints. Corkscrews get dropped and must survive impacts of other kinds, so brittle materials like glass or ceramic are unacceptable. Given these requirements, summarised in Table 5.5, what materials would you choose?

Figure 5.14 shows the appropriate chart: that in which Young's modulus, $E$, is plotted against density, $\rho$. The selection line for the index $M$ has a slope of 2, as explained in Section 5.3; it is positioned so that a small group of materials is left above it. They are the materials

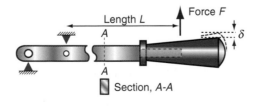

**Figure 5.13**   The corkscrew lever from Chapter 3. It must be adequately stiff and, for traveling, as light as possible.

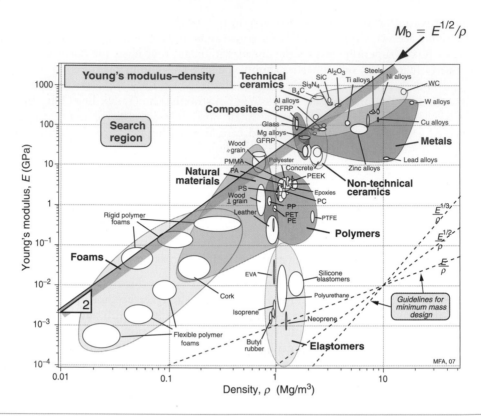

**Figure 5.14** Selection of materials for the lever. The objective is to make it as light as possible while meeting a stiffness constraint.

**Table 5.5** Design requirements for the corkscrew lever

| Function | • Lightweight lever, meaning light, stiff beam |
|---|---|
| Constraints | • Stiffness $S^*$ specified |
| | • Length $L$ |
| | • Section shape rectangular |
| Objective | • Minimise mass |
| Free variables | • Choice of material |
| | • Area $A$ of cross-section |

with the largest values of $M$, and it is these that are the best choice, provided they satisfy the other constraints. Three classes of materials lie above the line: woods, carbon-fiber reinforced polymers (CFRPs) and a number of ceramics. Ceramics are brittle and expensive, ruling them out. The recommendation is clear. Make the lever out of wood or—better—out of CFRP.

Table 5.6  Design requirements for floor beams

| Function | • Floor beam |
|---|---|
| Constraints | • Stiffness $S^*$ specified |
| | • Length $L$ specified |
| | • Section shape square |
| Objective | • Minimise material cost |
| Free variables | • Choice of material |
| | • Area $A$ of cross-section |

*Cost: structural materials for buildings*   The most expensive thing that most people ever buy is the house they live in. Roughly half the cost of a house is the cost of the materials of which it is made, and these are used in large quantities (family house: around 200 tonnes; large apartment block: around 20 000 tonnes). The materials are used in three ways: structurally to hold the building up; as cladding, to keep the weather out; and as 'internals', to insulate against heat and sound, and to provide comfort and decoration.

Consider the selection of materials for the structure (Figure 5.15). They must be stiff, strong and cheap: stiff, so that the building does not flex too much under wind loads or internal loading; strong, so that there is no risk of it collapsing; and cheap, because such a lot of material is used. The structural frame of a building is rarely exposed to the environment, and is not, in general, visible, so criteria of corrosion resistance or appearance are not important here. The design goal is simple: stiffness and strength at minimum cost. To be more specific: consider the selection of material for floor joists, focusing on stiffness. Table 5.6 summarises the requirements.

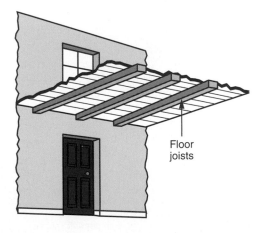

Floor
joists

Figure 5.15   The materials of a building are chosen to perform three different roles. Those for the structure are chosen to carry loads. Those for the cladding provide protection from the environment. Those for the interior control heat, light and sound. Here we explore structural materials.

The material index for a stiff beam of minimum mass, $m$, was developed earlier. The cost $C$ of the beam is just its mass, $m$, times the cost per kg, $C_m$, of the material of which it is made:

$$C = mC_m = AL\rho C_m$$

Proceeding as in Section 5.3, we find the index for a stiff beam of minimum cost to be:

$$M = \frac{E^{1/2}}{\rho C_m} \tag{5.31}$$

Figure 5.16 shows the relevant chart: modulus $E$ against relative cost per unit volume, $\rho C_m$ (the chart uses a *relative* cost $C_{v,R}$, defined in Chapter 4, in place of $C_m$ but this makes no difference to the selection). The shaded band has the appropriate slope for $M$; it isolates concrete, stone, brick, woods, cast irons and carbon steels.

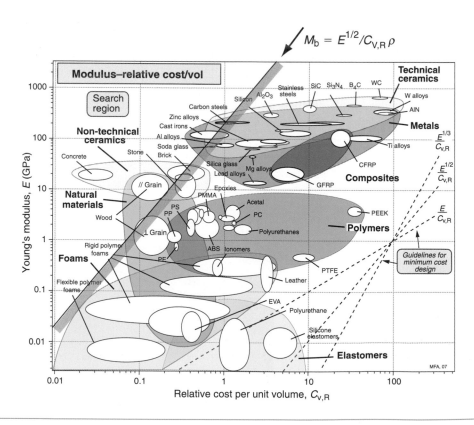

**Figure 5.16**   The selection of materials for stiff floor beams. The objective is to make them as cheap as possible while meeting a stiffness constraint.

Concrete, stone and brick have strength only in compression; the form of the building must use them in this way (walls, columns, arches). Wood, steel and reinforced concrete have strength both in tension and compression, and steel, additionally, can be given efficient shapes (I-sections, box sections, tubes) that can carry bending and tensile loads as well as compression, allowing greater freedom of the form of the building.

*Cushions and padding: the modulus of foams*   One way of manipulating the modulus is to make a material into a foam. Figure 4.26 showed an idealised foam structure: a network of struts of length $L$ and thickness $t$, connected at their mid-span to neighbouring cells. Cellular solids like this one are characterised by their *relative density*, which for the structure shown here (with $t << L$) is

$$\frac{\tilde{\rho}}{\rho_s} = \left(\frac{t}{L}\right)^2 \tag{5.32}$$

where $\tilde{\rho}$ is the density of the foam and $\rho_s$ is the density of the solid of which it is made.

If a compressive stress $\sigma$ is applied to a block of foam containing many cells, it is transmitted through the foam as forces $F$ pushing on edges that lie parallel to the direction of $\sigma$. The area of one cell face is $L^2$ so the force on one strut is $F = \sigma L^2$. This force bends the cell edge to which it connects, as on the right of Figure 4.23. Thus, the cell edge is just a beam, built-in at both ends, carrying a central force $F$. The bending deflection is given by equation (5.5):

$$\delta = \frac{FL^3}{C_1 E_s I} \tag{5.33}$$

where $E_s$ is the modulus of the solid of which the foam is made and $I = t^4/12$ is the second moment of area of the cell edge of square cross-section, $t \times t$. The compressive strain suffered by the cell as a whole is then $\varepsilon = 2\delta/L$. Assembling these results gives the modulus $\tilde{E} = \sigma/\varepsilon$ of the foam as

$$\frac{\tilde{E}}{E_s} \propto \left(\frac{\tilde{\rho}}{\rho_s}\right)^2 \tag{5.34}$$

Since $\tilde{E} = E_s$ when $\tilde{\rho} = \rho_s$, the constant of proportionality is 1, giving the result plotted earlier in Figure 4.27.

*Vibration: avoiding resonance when changing material*   Vibration, as the story at the start of this chapter relates, can be a big problem. Bridges are designed with sufficient stiffness to prevent wind-loads exciting their natural vibration frequencies (one, the Tacoma Straits bridge in the state of Washington, wasn't; its oscillations destroyed it). Auto-makers invest massively in computer simulation of new models to be sure that door and roof panels don't start thumping because the engine vibration hits a natural frequency. Even musical instruments, which rely on exciting natural frequencies, have problems with 'rogue' tones: notes that excite frequencies you didn't want as well as those you did.

Suppose you redesign your bridge, or car, or cello and, in a creative moment, decide to make it out of a new material. What will it do to the natural frequencies? Simple. Natural frequencies, $f$, as explained in Section 5.2, are proportional to $\sqrt{E/\rho}$. If the old material has a modulus $E_o$ and density $\rho_o$ and the new one has $E_n$, $\rho_n$, the change in frequency, $\Delta f$

$$\Delta f = \sqrt{\frac{E_n \rho_o}{E_o \rho_n}} \tag{5.35}$$

Provided this shift still leaves natural frequencies remote from those of the excitation, all is well. But if this is not so, a rethink would be prudent.

***Bendy design: part-stiff, part-flexible structures***   The examples thus far aimed at the design of components that did not bend too much—that is, they met a stiffness constraint. Elasticity can be used in another way: to design components that are strong but *not* stiff, arranging that they bend easily in a certain direction. Think, for instance, of a windscreen wiper blade. The frame to which the rubber squeegees are attached must adapt to the changing profile of the windscreen as it sweeps across it. It does so by flexing, maintaining an even pressure on the blades. This is a deliberately bendy structure.

Figure 5.17 shows two ways of making a spring-loaded plunger. The one on the left with a plunger and a spring involves sliding surfaces. The one on the right has none: it uses elastic bending to both locate and guide the plunger and to give the restoring force provided by the spring in the first design.

Exploiting elasticity in this way has many attractions. There is no friction, no wear, no need for lubrication, no precise clearances between moving parts. And in design with polymers there is another bonus; since there are no sliding surfaces it is often possible to mould the entire device as a single unit, reducing the part-count and doing away with the need for assembly. Reducing part-count is music to the ears of production engineers: it is cost-effective.

Figure 5.18 shows examples you will recognize: the 'living hinge' used on toothpaste tubes, moulded plastic boxes, clips and clothes pegs. Here the function of a rotational hinge is replaced by an elastic connecting strip, wide but thin. The width gives lateral registration; the thinness allows easy flexure. The traditional three-part hinge has been reduced to a single moulding.

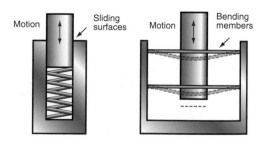

Figure 5.17   A sliding mechanism replaced by an elastic mechanism.

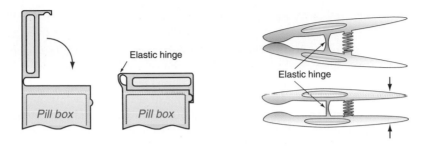

**Figure 5.18** Elastic or 'natural' hinges allowing flexure with no sliding parts.

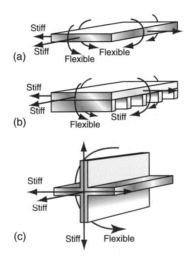

**Figure 5.19** The flexural degrees of freedom of three alternative section shapes. (a) Thin plates are flexible about any axis in the plane of the plate, but are otherwise stiff. (b) Ribbed plates are flexible about one in-plane axis but not in others. (c) Cruciform beams are stiff in bending but can be twisted easily.

There is much scope for imaginative design here. Figure 5.19 shows three section shapes, each allowing one or more degrees of elastic freedom, while retaining stiffness and strength in the other directions. By incorporating these into structures, parts can be allowed to move relative to the rest in controlled ways.

How do we choose materials for such elastic mechanisms? It involves a balance between stiffness and strength. Strength keeps appearing here—that means waiting until Chapter 7 for a full answer.

## **5.6** Summary and conclusions

Adequate stiffness is central to the design of structures that are *deflection limited*. The wing spar of an aircraft is an example: too much deflection compromises aerodynamic performance. Sports equipment is another: the feel of golf clubs, tennis rackets, skis and snowboards has much to do with stiffness. The sudden buckling of a drinking straw when you bend it is a stiffness-related problem—one that occurs, more disastrously, in larger structures. And when, as the turbine of an aircraft revs up, it passes through a speed at which vibration suddenly peaks, the cause is resonance linked to the stiffness of the turbine blades. Stiffness is influenced by the size and shape of the cross-section and the material of which it is made. The property that matters here is the elastic modulus *E*.

In selecting materials, adequate stiffness is frequently a *constraint*. This chapter explained how to meet it for various modes of loading and for differing *objectives*: minimising mass, or volume or cost. Simple modeling delivers expressions for the objective: for the mass, or for the material cost. These expressions contain material properties, either singly or in combination. It is these that we call the *material index*. Material indices measure the excellence of a material in a given application. They are used to rank materials that meet the other constraints.

Stiffness is useful, but lack of stiffness can be useful too. The ability to bend or twist allows *elastic mechanisms*: single components that behave as if they had moving parts with bearings. Elastic mechanisms have limitations, but—where practical—they require no assembly, they have no maintenance requirements and they are cheap.

The chapter ended by illustrating how indices are plotted onto material property charts to find the best selection. The method is a general one that we apply in later chapters to strength thermal, electrical, magnetic and optical properties.

## **5.7** Further reading

Ashby, M. F. (2005). *Materials Selection in Mechanical Design* (3rd ed.). Oxford, UK: Butterworth-Heinemann. Chapter 4. ISBN 0-7506-6168-2. (A more advanced text that develops the ideas presented here, including a much fuller discussion of shape factors and an expanded catalog of simple solutions to standard problems.)

Gere, J. M. (2006). *Mechanics of Materials* (6th ed.). Toronto, Canada: Thompson Publishing. ISBN 0-534-41793-0. (An intermediate level text on statics of structures by one of the fathers of the field; his books with Timoshenko introduced an entire generation to the subject.)

Hosford, W. F. (2005). *Mechanical Behavior of Materials*. Cambridge, UK: Cambridge University Press. ISBN 0-521-84670-6. (A text that nicely links stress–strain behavior to the micromechanics of materials.)

Jenkins, C. H. M., & Khanna, S. K. (2006). *Mechanics of Materials*. Boston, MA: USA: Elsevier Academic. ISBN 0-12-383852-5. (A simple introduction to mechanics, emphasising design.)

Riley, W. F., Sturges, L. D., & Morris, D. H. (2003). *Statics and Mechanics of Materials* (2nd ed.). Hoboken, NJ, USA: McGraw-Hill. ISBN 0-471-43446-9. (An intermediate level text on the stress, strain and the relationships between them for many modes of loading. No discussion of micromechanics—response of materials to stress at the microscopic level.)

Vable, M. (2002). *Mechanics of Materials*. Oxford, UK: Oxford University Press. ISBN 0-19-513337-4. (An introduction to stress—strain relations, but without discussion of the micromechanics of materials.)

Young, W. C. (1989). *Roark's Formulas for Stress and Strain* (6th ed.). New York, USA: McGraw-Hill. ISBN 0-07-100373-8. (This is the 'Yellow Pages' of formulae for elastic problems—if the solution is not here, it doesn't exist.)

# **5.8** Exercises

Exercise E5.1   Distinguish between tension, torsion, bending and buckling.

Exercise E5.2   Polymer ropes and lines for use on water are often designed to float, to aid in their retrieval and to avoid applying a downwards load to an object or person attached to them in the water. Excessive stretch is undesirable, so a lower limit of 0.5 GPa is also imposed on Young's modulus. Identify suitable polymers, using Figure 5.14.

Exercise E5.3   A beam is to be made by gluing and screwing together four wooden planks of length 2 m, thickness 20 mm and width 120 mm.

(a) Determine the second moment of area $I$ for the three configurations shown in cross-section in the figure below.

(a) Calculate the mid-span deflection $\delta$ of each of the beams when they are simply supported at their ends and loaded at mid-span with a force $F$ of 100N.

(c) Determine the stiffness $F/\delta$ for each of the beam configurations.

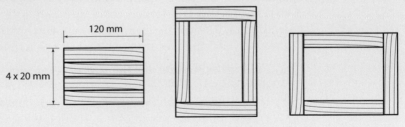

Exercise E5.4   What is meant by a *material index*?

Exercise E5.5   Your task is to design a lightweight tie of length $L$ with a circular cross-section of radius $R$. It has to carry an axial force $F$, without stretching by more than $\delta$. You will need to choose the material (with Young's modulus $E$ and density $\rho$ ) and the corresponding cross-section radius to suit your choice of material, that is, $R$ is a 'free variable'.

(a) Show that the extension of the tie is given by $\delta = FL/AE$, where $A = \pi R^2$.

(b) Rearrange this equation to find an expression for the radius $R$ of the tie that will carry the load without excessive deflection.

(c) If the length of the tie is $L = 0.3$m and the extension is not to exceed 0.1mm for a load of $F = 100$N, what value of $R$ is needed if the tie is made of: (i) PEEK; (ii) Butyl Rubber; (iii) Titanium; (iv) Copper? Use the material properties in the table below.

(d) Write an expression for the mass $m$ of the tie and determine the mass of each of the ties in (c). Which one would you choose?

(e) Substitute the expression for radius $R$ from (b) into the expression for mass $m$ from (d) to show that the material index to be *minimised* is $\rho/E$. Determine the value of the material index for the four ties.

(f) Comment on the relationship between the mass of each beam and its material index $\rho/E$.

(g) Examine Figure 5.14 to see how these three materials compare in terms of the reciprocal material index $E/\rho$.

| | Young's Modulus $E$ (GPa) | Density $\rho$ (kg/m³) | Radius $R$ (mm) | Mass $m$ (kg) | Material index $\rho/E$ |
|---|---|---|---|---|---|
| PEEK | 3.8 | 1300 | | | |
| Butyl Rubber | 0.0015 | 2400 | | | |
| Titanium alloy | 110 | 4600 | | | |
| Copper alloy | 120 | 8900 | | | |

Exercise E5.6    You are asked to design a lightweight cantilever beam of length $L$ with a square cross-section of width $b$ and depth $b$. The beam is to be built-in to a wall at one end and loaded at the free end with a point load $F$. The tip deflection is $\delta$. When designing the beam, you are free to specify the value of $b$ to suit your choice of material, that is, $b$ is a 'free variable'.

(a) Find an expression for the size of the beam $b$ that can carry $F$ with deflection $\delta$.

(b) If the length of the cantilever is $L = 1$ m and the deflection $\delta$ is not to exceed 5 mm for a load of $F = 1000$N, what value of $b$ is needed if the beam is made of: (i) Steel; (ii) Aluminium; (iii) CFRP? Use the material properties in the table below.

(c) Write an expression for the mass $m$ of the beam and determine the mass of each of the beams in (b). Which one would you choose?

(d) Substitute the expression for beam dimension $b$ from (a) into the expression for mass $m$ from (c) to show that the material index to be *minimised* is $\rho/E^{1/2}$. Determine the value of the material index for the three beams.

(e) Comment on the relationship between the mass of each beam and its material index $\rho/E^{1/2}$.

(f) Examine Figure 5.14 to see how these three materials compare in terms of the reciprocal material index $E^{1/2}/\rho$.

| | Young's Modulus $E$ (GPa) | Density $\rho$ (kg/m³) | Section size $b$ (mm) | Mass $m$ (kg) | Material index $\rho / E^{1/2}$ |
|---|---|---|---|---|---|
| Steel | 210 | 7800 | | | |
| Aluminium | 70 | 2700 | | | |
| CFRP | 100 | 1500 | | | |

**Exercise E5.7**    The objective in selecting a material for a panel of given in-plane dimensions for the casing of a portable computer is that of minimising the panel thickness $h$ while meeting a constraint on bending stiffness, $S^*$. What is the appropriate material index?

**Exercise E5.8**    A sailing enthusiast is seeking materials for lightweight panels to use in a sea-going yacht. The panels are of rectangular cross-section and will be loaded in bending, as shown in the figure. The span $L$ and width $b$ of the panels are fixed, but the depth $d$ may vary (up to a maximum specified value). The required stiffness is specified as a maximum allowable deflection $\delta$ under a given central load $W$.

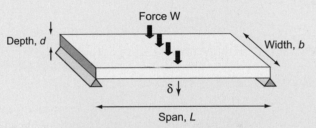

The central deflection for the simply supported span shown is given by:

$$\delta = \frac{W\,L^3}{48\,E\,I} \quad \text{where} \quad I = \frac{b\,d^3}{12}$$

The designer is interested in two scenarios: (i) minimum mass; (ii) minimum material cost.

(a) Show that the stiffness and geometric constraints lead to the relationship $E\,d^3 = $ constant, where $E$ is Young's modulus. Explain why the design specification leads to a minimum allowable value for $E$, and find an expression for this minimum value.

(b) Write down an expression for the first objective to be minimised (mass) and use the stiffness constraint to eliminate the free variable (thickness). Hence define the material performance index to *maximise* for a minimum mass design.

(c) On a Young's modulus—density property chart (Figure 5.14), how are materials with the best values of the performance index identified? Use this method to identify a short-list of candidate materials. Exclude ceramics and glasses—why is this?

(d) How does the short-list of materials in (c) change if the limit on depth leads to a requirement for a minimum required Young's modulus of 5 GPa?

(e) Modify your material performance index to describe the alternative objective of minimum material cost. Use the material property chart in Figure 5.16 to identify a short-list of materials, noting any that may be excluded because they have excessive thickness.

Exercise E5.9    Derive the material index for a torsion bar with a solid circular section. The length $L$ and the stiffness $S^*$ are specified, and the torsion bar is to be as light as possible. Follow the steps used in the text for the beam, but replace the bending stiffness $S^* = F/\delta$ with the torsional stiffness $S^* = T/(\theta/L)$ (equation 5.8), using the expression for $K$ given in Figure 5.2.

Exercise E5.10   The body of a precision instrument can be modelled as a beam with length $L$, second moment of area $I$, and mass per unit length $m_0$. It is made of a material with density $\rho$ and Young's modulus $E$. For the instrument to have the highest possible accuracy, it is desired to maximise the first natural frequency of vibration, which is given by equation (5.11), with the value of $C_2$ from Figure 5.6:

$$f = \frac{C_2}{2\pi}\sqrt{\frac{EI}{m_0 L^4}}$$

If the beam has a square cross-section $b \times b$, where the value of $b$ is free to be selected by the designer, and the stiffness $S$ is fixed by the design, show that the best material is one with a high value of the material index $E^{1/2}/\rho$.

Exercise E5.11   A material is required for a cheap column with a solid circular cross-section that must support a load $F^*$ without buckling. It is to have a height $L$. Write down an equation for the material cost of the column in terms of its dimensions, the price per kg of the material, $C_m$, and the material density $\rho$. The cross-section area $A$ is a free variable: eliminate it by using the constraint that the buckling load must not be less than $F_{crit}$ (equation 5.9). Hence read off the index for finding the cheapest tie. Plot the index on a copy of the appropriate chart and identify three possible candidates.

Exercise E5.12   The following table shows a selection of candidate materials for the light-weight, stiffness-limited design of a beam. The shape is initially constrained to be square, but the cross-sectional area can vary. Evaluate the relevant index for this problem, $E^{1/2}/\rho$, for each material. Which will be the lightest? This chapter showed that if the shape is allowed to vary, the mass can be improved by a factor of $\Phi^{-1/2}$, which is equivalent to increasing the material index by $\Phi^{1/2}$. Evaluate the modified index $(\Phi E)^{1/2}/\rho$, using the maximum shape factor for each material (as listed in Table 5.3). Does the ranking of the materials change?

| | Young's Modulus $E$ (GPa) | Density $\rho$ (kg/m$^3$) | Material index without shape $E^{1/2}/\rho$ | Material index with shape $(\Phi E)^{1/2}/\rho$ |
|---|---|---|---|---|
| Steel | 210 | 7800 | | |
| Al alloy | 70 | 2700 | | |
| GFRP | 25 | 1900 | | |
| Wood | 10 | 650 | | |

Exercise E5.13    Devise an elastic mechanism that, when compressed, shears in a direction at right angles to the axis of compression.

Exercise E5.14    Universal joints usually have sliding bearings. Devise a universal joint that could be moulded as a single elastic unit, using a polymer.

## 5.9 Exploring design with CES

Use Level 2, Materials, throughout the following exercises.

Exercise E5.15    Use a 'Limit' stage to find materials with modulus $E > 180$ GPa and price $C_m < 3$ \$/kg.

Exercise E5.16    Use a 'Limit' stage to find materials with modulus $E > 2$ GPa, density $\rho < 1000$ kg/m$^3$ and price $C_m < 3$ \$/kg.

Exercise E5.17    Make a bar chart of modulus, $E$. Add a tree stage to limit the selection to polymers alone. Which three polymers have the highest modulus?

Exercise E5.18    Repeat the material selection in Exercise E5.2 using CES. Examine the environmental resistance of the candidate materials and comment on any possible weaknesses for this application. Use the information for material applications to see if the materials identified are used in practice.

Exercise E5.19    (a) A component is made from brass (a copper alloy). Identify three alternative classes of alloy that offer higher Young's modulus for this component.

(b) Find materials with $E > 40$ GPa and $\rho < 2$ Mg/m$^3$ and identify the cheapest.

(c) Find metals and composites that are both stiffer *and* lighter than (i) steels; (ii) Ti alloys; (iii) Al alloys.

(d) Compare the specific stiffness, $E/\rho$, of steels, Ti alloys, Al alloys, Mg alloys, GFRP, CFRP and wood (parallel to the grain).

(e) Comment on the usefulness of the approaches in (c) and (d) for seeking improved performance in lightweight, stiffness-limited design.

Exercise E5.20    Make a chart showing modulus $E$ and density $\rho$. Apply a selection line of slope 1, corresponding to the index $E/\rho$ positioning the line such that six materials are left above it. Which are they and to which families do they belong?

Exercise E5.21    A material is required for a tensile tie to link the front and back walls of a barn to stabilise both. It must meet a constraint on stiffness and be as cheap as possible. To be safe the material of the tie must have a fracture toughness $K_{1c} > 18$ MPa.m$^{1/2}$ (defined in Chapter 8). The relevant index is

$$M = \frac{E}{C_m \rho}$$

Construct a chart of $E$ plotted against $C_m \rho$. Add the constraint of adequate fracture toughness, meaning $K_{1c} > 18$ MPa.m$^{1/2}$, using a 'Limit' stage. Then plot an appropriate selection line on the chart and report the three materials that are the best choices for the tie.

## 5.10  Exploring the science with CES Elements

Exercise E5.22    There is nothing that we can do to change the modulus or the density of the building blocks of all materials: the elements. We have to live with the ones we have. Make a chart of modulus $E$ plotted against the atomic number $A_n$ to explore the modulus across the periodic table. (Use a linear scale for $A_n$. To do so, change the default log scale to linear by double-clicking on the axis name to reveal the axis-choice dialog box and choose 'Linear'.) Which element has the highest modulus? Which has the lowest?

Exercise E5.23    Repeat Exercise E5.15, exploring instead the density $\rho$. Which solid element has the lowest density? Which has the highest?

# Chapter 6

# Beyond elasticity: plasticity, yielding and ductility

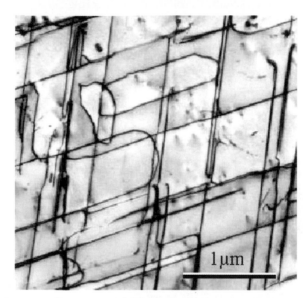

Dislocations in the intermetallic compound, Ni₃Al. (Image courtesy of C. Rentenberger and H. P. Karnthaler, Institute of Materials Physics, University of Vienna, Austria)

## **6.1** Introduction and synopsis

The verb 'to yield' has two seemingly contradictory meanings. To yield under force is to submit to it, to surrender. To yield a profit has a different, more comfortable connotation: to bear fruit, to be useful. The *yield strength*, when speaking of a material, is the stress beyond which it becomes plastic. The term is well chosen: yield and the plasticity that follows can be profitable—it allows metals to be shaped and it allows structures to tolerate impact and absorb energy. But the unplanned yield of the span of a bridge or of the wing spar of an aircraft or of the forks of your bicycle spells disaster.

   This chapter is about yield and plasticity. For that reason it is mainly (but not wholly) about metals: it is the plasticity of iron and steel that made them the structural materials on which the Industrial Revolution was built, enabling the engineering achievements of the likes of Telford[1] and Brunel[2]. The dominance of metals in engineering, even today, derives from their ability to be rolled, forged, drawn and stamped.

## **6.2** Strength, plastic work and ductility: definition and measurement

Yield properties and ductility are measured using the standard tensile tests introduced in Chapter 4, with the materials taken to failure. Figures 6.1–6.3 show the types of stress–strain behavior observed in different material classes. The *yield strength* $\sigma_y$ (or *elastic limit* $\sigma_{el}$)—units: MPa or $MN/m^2$—requires careful definition. For metals, the onset of plasticity is not always distinct so we identify $\sigma_y$ with the *0.2% proof stress*—that is, the stress at which the stress–strain curve for axial loading deviates by a strain of 0.2% from the linear elastic line as shown in Figure 6.1. It is the same in tension and compression. When strained beyond the yield point, most metals *work harden*, causing the rising part of the curve, until a maximum, the *tensile strength*, is reached. This is followed in tension by non-uniform deformation (*necking*) and fracture.

   For polymers, $\sigma_y$ is identified as the stress at which the stress–strain curve becomes markedly non-linear: typically, a strain of 1% (Figure 6.2). The behavior beyond yield depends on the temperature relative to the glass temperature $T_g$. Well below $T_g$ most polymers are brittle. As $T_g$ is approached, plasticity becomes possible until, at about $T_g$, thermoplastics exhibit *cold drawing*: large plastic extension at almost constant stress during which the molecules are pulled into alignment with the direction of straining, followed by

---

[1] Thomas Telford (1757–1834), Scottish engineer, brilliant proponent of the suspension bridge at a time when its safety was a matter of debate. Telford may himself have had doubts—he was given to lengthy prayer on the days that the suspension chains were scheduled to take the weight of the bridge. Most of his bridges, however, still stand.

[2] Isambard Kingdom Brunel (1806–1859), perhaps the greatest engineer of the Industrial Revolution (c. 1760–1860) in terms of design ability, personality, power of execution and sheer willingness to take risks—the *Great Eastern*, for example, was five times larger than any previous ship ever built. He took the view that 'great things are not done by those who simply count the cost'. Brunel was a short man and self-conscious about his height; he favored tall top hats to make himself look taller.

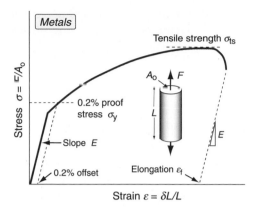

**Figure 6.1**   Stress–strain curve for a metal.

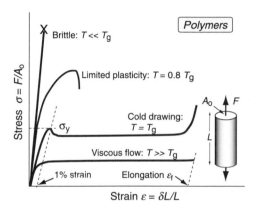

**Figure 6.2**   Stress–strain curve for a polymer.

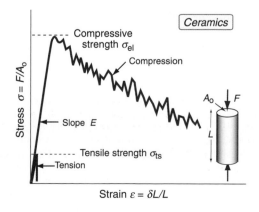

**Figure 6.3**   Stress–strain curve for a ceramic.

hardening and fracture when alignment is complete. At still higher temperatures, thermo-plastics become viscous and can be moulded; thermosets become rubbery and finally decompose.

The yield strength $\sigma_y$ of a polymer—matrix composite is best defined by a set deviation from linear elastic behavior, typically 0.5%. Composites that contain fibres (and this includes natural composites like wood) are a little weaker (up to 30%) in compression than tension because the fibres buckle on a small scale.

*Plastic strain*, $\varepsilon_{pl}$, is the permanent strain resulting from plasticity; thus it is the total strain $\varepsilon_{tot}$ minus the recoverable, elastic, part:

$$\varepsilon_{pl} = \varepsilon_{tot} - \frac{\sigma}{E} \tag{6.1}$$

The *ductility* is a measure of how much plastic strain a material can tolerate. It is measured in standard tensile tests by the *elongation* $\varepsilon_f$ (the tensile strain at break) expressed as a percentage (Figures 6.1 and 6.2). Strictly speaking, $\varepsilon_f$ is not a material property because it depends on the sample dimensions—the values that are listed in handbooks and in the CES software are for a standard test geometry—but it remains useful as an indicator of the ability of a material to be deformed.

In Chapter 4, the area under the elastic part of the stress—strain curve was identified as the elastic energy stored per unit volume ($\sigma_y^2/2E$). Beyond the elastic limit *plastic work* is done in deforming a material permanently by yield or crushing. The increment of plastic work done for a small permanent extension or compression $dL$ under a force $F$, per unit volume $V = AL_o$, is

$$dW_{pl} = \frac{FdL}{V} = \frac{F}{A_o}\frac{dL}{L_o} = \sigma d\varepsilon_{pl}$$

Thus, the plastic work per unit volume at fracture, important in energy-absorbing applications, is

$$W_{pl} = \int_0^{\varepsilon_f} \sigma \, d\varepsilon_{pl} \tag{6.2}$$

which is just the area under the stress—strain curve.

---

### Example 6.1

A metal rod with a yield strength $\sigma_y = 500$ MPa and a density $\rho = 8000$ kg/m$^3$ is uniformly stretched to a strain of 0.1 (10%). How much plastic work is absorbed per unit volume by the rod in this process? Assume that the material is elastic perfectly-plastic, that is, it deforms plastically at a constant stress $\sigma_y$. Compare this plastic work with the elastic energy calculated in Example 4.4.

*Answer.* Assuming that the material is 'elastic perfectly-plastic', that is, the stress equals the yield strength up to failure, the plastic energy per unit volume at failure is:

$$W_{pl} = \sigma_y \varepsilon_f = 500 \times 10^6 \times 0.1 = 50 \text{ MJ/m}^3$$

This plastic energy is 80 times the elastic strain energy of 625 kJ/m$^3$ in Example 4.4, but is still only 0.015% of the chemical energy stored in the same mass of gasoline.

Ceramics and glasses are brittle at room temperature (Figure 6.3). They do have yield strengths, but these are so enormously high that, in tension, they are never reached: the materials fracture first. Even in compression ceramics and glasses crush before they yield. To measure their yield strengths, special tests that suppress fracture are needed. We will return to the fracture-controlled strength of ceramics in Chapter 8. For now, it is useful to have a practical measure of the strength of ceramics to allow their comparison with other materials. The measure used here is the *compressive crushing strength*, and since it is not true yield even though it is the end of the elastic part of the stress strain curve, we call it the *elastic limit* and give it the symbol $\sigma_{el}$.

Tensile and compression tests are not always convenient: you need a large sample and the test destroys it. The hardness test (Figure 6.4) avoids these problems, although it has problems of its own. In it, a pyramidal diamond or a hardened steel ball is pressed into the surface of the material, leaving a tiny permanent indent, the size of which is measured with a microscope. The indent means that plasticity has occurred, and the resistance to it—a measure of

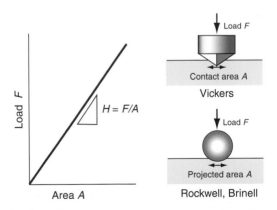

**Figure 6.4** The hardness test. The Vickers test uses a diamond pyramid; the Rockwell and Brinell tests use a steel sphere.

strength—is the load $F$ divided by the area $A$ of the indent projected onto a plane perpendicular to the load:

$$H = \frac{F}{A} \tag{6.3}$$

The indented region is surrounded by material that has not deformed, and this constrains it so that $H$ is larger than the yield strength $\sigma_y$; in practice it is about $3\sigma_y$. Strength, as we have seen, is measured in units of MPa, and since $H$ is a strength it would be logical and proper to measure it in MPa too. But things are not always logical and proper, and hardness scales are among those that are not. A commonly used scale, that of *Vickers*, symbol $H_v$, uses units of kg/mm$^2$, with the result that

$$H_v \approx \frac{\sigma_y}{3} \tag{6.4}$$

Figure 6.5 shows conversions to other scales.

The hardness test has the advantage of being non-destructive, so strength can be measured without destroying the component, and it requires only a tiny volume of material. But the information it provides is less accurate and less complete than the tensile test, so it is not used to provide critical design data.

***True stress and true strain*** The graphs of Figures 6.1–6.3 strictly show the *nominal stress–nominal strain* responses, that is, the area and length used are original values at the start

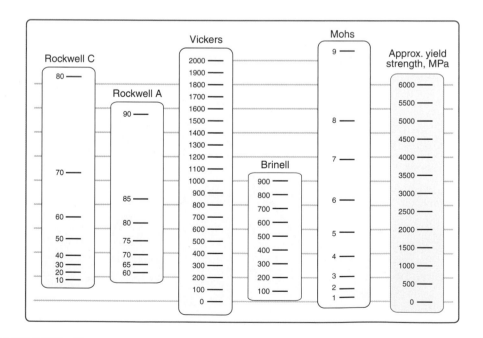

**Figure 6.5** Common hardness scales compared with the yield strength.

of the test. But once materials yield, they can change dimensions significantly. The *true stress*, $\sigma_t$, takes account of the current dimensions, so in tension, for example (equation 4.1)

$$\sigma_t = \frac{F}{A} \tag{6.5}$$

To relate this to the nominal stress, which we will refer to as $\sigma_n$, we note that in plastic deformation the volume of the sample is conserved, for reasons discussed later in this chapter, that is

$$\text{Volume} = A_o \, L_o = A \, L \tag{6.6}$$

where $A$ and $L$ are the current area and length of the test sample and were initially $A_o$ and $L_o$. Hence the nominal stress is

$$\sigma_n = \frac{F}{A_o} = \frac{F}{A}\left(\frac{L_o}{L}\right) \tag{6.7}$$

Returning to the definition of nominal strain, $\varepsilon_n$ (equation 4.3)

$$\varepsilon_n = \frac{\delta L}{L_o} = \left(\frac{L - L_o}{L_o}\right) = \left(\frac{L}{L_o}\right) - 1 \tag{6.8}$$

Combining with equations (6.5) and (6.7) gives the true stress as

$$\sigma_t = \sigma_n(1 + \varepsilon_n) \tag{6.9}$$

True strain, $\varepsilon_t$, is a bit more involved. An incremental change in length, $dL$, gives an incremental strain relative to the current length $L$. To find the cumulative true strain, we therefore integrate these increments over the full extension from $L$ to $L_o$, giving

$$\varepsilon_t = \int_{L_o}^{L} \frac{dL}{L} = \ln\left(\frac{L}{L_o}\right) = \ln\left(1 + \varepsilon_n\right) \tag{6.10}$$

To see how different the true and nominal quantities are in practice, let's do an example.

### Example 6.2

Find the true strain for the following values of the nominal strain during a tensile test on a metal: (a) the 0.2% offset strain, used to define the proof stress (Figure 6.1); (b) a strain of 20%, typical for the onset of failure at maximum stress. What is the ratio of true to nominal stress at these points?

*Answer.* The true strains are (a) $\ln(1+0.002) = 0.002$ (to 3 decimal places); (b) $\ln(1.2) = 0.182$.
The ratio of true to nominal stress is $(1 + \varepsilon_n)$, giving: (a) 1; (b) 1.2.

The example shows that for the elastic regime and small plastic strains, the difference between nominal and true stresses and strains is negligible. This explains the effectiveness of using nominal values to define key properties such as the yield stress. By the end of the tensile test in a metal, the discrepancy between the two is typically 20%.

What about compressive true stress and strain? Now the cross-sectional area increases with plastic deformation (to conserve volume). Equations (6.5) to (6.10) all hold, so long as care is taken with the sign of the applied force $F$ and the strain increments $dL$, which are negative. Figure 6.6 shows nominal and true stress-strain curves superimposed, for both compression testing, and for the stable part of a tensile test up to the tensile strength. As observed in Example 6.2, in tension the true stress is higher than nominal, whereas for strain the reverse is true. In compression, nominal stress exceeds true, and again the reverse for strain. Figure 6.6 shows one benefit of true stress-strain: the curves are identical in shape for tension and compression. The compressive side also extends to larger values of stress and strain, as the instability of necking in tension is avoided. The true stress-strain behaviour is most useful for *metal forming*, where plastic strains of the order of 50% or more are common. And some forming processes use multiple passes, with several plastic deformations one after another. Exercise E6.4 at the end of the chapter shows that in this case the total strain can simply be found by adding up the true strains, but this is not the case for nominal strains.

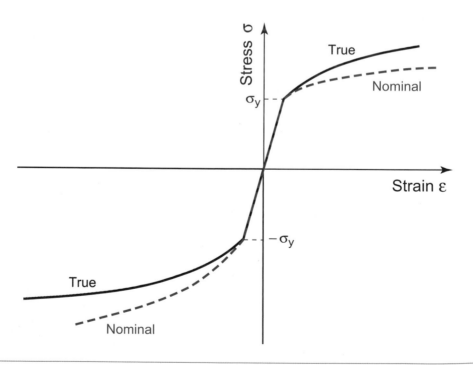

**Figure 6.6** Nominal and true stress-strain curves for a ductile metal, in tension and compression.

# **6.3** The big picture: charts for yield strength

Strength can be displayed on material property charts. Two are particularly useful: the strength–density chart and the modulus–strength chart.

***The strength–density chart*** Figure 6.7 shows the yield strength $\sigma_y$ or elastic limit $\sigma_{el}$ plotted against density $\rho$. The range of strength for engineering materials, like that of the modulus, spans about six decades: from less than 0.01 MPa for foams, used in packaging and energy-absorbing systems, to $10^4$ MPa for diamond, exploited in diamond tooling for machining and as the indenter of the Vickers hardness test. Members of each family again cluster together and can be enclosed in envelopes, each of which occupies a characteristic part of the chart.

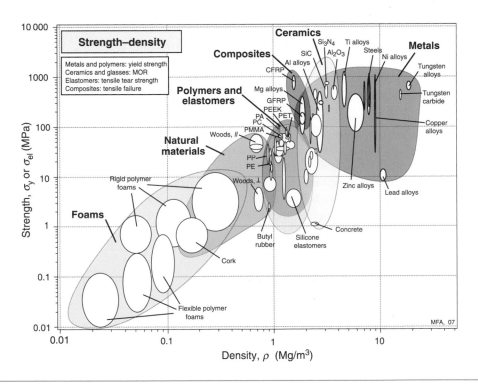

**Figure 6.7** The strength–density chart.

---

## Example 6.3

An unidentified metal from a competitor's product is measured to have a 'Rockwell A' hardness of 50 and a density of approximately 8800 kg/m$^3$. Estimate its Vickers hardness and its yield strength. Locate its position on the strength–density chart (Figure 6.7). What material is it likely to be?

*Answer.* From Figure 6.5, the Vickers hardness is approximately 510 and the yield strength is approximately 1600 MPa (which agrees reasonably with Equation 6.4). This is a strong metal. It falls near the top of the nickel alloys on the strength–density chart.

Comparison with the modulus–density chart (Figure 4.7) reveals some marked differences. The modulus of a solid is a well-defined quantity with a narrow range of values. The strength is not. The strength range for a given class of metals, such as stainless steels, can span a factor of 10 or more, while the spread in stiffness is at most 10%. Since density varies very little (Chapter 4), the strength bubbles for metals are long and thin. The wide ranges for metals reflect the underlying physics of yielding and present designers with an opportunity for manipulation of the strength by varying composition and process history. Both are discussed later in this chapter.

Polymers cluster together with strengths between 10 and 100 MPa. The composites CFRP and GFRP have strengths that lie between those of polymers and ceramics, as we might expect since they are mixtures of the two. The analysis of the strength of composites is not as straightforward as for modulus in Chapter 4, though the same bounds (with strength replacing modulus) generally give realistic estimates.

*The modulus–strength chart* Figure 6.8 shows Young's modulus, $E$, plotted against yield strength, $\sigma_y$ or elastic limit $\sigma_{el}$. This chart allows us to examine a useful material characteristic, the *yield strain*, $\sigma_y/E$, meaning the strain at which the material ceases to be linearly elastic. On log axes, contours of constant yield strain appear as a family of straight parallel lines, as shown in Figure 6.8. Engineering polymers have large yield strains, between 0.01 and 0.1; the values for metals are at least a factor of 10 smaller. Composites and woods lie on the 0.01 contour, as good as the best metals. Elastomers, because of their exceptionally low moduli, have values of $\sigma_y/E$ in the range 1 to 10, much larger than any other class of material.

This chart has many other applications, notably in selecting materials for springs, elastic diaphragms, flexible couplings and snap-fit components. We explore these in Chapter 7.

## **6.4** Drilling down: the origins of strength and ductility

*Perfection: the ideal strength* The bonds between atoms, like any other spring, have a breaking point. Figure 6.8 shows a stress–strain curve for a single bond. Here an atom is assumed to occupy a cube of side $a_o$ (as was assumed in Chapter 4) so that a force $F$ corresponds to a stress $F/a_o^2$. The force stretches the bond from its initial length $a_o$ to a new length $a$, giving a strain $(a - a_o)/a_o$. When discussing the modulus in Chapter 4 we focused on the initial, linear part of this curve, with a slope equal to the modulus, $E$. Stretched farther, the curve passes through a maximum and sinks to zero as the atoms lose communication. The peak is the bond strength—if you pull harder than this it will break. The same is true if you shear it rather than pull it.

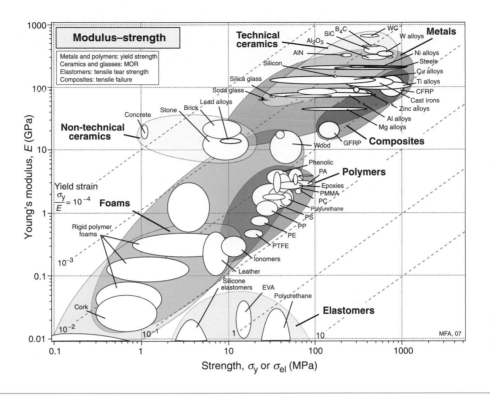

**Figure 6.8** The Young's modulus–strength chart. The contours show the strain at the elastic limit, $\sigma_y/E$.

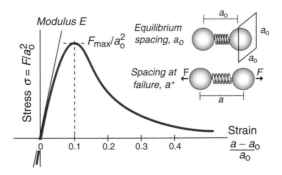

**Figure 6.9** The stress–strain curve for a single atomic bond (it is assumed that each atom occupies a cube of side $a_o$).

The distance over which inter-atomic forces act is small—a bond is broken if it is stretched to more than about 10% of its original length. So the force needed to break a bond is roughly:

$$F = \frac{Sa_o}{10} \tag{6.11}$$

where $S$, as before, is the bond stiffness. On this basis the *ideal strength* of a solid should therefore be roughly

$$\sigma_{\text{ideal}} = \frac{F_{\max}}{a_o^2} = \frac{S}{10a_o} = \frac{E}{10}$$

(remembering that $E = S_o/a_o$, equation (4.20))
  or

$$\frac{\sigma_{\text{ideal}}}{E} = \frac{1}{10} \tag{6.12}$$

This doesn't allow for the curvature of the force–distance curve; more refined calculations give a ratio of 1/15.

Figure 6.10 shows $\sigma_y/E$ for metals, polymers and ceramics. None achieves the ideal value of 1/10; most don't even come close. Why not? It's a familiar story: like most things in life, materials are imperfect.

***Crystalline imperfection: defects in metals and ceramics***   Crystals contain imperfections of several kinds. Figure 6.11 introduces the broad families, distinguished by their dimensionality. At the top left are *point defects*. All crystals contain *vacancies*, shown in (a): sites at which an

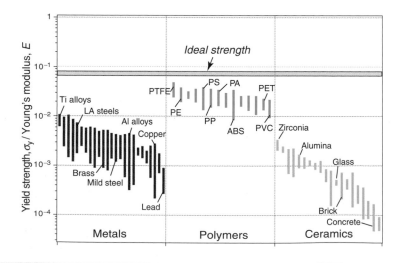

**Figure 6.10**   The ideal strength is predicted to be about *E*/15, where *E* is Young's modulus. The figure shows $\sigma_y/E$ with a shaded band at the ideal strength.

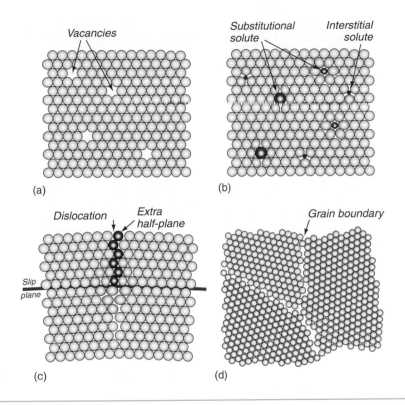

**Figure 6.11**  Defects in crystals. (a) Vacancies—missing atoms. (b) Foreign (solute) atom on interstitial and substitutional sites. (c) A dislocation—an extra half-plane of atoms. (d) Grain boundaries.

atom is missing. They play a key role in diffusion, creep and sintering (Chapter 13), but we don't need them for the rest of this chapter because they do not influence strength. The others do.

No crystal is totally, 100%, pure and perfect. Some impurities are inherited from the process by which the material was made; more usually they are deliberately added, creating *alloys*: a material in which a second (or third or fourth) element is dissolved. 'Dissolved' sounds like salt in water, but these are solid solutions. Figure 6.11(b) shows both a *substitutional solid solution* (the dissolved atoms replace those of the host) and an *interstitial solid solution* (the dissolved atoms squeeze into the spaces or 'interstices' between the host atoms). The dissolved atoms or solute rarely have the same size as those of the host material, so they distort the surrounding lattice. The red atoms here are substitutional solute, some bigger and some smaller than those of the host; the cages of host atoms immediately surrounding them, shown green, are distorted. If the solute atoms are particularly small, they don't need to replace a host atom; instead, they dissolve *interstitially* like the black atoms in the figure, again distorting the surrounding lattice. So solute causes local distortion; this distortion is one of the reasons that alloys are stronger than pure materials, as we shall see in a moment.

Now to the key player, portrayed in Figure 6.11(c): the *dislocation*. 'Dislocated' means 'out of joint' and this is not a bad description of what is happening here. The upper part of the crystal has one more double-layer of atoms than the lower part (the double-layer is needed to get the top-to-bottom registry right). It is dislocations that make metals soft and ductile. Dislocations distort the lattice—here the green atoms are the most distorted—and because of this they have elastic energy associated with them. If they cost energy, why are they there? To grow a perfect crystal just one cubic centimeter in volume from a liquid or vapour, about $10^{23}$ atoms have to find their proper sites on the perfect lattice, and the chance of this happening is just too small. Even with the greatest care in assembling them, all crystals contain point defects, solute atoms and dislocations.

Most contain yet more drastic defects, among them grain boundaries. Figure 6.11(d) shows such boundaries. Here three perfect, but differently oriented, crystals meet; the individual crystals are called *grains*, the meeting surfaces are *grain boundaries*. In this sketch the atoms of the three crystals have been given different colours to distinguish them, but they are all the same atoms. In reality grain boundaries form in pure materials (when all the atoms are the same) and in alloys (when the mixture of atoms in one grain may differ in chemical composition from those of the next).

Now put all this together. The seeming perfection of the steel of a precision machine tool or of the polished case of a gold watch is an illusion: they are riddled with defects. Imagine all the frames of Figure 6.11 superimposed and you begin to get the picture. Between them they explain diffusion, strength, ductility, electrical resistance, thermal conductivity and much more.

So defects in crystals are influential. For the rest of this section we focus on getting to know just one of them: the dislocation.

***Dislocations and plastic flow***    Recall that the strength of a perfect crystal computed from inter-atomic forces gives an 'ideal strength' around $E/15$ (where $E$ is the modulus). In reality the strengths of engineering materials are nothing like this big; often they are barely 1% of it. This was a mystery until halfway through the last century—a mere 60 years ago—when an Englishman, G. I. Taylor[3], and a Hungarian, Egon Orowan[4], realized that a 'dislocated' crystal could deform at stresses far below the ideal. So what is a dislocation, and how does it enable deformation?

Figure 6.12(a) shows how to make a dislocation. The crystal is cut along an atomic plane up to the line shown as $\perp - \perp$, the top part is slid across the bottom by one full atom spacing, and the atoms are reattached across the cut plane to give the atom configuration shown in Figure 6.12(b). There is now an extra half-plane of atoms with its lower edge along the $\perp - \perp$ line, the *dislocation line*—the line separating the part of the plane that has slipped from the

---

[3]  Geoffrey (G. I.) Taylor (1886—1975), known for his many fundamental contributions to aerodynamics, hydrodynamics and to the structure and plasticity of metals—it was he, with Egon Orowan, who realised that the ductility of metals implied the presence of dislocations. One of the greatest of contributors to theoretical mechanics and hydrodynamics of the 20th century, he was also a supremely practical man—a sailor himself, he invented (among other things) the anchor used by the Royal Navy.

[4]  Egon Orowan (1901—1989), Hungarian/US physicist and metallurgist, who, with G. I. Taylor, realised that the plasticity of crystals could be understood as the motion of dislocations. In his later years he sought to apply these ideas to the movement of fault lines during earthquakes.

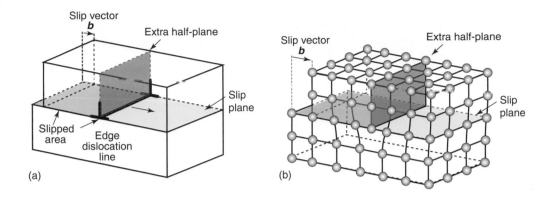

**Figure 6.12**   (a) Making a dislocation by cutting, slipping and rejoining bonds across a slip plane. (b) The atom configuration at an edge dislocation in a simple cubic crystal. The configurations in other crystal structures are more complex but the principle remains the same.

part that has not. This particular configuration is called an *edge* dislocation because it is formed by the edge of the extra half-plane, represented by the symbol $\perp$.

When a dislocation moves it makes the material above the slip plane slide relative to that below, producing a shear strain. Figure 6.13 shows how this happens. At the top is a perfect crystal. In the central row a dislocation enters from the left, sweeps through the crystal and

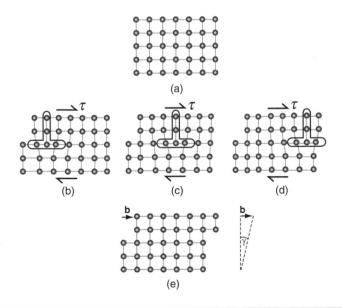

**Figure 6.13**   An initially perfect crystal is shown in (a). The passage of the dislocation across the slip plan, shown in the sequence (b), (c) and (d), shears the upper part of the crystal over the lower part by the slip vector **b**. When it leaves the crystal has suffered a shear strain γ.

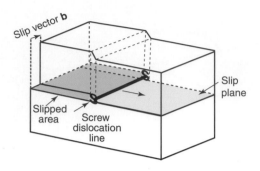

**Figure 6.14** A screw dislocation. The slip vector **b** is parallel to the dislocation line **S—S**.

exits on the right. By the end of the process the upper part has slipped by **b**, the slip vector (or Burger's vector) relative to the part below. The result is the shear strain γ shown at the bottom.

There is another way to make a dislocation in a crystal. After making the cut in Figure 6.12(a), the upper part of the crystal can be displaced *parallel* to the edge of the cut rather than *normal* to it, as in Figure 6.14. That too creates a dislocation, but one with a different configuration of atoms along its line—one more like a corkscrew than like a squashed worm—and for this reason it is called a *screw dislocation*. We don't need the details of its structure; it is enough to know that its properties are like those of an edge dislocation except that when it sweeps through a crystal (moving normal to its line), the lattice is displaced parallel to the dislocation line, not normal to it. All dislocations are either edge or screw or *mixed*, meaning that they are made up of little steps of edge and screw. The line of a mixed dislocation can be curved but every part of it has the same slip vector **b** because the dislocation line is just the boundary of a plane on which a fixed displacement **b** has occurred.

It is far easier to move a dislocation through a crystal, breaking and remaking bonds only along its line as it moves, than it is to simultaneously break all the bonds in the plane before remaking them. It is like moving a heavy carpet by pushing a fold across it rather than sliding the whole thing at one go. In real crystals it is easier to make and move dislocations on some planes than on others. The preferred planes are called *slip planes* and the preferred directions of slip in these planes are called *slip directions*. A slip plane is shown in gray and a slip direction as an arrow on the FCC and BCC unit cells of Figure 4.13.

Slip displacements are tiny—one dislocation produces a displacement of about $10^{-10}$ m. But if large numbers of dislocations traverse a crystal, moving on many different planes, the shape of a material changes at the macroscopic length scale. Figure 6.15 shows just two dislocations traversing a sample loaded in tension. The slip steps (here very exaggerated) cause the sample to get a bit thinner and longer. Repeating this millions of times on many planes gives the large plastic extensions observed in practice. Since none of this changes the average atomic spacing, the volume remains unchanged.

***Why does a shear stress make a dislocation move?*** Crystals resist the motion of dislocations with a friction-like resistance *f* per unit length—we will examine its origins in a moment. For yielding to take place the external stress must overcome the resistance *f*.

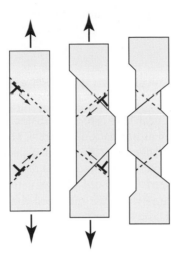

Figure 6.15 Dislocation motion causes extension, at constant volume.

Imagine that one dislocation moves right across a slip plane, traveling the distance $L_2$, as in Figure 6.16. In doing so, it shifts the upper half of the crystal by a distance $b$ relative to the lower half. The shear stress $\tau$ acts on an area $L_1L_2$, giving a shear force $F_s = \tau L_1 L_2$ on the surface of the block. If the displacement parallel to the block is $b$, the force does work

$$W = \tau L_1 L_2 b \tag{6.13}$$

This work is done against the resistance $f$ per unit length, or $f\,L_1$ on the length $L_1$, and it does so over a displacement $L_2$ (because the dislocation line moves this far against $f$), giving

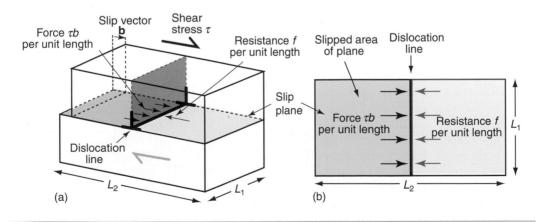

Figure 6.16 The force on a dislocation. (a) Perspective view. (b) Plan view of slip plane.

a total work against $f$ of $fL_1L_2$. Equating this to the work $W$ done by the applied stress $\tau$ gives

$$\tau b = f \tag{6.14}$$

This result holds for any dislocation—edge, screw or mixed. So, provided the shear stress $\tau$ exceeds the value $f/b$ it will make dislocations move and cause the crystal to shear.

*Line tension*    The atoms near the core of a dislocation are displaced from their proper positions, as shown by green atoms back in Figure 6.11(c), and thus they have higher potential energy. To keep the potential energy of the crystal as low as possible, the dislocation tries to be as short as possible—it behaves as if it had a *line tension*, $T$, like an elastic band. The tension can be calculated but it needs advanced elasticity theory to do it (the books listed under 'Further reading' give the analysis). We just need the answer. It is that the line tension, an energy per unit length (just as a surface tension is an energy per unit area), is

$$T = \frac{1}{2}Eb^2 \tag{6.15}$$

where $E$, as always, is Young's modulus. The line tension has an important bearing on the way in which dislocations interact with obstacles, as we shall see in a moment.

*The lattice resistance*    Where does the resistance to slip, $f$, come from? There are several contributions. Consider first the *lattice resistance*, $f_i$: the intrinsic resistance of the crystal structure to plastic shear. Plastic shear, as we have seen, involves the motion of dislocations. Pure metals are soft because the non-localised metallic bond does little to obstruct dislocation motion, whereas ceramics are hard because their more localised covalent and ionic bonds (which must be broken and reformed when the structure is sheared) lock the dislocations in place. When the lattice resistance is high, as in ceramics, further hardening is superfluous—the problem becomes that of suppressing fracture. On the other hand, when the lattice resistance $f_i$ is low, as in metals, the material can be strengthened by introducing obstacles to slip. This is done by adding alloying elements to give *solid solution hardening* ($f_{ss}$), precipitates or dispersed particles giving *precipitation hardening* ($f_{ppt}$), other dislocations giving what is called *work hardening* ($f_{wh}$) or grain boundaries introducing *grain-size hardening* ($f_{gb}$). These techniques for manipulating strength are central to alloy design. We look at them more closely in the next section.

*Plastic flow in polymers*    Figure 6.2 showed a range of stress-strain responses in polymers. What happens in a given polymer depends strongly on the temperature. In Chapter 4, the glass transition temperature $T_g$ was defined, being the temperature at which the inter-molecular hydrogen bonds melt. This is usually associated with a marked drop in the material stiffness, depending on whether the polymer is amorphous, semi-crystalline or cross-linked. Not surprisingly, the strength of a polymer depends strongly on whether the material is above or below the glass transition, and a number of different failure mechanisms can occur.

At low temperatures, meaning below about $0.75\ T_g$, polymers are elastic-brittle, like ceramics, and fail by propagation of a dominant flaw (see Chapter 8). Above this temperature, they become plastic. When pulled in tension, the chains slide over each other, unraveling, so that they become aligned with the direction of stretch, as in Figure 6.17(a), a process called

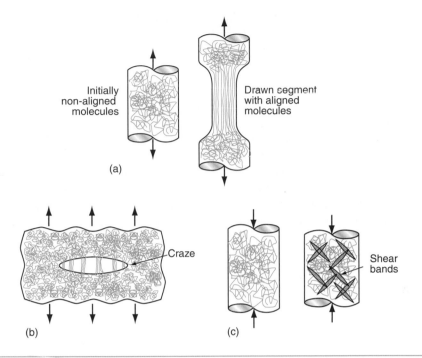

**Figure 6.17**   (a) Cold drawing—one of the mechanisms of deformation of thermoplastics. (b) Crazing—local drawing across a crack. (c) Shear banding.

*drawing*. It is harder to start drawing than to keep it going, so the zone where it starts draws down completely before propagating farther along the sample, leading to profiles like that shown in the figure. The drawn material is stronger and stiffer than before, by a factor of about 8, giving drawn polymers exceptional properties, but because you can only draw fibres or sheet (by pulling in two directions at once) the geometries are limited.

Many polymers, among them PE, PP and nylon, draw at room temperature. Others with higher glass temperatures, such as PMMA, do not, although they draw well at higher temperatures. At room temperature they *craze*. Small crack-shaped regions within the polymer draw down. Because the crack has a larger volume than the polymer that was there to start with, the drawn material ends up as ligaments that link the craze surfaces, as in Figure 6.17(b). Crazes scatter light, so their presence causes whitening, easily visible when cheap plastic articles are bent. If stretching is continued, one or more crazes develop into proper cracks, and the sample fractures.

While crazing limits ductility in tension, large plastic strains may still be possible in compression by shear banding (Figure 6.17(c)). Within each band, shear takes place with much the same consequences for the shape of the sample as shear by dislocation motion. Continued compression causes the number of shear bands to increase, giving increased overall strain.

## **6.5** Manipulating strength

***Strengthening metals***   The way to make crystalline materials stronger is to make it harder for dislocations to move. As we have seen, dislocations move in a pure crystal when the force $\tau b$ per unit length exceeds the lattice resistance $f_i$. There is little we can do to change this—it is an intrinsic property like the modulus $E$. Other strengthening mechanisms add to it, and here there is scope for manipulation. Figure 6.18 introduces them. It shows the view of a slip plane from the perspective of an advancing dislocation: each strengthening mechanism presents a new obstacle course. In the perfect lattice shown in (a) the only resistance is the intrinsic strength of the crystal; solution hardening, shown in (b), introduces atom-size obstacles to motion; precipitation hardening, shown in (c), presents larger obstacles; and in work hardening, shown in (d), the slip plane becomes stepped and threaded with 'forest' dislocations.

Obstacles to dislocation motion increase the resistance $f$ and thus the strength. To calculate their contribution to $f$, there are just two things we need to know about them: their spacing and their strength. Spacing means the distance $L$ between them in the slip plane. The number of obstacles touching unit length of dislocation line is then

$$N_L = \frac{1}{L}$$

Each individual obstacle exerts a *pinning force p* on the dislocation line—a resisting force per unit length of dislocation—so the contribution of the obstacles to the resistance $f$ is

$$f = \frac{p}{L}$$

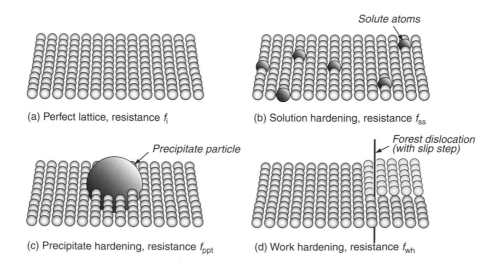

(a) Perfect lattice, resistance $f_i$

(b) Solution hardening, resistance $f_{ss}$

Solute atoms

(c) Precipitate hardening, resistance $f_{ppt}$

Precipitate particle

(d) Work hardening, resistance $f_{wh}$

Forest dislocation (with slip step)

Figure 6.18   A 'dislocation-eye' view of the slip plane across which it must move.

Thus, the added contribution to the shear stress $\tau$ needed to make the dislocation move is (from equation (6.14))

$$\Delta\tau = \frac{p}{bL} \qquad (6.16)$$

The pinning is an elastic effect—it derives from the fact that both the dislocation and the obstacle distort the lattice elastically even though, when the dislocation moves, it produces plastic deformation. Because of this $p$, for any given obstacle in any given material, scales as $Eb^2$, which has the units of force. The shear stress $\tau$ needed to force the dislocation through the field of obstacles then has the form

$$\tau = \alpha\frac{Eb}{L} \qquad (6.17)$$

where $\alpha$ is a dimensionless constant characterizing the obstacle strength.

Armed with this background we can explain strengthening mechanisms. We start with solid solutions.

*Solution hardening*   Solid solution hardening is strengthening by deliberate additions of impurities or, more properly said, by *alloying* (Figure 6.19(a)). The addition of zinc to copper makes the alloy brass—copper dissolves up to 30% zinc. The zinc atoms replace copper atoms to form a *random substitutional solid solution*. The zinc atoms are bigger that those of copper and, in squeezing into the copper lattice, they distort it. This roughens the slip plane, so to speak, making it harder for dislocations to move, thereby adding an additional resistance $f_{ss}$, opposing dislocation motion. The figure illustrates that the concentration of solute, expressed as an atom fraction, is on average:

$$c = \frac{b^2}{L^2}$$

where $L$ is the spacing of obstacles in the slip plane and $b$ is the atom size. Thus,

$$L = \frac{b}{c^{1/2}}$$

Plugging this into equation (6.17) relates the contribution of solid solution to the shear stress required to move the dislocation:

$$\tau_{ss} = \alpha E c^{1/2} \qquad (6.18)$$

$\tau_{ss}$ increases as the square root of the solute concentration. Brass, bronze and stainless steels, and many other metallic alloys, derive their strength in this way. They differ only in the extent to which the solute distorts the crystal, described by the constant $\alpha$.

*Dispersion and precipitate strengthening*   A more effective way to impede dislocations is to *disperse* small, strong particles in their path. One way to make such a microstructure is to disperse small solid particles of a high melting point compound into a liquid metal, and to cast it to shape, trapping the particles in place—it is the way that metal–matrix composites such as Al–SiC are made. An alternative is to form the particles *in situ* by a *precipitation* process. If a

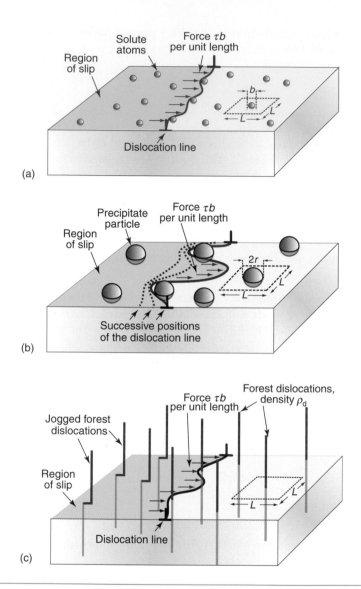

**Figure 6.19** (a) Solution hardening. (b) Precipitation or dispersion hardening. (c) Forest hardening (work hardening).

solute (copper, say) is dissolved in a metal (aluminum, for instance) at high temperature when both are molten, and the alloy is solidified and cooled to room temperature, the solute precipitates as small particles, much as salt will crystallize from a saturated solution when it is cooled. An alloy of aluminum containing 4% copper, treated in this way, gives very small, closely spaced precipitates of the hard compound $CuAl_2$. Copper alloyed with a little beryllium, similarly treated, gives precipitates of the compound CuBe. Most steels are strengthened by precipitates of carbides, obtained in this way. The precipitates give a large contribution to $f$.

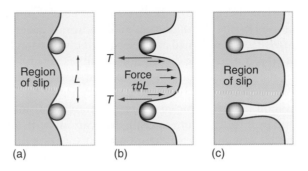

(a) (b) (c)

**Figure 6.20** Successive positions of a dislocation as it bypasses particles that obstruct its motion. The critical configuration is that with the tightest curvature, shown in (b).

Figure 6.19(b) shows how particles obstruct dislocation motion. If the particles are too strong for the dislocation to slice through them, the force $\tau b$ pushes the dislocation between them, bending it to a tighter and tighter radius against its line tension (equation (6.15)). The radius is at a minimum when it reaches half the particle spacing, $L$; after that it can expand under lower stress. It is a bit like blowing up a bicycle inner tube when the outer tire has a hole in it: once you reach the pressure that balloons the inner tube through the hole, the balloon needs a smaller pressure to get bigger still. The *critical configuration* is the semi-circular one: here the total force $\tau bL$ on one segment of length $L$ is just balanced by the force $2T$ due to the line tension (equation (6.16)), acting on either side of the bulge, as in Figure 6.20. The dislocation escapes when

$$\tau_{\text{ppt}} = \frac{2T}{bL} = \frac{Eb}{L} \tag{6.19}$$

The obstacles thus exert a resistance of $f_{\text{ppt}} = 2T/L$. Precipitation hardening is an effective way to increasing strength: precipitation-hardened aluminum alloys can be 15 times stronger than pure aluminum.

---

### Example 6.4

A polycrystalline aluminum alloy contains a dispersion of hard particles of diameter $10^{-8}$ m and an average centre-to-centre spacing of $6 \times 10^{-8}$ m, measured in the slip planes. The Young's modulus of aluminum is $E = 70$ GPa and $b = 0.286$ nm. Estimate the contribution of the hard particles to the yield strength of the alloy.

*Answer.* For the alloy, $L = (6 - 1) \times 10^{-8}$ m $= 5 \times 10^{-8}$ m, giving

$$\tau_{\text{ppt}} = \frac{Eb}{L} = \frac{70 \times 10^9 \times 0.286 \times 10^{-9}}{5 \times 10^{-8}} = 400 \text{ MPa}$$

Noting that $\sigma_y = 3\tau_y$ for polycrystals (see below), $\sigma_y = 1200$ MPa.

*Work hardening*    The rising part of the stress–strain curve of Figure 6.1 is caused by *work hardening*: it is caused by the accumulation of dislocations generated by plastic deformation. The *dislocation density*, $\rho_d$, is defined as the length of dislocation line per unit volume (m/m$^3$). Even in an annealed soft metal, the dislocation density is around $10^{10}$ m/m$^3$, meaning that a 1 cm cube (the size of a cube of sugar) contains about 10 km of dislocation line. When metals are deformed, dislocations multiply, causing their density to grow to as much as $10^{17}$ m/m$^3$ or more—100 million km per cubic centimeter. A moving dislocation now finds that its slip plane is penetrated by a forest of intersecting dislocations with an average spacing $L = \rho_d^{-1/2}$ (since $\rho_d$ is a number per unit area). Figure 6.19(c) suggests the picture. If a moving dislocation advances, it shears the material above the slip plane relative to that below, and that creates a little step called a *jog* in each forest dislocation. The jogs have potential energy—they are tiny segments of dislocation of length $b$—with the result that each exerts a pinning force $p = Eb^2/2$ on the moving dislocation. Assembling these results into equation (6.16) gives

$$\tau_{wh} = \frac{Eb}{2}\sqrt{\rho_d} \qquad\qquad (6.20)$$

The greater the density of dislocations, the smaller the spacing between them, and so the greater their contribution to $\tau_{wh}$.

All metals work harden. It can be a nuisance: if you want to roll thin sheet, work hardening quickly raises the yield strength so much that you have to stop and *anneal* the metal (heat it up to remove the accumulated dislocations) before you can go on—a trick known to blacksmiths for centuries. But it is also useful: it is a potent strengthening method, particularly for alloys that cannot be heat-treated to give precipitation hardening.

---

### Example 6.5

A cube of the aluminum in Example 6.4 has a side length 1 cm and a dislocation density of $10^8$ mm$^{-2}$. Assuming a square array of parallel dislocations, estimate: (a) the total length of dislocation in the sample; (b) the distance between dislocations; (c) the contribution of the dislocations to the strength of the crystal.

*Answer.*

(a) The dislocation density $\rho_d$ is the length of dislocation per unit volume. For a 1 cm cube, volume = 1000 mm$^3$. Dislocation density = $10^8$ mm$^{-2}$ (mm/mm$^3$), hence total length of dislocation = $10^8 \times 1000 = 10^{11}$ mm = 100 000 km!

(b) Assume that the dislocations form a uniform square array of side $d$. In this case, each dislocation sits at the centre of an area $d^2$. For a unit length of dislocation, the volume of material per dislocation is $1 \times d^2$, so the dislocation density is $\rho_d = 1/d^2$, hence

$$d = 1/\sqrt{\rho_d} = 1/\sqrt{10^8} = 10^{-4}\ \text{mm} = 100\ \text{nm}$$

(c) From equation (6.20)

$$\tau_{wh} = \frac{Eb}{2}\sqrt{\rho_d} = \frac{70 \times 10^9 \times 0.286 \times 10^{-9}}{2}\sqrt{10^8 \times 10^4} = 10 \text{ MPa}$$

Noting that $\sigma_y = 3\tau_y$ for polycrystals, $\sigma_y = 30$ MPa.

*Grain boundary hardening*   Almost all metals are polycrystalline, made up of tiny, randomly oriented crystals, or *grains*, meeting at grain boundaries like those of Figure 6.11(d). The grain size, $D$, is typically 10–100 µm. These boundaries obstruct dislocation motion. A dislocation in one grain—call it grain 1—can't just slide into the next (grain 2) because the slip planes don't line up. Instead, new dislocations have to nucleate in grain 2 with slip vectors that, if superimposed, match that of the dislocation in grain 1 so that the displacements match at the boundary. This gives another contribution to strength, $\tau_{gb}$, that is found to scale as $D^{-1/2}$, giving

$$\tau_{gb} = \frac{k_p}{\sqrt{D}} \tag{6.21}$$

where $k_p$ is called the Petch constant, after the man who first measured it. For normal grain sizes $\tau_{gb}$ is small and not a significant source of strength, but for materials that are microcrystalline ($D < 1$ µm) or nanocrystalline ($D$ approaching 1 nm) it becomes significant.

*Relationship between dislocation strength and yield strength*   To a first approximation the strengthening mechanisms add up, giving a shear yield strength, $\tau_y$, of

$$\tau_y = \tau_i + \tau_{ss} + \tau_{ppt} + \tau_{wh} + \tau_{gb} \tag{6.22}$$

Strong materials either have a high intrinsic strength, $\tau_i$ (like diamond), or they rely on the superposition of solid solution strengthening $\tau_{ss}$, precipitation $\tau_{ppt}$ and work hardening $\tau_{wh}$ (like high-tensile steels). Nanocrystalline solids exploit, in addition, the contribution of $\tau_{gb}$.

Before we can use this information, one problem remains: we have calculated the yield strength of one crystal, loaded in shear. We want the yield strength of a *polycrystalline material* in *tension*. To link them there are two simple steps. First, a uniform tensile stress $\sigma$ creates a shear stress on planes that lie at an angle to the tensile axis; dislocations will first move on the slip plane on which this shear stress is greatest. Figure 6.21 shows how this is calculated. A tensile force $F$ acting on a rod of cross-section $A$, if resolved parallel to a plane with a normal that lies at an angle $\theta$ to the axis of tension, gives a force $F \sin \theta$ in the plane. The area of this plane is $A/\cos \theta$, so the shear stress is

$$\tau = \frac{F \sin \theta}{A/\cos \theta} = \sigma \sin \theta \cos \theta$$

where $\sigma = F/A$ is the tensile stress. The value of $\tau$ is plotted against $\theta$ in the figure. The maximum lies at an angle of 45 degrees, when $\tau = \sigma/2$.

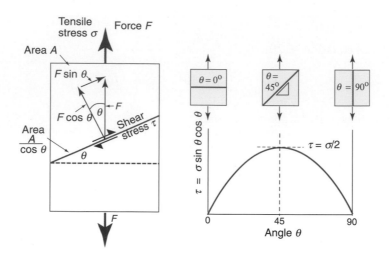

**Figure 6.21** The resolution of stress. A tensile stress $\sigma$ gives a maximum shear stress $\tau = \sigma/2$ on a plane at 45 degree to the tensile axis.

Second, when this shear stress acts on an aggregate of crystals, some crystals will have their slip planes oriented favorably with respect to the shear stress, others will not. This randomness of orientation jacks up the strength by a further factor of 1.5 (called the Taylor factor). Combining these results, the tensile stress to cause yielding of a sample that has many grains is approximately three times the shear strength of a single crystal:

$$\sigma_y \approx 3\tau_y$$

Thus, the superposition of strengthening mechanisms in equation (6.22) applies equally to the yield strength, $\sigma_y$.

***Strength and ductility of alloys*** Of all the properties that materials scientists and engineers have sought to manipulate, the strength of metals and alloys is probably the most explored. It is easy to see why: Table 6.1 gives a small selection of the applications of metals and their alloys. Their importance in engineering design is enormous. The hardening mechanisms are often used in combination. This is illustrated graphically for copper alloys in Figure 6.22. Good things, however, have to be paid for. Here the payment for increased strength is, almost always, loss of ductility, so the elongation $\varepsilon_f$ is reduced. The material is stronger but it cannot be deformed as much without fracture. This general trend is evident in Figure 6.23, which shows the nominal stress-strain curves for a selection of engineering alloys.

***Strengthening polymers*** In non-crystalline solids the dislocation is not a helpful concept. We think instead of some unit step of the flow process: the relative slippage of two segments of a polymer chain, or the shear of a small molecular cluster in a glass network. Their strength has the same origin as that underlying the lattice resistance: if the unit step involves breaking strong bonds (as in an inorganic glass), the materials will be strong and brittle, as ceramics are. If it

**Table 6.1** Metal alloys with typical applications, indicating the strengthening mechanisms used

| Alloy | Typical uses | Solution hardening | Precipitation hardening | Work hardening |
|---|---|---|---|---|
| Pure Al | Kitchen foil | | | ✔✔✔ |
| Pure Cu | Wire | | | ✔✔✔ |
| Cast Al, Mg | Automotive parts | ✔✔✔ | ✔ | |
| Bronze (Cu–Sn), Brass (Cu–Zn) | Marine components | ✔✔✔ | ✔ | ✔✔ |
| Non-heat-treatable wrought Al | Ships, cans, structures | ✔✔✔ | | ✔✔✔ |
| Heat-treatable wrought Al | Aircraft, structures | ✔ | ✔✔✔ | ✔ |
| Low-carbon steels | Car bodies, structures, ships, cans | ✔✔✔ | | ✔✔✔ |
| Low alloy steels | Automotive parts, tools | ✔ | ✔✔✔ | ✔ |
| Stainless steels | Pressure vessels | ✔✔✔ | ✔ | ✔✔✔ |
| Cast Ni alloys | Jet engine turbines | ✔✔✔ | ✔✔✔ | |

Symbols: ✔✔✔ = Routinely used. ✔ = Sometimes used.

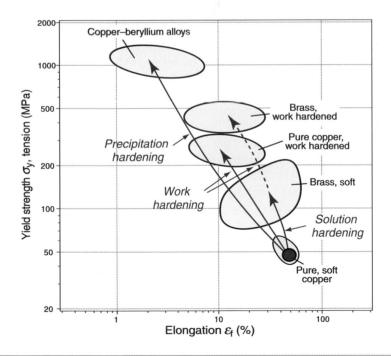

**Figure 6.22** Strengthening mechanisms and the consequent drop in ductility, here shown for copper alloys. The mechanisms are frequently combined. The greater the strength, the lower the ductility (the elongation to fracture, $\varepsilon_f$).

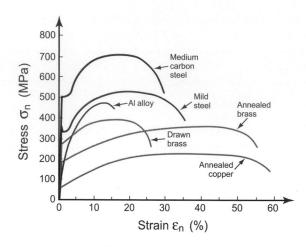

**Figure 6.23** Nominal stress-strain curves for a selection of engineering alloys. Stronger alloys tend to have lower ductility.

only involves the rupture of weak bonds (the Van der Waals bonds in polymers, for example), it will be weak. Polymers too must therefore be strengthened by impeding the slippage of segments of their molecular chains. This is achieved by *blending*, by *drawing*, by *cross-linking* and by *reinforcement* with particles, fibres or fabrics.

A blend is a mixture of two polymers, stirred together in a sort of industrial food-mixer. The strength and modulus of a blend are just the average of those of the components, weighted by volume fraction (a rule of mixtures again). If one of these is a low molecular weight hydro-carbon, it acts as a plasticiser, reducing the modulus and giving the blend a leather-like flexibility.

Drawing is the deliberate use of the molecule-aligning effect of stretching, like that sketched in Figure 6.17(a), to greatly increase stiffness and strength in the direction of stretch. Fishing line is drawn nylon, Mylar film is a polyester with molecules aligned parallel to the film, and geotextiles, used to restrain earth banks, are made from drawn polypropylene.

Cross-linking, sketched in Figures 4.19 and 4.20, creates strong bonds between molecules that were previously linked by weak Van der Waals forces. Vulcanized rubber is rubber that has been cross-linked, and the superior strength of epoxies derives from cross-linking.

Reinforcement is possible with particles of cheap fillers—sand, talc or wood dust. Far more effective is reinforcement with fibres—usually glass or carbon—either continuous or chopped, as explained in Chapter 4.

## 6.6 Summary and conclusions

Load-bearing structures require materials with reliable, reproducible strength. There is more than one measure of strength. *Elastic design* requires that no part of the structure suffers plastic deformation, and this means that the stresses in it must nowhere exceed the *yield strength*, $\sigma_y$, of ductile materials or the *elastic limit* of those that are not ductile. *Plastic design*, by contrast, allows some parts of the structure to deform plastically so long as the structure as a whole does not collapse. Then two further properties become relevant: the *ductility*, $\varepsilon_f$, and the tensile strength, $\sigma_{ts}$, which are the maximum strain and the maximum stress the material can tolerate before fracture. The tensile strength is generally larger than the yield strength because of *work hardening*.

Charts plotting strength, like those plotting modulus, show that material families occupy different areas of material property space, depending on the strengthening mechanisms on which they rely. Crystal defects—particularly *dislocations*—are central to the understanding of these. It is the motion of dislocations that gives plastic flow in crystalline solids, giving them unexpectedly low strengths. When strength is needed it has to be provided by the strengthening mechanisms that impede dislocation motion.

First among these is the *lattice resistance*—the intrinsic resistance of the crystal to dislocation motion. Others can be deliberately introduced by alloying and heat treatment. *Solid solution hardening, dispersion* and *precipitation hardening, work hardening* and *grain boundary hardening* add to the lattice resistance. The strongest materials combine them all.

Non-crystalline solids—particularly polymers—deform in a less organised way by the pulling of the tangled polymer chains into alignment with the direction of deformation. This leads to *cold drawing* with substantial plastic strain and, at lower temperatures, to *crazing*. The stress required to do this is significant, giving polymers a considerable intrinsic strength. This can be enhanced by blending, cross-linking and reinforcement with particles or fibres to give the engineering polymers we use today.

## **6.7** Further reading

Ashby, M. F., & Jones, D. R. H. (2006). *Engineering Materials* (Vols. I–II). Oxford, UK: Butterworth-Heinemann. ISBN 7-7506-6380-4 and ISBN 0-7506-6381-2. (An introduction to mechanical properties and processing of materials.)

Cottrell, A. H. (1953). *Dislocations and Plastic Flow in Crystals*. Oxford, UK: Oxford University Press. (Long out of print but worth a search: the book that first presented a coherent picture of the mechanisms of plastic flow and hardening.)

Friedel, J. (1964). *Dislocations*. Reading. MA, USA: Addison-Wesley. Library of Congress No. 65-21133. (A book that, with that of Cottrell, first established the theory of dislocations.)

Hertzberg, R. W. (1989). *Deformation and Fracture of Engineering Materials* (3rd ed.). New York, USA: Wiley. ISBN 0-471-61722-9. (A readable and detailed coverage of deformation, fracture and fatigue.)

Hull, D., & Bacon, D. J. (2001). *Introduction to Dislocations* (4th ed.). Oxford, UK: Butterworth-Heinemann. ISBN 0-750-064681-0. (An introduction to dislocation mechanics.)

Hull, D., & Clyne, T. W. (1996). *An Introduction to Composite Materials* (2nd ed.). Cambridge, UK: Cambridge University Press. ISBN 0-521-38855-4. (A concise and readable introduction to composites that takes an approach that minimises the mathematics and maximises the physical understanding.)

Young, R. J. (1981). *Introduction to Polymers*. London, UK: Chapman & Hall. ISBN 0-412-22180-2. (A good starting point for more information on the chemistry, structure and properties of polymers.)

# 6.8 Exercises

Exercise E6.1   Sketch a stress-strain curve for a typical metal. Mark on it the yield strength, $\sigma_y$, the tensile strength $\sigma_{ts}$ and the ductility $\varepsilon_f$. Indicate on it the work done per unit volume in deforming the material up to a strain of $\varepsilon < \varepsilon_f$ (pick your own strain $\varepsilon$).

Exercise E6.2   On a sketch of a typical stress-strain curve for a ductile metal, distinguish between the maximum elastic energy stored per unit volume and the total energy dissipated per unit volume in plastic deformation. Give examples of engineering applications in which a key material characteristic is (i) maximum stored elastic energy per unit volume; (ii) maximum plastic energy dissipated per unit volume.

Exercise E6.3   (a) Write down expressions for the nominal strain and the true strain for the following situations:
　　　　　(i) in a tensile test where the length of the specimen increases from $\ell_o$ to $\ell$;
　　　　　(ii) in a compression test where the height of the specimen decreases from $h_o$ to $h$.
　　　　(b) In what circumstances can the true strain be approximated by the nominal strain?

Exercise E6.4   In a two-stage elongation of a specimen, the length is first increased from $\ell_o$ to $\ell_1$ and then from $\ell_1$ to $\ell_2$. Write down the true strains $\varepsilon_1$ and $\varepsilon_2$, respectively, for each of the two elongation processes considered separately. Show that the true strain for the overall elongation $\ell_o$ to $\ell_2$ is given by the sum of the true strains for each of the two separate processes. Is the same true for the nominal strains?

Exercise E6.5   In a 'tandem' rolling process, a wide metal strip passes continuously between four sets of rolls, each of which apply a reduction in thickness. The strip elongates without getting wider and therefore accelerates. For an incoming thickness $h_{in} = 25$ mm moving at speed $v_{in} = 5$ cm.s$^{-1}$, and an exit thickness $h_{out} = 3.2$ mm, calculate (i) the overall true strain applied to the material; (ii) the exit speed $v_{out}$. State any assumptions made.

Exercise E6.6   Explain briefly why the material property bubbles for metals are very elongated on the Young's Modulus−Strength chart (Figure 6.8). Comment on whether the same is true for the polymers. Why is this?

Exercise E6.7   Use the yield strength−density chart or the yield strength−modulus chart (Figures 6.7 and 6.8) to find:
1. The metal with the lowest strength.
2. The approximate range of strength of the composite GFRP.

3. Whether there are any polymers that are stronger than wood measured parallel to the grain.
4. How the strength of GFRP compares with that of wood.
5. Whether elastomers, that have moduli that are far lower than polymers, are also far lower in strength.

Exercise E6.8    What is meant by the ideal strength of a solid? Which material class most closely approaches it?

Exercise E6.9    (a) Explain briefly the three main microstructural mechanisms by which metals are hardened. Identify which mechanisms account for the following increases in yield stress:
   (i) Pure annealed aluminium: 25 MPa; cold-rolled non-heat-treatable Al-Mn-Mg alloy: 200 MPa.
   (ii) Pure annealed copper: 20 MPa; cast 60–40 Brass (60% Cu, 40% Zn): 105 MPa.
   (iii) Pure annealed iron: 140 MPa; quenched and tempered medium carbon steel: 550 MPa.
(b) For each of the hardening mechanisms in part (a), give other examples of important engineering alloys that exploit these mechanisms.

Exercise E6.10    An aluminium alloy used for making cans is cold-rolled into a strip of thickness 0.3 mm and width 1m. It is coiled around a drum of diameter 15 cm, and the outer diameter of the coil is 1 m. In the cold-rolled condition, the dislocation density is approximately $10^{15}$ m$^{-2}$. Estimate:
   (i) The mass of aluminium on the coil.
   (ii) The total length of strip on the coil.
   (iii) The total length of dislocation in the coiled strip.

Exercise E6.11    The lattice resistance of copper, like that of most FCC metals, is small. When 10% of nickel is dissolved in copper to make a solid solution, the strength of the alloy is 150 MPa. What would you expect the strength of an alloy with 20% nickel to be?

Exercise E6.12    A metal matrix composite consists of aluminum containing hard particles of silicon carbide (SiC) with a mean spacing of 3 μm. The composite has a strength of 180 MPa. If a new grade of the composite with a particle spacing of 2 μm were developed, what would you expect its strength to be?

Exercise E6.13    (a) Explain briefly how impeding the motion of dislocations with dispersed precipitates in a metal can raise the yield stress.
(b) A straight segment of dislocation line of length $L$ is pinned at two particles in a crystal. The application of a shear stress $\tau$ causes the dislocation to bow outwards as shown in the following figure, decreasing its radius of curvature as it does so. The force per unit length acting on the

dislocation is $\tau b L$, where $b$ is the Burger's vector. Given that the dislocation line tension $T = Gb^2/2$, show that the stress just large enough to fully bow out the dislocation segment into a semi-circle is given by $\tau = Gb/L$

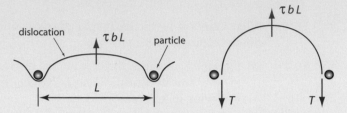

(c) Estimate the tensile yield stress of aluminium containing an array of equally spaced second-phase precipitates, where the volume fraction of the second phase is 5% and the precipitate diameter is 0.1 μm. The Burger's vector for aluminium is 0.29 nm.

(d) If incorrect heat treatment of the aluminium alloy results in a non-uniform distribution of precipitates, explain the effect this would have on the yield stress.

Exercise E6.14   Nano-crystalline materials have grain sizes in a range 0.01 to 0.1 μm. If the contribution of grain boundary strengthening in an alloy with grains of 0.1 μm is 20 MPa, what would you expect it to be if the grain size were reduced to 0.01 μm?

Exercise E6.15   The yield strength $\sigma_y$ of plain carbon steel is moderately dependent on the grain size $d$, and the relation can be described by the equation:

$$\sigma_y = \sigma_o + k\sqrt{\frac{1}{d}}$$

where $\sigma_o$ and $k$ are material constants. The yield strength is 622 MPa for a grain size of 180 μm and 663 MPa for a grain size of 22 μm.

(a) Calculate the yield strength of the steel for a grain size of 11μm.

(b) Explain briefly the physical significance of the constant $\sigma_o$.

Exercise E6.16   From the uniaxial stress-strain responses for various metals in Figure 6.23:

(a) Find the yield stress (or if appropriate a 0.2% proof stress), the tensile strengths and the ductility of each alloy.

(b) Account for the differences in strength, comparing annealed copper with each of the other alloys in turn.

Exercise E6.17 From the uniaxial stress-strain responses for various polymers in the following figure:
(a) Estimate the tensile strengths and ductility of each material.
(b) Account for the main differences between the shapes of the graphs for PMMA, ABS and Polypropylene.

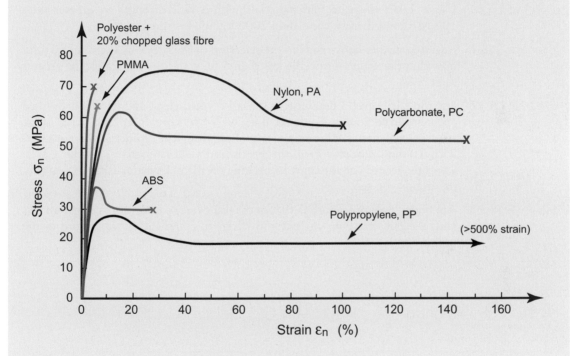

Exercise E6.18 Polycarbonate, PC (yield strength 70 MPa) is blended with polyester, PET (yield strength 50 MPa) in the ratio of 30% to 70%. If the strength of blends follows a rule of mixtures, what would you expect the yield strength of this blend to be?

## 6.9 Exploring design with CES

Use Level 2, Materials, throughout exercises.

| | |
|---|---|
| Exercise E6.19 | Find, by opening the records, the yield strengths of copper, brass (a solid solution of zinc in copper) and bronze (a solid solution of tin in copper). Report the mean values of the ranges that appear in the records. What explains the range within each record, since the composition is not a variable? What explains the differences in the mean values, when composition is a variable? |
| Exercise E6.20 | Use a 'Limit' stage to find materials with a yield strength $\sigma_y$ greater than 100 MPa and density $\rho$ less than 2000 kg/m$^3$. List the results. |
| Exercise E6.21 | Add two further constraints to the selection of the previous exercise. Require now that the material price be less than 5 \$/kg and the elongation be greater than 5%. |
| Exercise E6.22 | Use the CES Level 3 database to select Polypropylene and its blended, filled and reinforced grades. To do so, open CES Edu Level 3, apply a 'Tree' stage selecting Polymers—Thermoplastics—Polypropylene (folder). Make a chart with Young's modulus $E$ on the $x$-axis and yield strength $\sigma_y$ on the $y$-axis. Label the records on the chart by clicking on them. Explain, as far as you can, the trends you see. |
| Exercise E6.23 | Apply the same procedure as that of the last exercise to explore copper and its alloys. Again, use your current knowledge to comment on the origins of the trends. |

## 6.10 Exploring the science with CES Elements

| | |
|---|---|
| Exercise E6.24 | The elastic (potential) energy per unit length of a dislocation is 0.5 $Eb^2$ J/m. Make a bar chart of the energy stored in the form of dislocations, for a dislocation density of 10$^{14}$ m/m$^3$. Assume that the magnitude of Burger's vector, **b**, is the same as the atomic diameter. (You will need to use the 'Advanced' facility in the axis-choice dialog box to make the function.) How do the energies compare with the cohesive energy, typically $5 \times 10^4$ MJ/m$^3$? |
| Exercise E6.25 | Work hardening causes dislocations to be stored. Dislocations disrupt the crystal and have potential energy associated with them. It has been suggested that sufficient work hardening might disrupt the crystal so much that it becomes amorphous. To do this, the energy associated with the dislocations would have to be about equal to the heat of fusion, since this is the difference |

in energy between the ordered crystal and the disordered liquid. The energy per unit length of a dislocation is $0.5 \, Eb^2$ J/m. Explore this in the following way:

(a) Calculate and plot the energy associated with a very high dislocation density of $10^{17}$ m/m$^3$ for the elements, i.e. plot a bar chart of $0.5 \times 10^{17} Eb^2$ on the $y$-axis using twice atomic radius as equal to Burger's vector **b**. Remember that you must convert GPa into kPa and atomic radius from nm to m to get the energy in kJ/m$^3$.

(b) Now add, on the $x$-axis, the heat of fusion energy. Convert it from kJ/mol to kJ/m$^3$ by multiplying $H_c$ by 1000/molar volume, with molar volume in m$^3$/kmol (as it is in the database). What, approximately, is the ratio of the dislocation energy to the energy of fusion? Would you expect this very high dislocation density to be enough to make the material turn amorphous?

# Chapter 7
# Bend and crush: strength-limited design

Elastic design, avoiding plasticity, ensures that the cabin of the car does not deform in a crash. Plasticity absorbs the energy of impact, and allows metals to be shaped and polymers to be moulded. (Image of crash testing courtesy AutoNews; image of hot rolling courtesy of Tanis Inc., Delafield, WI)

## 7.1 Introduction and synopsis

Stiffness-limited design, described in Chapter 5, is design to avoid excessive elastic deflection. Strength-limited design, our concern here, is design to avoid plastic collapse. That generally means design to avoid yield, arranging that the component remains elastic throughout, when it is called *elastic design*. Elastic design is not always possible or necessary: *local* yielding may be permissible provided *general* yield is avoided.

That is half the picture. The other half is design to permit controlled plastic collapse. The safety of modern cars relies on the front of the car absorbing the kinetic energy in a collision by plastic deformation. And the manufacturing processes of metal rolling, forging, extrusion and pressing use plastic flow. Here, strains are large, elastic deformation irrelevant and the focus is on the forces and work necessary to achieve a prescribed change of shape (see chapter cover picture).

Plasticity problems are solved in more than one way. When yield is to be avoided, we analyse the elastic state of stress in a component and make sure that this nowhere exceeds the yield strength. Full plasticity, by contrast, requires general yield. Then the mechanism of plastic collapse must be identified and the collapse load calculated by requiring that yield does occur where it needs to.

Chapter 5 introduced elastic solutions for common modes of loading: tension, compression, bending, torsion or internal pressure. Not surprisingly there are equivalent results for plastic design, and we start with these. As before, you don't need to know how to derive them, just where to find them and how to use them. We use them to develop material indices for strength-limited design and apply them via case studies.

## 7.2 Standard solutions to plastic problems

*Yielding of ties and columns*    A tie is a rod loaded in tension, a column is a rod loaded in compression. The state of stress within them is uniform, as was shown in Figure 5.1(a). If this stress, $\sigma$, is below the yield strength $\sigma_y$ the component remains elastic; if it exceeds $\sigma_y$, it yields. Yield in compression is only an issue for short, squat columns. Slender columns and panels in compression are more likely to buckle elastically first (Chapter 5).

*Yielding of beams and panels*    The stress state in bending was introduced in Chapter 5. A bending moment $M$ generates a linear variation of longitudinal stress $\sigma$ across the section (Figure 7.1(a)) defined by

$$\frac{\sigma}{y} = \frac{M}{I} = E\kappa \tag{7.1}$$

where $y$ is the distance from the neutral axis, and the influence of the cross-section shape is captured by $I$, the second moment of area.

For elastic deflection, we were interested in the last term in the equation—that containing the curvature $\kappa$. For yielding, it is the first term. The maximum longitudinal stress $\sigma_{max}$ occurs at the surface (Figure 7.1(a)), at the greatest distance $y_m$ from the neutral axis

$$\sigma_{max} = \frac{My_m}{I} = \frac{M}{Z_e} \tag{7.2}$$

The quantity $Z_e = I/y_m$ is called the *elastic section modulus* (not to be confused with the elastic modulus of the material, $E$). If $\sigma_{max}$ exceeds the yield strength $\sigma_y$ of the material of the beam, small zones of plasticity appear at the surface where the stress is highest, as in Figure 7.1(b). The beam is no longer elastic and, in this sense, is damaged even if it has not

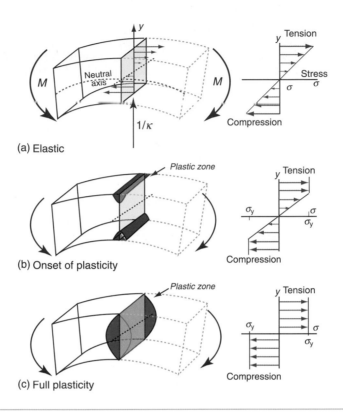

**Figure 7.1** A beam loaded in bending. The stress state is shown on the right for purely elastic loading (a), the onset of plasticity (b) and full plasticity (c).

failed completely. If the moment is increased further, the linear profile is truncated—the stress near the surface remains equal to $\sigma_y$ and plastic zones grow inward from the surface. Although the plastic zone has yielded, it still carries load. As the moment increases further the plastic zones grow until they penetrate through the section of the beam, linking to form a *plastic hinge* (Figure 7.1(c)). This is the maximum moment that can be carried by the beam; further increase causes it to collapse by rotating about the plastic hinge.

### Example 7.1

Calculate the maximum 'extreme fibre' stress in the $300 \times 25 \times 1$ mm stainless steel ruler in Examples 5.1 and 5.2, when it is mounted as a cantilever with the X-X axis horizontal with 250 mm protruding and a weight of 10 N is hung from its end. Compare this bending stress with the yield stress of the material, and with the direct stress when a 10 N force is used to pull the ruler axially.

*Answer.* The maximum bending moment $M$ in the cantilever (at its base—adjacent to the support) is $M = WL = 0.25 \times 10 = 2.5$ Nm. Using equation (7.2) with $I_{XX} = 2.1$ mm$^4$ (from Example 5.1), and $y_m = 0.5$ mm at the extreme fibre of the beam,

$$\sigma = \frac{My_m}{I_{XX}} = \frac{2.5 \times 0.0005}{2.1 \times 10^{-12}} = 595 \text{ MPa}$$

The yield strength of stainless steels lies in the range 200 to 1000 MPa. So the ruler would have to be made of quite a strong stainless steel to withstand this load without yielding.

When loaded by the same force in tension, the stress would be

$$\sigma = \frac{F}{wt} = \frac{10}{25 \times 1} = 0.4 \text{ MPa}$$

This stress is far less than the strength of most engineering materials (see Figure 6.7).

Figure 7.2 shows simply supported beams loaded in bending. In the first, the maximum moment $M$ is $FL$, in the second it is $FL/4$ and in the third $FL/8$. Plastic hinges form at the positions indicated in red when the maximum moment reaches the moment for collapse. This failure moment, $M_f$, is found by integrating the moment caused by the constant stress distribution over the section (as in Figure 7.1(c), compression one side, tension the other)

$$M_f = \int_{\text{section}} b(y)|y|\sigma_y \, dy = Z_p \sigma_y \tag{7.3}$$

where $Z_p$ is the *plastic section modulus*. So two new functions of section shape have been defined for failure of beams: one for first yielding, $Z_e$, and one for full plasticity, $Z_p$. In both cases the moment required is simply $Z\sigma_y$. Values for both are listed in Figure 7.3. The

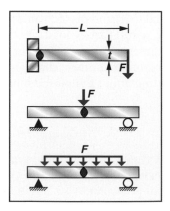

**Figure 7.2** The plastic bending of beams.

| Section shape | Area $A$ m² | Elastic section modulus $Z_e$ m³ | Plastic section modulus $Z_p$ m³ |
|---|---|---|---|
| 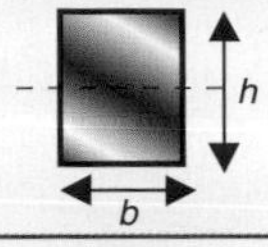 $h$ $b$ | $bh$ | $\dfrac{bh^2}{6}$ | $\dfrac{bh^2}{4}$ |
| 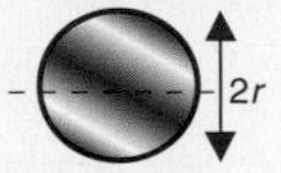 $2r$ | $\pi r^2$ | $\dfrac{\pi}{4} r^3$ | $\dfrac{\pi}{3} r^3$ |
| 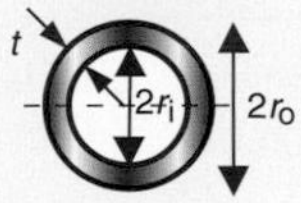 $t$ $2r_i$ $2r_o$ | $\pi(r_o^2 - r_i^2)$ $\approx 2\pi r t$ | $\dfrac{\pi}{4 r_o}(r_o^4 - r_i^4)$ $\approx \pi r^2 t$ | $\dfrac{\pi}{3}(r_o^3 - r_i^3)$ $\approx \pi r^2 t$ |

**Figure 7.3** The area $A$, section modulus $Z_e$ and fully plastic modulus $Z_p$ for three simple sections.

ratio $Z_p/Z_e$ is always greater than 1 and is a measure of the safety margin between initial yield and collapse. For a solid rectangle, it is 1.5, meaning that the collapse load is 50% higher than the load for initial yield. For efficient shapes, like tubes and I-beams, the ratio is much closer to 1 because yield spreads quickly from the surface to the neutral axis.

*Yielding of shafts* We saw in Chapter 5 that a torque, $T$, applied to the ends of a shaft with a uniform circular section, and acting in the plane normal to the axis of the bar as in Figure 7.4, produces a shear stress that increases linearly with distance $r$ from the central axis:

$$\tau = \frac{Tr}{K} = \frac{G\theta r}{K} \tag{7.4}$$

where $K$ is the *polar second moment of area*. The resulting elastic deformation was described by the angle of twist per unit length $\theta/L$. Failure occurs when the maximum surface stress exceeds the yield strength $\sigma_y$ of the material. The maximum shear stress, $\tau_{max}$, is at the surface and has the value

$$\tau_{max} = \frac{TR}{K} \tag{7.5}$$

where $R$ is the radius of the shaft. From Chapter 6, the yield stress in shear, $k$, is half the tensile yield stress, so first yield occurs when $\tau_{max} = \sigma_y/2$. When the torque is increased further, plasticity spreads inward. The maximum torque that the shaft can carry occurs when $\tau = k$ over the whole section. Any greater torque than this causes the shaft to collapse in torsion by unrestrained rotation. For example, for a solid circular section, the collapse torque is

$$T = \frac{2}{3}\pi R^3 k \tag{7.6}$$

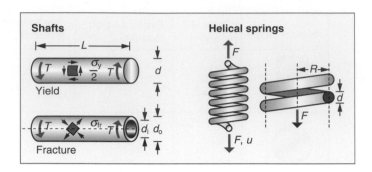

**Figure 7.4** Elastic torsion of shafts. The stress in the shaft depends on the torque *T* and the polar moment of area *K*. Helical springs are a special case of torsional loading.

---

### Example 7.2

The solid brass shaft in Example 5.3 has a yield strength in tension of 300 MPa. Calculate the torque at first yield and the torque at plastic collapse (that is, it is plastic right through).

*Answer.* At first yield, using equation (7.5) and noting that $J = K$ for circular sections (see equation (5.7)):

$$T = \frac{(\sigma_y/2)K}{R} = \frac{(150 \times 10^6)(0.98 \times 10^{-9})}{0.005} = 29.4 \text{ Nm}$$

At plastic collapse, using equation (7.6) gives

$$T = \frac{2}{3}\pi R^3 k = \frac{2}{3}\pi(0.005)^3(150 \times 10^6) = 39.3 \text{ Nm}$$

---

Helical springs are a special case of torsional loading (Figure 7.4): when the spring is loaded axially, the individual turns twist. It is useful to know the spring stiffness, *S*. If the spring has *n* turns of wire of shear modulus *G*, each of diameter *d*, wound to give a spring of radius *R*, the stiffness is

$$S = \frac{F}{u} = \frac{Gd^4}{64nR^3}$$

where *F* is the axial force applied to the spring and *u* is its extension. The elastic extension is limited by the onset of plasticity. This occurs at the force

$$F_{\text{crit}} = \frac{\pi}{32}\frac{d^3\sigma_y}{R} \tag{7.7}$$

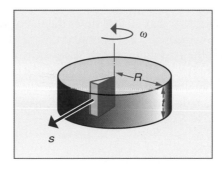

**Figure 7.5** Spinning disks, as in flywheels and gyroscopes, carry radial tensile stress caused by centrifugal force.

*Spinning disks (flywheels)* Spinning disks or rings store kinetic energy $U$ (Figure 7.5). Centrifugal forces generate a radial tensile stress in the disk that reaches a maximum value $\sigma_{max}$. Analysis of a disk of density $\rho$, radius $R$ and thickness $t$, rotating at an angular velocity $\omega$ radians/second, gives the kinetic energy and the maximum stress (when Poisson's ratio is taken as 1/3) as

$$U = \frac{\pi}{4}\rho t\omega^2 R^4 \quad \text{and} \quad \sigma_{max} = 0.42\rho\omega^2 R^2 \tag{7.8}$$

The disk yields when $\sigma_{max}$ exceeds $\sigma_y$, and this defines the maximum allowable $\omega$ and limits the inertial energy storage.

---

### Example 7.3

A flywheel has a radius of 150 mm and a thickness of 50 mm. It is made of CFRP with a strength of 700 MPa and a density of 1600 kg/m$^3$. What is its burst speed (the speed at which the material at the centre reaches its failure strength)? What is the maximum amount of kinetic energy it can store? How does this value compare with the chemical energy stored in the same mass of gasoline?

*Answer.* When the stress at the centre reaches the yield strength, equation (7.8) gives:

$$\omega = \sqrt{\frac{\sigma_y}{0.42\rho R^2}} = \sqrt{\frac{700 \times 10^6}{0.42 \times 1600 \times 0.15^2}} = 6800 \text{ rad/s } (65\,000 \text{ rpm})$$

At this speed, equation (7.8) gives the maximum possible kinetic energy as

$$U = \frac{\pi}{4} \times 1600 \times 0.05 \times 6800^2 \times 0.15^4 = 1.5 \text{ MJ}$$

The mass of the flywheel is $m = \pi R^2 t\rho = \pi \times 0.15^2 \times 0.05 \times 1600 = 5.7$ kg. Using the calorific value of gasoline of 43 000 kJ/kg (Example 4.4), the equivalent chemical energy is 43 000 × 5.7 = 245 MJ; that is, 160 times the maximum kinetic energy stored in the flywheel. Flywheel-powered cars are impractical. The flywheel would have to weigh at least 100 times more than a full gas tank for an equivalent range.

Flywheel energy storage *is*, however, practical for regenerative braking systems, which store energy during deceleration of a vehicle and re-insert it during acceleration. This is the principle of the Kinetic Energy Recovery Systems (KERS), introduced into Formula I racing cars in 2009.

*Contact stresses*    Contact stress analysis is important in design of rolling and sliding contacts as in bearings, gears and railway track. Yielding at contacts is closely linked to failure by wear and fatigue. When surfaces are placed in contact they touch at a few discrete points. If the surfaces are loaded, the contacts flatten elastically and the contact areas grow (Figure 7.6). The stress state beneath the contact is complex, first analysed by the very same Hertz[1] for whom the unit of frequency is named. Consider a sphere of radius $R$ pressed against a flat surface with a load $F$. Both sphere and surface have Young's modulus $E$. While the contact is elastic (and again assuming Poisson's ratio $v = 1/3$), the radius of the contact area is

$$a \approx 0.7 \left( \frac{FR}{E} \right)^{1/3} \tag{7.9}$$

and the relative displacement of the two bodies is

$$u \approx - \left( \frac{F^2}{E^2 R} \right)^{1/3} \tag{7.10}$$

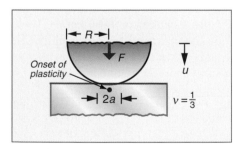

**Figure 7.6**  Contact stresses are another form of stress concentration. When elastic, the stresses and displacement of the surfaces toward each other can be calculated.

[1] Heinrich Rudolph Hertz (1821—1894), German physicist, discoverer of radio waves and how to generate them and inventor of the transmitter that started the radio age. In his spare time he dabbled in mechanics.

For failure in contact, we need the maximum value of the shear stress, since it is this that causes first yield. It is beneath the contact at a depth of about $a/2$, and has the value

$$\tau_{max} = \frac{F}{2\pi a^2} \tag{7.11}$$

If this exceeds the shear yield strength $k = \sigma_y/2$, a plastic zone appears beneath the center of the contact.

---

### Example 7.4

A steel ball bearing of radius 5 mm is pressed with a force of 10 N onto a steel ball-race. Both bearing and race have a Young's modulus of 210 GPa and yield strength of 400 MPa. Calculate the contact radius and the relative displacement of the two bodies. Compare the maximum stress beneath the contact patch with the strength of the surface and with the stress when the load of 10 N is uniformly distributed over a circular section radius 5 mm.

*Answer.* The contact radius is equation (7.9)

$$a \approx 0.7\left(\frac{FR}{E}\right)^{1/3} = 0.7\left(\frac{10 \times 0.005}{210 \times 10^9}\right)^{1/3} = 0.044 \text{ mm} = 44 \text{ μm}$$

The relative displacement is equation (7.10)

$$u \approx \left(\frac{F^2}{E^2 R}\right)^{1/3} = \left(\frac{10^2}{(210 \times 10^9)^2 \times 0.005}\right)^{1/3} = 0.78 \text{ μm}$$

The maximum shear stress in the ball race is equation (7.11)

$$\tau_{max} = \frac{F}{2\pi a^2} = \frac{10}{2\pi(9.4 \times 10^{-5})^2} = 180 \text{ MPa}$$

This is a significant proportion of the yield strength of the surface, indicating that fatigue failure due to cyclic loading may be an issue of concern (Chapter 9). For comparison, a load of 10 N applied to an area of radius 5 mm would generate a very low stress of

$$\frac{F}{\pi a^2} = \frac{10}{\pi(5 \times 10^{-3})^2} = 0.13 \text{ MPa}$$

---

*Stress concentrations*   Holes, slots, threads and changes in section concentrate stress locally (Figure 7.7). Yielding will therefore start at these places, though as the bulk of the component is still elastic this initial yielding is not usually catastrophic. The same cannot be said for fatigue (Chapter 9), where stress concentrations often are implicated as the origins of failure.

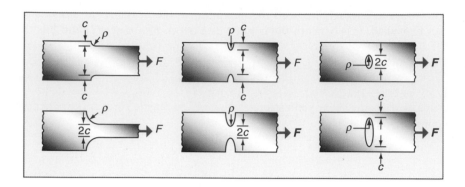

**Figure 7.7** Stress concentrations. The change of section concentrates stress most strongly where the curvature of the surface is greatest.

We define the *nominal stress* in a component $\sigma_{nom}$ as the load divided by the cross-section, ignoring features that cause the stress concentration. The *maximum local stress* $\sigma_{max}$ is then found approximately by multiplying the nominal stress $\sigma_{nom}$ by a *stress concentration factor* $K_{sc}$, where

$$K_{sc} = \frac{\sigma_{max}}{\sigma_{nom}} = 1 + \alpha \left(\frac{c}{\rho_{sc}}\right)^{1/2} \tag{7.12}$$

Here $\rho_{sc}$ is the minimum radius of curvature of the stress-concentrating feature and $c$ is a characteristic dimension associated with it: either the half-thickness of the remaining ligament, the half-length of a contained notch, the length of an edge notch or the height of a shoulder, whichever is least (Figure 7.7). The factor $\alpha$ is roughly 2 for tension, but is nearer 1/2 for torsion and bending. Though inexact, the equation is an adequate working approximation for many design problems. More accurate stress concentration factors are tabulated in compilations such as Roark (see Further reading at the end of this chapter).

As a simple example, consider a circular hole in a plate loaded in tension. The radius of curvature of the feature is the hole radius, $\rho_{sc} = R$, and the characteristic dimension is also the radius, $c = R$. From equation (7.12), the local stress next to the hole is thus three times the nominal tensile stress—however small the hole. Local yielding therefore occurs when the nominal stress is only $\sigma_y/3$.

---

### Example 7.5

The skin of an aircraft is made of aluminum with a yield strength of 300 MPa and a thickness of 8 mm. The fuselage is a circular cylinder of radius 2 m, with an internal pressure that is $\Delta p = 0.5$ bar ($5 \times 10^4$ N/m$^2$) greater than the outside pressure. A window in the side of the aircraft is nominally square (160 mm $\times$ 160 mm). It has corners with a radius of curvature of $\rho = 5$ mm.

Calculate the nominal hoop stress in the aircraft wall due to pressure and estimate the peak stress near the corner of the window. Ignore the effects of longitudinal stress in the tube. Note that the hoop stress in the wall of a circular pressure vessel of radius $R$ and thickness $t$ with internal pressure $\Delta p$ is given by $\sigma = \Delta p R/t$. If the maximum allowable stress is 100 MPa, is 5 mm a suitable radius for the corners of the window?

*Answer.* The nominal 'hoop' stress is:

$$\sigma = \frac{\Delta p R}{t} = \frac{5 \times 10^4 \times 2}{0.008} = 12.5 \text{ MPa}$$

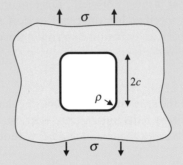

The sketch shows how the stress concentration factor for the window geometry is equivalent to one of the standard cases shown in Figure 7.7. Taking $c$ as half the width of the window (80 mm) and $\rho_{sc} = 5$ mm, the stress concentration factor is:

$$K_{sc} \approx 1 + \alpha \left( \frac{c}{\rho_{sc}} \right)^{1/2} = 1 + 2 \left( \frac{80}{5} \right)^{1/2} = 9$$

Consequently the peak stress near the corner of the window is approximately $9 \times 12.5 =$ 112 MPa, which is greater than the allowable value, risking fatigue failure after a small number of pressurisations. Such was the fate of the de Havilland Comet—the world's first pressurised commercial jet airliner, which first flew in 1949. Hence the large corner radii of aircraft windows today.

# 7.3 Material indices for yield-limited design

*Minimising weight: a light, strong tie-rod* Many structures rely on tie members that must carry a prescribed tensile load without yielding—the cover picture of Chapter 4 showed two—often with the requirement that they be as light as possible. Consider a design that calls for a cylindrical tie-rod of given length $L$ that must carry a tensile force $F$ as in Figure 5.7(a),

Table 7.1 Design requirements for the light tie

| Function | • Tie-rod |
|---|---|
| Constraints | • Tie must support tensile load $F$ without yielding |
| | • Length $L$ specified |
| Objective | • Minimise the mass $m$ of the tie |
| Free variables | • Choice of cross-section area, $A$ |
| | • Choice of material |

with the constraint that it must not yield, but remain elastic. The objective is to minimise its mass. The length $L$ is specified but the cross-section area $A$ is not (Table 7.1).

As before, we first seek an equation describing the quantity to be maximised or minimised. Here it is the mass $m$ of the tie

$$m = AL\rho \tag{7.13}$$

where $A$ is the area of the cross-section and $\rho$ is the density of the material of which it is made. We can reduce the mass by reducing the cross-section, but there is a constraint: the section area $A$ must be sufficient to carry the tensile load $F$ without yielding, requiring that

$$\frac{F}{A} \leq \sigma_y \tag{7.14}$$

where $\sigma_y$ is the yield strength. Eliminating $A$ between these two equations gives

$$m \geq FL\left(\frac{\rho}{\sigma_y}\right) \tag{7.15}$$

The lightest tie that will carry $F$ safely is that made of the material with the smallest value of $\rho/\sigma_y$. We could define this as the material index of the problem, seeking a minimum, but as in Chapter 5 we will invert it, seeking materials with the largest values of

$$M_t = \frac{\sigma_y}{\rho} \tag{7.16}$$

This index, the *specific strength*, is plotted as a line of slope 1 in the chart of Figure 7.8. A particular value of $\sigma_y/\rho$ is identified, passing through the high-strength end of several major alloy systems—nickel alloys, high-strength steels and aluminum and magnesium alloys. Titanium alloys are significantly better than the other metals; CFRP is better still. Ceramics and glasses have high values of $M_t$ but are impractical as structural ties because of their brittleness.

Tension, then, is straightforward. The real problem in elastic design is seldom tension; it is *bending*. We therefore revisit the bending of panels and beams, applying a constraint on strength rather than stiffness.

***Minimising weight: light, strong panels***  Figure 5.7(b) showed a panel supported at its edges, carrying a specified central load. The width $b$ and span $L$ are fixed, but we are free to choose the thickness $h$. The objective is to minimise the mass. Reducing the thickness reduces the mass, but it must be sufficient for the maximum stress to be below the elastic limit (Table 7.2).

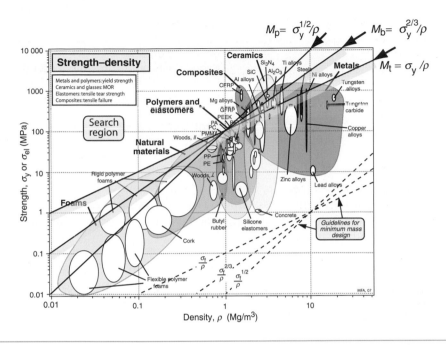

**Figure 7.8**  The strength–density chart with the indices $\sigma_{y}/\rho$, $\sigma_{y}^{2/3}/\rho$ and $\sigma_{y}^{1/2}/\rho$ plotted on it.

**Table 7.2**  Design requirements for the light panel

| Function | • Panel in bending |
|---|---|
| Constraints | • Panel must support bending load $F$ without yielding |
| | • Width $b$ and span $L$ specified |
| Objective | • Minimise the mass $m$ of the panel |
| Free variables | • Choice of thickness $h$ |
| | • Choice of material |

The procedure is much as before—set up an equation for the mass; find an expression for the maximum stress (noting that for a rectangular section, $I = bh^3/12$); use this constraint to eliminate the free variable $h$ and read off the material index. The analysis itself is left for the Exercises; the result is

$$M_{\mathrm{p}} = \frac{\sigma_{y}^{1/2}}{\rho} \tag{7.17}$$

This index also is shown by a shaded guideline in Figure 7.8. Now all the light alloys (Mg, Al and Ti) outperform steel, as do GFRP and wood. CFRP still leads the way.

Table 7.3 Design requirements for the light strong beam

| | |
|---|---|
| Function | • Beam in bending |
| Constraints | • Beam must support bending load *F* without yielding |
| | • Span *L* specified, section shape square |
| Objective | • Minimise the mass *m* of the beam |
| Free variables | • Area *A* (or square section dimension *b*) |
| | • Choice of material |

*Light, strong beams: the effect of shape* In beam design we are free to choose the shape as well as the dimensions of the cross-section. First we consider beams of prescribed shape, with freedom to change their size in a self-similar way. Then we explore how much better we can make the design by using efficient shapes.

Start with the beam of square section $A = b \times b$, which may vary in size, loaded in bending over a span of fixed length $L$ with a central load $F$ (Figure 5.7(c) and Table 7.3).

The analysis is similar to the panel (see Exercises), with the modified second moment of area, $I = b^4/12 = A^2/12$. The resulting material index is

$$M_b = \frac{\sigma_y^{2/3}}{\rho} \tag{7.18}$$

The chart in Figure 7.8 shows that the slope of this index lies between the other two, so the competition between metals, wood and composites changes again, with CFRP still on top.

But, you will say, no one uses solid, square-section beams for minimum mass design—and you are right. If you want to support bending loads it is better to choose a shape that uses less material to provide the same strength. Wooden floor-joists in houses are typically twice as deep as they are wide; standard steel and aluminum beams have an I-section or a box section; space frames, commonly, are made from tubes. Does this change the index? Well, yes and no. Let's start with the no.

Equation (7.18) was derived by analysing a square beam, but the result holds for any *self-similar* shape, meaning one in which all dimensions remain in proportion as the size is varied. However, shaping a given cross-sectional area from a solid square beam into a tube or an I-beam increases the second moment of area, without changing the cross-sectional area or the mass. In Chapter 5, it was shown that this provides greater stiffness with no mass penalty. It also makes the beam stronger. Equation (7.2) shows that a bending moment gives a maximum stress determined by the elastic section modulus, $Z_e = I/y_m$. Consequently the gain in strength by increasing $I$ is not quite as great as the gain in stiffness—shaping the section to an I-beam increases $I$, but usually makes the beam deeper too, increasing $y_m$. As before, we can define a 'shape factor' $\phi_B^y$ for strength—the ratio of $Z_e$ for the shaped section to $Z_e$ for the same area of material in a solid square section:

$$\phi_B^y = \frac{Z_e^{\text{shaped}}}{Z_e^{\text{solid}}}$$

**Table 7.4**  The effect of shaping on strength and mass of beams in different structural materials

| Material | Maximum failure shape factor (failure moment relative to solid square beam) $\phi_B^y$ | Mass ratio by shaping (relative to solid square beam) |
| --- | --- | --- |
| Steels | 13 | 0.18 |
| Al alloys | 10 | 0.22 |
| Composites (GFRP, CFRP) | 9 | 0.23 |
| Wood | 3 | 0.48 |

Recall from Chapter 5 that materials are not all equally easy to shape; if they were, all could be given the same efficient shape and the index in equation (7.18) would be sufficient. Table 7.4 gives typical maximum values of the shape factor for strength, for a range of competing materials.

As was the case for stiffness-limited design in Chapter 5, shaping also enables a lighter structure to be made with the same strength, that is, less material is needed to carry the same load in bending without failure. The right-hand column of Table 7.4 shows the mass saving, relative to a square beam, that we might achieve by shaping. In this case the mass ratio (shaped to square) is given by $\Phi^{-2/3}$. This more complex dependence arises from the way $Z_e (= I/y_m)$ relates to the cross-section area $A$. Metals and composites can all be improved significantly (though again, metals do a little better), but wood has more limited potential. Thus, when comparing materials for light, strong beams using the index in equation (7.18), the performance of wood should not be over-estimated; other materials allow more efficient shapes to be made.

*Minimising material cost or volume*   When the objective is to minimise cost rather than weight, the indices change exactly as before. The objective function for the material cost $C$ of the tie, panel or beam becomes

$$C = mC_m = ALC_m\rho \tag{7.19}$$

where $C_m$ is the cost per kg of the material. This leads to indices that are just those of equations (7.16)–(7.18) with $\rho$ replaced by $C_m\rho$.

If instead the objective is to minimise volume, density is no longer relevant. The indices for ties, panels and beams are then the same as those for minimum mass with $\rho$ deleted.

# 7.4 Case studies

Plasticity problems are of two types. In the first, yield is avoided everywhere so that the entire component remains elastic, meeting the condition $\sigma < \sigma_y$ everywhere. Limited local plastic flow at stress concentrations may be allowed provided that, once it has happened, the condition $\sigma < \sigma_y$ is met everywhere (bedding down). In the second, full plasticity is the aim. Manufacturing processes such as metal forging and extrusion are well-managed plasticity; crash barriers and packaging, too, rely on full plasticity to absorb energy.

First a cautionary note. Carrying loads safely is not just a question of strength, but also of toughness—the resistance of the material to fracture. Strength and toughness are not the same thing—we explain why in the next chapter. For now, it is enough to know that some materials may appear to be good options for strength-limited design, but that they are impractical because they are too brittle in tension or in shock loading.

*Corkscrew levers again: strength* The lever of the corkscrew of Figure 5.13 and described in Section 5.5 is loaded in bending. It needs some stiffness, but if it flexes slightly, no great harm is done. If, however, it yields, bending permanently before it extracts the cork, the user will not be happy. So it must also meet a strength constraint.

The cross-section is rectangular. As in Chapter 5 we make the assumption of self-similarity, meaning that we are free to change the scale of the section but not its shape. Then the criterion for selection is that of the index of equation (7.18), $M_b = \sigma_y^{2/3}/\rho$. The index is plotted in Figure 7.8, isolating materials for light, strong beams. The selection is almost the same as that for stiffness: CFRP, magnesium and aluminum alloys are the best choice.

But what about the stress-concentrating effect at the holes? A stress concentration factor $K_{sc}$ means that yield starts when the nominal stress exceeds $\sigma_y/K_{sc}$. Changing the scale of the part does not change $K_{sc}$ because it depends only on the shape of the defect—the ratio $\rho/c$ (its value for a circular hole is 3, regardless of scale)—with the result that the index remains unchanged and the selection remains valid.

*Elastic hinges and couplings* Nature makes much use of elastic hinges: skin, muscle, cartilage all allow large, recoverable deflections. Man, too, designs with *flexural* and *torsional hinges*: ligaments that connect or transmit load between components while allowing limited relative movement between them by deflecting elastically (Figures 5.18 and 5.19). Which materials make good hinges?

Consider the hinge for the lid of a box. The box, lid and hinge are to be moulded in one operation—there are no separate screws or pins. The hinge is a thin ligament of material that flexes elastically as the box is closed, as in Figure 5.18, but it carries no significant axial loads. Then the best material is the one that (for given ligament dimensions) bends to the smallest radius without yielding or failing (Table 7.5). When a ligament of thickness $t$ is bent elastically to a radius $R$, the surface strain is

$$\varepsilon = \frac{t}{2R} \tag{7.20}$$

Table 7.5  Design requirements for elastic hinges

| Function | • Elastic hinge |
|---|---|
| Constraint | • No failure, meaning $\sigma < \sigma_y$ throughout the hinge |
| Objectives | • Maximise elastic flexure |
| Free variables | • Choice of material |

and—since the hinge is elastic—the maximum stress is

$$\sigma = E\frac{t}{2R}$$

This must not exceed the yield or failure strength $\sigma_y$. Thus, the minimum radius to which the ligament can be bent without damage is

$$R \geq \frac{t}{2}\left[\frac{E}{\sigma_y}\right] \tag{7.21}$$

The best material is the one that can be bent to the smallest radius—that is, the one with the greatest value of the index

$$M = \frac{\sigma_y}{E} \tag{7.22}$$

Here we need the $\sigma_y - E$ chart (Figure 7.9). Candidates are identified by using the guideline of slope 1; a line is shown at the position $M = \sigma_y/E = 3 \times 10^{-2}$. The best choices for the hinge are all polymeric materials. The short-list includes polyethylene, polypropylene, nylon and,

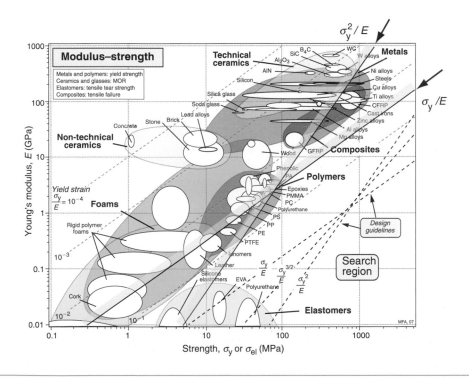

**Figure 7.9** Materials for elastic hinges and springs. Polymers are the best choice for the former. High-strength steel, CFRP and certain polymers and elastomers are the best choice for the latter.

best of all, elastomers, though these may be too flexible for the body of the box itself. Cheap products with this sort of elastic hinge are generally moulded from polyethylene, polypropylene or nylon. Spring steel and other metallic spring materials (like phosphor bronze) are possibilities: they combine usable $\sigma_y/E$ with high $E$, giving flexibility with good positional stability (as in the suspensions of relays).

*Materials for springs*    Springs come in many shapes (Figure 7.10) and have many purposes: axial springs (a rubber band, for example), leaf springs, helical springs, spiral springs, torsion bars. All depend on storing elastic energy when loaded, releasing it when unloaded again. The stored energy per unit volume in a material carrying a tensile stress $\sigma$ was derived in Chapter 4—it is given by $\sigma^2/2E$. If the spring yields it deforms permanently and ceases to fulfill its function, so the maximum value of $\sigma$ must not exceed $\sigma_y$, when the stored energy is $\sigma_y^2/2E$ per unit volume. The best material for a spring of minimum volume (Table 7.6) is therefore that with the greatest value of

$$M = \frac{\sigma_y^2}{E} \tag{7.23}$$

Though it is less obvious, the index is the same for leaf springs, torsional spring and coil springs—the best choice for one is the best choice for all.

The choice of materials for springs of minimum volume is shown in Figure 7.9. A family of lines of slope 2 link materials with equal values of $\sigma_y^2/E$; those with the highest values of $M$ lie toward the bottom right. The heavy line is one of the family; it is positioned so that a subset of materials is left exposed. The best choices are a *high-strength steel* lying near the top end of the line. Other materials are suggested too: *CFRP* (now used for truck springs), *titanium alloys* (good but expensive) and *nylon* (children's toys often have nylon springs), and, of course,

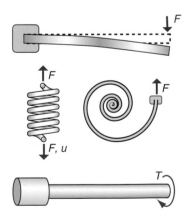

**Figure 7.10**   Springs: leaf, helical, spiral and torsion bar. Springs store energy. The best material for a spring, regardless of its shape or the way it is loaded, is that of a material with a large value of $\sigma_y^2/E$.

Table 7.6  Design requirements for springs

| Function | • Elastic spring |
| --- | --- |
| Constraint | • No failure, meaning stress below yield throughout the spring |
| Objectives | • Maximum stored elastic energy per unit volume |
| Free variables | • Choice of material |

*elastomers*. Note how the procedure has identified a candidate from almost every class of materials: metals, polymers, elastomers and composites.

*Full plasticity: metal forming*    Metal forming by plasticity is an ancient practice. Once metals could be extracted and cast, it did not take long to discover that they could be shaped by beating and bending them, and that their properties improved too. Almost any metal part that is not cast directly to shape will undergo some plastic forming in the solid state: standard sizes of plate, sheet, tube and I-sections, as well as other prismatic shapes like railway track, are shaped by rolling or extrusion; 3D shapes like tooling, gears and cranks are forged; and rolled sheet is shaped further by stamping (car body panels) and deep drawing (beverage cans). Figure 7.11 shows schematics of some of these processes, characterised in most cases by compressive loading (to avoid early failure in tension) and very large plastic strains. We revisit metal-forming processes in Chapters 13, 18 and 19, but here we can introduce the idea of the power input needed to shape a metal.

Figure 7.11(a) shows the rolling of plate with an initial thickness $t_o$. The plate emerges from the rolls with a lesser thickness $t_1$, a reduction of $\Delta t = t_o - t_1$. A lower bound for the torque $T$ and power $P$ required to do this is found from the plastic work, $\sigma_y \varepsilon_{pl}$ per unit volume that it takes to produce a plastic strain $\varepsilon_{pl}$ of $\Delta t / t_o$. If the rolls rotate through $\Delta\theta$, a length of approximately $R\Delta\theta$ is fed into the 'bite' (and thus of volume $V = R\Delta\theta t_o$ per unit width), where it is compressed to the emerging thickness $t_1$. Equating the work done by the pair of rolls, $2T\Delta\theta$, to the plastic work $V\sigma_y\varepsilon_{pl}$ gives the torque per roll

$$T = \frac{1}{2}R\sigma_y\Delta t \tag{7.24}$$

The power $P$ is just the torque $T$ times the angular velocity $\omega$ radians per second, giving

$$P = 2\,T\,\omega = R\omega\,\sigma_y\,\Delta t \tag{7.25}$$

The torque and power increase with $\sigma_y$, so hot-rolling takes less power than cold-rolling because $\sigma_y$ is smaller and work hardening (the increase of yield strength with strain) is negligible. We come back to the effect of temperature on yielding in Chapter 13.

The values of $T$ and $P$ calculated here are lower bound estimates. Most metal-working operations involve *redundant work* (plastic work that is not strictly necessary but happens because of the geometry of metal flow imposed by the machine). And there is always some friction in forming, as the compressed metal slides against the rolls or dies, which also increases the forces and power that are needed.

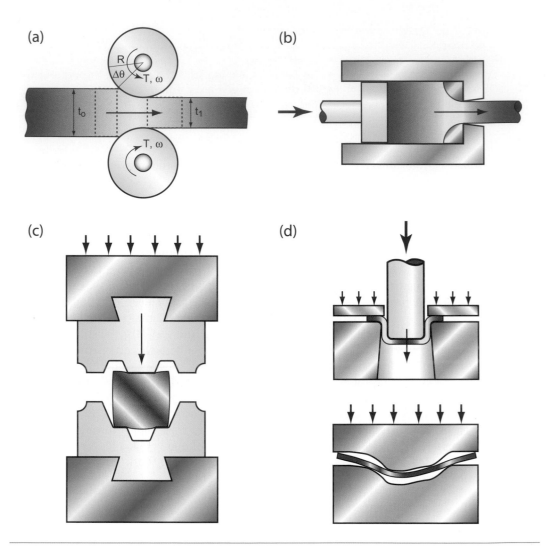

**Figure 7.11** Metal forming processes: (a) rolling; (b) extrusion; (c) forging (closed die); (d) sheet forming (deep drawing, stamping).

### Example 7.6

A rolling mill has a pair of rolls of radius 100 mm that rotate at an angular velocity of 0.5 rad/s. A steel strip passes through the rolls and its thickness is reduced from 25 mm to 18 mm. (a) Calculate the nominal plastic strain in the strip. (b) Calculate the power required to drive the rolls if the steel is: (i) at room temperature, when $\sigma_y = 480$ MPa; (ii) at 600 °C, when $\sigma_y = 150$ MPa.

*Answer.*

(a) The plastic strain is $\Delta t / t_{\mathrm{o}} = (25 - 18)/25 = 0.28$ (28%).
(b) (i) From equation (7.25), the necessary power for cold-rolling is at least

$$P - R\,\omega\,\sigma_{\mathrm{y}}\,\Delta t - 0.1 \times 0.5 \times \left(480 \times 10^6\right) \times \left(7 \times 10^{-3}\right) = 170\,\mathrm{kW}.$$

   (ii) The power scales with yield strength, so hot-rolling at 600 °C requires a power of at least $(150/480) \times 170 = 50$ kW.

# 7.5 Summary and conclusions

Elastic design is design to avoid yield. That means calculating the maximum stress, $\sigma$, in a loaded component and ensuring that it is less than $\sigma_{\mathrm{y}}$, the yield strength. Standard solutions give this maximum stress for panels, beams, torsion bars, spinning disks and many other components in terms of their geometry and the loads applied to them. From these, indices are derived to guide material choice. The indices depend on objectives such as minimising mass or cost, much as they did for stiffness—limited design in Chapter 5.

Sometimes, however, controlled plasticity is the aim. Then the requirement is that the stress must exceed the yield strength over the entire section of the component. Metal-forming operations such as forging, rolling and deep drawing rely on full plasticity. The forces and power they require scale with the yield strength of the material being shaped. Thus, soft metals such as lead, or aluminium alloys and low-carbon steel are easy to shape by plastic deformation, particularly when hot (Chapter 13). Some of the hardest alloys, like tool steels and the nickel-based super-alloys used in gas turbines, have such high yield strengths that they have to be shaped in other ways.

# 7.6 Further reading

Ashby, M. F. (2005). *Materials Selection in Mechanical Design* (3rd ed.). Oxford, UK: Butterworth-Heimann. ISBN 0-7506-6168-2. (Appendix A of this text is an expanded catalog of simple solutions to standard problems.)

Gere, J. M. (2006). *Mechanics of Materials* (6th ed.). Toronto, Canada: Thompson Publishing. ISBN 0-534-41793-0. (An intermediate level text on statics of structures by one of the fathers of the field; his books with Timoshenko introduced an entire generation to the subject.)

Hosford, W. F. (2005). *Mechanical Behavior of Materials*. Cambridge, UK: Cambridge University Press. ISBN 0-521-84670-6. (A text that nicely links stress—strain behaviour to the micromechanics of materials.)

Jenkins, C. H. M., & Khanna, S. K. (2006). *Mechanics of Materials*. Boston, MA, USA: Elsevier Academic. ISBN 0-12-383852-5. (A simple introduction to mechanics, emphasising design.)

Riley, W. F., Sturges, L. D., & Morris, D. H. (2003). *Statics and Mechanics of Materials* (2nd ed.). Hoboken, NJ, USA: McGraw-Hill. ISBN 0-471-43446-9. (An intermediate level text on the stress, strain and the relationships between them for many modes of loading. No discussion of micromechanics—response of materials to stress at the microscopic level.)

Vable, M. (2002). *Mechanics of Materials*. Oxford, UK: Oxford University Press. ISBN 0-19-513337-4. (An introduction to stress–strain relations, but without discussion of the micromechanics of materials.)

Young, W. C. (1989). *Roark's Formulas for Stress and Strain* (6th ed.). New York, USA: McGraw-Hill. ISBN 0-07-100373-8. (This is the 'Yellow Pages' of formulae for elastic problems—if the solution is not here, it doesn't exist.)

## 7.7 Exercises

Exercise E7.1   What is meant by the *elastic section modulus*, $Z_e$? A beam carries a bending moment $M$. In terms of $Z_e$, what is the maximum value that $M$ can take without initiating plasticity in the beam?

Exercise E7.2   (a) Calculate the elastic bending moment $M_e$ that causes first yield in a beam of square cross-section $b \times b$ with tensile yield strength $\sigma_y$. Calculate the fully plastic moment $M_p$ in the same beam. What is the ratio of $M_p/M_e$?

(b) Calculate the elastic torque $T_e$ that causes first yield in a shaft of radius $R$ with shear yield strength $\sigma_y/2$. Calculate the fully plastic torque $T_p$ in the same shaft. What is the ratio of $T_p/T_e$?

(c) Comment on the safety margin between initial yield and full plasticity in each case.

Exercise E7.3   A plate with a rectangular section 500 mm by 15 mm carries a tensile load of 50 kN. It is made of a ductile metal with a yield strength of 50 MPa. The plate contains an elliptical hole of length 100 mm and a minimum radius of 1 mm, oriented as shown in the diagram. What is:

(a) The nominal stress?

(b) The maximum stress in the plate?

Will the plate start to yield? Will it collapse completely?

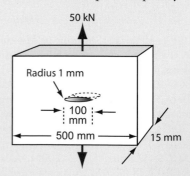

Exercise E7.4   A tie-rod made of a strip of metal of thickness $t$ has width $b$ that reduces to width $a$ with a radius of $\rho$, as shown in the figure ahead. The material has a yield stress $\sigma_y$.

(a) Find an expression for the value of $a$ that would be needed to avoid yield at the nominal stress level in the thinner section, with a safety factor of $S$.

(b) The applied load is expected to reach the stress level in (a). Find an expression for the value of the fillet radius $\rho$ that would ensure that the maximum stress in the tie-rod is less than the yield stress.

(c) Determine the value of $\rho$ if $a = 30$ mm, $b = 40$ mm, and $S = 3$.

(d) What is the real safety factor of the part with this value of $\rho$?

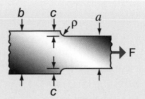

Exercise E7.5    Your task is to design a lightweight tie of length $L$ with a circular cross-section of radius $R$. It has to carry an axial force $F$, without tensile failure. You will need to choose the material (with Young's modulus $E$ and density $\rho$) and the corresponding cross-section radius to suit your choice of material (i.e. $R$ is a 'free variable'). Use a safety factor of 2 throughout.

(a) Find an expression for the radius $R$ of the tie that will carry the load without failure at stress $\sigma_f$.

(b) If the length of the tie is $L = 0.3$ m and the load is $F = 1000$ N, what value of $R$ is needed if the tie is made of (i) PEEK; (ii) Butyl Rubber; (iii) Ti alloy; (iv) Cu alloy? Use the material properties in the table below.

(c) Write an expression for the mass $m$ of the tie and determine the mass for each of the materials in (b). Which one would you choose?

(d) Substitute the expression for radius $R$ from (a) into the expression for mass $m$ from (c) to show that the material index to be *minimised* is $\rho/\sigma_f$. Determine the value of the material index for the four ties.

(e) Comment on the relationship between the mass of each tie and its material index $\rho/\sigma_f$.

(f) Examine Figure 7.8 to see how these three materials compare in terms of the reciprocal material index $\sigma_f/\rho$.

| | Failure stress $\sigma_f$ (MPa) | Density $\rho$ (kg/m³) | $R$ (mm) | $m$ (kg) | $\rho/\sigma_f$ |
|---|---|---|---|---|---|
| PEEK | 80 | 1300 | | | |
| Butyl Rubber | 2.5 | 2400 | | | |
| Ti alloy | 1200 | 4600 | | | |
| Cu alloy | 150 | 8900 | | | |

Exercise E7.6    You are asked to design a lightweight cantilever beam of length $L$ with a square cross-section of width $b$ and depth $b$. The beam is to be built-in to a wall at one end and loaded at the free end with a point load $F$ without exceeding the elastic limit. When designing the beam, you are free to specify the value of $b$ to suit your choice of material (i.e. $b$ is a 'free variable').

(a) Find an expression for the size of the beam $b$ that can carry $F$ without exceeding the elastic limit (failure stress) $\sigma_f$.

(b) If the length of the cantilever is $L = 1$ m, the load is $F = 1000$ N and the safety factor is 2, what value of $b$ is needed if the beam is made of (i) low-alloy steel; (ii) Al alloy; (iii) CFRP? Use the material properties in the table below.

(c) Write an expression for the mass $m$ of the beam and determine the mass for each of the materials in (b). Which one would you choose?

(d) Substitute the expression for beam dimension $b$ from (a) into the expression for mass $m$ from (c) to show that the material index to be *minimised* is $\rho/\sigma_f^{2/3}$. Determine the value of the material index for the three beams.

(e) Comment on the relationship between the mass of each beam and its material index $\rho/\sigma_f^{2/3}$.

(f) Examine Figure 7.8 to see how these three materials compare in terms of the reciprocal material index $\sigma_f^{2/3}/\rho$.

| | Failure stress $\sigma_f$ (MPa) | Density $\rho$ (kg/m$^3$) | $b$ (mm) | $m$ (kg) | $\rho/\sigma_f^{2/3}$ |
|---|---|---|---|---|---|
| Low-alloy steel | 900 | 7800 | | | |
| Al Alloy | 300 | 2600 | | | |
| CFRP | 700 | 1500 | | | |

**Exercise E7.7**  Derive the index for selecting materials for a light, strong panel, equation (7.17), following the steps outlined above the equation in the text.

**Exercise E7.8**  Derive the index for selecting materials for a light, strong beam with a square cross-section, equation (7.18).

**Exercise E7.9**  Derive an index for selecting materials for a panel that meets a constraint on bending strength and is as thin as possible.

**Exercise E7.10**  Derive an index for selecting materials for a panel that meets a constraint on bending strength and is as cheap as possible.

**Exercise E7.11**  A low temperature furnace operating at 250 °C uses solid shelves of rectangular cross-section for supporting components during heat treatment. The shelves are simply supported at the edges with the components located towards the centre, where the temperature is uniform. The designer wishes to minimise the mass, for ease of automated removal of the shelves, but failure of the shelves in bending must be avoided. The width $b$ and length $L$ of the shelves are fixed, but the thickness $d$ can vary, up to a specified limit.

(a) For a specified load W applied at mid-span of the shelves, the maximum stress in bending is given by:

$$\sigma_{max} = \frac{3\,WL}{2bd^2}$$

Show that the performance index to be maximised for minimum mass is $M = \sigma_f^{1/2}/\rho$, where $\sigma_f$ is the material strength, and $\rho$ is the density. Explain why there is also a lower limit on the material strength.

(b) Use a Strength–Density property chart to identify a short-list of candidate materials (excluding ceramics and glasses—why is this?). Eliminate materials that cannot operate at the required temperature; to do this, look ahead at the property chart in Figure 13.7, which indicates the usual upper limit on the service temperature of materials. Check whether the remaining materials would have a problem with the thickness limit, for which a minimum strength of 200 MPa is suggested.

Exercise E7.12  A cable of cross-sectional area $A$ and material density $\rho$ is suspended over a fixed span $L$. The maximum allowable sag $\delta$ is specified, and the tensile stress in the cable must not exceed 0.8 of the yield stress, $\sigma_y$. It may be assumed that the sag is small compared to the span, in which case the tension in the cable is given by

$$T = \frac{wL^2}{8\delta}$$

where $w$ is the weight per unit length. Show that this design specification places a lower limit on the specific strength of the material used for the cable. Does this result depend on the cross-sectional area?

Exercise E7.13  You are to design a set of 'rising bollards' (that is, columns that rise up from beneath the road to provide a barrier to prevent cars from passing). The bollards have a specified height $l$ and may be assumed to be built-in at the ground (i.e. cantilevered). A solid cylindrical cross-section has been chosen, but the radius $R$ may be varied. In service the bollards are vertical and must support a given maximum load $W$, applied horizontally at mid-height, without failure. To minimise the power requirements for lifting the bollards, they must be as light as possible.

(a) Show that the maximum stress in a bollard is given by

$$\sigma = \frac{2Wl}{\pi R^3}$$

(b) Derive a performance index, including only material properties, which should be minimised for the given design specification. Hence select the best two materials for the bollards from those given in the table below.

(c) A bollard of height 0.6 m is required to resist a load of 40 kN at mid-height without failure. An upper limit of 50 mm is imposed on the radius of the bollard. Show that this sets a lower limit on the failure stress of the material, and if necessary, revise your material selection from the table.

(d) Briefly outline two factors, other than strength, that might eliminate either of your chosen materials.

| Material | Density (Mg/m$^3$) | Failure stress (MPa) |
|---|---|---|
| CFRP | 1.6 | 300 |
| Wood | 0.6 | 100 |
| Mild steel | 7.6 | 350 |
| Al alloy | 2.7 | 350 |
| Concrete | 2.1 | 80 |

Exercise E7.14    A centrifuge has a rotor that can be idealised as a uniform disk. It has a diameter of 200 mm and is made of a material of strength 450 MPa and density 7900 kg/m$^3$. Use equation (7.8) to see how fast it can be spun (in radians per second) before the stresses it carries exceed its yield strength.

Exercise E7.15    There is a plan to build an energy storage system for the London Underground rail system. The objective is to slow trains down using 'regenerative braking' by running their electric motors 'in reverse' as electricity generators. The electricity generated by the trains will be injected back into the rail system's power grid. A flywheel (located in a hole in the ground in central London) will be used to store the surplus electrical energy and then supply it back to the next accelerating train. The goal is to store at least the kinetic energy of a single train of mass 100 tonnes, decelerated to rest from a speed of 60 km/h. The flywheel is a spinning disc with thickness $t = 1$ m and radius $R = 4$ m.

(a) How much energy must be stored in the flywheel (assuming that all energy conversions are 100% efficient)?
(b) Find the maximum speed $\omega$ for bursting of flywheels made of low-alloy steel, GFRP and CFRP, using the property values in the table below and a safety factor of $S = 2$.
(c) Determine the corresponding stored energy $U$ in each case, and compare this with the energy recovered from stopping the train.
(d) Calculate the mass and the material cost of the flywheels. Which material would you choose?

| | Failure stress $\sigma_f$ (MPa) | Density $\rho$ (kg/m$^3$) | Material cost $C_m$ (GBP/kg) | $\omega$ (rad/sec) | $U$ (MJ) | Mass $m$ (kg) | Cost (GBP) |
|---|---|---|---|---|---|---|---|
| Low-alloy steel | 900 | 7800 | 0.6 | | | | |
| GFRP | 190 | 1850 | 12 | | | | |
| CFRP | 700 | 1600 | 25 | | | | |

Exercise E7.16    A material is sought for a high-speed centrifuge. The objective is to achieve as high an angular velocity $\omega$ of the centrifuge disk as possible. The constraint is that the stress created by the centripetal force must not exceed the yield strength of the material of which it is made. Derive an index to guide the choice of material to allow the maximum $\omega$.

**Exercise E7.17**  The engine of a car is mounted on four shear bolts designed to fail in shear in a front-end collision, detaching the engine from the car if the deceleration exceeds 10g. Assume that all four bolts carry the same load. The mass of the engine is 80 kg. Because of space limitations, the maximum diameter of each shear bolt cannot exceed 5 mm.

(a) Calculate the constraint on the yield strength $\sigma_y$ of the material of the shear bolt, assuming that the shear strength is $\sigma_y/2$ and that the mounting is done in such a way that the shear can take place without friction.

(b) If, to save weight, the shear bolts have to be as light as possible, what metal (of those identified on the chart of Figure 7.8) would you choose to make them?

**Exercise E7.18**  Valve springs for high-performance automobile engines must be light to minimise inertial loads, since part of their mass moves with the valves. At high engine speeds the valves, if heavy, bounce out of contact with the valve itself ('valve bounce'), impeding the flow of gas into and out of the combustion chamber. Derive the index for light springs to guide material choice for the springs, noting also that engine temperatures can reach 200 °C.

**Exercise E7.19**  When the new hospital in Cambridge was first opened, all the ward doors had elastic hinges, as in the figure, so that they could be pushed both ways and close automatically. Consider such a hinge as a thin ligament of material that flexes elastically.

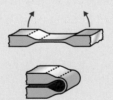

(a) Explain briefly why the best material for the elastic hinge (of a given design) would be one that bends to the smallest radius without yielding.

(b) Consider a section of beam of thickness $t$, bent elastically into an arc of radius R (as in Figure 7.1(a)). Show that the maximum strain $\varepsilon$ at the surface is given by $\varepsilon = t/2R$. Hence write down an expression for the maximum stress $\sigma$ at the surface in terms of the Young's modulus $E$, $t$ and $R$.

(c) Deduce the minimum radius $R$ of the ligament that can be formed without the material yielding, in terms of $E$, $t$ and the yield stress $\sigma_y$. Hence propose a material performance index for minimum radius.

(d) Using the material selection chart in Figure 7.9, identify the best choices of material for the elastic hinge. Make a short-list of materials and from it select the best material, giving reasons for discarding others.

(e) Some six months following the opening of the hospital, many of the elastic hinges had failed. Suggest other factors that should have been taken into account (Chapters 9 and 10 will explore issues that could be relevant here).

Exercise E7.20    You have been asked to design the pressure hull for a deep-sea submersible vehicle capable of descending to the bottom of the Pacific Ocean. The hull is to be a thin-walled sphere with a specified radius $r$ (equal to 1 m) and a uniform wall thickness $t$ (which must be chosen). The design pressure at the bottom of the ocean is 200 MPa. The sphere can fail by one of the following two mechanisms:

(i)  External-pressure buckling at a pressure given by

$$p_b = 0.3E\left(\frac{t}{r}\right)^3$$

(ii)  Compressive failure at a pressure given by

$$p_f = 2\sigma_f\left(\frac{t}{r}\right).$$

$E$ is the Young's modulus and $\sigma_f$ is the failure stress.

(a)  Derive material performance indices that must be maximised in order to achieve the miniumum mass:
   (i)  in the case of external-pressure buckling;
   (ii)  in the case of compressive failure.
(b)  For each material listed in the table below, calculate the mass and the wall thickness of the hull for each failure mechanism at the design pressure.
(c)  Hence determine the limiting failure mechanism for each material.
(d)  Which is the best material for the hull?

| Material | Modulus $E$ (GPa) | Strength $\sigma_f$ (MPa) | Density $\rho$ (kg/m$^3$) | Thickness $t$ and mass $m$ (buckling) | Thickness $t$ and mass $m$ (compression) | Failure mechanism |
|---|---|---|---|---|---|---|
| Alumina | 390 | 5000 | 3900 | | | |
| Glass | 70 | 2000 | 2600 | | | |
| Alloy steel | 210 | 2000 | 7800 | | | |
| Ti alloy | 120 | 1200 | 4700 | | | |
| Al alloy | 70 | 500 | 2700 | | | |

Exercise E7.21    The table ahead shows a selection of candidate materials for lightweight, strength-limited design of a beam. The shape is initially constrained to be square, but the cross-sectional area can vary. Evaluate the relevant index for this problem, $\sigma_f^{2/3}/\rho$, for each material. Which will be the lightest? This chapter showed that if the shape is allowed to vary, the mass can be improved by a factor of $\Phi^{-2/3}$, which is equivalent to increasing the material index by $\Phi^{2/3}$. Evaluate the modified index $(\Phi\sigma_f)^{2/3}/\rho$, using the maximum shape factor for each material (as listed in Table 7.4). Does the ranking of the materials change?

| | Elastic limit $\sigma_f$ (MPa) | Density $\rho$ (kg/m³) | Material index without shape $\sigma_f^{2/3}/\rho$ | Material index with shape $(\Phi\sigma_f)^{2/3}/\rho$ |
|---|---|---|---|---|
| Steel | 600 | 7800 | | |
| Al alloy | 400 | 2700 | | |
| GFRP | 150 | 1900 | | |
| Wood | 70 | 650 | | |

# 7.8 Exploring design with CES

Use Level 2 unless otherwise indicated.

| | |
|---|---|
| Exercise E7.22 | Use CES Level 2 to answer the following: |

(a) Find materials for which the failure strength (elastic limit) $\sigma_f > 300$ MPa, and the density $\rho < 3$ Mg/m³. Eliminate materials with poor ductility (e.g. apply a minimum elongation of 5%) and identify the cheapest.

(b) Compare the specific strength, $\sigma_f/\rho$, of steels, Ti alloys, Al alloys, Mg alloys and CFRP (taking the highest strength in each range for each material).

(c) Find metals and composites that are both stiffer *and* stronger than the highest strength Al alloys.

(d) State the performance index for maximum elastic energy storage per unit volume. Use this index to find materials that are suitable for efficient springs. Can you identify spring applications using these materials?

| | |
|---|---|
| Exercise E7.23 | Use the Search facility in CES to search for materials that are used for *springs*. Report what you find. |
| Exercise E7.24 | Use the Search facility in CES to search for materials that are used for *light springs*. Report what you find. |
| Exercise E7.25 | Make a property chart with $\sigma_y$ on one axis and $E$ on the other. Use it to select materials for springs, using the index $\sigma_y^2/E$ derived in the text. Which three metals emerge as the best metallic choices? |
| Exercise E7.26 | A material is required for a spring that must be as light as possible. To be stiff enough it must also have a Young's modulus $E > 20$ GPa. Make a bar chart with the index $\sigma_y^2/E\rho$ for selecting light springs (you will need to use the Advanced facility in the axis-choice dialog box to do this). Add a Limit stage to apply the constraint $E > 20$ GPa. Hence find the two materials that are the best choices for this application. |

Exercise E7.27    Exercise E7.18 describes the requirements for valve springs for high-performance engines. Apply the index derived there, $\sigma_y^2/\rho E$, by making an appropriate chart and plotting an appropriate selection line on it. Engines are hot: add a Limit stage on maximum service temperature of 250 °C. Hence select metals for this application.

Exercise E7.28    The designer in Exercise E7.18 is concerned about the cost and asks for an alternative short-list based on minimum material cost. Modify your material performance index accordingly, and use a suitable material property chart in CES Level 2, to identify a short-list of materials (applying the same secondary constraints as before).

Exercise E7.29    Abrasives have high hardness, $H$. Make a bar chart of hardness and identify the four materials with the highest values. They are prime choices for abrasive wheels and pastes.

Exercise E7.30    Crash barriers, auto fenders and other protective structures rely on absorbing kinetic energy by plastic deformation. The energy $W_{pl}$ absorbed (per unit volume) in deforming a material to fracture (the area under the stress-strain curve) can be estimated approximately in CES as

$$W_{pl} = \frac{1}{2}\left(\sigma_y + \sigma_{ts}\right)\varepsilon_f$$

Make a chart with density on the x-axis and $W_{pl}$ on the y-axis by using the 'Advanced' facility in the axis-selection dialog box. Use a box selection to find the three materials that absorb the most energy. Rank them by price, using a Graph stage to plot price.

Exercise E7.31    If the crash barrier of Exercise E7.30 is part of a vehicle, fuel is saved if the vehicle is light. We then want the materials with the largest value of $W_{pl}/\rho$, where $\rho$ is the density. These are found by using a selection line of slope 1 on the chart made in the last exercise and selecting the materials above the line. In this application the materials must also have adequate stiffness so that they do not bend elastically too much, requiring a Young's modulus of at least 2.5 GPa. Apply this requirement using a Limit stage, return to the chart and move the selection line until only three materials remain. Comment on the choice.

Exercise E7.32    Now explore what Level 3 of the database can do. Leave all the selection lines and criteria in Exercise E7.31. Go to File > Change Database > CES Edu Level 3 > Open. A dialog box asks if you wish apply this to your current project. Click Yes. Then list the materials that now appear in the results window. Comment on the choice.

Exercise E7.33    This chapter showed that the power required to roll a metal is proportional to its yield strength. Make a bar chart of yield strength, $\sigma_y$ for metals. Open the record for low-carbon steel, find the range of its yield strength and take the average. Normalise all the yield strengths in the database by dividing them by this value, using the 'Advanced' option in the dialog box for selecting the axes, so that low-carbon steel now lies at the value 1 on the bar chart. Use the chart to read off how much less power is required to roll (a) commercially pure zinc and (b) commercially pure lead.

# Chapter 8
# Fracture and fracture toughness

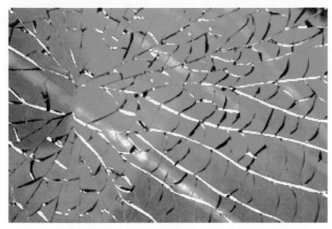

Ductile and brittle fracture. (Image of bolt courtesy of Boltscience; www.boltscience.com)

# **8.1** Introduction and synopsis

It is easy to set a value on the engineering science that enables success, that makes things happen, but much harder to value engineering science that prevents failure, that *stops* things happening. One of the great triumphs of recent engineering science has been the development from the 1960s onward of a rigorous mechanics of material fracture. We have no numbers for the money and lives it has saved by preventing failures; all we know is that, by any measure, it is enormous. This chapter is about the ways in which materials fail when loaded progressively, and design methods to ensure that fracture won't happen unless you want it to.

Sometimes, of course, you do. Aircraft engines are attached to the wing by shear-bolts, designed to fail and shed the engine if it suddenly seizes. At a more familiar level, peel-top cans, seals on food containers and many other safety devices rely on controlled tearing or fracture. And processes like machining and cutting use a combination of plasticity and fracture.

We start by distinguishing *strength* from *toughness*. Toughness—resistance to fracture—requires a new material property, the *fracture toughness* developed in Section 8.3, to describe it. This new property is explored in Section 8.4 using charts like those we have already seen for modulus and strength. The underlying science mechanisms (Section 8.5) give insight into ways in which toughness can be manipulated (Section 8.6). The chapter ends in the usual way with a summary, suggestions for Further reading and Exercises.

# **8.2** Strength and toughness

**Strength and toughness? Why both? What's the difference?** *Strength*, when speaking of a material, is its resistance to plastic flow. Think of a sample loaded in tension. Increase the stress until dislocations sweep right across the section, meaning the sample just yields, and you measure the initial *yield strength*. Strength generally increases with plastic strain because of work hardening, reaching a maximum at the *tensile strength*. The area under the whole stress–strain curve up to fracture is the *work of fracture*. We've been here already—it was the subject of Chapter 5.

*Toughness* is the resistance of a material to the propagation of a crack. Suppose that the sample of material contained a small, sharp crack, as in Figure 8.1(a). The crack reduces the cross-section $A$ and, since stress $\sigma$ is $F/A$, it increases the stress. But suppose the crack is small, hardly reducing the section, and the sample is loaded as before. A tough material will yield, work harden and absorb energy as before—the crack makes no significant difference. But if the material is *not* tough (defined in a moment) then the unexpected happens; the crack suddenly propagates and the sample fractures at a stress that can be far below the yield strength. Design based on yield is common practice. The possibility of fracture at stresses below the yield strength is really bad news. And it has happened, on spectacular scales, causing boilers to burst, bridges to collapse, ships to break in half, pipelines to split and aircraft to crash. We get to that in Chapter 10.

So what is the material property that measures the resistance to the propagation of a crack? And just how concerned should you be if you read in the paper that cracks have been detected

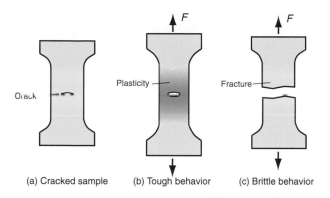

(a) Cracked sample    (b) Tough behavior    (c) Brittle behavior

**Figure 8.1**  Tough and brittle behavior. The crack in the tough material, shown in (b), does not propagate when the sample is loaded; that in the brittle material propagates without general plasticity, and thus at a stress less than the yield strength.

in the track of the railway on which you commute or in the pressure vessels of the nuclear reactor of the power station a few miles away? If the materials are tough enough you can sleep in peace. But what is 'tough enough'?

This difference in material behaviour, once pointed out, is only too familiar. Buy a CD, a pack of transparent folders or even a toothbrush: all come in perfect transparent packaging. Try to get them out by pulling and you have a problem: the packaging is strong. But nick it with a knife or a key or your teeth and suddenly it tears easily. That's why the makers of shampoo sachets do the nick for you. What they forget is that the polymer of the sachet becomes tougher when wet, and that soapy fingers can't transmit much force. But they had the right idea.

***Tests for toughness***  If you were asked to devise a test to characterise toughness, you might dream up something like those of Figure 8.2: notch the material, then yank it or whack it till it

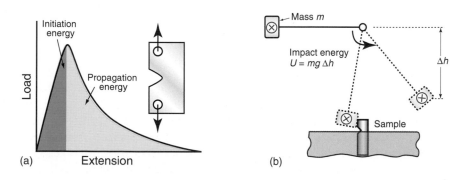

**Figure 8.2**  (a) The tear test. (b) The impact test. Both are used as acceptance tests and for quality control, but neither measures a true material property.

breaks, measuring the energy to do so. Tests like these (there are many variants) in fact are used for ranking and as an acceptance procedure when taking delivery of a new batch of material. The problem is that they do not measure a true material property, meaning one that is independent of the size and shape of the test sample, so the energy measurements do not help with design. To get at the real, underlying, material properties we need the ideas of *stress intensity* and *fracture toughness*.

## 8.3 The mechanics of fracture

*Stress intensity* $K_1$ *and fracture toughness* $K_{1c}$    Cracks and notches concentrate stress. For notches we defined (in Chapter 7) a 'stress concentration factor', which tells us how much greater the peak local stress is compared to the remote stress, for changes in cross-section with a well-defined size and radius, such as a circular hole. Now we consider how cracks affect the stress field. Figure 8.3 shows a remote stress $\sigma$ applied to a cracked material. We can envisage 'lines of force' that are uniformly spaced in the remote region but bunched together around the crack, giving a stress that rises steeply as the crack tip is approached. Referring to equation (7.12), we can see that the stress concentration factor for a notch does not help here—it is relevant only for features with a finite radius of curvature. Cracks are sharp: the radius at the tip is essentially zero, giving (notionally) an infinite stress by equation (7.12). A different approach is needed for cracks. Analysis of the elastic stress field ahead of a sharp crack of length $c$ shows that the local stress at a distance $r$ from its tip, caused by a remote uniform tensile stress $\sigma$, is

$$\sigma_{\text{local}} = \sigma\left(1 + Y\sqrt{\frac{\pi c}{2\pi r}}\right) \tag{8.1}$$

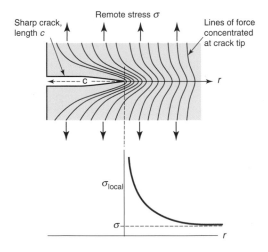

**Figure 8.3**  Lines of force in a cracked body under load; the local stress is proportional to the number of lines per unit length, increasing steeply as the crack tip is approached.

where $Y$ is a constant with a value near unity that depends weakly on the geometry of the cracked body. Far from the crack, where $r \gg c$, the local stress falls to the value $\sigma$; near the tip, where $r \ll c$, it rises steeply (as shown in Figure 8.3) as

$$\sigma_{local} = Y \frac{\sigma \sqrt{\pi c}}{\sqrt{2\pi r}} \tag{8.2}$$

So for any given value of $r$ the local stress scales as $\sigma \sqrt{\pi c}$, which therefore is a measure of the 'intensity' of the local stress field (the inclusion of the $\pi$ is a convention used universally). This quantity is called the *mode 1 stress intensity factor* (the 'mode 1' means tensile loading perpendicular to the crack), and given the symbol $K_1$ (with units of MPa.m$^{1/2}$):

$$K_1 = Y\sigma \sqrt{\pi c} \tag{8.3}$$

The stress intensity factor is thus a measure of the elastic stress field near the tip of a sharp crack, equation (8.2). Note that as $r \to 0$, this equation predicts an infinite stress. In practice the material will yield in a small contained zone at the crack tip (or some other localised form of damage will occur, such as micro-cracking); this is discussed further below. The important point for now is that the loading on the crack tip region that drives potential failure is an elastic stress field that scales with $K_1$. As a result, for reasons explored further below, cracks propagate when the stress intensity factor exceeds a critical value. This critical value is called the *fracture toughness*, $K_{1c}$.

Figure 8.4 shows two sample geometries used to measure $K_{1c}$ (there are others, described in Chapter 10). A sample containing a sharp crack of length $c$ (if a surface crack) or $2c$ (if a contained crack) is loaded, recording the tensile stress $\sigma^*$ at which it suddenly propagates. It is essential that the crack be sharp—not an easy thing to achieve—because if it is not, the part of the stress field with the highest stresses, where the lines of force in Figure 8.3 are closest together, is changed. There are ways of making sharp cracks for doing this test: growing a

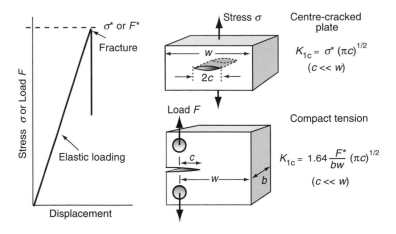

**Figure 8.4** Measuring fracture toughness, $K_{1c}$. Two test configurations are shown here; others are described in Chapter 10.

crack by cyclic loading is one, a process called *fatigue* that is described in the next chapter. Indeed, fracture mechanics based on the stress intensity factor is particularly relevant to design against fracture due to cracks that developed by fatigue. The value of $Y$ for the centre-cracked plate in Figure 8.4 is 1, provided $c << w$; for other geometries, there is a small correction factor. Then the quantity $K_{1c}$ is given by

$$K_{1c} = K_1 = Y\sigma^*\sqrt{\pi c} \approx \sigma^*\sqrt{\pi c} \tag{8.4}$$

Fracture toughness is a *material property*, and this means two things. The first is that its value is independent of the way it is measured: tests using different geometries, if properly conducted, give the same value of $K_{1c}$ for any given material (Figure 8.4). The second is that it can be used for design, in ways described in Chapter 10. For now, we note two important ways that we could use equation (8.4): to find the failure stress, if we know that there is a crack of a given size present; and the reverse, to find the 'critical crack length', being the maximum crack size that we can tolerate without failure, for a given stress.

---

### Example 8.1

The stainless steel and polystyrene rulers in Example 5.2 are loaded as cantilevers of length 250 mm. Both have sharp transverse scratches of depth 0.2 mm, near the base of the cantilever on the tensile side. For this surface crack geometry, assume the factor $Y$ in equation (8.4) is 1.1. The fracture toughness $K_{1c}$ of steel and polystyrene are 80 and 1 MPa.m$^{1/2}$, respectively. Calculate the stress needed to cause fast fracture in each case. Will the cantilevers fail by yielding or by fast fracture?

*Answer.* From equation (8.4), the stress at fast fracture for stainless steel is

$$\sigma^* = \frac{K_{1c}}{1.1\sqrt{\pi c}} = \frac{80 \times 10^6}{1.1\sqrt{\pi \times 2 \times 10^{-4}}} = 2.9 \text{ GPa}$$

This is far in excess of its yield strength—meaning that the ruler will yield before it fractures. Conversely, for polystyrene the fracture stress is

$$\sigma^* = \frac{K_{1c}}{1.1\sqrt{\pi.c}} = \frac{1 \times 10^6}{1.1\sqrt{\pi \times 2 \times 10^{-4}}} = 36 \text{ MPa}$$

which is less than its yield strength—so the ruler will fracture before it yields.

---

### Example 8.2

A compact tension specimen, as in Figure 8.4, is manufactured from an Al alloy with dimensions $w = 50$ mm and $b = 20$ mm, with a sharp pre-crack of length $c = 10$ mm. The maximum load capacity available on the test machine is 50 kN. Is the machine capacity sufficient to fracture the alloy, if its fracture toughness is 25 MPa.m$^{1/2}$?

*Answer.* From Figure (8.4), the stress intensity factor is

$$K_{1c} = 1.64 \, \frac{F^*}{b \, w} \, \sqrt{\pi \, c}$$

So for $K_{1c} = 25$ MPa. m$^{1/2}$, the failure load will be:

$$F^* = \frac{25 \times 10^6 \times 20 \times 10^{-3} \times 50 \times 10^{-3}}{1.64 \times \sqrt{\pi} \times 10 \times 10^{-3}} = 86\text{kN}.$$

So the test machine is not strong enough—we need to use either a larger machine, or reduce the sample thickness (though there are limits on this for a $K_{1c}$ test to be considered valid).

*Energy release rate G and toughness G$_c$*   When a sample fractures, a new surface is created. Surfaces have energy, the *surface energy* $\gamma$, with units of joules[1] per square meter (typically $\gamma = 1$ J/m$^2$). If you fracture a sample across a cross-section area $A$ you make an area $2A$ of new surface, requiring an energy of at least $2A\gamma$ joules to do so. Consider first the question of the *necessary condition for fracture*. It is that sufficient external work be done, or elastic energy released, to at least supply the surface energy, $\gamma$ per unit area, of the two new surfaces that are created. We write this as

$$G \geq 2\gamma \tag{8.5}$$

where $G$ is called the *energy release rate*. In practice, it takes much more energy than $2\gamma$ because of plastic deformation round the crack tip. But the argument still holds: growing a crack costs energy $G_c$ J/m$^2$ for the two surfaces—a sort of 'effective' surface energy, replacing $2\gamma$. It is called, confusingly, the *toughness* (or the *critical strain energy release rate*). This toughness $G_c$ is related to the *fracture toughness* $K_{1c}$ in the following way.

   Think of a slab of material of unit thickness carrying a stress $\sigma$. The elastic energy stored in it (Chapter 4) is

$$U_v = \frac{1}{2} \frac{\sigma^2}{E} \tag{8.6}$$

per unit volume. Now put in a crack of length $c$, as in Figure 8.5. The crack relaxes the stress in a half-cylinder of radius about $c$—the reddish half-cylinder in the figure—releasing the energy it contained:

$$U(c) = \frac{1}{2} \frac{\sigma^2}{E} \cdot \frac{1}{2} \pi c^2 \tag{8.7}$$

---

[1] James Joule (1818–1889), English physicist, did not work on fracture or on surfaces, but his demonstration of the equivalence of heat and mechanical work linked his name to the unit of energy.

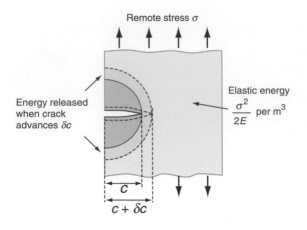

**Figure 8.5** The release of elastic energy when a crack extends.

Suppose now that the crack extends by $\delta c$, releasing the elastic energy in the yellow segment. This energy must pay for the extra surface created, and the cost is $G_c \delta c$. Thus, differentiating the last equation, the condition for fracture becomes

$$\delta U = \frac{\sigma^2 \pi c}{2E} \delta c = G_c \delta c \qquad (8.8)$$

But $\sigma^2 \pi c$ is just $K_{1c}^2$, so from equation (8.4), taking $Y = 1$,

$$\frac{K_{1c}^2}{2E} = G_c \qquad (8.9)$$

This derivation is an approximate one. A more rigorous (but much more complicated) one shows that the form of equation (8.9) is right, but that it is too small by exactly a factor of 2. Thus, correctly, the result we want (taking the square root) is:

$$K_{1c} = \sqrt{EG_c} \qquad (8.10)$$

Toughness $G_c$ is also therefore a material property, and is perhaps more obviously related to the physics of fracture: a tough material requires more energy to be dissipated when a crack propagates. But quantifying the energy release rate $G$ for a loaded crack is difficult, so for design it is preferable to use $K = K_{1c}$ as the failure criterion. The elastic stress field is something we will be analysing anyway, and the crack length is a physical quantity that can be directly observed and measured. The physical equivalence of the two criteria, expressed by equation (8.10), is straightforward: crack growth is driven by the release of elastic stored energy ($G$), and this scales with the crack tip elastic stress field ($K$).

***The crack tip plastic zone*** The intense stress field at the tip of a crack generates a *process zone*: a plastic zone in ductile solids, a zone of micro-cracking in ceramics, a zone of delamination, debonding and fibre pull-out in composites. Within the process zone, work is done against plastic and frictional forces; it is this that accounts for the difference between the measured fracture energy $G_c$ and the true surface energy $2\gamma$. We can estimate the size of a plastic zone that forms at the crack tip as follows. The stress rises as $1/\sqrt{r}$ as the crack tip is approached (equation (8.2)). At the point where it reaches the yield strength $\sigma_y$ the material yields (Figure 8.6) and—except for some work hardening—the stress cannot climb higher than this. The distance from the crack tip where $\sigma_{local} - \sigma_y$ is found by setting equation (8.2) equal to $\sigma_y$ and solving for $r$. But the truncated part of the elastic stress field is redistributed, making the plastic zone larger. The analysis of this is complicated but the outcome is simple: the radius $r_y$ of the plastic zone, allowing for stress redistribution, is twice the value found from equation (8.2), giving

$$r_y = 2\left(\frac{\sigma^2 \pi c}{2\pi \sigma_y^2}\right) = \frac{K_I^2}{\pi \sigma_y^2} \tag{8.11}$$

(taking $Y = 1$). Note that the size of the zone shrinks rapidly as $\sigma_y$ increases: cracks in soft metals have large plastic zones; those in ceramics and glasses have small zones or none at all.

The property $K_{1c}$ has well-defined values for brittle materials and for those in which the plastic zone is small compared to all dimensions of the test sample so that most of the sample is elastic. When this is not so, a more complex characterisation is needed. In very ductile materials the plastic zone size exceeds the width of the sample; then the crack does not propagate at all—the sample simply yields.

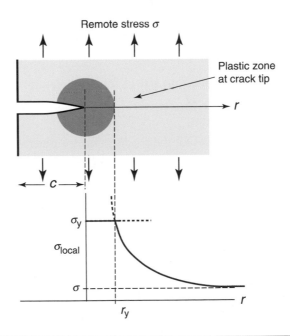

**Figure 8.6** A plastic zone forms at the crack tip where the stress would otherwise exceed the yield strength $\sigma_y$.

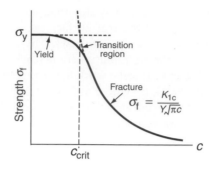

**Figure 8.7** The transition from yield to fracture at the critical crack length $c_{crit}$.

When cracks are small, materials yield before they fracture; when they are large, the opposite is true. But what is 'small'? Figure 8.7 shows how the tensile failure stress varies with crack size. When the crack is small, this stress is equal to the yield stress; when large, it falls off according to equation (8.4), which we write (taking $Y = 1$ again) as

$$\sigma_f = \frac{K_{lc}}{\sqrt{\pi c}} \qquad (8.12)$$

The transition from yield to fracture is smooth, as shown in the figure, but occurs around the intersection of the two curves, when $\sigma_f = \sigma_y$, giving the transition crack length

$$c_{crit} = \frac{K_{lc}^2}{\pi \sigma_y^2} \qquad (8.13)$$

This is the same as the plastic zone size at fracture (equation (8.11)), when $K_1 = K_{1c}$.

---

### Example 8.3

Estimate the plastic zone sizes for the cracks in the stainless steel and polystyrene rulers in Example 8.1.

*Answer.* The plastic zone size is given by

$$r_y = \frac{K_{lc}^2}{Y^2 \pi \sigma_y^2}$$

For stainless steel this is $\dfrac{80^2}{1.1^2 \pi 600^2} = 4.7$mm—much greater than the crack depth of 0.2 mm and greater than the thickness of the ruler. For polystyrene it is $\dfrac{1^2}{1.1^2 \pi 50^2} = 0.1$mm, which is less than the crack depth. So the stainless steel ruler yields and the polystyrene one fractures.

---

Table 8.1 lists the range of values for the main material classes. These crack lengths are a measure of the *damage tolerance* of the material. Tough metals are able to contain large cracks

Table 8.1 Approximate crack lengths for transition between yield and fracture

| Material class | Transition crack length, $c_{crit}$ (mm) |
| --- | --- |
| Metals | 1–1000 |
| Polymers | 0.1–10 |
| Ceramics | 0.01–0.1 |
| Composites | 0.1–10 |

but still yield in a predictable, ductile, manner. Ceramics (which always contain small cracks) fail in a brittle way at stresses far below their yield strengths. Glass can be used structurally, but requires careful treatment to prevent surface flaws developing. Polymers are perceived as tough, due to their resistance to impact when they are not cracked. But the table shows that defects less than 1 mm can be sufficient to cause some polymers to fail in a brittle manner.

## **8.4** Material property charts for toughness

*The fracture toughness–modulus chart* The fracture toughness $K_{1c}$ is plotted against modulus $E$ in Figure 8.8. The range of $K_{1c}$ is large: from less than 0.01 to over 100 MPa.m$^{1/2}$. At the lower end of this range are brittle materials, which, when loaded, remain elastic until

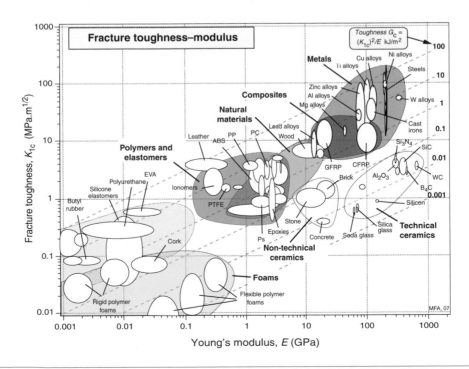

Figure 8.8 A chart of fracture toughness $K_{1c}$ and modulus $E$. The contours show the toughness, $G_c$.

they fracture. For these, linear elastic fracture mechanics works well, and the fracture toughness itself is a well-defined property. At the upper end lie the super-tough materials, all of which show substantial plasticity before they break. For these the values of $K_{1c}$ are approximate but still helpful in providing a ranking of materials. The figure shows one reason for the dominance of metals in engineering; they almost all have values of $K_{1c}$ above 15 MPa.m$^{1/2}$, a value often quoted as a minimum for conventional design.

The log scales of Figure 8.8 allow us to plot contours of *toughness*, $G_c$, the apparent fracture surface energy (since $G_c \approx K_{1c}^2/E$). The diagonal broken lines on the chart show that the values of the toughness start at $10^{-3}$ kJ/m$^2$ (about equal to the surface energy, $\gamma$) and range through almost five decades to over 100 kJ/m$^2$. On this scale, ceramics ($10^{-3} - 10^{-1}$ kJ/m$^2$) are much lower than polymers ($10^{-1} - 10$ kJ/m$^2$); this is part of the reason polymers are more widely used in engineering than ceramics, a point we return to in Chapter 10.

*The fracture toughness–strength chart*  Strength-limited design relies on the component yielding before it fractures. This involves a comparison between strength and toughness—Figure 8.9 shows them on a property chart. Metals are both strong and tough—that is why they have become the workhorse materials of mechanical and structural engineering.

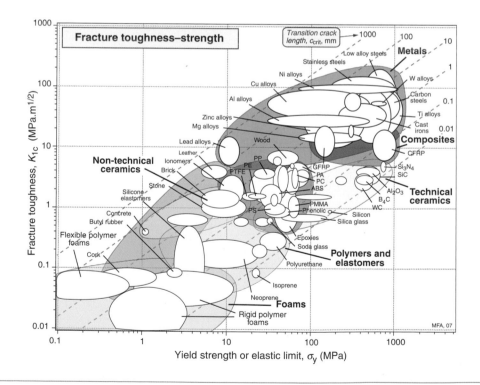

**Figure 8.9**  A chart of fracture toughness $K_{1c}$ and yield strength $\sigma_y$. The contours show the transition crack size, $c_{crit}$.

The stress at which fracture occurs depends on both $K_{1c}$ and the crack length $c$ (equation (8.12)). The transition crack length $c_{crit}$ at which ductile behaviour is replaced by brittle is given by equation (8.13). It is plotted on the chart as broken lines labeled 'Transition crack length'. The values vary enormously, from near-atomic dimensions for brittle ceramics and glasses to almost a meter for the most ductile of metals like copper or lead. Materials toward the bottom right have high strength and low toughness; they *fracture before they yield*. Those toward the top left do the opposite: they *yield before they fracture*.

The diagram has application in selecting materials for the safe design of load-bearing structures (Chapter 10). The strength–fracture toughness chart is also useful for assessing the influence of composition and processing on properties.

## **8.5** Drilling down: the origins of toughness

***Surface energy*** The surface energy of a solid is the energy it costs to make it. It is an energy per unit area, units $J/m^2$. Think of taking a 1 m cube of material and cutting it in half to make two new surfaces, one above and one below, as in Figure 8.10. To do so we have to provide the cohesive energy associated with the bonds that previously connected across the cut. The atoms are bonded on all sides so the surface atoms lose one-sixth of their bonds when the cut is made. This means that we have to provide one-sixth of the cohesive energy $H_c$ (an energy per unit volume) to a slice $4r_0$ thick, were $r_0$ is the atom radius, thus to a volume $4r_0$ $m^3$. So the surface energy should be:

$$2\gamma = \frac{1}{6}H_c \cdot 4r_0 \quad \text{or} \quad \gamma = \frac{1}{3}H_c \cdot r_0$$

with $H_c$, typically, $3 \times 10^{10}$ $J/m^3$ and $r_0$ typically $10^{-10}$ m, so surface energies are around 1 $J/m^2$.

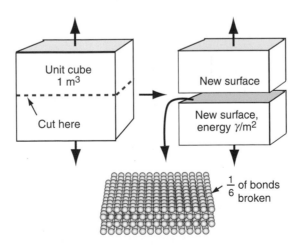

**Figure 8.10** When new surface is created as here, atomic bonds are broken, requiring some fraction of the cohesive energy, $H_c$.

The toughness $G_c$ cannot be less than $2\gamma$. The chart of Figure 8.8 shows contours of $G_c$; for most materials its value is hundreds of times larger than $2\gamma$. Where is the extra energy going? The answer is: into plastic work. We will examine that in more detail in a moment. First, let's examine cleavage fracture.

*Brittle 'cleavage' fracture*   Brittle fracture is characteristic of ceramics and glasses. These have very high yield strengths, giving them no way to relieve the crack tip stresses by plastic flow. This means that, near the tip, the stress reaches the ideal strength (about $E/15$, Chapter 6). That is enough to tear the atomic bonds apart, allowing the crack to grow as in Figure 8.11. And since $K_1 = \sigma\sqrt{\pi c}$, an increase in $c$ means an increase in $K_1$, causing the crack to accelerate until it reaches the speed of sound—that is why brittle materials fail with a bang. Some polymers are brittle, particularly the amorphous ones. The crack tip stresses unzip the weak Van der Waals bonds between the molecules.

*Tough 'ductile' fracture*   To understand how cracks propagate in ductile materials, think first of pulling a sample with no crack, as in Figure 8.12. Ductile metals deform plastically when loaded above their yield strength, work hardening until the tensile strength is reached. Thereafter, they weaken and fail. What causes the weakening? If ultra-pure, the metal may simply thin down until the cross-section goes to zero. Engineering alloys are not ultra-pure; almost all contain inclusions—small, hard particles of oxides, nitrides, sulfides and the like. As the material—here shown as a test specimen—is stretched, it deforms at first in a uniform way, building up stress at the inclusions, which act as stress concentrations. These either

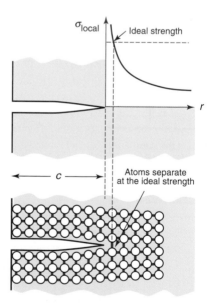

**Figure 8.11**   Cleavage fracture. The local stress rises as $1/\sqrt{r}$ toward the crack tip. If it exceeds that required to break inter-atomic bonds (the 'ideal strength') they separate, giving a cleavage fracture. Very little energy is absorbed.

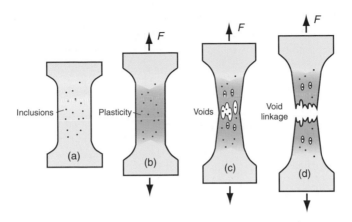

**Figure 8.12** Ductile fracture. Plasticity, shown in brown, concentrates stress on inclusions that fracture or separate from the matrix, nucleating voids that grow and link, ultimately causing fracture.

separate from the matrix or fracture, nucleating tiny holes. The holes grow as strain increases, linking and weakening the part of the specimen in which they are most numerous until they finally coalesce to give a *ductile fracture*. Many polymers, too, are ductile. They don't usually contain inclusions because of the way in which they are made. But when stretched they *craze*—tiny cracks open up in the most stretched regions, whitening them if the polymer is transparent, simply because the Van der Waals bonds that link their long chains to each other are weak and pull apart easily. The details differ but the results are the same: the crazes nucleate, grow and link to give a ductile fracture.

Return now to the cracked sample, shown in Figure 8.13. The stress still rises as $1/\sqrt{r}$ as the crack tip is approached, but at the point that it exceeds the yield strength $\sigma_y$ the material yields and a plastic zone develops. Within the plastic zone the same sequence as that of Figure 8.12 takes place: voids nucleate, grow and link to give a *ductile fracture*. The crack advances and the process repeats itself. The plasticity blunts the crack and the stress-concentrating effect of a blunt crack is less severe than that of a sharp one, so that at the crack tip itself the stress is just sufficient to keep plastically deforming the material there. This plastic deformation absorbs energy, increasing the toughness $G_c$.

***The ductile-to-brittle transition*** A cleavage fracture is much more dangerous than one that is ductile: it occurs without warning or any prior plastic deformation. At low temperatures some metals and all polymers become brittle and the fracture mode switches from one that is ductile to one of cleavage—in fact only those metals with an FCC structure (copper, aluminum, nickel and stainless steel, for example) remain ductile to the lowest temperatures. All others have yield strengths that increase as the temperature falls, with the result that the plastic zone at any crack they contain shrinks until it becomes so small that the fracture mode switches, giving a *ductile-to-brittle transition*. For some steels that transition temperature is as high as 0 °C (though for most it is considerably lower), with the result that steel ships, bridges and oil rigs are more likely to fail in winter than in summer. Polymers, too, have a

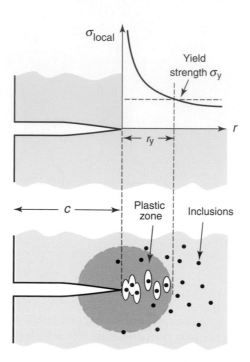

**Figure 8.13** If the material is ductile a plastic zone forms at the crack tip. Within it voids nucleate, grow and link, advancing the crack in a ductile mode, absorbing energy in the process.

ductile-to-brittle transition, a consideration in selecting those that are to be used in freezers and fridges.

***Embrittlement of other kinds*** Change of temperature can lead to brittleness; so, too, can chemical segregation. When metals solidify, the grains start as tiny solid crystals suspended in the melt, and grow outward until they impinge to form grain boundaries. The boundaries, being the last bit to solidify, end up as the repository for the impurities in the alloy. This grain boundary segregation can create a network of low-toughness paths through the material so that, although the bulk of the grains is tough, the material as a whole fails by brittle intergranular fracture (Figure 8.14). The locally different chemistry of grain boundaries causes other problems, such as corrosion (Chapter 17)—one way in which cracks can appear in initially defect-free components.

## **8.6** Compressive and tensile failure of ceramics

Ceramics are brittle materials in tension, with a low fracture toughness (Figures 8.9 and 8.11). But provided we can control the size of the defects, ceramics can have useful tensile strength. The difficulty is that failure is not now characterised by a well-defined material property, like a yield stress, but has statistical variability (reflecting the distribution of the underlying cracks in

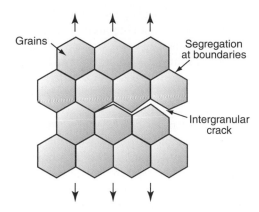

**Figure 8.14** Chemical segregation can cause brittle intergranular cracking.

the material). Furthermore, in compression their strength is typically 10 to 15 times greater than their tensile strength—something that we exploit to the full in civil engineering, building with stone, bricks and concrete. Failure is still controlled by cracking, however, but now as the collective growth of many defects. So the way we design with ceramics doesn't use a fracture toughness criterion, and both compressive and tensile failure processes in ceramics need some further explanation.

*Compressive failure of ceramics*   Figure 8.15 shows ceramic samples loaded in tension and in compression. Ceramics contain a distribution of micro-cracks, usually of a size related to the solid particles that were compacted to make the component (Chapter 19). In tension, the 'worst flaw' (in terms of both size and orientation to the applied stress) propagates to failure (Figure 8.15(a)). But in compression something rather different happens. The cracks can still

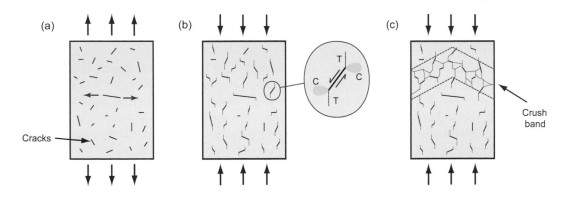

**Figure 8.15**   Ceramic failure mechanisms: (a) tensile failure from the largest flaw; (b) stable growth of wing cracks in compression, driven by the shear acting on inclined cracks generating crack tip regions of compression (C) and tension (T); (c) final compressive failure by collapse of a crush band of fragmented material.

extend, but in a stable manner, as illustrated in Figure 8.15(b). Any crack that is inclined to the compressive stress axis is subjected to a combination of compression and shear. Shear stresses applied to a sharp crack also generate a crack tip stress field (which we now call 'mode 2'), and part of this field is tensile. The cracks extend at right angles to this induced tension, developing 'wing cracks' that grow, rather surprisingly, *parallel* to the applied compression. In this configuration, crack growth is stable—the crack only continues to extend if the stress is increased. Multiple cracking therefore occurs, with cracks growing different amounts depending on their size and orientation. Eventually, the extent of damage to the material is such that an overall sample instability occurs, with the material fragmenting into a band of crushed material that shears away (Figure 8.15(c)). Compression tests on cylindrical blocks of concrete usually end up as a conical remnant of the sample surrounded by rubble. This progressive failure mechanism also explains the serrated stress-strain response that was shown in Figure 6.3. It explains why the peak compressive stress required for failure is so much greater than that needed in tension to propagate the single worst defect.

*Tensile failure of ceramics—Weibull statistics* Look again at the tensile sample in Figure 8.15(a). If we first sliced it vertically into three identical pieces, the middle sample would fail at the same stress as before, since it would contain the critical flaw, and the criterion $K = K_{1c}$ is reached for this crack at the same stress. But the other two pieces will survive to a higher stress, since we need to seek out the worst flaw in each of these pieces, and by definition they are less of a problem than the one in the middle. On average, therefore, the strength of the material appears to be greater as the sample size decreases, but we also have a statistical probability of failure at a given stress. This is the central difficulty of designing with ceramics in tension. We cannot guarantee that failure will not occur, but must live with a probability of failure, and we need a design formula that scales with the volume of the component. Weibull[2] statistics manages both for us.

First consider a sample of given reference volume, $V_o$ (such as one of the slices in the previous discussion), loaded in uniaxial tension. The statistical distribution in flaw size and orientation leads to a probability of survival $P_s$ of the form

$$P_s(V_o) = \exp\left\{-\left(\frac{\sigma}{\sigma_o}\right)^m\right\} \qquad (8.14)$$

where $\sigma_o$ is a reference stress for this volume—it is the stress at which there is a survival probability equal to 1/e. This *Weibull distribution* is illustrated in Figure 8.16. The parameter $m$, the *Weibull modulus*, controls the sensitivity of the probability to stress: low values give a wide variability, high values give much greater confidence in survival below the reference stress. Values for $\sigma_o$ and $m$ are determined empirically. By testing a large number of identical samples and ranking their failure stresses, a probability can be assigned to survival above any

---

[2] Waloddi Weibull (1887–1979) was a Swedish engineer and mathematician whose technical interests were stimulated by his early naval career with the coast guard, where he developed the use of sound waves from explosive charges to investigate the nature of the seabed, something the oil industry can thank him for. The probability distribution that bears his name had previously been associated with describing the statistics of particle-size distributions.

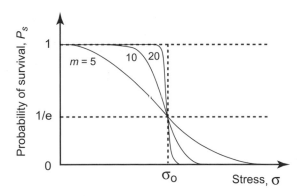

**Figure 8.16** The Weibull distribution for the survival probability of a ceramic loaded in tension. The variability is captured by the Weibull modulus, $m$.

stress level (from the number of samples that survived as a fraction of the total). A suitable log-log plot then delivers the material parameters.

The Weibull equation is readily adapted to cope with sample volumes other than $V_o$ with the following thought experiment. Consider a tensile sample that is $n$ times longer than the reference sample, such that its volume is $V = nV_o$. For this sample to survive a given stress $\sigma$, we need all $n$ sections of volume $V_o$ to survive. In these circumstances, the probabilities are multiplied together, that is, the overall survival probability is $P_s(V_o)$ multiplied by itself n times, while noting that $n = V/V_o$

$$P_s(V) = [P_s(V_o)]^n = [P_s(V_o)]^{V/V_o} \tag{8.15}$$

This is often referred to as a 'weakest link theory': ceramic failure is like the failure of a chain made up of many links, for which the strength of the chain is the same as that of the link that fails first. Substituting from equation (8.14) we get the more general form of Weibull equation for uniform tensile loading of a volume $V$:

$$P_s(V) = \left[\exp\left\{-\left(\frac{\sigma}{\sigma_o}\right)^m\right\}\right]^{V/V_o} = \exp\left\{-\frac{V}{V_o}\left(\frac{\sigma}{\sigma_o}\right)^m\right\} \tag{8.16}$$

Given the survival probability for loading of one volume, it is easy enough to scale to the failure of a different volume.

---

### Example 8.4

A ceramic cylindrical sample of length 50 mm and radius 5 mm has a survival probability $P$ when loaded with a tensile stress of 100 MPa. A second batch of this ceramic has a square section of 15 mm and a length of 80 mm. At what tensile stress do you expect to find the same probability of survival, if the Weibull modulus for the material $m = 10$?

*Answer.* Sample volume $V_1 = 50 \times \pi \times 5^2 = 3927$ mm$^3$, and sample volume $V_2 = 80 \times 15 \times 15 = 18{,}000$ mm$^3$.

From equation (8.16), for $P_s(V_1) = P_s(V_2) = P$:

$$\frac{V_1}{V_o}\left(\frac{\sigma_1}{\sigma_o}\right)^m = \frac{V_2}{V_o}\left(\frac{\sigma_2}{\sigma_o}\right)^m \quad \text{and} \quad \text{hence} \quad \frac{V_1}{V_2} = \left(\frac{\sigma_2}{\sigma_1}\right)^m. \quad \text{So} \quad \sigma_2 = 100\left(\frac{3927}{18000}\right)^{1/10}$$

$$= 85.9 \text{ MPa}$$

As expected, for the same probability of failure, the larger volume can withstand a lower stress.

Note that in Example 8.4 we didn't need to specify the probability—the same scaling of stress with volume applies whatever the probability. And we didn't need to specify $V_o$ and $\sigma_o$, which are to an extent arbitrary—often it is simplest to assign one of the test volumes to be $V_o$, and find the relevant $\sigma_o$ for that volume using the Weibull equation itself, as in the example. The reference stress $\sigma_o$ is not a material property, since its value is associated with a particular test volume.

A final refinement: what happens if the stress state isn't uniform, for example in bending? Figure 8.17 compares the same sample subjected to tension and pure bending (i.e. with a constant bending moment $M$ along the length). The worst flaw controls failure in tension. In bending, this flaw is not even loaded in tension—half of the sample is in compression and can effectively be ignored (as the failure strength is so much higher in compression). So there is already a volume effect at work, but in addition the stress varies linearly across the section (recall Figure 5.1 for elastic bending). The coupling between flaw size and stress is therefore

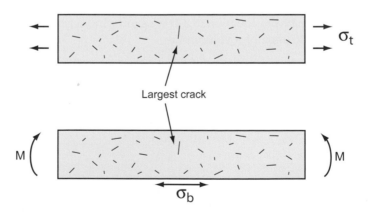

**Figure 8.17** Failure of ceramic samples in tension and pure bending. The largest crack controls failure in tension; in bending, the stress variation across the sample means that the maximum stress $\sigma_b$ is greater than $\sigma_t$.

more complicated, since the critical flaw that governs failure depends both on its size and where it is. Since each element of the sample volume sees a different stress, we need to use an integral form of the Weibull equation:

$$P_s(V) = \exp\left\{ -\frac{1}{V_o\,\sigma_o^m} \int\limits_V \sigma''' \, dV \right\}$$ (8.17)

This does the same job as the weakest link theory for uniform stress, but with each infinitesimal volume element $dV$ associated with its local stress. For bending tests, it is common to characterise the strength by the maximum bending stress $\sigma_b$ produced by the applied bending moment (Figure 8.17). Applying equations (8.16) and (8.17) to the two scenarios in the figure allows this strength in bending to be compared with the strength of the same sample in tension, $\sigma_t$, for the same probability of survival. The ratio of these strengths is given by

$$\frac{\sigma_b}{\sigma_t} = [2\,(m\,+\,1)]^{\,1/m}$$ (8.18)

For typical values of $m$ (3–20), this ratio is greater than 1: as expected, the bending strength (or 'modulus of rupture') is greater than in tension, due to the way the linear stress gradient samples the volume. We'll explore by how much in the Exercises. In the limit of large $m$, the ratio tends to 1, which is equivalent to saying there is a characteristic unique failure stress at which the survival probability drops directly from 1 to 0 (Figure 8.16), and failure always initiates at the surface in bending at the same stress as it would in tension.

## **8.7** Manipulating properties: the strength–toughness trade-off

*Metals*   It is not easy to make materials that are both strong and tough. The energy absorbed by a crack when it advances, giving toughness, derives from the deformation that occurs in the plastic zone at its tip. Equation (8.11) showed that increasing the yield strength causes the zone to shrink and the smaller the zone, the smaller the toughness. Figure 8.18 shows strength and fracture toughness for wrought and cast aluminum alloys, indicating the strengthening mechanisms. All the alloys have a higher strength and lower toughness than pure aluminum. The wrought alloys show a slight drop in toughness with increasing strength; for the cast alloys the drop is much larger because of intergranular fracture.

Separation within the plastic zone, allowing crack advance, results from the growth of voids that nucleate at inclusions (Figure 8.13). Toughness is increased with no loss of strength if the inclusions are removed, delaying the nucleation of the voids. 'Clean' steels, superalloys and aluminum alloys, made by filtering the molten metal before casting, have significantly higher toughness than those made by conventional casting methods.

*Polymers and composites*   The microstructures and properties of polymers are manipulated by cross-linking and by adjusting the molecular weight and degree of crystallinity. Figures 8.8 and 8.9 showed that the strengths of polymers span around a factor of 5, whereas fracture toughness spans a factor of 20. More dramatic changes are possible by *blending*, by adding

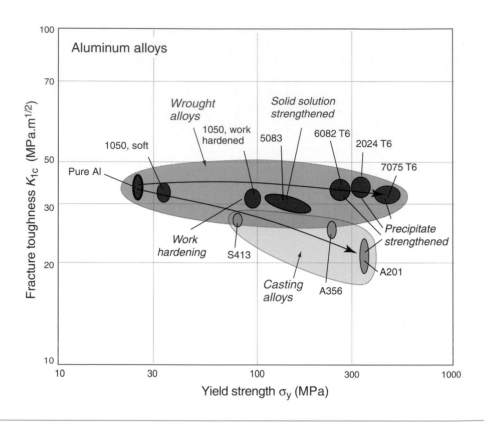

**Figure 8.18** The strength and fracture toughness of wrought and cast aluminum alloys.

*fillers* and by *reinforcement* with chopped or continuous fibres to form composites. Figure 8.19 shows how these influence the modulus $E$ and fracture toughness $K_{1c}$ of polypropylene (PP). Blending or co-polymerization with elastomers such as EPR or EDPM ('impact modifiers') reduces the modulus but increases the fracture toughness $K_{1c}$ and toughness $G_c$. Filling with cheap powdered glass, talc or calcium carbonate more than doubles the modulus, but at the expense of some loss of toughness.

Reinforcement with glass or carbon fibres most effectively increases both modulus and fracture toughness $K_{1c}$, and in the case of glass, the toughness $G_c$ as well. How is it that a relatively brittle polymer ($K_{1c} \approx 3$ MPa.m$^{1/2}$) mixed with even more brittle fibres (glass $K_{1c} \approx 0.8$ MPa.m$^{1/2}$) can give a composite $K_{1c}$ as high as 10 MPa.m$^{1/2}$? The answer is illustrated in Figure 8.20. The fine fibres contain only tiny flaws and consequently have high strengths. When a crack grows in the matrix, the fibres remain intact and bridge the crack. This promotes multiple cracking—each contributing its own energy and thereby raising the overall dissipation. When the fibres do break, the breaks are statistically distributed, leaving ligaments of fibre buried in the matrix. Fibre pull-out, as the cracks open up, dissipates more energy by friction. Composites are processed to control adhesion between the fibres and the matrix to maximise the toughening by these mechanisms.

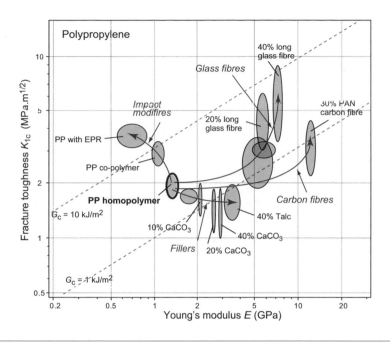

**Figure 8.19** The fracture toughness and modulus of polypropylene, showing the effect of fillers, impact modifiers and fibres.

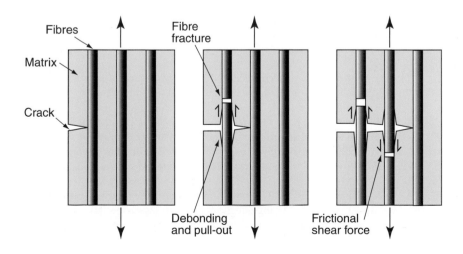

**Figure 8.20** Toughening by fibres. The pull-out force opposes the opening of the crack.

## **8.8** Summary and conclusions

Toughness is the resistance of a material to the propagation of a crack. Tough materials are forgiving: they tolerate the presence of cracks, they absorb impact without shattering and, if overloaded, they yield rather than fracture. The dominance of steel as a structural material derives from its unbeatable combination of low cost, high stiffness and strength and high toughness.

Toughness is properly measured by loading a pre-cracked sample with one of a number of standard geometries, measuring the load at which the crack first propagates. From this is calculated the fracture toughness, $K_{1c}$, as the value of the stress intensity, $K_1$, at which the crack advances. Values of $K_{1c}$ above about 15 MPa.m$^{1/2}$ are desirable for damage-tolerant design—design immune to the presence of small cracks. The charts of Figures 8.8 and 8.9 show that most metals and fibre-reinforced composites meet this criterion, but polymers and ceramics fall below it. We shall see in Chapter 10 that some designs require a high value not of $K_{1c}$ but of the toughness $G_c = K_{1c}^2/E$, also shown in the chart of Figure 8.8. Polymers do well by this criterion and it is in applications that require it that they become a good choice. Ceramics are poor by this criterion too, making design with them much more difficult—at least in tension; in compression, they have considerable strength.

High toughness means that crack advance absorbs energy. It does so in the plastic zone that forms at the crack tip; the larger this becomes, the higher is the fracture toughness. As the material within the plastic zone deforms, voids nucleate at inclusions and link up, advancing the crack; the cleaner the material, the fewer are the inclusions, contributing further to high toughness. The higher the yield strength, the smaller is the zone until, as in ceramics with their very high yield strengths, the zone becomes vanishingly small, the stresses approach the ideal strength and the material fails by brittle cleavage fracture.

This is the background we need for design. That comes in Chapter 10, after we have examined fracture under cyclic loads.

## **8.9** Further reading

Broek, D. (1981). *Elementary Engineering Fracture Mechanics* (3rd ed.). Boston, USA: Martinus Nijhoff. ISBN 90-247-2580-1. (A standard, well-documented introduction to the intricacies of fracture mechanics.)

Ewalds, H. L., & Wanhill, R. J. H. (1984). *Fracture Mechanics*. London, UK: Edward Arnold. ISBN 0-7131-3515-8. (An introduction to fracture mechanics and testing for both static and cyclic loading.)

Hertzberg, R. W. (1989). *Deformation and Fracture of Engineering Materials* (3rd ed.). New York, USA: Wiley. ISBN 0-471-63589-8. (A readable and detailed coverage of deformation, fracture and fatigue.)

Kinloch, A. J., & Young, R. J. (1983). *Fracture Behavior of Polymers*. London, UK: Elsevier Applied Science. ISBN 0-85334-186-9. (An introduction both to fracture mechanics as it is applied to polymeric systems and to the fracture behaviour of different classes of polymers and composites.)

Knott, J. F. (1973). *The Mechanics of Fracture*. London, UK: Butterworths. ISBN 0-408-70529-9. (One of the first texts to present a systematic development of the mechanics of fracture; dated, but still a good introduction.)

Tada, H., Paris, G., & Irwin, G. R. (2000). *The Stress Analysis of Cracks Handbook* (3rd ed.). ISBN 1-86058-304-0. (Here we have another 'Yellow Pages', like Roark for stress analysis of uncracked bodies—this time of stress intensity factors for a great range of geometries and modes of loading.)

## 8.10 Exercises

| | |
|---|---|
| Exercise E8.1 | What is meant by stress intensity factor? How does this differ from the stress concentration factor? |
| Exercise E8.2 | What is meant by toughness? How does it differ from strength, and from fracture toughness? |
| Exercise E8.3 | Why does a plastic zone form at the tip of a crack when a cracked body is loaded in tension? |
| Exercise E8.4 | Why is there a transition from ductile to brittle behaviour at a transition crack length, $c_{crit}$ ? |
| Exercise E8.5 | Compare the critical crack length $c_{crit}$ of a Titanium alloy with $\sigma_y = 550$ MPa and $K_{1c} = 40$ MPam$^{1/2}$ with that of Silicon Carbide, with $\sigma_y = 490$ MPa and $K_{1c} = 4$ MPam$^{1/2}$. Comment on the ductility of the two materials. |
| Exercise E8.6 | A tensile sample of width 10 mm contains an internal crack of length 0.3 mm. When loaded in tension the crack suddenly propagates when the stress reaches 450 MPa. What is the fracture toughness $K_{1c}$ of the material of the sample? If the material has a modulus $E$ of 200 GPa, what is its toughness $G_c$? (Assume the geometric factor Y = 1). |
| Exercise E8.7 | The aircraft designer in Example 7.5 changes the design so that the window has a corner radius of 20 mm instead of 5 mm.<br><br>(a) What is the new stress concentration factor at the corner of the window, and what is the maximum stress?<br>(b) Now suppose there is a small sharp defect near the corner of the window, such that the effective length of the crack in the fuselage is the size of the window itself. If the fracture toughness of the fuselage material is 25 MPa.m$^{1/2}$ and the window is of length 160 mm, what stress would cause fast fracture? Is the design safe? |
| Exercise E8.8 | Two long wooden beams of square cross-section $t \times t = 0.1$ m are butt-jointed using an epoxy adhesive. The adhesive was stirred before application, trapping air bubbles that, under pressure in forming the joint, deformed to flat, penny-shaped cracks of diameter 2 mm (see figure below). For these cracks, assume a geometry factor Y = 1 in the expression for stress intensity factor. The fracture toughness of the epoxy $K_{1c} = 1.3$ MPa.m$^{1/2}$. |

(a) If the beams are loaded in axial tension, what is the load $F_t$ that leads to fast fracture?

(b) The beams are now loaded in three-point bending, with a central load $F_b$ and a total span $L = 2$ m.

Show that the maximum tensile stress at the surface of the beam is given by $\sigma_b = 3F_b\,L/2t^3$. Assuming that there is a bubble close to the beam surface, find the maximum load that the beam can support without fast fracture.

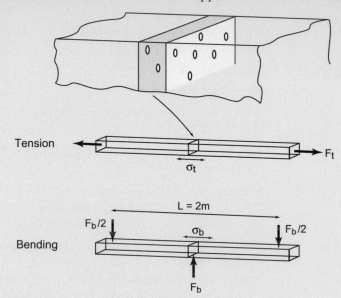

Exercise E8.9   Use the $K_{1c} - E$ chart of Figure 8.8 to establish whether:

1. CFRP has a higher fracture toughness $K_{1c}$ than aluminum alloys.
2. Polypropylene (PP) has a higher toughness $G_c$ than aluminum alloys.
3. Polycarbonate (PC) has a higher fracture toughness $K_{1c}$ than glass.

Exercise E8.10   Find epoxy, soda glass and GFRP (epoxy reinforced with glass fibres) on the chart of Figure 8.8 and read off an approximate mean value for the toughness $G_c$ for each. Explain how it is that the toughness of the GFRP is so much larger than that of either of its components.

Exercise E8.11   Use the chart of Figure 8.8 to compare the fracture toughness, $K_{1c}$, of the two composites GFRP and CFRP. Do the same for their toughness, $G_c$. What do the values suggest about applications they might best fill?

Exercise E8.12   Use the $K_{1c} - \sigma_y$ chart of Figure 8.9 to find:

1. The range of transition crack sizes for stainless steel.
2. The range of transition crack sizes for polycarbonate (PC).
3. The range of transition crack sizes for silicon nitride (Si$_3$N$_4$).

Exercise E8.13  The cylindrical ceramic samples from Example 8.4 (50 mm long and 5 mm radius) were found to have a survival probability of 50% when the applied stress was 120 MPa. The Weibull modulus of the ceramic is $m = 10$. Find the reference stress $\sigma_0$ for this volume of material. What stress can be carried with a probability of failure lower than 1%?

Exercise E8.14  The strengths of a square ceramic beam loaded in pure bending and in tension were compared in this chapter, equation (8.18). Plot how the strength ratio varies with the value of the Weibull modulus, $m$, and explain the shape of the curve.

Exercise E8.15  Square section silicon nitride beams were loaded in pure bending. When the maximum bending stress was below 500 MPa, 50% of the beams had broken. Use equation (8.18) to find the tensile stress applied to the same samples that gives the same probability of failure. The Weibull modulus for silicon nitride is $m = 10$.

## 8.11 Exploring design with CES

Use the Level 2 database unless otherwise stated.

Exercise E8.16  Use the 'Advanced' facility to make a bar chart showing $\sigma_f = K_{1c}/\sqrt{\pi c}$ for an internal crack of length $2c = 1$ mm plotted on the y-axis (i.e. using the CES parameter 'Fracture toughness' for $K_{1c}$).

Which materials have the highest values? Add an axis of density, $\rho$. Use the new chart to find the two materials with highest values of $\sigma_f/\rho$.

Exercise E8.17  Suppose that the resolution limit of the non-destructive testing (NDT) facility available to you is 1 mm, meaning that it can detect cracks of this length or larger. You are asked to explore which materials will tolerate cracks equal to or smaller than this without brittle fracture. Make a bar chart with $\sigma_f = K_{1c}/\sqrt{\pi c}$ for an internal crack of length $2c = 1$ mm plotted on the y-axis, as in the previous exercise. Add yield strength $\sigma_y$ on the x-axis. The material will fracture in tension if $\sigma_f < \sigma_y$, and it will yield, despite being cracked, if $\sigma_f > \sigma_y$. Plot a line of slope 1 along which $\sigma_f = \sigma_y$, either on a printout of the chart or in CES (using the line selection tool). All the materials above the line will yield, all those below will fracture. Do age-hardening aluminum alloys lie above the line? Does CFRP?

Exercise E8.18  Find data for PVCs in the Level 3 database and make a plot like that of Figure 8.19 showing how fillers, blending and fibres influence modulus and toughness.

## 8.12 Exploring the science with CES Elements

Exercise E8.19    Explore the origins of surface energy, $\gamma$ J/m$^2$. The toughness, $G_c$, cannot be less than $2\gamma$ because two new surfaces are created when a material is fractured. What determines $\gamma$ and how big is it? The text explained how bonds are broken and atoms separated when a new surface is created. It was shown in Section 8.4 that

$$\gamma \approx \frac{1}{3} H_c \cdot r_o$$

where $H_c$ is the cohesive energy in J/m$^3$ and $r_o$ is the atomic radius in m. To convert $H_c$ in kJ/mol (as it is in the database) into these units, multiply it by $10^6$/molar volume, and to convert $r_o$ from nm into m multiply by $10^{-9}$. All this can be done using the 'Advanced' facility in the axis-choice dialog box.

     Make a chart with $\gamma$ calculated in this way on the $x$-axis and the measured value, 'Surface energy, solid' on the $y$-axis, and see how good the agreement is.

Exercise E8.20    If you pull on an atomic bond, it breaks completely at a strain of about 0.1. If the atom spacing is $a_o$, then breaking it requires a displacement $\delta = 0.1 a_o$ and doing so creates two new surfaces, each of area $a_o^2$. If the bond stiffness is $S$, then the work done to stretch the bonds to breaking point, and hence to create the new surfaces, per unit area, is

$$2\gamma \approx \frac{1}{a_o^2}\left(\frac{1}{2} S\delta\right) = \frac{E a_o}{20} = \frac{1}{10} E r_o$$

(using equation (4.20) for $S$), where $r_o$ is the atomic radius. Use the CES Elements database to explore whether real surface energies can be explained in this way. (Watch out for the units.)

Exercise E8.21    Observe the general magnitudes of surface energies $\gamma$: they are about 1.5 J/m$^2$. Thus, the minimum value for $G_c$ should be about $2\gamma$ or 3 J/m$^2$. Return to the CES Edu Level 3 database and find the material with the lowest value of $G_c$, which you can calculate as $K_{1c}^2/E$. Is it comparable with $2\gamma$? Limit the selection to metals and alloys, polymers, technical ceramics and glasses only, using a 'Tree' stage (materials such as foam have artificially low values of $G_c$ because they are mostly air).

# Chapter 9

# Shake, rattle and roll: cyclic loading, damage and failure

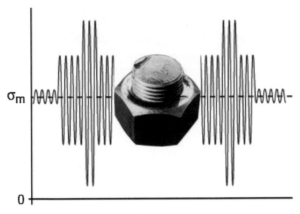

A bolt that has failed by fatigue. The fatigue crack initiated at the root of a thread, which acts as a notch, concentrating stress. (Image courtesy Bolt Science: www.boltscience.com)

## **9.1** Introduction and synopsis

German engineering was not always what it is today. The rapidly expanding railway system of the mid-19th century was plagued, in Germany and elsewhere, by accidents, many catastrophic, caused by the failure of the axles of the coaches. The engineer August Wöhler[1] was drafted in to do something about it. It was his systematic tests that first established the characteristics of what we now call *fatigue*.

Repetition is tiring, the cause of many human mistakes and accidents. Materials, too, grow tired if repeatedly stressed, with failure as a consequence. This chapter is about the energy that is dissipated, and the damage and failure that can result, when materials are loaded in a cyclic, repetitive way. Even when the amplitude of the cycles is very small, some energy dissipation or *damping* occurs. Larger amplitudes cause the slow accumulation of damage, a little on each cycle, until a critical level is reached at which a crack forms. Continued cycling causes the crack to grow until the component suddenly fails. Fatigue failure is insidious—there is little sign that anything is happening until, bang, it does. So when the clip breaks off your pen or your office chair collapses, it is probably fatigue that is responsible (cover picture).

We start with cyclic loading at very small amplitudes and the energy loss or damping that goes with it. We then turn to the accumulation of damage and cracking that is associated with fatigue loading proper.

## **9.2** Vibration and resonance: the damping coefficient

Bells, traditionally, are made of bronze. They can be (and sometimes are) made of glass, and they could (if you could afford it) be made of silicon carbide. Metals, glasses and ceramics all, under the right circumstances, have low material damping or 'internal friction', an important material property when structures vibrate. By contrast, leather, wood (particularly when green or wet) and most foams, elastomers and polymers have high damping, useful when you want to kill vibration.

We are speaking here of elastic response. Until now we have thought of the elastic part of the stress—strain curve as linear and completely reversible so that the elastic energy stored on loading is completely recovered when the load is removed. No material is so perfect—some energy is always lost in a load—unload cycle. If you load just once, you might not notice the loss, but in vibration at, say, acoustic frequencies, the material that is loaded between 20 and 20,000 times per second, as in the stress cycle of Figure 9.1(a). Then the energy loss becomes only too obvious.

The *mechanical loss coefficient* or *damping coefficient*, $\eta$ (a dimensionless quantity), measures the degree to which a material dissipates vibrational energy. If an elastic material is loaded, energy is stored (Chapter 5). If it is unloaded, the energy is returned—it is how springs work. But materials have ways of cheating on this: the amount of energy returned is slightly less. The difference is called the *loss coefficient*, $\eta$. It is the fraction of the stored elastic energy

---

[1] August Wöhler (1819—1914), German engineer, and from 1854 to 1889 Director of the Prussian Imperial Railways. It was Wöhler's systematic studies of metal fatigue that first gave insight into design methods to prevent it.

that is not returned on unloading. If you seek materials for bells, you choose those with low $\eta$, but if you want to damp vibration, then you choose those with high.

## 9.3 Fatigue

*The problem of fatigue*    Low-amplitude vibration causes no permanent damage in materials. Increase the amplitude, however, and the material starts to suffer fatigue. You will, at some time or other, have used fatigue to flex the lid off a sardine can or to break an out-of-date credit card in two. The cyclic stress hardens the material and causes damage in the form of dislocation tangles to accumulate, from which a crack nucleates and grows until it reaches the critical size for fracture. Anything that is repeatedly loaded (like an oil rig loaded by the waves or a pressure vessel that undergoes pressure cycles), or rotates under load (like an axle), or reciprocates under load (like an automobile connecting rod), or vibrates (like the rotor of a helicopter) is a potential candidate for fatigue failure.

Figure 9.2 shows in a schematic way how the stress on the underside of an aircraft wing might vary during a typical flight, introducing some important practicalities of fatigue. First, the amplitude of the stress varies as the aircraft takes off, cruises at altitude, bumps its way through turbulence and finally lands. Second, on the ground the weight of the engines (and the wings themselves) bends them downward, putting the underside in compression. Once in flight, aerodynamic lift bends them upward, putting the underside in tension. The amplitude of the stress cycles and their mean values change during flight. Fatigue failure depends on both. And of course it depends on the total number of cycles. Aircraft wings bend to and fro at a frequency of a few hertz. In a trans-Atlantic flight, tens of thousands of loading cycles take place; in the lifetime of the aircraft, it is millions. For this reason, fatigue testing needs to apply tens of millions of fatigue cycles to provide meaningful design data.

The food can and credit card are examples of *low-cycle fatigue*, meaning that the component survives for only a small number of cycles. It is typical of cycling at stresses above the yield stress, $\sigma_y$, like that shown in Figure 9.1(c). More significant in engineering terms is *high-cycle fatigue*: here the stresses remain generally elastic and may be well below $\sigma_y$, as in cycle (b) of Figure 9.1; cracks nonetheless develop and cause failure, albeit taking many more cycles to do so, as in Wöhler's railway axles.

In both cases we are dealing with components that are initially undamaged, containing no cracks. In these cases, most of the fatigue life is spent generating the crack. Its growth to failure occurs only at the end. We call this *initiation-controlled* fatigue. Some structures contain cracks right from the word go, or are so safety-critical (like aircraft) that they are assumed to have small cracks. There is then no initiation stage—the crack is already there—and the fatigue life is *propagation-controlled*, that is, it depends on the rate at which the crack grows. A different approach to design is then called for—we return to this later.

*High-cycle fatigue and the S−N curve*    Figure 9.3 shows how fatigue characteristics are measured and plotted. A sample is cyclically stressed with amplitude $\Delta\sigma/2$ about a mean value $\sigma_m$, and the number of cycles to cause fracture is recorded. The data are presented as $\Delta\sigma - N_f$ ('S−N')

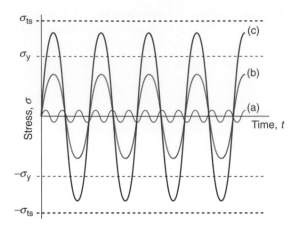

**Figure 9.1**   Cyclic loading. (a) Very low amplitude acoustic vibration. (b) High-cycle fatigue: cycling well below general yield, $\sigma_y$. (c) Low-cycle fatigue: cycling above general yield (but below the tensile strength $\sigma_{ts}$).

curves, where $\Delta\sigma$ is the peak-to-peak range over which the stress varies and $N_f$ is the number of cycles to failure. Most tests use a sinusoidally varying stress with an amplitude $\sigma_a$ of

$$\sigma_a = \frac{\Delta\sigma}{2} = \frac{\sigma_{max} - \sigma_{min}}{2} \tag{9.1}$$

and a mean stress $\sigma_m$ of

$$\sigma_m = \frac{\sigma_{max} + \sigma_{min}}{2} \tag{9.2}$$

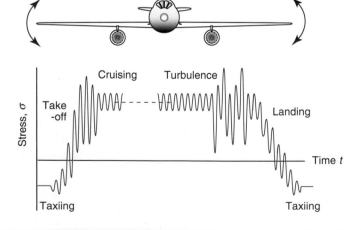

**Figure 9.2**   Schematic of stress cycling on the underside of a wing during flight.

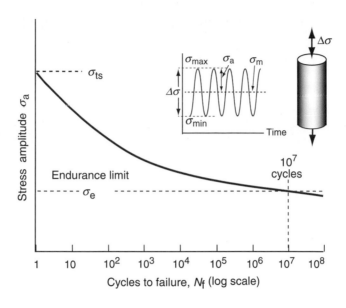

**Figure 9.3** Fatigue strength at $10^7$ cycles, the endurance limit, $\sigma_e$.

all defined in the figure. Fatigue data are usually reported for a specified $R$-value:

$$R = \frac{\sigma_{min}}{\sigma_{max}} \tag{9.3}$$

An $R$-value of $-1$ means that the mean stress is zero; an $R$-value of $0$ means the stress cycles from 0 to $\sigma_{max}$. For many materials there exists a *fatigue* or *endurance limit*, $\sigma_e$ (units: MPa). It is the stress amplitude $\sigma_a$, about zero mean stress, below which fracture does not occur at all, or occurs only after a very large number ($N_f > 10^7$) of cycles. Design against high-cycle fatigue is therefore very similar to strength-limited design, but with the maximum stresses limited by the endurance limit $\sigma_e$ rather than the yield stress $\sigma_y$.

Experiments show that the high-cycle fatigue life is approximately related to the stress range by what is called Basquin's law:

$$\Delta \sigma N_f^b = C_1 \tag{9.4}$$

where $b$ and $C_1$ are constants; the value of $b$ is small, typically 0.07 and 0.13. Dividing $\Delta \sigma$ by the modulus $E$ gives the strain range $\Delta \varepsilon$ (since the sample is elastic):

$$\Delta \varepsilon = \frac{\Delta \sigma}{E} = \frac{C_1/E}{N_f^b} \tag{9.5}$$

or, taking logs,

$$\log \left( \Delta \varepsilon \right) = -b \log \left( N_f \right) + \log \left( C_1/E \right)$$

This is plotted in Figure 9.4, giving the right-hand, high-cycle fatigue part of the curve with a slope of $2b$.

---

### Example 9.1

The fatigue life of a component obeys Basquin's law, equation (9.4), with $b = 0.1$. The component is loaded cyclically with a sinusoidal stress of amplitude 100 MPa (stress *range* of 200 MPa) with zero mean, and has a fatigue life of 200 000 cycles. What will be the fatigue life if the stress amplitude is increased to 120 MPa (stress *range* = 240 MPa)?

*Answer.* From equation (9.4), $C_1 = \Delta\sigma_1 N_1^b = \Delta\sigma_2 N_2^b$, consequently

$$N_2 = N_1\left(\frac{\Delta\sigma_1}{\Delta\sigma_2}\right)^{1/b} = 200\,000\left(\frac{200}{240}\right)^{10} = 32300 \text{ cycles}$$

This is an 84% reduction in fatigue life for a 20% increase in stress amplitude. Fatigue is very sensitive to stress level.

---

*Low-cycle fatigue* In low-cycle fatigue the peak stress exceeds yield, so at least initially (before work hardening raises the strength), the entire sample is plastic. Basquin is no help to us here; we need another empirical law, this time that of Dr Lou Coffin:

$$\Delta\varepsilon^{pl} = \frac{C_2}{N_f^c} \tag{9.6}$$

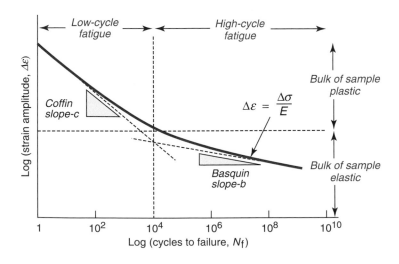

Figure 9.4  The low- and high-cycle regimes of fatigue and their empirical description of fatigue.

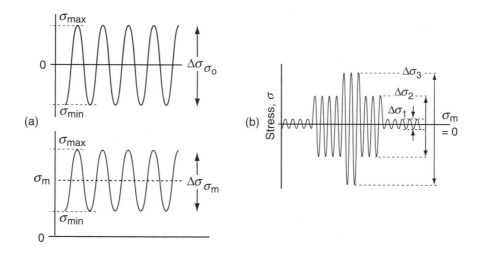

**Figure 9.5** (a) The endurance limit refers to a zero mean stress. Goodman's law scales of the stress range to a mean stress $\sigma_m$. (b) When the cyclic stress amplitude changes, the life is calculated using Miner's cumulative damage rule.

where $\Delta\varepsilon^{pl}$ means the plastic strain range—the total strain minus the (usually small) elastic part. For our purposes we can neglect that distinction and plot it in Figure 9.4 as well, giving the left-hand branch. Coffin's exponent, $c$, is much larger than Basquin's; typically it is 0.5.

*High-cycle fatigue: mean stress and variable amplitude* These laws adequately describe the fatigue failure of uncracked components cycled at constant amplitude about a mean stress of zero. But as we saw, real loading histories are often much more complicated (Figure 9.2). How do we make some allowance for variations in mean stress and stress range? Here we need yet more empirical laws, courtesy this time of Goodman and Miner. Goodman's rule relates the stress range $\Delta\sigma_{\sigma_m}$ for failure under a mean stress $\sigma_m$ to that for failure at zero mean stress $\Delta\sigma_{\sigma_o}$:

$$\Delta\sigma_{\sigma_m} = \Delta\sigma_{\sigma_o}\left(1 - \frac{\sigma_m}{\sigma_{ts}}\right) \tag{9.7}$$

where $\sigma_{ts}$ is the tensile strength, giving a correction to the stress range. Increasing $\sigma_m$ causes a small stress range $\Delta\sigma_{\sigma_m}$ to be as damaging as a larger $\Delta\sigma_{\sigma_o}$ applied with zero mean (Figure 9.5(a)). The corrected stress range may then be plugged into Basquin's law.

### Example 9.2

The component in Example 9.1 is made of a material with a tensile strength $\sigma_{ts} = 200$ MPa. If the mean stress is 50 MPa (instead of zero), and the stress amplitude is 100 MPa, what is the new fatigue life?

*Answer.* Using Goodman's rule, equation (9.7), the equivalent stress amplitude for zero mean stress is:

$$\Delta\sigma_{\sigma_0} = \frac{\Delta\sigma_{\sigma_m}}{\left(1 - \dfrac{\sigma_m}{\sigma_{ts}}\right)} = \frac{100}{\left(1 - \dfrac{50}{200}\right)} = 133 \text{ MPa}$$

So, following the approach in Example 9.1, Basquin's law gives the new fatigue life to be

$$N_2 = N_1\left(\frac{\Delta\sigma_1}{\Delta\sigma_2}\right)^{1/b} = 200\,000\left(\frac{200}{266}\right)^{10} = 11550 \text{ cycles}$$

The variable amplitude problem can be addressed approximately with Miner's *rule of cumulative damage*. Figure 9.5(b) shows an idealised loading history with three stress amplitudes (all about zero mean). Basquin's law gives the number of cycles to failure if each amplitude was maintained throughout the life of the component. So if $N_1$ cycles are spent at stress amplitude $\Delta\sigma_1$ a fraction $N_1/N_{f1}$ of the available life is used up, where $N_{f1}$ is the number of cycles to failure at that stress amplitude. Miner's rule assumes that damage accumulates in this way at each level of stress. Then failure will occur when the sum of the damage fractions reaches 1—that is, when

$$\sum_{i=1}^{n} \frac{N_i}{N_{f,i}} = 1 \tag{9.8}$$

Example 9.3

The component in Examples 9.1 and 9.2 is loaded for $N_1 = 5\,000$ cycles with a mean stress of 50 MPa and a stress amplitude of 100 MPa (as per Example 9.2). It is then cycled about zero mean for $N_2$ cycles with the same stress amplitude (as per Example 9.1) until it breaks. Use Miner's rule, equation (9.8), to determine $N_2$.

*Answer.* From equation (9.8)

$$\frac{5000}{11300} + \frac{N_2}{200000} = 1$$

So $N_2 = 200000\left(1 - \frac{5000}{11300}\right) = 112000$. Approximately half of the fatigue life is used up at each load level.

Goodman's law and Miner's rule are adequate for preliminary design, but they are approximate; in safety-critical applications, tests replicating service conditions are essential. It is for this reason that new models of cars and trucks are driven over rough 'durability tracks' until they fail—it is a test of fatigue performance. The discussion so far has focused on initiation-controlled fatigue failure of uncracked components. Now it is time to look at those containing cracks.

***Fatigue loading of cracked components*** In fabricating large structures like bridges, ships, oilrigs, pressure vessels and steam turbines, cracks and other flaws cannot be avoided. Cracks in castings appear because of differential shrinkage during solidification and entrapment of oxide and other inclusions. Welding, a cheap, widely used joining process, can introduce both cracks and internal stresses caused by the intense local heating. If the cracks are sufficiently large that they can be found it may be possible to repair them, but finding them is the problem. All non-destructive testing (NDT) methods for detecting cracks have a resolution limit; they cannot tell us that there are no cracks, only that there are none longer than the resolution limit, $c_{lim}$. Thus, it is necessary to assume an initial crack exists and design the structure to survive a given number of loadings. So how is the propagation of a fatigue crack characterised?

Fatigue crack growth is studied by cyclically loading specimens containing a sharp crack of length $c$ like that shown in Figure 9.6. We define the cyclic stress intensity range, $\Delta K$, using equation (8.4), as

$$\Delta K = K_{max} - K_{min} = \Delta\sigma\sqrt{\pi c} \tag{9.9}$$

The range $\Delta K$ increases with time under constant cyclic stress because the crack grows in length: the growth per cycle, $dc/dN$, increases with $\Delta K$ in the way shown in Figure 9.7. The rate is zero below a threshold cyclic stress intensity $\Delta K_{th}$, useful if you want to make sure it does not grow at all. Above it, there is a steady-state regime described by the Paris law:

$$\frac{dc}{dN} = A\Delta K^m \tag{9.10}$$

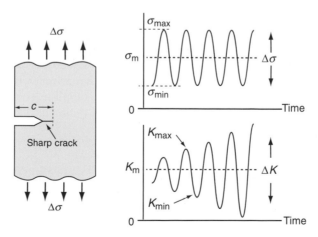

**Figure 9.6** Cyclic loading of a cracked component. A constant stress amplitude $\Delta\sigma$ gives an increasing amplitude of stress intensity, $\Delta K = \Delta\sigma\sqrt{\pi c}$, as the crack grows in length.

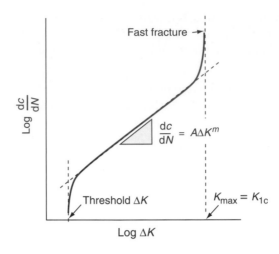

**Figure 9.7** Crack growth during cyclic loading.

where $K$ and $m$ are constants. At high $\Delta K$ the growth rate accelerates as the maximum applied $K$ approaches the fracture toughness $K_{1c}$. When it reaches $K_{1c}$ the sample fails in a single load cycle.

Safe design against fatigue failure in potentially cracked components means calculating the number of loading cycles that can safely be applied without the crack growing to a dangerous length. We return to this in Chapter 10.

---

### Example 9.4

The polystyrene ruler in Example 8.1 is to be loaded as a cantilever with a cyclic force that varies from 0 to 10 N. (a) What depth of transverse crack is needed at the base of the cantilever for fast fracture to occur when the end force is 10 N? (b) If the ruler has an initial transverse scratch of depth $c_i = 0.1$ mm, how many cycles of the force will it take before the ruler breaks? The Paris law constants in equation (9.10) are $m = 4$ and $A = 5 \times 10^{-6}$ when $\Delta\sigma$ is in MPa. (Assume that the scratch is shallow so that the stress at the crack tip is the same as the stress at the extreme fibre of the beam.)

*Answer.*

(a) The stress in the extreme fibre of the bending beam is given by equation (7.1)

$$\sigma = \frac{Mt/2}{I}, \text{ with } M = FL \text{ and } I = \frac{wt^3}{12}$$

Thus

$$\sigma = \frac{6\,FL}{wt^2} = \frac{6 \times 10 \times 0.25}{0.025 \times 0.0046^2} = 28.4 \text{ MPa}$$

From equation (8.4), the critical crack length (depth) at fast fracture is

$$c^* = \frac{K_{1c}^2}{\pi Y^2 \sigma^2} = \frac{(10^6)^2}{\pi 1.1^2 (28.4 \times 10^6)^2} = 0.33 \text{ mm}$$

(b) From equation (9.10), the range of stress intensity factor is $\Delta K = Y\Delta\sigma\sqrt{\pi c}$, where $\Delta\sigma = \sigma_{max} - \sigma_{min}$. In this case, $\Delta\sigma = \sigma_{max}$ because $\sigma_{min} = 0$. The crack starts at length $c_i$ and grows steadily according to the Paris law, equation (9.10), $dc/dN = A\Delta K^m$, until it reaches the critical length $c^*$, when the ruler breaks.

Combining equations (9.9) and (9.10) gives:

$$\frac{dc}{dN} = A\left(Y\sigma_{max}\sqrt{\pi c}\right)^m$$

from which

$$\int_0^{N_f} dN = \frac{1}{AY^m \sigma_{max}^m \pi^{m/2}} \int_{c_f}^{c^*} \frac{dc}{c^{m/2}}$$

where $N_f$ is the number of cycles to failure. Integrating this gives:

$$N_f = \frac{1}{AY^m \sigma_{max}^m \pi^{m/2}(1 - m/2)}\left[(c^*)^{1-m/2} - (c_i)^{1-m/2}\right]$$

$$= \frac{1}{5 \times 10^{-6} \times 1.1^4 \times 28.4^4 \pi^2(-1)}\left[\frac{1}{0.00033} - \frac{1}{0.0001}\right] = 148$$

## 9.4 Charts for endurance limit

The most important single property characterising fatigue strength is the endurance limit, $\sigma_e$, at $10^7$ cycles and zero mean stress (an $R$-value of $-1$). Given this and the ability to scale it to correct for mean stress, and to sum contributions when stress amplitude changes (equations (9.7) and (9.8)), enables design to cope with high-cycle fatigue.

Not surprisingly endurance limit and strength are related. The strongest correlation is with the tensile strength $\sigma_{ts}$, shown in the chart of Figure 9.8. The data for metals and polymers cluster around the line

$$\sigma_e \approx 0.33\sigma_{ts}$$

shown on the chart. For ceramics and glasses

$$\sigma_e \approx 0.9\sigma_{ts}$$

In the next section we examine why.

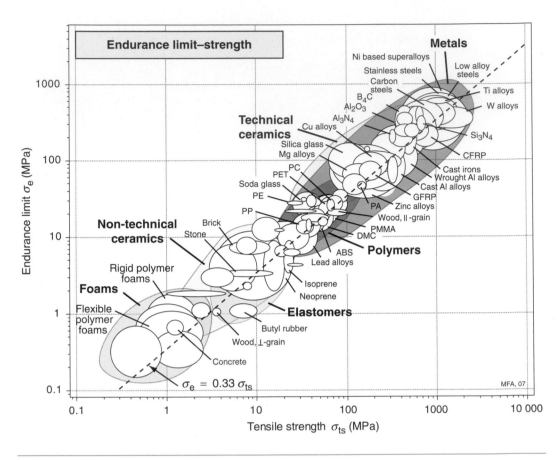

**Figure 9.8** The endurance limit plotted against the tensile strength. Almost all materials fail in fatigue at stresses well below the tensile strength.

## 9.5 Drilling down: the origins of damping and fatigue

*Material damping: the mechanical loss coefficient* There are many mechanisms of material damping. Some are associated with a process that has a specific time constant; then the energy loss is centred about a characteristic frequency. Others are frequency independent; they absorb energy at all frequencies. In metals a large part of the loss is caused by small-scale dislocation movement: it is high in soft metals like lead and pure aluminum. Heavily alloyed metals like bronze and high-carbon steels have low loss because the solute pins the dislocations; these are the materials for bells. Exceptionally high loss is found in some cast irons, in manganese—copper alloys and in magnesium, making them useful as materials to dampen vibration in machine tools and test rigs. Engineering ceramics have low damping because the dislocations in them are immobilised by the high lattice resistance (which is why they are hard). Porous ceramics, on the other hand, are filled with cracks, the surfaces of which rub, dissipating energy, when the material is loaded. In polymers, chain segments slide against each other when loaded; the relative motion dissipates energy. The ease with which they slide depends on the ratio of the temperature $T$ to the glass temperature, $T_g$, of the polymer. When $T/T_g < 1$, the secondary bonds are 'frozen', the modulus is high and the damping is relatively low. When $T/T_g > 1$, the secondary bonds have melted, allowing easy chain slippage; the modulus is low and the damping is high.

*Fatigue damage and cracking*   A perfectly smooth sample with no changes of section, and containing no inclusions, holes or cracks, would be immune to fatigue provided neither $\sigma_{max}$ nor $\sigma_{min}$ exceeds its yield strength. But that is a vision of perfection that is unachievable. Blemishes, small as they are, can be deadly. Rivet holes, sharp changes in section, threads, notches and even surface roughness concentrate stress in the way described in Chapter 7, Figure 7.7. Even though the general stress levels are below yield, the locally magnified stresses can lead to reversing plastic deformation. Dislocation motion is limited to a small volume near the stress concentration, but that is enough to cause damage that finally develops into a tiny crack.

In high-cycle fatigue, once a crack is present it propagates in the way shown in Figure 9.9(a). During the tensile part of a cycle a tiny plastic zone forms at the crack tip, stretching it open and thereby creating a new surface. On the compressive part of the cycle the crack closes again and the newly formed surface folds forward, advancing the crack. Repeated cycles make it inch

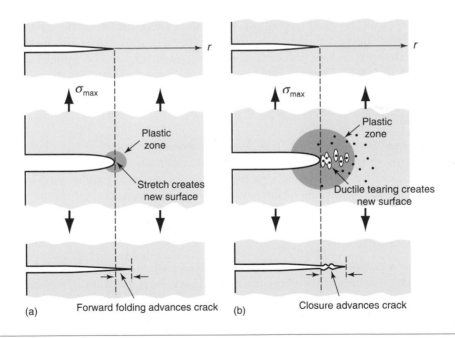

(a)   Forward folding advances crack   (b)   Closure advances crack

**Figure 9.9**   (a) In high-cycle fatigue a tiny zone of plasticity forms at the crack tip on each tension cycle; on compression the newly formed surface folds forward. (b) In low-cycle fatigue the plastic zone is large enough for voids to nucleate and grow within it. Their coalescence further advances the crack.

forward, leaving tiny ripples on the crack face marking its position on each cycle. These 'striations' are characteristic of a fatigue failure and are useful, in a forensic sense, for revealing where the crack started and how fast it propagated.

In low-cycle fatigue the stresses are higher and the plastic zone larger, as in Figure 9.9(b). It may be so large that the entire sample is plastic, as it is when you flex the lid of a tin to make it break off. The largest strains are at the crack tip, where plasticity now causes voids to nucleate, grow and link, just as in ductile fracture (Chapter 8).

## **9.6** Manipulating resistance to fatigue

Fatigue life is enhanced by choosing materials that are strong, making sure they contain as few defects as possible, and giving them a surface layer in which the internal stresses are compressive.

***Choosing materials that are strong*** The chart of Figure 9.8 established the close connection between endurance limit and tensile strength. Most design is based not on tensile strength but on yield strength, and here the correlation is less strong. We define the fatigue ratio, $F_r$, as

$$F_r = \frac{\text{Endurance limit } \sigma_e}{\text{Yield strength } \sigma_y}$$

Figure 9.10 shows data for $F_r$ for an age-hardening aluminum alloy and a low alloy steel. Both can be heat-treated to increase their yield strength, and higher strength invites higher design stresses. The figure, however, shows that the fatigue ratio $F_r$ falls as strength $\sigma_y$ increases, meaning that the gain in endurance limit is considerably less than that in yield strength. In propagation-controlled fatigue, too, the combination of higher stress and lower fracture toughness implies that smaller crack sizes will propagate, causing failure (and the resolution of inspection techniques may not be up to the job).

***Making sure they contain as few defects as possible*** Mention has already been made of clean alloys—alloys with carefully controlled compositions that are filtered when liquid to remove unwanted particles that, in the solid, can nucleate fatigue cracks. Non-destructive testing using X-ray imaging or ultrasonic sensing detects dangerous defects, allowing the part to be rejected. And certain secondary processes like hot isostatic pressing (HIPing), which heats the component under a large hydrostatic pressure of inert argon gas, can seal cracks and collapse porosity.

***Providing a compressive surface stress*** Cracks only propagate during the tensile part of a stress cycle; a compressive stress forces the crack faces together, clamping it shut. Fatigue cracks frequently start from the surface, so if a thin surface layer can be given an internal stress that is compressive, any crack starting there will remain closed even when the average stress across the entire section is tensile. This is achieved by treatments that plastically compress the surface by shot peening (Figure 9.11): leaf-springs for cars and trucks are 'stress peened', meaning that they are bent to a large deflection and then treated in the way shown in the figure; this increases their fatigue life by a factor of 5. A similar outcome is achieved by sandblasting

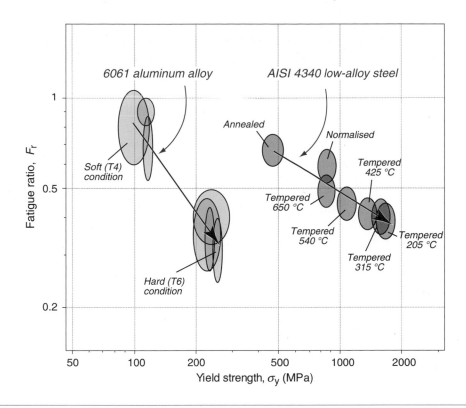

**Figure 9.10** The drop in fatigue ratio with increase in yield strength for an aluminum alloy and a steel, both of which can be treated to give a range of strengths.

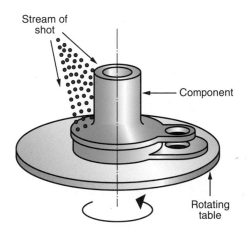

**Figure 9.11** Shot peening, one of several ways of creating compressive surface stresses.

or burnishing (local deformation with a smooth, polished tool), or diffusing atoms into the surface, expanding a thin layer, which, because it is bonded to the more massive interior, becomes compressed.

## 9.7 Summary and conclusions

Static structures like the Eiffel Tower or the Golden Gate Bridge stand for centuries; those that move have a much shorter life span. One reason for this is that materials are better at supporting static loads than loads that fluctuate. The long-term cyclic load a material can tolerate, its endurance limit $\sigma_e$, is barely one-third of its tensile strength, $\sigma_{ts}$. This sensitivity to cyclic loading—fatigue—was unknown until the mid-19th century, when a series of major industrial disasters made it the subject of intense study.

Empirical rules describing fatigue failure and studies of the underlying mechanisms have made fatigue failures less common, but they still happen, notably in rail track, rolling stock, engines and airframes. It is now known that cycling causes damage that slowly accumulates until a crack nucleates, grows slowly and suddenly runs unstably. This is pernicious behavior: the cracks are almost invisible until final failure occurs without warning.

The rules and data for their empirical constants, described in this chapter, give a basis for design to avoid fatigue failure. The scientific studies provide insight that guides the development of materials with greater resistance to fatigue failure. Surface treatments to inhibit crack formation are now standard practice. We return to these in the next chapter, in which the focus is on design issues.

## 9.8 Further reading

Hertzberg, R. W. (1989). *Deformation and Fracture of Engineering Materials* (3rd ed.). New York, USA: Wiley. ISBN 0-471-63589-8. (A readable and detailed coverage of deformation, fracture and fatigue.)

Suresh, S. (1998). *Fatigue of Materials* (2nd ed.). Cambridge, UK: Cambridge University Press. ISBN 0-521-57847-7. (The place to start for an authoritative introduction to the materials science of fatigue in both ductile and brittle materials.)

Tada, H., Paris, G., & Irwin, G. R. (2000). *The Stress Analysis of Cracks Handbook* (3rd ed.). ISBN 1-86058-304-0. (Here we have another 'Yellow Pages', like Roark for stress analysis of uncracked bodies—this time of stress intensity factors for a great range of geometries and modes of loading.)

## 9.9 Exercises

Exercise E9.1     What is meant by the mechanical loss coefficient, $\eta$, of a material? Give examples of designs in which it would play a role as a design-limiting property.

Exercise E9.2　Distinguish between 'low-cycle' and 'high-cycle' fatigue. Find examples of engineering components that may fail by high-cycle fatigue. What is meant by the endurance limit, $\sigma_e$, of a material?

Exercise E9.3　What is the fatigue ratio? If the tensile strength $\sigma_{ts}$ of an alloy is 900 MPa, what, roughly, would you expect its endurance limit $\sigma_e$ to be?

Exercise E9.4　The figure shows an $S - N$ curve for AISI 4340 steel, hardened to a tensile strength of 1800 MPa.

(a) What is the endurance limit?
(b) If cycled for 100 cycles at an amplitude of 1200 MPa and a zero mean stress, will it fail?
(c) If cycled for 100,000 cycles at an amplitude of 900 MPa and zero mean stress, will it fail?
(d) If cycled for 100,000 cycles at an amplitude of 800 MPa and a mean stress of 300 MPa, will it fail?

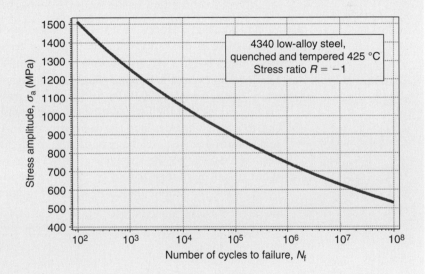

Exercise E9.5　The high-cycle fatigue life, $N_f$, of an aluminium alloy is described by Basquin's law:

$$\frac{\Delta\sigma}{2} = \sigma_a = 480\left(N_f\right)^{-0.12}$$

(stress in MPa). How many cycles will the material tolerate at a stress amplitude $\sigma_a$ of $\pm70$ MPa and zero mean stress? How will this change if the mean stress is 10 MPa? What if the mean stress is $-10$ MPa? The tensile strength of the alloy is 200 MPa.

Exercise E9.6    The low-cycle fatigue of an aluminium alloy is described by Coffin's law:

$$\Delta \varepsilon^{pl} = \frac{0.2}{N_f^{0.5}}$$

How many cycles will the material tolerate at a plastic strain amplitude $\Delta \varepsilon^{pl}$ of 2%?

Exercise E9.7    A material with a tensile strength $\sigma_{ts} = 350$ MPa is loaded cyclically about a mean stress of 70 MPa. If the stress range that will cause fatigue fracture in $10^5$ cycles under zero mean stress is ±60 MPa, what stress range about the mean of 70 MPa will give the same life?

Exercise E9.8    A component made of the AISI 4340 steel with a tensile strength of 1800 MPa and the S–N curve shown in Exercise E9.4 is loaded cyclically between 0 and 1200 MPa. What is the $R$ value and the mean stress, $\sigma_m$? Use Goodman's rule to find the equivalent stress amplitude for an $R$-value of −1, and read off the fatigue life from the S–N curve.

Exercise E9.9    Some uncracked bicycle forks are subject to fatigue loading. Approximate S–N data for the material used are given in the figure, for zero mean stress. This curve shows a 'fatigue limit': a stress amplitude below which the life is infinite.

(a) The loading cycle due to road roughness is assumed to have a constant stress range $\Delta \sigma$ of 1200 MPa and a mean stress of zero. How many loading cycles will the forks withstand before failing?

(b) Due to a constant rider load the mean stress is 100 MPa. Use Goodman's rule to estimate the percentage reduction in lifetime associated with this mean stress. The tensile strength $\sigma_{ts}$ of the steel is 1100 MPa.

(c) What practical changes could be made to the forks to bring the stress range below the fatigue limit and so avoid fatigue failure?

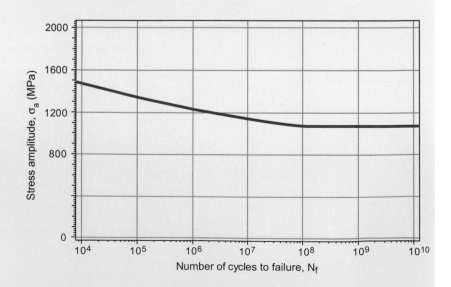

Exercise E9.10 An aluminium alloy for an airframe component was tested in the laboratory under an applied stress that varied sinusoidally with time about a mean stress of zero. The alloy failed under a stress range $\Delta\sigma$ of 280 MPa after $10^5$ cycles. Under a stress range of 200 MPa, the alloy failed after $10^7$ cycles. Assume that the fatigue behaviour of the alloy can be represented by:

$$\Delta\sigma \; N_f^b = C$$

where $b$ and C are material constants.

(a) Find the number of cycles to failure for a component subject to a stress range of 150 MPa.

(b) An aircraft using the airframe components has encountered an estimated $4 \times 10^8$ cycles at a stress range of 150 MPa. It is desired to extend the life of the airframe by another $4 \times 10^8$ cycles by reducing the performance of the aircraft. Use Miner's rule to find the decrease in the stress range needed to achieve this additional life.

Exercise E9.11 A medium-carbon steel was tested to obtain high-cycle fatigue data for the number of cycles to failure $N_f$ in terms of the applied stress range $\Delta\sigma$ (peak-to-peak). As the test equipment was only available for a limited time, the tests had to be accelerated. This was achieved by testing the specimen using the loading history shown schematically in the figure. A first set of $N_1$ cycles was applied with a stress range $\Delta\sigma_1$, followed by a second set of $N_2$ cycles with a stress range $\Delta\sigma_2$. The table shows the test programme. In every test the mean stress $\sigma_m$ was held constant at 150 MPa, and each test was continued until specimen failure. A separate tensile test gave a tensile strength for the steel of 600 MPa.

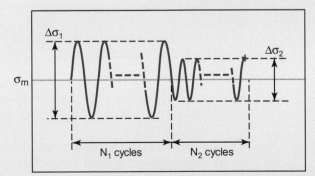

| Test | $\Delta\sigma_1$ (MPa) | $N_1$ | $\Delta\sigma_2$ (MPa) | $N_2$ |
|------|------------------------|-------|------------------------|-------|
| 1 | 630 | $10^4$ | — | — |
| 2 | 630 | $5 \times 10^3$ | 460 | $5 \times 10^5$ |
| 3 | 630 | $5 \times 10^3$ | 510 | $1.2 \times 10^5$ |
| 4 | 630 | $2.5 \times 10^3$ | 560 | $4.4 \times 10^4$ |

(a) Using Goodman's rule, show that the stress *range* that would give failure in $10^4$ cycles with *zero* mean stress is 840 MPa.

(b) Use Miner's rule to find the expected number of cycles to failure for each of the stress ranges $\Delta\sigma_2$ in the table, with the mean stress of 150 MPa.

(c) Convert the $\Delta\sigma_2 - N_f$ data obtained to the equivalent data for zero mean stress.

(d) Plot a suitable graph to show that all the fatigue life data for zero mean stress are consistent with Basquin's law for high-cycle fatigue $\Delta\sigma \, N_f^b = C_1$, and find the constant $b$.

**Exercise E9.12** The leaf spring of a heavy vehicle suspension is 1.2 m long, 75 mm wide and 30 mm thick. A central thickened section is 150 mm long and 34 mm thick with a fillet of radius 5 mm, as shown in the figure. It is supported by a pin joint at one end and a 'shackle' at the other. This can be assumed to be equivalent to 'simple' supports. The spring is deflected sideways in a testing machine using a hydraulic actuator attached to the central section. The actuator causes the spring to oscillate cyclically with an amplitude of ± 50 mm. The material is AISI 4340 steel, for which the Young's Modulus is 210 GPa and the S–N curve is provided in Example E9.4. You will need to use elastic solutions for the deflection and stresses in beams under three-point bending, from Chapter 5, and for the stress concentration factor, from Chapter 7.

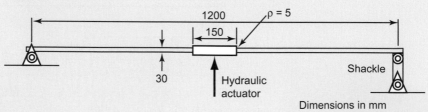

(a) Find the load amplitude applied by the actuator to give the specified deflection (ignore the stiffening effect of the central thickened section).

(b) Find the bending moment and maximum surface stress on the thinner part of the spring, at the location of the fillet (initially ignoring the change in thickness). What stress concentration factor should then be applied to the stress amplitude for this location?

(c) If the spring is initially cycled about its unstressed (equilibrium) position, estimate the number of cycles of loading that it will last before it fails by fatigue at the fillet.

(d) If the test is performed with a mean load of 30 kN, find the mean stress (including the stress concentration), and hence use Goodman's rule to estimate the number of cycles of loading that the spring will last before failure.

Exercise E9.13  A material has a threshold cyclic stress intensity $\Delta K_{th}$ of 2.5 MPa.m$^{1/2}$. If it contains an internal crack of length 1 mm, will it be safe (meaning no failure) if subjected to continuous cyclic range of tensile stress $\Delta\sigma$ of 50 MPa?

Exercise E9.14  A plate of width $a = 50$ mm contains a sharp, transverse edge-crack of length $c = 5$ mm. The plate is subjected to a cyclic axial stress $\sigma$ that varies from 0 to 50 MPa. The Paris law constants (equation (9.10)) are $m = 4$ and $A = 5 \times 10^{-9}$, where $\sigma$ is in MPa. The fracture toughness of the material is $K_{1c} = 10$ MPa.m$^{1/2}$.

(a) At what 'critical' crack length will fast fracture occur? (Assume the geometry factor in the expression for $K$ is $Y = 1.1$)

(b) How many load cycles will it take for the crack to grow to the critical crack length?

Exercise E9.15  Components that are susceptible to fatigue are sometimes surface treated by 'shot peening'. Explain how the process works, and why it is beneficial to fatigue life.

# 9.10 Exploring design with CES

Exercise E9.16  Make a bar chart of Mechanical loss coefficient, $\eta$. Low loss materials are used for vibrating systems where damping is to be minimised—bells, high-frequency relays and resonant systems. High loss materials are used when damping is desired—sound-deadening cladding for buildings, cars and machinery, for instance. Use the chart to find:

(a) The metal with the lowest loss coefficient.

(b) The metal with the highest loss coefficient.

Do their applications include one or more of those listed above?

Exercise E9.17  Use the 'Search' facility to search materials that are used for:

(a) Bells.

(b) Cladding.

Exercise E9.18   Use a 'Limit' stage, applied to the *Surface treatment* data table, to find surface treatment processes that enhance fatigue resistance. To do this:
(a) Change the selection table to Process universe Level 2 Surface treatment, open a 'Limit' stage, locate Function of treatment and click on Fatigue resistance > Apply. Copy and report the results.
(b) Repeat, using the Level 3 Surface treatment data table.

Exercise E9.19   Explore the relationship between fatigue ratio and strength for a heat-treatable low alloy steel AISI 4340. The endurance limit $\sigma_e$ is stored in the database under the heading 'Fatigue strength at $10^7$ cycles'.
(a) Plot the fatigue ratio $\sigma_e/\sigma_y$ against the yield strength $\sigma_y$.
(b) Plot the fatigue ratio $\sigma_e/\sigma_{ts}$ against the tensile strength $\sigma_{ts}$.
Use Level 3 of the database, apply a 'Tree' stage to isolate the folder for the low alloy steel, AISI 4340, then make the two charts, hiding all the other materials. How do you explain the trends?

# Chapter 10
# Keeping it all together: fracture-limited design

Fractured pipe. (Image courtesy Prof. Robert Akid, School of Engineering, Sheffield Hallam University, Sheffield, UK)

## **10.1** Introduction and synopsis

It is very hard to build a structure that is completely without cracks. As explained in Chapter 8, cracks caused by shrinkage in casting and welding, by the cracking of inclusions during rolling or just caused by careless machining are commonplace. And even if there are no cracks to start with, cyclic loading (Chapter 9) and corrosion (Chapter 17) can introduce them later.

This creates the need for design methods to deal with cracked structures. The idea of the *tensile stress intensity* $K_1$ caused by a crack was introduced in Chapter 8. It depends on crack length, component geometry and the way the component is loaded. We start with standard solutions for the stress intensity $K_1$ associated with generic configurations—there are others, but this is enough to get started. Cracks will not propagate if this $K_1$ is kept below the fracture toughness $K_{1c}$ of the material of the structure.

This might suggest that the best material to resist fracture is the one with the highest $K_{1c}$, and in load-limited design it is. But sometimes the requirement is not to carry a given load without failure, but to store a given energy (springs) or allow a given deflection (elastic couplings) without failure. Then the best choice of material involves combinations of $K_{1c}$ and Young's modulus $E$, as we will see in Section 10.3.

Cyclic loading, too, causes fracture. Chapter 9 explained how accumulating fatigue damage causes cracks to form and grow, a little on each cycle, until the length is such that the stress intensity exceeds $K_{1c}$, when fast fracture follows. Stress intensities in cyclic loading are given by the same standard solutions as for static loading. Cracks will not propagate if the cyclic stress intensity range, $\Delta K$, is below the threshold range $\Delta K_{th}$.

The chapter concludes with case studies illustrating how these ideas enable *no-fail* or *fail-safe* design.

## **10.2** Standard solutions to fracture problems

As we saw in Chapter 8, sharp cracks concentrate stress in an elastic body. To summarise, the local stress falls off as $1/r^{1/2}$ with radial distance $r$ from the crack tip. A tensile stress $\sigma$, applied normal to the plane of a crack of length $2c$ contained in an infinite plate, gives rise to a local stress field $\sigma_1$ that is tensile in the plane containing the crack and given by

$$\sigma_1 = \frac{Y\sigma\sqrt{\pi c}}{\sqrt{2\pi r}} \tag{10.1}$$

where $r$ is measured from the crack tip and $Y$ is a constant. The mode 1 *stress intensity factor*, $K_1$, is defined as

$$K_1 = Y\sigma\sqrt{\pi c} \tag{10.2}$$

The crack propagates when

$$K_1 = Y\sigma\sqrt{\pi c} = K_{1c} \tag{10.3}$$

where $K_{1c}$ is the fracture toughness, a material property.

Equation (10.3), illustrated in Figure 10.1, summarises the key factors. In the design of a component to avoid fracture, we can manipulate the geometry and loads, we can monitor it

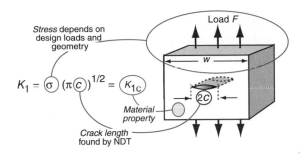

**Figure 10.1** The key players in fracture-limited design: load and geometry, crack length and material

using non-destructive testing (NDT) to be sure it contains no crack larger than an acceptable value, and we can choose materials with adequate fracture toughness.

Expressions for $K_1$ for various geometries and modes of loading are given in Figure 10.2. When the stress is uniform and the crack length is small compared with all specimen dimensions, $Y$ is equal to or close to 1. As the crack extends in a uniformly loaded component, it interacts with the free surfaces, requiring correction factors. If, in addition, the stress field is non-uniform (as it is in an elastically bent beam), $Y$ differs from 1. More accurate approximations and other less common loading geometries can be found in the references listed in Further Reading.

## **10.3** Material indices for fracture-safe design

Among mechanical engineers there is a rule of thumb already mentioned: avoid materials with a fracture toughness $K_{1c}$ less than 15 MPa.m$^{1/2}$. The $K_{1c}$–$E$ chart of Figure 8.8 shows that almost all metals pass: most have values of $K_{1c}$ in the range 20–100 in these units. Just a few fail: white cast iron and some powder-metallurgy products have values as low as 10 MPa.m$^{1/2}$. Ordinary engineering ceramics have values in the range 1–6 MPa.m$^{1/2}$; mechanical engineers view them with deep suspicion. But engineering polymers have even smaller values of $K_{1c}$ in the range 0.5–3 MPa.m$^{1/2}$ and yet engineers use them all the time. What is going on here?

In a *load-limited* design—the structural member of a bridge or the wing spar of an aircraft, for instance (Figure 10.3(a) and (b))—the part will fail in a brittle way if the stress exceeds that given by equation (10.3). To maximise the load we want materials with highest values of

$$M_1 = K_{1c} \tag{10.4}$$

But not all designs are load limited; some are *energy limited*, others are *deflection limited*. Then the criterion for selection changes. Consider, then, the other two scenarios sketched in Figure 10.3.

***Energy-limited design*** Springs and containment systems for turbines and flywheels (to catch the bits if they disintegrate) are *energy* limited. Take the spring (Figure 10.3(c)) as an example. The elastic energy $U_e$ stored in it is the integral over the volume of the energy per unit volume, $U_e$, where

$$U_e = \frac{1}{2}\sigma\varepsilon = \frac{1}{2}\frac{\sigma^2}{E}$$

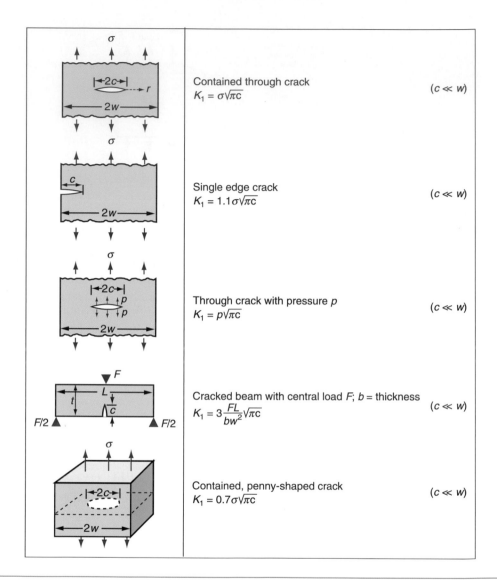

**Figure 10.2**  Stress intensities $K_1$ associated with cracks in standard geometries

The stress is limited by the fracture stress of equation (10.3) so that—if 'failure' means 'fracture'—the maximum energy per unit volume that the spring can store is

$$U_e^{max} = \frac{Y^2}{2\pi c}\left(\frac{K_{lc}^2}{E}\right)$$

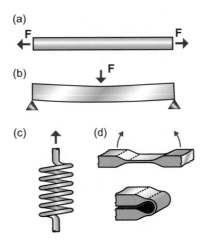

**Figure 10.3** (a, b) Load-limited design. (c) Energy-limited design. (d) Displacement-limited design. Each must perform its function without fracturing

For a given initial flaw size $c$, the energy is maximised by choosing materials with large values of

$$M_2 = \frac{K_{1c}^2}{E} = G_c \tag{10.5}$$

where $G_c$ is the toughness, defined in Chapter 8.

*Displacement-limited design*    There is a third scenario: that of *displacement*-limited design (Figure 10.3(d)). Snap-on bottle tops, elastic hinges and couplings are displacement limited: they must allow sufficient elastic displacement to permit the snap-action or flexure without failure, requiring a large failure strain $\varepsilon_f$. The strain is related to the stress by Hooke's law, $\varepsilon = \sigma/E$, and the stress is limited by the fracture equation (10.3). Thus, the failure strain is

$$\varepsilon_f = \frac{C}{\sqrt{\pi c_{max}}} \frac{K_{1c}}{E}$$

The best materials for displacement-limited design are those with large values of

$$M_3 = \frac{K_{1c}}{E} \tag{10.6}$$

*Plotting indices on charts*    Figure 10.4 shows the $K_{1c}$–$E$ chart again. It allows materials to be compared by values of fracture toughness, $M_1$, by toughness, $M_2$, and by values of the deflection-limited index, $M_3$. As the engineer's rule of thumb demands, almost all metals have values of $K_{1c}$ that lie above the 15 MPa.m$^{1/2}$ acceptance level for load-limited design, shown as a horizontal selection line in the figure. Polymers and ceramics do not.

The line showing $M_2$ in Figure 10.4 is placed at the value 1 kJ/m$^2$. Materials with values of $M_2$ greater than this have a degree of shock resistance with which engineers feel comfortable

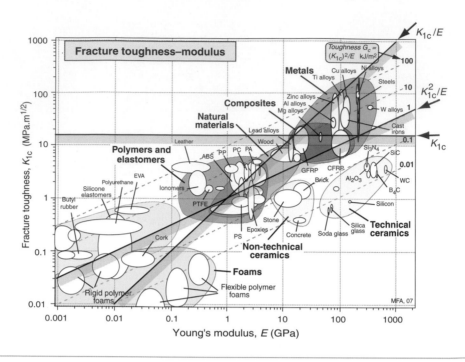

**Figure 10.4** The $K_{1c}$–$E$ chart showing the three indices $M_1$, $M_2$ and $M_3$

(another rule of thumb). Metals, composites and many polymers qualify; ceramics do not. When we come to deflection-limited design, the picture changes again. The line shows the index $M_3 = K_{1c}/E$ at the value $10^{-3}$ m$^{1/2}$. It illustrates why polymers find such wide application: when the design is deflection limited, polymers—particularly nylons (PA), poly-carbonate (PC) and ABS—are better than the best metals.

The figure gives further insights. Mechanical engineers' love of metals (and, more recently, of composites) is inspired not merely by the appeal of their $K_{1c}$ values. Metals are good by all three criteria ($K_{1c}$, $K_{1c}^2/E$ and $K_{1c}/E$). Polymers have good values of $K_{1c}/E$ and are acceptable by $K_{1c}^2/E$. Ceramics are poor by *all three* criteria. Herein lie the deeper roots of the engineers' distrust of ceramics.

## 10.4 Case studies

*Forensic fracture mechanics: pressure vessels*   An aerosol or fizzy beverage can is a pressure vessel. So, too, is a propane gas cylinder, the body of an airliner, the boiler of a power station, and the containment of a nuclear reactor. Their function is to contain a gas under pressure—$CO_2$, propane, air, steam. Failure can be catastrophic. Think of the bang that a party balloon can make, then multiply it by $10^{14}$—yes, really—and you begin to get an idea of what happens when a large pressure vessel explodes. When working with pressure, safety is an issue, as the following example illustrates.

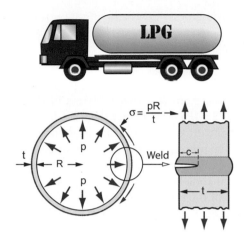

**Figure 10.5** A cylindrical pressure vessel with a cracked weld

A truck-mounted propane tank (Figure 10.5) was filled, then driven to the driver's home where he parked it in the sun and went inside to have lunch, leaving the engine running. As he was drinking his coffee the tank exploded causing considerable property damage and one death. The tank was made of rolled AISI 1030 steel plate formed into a cylinder, with a longitudinal seam weld and two domes joined to the rolled plate by circumferential welds. The failure occurred through the longitudinal weld, causing the tank to burst (see chapter opening picture). Subsequent examination showed that the weld had contained a surface crack of depth 10 mm that had been there for a long time, as indicated by discoloration and by 'striations' showing that it was growing slowly by fatigue each time the tank was emptied and refilled. At first sight it appeared that the crack was the direct cause of the failure.

That's where fracture mechanics comes in. Measurements on a section of the tank wall gave a fracture toughness of 45 MPa.m$^{1/2}$. The specification for the tank is listed in Table 10.1. It was designed for a working pressure of 1.4 MPa, limited for safety at 1.5 MPa by a pressure release valve.

The tensile stress in the wall of a thin-walled cylindrical pressure vessel of radius $R$ and wall thickness $t$ containing a pressure $p$ (Figure 10.5) was given in Chapter 4. With the values listed above the working stress in the tank should have been

$$\sigma = \frac{pR}{t} = 84 \text{ MPa} \tag{10.7}$$

**Table 10.1** Tank specification

| | |
|---|---|
| Working design pressure, $p$ | 1.4 MPa |
| Wall thickness, $t$ | 14 mm (thicker at weld) |
| Outer diameter, $2R$ | 1680 mm |
| Length, $L$ | 3710 mm |

A plate with a fracture toughness 45 MPa.m$^{1/2}$ containing a surface crack of depth 10 mm will fail at the stress

$$\sigma = \frac{K_{1c}}{Y\sqrt{\pi c}} = 254 \text{ MPa (assuming } Y \approx 1).$$

This is three times higher than the expected stress in the tank wall. The pressure needed to generate this stress is 4.2 MPa—far higher than the limit set by the safety valve of 1.5 MPa.

At first it appears that the calculations cannot be correct since the tank had a relief valve set just above the working pressure, so the discrepancy prompted the investigators to look further. An inspection of the relief valve showed rust and corrosion, rendering it inoperative. Further tests confirmed that heat from the sun and from the exhaust system of the truck raised the temperature of the tank, vaporising the liquefied gas and driving the pressure up to the value of 4.2 MPa needed to make the crack propagate. The direct cause of the failure was the jammed pressure release valve. The crack would not have propagated at the normal operating pressure.

This story indicates that we need a way to design such that there is warning before cracks propagate unstably. In other words, we need *fail-safe design*.

*Fail-safe design* If structures are made by welding or riveting, it is wise to assume that cracks are present—either because the process can lead to various forms of cracking, or because stress concentrations and residual stresses around the joints accelerate the formation of fatigue cracks. A number of techniques exist for detecting cracks and measuring their length, without damaging the component or structure; this is called *non-destructive testing* (NDT). X-ray imaging is one—alloy wheels for cars are checked in this way. Ultrasonic testing is another: sound waves are reflected by a crack and can be analysed to determine its size. Surface cracks can be revealed with fluorescent dyes. When none of these methods can be used, then proof testing—filling the pressure vessel with water and pressurising it to a level above the planned working pressure—demonstrates that there are no cracks large enough to propagate during service. All these techniques have a resolution limit, or a crack size $c_{\lim}$ below which detection is not possible. They do not demonstrate that the component or structure is crack-free, only that there are no cracks larger that the resolution limit. Proof testing is the safest, though it is expensive, since it does guarantee a maximum crack size that could be present. The other techniques carry a risk of operator error, with cracks being overlooked.

The first condition, obviously, is to design in such a way that the stresses are everywhere less than that required to make a crack of length $c_{\lim}$ propagate. Applying this condition gives the allowable pressure:

$$p \leq \frac{t}{R} \frac{K_{1c}}{\sqrt{\pi c_{\lim}}} \tag{10.8}$$

The largest pressure (for a given $R$, $t$ and $c_{\lim}$) is carried by the material with the greatest value of

$$M_1 = K_{1c} \tag{10.9}$$

—our load-limited index of Section 10.3.

But this design is not fail-safe. If the inspection is faulty, or if, by some mechanism such as fatigue, a crack of length greater than $c_{\text{lim}}$ appears, catastrophe follows. Greater security is obtained by requiring that the crack will not propagate even if the stress is sufficient to cause general yield, for then the vessel will deform stably in a way that can be detected. This condition, called the *yield-before-break* criterion, is expressed by setting the fracture stress σ equal to the yield stress $\sigma_y$, giving

$$c_{max} \leq \frac{1}{Y^2}\frac{1}{\pi}\left[\frac{K_{1c}}{\sigma_y}\right]^2 \tag{10.10}$$

Using this criterion, the tolerable crack size, and thus the integrity of the vessel, is maximised by choosing a material with the largest value of

$$M_4 = \frac{K_{1c}}{\sigma_y} \tag{10.11}$$

Large pressure vessels cannot always be X-rayed or ultrasonically tested; and proof testing may be something we wish to conduct infrequently, due to the cost. But since cracks can grow slowly because of corrosion or cyclic loading, a single examination at the beginning of service life is not sufficient. Then safety can be ensured by arranging that a crack that is long enough to penetrate all the way from the inner to the outer surface of the vessel is still stable. The vessel will leak, but this is not catastrophic and can be detected. This condition, called the *leak-before-break* criterion, is achieved by setting $c = t$ in equation (10.3), assuming the crack starts as a crack in one face and grows through the thickness to the other face, giving

$$\sigma = \frac{K_{1c}}{Y\sqrt{\pi t}} \tag{10.12}$$

### Example 10.1

A cylindrical steel pressure vessel of 5 m diameter is to hold a pressure $p$ of 4 MPa. The fracture toughness is 110 MPa.m$^{1/2}$. What is the greatest wall thickness that can be used while ensuring that the vessel will fail by leaking rather than fracture, should a crack grow within the wall? Assume a semi-circular surface crack of length $c$, for which $K \approx \sigma\sqrt{\pi c}$, assuming that the crack is orientated normal to the maximum stress $\sigma$ in the wall.

*Answer.* For the vessel to leak, the crack length $c$ has to reach the wall thickness $t$ without fracture. Consequently, $c = t$ and $K_{1c} = \sigma\sqrt{\pi t}$. The greatest hoop stress in the vessel wall drives the crack, so $\sigma = \frac{p\,R}{t}$. Combining these gives an equation for the maximum wall thickness for which the crack will penetrate the wall without fast fracture, with $R = 2.5$m:

$$t = \frac{\pi\, p^2\, R^2}{K_{1c}^2} = \frac{\pi \times 4^2 \times 2.5^2}{110^2} = 0.026 \text{ m} = 26 \text{ mm}$$

The wall thickness $t$ of the pressure vessel must also contain the pressure $p$ without yielding. From equation (10.7), the maximum pressure that can be carried is when the stress $\sigma = \sigma_y$ giving

$$\frac{p\,R}{t} = \sigma_y$$

The pressure that can be carried increases with the wall thickness, but there is an upper limit on the thickness given by the leak-before-break criterion, equation (10.12). Substituting for thickness into the previous equation (again with $\sigma = \sigma_y$) gives

$$p \leq \frac{1}{Y^2\,\pi\,R}\left(\frac{K_{1c}^2}{\sigma_y}\right)$$

So the pressure is carried most safely by the material with the greatest value of

$$M_5 = \frac{K_{1c}^2}{\sigma_y} \tag{10.13}$$

Both $M_4$ and $M_5$ could be made large by using a material with a low yield strength, $\sigma_y$: lead, for instance, has high values of both $M_4$ and $M_5$, but you would not choose it for a pressure vessel. That is because the vessel wall must also be reasonably thin, for both acceptable mass and economy of material. From equation (10.7), thickness is inversely proportional to the yield strength, $\sigma_y$, so we also seek a reasonably high value of

$$M_6 = \sigma_y \tag{10.14}$$

narrowing further the choice of material.

All of these criteria can be explored using the $K_{1c} - \sigma_y$ chart introduced in Chapter 8 (Figure 10.6). The indices $M_1, M_4, M_5$ and $M_6$ can be plotted onto it as lines of slope 0, 1, 1/2 and as lines that are vertical. Take 'leak-before-break' as an example. A diagonal line corresponding to a constant value of $M_5 = K_{1c}^2/\sigma_y$ links materials with equal performance; those above the line are better. The line shown in the figure excludes everything but the toughest steels, copper, nickel and titanium alloys. A second selection line for $M_6$ at $\sigma_y = 200$ MPa narrows the selection further. A more detailed treatment to minimise mass would seek to maximise the specific strength, $\sigma_y/\rho$. Exercises at the end of this chapter develop further examples of the use of the others.

Boiler failures used to be commonplace—there are even songs about it. Now they are rare, though when safety margins are pared to a minimum (rockets) or maintenance is neglected (the propane tank) pressure vessels still occasionally fail. This relative success is one of the major contributions of fracture mechanics to engineering practice.

*Materials to resist fatigue: con-rods for high-performance engines*   The engine of a family car is designed to tolerate speeds up to about 6000 rpm—100 revolutions per second. That of a Formula 1 racing car revs to about three times that. Comparing an F1 engine to that of a family car is like comparing a Rolex watch to an alarm clock, and their cost reflects this: about $200 000 per engine. Performance, here, is everything.

The connecting rods of a high-performance engine are critical components: if one fails the engine self-destructs. Yet to minimise inertial forces and bearing loads, each must weigh as

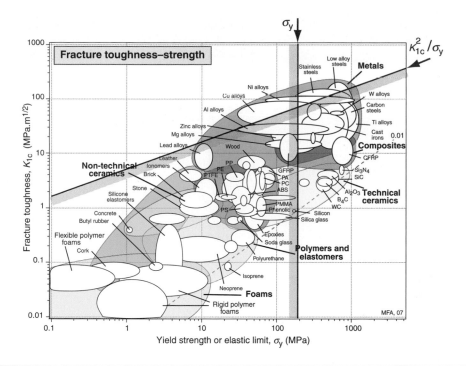

**Figure 10.6** Selecting the best materials for leak-before-break design. The additional requirement of high yield strength is also shown. The best choices are the materials in the small area at the top right

little as possible. This implies the use of light, strong materials, stressed near their limits. When minimising cost not maximising performance is the objective, con-rods are made from cast iron. But here we want performance. What, then, are the best materials for such con-rods?

An F1 engine is not designed to last long—about 30 hours—but 30 hours at around 15 000 rpm means about $3 \times 10^6$ cycles of loading and that means high-cycle fatigue. For simplicity, assume that the shaft has a uniform section of area $A$ and length $L$ (Figure 10.7) and that it carries a cyclic load $\pm F$. Its mass is

$$m = AL\rho \tag{10.15}$$

where $\rho$ is the density. The fatigue constraint requires that

$$\frac{F}{A} \leq \sigma_e$$

where $\sigma_e$ is the endurance limit of the material of which the con-rod is made. Using this to eliminate $A$ in equation (10.15) gives an equation for the mass:

$$m \geq F\ell \left( \frac{\rho}{\sigma_e} \right)$$

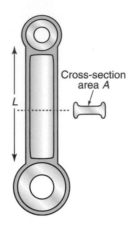

**Figure 10.7** A connecting rod

containing the material index

$$M = \frac{\sigma_e}{\rho} \tag{10.16}$$

Materials with high values of this index are identified by creating a chart with $\sigma_e$ and $\rho$ as axes, applying an additional, standard constraint that the fracture toughness exceeds 15 MPa.m$^{1/2}$ (Figure 10.8). Materials near the top-left corner are attractive candidates: high-strength magnesium, aluminum and ultra-high-strength steels; best of all are titanium alloys and CFRP. This last material has been identified by others as attractive in this application. To go further we need S–N curves for the most attractive candidates. Figure 10.9 shows such a curve for the most widely used of titanium alloys: the alloy Ti–6Al–4V. The safe stress amplitude at $R = -1$ for a design life of $2.5 \times 10^6$ cycles can be read off: it is 620 MPa.

If the selection is repeated using the much larger CES Level 3 database, really exotic materials emerge: aluminum reinforced with SiC, boron or $Al_2O_3$ fibres, beryllium alloys and a number of high-performance carbon-reinforced composites. A con-rod made of CFRP sounds a difficult thing to make, but at least three prototypes have been made and tested. They use a compression strut (a CFRP tube) with an outer wrapping of filament-wound fibres to attach the bearing housings (made of titanium or aluminum) to the ends of the strut.

*Rail cracking* At 12.23 p.m. on 7 October 2000, the 12.10 express from King's Cross (London) to Leeds (in the north of England) entered a curve just outside Hatfield at 115 mph—the maximum permitted for that stretch of line—and left the track. Four people died. More are killed on British roads every day, but the event brought much of the railway system, and through this, the country, to a near halt. It was clear almost immediately that poor maintenance and neglect had allowed cracks to develop in much of the country's track; it was an extreme case of this cracking that caused the crash.

Rail-head cracking is no surprise—rails are replaced regularly for that reason. The cracks develop because of the extreme contact stresses—up to 1 GPa—at the point where the wheels contact the rail (Figure 10.10(a)). This deforms the rail surface, and because the driving wheels

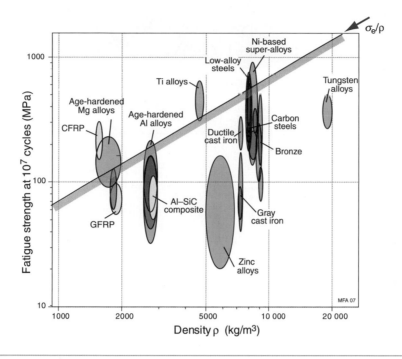

**Figure 10.8** Endurance limit and density for high-strength metals and composites. (All materials with $K_{1c} < 15$ MPa.m$^{1/2}$ were screened out before making this chart)

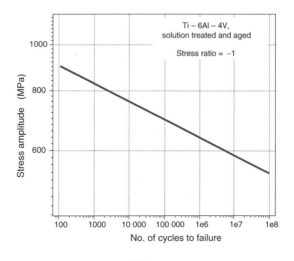

**Figure 10.9** An S–N curve for the titanium alloy Ti–6Al–4V

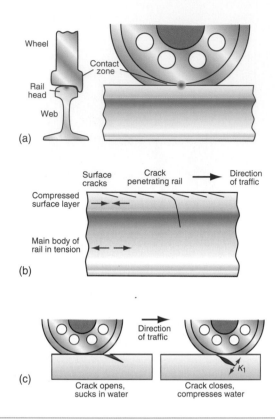

**Figure 10.10** Rail cracking. (a) The rolling contact. (b) The surface cracks, one of which has penetrated the part of the rail that is in tension. (c) The mechanism by which the crack advances

also exert a shear traction, the steel surface is smeared in a direction opposite to that of the motion of the train, creating a very heavily sheared surface layer. Cracks nucleate in this layer after about 60 000 load cycles, and because of the shearing they lie nearly parallel to the surface, as in Figure 10.10(b). The deformation puts the top of the rail into compression, balanced by tension in the main web. Welded rails are also pre-tensioned to compensate for thermal expansion on a hot day, so this internal stress adds to the pre-tension that was already there.

The mystery is how these harmless surface cracks continue to grow when the stress in the rail head is compressive. One strange observation gives a clue: it is that the cracks propagate much faster when the rails are wet than when they are dry—as they remain in tunnels, for instance. Research following the Hatfield crash finally revealed the mechanism, illustrated in Figure 10.10(c). As a wheel approaches, a surface crack is forced open. If the rail is wet, water (shown in red) is drawn in. As the wheel passes the mouth of the crack, it is forced shut, trapping the water, which is compressed to a high pressure. Figure 10.2 lists the stress intensity of a crack containing an internal pressure $p$:

$$K_1 = p\sqrt{\pi c} \tag{10.17}$$

So although the rail head is compressed, at the tip of the crack there is tension—enough to drive the crack forward. The crack inches forward and downward, driven by this hydraulic pressure, and in doing so it grows out of the compression field of the head of the rail and into the tensile field of the web. It is at this point that it suddenly propagates, fracturing the rail.

The answer—looking at equation (10.17)—is to make sure the surface cracks do not get too long; $K_1$ for a short pressurised crack is small, for a long one it is large. Proper rail maintenance involves regularly grinding the top surface of the track to remove the cracks, and when this has caused significant loss of section, replacing the rails altogether.

*Fatigue crack growth: living with cracks*    The crack in the LPG tank of the earlier case study showed evidence of 'striations': ripples marking the successive positions of the crack front as it slowly advanced, driven by the cyclic pressurising of the tank. The cause of the explosion was over-pressure caused by poor maintenance, but suppose maintenance had been good and no over-pressure had occurred. How long would it have lasted before fatigue grew the crack right through the wall and it leaked? First we had better check that the material had been correctly chosen to ensure leak-before-break. To calculate the critical crack length, $c^*$, we invert equation (10.3) giving:

$$c^* = \frac{K_{1c}^2}{Y^2 \pi \sigma^2}$$
(10.18)

With $K_{1c} = 45$ MPa.m$^{1/2}$, $\sigma = 84$ MPa (the working stress) and $Y = 1$, the answer is $c^* = 91$ mm. This is much greater than the wall thickness of 14 mm, so leak-before-break is satisfied (assuming the pressure release system works). So how long would it take the initial crack that was identified in the failure, of length 10 mm, to grow to 14 mm? A week? A month? 20 years?

Fatigue-crack growth where there is a pre-existing crack was described in Chapter 9. Its rate is described by the Paris' law equation

$$\frac{dc}{dN} = A\Delta K^m$$
(10.19)

in which the cyclic stress intensity range, $\Delta K$, is

$$\Delta K = K_{\max} - K_{\min} = \Delta\sigma\sqrt{\pi c}$$

Note that when using this equation, care is needed with the value of the constant $A$, which depends on the units of $c$ and $\Delta K$, and on the value of the exponent m. Substituting for $\Delta K$ in equation (10.19) gives

$$\frac{dc}{dN} = A\,\Delta\sigma^m(\pi c)^{m/2}$$
(10.20)

The residual life, $N_R$, is found by integrating this between $c = c_i$ (the initial crack length) and (in this case) $c = t$, the value at which leaking will occur:

$$N_R = \int_0^{N_R} dN = \frac{1}{A\,\pi^{m/2}\Delta\sigma^m} \int_{c_i}^{t} \frac{dc}{c^{m/2}}$$
(10.21)

For the steel of which the tank is made we will take a typical crack-growth exponent $m = 4$ and a value of $A = 2.5 \times 10^{-14}$ (for which $\Delta\sigma$ must be in MPa). Taking the initial crack length $c_i = 10$ mm and $t = 14$ mm, the integral gives

$$N_R = \frac{1}{A \, \pi^2 \Delta\sigma^4} \left( \frac{1}{c_i} - \frac{1}{t} \right) = 2.3 \times 10^6 \text{ cycles.}$$

So assuming that the tank is pressurised once per day, it will last forever.

This sort of question always arises in assessing the safety of a large plant that (because of welds) must be assumed to contain cracks: the casing of steam turbines, chemical engineering equipment, boilers and pipe-work. The cost of replacement of a large turbine casing is considerable, and there is associated down-time. It does not make sense to take it out of service if, despite the crack, it is perfectly safe. *Proof testing* is a technique used to guarantee a number of cycles, as follows. The vessel is over-pressurised above the working pressure, using a liquid rather than gas (since the stored energy in compressed liquid is much lower than in a gas at the same pressure, so that failure of the vessel will not lead to an explosion). The proof stress in the wall must be high enough to detect crack lengths below the wall thickness. There may be no cracks at all, but if the vessel survives the proof test we can guarantee that there are no cracks present above the critical crack length for this (higher) stress level—this is then used as $c_i$ in equation (10.21). Note that the guaranteed residual life is increased if the initial crack length in the integral is smaller. This means increasing the proof stress, but there is a practical limit on this: it cannot exceed the yield stress of the material. However, after the specified number of safe pressurisation cycles, the proof test can be repeated and the same lifetime is guaranteed all over again.

---

### Example 10.2

The cylindrical steel pressure vessel in Example 10.1 is built with a wall thickness of 15 mm. It is to be proof tested to guarantee 2000 safe cycles of pressurisation. The rate of crack growth is given by $da/dN = A(\Delta K)^4$ where $A = 3 \times 10^{-14}$ (with $a$ in metres, and $\Delta K$ in units of MPa.m$^{1/2}$).

Use the integral in equation (10.21) to find the maximum initial crack size that will require 2000 cycles to grow to the wall thickness (and failure by leaking). Hence calculate the pressure required in a proof test to ensure that such a crack does not exist. Use the chart in Figure 10.6 to see if the applied stress looks reasonable relative to the yield stress of steels with fracture toughness of 110 MPa.m$^{1/2}$.

*Answer.* The stress amplitude in the wall during normal pressurisation cycles is

$$\Delta\sigma = \frac{pR}{t} = \frac{4 \times 2.5}{0.015} = 667 \text{ MPa}$$

From equation (10.21) with a Paris' law exponent $m = 4$:

$$N_R = \frac{1}{A \, \pi^2 \Delta\sigma^4} \left( \frac{1}{c_i} - \frac{1}{t} \right) \Rightarrow 2000 = \frac{1}{3 \times 10^{-14} \times \pi^2 \times 667^4} \left( \frac{1}{c_i} - \frac{1}{t} \right) \Rightarrow c_i = 5.4 \text{ mm.}$$

The proof stress required to fracture the steel with this crack length is

$$\sigma_{proof} = \frac{K_{1c}}{Y\sqrt{\pi\,c_i}} = \frac{110}{1 \times \sqrt{\pi \times 0.0054}} = 845 \text{ MPa}.$$

The corresponding pressure is

$$p_{proof} = \frac{\sigma_{proof}\,t}{R} = \frac{845 \times 0.015}{2.5} = 5.1 \text{ MPa}$$

From Figure 10.6, the yield strength of (low alloy) steels with fracture toughness of 110 MPa.m$^{1/2}$ is in the range 500–1000 MPa. So the proof stress will be all right provided a sufficiently high strength variant is used. If not, the proof stress must be reduced, giving a larger starting crack length and a shorter guaranteed number of cycles.

*Designing for fracture* Manufacturers who distribute their products packaged in toughened envelopes, sheathed in plastic, or contained in aluminum or steel cans, need to design the containment so that the purchaser can get at the contents. As you will know, not all do—even getting the wrapper off a newly purchased CD is a pain. And there is a more serious dimension. Producers of foods, drugs, chemicals, even washing powder, now package their products in 'break to access' packaging so that the purchaser knows that it has not been tampered with. How do you arrange protection and yet enable fracture? One part of the answer is to choose the right material; the other is to provide stress concentrations to focus stress on the break-lines or to use adhesives that have high shear strength but low peel resistance.

Start with material: materials with low ductility, in thin sheets, tear easily. If 'opening' means pulling in such a way as to tear (as it often does), then the first step is a to choose a material with adequate stiffness, strength and durability, but low ductility. The lids of top-opening drink cans are made of a different alloy than the can itself for exactly this reason.

Next, stress concentration. This means reducing the section by grooving or serrating the package locally along the line where tearing is wanted. Toilet paper, as we all know, hardly ever tears along the perforations, but it was the right idea. A sardine can is made of low-ductility aluminum alloy with a groove with a sharp radius of curvature along the tear line to provide a stress concentration factor (Chapter 7) of

$$K_{sc} = \left(1 + \frac{1}{2}\sqrt{\frac{c}{\rho}}\right) \tag{10.22}$$

where $c$ is the groove depth and $\rho$ its root radius and the factor (1/2) appears because the loading is shear rather than tension. A 0.2 mm groove with root radius of 0.02 mm gives a local stress that is 2.5 times higher than that elsewhere, localising the tearing at the groove.

The peel-strip of a CD wrapper is not a groove; it is an additional *thicker* strip. How does this apparent reinforcement make it easier to tear open the package? Figure 7.7 provides the answer: it is because *any* sudden change of section concentrates stress. If the strip thickness is $c$ and the radius where it joins the wrapping is $\rho$, the stress concentration is still given by equation (10.22).

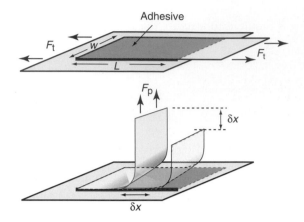

**Figure 10.11** Peeling of an adhesive bond

The alternative to tearing is adhesive peeling. Figure 10.11 shows an adhesive joint before and during peeling. The adhesive has shear strength $\sigma_s^*$ and toughness $G_c^*$. Adhesively bonded packaging must accept in-plane tension, since to protect the content it must support the mass of its contents and handling loads. The in-plane pull force $F_t$ that the joint can carry without failing is

$$F_t = \sigma_s^* A = \sigma_s^* w L$$

where $A = wL$ is the area of the bonded surface. To open it, a peel force $F_p$ is applied. The lower part of Figure 10.11 shows how $F_p$ does work when the joint is peeled back by a distance $\delta x$, creating new surface of area $w\delta x$. This requires an energy $G_c^* w\delta x$ (since $G_c^*$, the toughness, is the energy to create unit area of new surface). The work done by $F_p$ must provide this energy, giving

$$F_p\, \delta x = G_c^* w\, \delta x$$

Thus, $F_p = w\, G_c^*$. The ratio of the peel force to the tensile force is

$$\frac{F_p}{F_t} = \frac{G_c^*}{\sigma_s^* L}$$

Adhesive joints are designed to have a particular value for this ratio. The choice of adhesive sets the values of $G_c^*$ and $\sigma_s^*$, allowing the length $L$ to be chosen to give tensile strength with ease of peeling.

## 10.5 Summary and conclusions

Elastic deformation is recoverable. Plastic deformation is gradual and detectable before really bad things happen. Failure by fast fracture is none of these. It has been the cause of many great engineering disasters: collapsed bridges, burst boilers, rail accidents, aircraft crashes.

A component will fail by fast fracture if the stress intensity $K_1$ at any crack-like defect it contains exceeds the fracture toughness $K_{1c}$ of the material of which it is made. The understanding of fast fracture and the development of design methods to deal with it are relatively new—before 1950 little was known. Both are now on a solid basis. This chapter introduced them, describing the ways in which $K_1$ is calculated, how $K_{1c}$ is measured and the various scenarios in which it is relevant. It ended with examples of their application.

Much the same is true of failure by fatigue. A seemingly healthy component, one that has served its purpose well, fails without warning because an initially small and harmless crack has grown until it reached the size at which fast fracture takes over. Small components may, initially, be crack free, but cyclic loading can lead to crack nucleation and subsequent growth; design is then based on the S–N curve of the material. Larger structures, particularly those that are welded, cannot be assumed to be crack free; then the life is estimated from the predicted rate of crack growth, integrating it to find the number of cycles to grow the crack to a dangerous size.

Fracture, however, is not always bad. Enabling fracture allows you to get at the contents of shrink-wrapped packages, to access food and drugs in tamper-proof containers, and to have the convenience of drinks in pop-top cans. Here the trick is to design-in stress concentrations that create locally high stresses along the desired fracture path, and to choose a material with adequate strength but low ductility.

## 10.6 Further reading

Broek, D. (1981). *Elementary Engineering Fracture Mechanics* (3rd ed.). Boston, USA: Martinus Nijhoff. ISBN 90-247-2580-1. (A standard, well-documented introduction to the intricacies of fracture mechanics.)

Ewalds, H. L., & Wanhill, R. J. H. (1984). *Fracture Mechanics*. London, UK: Edward Arnold. ISBN 0-7131-3515-8. (An introduction to fracture mechanics and testing for both static and cyclic loading.)

Hertzberg, R. W. (1989). *Deformation and Fracture of Engineering Materials* (3rd ed.). New York, USA: Wiley. ISBN 0-471-63589-8. (A readable and detailed coverage of deformation, fracture and fatigue.)

Hudson, C. M., & Rich, T. P. (Eds.). (1986). *Case Histories Involving Fatigue and Fracture Mechanics*. USA: American Society for Testing and Materials. ASTM STP 918. ISBN 0-8031-0485-5. (A compilation of case studies of fatigue failures, many of them disastrous, analysing what went wrong.)

Suresh, S. (1998). *Fatigue of Materials* (2nd ed.). Cambridge, UK: Cambridge University Press. ISBN 0-521-57847-7. (The place to start for an authoritative introduction to the materials science of fatigue in both ductile and brittle materials, with case studies.)

Tada, H., Paris, G., & Irwin, G. R. (2000). *The Stress Analysis of Cracks Handbook* (3rd ed.). ISBN 1-86058-304-0. (A comprehensive catalog of stress intensity factors for a great range of geometries and modes of loading.)

## 10.7 Exercises

Exercise E10.1    Supersonic wind tunnels store air under high pressure in cylindrical pressure vessels—the pressure, when released, produces hypersonic rates of flow. The pressure vessels are routinely proof tested to ensure that they are safe. If such a cylinder, of diameter 400 mm and wall thickness 20 mm, made of a steel with a fracture toughness of 42 MPa.m$^{1/2}$, survives a proof test to 40 MPa (400 atmospheres), what is the length of the largest crack it might contain?

**Exercise E10.2** A thin-walled spherical pressure vessel of radius $R = 1$ m is made of a steel with a yield stress of 600 MPa and a fracture toughness $K_{1c}$ of 100 MPa.m$^{1/2}$. The vessel is designed for an internal working pressure $p = 20$ MPa.

(a) Find a suitable wall thickness if the maximum stress is to be kept below 60% of the yield stress. The stress in the wall of a uniform spherical pressure vessel is given by $\sigma = pR/2t$.

(b) The pressure vessel is inspected by a non-destructive method, and an embedded penny-shaped crack of length $2c = 10$ mm is found in the wall (i.e., from Figure 10.2, the geometry factor $Y = 0.7$ in the expression for $K$). Will the pressure vessel survive at the working pressure, and in a proof test at 90% of the yield stress?

**Exercise E10.3** You are asked to select a polymer to make a flexible coupling. The polymer must have a modulus greater than 2 GPa. The objective is to maximise the available flexure without fracture. Use the chart of Figure 10.4 to identify two good choices to meet these requirements. Are there any metals that are as good? If the flexible coupling is a door hinge, what other material behaviour should be considered to avoid premature failure in service?

**Exercise E10.4** Crash barriers like car fenders must absorb energy without total fracture. The most effective are those that deform plastically, absorbing energy in plastic work, but they are not reusable. Fenders that remain elastic spring back after impact. For practical reasons the material must have a modulus greater than 10 GPa. Use the chart of Figure 10.4 to find non-metallic materials for elastic fenders, assuming that the overriding consideration is that the displacement before fracture is as great as possible (the constraint on modulus ensures that it absorbs enough energy).

**Exercise E10.5** Materials with high toughness $G_c$ generally have high modulus. Sometimes, however, the need is for high toughness with low modulus, so that the component has some flexibility. Use the chart of Figure 10.4 to find the material (from among those on the chart) that has a modulus less than 0.5 GPa and the highest toughness $G_c$. List applications of this material that you think exploit this combination of properties.

**Exercise E10.6** If you want to support a precision optical system (laser metrology equipment, for instance) on a stable platform, you put it on a granite slab supported on end plinths to bring it to working height. (Granite can be ground to a flat surface and is thermally very stable and hard-wearing.) The granite chosen for one such table has a fracture toughness of 0.9 MPa.m$^{1/2}$ and is known, from non-destructive testing (NDT) procedures, to contain internal cracks up to 5 mm in length. If the table is 2 m long and 1 m deep, simply supported at its ends and must carry a uniformly distributed load of 2000 N on its upper surface (as in Figure 7.2), what is the minimum thickness the slab must have? Include the self-weight of the slab in the analysis. Assume that at least one of the cracks will lie in the part of the beam that carries the highest tensile stress—that is, at the lower surface. (The density of granite is 2700 kg/m$^3$).

Exercise E10.7    The figure below shows, in (a), a cylindrical tie-rod with a diameter 20 mm. The plan is to use it to carry a cyclic load with a stress range ±200 kN. The figure also shows the S–N curve of the material of which it is to be made. Will it survive without failure for at least 10$^5$ cycles?

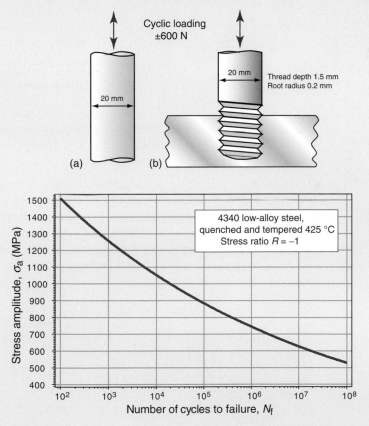

Exercise E10.8    The component of the previous exercise was made and tested. It failed in less than 10$^5$ cycles. A post-mortem revealed that it had fractured at the root of a threaded end, shown in (b) in the figure. The threads have a depth of 2 mm and a root radius of 0.1 mm. Given this additional information, how many cycles would you expect it to survive?

Exercise E10.9    The figure shows a component to be made from the high-strength aerospace alloy Ti−6Al−4V. It will be loaded cyclically at ± 210 MPa. How long will it last?

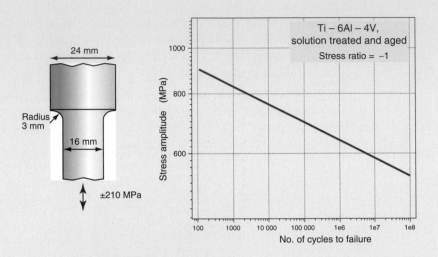

Exercise E10.10   Discuss the factors that determine whether a pressure vessel will fail gradually by yielding, leaking or by fast fracture, when subjected to repeated applications of internal pressure.

Exercise E10.11   Consider three cylindrical pressure vessels made of the aluminium alloys in the table below. Each has a wall thickness of $t = 14$ mm, and a radius of $R = 840$ mm. The vessels are each periodically emptied and filled to an internal pressure of $p = 2$ MPa. Cracks can grow through the wall by fatigue, driven by the cycle of stress on pressurisation. Determine the failure mode of each vessel: yield / leak / fracture. Comment on which material would be best for this application.

| Material | Yield stress, $\sigma_y$ (MPa) | Fracture toughess, $K_{1c}$ (MPa.m$^{1/2}$) |
|---|---|---|
| Aluminium 2024 T4 | 300 | 40 |
| Aluminium 6061 T4 | 110 | 35 |
| Aluminium 5454 H111 | 140 | 20 |

Exercise E10.12   (a) Use the Paris' law for crack growth rate (equation (10.19)) to show that the number of cycles to failure $N_f$ of a component cycled through a uniform stress range $\Delta\sigma$, and containing an initial crack of length $c_i$, is:

$$N_f = \frac{1}{A\, Y^m \Delta\sigma^m \pi^{m/2}(1 - m/2)}\left[(c^*)^{1-m/2} - (c_i)^{1-m/2}\right]$$

where $c^*$ is the critical crack length, $Y$ is a geometric factor, $A$ and $m$ are constants in the Paris law (with $m > 2$).

(b) A die fabricated from a low alloy steel is to withstand tensile hoop stresses that cycle between 0 and 400 MPa. Prior to use it has been determined that the length of the largest crack in the steel is 1 mm. Determine the minimum number of times that the die may be used before failing by fatigue, assuming $n = 3$, $A = 1.0 \times 10^{-12}$ (with $\Delta\sigma$ in MPa), and $Y = 1.13$. The fracture toughness of the steel is 170 MPa.m$^{1/2}$.

Exercise E10.13 A cylindrical steel pressure vessel of 7.5 m diameter and 40 mm wall thickness is to operate at a working pressure of 5.1 MPa. The design assumes that small thumbnail-shaped flaws in the inside wall will gradually extend through the wall by fatigue. Assume for this crack geometry that $K = \sigma\sqrt{\pi c}$, where $c$ is the length of the edge-crack and $\sigma$ is the hoop stress in the vessel. The fracture toughness of the steel is 200 MPa.m$^{1/2}$ and the yield stress is 1200 MPa.

(a) Would you expect the vessel to fail in service by leaking (when the crack penetrates the thickness of the wall) or by fast fracture?

(b) The vessel will be subjected to 3000 pressurisation cycles during its service life. The growth of flaws by fatigue is given by equation (10.19) with $m = 4$, $A = 2.44 \times 10^{-14}$ MPa$^{-4}$m$^{-1}$, and stress in MPa. Find the critical crack length for fracture, and check that the vessel will leak before fracture. Determine the initial crack length $c_i$ that would grow to failure in 3000 cycles. Hence find the pressure to which the vessel must be subjected in a proof test to guarantee against failure in the service life. Check that this test will not yield the vessel.

Exercise E10.14 Consider the design of a bicycle crank. It is assumed that the design is governed by fatigue failure, and that to avoid this the maximum stress amplitude must be kept below the endurance limit $\sigma_e$ of the material. The crank is modelled as a beam in bending of square cross-section $a \times a$, where $a$ is free to vary. The length of the beam and the amplitude of the applied cyclic moment are fixed.

(a) Derive an expression for the material index which should be maximised to minimise the mass. Hence choose a short-list of materials using the property chart, Figure 10.8, commenting on your choice. Use the most advantageous material properties within the ranges shown on the chart.

(b) An existing crank made of the best performing Al alloy weighs 0.2 kg. A cyclist is prepared to pay extra for a lighter component made of titanium, but only if it saves enough weight. Use the property chart with the material index from (a) to find the ratio of the mass of the best performing Ti alloy to that of the best Al alloy. Hence estimate the mass of the corresponding titanium component.

(c) The as-manufactured costs per kg of aluminium and titanium are 10 and 40 £/kg, respectively. Estimate the costs of the two cranks. How much value (in £) would the cyclist need to associate with each kg of weight saved, in order to go ahead with the switch from aluminium to titanium?

Exercise E10.15 An adhesive has a toughness $G_c = 100$ J/m$^2$ and a shear strength $\sigma_s = 0.1$ MPa. What must the dimensions of the bonded area of a lap joint be if it is to carry an in-plane tensile $F_t = 100$ N but allow peeling at a force $F_p = 5$ N?

## 10.8 Exploring design with CES

Use Level 2 Materials for all selections.

Exercise E10.16 Use the 'Search' facility in CES to find materials that are used for:
(a) Pressure vessels.
(b) Connecting rods.
(c) Rail track.
(Search using the singular—e.g., pressure vessel—since that will find the plural too.)

Exercise E10.17 You are asked, as in Exercise E10.3, to select a polymer to make a flexible coupling. The polymer must have a modulus $E$ greater than 2 GPa. The objective is to maximise the available flexure without fracture, and that means materials with high $K_{1c}/E$. Use a 'Limit' stage to impose the constraint on $E$, then make use of a 'Graph' stage to make a chart for $K_{1c}$ and $E$, put on an appropriate selection line and move it until only three materials remain in the Results window. What are they? Rank them by price.

Exercise E10.18 You are asked to recommend materials that have yield strength above 500 MPa and perform best in a design based on the leak-before-break criterion. Construct an appropriate limit stage and chart, put on the necessary selection line and list the three materials you would recommend.

Exercise E10.19 Repeat the previous selection, applying instead a yield-before-break selection criterion.

Exercise E10.20 A material is sought for a high-performance con-rod, requiring that the index has a high value of the index $\sigma_e/\rho$, where $\sigma_e$ is the endurance limit (the fatigue strength at $10^7$ cycles). It must have enough toughness to tolerate stress concentrations, requiring that $K_{1c} > 15$ MPa.m$^{1/2}$. Make the appropriate selection stages and list the three materials that best meet the criteria.

# Chapter 11
# Rub, slither and seize: friction and wear

A disc brake. (Image courtesy Paul Turnbull, SEMTA, Watford, UK: www.gcseinenginering.com)

## **11.1** Introduction and synopsis

God, it is said, created materials, but their surfaces are the work of the devil. They are certainly the source of many problems. When surfaces touch and slide, there is friction; and where there is friction, there is wear. Tribologists—the collective noun for those who study friction and wear—are fond of citing the enormous cost, through lost energy and worn equipment, for which these two phenomena are responsible. It is certainly true that, if friction could be eliminated, the efficiency of engines, gearboxes, drive trains and the like would increase enormously; and if wear could be eradicated, they would also last much longer. But before accepting this negative image, we should remember that, without wear, pencils would not write on paper or chalk on blackboards; and without friction, we would slither off the slightest incline. Friction plays a significant role in shaping processes too, often leading to increases in forming loads and unwanted heating. But there are also welding processes that exploit frictional heating to make joints without having to melt the components.

Tribological properties are not attributes of one material alone, but of one material sliding on another with—almost always—a third in between. The number of combinations is far too great to allow choice in a simple, systematic way. The selection of materials for bearings, drives and sliding seals relies heavily on experience. This experience is captured in reference sources (for which, see 'Further reading'); in the end it is these that must be consulted. But it does help to have a feel for the magnitude of friction coefficients and wear rates, and an idea of how these relate to material class. This chapter provides it.

## **11.2** Tribological properties

When two surfaces are placed in contact under a normal load $F_n$ and one is made to slide over the other, a force $F_s$ opposes the motion. This force $F_s$ is proportional to $F_n$ but does not depend on the area of the surface, facts discovered by none other than Leonardo da Vinci[1], then forgotten, then rediscovered 200 years later by Amontons[2], whose name they carry. The *coefficient of friction $\mu$* is defined as in Figure 11.1.

$$\mu = \frac{F_s}{F_n} \tag{11.1}$$

---

[1] Leonardo da Vinci (1452–1519), Renaissance painter, architect, engineer, mathematician, philosopher and prolific inventor.
[2] Guillaume Amontons (1663–1705), French scientific instrument maker and physicist, perfector of the *clepsydra* or water clock—a device for measuring time by letting water flow from a container through a tiny hole.

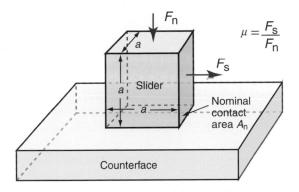

**Figure 11.1**  The definition of the coefficient of friction, $\mu$. Here $F_n$ is the force normal to the interface and $F_s$ the force parallel to the interface required to maintain sliding.

---

### Example 11.1

Wood, sliding on concrete, has a coefficient of friction of 0.4. How much force is required to slide a wooden palette carrying equipment weighing 300 kg across concrete?

*Answer.* The normal force $F_n$ is the weight times the acceleration due to gravity (9.81 m/s$^2$), giving $F_n = 2943$ N. The sliding force $F_s = \mu F_n = 1177$ N.

---

When surfaces slide, they wear. Material is lost from both surfaces, even when one is much harder than the other. The *wear rate*, $W$, is conventionally defined as

$$W = \frac{\text{Volume of material removed from contact surface}\,(\text{m}^3)}{\text{Distance slid (m)}} \tag{11.2}$$

and thus has units of m$^2$. A more useful quantity, for our purposes, is the wear rate per unit area of surface, called the *specific wear rate*, $\Omega$:

$$\Omega = \frac{W}{A_n} \tag{11.3}$$

which is dimensionless. It increases with bearing pressure $P$ (the normal force $F_n$ divided by $A_n$), such that

$$\Omega = k_a \frac{F_n}{A_n} = k_a P \tag{11.4}$$

where $k_a$, the *Archard wear constant*, has units of (MPa)$^{-1}$. It is a measure of the propensity of a sliding couple for wear: high $k_a$ means rapid wear at a given bearing pressure.

### Example 11.2

A dry sliding bearing oscillates with an amplitude of $a = 2$ mm at a frequency $f = 1$ Hz under a normal load of $F_n = 10$ N. The nominal area of the contact is $A_n = 100$ mm$^2$. The Archard wear constant for the materials of the bearing is $k_a = 10^{-8}$ MPa$^{-1}$. If the bearing is used continuously for one year, how much will the surfaces have worn down?

*Answer.* Define $\Delta x$ as the distance the surfaces have worn down. Then

$$\Delta x = \frac{\text{Volume removed}}{\text{Nominal contact area}} = \Omega \cdot \text{Distance slid} = k_a \frac{F_n}{A_n} \cdot \text{Distance slid}$$

The distance slid per cycle is $4a$. One year is $t = 3.15 \times 10^7$ s. The distance slid is $4a \times f \times t = 2.5 \times 10^5$ m giving a wear $\Delta x = 2.5 \times 10^{-4}$ m $= 0.25$ mm.

## 11.3 Charting friction and wear

***The coefficient of friction*** Approximate values for the coefficient of friction for dry—that is, unlubricated—sliding of materials on a steel counterface are shown in Figure 11.2. The values depend on the counterface material, but the range is much the same: typically,

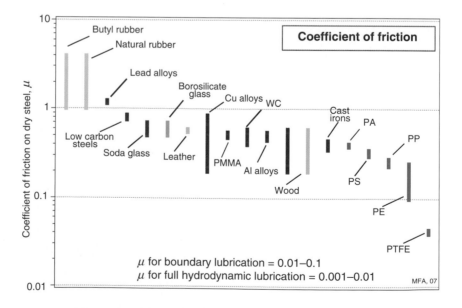

Figure 11.2    The coefficient of friction $\mu$ of materials sliding on an unlubricated steel surface.

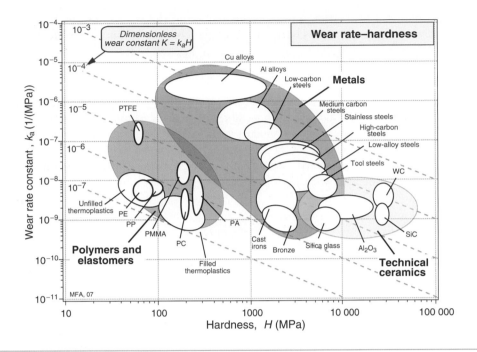

**Figure 11.3** The normalised wear rate for lubricated sliding, $k_a$ plotted against hardness $H$, here expressed in MPa rather than Vickers ($H$ in MPa = 10 $H_v$). The chart gives an overview of the way in which common engineering materials behave.

$\mu \approx 0.5$. Certain combinations show much higher values, either because they seize when rubbed together (a soft metal rubbed on itself in a vacuum with no lubrication, for instance) or because one surface has a sufficiently low modulus that it conforms to the rough surface of the other (rubber on rough concrete). At the other extreme are sliding combinations with exceptionally low coefficients of friction, such as PTFE, or bronze bearings loaded with graphite, sliding on polished steel. Here the coefficient of friction falls as low as 0.04, though this is still high compared with friction for lubricated surfaces, as noted at the bottom of the diagram.

***The wear rate–hardness chart***   Figure 11.3 shows the Archard wear constant $k_a$ plotted against hardness, here expressed in MPa. (Hardness in MPa $\approx$ 10 $\times$ Vickers hardness, $H_V$.) The bearing pressure $P$ is the quantity specified by the design. The ability of a surface to resist a static contact pressure is measured by its hardness, so the *maximum* bearing pressure $P_{max}$ is just the hardness $H$ of the softer surface. Thus, the wear rate of a bearing surface can be written:

$$\Omega = k_a P = k_a \left( \frac{P}{P_{max}} \right) H$$

Two material properties appear in this equation: the wear constant $k_a$ and the hardness, $H$—they are the axes of the chart. The diagonal contours show the product

$$K = k_a H \tag{11.5}$$

The best materials for bearings for a given bearing pressure $P$ are those with the lowest value of $k_a$—that is, those nearest the bottom of the diagram. On the other hand, an efficient bearing, in terms of size or weight, will be loaded to a safe fraction of its maximum bearing pressure—that is, to a constant value of $P/P_{max}$—and for these, materials with the lowest values of the product $k_a H$ are best. Materials of a given class (metals, for instance) tend to lie along a downward sloping diagonal across the figure, reflecting the fact that low wear rate is associated with high hardness.

---

### Example 11.3

A newly developed polymer composite hardened with nano-scale ceramic particles is found to have a hardness of 100 Vickers. Extrapolate the data for polymers on the Chart of Figure 11.3 to estimate approximately the Archard wear constant you might expect it to have.

*Answer.* A hardness of 100 Vickers = 1000 MPa. An extrapolation of the polymers data to 1000 MPa gives an Archard wear constant of about $10^{-10}$ MPa$^{-1}$.

---

## 11.4 The physics of friction and wear[3]

*Friction*   Surfaces, no matter how meticulously honed and polished, are never perfectly flat. The *roughness*, defined in Chapter 18, depends on how the surface was made: a sand-casting has a very rough surface, while one that is precision-machined has a much smoother one. But even a mirror finish is not perfectly smooth. Magnified vertically, surfaces look like Figure 11.4(a)—an endless range of asperity peaks and troughs. So when two surfaces are placed in contact, no matter how carefully they have been crafted, they touch only at points where asperities meet, as in Figure 11.4(b). The load $F_n$ is supported solely by the contacting asperities. The real contact area $A_r$ is only a tiny fraction of the apparent, nominal, area $A_n$.

What is this fraction? When they first touch the asperities deform elastically. But even small loads cause large contact stresses, enough to cause plastic deformation as in Figure 11.5(a). The contact points flatten, forming junctions with a total area $A_r$, as in Figure 11.5(b). The contacts can carry a greater stress than the uniaxial yield stress, as the deformation of the

---

[3] We owe our understanding of friction and wear to the work of many engineers and scientists, above all to the remarkably durable and productive collaboration between the Australian Philip Bowden (1903—1968) and David Tabor (1913—2005), English of Lithuanian descent, between 1936 and 1968 in Melbourne and Cambridge.

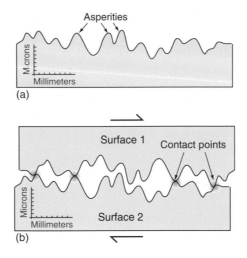

(a)

(b)

**Figure 11.4** (a) The profile of a surface, much magnified vertically. (b) Two surfaces in contact touch only at asperities.

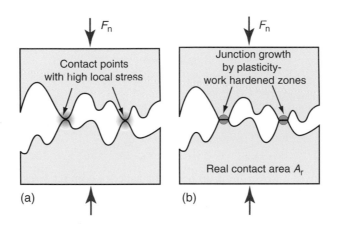

(a)

(b)

**Figure 11.5** When surfaces first touch, high contact stresses appear, as in (a). These cause plasticity and junction growth until the junction area $A_r$ can just support the load $F_n$ without further plasticity, as in (b).

asperities is constrained by the surrounding material—the peaks of Figure 11.5 are exaggerated for clarity. Localised contacts on a flat surface resemble hardness indentations (Chapter 6) such that the total load transmitted across the surface is approximately

$$F_n = A_r H \tag{11.6}$$

where the hardness $H \approx 3\,\sigma_y$ and where $\sigma_y$ is the yield strength. Thus a lower limit on the real contact area is

$$A_r \approx \frac{F_n}{3\,\sigma_y} \tag{11.7}$$

To see how small this is, return to the cube of Figure 11.1. Its mass is $m = \rho\,a^3$ so the force needed to support it is $F_n = \rho\,g\,a^3 = \rho\,g\,a\,A_n = 3\,A_r\,\sigma_y$ where $A_n = a^2$ is the nominal contact area and $g$ is the acceleration due to gravity (9.81 m/s$^2$). Thus the ratio of real to nominal contact areas is

$$\frac{A_r}{A_n} = \frac{\rho\,g\,a}{3\,\sigma_y}$$

Thus a 100 mm cube of ordinary steel (density 7900 kg/m$^3$, yield strength 250 MPa) weighing a hefty 7.9 kg, contacts any hard surface on a fraction of 1% of its nominal area—in this example, around 0.1 mm$^2$.

Now think of sliding one surface over the other. If the junctions weld together (as they do when surfaces are clean) it will need a shear stress $F_s$ equal to the shear yield strength $k$ of the material to shear them or, if the materials differ, one that is equal to the shear strength of the softer material. Thus for unlubricated sliding

$$\frac{F_s}{A_r} \approx k$$

or, since $k \approx \sigma_y/2$,

$$F_s \approx \frac{1}{2} A_r\,\sigma_y \tag{11.8}$$

Dividing this by equation (11.6) gives an estimate for the coefficient of friction

$$\mu = \frac{F_s}{F_n} \approx 0.17 \tag{11.9}$$

This estimate gives a lower limit on $\mu$—if the asperity contact stress is nearer to $\sigma_y$ than to $H$, the estimate for $\mu$ would be $\approx 0.5$. This is consistent with the spread in Figure 11.2.

So far we have spoken of *sliding* friction: $F_s$ is the force to maintain a steady rate of sliding. If surfaces are left in static contact, the junctions tend to grow by creep, and the bonding between them becomes stronger so that the force to start sliding is larger than that to maintain it. This means that the coefficient of friction to start sliding, called the static coefficient, $\mu_s$ is larger than that to sustain it. This can result in the stick-slip behaviour that leads to vibration in brakes, and is also the way that the bow-hair causes a violin string to resonate.

**Wear** We distinguish two sorts of sliding wear. In *adhesive wear*, characteristic of wear between the same or similar materials (copper on aluminum, for example), asperity tips, stuck together, shear off to give wear damage. In *abrasive wear*, characteristic of wear when one surface is much harder than the other (steel on plastic, say), the asperity tips of the harder material plough through the softer one, abrading it like sandpaper.

Figure 11.6 shows the way in which adhesion between surfaces causes wear fragments to be torn from the surface. To minimise the rate of wear we need to minimise the size of the fragments; that means minimising the area of contact. Since $A_r = F_n/\sigma_y$, reducing the load or increasing $\sigma_y$ (that is, the surface hardness) reduces wear.

Figure 11.7 shows abrasive wear. The asperities of the harder surface slice off segments of the softer one, like grating cheese. More commonly the problem is one of contamination: hard particles that sneak into the surface—dust, sand, oxidised wear particles—embed themselves in one surface and plough through the other. It is to prevent this that cars have air and oil filters.

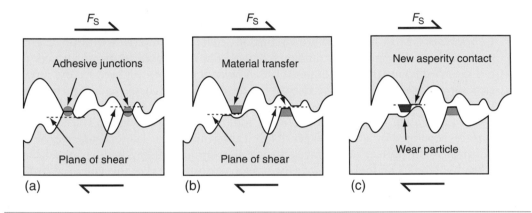

**Figure 11.6**  Adhesive wear: adhesion at the work-hardened junctions causes shear-off of material when the surfaces slide.

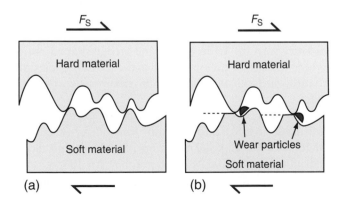

**Figure 11.7**  Abrasive wear: asperities of the harder material, or embedded hard particles, plough through the softer one, grinding off wear particles.

## **11.5** Friction in design and metal processing

*Bearings* allow two components to move relative to each other in one or two dimensions while constraining them in the others. The key criteria are low friction coupled with the ability to support the bearing load $F_n$ without yielding, overheating or wearing excessively. *Brakes* and *clutches*, by contrast, rely on high friction, but there is more to their design than that, as we shall see in a moment. But first, bearings.

Dry sliding, as we have seen, generally leads to friction with a coefficient $\mu$ of about 0.5. If the bearings of the drive-train of your car had a $\mu$ of this magnitude, negligible power would reach the wheels—almost all would be lost as heat in the bearing surfaces. In reality, about 15% of the power of the engine is lost in friction because of two innovations: *lubrication* and the replacement of sliding by *rolling bearings*.

*Lubrication*   Most lubricants are viscous oils containing additives with polar groups (fatty acids with $-OH^-$ groups not unlike ordinary soap) that bind to metal surfaces to create a chemisorbed boundary layer (Figure 11.8). A layer of the lubricant between the surfaces allows easy shear, impeded only by the viscosity of the oil, giving a coefficient of fraction in the range 0.001−0.01. When the sliding surfaces are static, the bearing load $F_n$ tends to squeeze the oil out of the bearing, but the bonding of the additive is strong enough to resist, leaving the two surfaces separated by a double-layer of lubricating molecules. This *boundary-layer lubrication* is sufficient to protect the surfaces during start-up and sliding at low loads, but it is not something to rely on for continuous use or high performance. For this you need *hydrodynamic lubrication*.

The viscosity of the oil is both a curse and a blessing—a curse because it exerts a drag (and thus friction) on the sliding components; a blessing because, without it, hydrodynamic lubrication would not be possible. Figure 11.9 explains this (an explanation first provided by Reynolds[4]). Sliding bearings are designed so that the surfaces are slightly inclined. Oil is dragged along the wedge-shaped region between them from the wide to the narrow end, as in Figure 11.9(a). The rise in pressure as the oil is compressed forces the surfaces apart against the force of the normal load carried by the bearing. The maximum pressure is reached just behind the centre of the slider, where the color is most intense in the figure. The same principle governs the self-alignment of a rotating journal bearing, as in Figure 11.9(b). Again, the lubricant is dragged by viscosity into the wide end of the wedge-shaped region and swept toward the narrow end, where the rise of pressure pushes the journal back onto the axis of the bearing. The result is that the surfaces never touch but remain separated by a thin film of oil or other lubricating fluid.

At high temperatures, hydrocarbon lubricants decompose. Solid lubricants—PTFE, molybdenum disulfide ($MoS_2$) and graphite (one form of carbon)—extend the operating temperature to 500 °C. PTFE, good to about 250 °C, has an intrinsically low coefficient of friction. Moly-disulfide (up to 300 °C) and graphite (to 500 °C) rely on their lamellar (layer-like) crystal structure, able to shear easily on certain crystal planes, to provide coefficients of

---

[4] Osborne Reynolds (1842–1912), Northern Irishman, brilliant mathematician and engineer but awful lecturer, known for his contributions to lubrication, turbulence and for his celebrated Number—the dimensionless group describing the transition from laminar to turbulent flow.

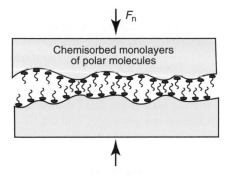

**Figure 11.8** The formation of a boundary layer of chemisorbed, polar molecules; the red dot represents the polar group, the tail the hydrocarbon chain attached to it.

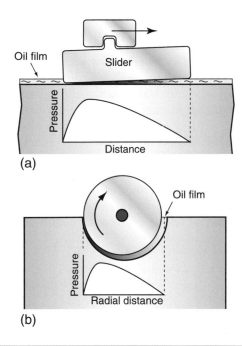

**Figure 11.9** The distribution of pressure under a sliding and a rotating bearing with hydrodynamic lubrication.

friction well below 0.1. All these can be applied as lubricants or incorporated into the pores of porous bearings (made by sintering powdered metal like bronze and cast iron) to give self-lubricating bearings.

That's the background. So what materials are good for bearings? That depends—plane bearings or roller bearings?

*Plane bearings*   The surfaces of sliding bearings, if properly lubricated, do not touch. Then almost any material that meets the other constraints of the design will do—steel, cast iron, brass, bronze. The key material properties are strength (to carry the bearing loads), corrosion resistance and conformability—the ability to adapt to slight misalignment. Full hydrodynamic lubrication breaks down under static loads or when sliding velocities are small, and long-term reliance on boundary lubrication is risky. Bearing materials are designed to cope with this.

When soft metals slide over each other (lead on lead, for instance), the junctions are weak but their area of contact is large so $\mu$ is large. When hard metals slide (e.g., steel on steel), the junctions are small but they are strong, and again $\mu$ is large. The way to get low friction with high strength is to support a soft metal with a hard one, giving weak junctions of small area. White-metal bearings, for example, consist of soft lead or tin supported in a matrix of a stronger material; leaded-bronze bearings consist of a matrix of strong, corrosion-resistant bronze containing particles of soft lead that smear out to give an easily sheared surface film when asperities touch; polymer-impregnated porous bearings are made by partly sintering bronze and then forcing a polymer (usually PTFE) into its pores. Bearings like these are not designed to run dry, but if they do, the soft component saves them, at least for a while.

*Rolling bearings*   In rolling element bearings the load is carried by a set of balls or rollers trapped between an inner and an outer *race* by a *cage*, as in Figure 11.10. When the loading is radial the load is carried at any instant by half the balls or rollers, but not equally: the most heavily loaded one carries about $5/Z$ times the applied load, where $Z$ is the total number of balls or rollers in the bearing. Its contact area with the race is small, so the contact stresses are high.

The main requirement is that the deformation of the components of the bearing remain at all times elastic, so the criterion for material choice is that of high hardness. That usually means a high-carbon steel or a tool steel. If the environment is corrosive or the temperature high, stainless steel or ceramic (silicon carbide or silicon nitride) is used. When the loads are very light the races can be made from thermoplastic—polypropylene or acetal—with the advantage that no lubrication is needed (cheap bike pedals sometimes have such bearings).

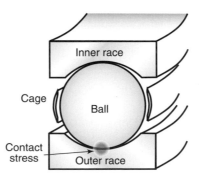

**Figure 11.10**   A rolling-contact bearing. The contact stresses are high, requiring materials with high hardness and resistance to fatigue failure.

When rolling element bearings fail, it is usually because of fatigue—the repeated loading and unloading of ball and race as the bearing rotates causes cracks to nucleate, splitting the balls or pitting the races. Inclusions in the material are sources of fatigue cracks, so steels for rolling bearings are processed to be as clean and inclusion free as possible by processing under vacuum.

*High friction: materials for brakes and clutches*   Applying friction sounds easy—just rub. Brakes for horse-drawn carriages did just that—spoon-shaped blocks of wood, sometimes faced with leather (bigger $\mu$), were wound down with a screw to rub on the rim of the carriage wheel. But brake materials today do much more.

The friction coefficient of materials for brake linings is—surprisingly—not special: $\mu$ is about 0.4. The key thing is not $\mu$ but the ability to apply it without vibration, wear or fade. That is not easy when you reckon that, to stop a one-tonne car from 100 kph (60 mph) requires that you find somewhere to put about 0.4 MJ. That's enough (if you don't lose any) to heat 1 kg of steel to 800 °C. Brake materials must apply friction at temperatures that can approach such values, and they must do so consistently and without the stick-slip that causes 'brake squeal'—not acceptable in even the cheapest car. That means materials that can tolerate heat, conduct it away to limit temperature rise and—curiously—lubricate, since it is this that quenches squeal.

For most of the history of the automobile, brake linings were made of compacted asbestos (able to tolerate 2000 °C) mixed with brass or copper particles or wires for heat conduction and strength. When it became apparent that asbestos dust—the product of brake wear—was perhaps not so good for you, new materials appeared. They are an example of hybrid material design methods: no one material can do what you want, so we synthesise one that does.

Today's materials are a synthesis: a matrix, an abrasive additive to increase friction, a reinforcement—one to help conduct heat if possible—and a lubricant to suppress vibration. Intimate details are trade secrets, but here is the idea. The brake pads on your bike are made up of a synthetic rubber with particles of a cheap silicate to reduce pad wear and give wet friction. Those on your car have a phenolic matrix with particles of silicates, silicon carbide (sandpaper grit) to control friction and graphite or $MoS_2$ as a lubricant. Those on a military jet, a 747 or an F1 car are carbon or ceramic—here high temperature is the problem; replacement every 10 days is acceptable if they stand the heat (try calculating how much aluminum you could melt with the energy dissipated in bringing a 200-tonne 747 to rest from 200 kph).

*Waging war on wear*   As explained earlier, the rate of wear increases with the bearing pressure $P$ (equation (11.4)). It also increases with sliding velocity $v$ because the faster the sliding, the more work is done against friction and it all turns into heat. A given bearing material has an acceptable $P-v$ envelope within which it is usable; venture outside it and seizure or catastrophic wear awaits you (Figure 11.11(a)). To extend the envelope upward, we select materials with higher strength; to extend it laterally, we choose materials with higher thermal conductivity and melting point—or lower friction.

Lubrication, of course, reduces friction. Figure 11.11(b) shows that this does not increase the admissible bearing pressure—that is a static property of the material—but it does increase the permissible sliding velocity. Boundary lubrication expands the envelope a little; hydrodynamic lubrication expands it much more—but note its drop-off at low velocities because there is not enough sliding speed to build up the pressure profiles of Figure 11.9.

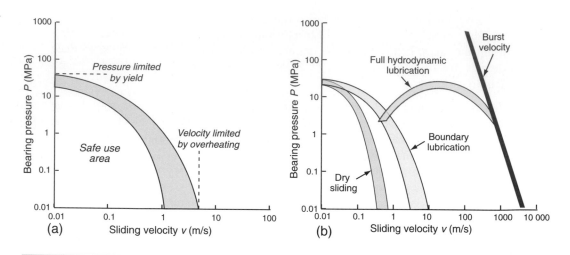

**Figure 11.11** (a) A $P–v$ curve for a bearing material. (b) $P–v$ envelopes for various lubricated and unlubricated sliding, ultimately limited by the disintegration of the bearing under centrifugal force at very high velocities.

At very high rotational velocities, approached in ultra-centrifuges, gyroscopic navigation systems and super-flywheels for energy storage, there is an ultimate cut-off. It is the speed at which centrifugal forces exceed the strength of the material and the whole system disintegrates.

Hard materials have low wear rates but they tend to be brittle. Soft ones are tougher but they wear rapidly. Wear is a surface property, toughness a property of the bulk. So the way to fight wear is to take a tough material and give it a hard surface coating, using techniques of surface treatment. High-carbon steels are surface hardened by rapidly heating with a flame, an electron beam or a laser beam and then quenching the surface. *Carburising, nitriding* and *boriding* give wear resistance to components like crankshafts or ball races by diffusing carbon, nitrogen or boron atoms into a thin surface layer, where they react to give particles of carbides, nitrides or borides. Hard, corrosion-resistant layers of alloys rich in tungsten, cobalt or chromium can be sprayed onto surfaces, but a refinishing process is then necessary to restore dimensional precision. Hard ceramic coatings of alumina ($Al_2O_3$) or of tungsten or titanium carbide (WC, TiC) can be applied by plasma spraying to give both wear resistance and resistance to chemical attack, and chemical and physical vapour deposition methods allow surface coatings of other metals and ceramics, including diamond-like carbon.

Records for all these treatments (and more) are contained in the CES software under the heading 'Surface treatment'. The software allows selection of coatings to meet given design requirements, of which wear resistance is one of the more important. We return to these in Chapter 18 on selecting processes.

***Friction in metal processing*** When blocks of material slide over one another, work must be done by the applied forces to overcome friction. Metal forming processes often involve metal sliding past a tool (refer to Figure 11.7). In rolling, friction is essential for the process to work at all: it is friction at entry to the roll bite that draws the material in. In extrusion, the billet is forced down the cylindrical container and squeezed through the shaped hole in the die—friction accounts for a substantial share of the load applied to the ram. And in die forging and sheet forming, the metal slides over the tooling as it deforms to adopt the shape of the dies. Lubrication can help in cold rolling and forging, but this is not generally possible in hot forming. So some of the benefit of lowering the yield stress, by forming at high temperature, will be lost due to friction. Friction not only increases the loads needed in forming, but it also sets a limit on the section of a component that can be rolled or forged: as the ratio of width to thickness increases, it becomes impossible to squash the metal plastically at all, a processing limit we will look at again in Chapter 18.

The work done against friction is dissipated as heat at the sliding surfaces, giving a non-uniform distribution of temperature in the workpiece. Since temperature controls the underlying development of microstructure (Chapter 19), here is yet another complicating factor to worry about. As an example, friction against the die in extrusion can generate enough surface heating to locally melt the emerging product, giving a streaky surface finish and increased scrap-rate. But there is one class of processes in which frictional heating is deliberately exploited to soften the metal locally—friction welding processes. Figure 11.12 shows two common examples. In rotary friction welding, one component is rotated against another at high speed under an axial load. The contact area rapidly heats up, and the oxide on the original surfaces is broken up and dispersed, giving a strong metal-metal bond. After a short period the rotation is stopped, and a final axial forging force is applied, leaving a flash of expelled material that is usually machined off. In friction stir welding, a rotating tool consisting of a shoulder and profiled pin is pressed into the interface between two plates, and then traversed along the joint line. The material is heated by friction and plastic deformation, and the metal is swirled round the pin to give a strong metal joint. As melting is avoided, the process has proved

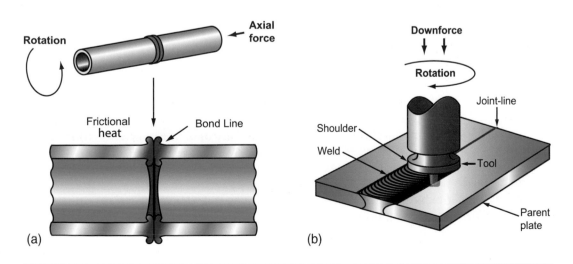

Figure 11.12   (a) Rotary friction welding; (b) Friction stir welding.

particularly effective for joining alloys that are traditionally more difficult to join by fusion welding, such as age-hardened aluminium alloys.

We can estimate how much power is needed in friction welding as follows. Consider rotary friction welding with the tubes in Figure 11.12(a) being replaced by solid cylinders of radius $R$, rotating at a speed $\omega$ (radians per sec) under an axial force $F$. On initial contact, the surfaces rub against each other with a coefficient of friction $\mu$, giving a frictional force acting circumferentially. Hence a torque $T$ must be applied to overcome this frictional resistance. Each point on the contact area contributes to the torque via the local circumferential force multiplied by the radial distance from the centre. Integrating over the contact area gives the torque

$$T = \frac{2}{3}\, \mu\, F\, R \tag{11.10}$$

The power $q$ is given by

$$q = T\,\omega$$

Typical values for friction welding indicate a power per unit area of order $1\,\text{kW/cm}^2$ (see Exercises). Transient thermal analysis, discussed in the next chapter, would tell us that with this intensity of heating the contacts reach hot forming temperatures (i.e. over 0.5 times the melting temperature) within seconds. The asperities quickly flatten completely, and the ratio $A_r/A_n$ rises to unity. This condition is referred to as 'sticking friction', with the shear stress over the whole contact area being equal to $k$, the (hot) shear yield stress. The torque then becomes

$$T = \frac{2\,\pi}{3}\, k\, R^3 \tag{11.11}$$

The torque (and thus the power) are lower in the hot condition, due to the reduced value of $k$, and the temperature stabilises when the heat input is balanced by thermal conduction along the cylinders away from the welding zone. At this point the rotation is stopped, and the joint is finally forged together.

## 11.6 Summary and conclusions

When surfaces slide, there is friction and there is wear. Eliminate all friction and a spinning wheel would spin forever. Eliminate all wear and it would last forever too. Both are about as probable as living forever while eating nothing. So we have to learn to live with friction and wear: to use them when they are useful, to minimise the problems they cause when they are a nuisance.

The remarkable thing about friction is that, if you slide one surface over another, the force to do so depends only on the normal force pushing the surfaces together, not on the area of the surface. That is because surfaces are not perfectly smooth; they touch only at points (flattened a little by the pressure)—just enough of them to support the normal force. Changing the area does not change the number of contact points. The sliding force is the force to shear the contact points, so it, too, is independent of area.

The points of contact are where wear happens. When surfaces slide the points shear off, becoming wear particles. This process can be useful—it helps surfaces bed down and conform by what is called 'run-in'. But if the wear particles or particles from elsewhere (dust, grit) get between the surfaces, the grinding action continues and the surfaces wear.

Soft materials—lead, polyethylene, PTFE, graphite—shear easily. Impregnating a sliding material with one of these reduces friction—skis and snow boards have polyethylene surfaces for just this reason. Better still is hydrodynamic lubrication: floating the sliding surfaces apart by dragging fluid between them. Flotation requires that the surfaces slide; if they stop the fluid is squeezed out. Clever lubricants overcome this by bonding to the surfaces, leaving a thin boundary layer that is enough to separate the surfaces until hydrodynamic lubrication gets going.

In metal forming processes, friction is usually a problem, increasing the loads needed, limiting the shapes that can be made, or causing local surface heating and non-uniform microstructures. But friction welding is an exception, exploiting frictional heating directly to join alloys without having to melt them.

Friction, wear and corrosion cause more damage to mechanisms than anything else. They are the ultimate determinants of the life of products—or they would be, if we kept them until they wore out. But in today's society, many products are binned before they are worn out; many when they still work perfectly well. More on that subject in Chapter 20.

## 11.7 Further reading

Bowden, F. P., & Tabor, D. (1950, 1964). *The Friction and Lubrication of Solids, Parts 1 and 2*. Oxford, UK: Oxford, University Press. ISBN 0-19-851204-X. (Two volumes that establish the cornerstones of tribology: contact between solids; friction of metals and non-metals; frictional temperature rise; boundary lubrication and other chemical effects; adhesion; sliding wear; and hardness.)

Hutchings, I. M. (1992). *Tribology: Friction and Wear of Engineering Materials*. London, UK: Arnold. ISBN 0-340-56184-X. (An introduction to the mechanics and materials science of friction, wear and lubrication.)

Neale, M. J. (1995). *The Tribology Handbook* (2nd ed.). Oxford, UK: Butterworth-Heinemann. ISBN 0-7506-1198-7. (A tribological bible: materials, bearings, drives, brakes, seals, lubrication failure and repair.)

## 11.8 Exercises

Exercise E11.1    Define the coefficient of friction. Explain why it is independent of the area of contact of the sliding surfaces.

Exercise E11.2    Historically, woods, both resinous wood from the pine family and hardwoods like *lignum vitae* (which is so dense that it sinks in water), were used for bearings—even clocks had hardwood bearings. Why? Think of your own ideas, then research the Web.

Exercise E11.3    Give examples, based on your experience in sport (tennis, golf, swimming, skiing, rock climbing, hang-gliding, etc.) of instances in which friction is wanted and when it is not.

Exercise E11.4    Now a more challenging exercise: give examples, again based on your experience in sport, of instances in which wear is desirable and in which it is not.

Exercise E11.5    What are the characteristics of materials that are a good choice for use as brake pads?

Exercise E11.6    A prototype disk brake for a bicycle is to be mounted on the hub of the rear wheel, and consists of a pair of pads of area 5 cm × 1 cm that are pressed against opposing faces of the steel disk. The pads are aligned with their longer dimension in the tangential rotation direction, with the pad centres at a distance of 10 cm from the centre of the disk. The brake is to be tested on a rotating machine to simulate the loading and wear in service. The disk rotates at a suitable constant speed, and the brakes are applied for 10 seconds every minute, with each pad applying a normal load of 1 kN.

(a) Find a suitable rotation speed (in revolutions per minute, rpm) if the test is to simulate braking a bicycle travelling at a speed of 20 km/hour, with a wheel diameter (including the tyre) of 70 cm.

(b) When the brakes are applied, the machine registers an increase in torque of 30 Nm. Find the friction force that is being applied to each pad, and hence estimate the friction coefficient.

(c) Estimate the depth of material removed per hour of testing, if the Archard wear constant for the materials is $k_a = 2 \times 10^{-8}$ MPa$^{-1}$.

(d) What other factors would influence the wear rate in service?

Exercise E11.7    A bronze statue weighing 4 tonnes with a base of area 0.8 m$^2$ is placed on a granite museum floor. The yield strength of the bronze is 240 MPa. What is the true area of contact, $A_r$, between the base and the floor?

Exercise E11.8    The statue of the previous example was moved to a roof courtyard with a lead floor. The yield strength of the lead is 6 MPa. What now is the true area of contact?

Exercise E11.9    How would you measure the true area of contact $A_r$ between two surfaces pressed into contact? Use your imagination to devise ways.

Exercise E11.10   A 30 mm diameter plane bearing of length 20 mm is to be made of a material with the $P - v$ characteristics shown in Figure 11.11(b). If the bearing load is 300 N and the maximum rotation rate is 500 rpm, is it safe to run it dry? If not, is boundary lubrication sufficient? (Remember that the bearing pressure is the load divided by the projected area normal to the load.)

Exercise E11.11   A rotary friction weld is to be made between two solid aluminium cylinders of diameter 2 cm. The rotation speed is 600 rpm, an initial axial force of 50 kN is applied, and the increase in the measured machine torque is 60 Nm. Estimate the friction coefficient and heating rate on first contact. After a few seconds, the contact has heated up and sticking friction applies over the whole contact area, with a shear yield stress $k = 15$ MPa. Find the new machine torque and the power being dissipated at the contact.

## **11.9** Exploring design with CES

Use Level 2, Materials, throughout exercises.

Exercise E11.12   Use the 'Search' facility to find materials that are used as ball bearings. (Search on the singular—ball bearing—since that picks up the plural as well.)

Exercise E11.13   Use the 'Search' facility to find materials that are used for brake pads. Do the same for brake discs.

Exercise E11.14   Use a 'Limit' selection stage applied to the Surface Treatment data table to find processes that enhance wear resistance. (To do this, click 'Select' in the main toolbar, then 'Selection Data', just below it. Select Level 2 Surface treatment in the dialog box. Open a 'Limit' stage, open Function of treatment, and click Wear resistance.) Explore the record for laser-based methods. What are its typical uses?

Exercise E11.15   Follow the same procedure as that of the previous example to search for processes used to control friction. Explore the record for grinding and mechanical polishing. What are grinding wheels made of?

# Chapter 12
# Agitated atoms: materials and heat

A heat exchanger. (Image courtesy of the International Copper Association)

## **12.1** Introduction and synopsis

Thermal properties quantify the response of materials to heat. Heat, until about 1800, was thought to be a physical substance called 'caloric' that somehow seeped into things when they were exposed to flame. It took the American Benjamin Thompson[1], backed up by none other than the formidable Carnot[2], to suggest what we now know to be true: that heat is atoms or molecules in motion. In gases, they are flying between occasional collisions with each other. In solids, by contrast, they vibrate about their mean positions; the higher the temperature, the greater the amplitude of vibrations. From this perception emerges all of our understanding of thermal properties of solids: their heat capacity, expansion coefficient, conductivity, even melting.

Heat affects mechanical and physical properties too. As temperature rises, materials expand, the elastic modulus decreases, the strength falls and the material starts to creep, deforming slowly with time at a rate that increases as the melting point is approached until, on melting, the solid loses all stiffness and strength. This we leave for Chapter 13.

Thermal design, the first theme of this chapter, is design to cope properly with the effects of heat, and in some cases to exploit them. The chapter opening page shows an example: a copper heat exchanger, designed to transfer heat efficiently between the two circulating fluids. The thermal properties of materials also underpin the second theme of the chapter: manufacturing. Most processes involve heating the material up to make it easier to work with, but the thermal cycles imposed have all sorts of additional consequences for achieving the target properties. So the design and selection of the manufacturing process are closely bound up with the design of the component.

## **12.2** Thermal properties: definition and measurement

Two temperatures, the *melting temperature*, $T_m$, and the *glass transition temperature*, $T_g$ (units for both: Kelvin[3], K, or Centigrade, °C), are fundamental points of reference because they relate directly to the strength of the bonds in the solid. Pure crystalline solids have a sharp melting point, $T_m$, abruptly changing state from solid to low-viscosity liquid. Alloys usually melt over a temperature range—but the onset and completion of melting are still at precise temperatures. Non-crystalline molecular materials such as thermoplastics have a more gradual transition from true solid to very viscous liquid, with the onset of the change being characterised by the glass transition temperature $T_g$. It is helpful in engineering design

---

[1] Benjamin Thompson (1753–1814), later Lord Rumford, while in charge of the boring of cannons at the Watertown Arsenal near Boston, Mass., noted how hot they became and formulated the 'mechanical equivalent of heat'—what we now call the heat capacity.

[2] Sadi Nicolas Léonhard Carnot (1796–1832), physicist and engineer, formulator of the Carnot cycle and the concept of entropy, the basis of the optimisation of heat engines. He died of cholera, a fate most physicists today, mercifully, are spared.

[3] William Thompson, Lord Kelvin (1824–1907), Scottish mathematician who contributed to many branches of physics; he was known for a self-confidence that led him to claim (in 1900) that there was nothing more to be discovered in physics—he already knew it all.

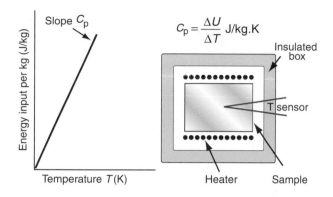

**Figure 12.1**  Measuring heat capacity, $C_p$. Its units are J/kg.K.

to define two further reference temperatures: the *maximum* and *minimum service temperatures*, $T_{max}$ and $T_{min}$ (units for both: K or °C), established empirically from practical experience. The first tells us the highest temperature at which the material can be used continuously without oxidation, chemical change or excessive distortion becoming a problem. The second is the temperature below which the material becomes brittle or otherwise unsafe to use.

It costs energy to heat a material up. The energy to heat 1 kg of a material by 1 K is called the *heat capacity* or *specific heat*, and since the measurement is usually made at constant pressure (atmospheric pressure) it is given the symbol $C_p$. Heat is measured in Joules[4], symbol J, so the units of specific heat are J/kg · K. When dealing with gases, it is more usual to measure the heat capacity at constant volume (symbol $C_v$), and for gases this differs from $C_p$. For solids the difference is so slight that it can be ignored, and we shall do so here. $C_p$ is measured by the technique of calorimetry, which is also the standard way of measuring the glass temperature $T_g$. Figure 12.1 shows how, in principle, this is done. A measured quantity of energy (here, electrical energy) is pumped into a sample of material of known mass. The temperature rise is measured, allowing the energy/kg.K to be calculated. Real calorimeters are more elaborate than this, but the principle is the same.

Most materials expand when they are heated (Figure 12.2). The thermal strain per degree of temperature change is measured by the *linear thermal expansion coefficient*, $\alpha$. It is defined by

$$\alpha = \frac{1}{L}\frac{dL}{dT} \tag{12.1}$$

where $L$ is a linear dimension of the body. If the material is anisotropic it expands differently in different directions, and two or more coefficients are required. Since strain is dimensionless, the units of $\alpha$ are $K^{-1}$ or, more conveniently, 'microstrain/K', that is, $10^{-6}\ K^{-1}$.

---

[4] James Joule (1818–1889), English physicist, who, motivated by theological belief, sought to establish the unity of the forces of nature. His demonstration of the equivalence of heat and mechanical work did much to discredit the caloric theory.

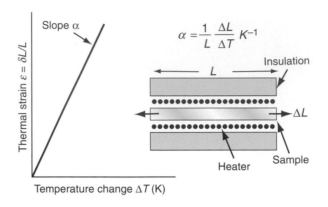

**Figure 12.2** Measuring the thermal expansion coefficient, $\alpha$. Its units are 1/K or, more usually, $10^{-6}$/K (microstrain/K).

---

### Example 12.1

The volume of water in the earth's oceans is $V = 1.37 \times 10^9$ km$^3$. The oceans cover an area $A = 3.61 \times 10^8$ km$^2$ (71% of the earth's surface). Thus the average depth is 3795 m. The average surface temperature is 17 °C, but the deeper water is colder—we take the average to be 10 °C. The volumetric coefficient of thermal expansion of water at 10 °C, $\alpha_v = 8.8 \times 10^{-5}$ K$^{-1}$. If the earth's oceans warmed by 2 °C, by how much would sea level rise?

*Answer.* The volume change caused by the temperature rise is $\Delta V = \alpha_v . V . \Delta T = 2.4 \times 10^8$ m$^3$. Dividing this by the surface area of the oceans gives a rise of 0.66 m.

---

### Example 12.2

A steel rail of square cross-section 25 mm $\times$ 25 mm is rigidly clamped at two points a distance $L = 2$ m apart. If the rail is stress-free at 20 °C, at what temperature will thermal expansion cause it to buckle elastically? (The thermal expansion coefficient of steel is $\alpha = 12 \times 10^{-6}$/K.)

*Answer.* The elastic buckling load of a column clamped in this way is given by equation (5.9) and Figure 5.5:

$$F_{\text{crit}} = \frac{4\pi^2 EI}{L^2}$$

Here $E$ is Young's modulus and $I = \dfrac{t^4}{12} = 3.26 \times 10^{-8}$ m$^4$ is the second moment of area of the rail. When the clamped rail is heated by $\Delta T$ it expands, causing a stress $\sigma = E\alpha\Delta T$ to appear in it. The clamps there exert an axial load $F = \sigma A = E A \alpha \Delta T$ on the rail, where $A = 6.25 \times 10^{-4}$ m$^2$ is the area of its cross-section. Equating this to $F_{crit}$ and solving for $\Delta T$ gives

$$\Delta T = \frac{4\pi^2 I}{\alpha A L^2} = 43°C$$

Power is measured in Watts[5]; 1 Watt (W) is 1 Joule/second. The rate at which heat is conducted through a solid at steady-state (meaning that the temperature profile does not change with time) is measured by the *thermal conductivity*, $\lambda$ (units: W/m.K). Figure 12.3 shows how it is measured: by recording the heat flux $q$ (W/m$^2$) flowing through the material from a surface at higher temperature $T_1$ to a lower one at $T_2$ separated by a distance $x$. The conductivity is calculated from Fourier's[6] law:

$$q = -\lambda \frac{dT}{dx} = \lambda \frac{(T_1 - T_2)}{x} \tag{12.2}$$

where $q$ is the heat flux per unit area (units: W/m$^2$).

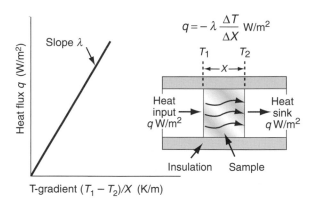

**Figure 12.3** Measuring the thermal conductivity, $\lambda$. Its units are W/m·K.

[5] James Watt (1736–1819), instrument maker and inventor of the condenser steam engine (the idea came to him while 'walking on a fine Sabbath afternoon'), which he doggedly developed. Unlike so many of the characters footnoted in this book, Watt, in his final years, was healthy, happy and famous.

[6] Baron Jean Baptiste Joseph Fourier (1768–1830), mathematician and physicist; he nearly came to grief during the French Revolution but survived to become one of the savants who accompanied Napoleon Bonaparte in his conquest of Egypt.

Thermal conductivity, as we have said, governs the flow of heat through a material at steady-state. The property governing transient heat flow (when temperature varies with time) is the *thermal diffusivity, a* (units: m²/s). The two are related by

$$a = \frac{\lambda}{\rho C_p} \tag{12.3}$$

where $\rho$ is the density and $C_p$ is, as before, the heat capacity. The thermal diffusivity can be measured directly by measuring the time it takes for a temperature pulse to traverse a specimen of known thickness when a heat source is applied briefly to the one side; or it can be calculated from $\lambda$ (and $\rho C_p$) via equation (12.3).

---

### Example 12.3

A material has a thermal conductivity $\lambda = 0.3$ W/m.K, a density $\rho = 1200$ kg/m³ and a specific heat $C_p = 1400$ J/kg.K. What is its thermal diffusivity?

*Answer.* The thermal diffusivity is $a = \lambda/\rho C_p = 1.79 \times 10^{-7}$ m²/s.

---

## 12.3 The big picture: thermal property charts

All materials have a thermal expansion coefficient $\alpha$, a specific heat $C_p$ and a thermal conductivity $\lambda$ (and hence a thermal diffusivitiy, $a$). Three charts give an overview of these thermal properties and their relationship to strength. The charts are particularly helpful in choosing materials for applications that use these properties, or where failure may be a consequence of poor thermal design. Certain thermal properties, however, are shown only by a few special materials: shape-memory behaviour is an example. There is no point in making charts for these because so few materials have them, but they are still of engineering interest. Examples that use them are described near the end of Section 12.6.

*Thermal expansion, α, and thermal conductivity λ* Figure 12.4 maps thermal expansion, $\alpha$, and thermal conductivity, $\lambda$. Metals and technical ceramics lie toward the lower right: they have high conductivities and modest expansion coefficients. Polymers and elastomers lie at the upper left: their conductivities are 100 times less and their expansion coefficients 10 times greater than those of metals. The chart shows contours of $\lambda/\alpha$, a quantity important in designing against thermal distortion—a subject we return to in Section 12.6. An extra material, Invar (a nickel alloy), has been added to the chart because of its uniquely low expansion coefficient at and near room temperature, a consequence of a trade-off between normal expansion and a contraction associated with a magnetic transformation.

*Thermal conductivity, λ, and thermal diffusivity, a* Figure 12.5 shows the room temperature values of thermal conductivity, $\lambda$, and thermal diffusivity, $a$, with contours

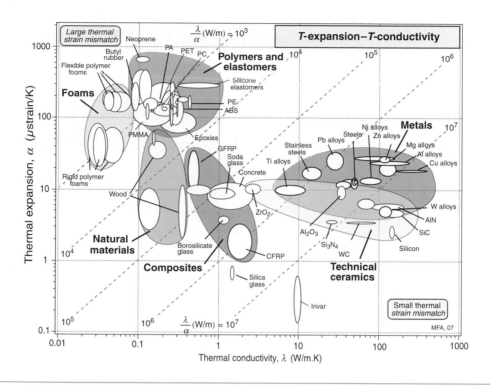

**Figure 12.4** The linear expansion coefficient, $\alpha$, plotted against the thermal conductivity, $\lambda$. The contours show the thermal distortion parameter $\lambda/\alpha$.

of volumetric heat capacity $\rho C_p$, equal to the ratio of the two, $\lambda/\alpha$ (from equation (12.3)). The data span almost five decades in $\lambda$ and $a$. Solid materials are strung out along the line

$$\rho C_p \approx 3 \times 10^6 \text{ J/m}^3 \cdot \text{K} \tag{12.4}$$

meaning that the heat capacity per unit volume, $\rho C_p$, is almost constant for all solids, something to remember for later. As a general rule, then,

$$\lambda \approx 3 \times 10^6 a \tag{12.5}$$

($\lambda$ in W/m.K and $a$ in m$^2$/s). Some materials deviate from this rule: they have lower-than-average volumetric heat capacity. The largest deviations are shown by porous solids: foams, low-density firebrick, woods and the like. Because of their low density they contain fewer atoms per unit volume and, averaged over the volume of the structure, $\rho C_p$ is low. The result is that, although foams have low *conductivities* (and are widely used for insulation because of this), their *thermal diffusivities* are not necessarily low. This means that they don't transmit much heat, but they do change temperature quickly.

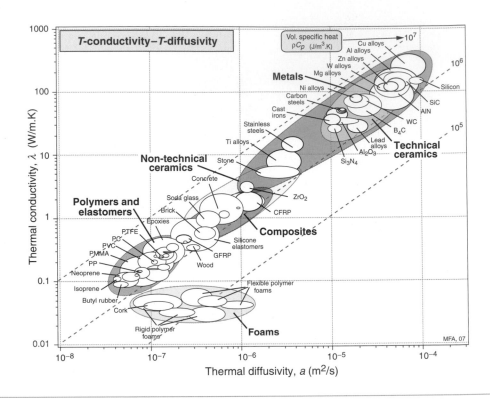

**Figure 12.5**  The thermal conductivity, $\lambda$, plotted against the thermal diffusivity, $a = \lambda/\rho C_p$. The contours show the specific heat per unit volume $\rho C_p$.

---

### Example 12.4

A locally available building material in a third-world country has a density that you find by weighing a cube of known dimensions to be $\rho = 2300$ kg/m³. To judge the thermal behaviour of the shelter you are proposing to build you wish to know its heat capacity $C_p$. What would you estimate it to be?

*Answer.*  The chart of Figure 12.5 shows that, for solids, the quantity $\rho C_p \approx 3 \times 10^6$ J/m³.K so the heat capacity of the material is $C_p \approx 1300$ J/kg.K.

---

***Thermal conductivity $\lambda$ and yield strength $\sigma_y$***  The final chart, Figure 12.6, shows thermal conductivity, $\lambda$ and strength, $\sigma_y$. Metals, particularly the alloys of copper, aluminum and nickel, are both strong and good conductors, a combination of properties we seek for applications such as heat exchangers—one we return to in Section 12.6.

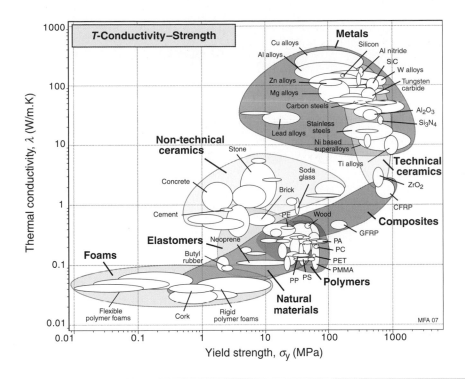

**Figure 12.6** Thermal conductivity $\lambda$ and strength $\sigma_y$.

## **12.4** Drilling down: the physics of thermal properties

*Heat capacity* We owe our understanding of heat capacity to Albert Einstein[7] and Peter Debye[8]. Heat, as already mentioned, is atoms in motion. Atoms in solids vibrate about their mean positions with an amplitude that increases with temperature. Atoms in solids can't vibrate independently of each other because they are coupled by their inter-atomic bonds; the vibrations are like standing elastic waves. Figure 12.7 shows how, along any row of atoms, there is the possibility of one longitudinal mode and two transverse modes, one in the plane of the page and one normal to it. Some of these have short wavelengths and high energy, others long wavelengths and lower energy (Figure 12.8). The shortest possible wavelength, $\lambda_1$, is just twice the atomic spacing; the other vibrations have wavelengths that are longer. In a solid with $N$ atoms there are $N$ discrete wavelengths, and each has a longitudinal mode and two transverse modes, $3N$ modes in all. Their amplitudes are such that, on average, each has energy

[7] Albert Einstein (1879–1955), Patent Officer, physicist and campaigner for peace, one of the greatest scientific minds of the 20th century; he was forced to leave Germany in 1933, moving to Princeton, NJ, where his influence on US defence policy was profound.

[8] Peter Debye (1884–1966), Dutch physicist and Nobel Prize winner, did much of his early work in Germany until in 1938, harassed by the Nazis, he moved to Cornell, NY, where he remained until his death.

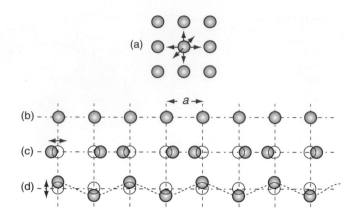

**Figure 12.7** (a) An atom vibrating in the 'cage' of atoms that surround it with three degrees of freedom. (b) A row of atoms at rest. (c) A longitudinal wave. (d) One of two transverse waves (in the other the atoms oscillate normal to the page).

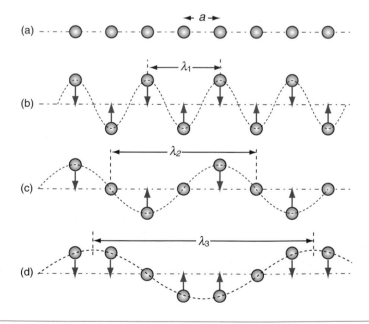

**Figure 12.8** Thermal energy involves atom vibrations. There is one longitudinal mode of vibration and two transverse modes, one of which is shown here. The shortest meaningful wavelength, $\lambda_1 = 2a$, is shown in (b).

$k_B T$, where $k_B$ is Boltzmann's constant[9], $1.38 \times 10^{-23}$ J/K. If the volume occupied by an atom is $\Omega$, then the number of atoms per unit volume is $N = 1/\Omega$ and the total thermal energy per unit volume in the material is $3k_B T/\Omega$. The heat capacity per unit volume, $\rho C_p$, is the *change* in this energy per Kelvin change in temperature, giving:

$$\rho C_p = \frac{3k_B}{\Omega} \ \text{J/m}^3\text{K} \tag{12.6}$$

Atomic volumes do not vary much; all lie within a factor of 3 of the value $2 \times 10^{-29}/\text{m}^3$, giving a volumetric heat capacity

$$\rho C_p \approx 2 \times 10^6 \ \text{J/m}^3\text{K} \tag{12.7}$$

The chart of Figure 12.5 showed that this is indeed the case.

*Thermal expansion* If a solid expands when heated (and almost all do) it must be because the atoms are moving farther apart. Figure 12.9 shows how this happens. It looks very like an earlier one, Figure 4.21, but there is a subtle difference: the force–spacing curve, a straight line in the earlier figure, is not in fact quite straight; the bonds become stiffer when the atoms are pushed together and less stiff when they are pulled apart. Atoms vibrating in the way described earlier oscillate about a mean spacing that increases with the amplitude of oscillation, and thus with increasing temperatures. So thermal expansion is a non-linear effect; if the bonds between atoms were linear springs, there would be no expansion.

The stiffer the springs, the steeper is the force–displacement curve and the narrower is the energy well in which the atom sits, giving less scope for expansion. Thus, materials with high modulus (stiff springs) have low expansion coefficient, those with low modulus (soft springs) have high expansion—indeed, to a good approximation

$$\alpha \approx \frac{1.6 \times 10^{-3}}{E} \tag{12.8}$$

($E$ in GPa, $\alpha$ in K$^{-1}$). It is an empirical fact that all crystalline solids expand by about the same amount on heating from absolute zero to their melting point: it is about 2%. The expansion coefficient is the expansion per degree Kelvin, meaning that

$$\alpha \approx \frac{0.02}{T_m} \tag{12.9}$$

For example, tungsten, with a melting point of around 3330 °C (3600 K) has $\alpha = 5 \times 10^{-6}$/K, while lead, with a melting point of about 330 °C (600 K, six times lower) expands six times more ($\alpha = 30 \times 10^{-6}$/K). Equation (12.9) (with $T_m$ in Kelvin, of course) is a remarkably good approximation for the expansion coefficient. Note in passing that equations (12.8) and (12.9)

---

[9] Ludwig Boltzmann (1844–1906), born and worked in Vienna at a time when that city was the intellectual centre of Europe; his childhood interest in butterflies evolved into a wider interest in science, culminating in his seminal contributions to statistical mechanics.

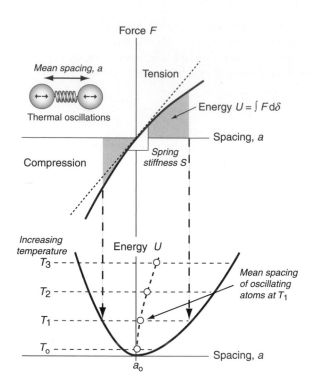

**Figure 12.9** Thermal expansion results from the oscillation of atoms in an unsymmetrical energy well.

indicate that $E$ is expected to scale with $T_m$—try exploring how well these approximations hold using the CES Elements database.

***Thermal conductivity*** Heat is transmitted through solids in three ways: by thermal vibrations; by the movement of free electrons in metals; and, if they are transparent, by radiation. Transmission by thermal vibrations involves the propagation of *elastic waves*. When a solid is heated the heat enters as elastic wave packets or *phonons*. The phonons travel through the material and, like any elastic wave, they move with the speed of sound, $c_0 \left( c_0 = \sqrt{E/\rho} \right)$. If this is so, why does heat not diffuse at the same speed? It is because phonons travel only a short distance before they are scattered by the slightest irregularity in the lattice of atoms through which they move, even by other phonons. On average they travel a distance called the *mean free path*, $\ell_m$, before bouncing off something, and this path is short: typically less than 0.01 μm ($10^{-8}$ m).

We calculate the conductivity by using a *net flux model*, much as you would calculate the rate at which cars accumulate in a car park by counting the rate at which they enter and subtracting the rate at which they leave. Phonon conduction can be understood in a similar

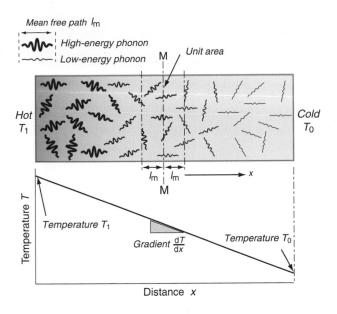

**Figure 12.10**  The transmission of heat by the motion of phonons.

way, as suggested by Figure 12.10. Here a rod with unit cross-section carries a uniform temperature gradient $dT/dx$ between its ends. Phonons within it have three degrees of freedom of motion (they can travel in the $\pm x$, the $\pm y$ and the $\pm z$ directions). Focus on the midplane M–M. On average, one-sixth of the phonons are moving in the $+x$ direction; those within a distance $\ell_m$ of the plane will cross it from left to right before they are scattered, carrying with them an energy $\rho C_p(T + \Delta T)$ where $T$ is the temperature at the plane M–M and $\Delta T = (dT/dx)\ell_m$. Another one-sixth of the phonons move in the $-x$ direction and cross M–M from right to left, carrying an energy $\rho C_p(T - \Delta T)$. Thus, the energy flux $q$ J/m².s across unit area of M–M per second is

$$q = -\frac{1}{6}\rho C_p c_o \left( T + \frac{dT}{dx}\ell_m \right) + \frac{1}{6}\rho C_p c_o \left( T - \frac{dT}{dx}\ell_m \right)$$

$$= -\frac{1}{3}\rho C_p \ell_m c_o \frac{dT}{dx}$$

Comparing this with the definition of thermal conductivity (equation (12.2)), we find the conductivity to be

$$\lambda = \frac{1}{3}\rho C_p \ell_m c_o \qquad (12.10)$$

Sound waves, which, like phonons, are elastic waves, travel through the same bar without much scattering. Why do phonons scatter so readily? It's because waves are scattered most by obstacles with a size comparable with their wavelengths—that is how diffraction gratings work. Audible sound waves have wavelengths of metres, not microns; they are scattered by buildings (which is why you can't always tell where a police siren is coming from) but not by the atom-scale obstacles that scatter phonons, with wavelengths starting at two atomic spacings (Figure 12.8).

Elastic waves contribute little to the conductivity of pure metals such as copper or aluminum because the heat is carried more rapidly by the free electrons. Equation (12.10) still applies but now $C_p$, $c_o$ and $\ell_m$ become the thermal capacity, the velocity and the mean free path of the electrons. Free electrons also conduct electricity, with the result that metals with high electrical conductivity also have high thermal conductivity (the Wiedemann–Franz law), as will be illustrated in Chapter 14.

---

### Example 12.5

A sample of silicon has modulus $E = 1500$ GPa, density $\rho = 2300$ kg/m$^3$ and specific heat capacity $C_p = 680$ J/kg.K. If its thermal conductivity is $\lambda = 145$ W/m.K what is the mean free path $\ell_m$ of its phonons?

*Answer.* The thermal conductivity is $\lambda = \frac{1}{3}\rho C_p \ell_m c_o$ with $c_o = \sqrt{E/\rho} = 8.08 \times 10^3$ m/s. Thus

$$\ell_m = \frac{3\lambda}{\rho C_p \sqrt{E/\rho}} = 0.034 \ \mu m$$

---

## 12.5 Manipulating thermal properties

*Thermal expansion* Expansion, as already mentioned, can generate thermal stress and cause distortion, undesirable in precise mechanisms and instruments. The expansion coefficient $\alpha$, like modulus or melting point, depends on the stiffness and strength of atomic bonds. There's not much you can do about these, so the expansion coefficient, normally, is beyond our control. Some few materials have exceptionally low expansion—borosilicate glass (Pyrex), silica and carbon fibres, for example—but they are hard to use in engineering structures. There is one exception: the family of alloys called Invars with very low expansion. They achieve this by the trick of canceling the thermal expansion (which is, in fact, happening) with a contraction caused by the gradual loss of magnetism as the material is heated.

*Thermal conductivity and heat capacity* Equation (12.10) describes thermal conductivity. Which of the terms it contains can be manipulated and which cannot? The volumetric heat capacity $\rho C_p$, as we have seen, is almost the same for all solid materials. The sound velocity $c_o \approx \sqrt{E/\rho}$ depends on two properties that are not easily changed. That leaves the mean free path, $\ell_m$, of the phonons or, in metals, of the electrons. Introducing *scattering centres* such as impurity atoms or finely dispersed particles achieves this. Thus, the conductivity of pure iron is 80 W/m.K, while that of stainless steel—iron with up to 30% of nickel and chromium—is

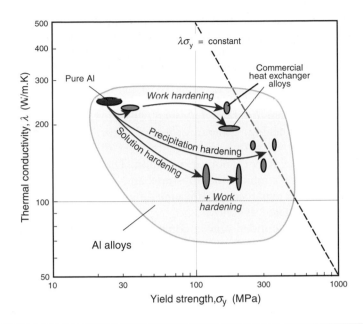

**Figure 12.11**  Thermal conductivity and strength for aluminum alloys. Good materials for heat exchangers have high values of $\lambda\sigma_y$, indicated by the dashed line.

only 18 W/m.K. Certain materials can exist in a glassy state, and here the disordered nature of the glass makes every molecule a scattering centre and the mean free path is reduced to a couple of atom spacings; the conductivity is correspondingly low.

Chapter 6 showed how strength is increased by defects such as solute atoms, precipitates and work hardening. Some of these reduce the thermal conductivity. Figure 12.11 shows a range of aluminum alloys hardened by each of the three main mechanisms: solid solution, work hardening and precipitation hardening. Work hardening strengthens significantly without changing the conductivity much, while solid solution and precipitation hardening introduce more scattering centres, giving a drop in conductivity. The chart guides material selection for heat exchangers, for which the performance index in the figure is derived in a case study later in the chapter.

We are not, of course, restricted to fully dense solids, making density $\rho$ a possible variable after all. In fact, the best thermal insulators are porous materials—low-density foams, insulating fabric, cork, insulating wool etc.—that take advantage of the low conductivity of still air trapped in the pores ($\lambda$ for air is 0.02 W/m.K).

## 12.6 Design and manufacture: using thermal properties

*Managing thermal stress*    Most structures, small or large, are made of two or more materials that are clamped, welded or otherwise bonded together. This causes problems when temperatures change. Railway track will bend and buckle in exceptionally hot weather (steel, high $\alpha$, clamped to mother earth with a much lower $\alpha$). Overhead transmission lines sag for the same reason—a problem for high-speed electric trains. Bearings seize, doors jam. Thermal distortion is a particular problem in equipment designed for precise measurement or registration like that used to make masks for high-performance computer chips, causing loss of accuracy when temperatures change.

All of these derive from *differential thermal expansion*, which, if constrained (clamped in a way that stops it happening) generates *thermal stress*. As an example, many technologies involve coating materials with a thin surface layer of a different material to impart wear resistance, or resistance to corrosion or oxidation. The deposition process often operates at a high temperature. On cooling, the substrate and the surface layer contract by different amounts because their expansion coefficients differ and this puts the layer under stress. This residual stress is calculated as follows.

Think of a thin film bonded onto a component that is much thicker than the film, as in Figure 12.12(a). First imagine that the layer is detached, as in Figure 12.12(b). A temperature drop of $\Delta T$ causes the layer to change in length by $\delta L_1 = \alpha_1 L_0 \Delta T$. Meanwhile, the substrate to which it was previously bonded contracts by $\delta L_2 = \alpha_2 L_0 \Delta T$. If $\alpha_1 > \alpha_2$ the surface layer shrinks more than the substrate. If we want to stick the film back on the much-more-massive substrate, covering the same surface as before, we must stretch it by the strain

$$\varepsilon = \frac{\delta L_1 - \delta L_2}{L_0} = \Delta T(\alpha_1 - \alpha_2)$$

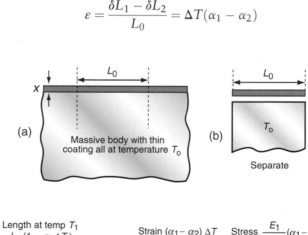

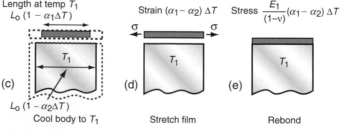

---

**Figure 12.12**  Thermal stresses in thin films arise on cooling or heating when their expansion coefficients differ. Here that of the film is $\alpha_1$ and that of the substrate, a massive body, is $\alpha_2$.

This requires a stress in the film of

$$\sigma_1 = \frac{E_1}{(1-v)}(\alpha_1 - \alpha_2)\Delta T \qquad (12.11)$$

where the Poisson's ratio term enters due to the biaxial stress state in the film. The stress can be large enough to crack the surface film. The pattern of cracks seen on glazed tiles arises in this way.

So if you join dissimilar materials you must expect thermal stress when they are heated or cooled. The way to avoid it is to avoid material combinations with very different expansion coefficients. The $\alpha$–$\lambda$ chart of Figure 12.4 has expansion $\alpha$ as one of its axes: benign choices are those that lie close together on this axis, dangerous ones are those that lie far apart.

But avoiding materials with $\alpha$ mismatch is not always possible. Think of joining glass to metal—Pyrex to stainless steel, say—a common combination in high vacuum equipment. Their expansion coefficients can be read from the $\alpha$–$\lambda$ chart of Figure 12.4: for Pyrex (borosilicate glass), $\alpha = 4 \times 10^{-6}$/K; for stainless steel, $\alpha = 20 \times 10^{-6}$/K: a big mismatch. Vacuum equipment has to be 'baked out' to desorb gases and moisture, requiring that it be heated to about 150 °C, enough for the mismatch to crack the Pyrex. The answer is to grade the joint with one or more materials with expansion that lies between the two: first join the Pyrex to a glass of slightly higher expansion and the stainless to a metal of lower expansion—the chart suggests soda glass for the first and a nickel alloy for the second—and then diffusion-bond these to each other, as in Figure 12.13. The graded joint spreads the mismatch, lowering the stress and the risk of damage.

There is an alternative, although it is not always practical. It is to put a compliant layer—rubber, for instance—in the joint; it is the way windows are mounted in cars. The difference in expansion is taken up by distortion in the rubber, transmitting very little of the mismatch to the glass or surrounding steel.

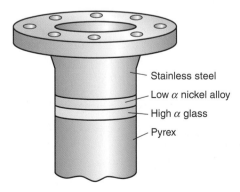

Stainless steel
Low $\alpha$ nickel alloy
High $\alpha$ glass
Pyrex

**Figure 12.13** A graded joint. The high expansion stainless steel is bonded to a nickel alloy of lower expansion, the low expansion Pyrex (a borosilicate glass) is bonded to a glass with higher expansion, and the set is assembled to give a stepwise graded expansion joint.

***Thermal sensing and actuation***   Thermal expansion can also be used to *sense* (to measure temperature change) and to *actuate* (to open or close valves or electrical circuits, for instance). The direct thermal displacement $\delta = \alpha L_o \Delta T$ is small; it would help to have a way to magnify it. Figure 12.14 shows one way to do this—the *bi-material strip*. Two materials, now deliberately chosen to have different expansion coefficients $\alpha_1$ and $\alpha_2$, are bonded together in the form of a bi-material strip, of thickness $2t$, as in Figure 12.14(a). When the temperature changes by $\Delta T = (T_1 - T_o)$, one expands more than the other, causing the strip to bend. The mid-thickness planes of each strip (dotted in Figure 12.14(b)) differ in length by $L_o(\alpha_2 - \alpha_1)\Delta T$. Simple geometry then shows that

$$\frac{R + t}{R} = 1 + (\alpha_2 - \alpha_1)\,\Delta T$$

from which

$$R = \frac{t}{(\alpha_2 - \alpha_1)\,\Delta T} \tag{12.12}$$

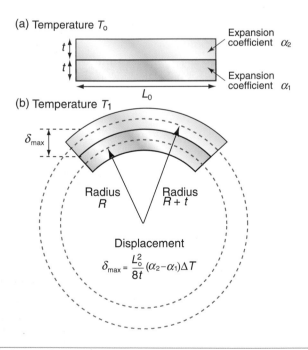

(a) Temperature $T_o$

(b) Temperature $T_1$

**Figure 12.14**   The thermal response of a bi-material strip when the expansion coefficients of the two materials differ. The configuration magnifies the thermal strain.

The resulting upward displacement of the centre of the bi-material strip (assumed to be thin) is

$$\delta_{max} = \frac{L_o^2}{8t}(\alpha_2 - \alpha_1)\Delta T \qquad (12.13)$$

A large aspect ratio $L_o/t$ produces a large displacement, and one that is linear in temperature.

*Managing thermal distortion*   Thermal expansion can lead to distortion in a single material too. A *temperature gradient* across a component will lead to differential thermal expansion. If the component is free to change shape, different amounts of expansion in different locations will lead to distortion. For example, an originally flat plate with a temperature gradient imposed across it will bend into a curve. Thermal distortion due to thermal gradients is a problem in precision equipment in which heat is generated by the electronics, motors, actuators or sensors that are necessary for its operation: different parts of the equipment function at different temperatures and so expand by different amounts. The answer is to make the equipment from materials with low expansion $\alpha$ (because this minimises the expansion difference in a given temperature gradient) and with high conductivity $\lambda$ (because this spreads the heat further, reducing the steepness of the gradient). It can be shown that the best choice is that of materials with high values of the ratio $\lambda/\alpha$, just as our argument suggests. The $\alpha - \lambda$ chart of Figure 12.4 guides selection for this too. The diagonal contours show the quantity $\lambda/\alpha$. Materials with the most attractive values lie towards the bottom right. Copper, aluminum, silicon, aluminum nitride and tungsten are good—they distort the least; stainless steel and titanium are much less good.

*Thermal shock resistance*   Thermal gradients in single materials also generate thermal stress, particularly when abrupt temperature changes are applied at the surface of a component—such as in quenching a hot component in water. It takes some time for heat to conduct out of the interior (we will look at this in more detail shortly), but right after quenching the full temperature difference will be retained between surface and interior. Suppose a chunk of material is suddenly cooled by $\Delta T$ (by dropping it hot into a bath of cold water, for example). The surface, almost immediately, adjusts to the temperature of the bath. It wants to contract by a strain $\varepsilon = \alpha \Delta T$. Since the surface is stuck to the interior, it is constrained, and thermal stresses appear in it just as they did in the thin film of Figure 12.12. If the bulk of the interior doesn't strain at all (as will be the case when only a thin layer at the surface has been cooled), the stress induced is

$$\sigma = \frac{E}{(1-\nu)} \alpha \Delta T \qquad (12.14)$$

Superficially this looks the same as equation (12.11), but here there is one material with a temperature difference of $\Delta T$ between surface and interior; in the previous case the whole component was cooled uniformly by an amount $\Delta T$, and the stress comes from the difference in expansion coefficients. In ductile materials, the stress can be sufficient to yield the interior, giving a final misfit in the strain and a state of *residual stress*, even after the whole component has cooled down. In brittle materials, the stress induced in a quenched material can cause fracture. The ability of a material to resist this, its *thermal shock resistance* $\Delta T_s$ (units: K or °C), is the maximum sudden change of temperature to which such a material can be subjected without damage. Safety critical components that must resist thermal shocks may be subjected to a proof test during manufacture. The components are heated and lowered rapidly into cold water; if they break, then they would not have been safe to put into service—better to find this out in the safety of the factory.

---

### Example 12.6

The glass proposed for the doors of a stove is known to be susceptible to fracture at a tensile stress around 30 MPa. Find a value for the corresponding thermal shock resistance for the glass, and comment on the suitability of the glass for this application. For the glass, $E = 62$ GPa, $\nu = 0.2$ and $\alpha = 4 \times 10^{-6}$ K$^{-1}$.

*Answer.* From equation (12.14)

$$\Delta T = \frac{\sigma(1 - \nu)}{E\alpha} = \frac{30 \times (1 - 0.2)}{62 \times 10^3 \times 4 \times 10^{-6}} = 97\,°\mathrm{C}$$

Oven doors can be heated to temperatures of 200 °C or more, and could inadvertently be splashed with cold water. The thermal shock resistance of this glass does not appear to be sufficient. Glasses with a lower expansion coefficient are not available, but higher strength might be achievable. In practice there are heat treatments available to strengthen glass, exploiting residual stresses generated by controlled cooling during processing: the surface is put into residual compression, balanced by internal tension.

---

***Thermal problems in manufacturing*** Manufacturing with metals spans a diverse range of primary shaping processes, such as casting and forging, followed by secondary heat treatment and surface engineering, then assembly, often by welding (see Chapter 2). The goals are to achieve the desired shapes and to assemble the parts, while simultaneously developing the internal microstructure on which the material properties depend. Polymers and ceramics, too, have their own characteristic processes and microstructural objectives. It's a complicated business, deserving its own discussion, which we'll address in Chapter 19. What virtually all processes have in common, though, is some aspect of *temperature history*, with heating and cooling rates varying with time and position in the component. Apart from controlling the evolution of microstructure, the thermal response of components during manufacturing influences other aspects of producing acceptable quality, such as distortion and residual stress (a particular problem in heat treatment and welding, often linked to later failure in service). Simply knowing how long something will take to heat up or cool down is useful in running a factory. In short, understanding the thermal behaviour of products during processing is an important part of materials engineering and design.

***Transient heat flow problems*** The thermal conductivity was defined via Figure 12.3 and equation (12.2). Here the heat flow was at *steady-state*, with uniform external temperatures and a temperature gradient that doesn't change with time. Sometimes this scenario is important in design and manufacture: for example, in the operation of a heat exchanger, or minimising the heat loss from a furnace. Often, however, the temperature history is more complex, with the temperature everywhere changing continuously—imagine quenching a large red hot steel plate into water, for example (Figure 12.15). We call this a *transient heat flow* problem: cooling progresses by the process of heat diffusion, discussed earlier, and while it is happening the temperature everywhere keeps evolving. In a large plate, the heat conduction is one-dimensional in the through-thickness direction—heat always seeks out the shortest path between the interior and the outside world. The simplification of 1D heat flow is a good approximation in many situations. Even then, a full mathematical treatment can get complicated, but fortunately we can extract a simple and useful rule-of-thumb to guide our understanding.

We start from the differential equation governing the temperature $T$ in terms of position $x$ and time $t$, when heat is being conducted in one dimension (the $x$-direction)

$$\frac{\partial^2 T}{\partial x^2} = \frac{1}{a}\frac{\partial T}{\partial t} \qquad (12.15)$$

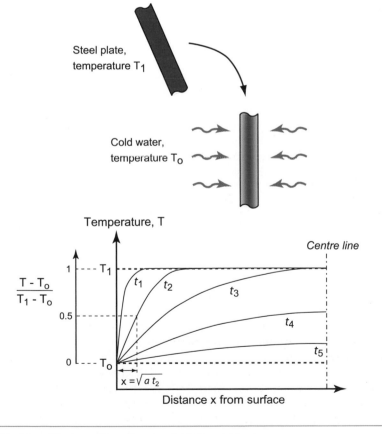

Figure 12.15 The diffusion of heat into an initially cold body.

(where the curly '$\partial$' signifies that this is a partial differential equation, with temperature varying with two parameters, $x$ and $t$). This equation can be derived directly from Fourier's Law for steady-state conduction, equation (12.2), applied to the instantaneous heat flows through a very small volume of material, of length $\delta x$. Rather than proving this in full, here is a simple thought experiment. Figure 12.15 shows that the thermal gradient, $\partial T/\partial x$, changes with position $x$. So for a small volume of length $\delta x$, it would have a slightly different gradient on one side compared to the other. Equation (12.2) tells us that the amount of heat flowing (per unit area) is proportional to the thermal gradient. So in a given time interval, $\delta t$, more heat enters the little volume than leaves it, and it will get hotter—there will be a small change in temperature $\delta T$. So the variation in the amount of heat being transferred from point to point governs the rate of change of temperature at that location—which is essentially what equation (12.15) says. Notice that the only property in the equation is the thermal diffusivity, $a$, that we encountered earlier:

$$a = \frac{\lambda}{\rho C_p} \tag{12.16}$$

with units of $m^2/s$. The bigger the conductivity, $\lambda$, the higher the diffusivity, $a$, but why is $\rho C_p$ on the bottom? If this quantity (the heat capacity per unit volume) is large, more heat has to diffuse in or out of unit volume to change the temperature. The more heat that has to diffuse, the longer it takes to do so, so that the diffusivity $a$ is reduced. So the thermal diffusivity captures the effect of both conductivity and heat capacity in transient thermal problems—it is the key property. We'll show why with an example.

In the early stages of quenching the plate in Figure 12.15, heat has only had time to diffuse from a short depth out to the surface—the two sides cool independently, and the cooling of the plate is the same is if it were semi-infinite in the $x$-direction (i.e. times well below $t_3$ in Figure 12.15). The simplest mathematical solution to equation (12.15) assumes that the surface is immediately cooled to the temperature of the bath, $T_o$, and the temperature is then given by

$$\frac{T - T_o}{T_1 - T_o} = \mathrm{erf}\left(\frac{x}{2\sqrt{a\,t}}\right) \tag{12.17}$$

The function erf $(X)$ is called the *error function*. It has exactly the shape of the temperature curves for short times $t_1$ and $t_2$ in Figure 12.15, rising from 0 to a value of 1 as the argument $X$ becomes large—as indicated by the dimensionless temperature axis to the left of the figure. The error function crops up in statistics as well as heat flow, and its values are tabulated in handbooks, or it is available as a standard function in spreadsheets and computer programs. The steepest part of the erf $(X)$ curve is close to being linear and equal to its own argument, $X$, so we can use the approximation erf $(X) \approx X$ (provided $X$ is less than about 0.7). The temperature below the surface is then

$$\frac{T - T_o}{T_1 - T_o} = \left(\frac{x}{2\sqrt{a\,t}}\right) \tag{12.18}$$

So the temperature at a given depth $x$ falls gradually with time. It would be useful to have a rough idea of how far heat has diffused in a given time, or alternatively, how long the component is taking to cool down, at a given depth. A suitable question might be to ask, 'How long does it take for the temperature to fall halfway between the initial and final temperatures?'—that is, to a dimensionless temperature of 0.5, as shown in Figure 12.15. In this case, equation (12.18) becomes

$$\frac{T - T_o}{T_1 - T_o} = 0.5 = \left(\frac{x}{2\sqrt{a\,t}}\right) \quad \text{so } x = \sqrt{a\,t} \tag{12.19}$$

This is a particularly useful little 'rule-of-thumb', and although derived for quenching a plate, it turns out that this formula gives us a characteristic length scale of heat diffusion in all transient problems. It is approximate, but the equation $x = \sqrt{a\,t}$ is perfectly adequate for first estimates of heating and cooling times (or distances). If the component is a more three-dimensional shape, only a modest numerical correction is needed to account for differences in the geometry of the problem, so long as we pick the shortest dimension of the part, which will dominate the heat flow.

Consider now the later stages of cooling (times $t_4$ to $t_5$ in the figure). At the mid-thickness of the plate (of thickness $l$), the solution to equation (12.15) is

$$\frac{T - T_o}{T_1 - T_o} = \frac{4}{\pi} \exp\left(-\frac{\pi^2 a\,t}{l^2}\right) \tag{12.20}$$

Let's see how well our rule-of-thumb estimates the cooling time.

### Example 12.7

A large plate of steel of thickness 100 mm is quenched from a uniform temperature of 920 °C, into a bath of cold water at 20 °C, and agitated to maintain this temperature at the surface. After what time has the centre of the plate cooled to a temperature of 470 °C? Compare this with a rough estimate for the cooling time, using $x = \sqrt{a\,t}$, assuming that on average heat has to diffuse a characteristic distance of between one quarter and one half of the plate thickness. The thermal diffusivity of the steel is $a = 9 \times 10^{-6}$ m$^2$/s.

*Answer.* The dimensionless temperature ratio is

$$\frac{T - T_o}{T_1 - T_o} = \frac{920 - 20}{470 - 20} = 0.5$$

Substituting into equation (12.20) and solving for time $t$:

$$0.5 = \frac{4}{\pi} \exp\left(-\frac{\pi^2 \times 9 \times 10^{-6}t}{(100 \times 10^{-3})^2}\right) \quad \text{so } t = 105 \text{ s.}$$

Using the rule-of-thumb, $x = \sqrt{a\,t}$, with distances of $x \approx (l/4)$ and $x \approx (l/2)$:

$$t \approx (l/4)^2/a \approx (0.025)^2/\,9 \times 10^{-6} = 69 \text{ s and } t \approx (l/2)^2/a \approx (0.05)^2\,/\,9 \times 10^{-6} = 278 \text{ s.}$$

The rule-of-thumb estimates of the cooling time span the exact result, and give the right timescale within a factor of 2 or so.

*Conduction with strength: heat exchangers*   A heat exchanger transfers heat from one fluid to another while keeping the fluids physically separated. You find them in central heating systems, power-generating plants, and cooling systems in car and marine engines. Heat is transferred most efficiently if the material separating the fluids is kept thin and if the surface area for heat conduction is large—a good solution is a tubular coil. At least one of the fluids is under pressure, so the tubes must be strong enough to withstand it. Here there is a conflict in the design: thinner walls conduct heat faster, thicker walls are stronger. How is this trade-off optimised by choosing the tube material?

Consider the idealised heat exchanger made of thin-walled tubes shown in Figure 12.16. The tubes have a given radius $R$, a wall thickness $t$ (which may be varied) and must support an internal pressure of $p$, with a temperature difference of $\Delta T$ between the fluids. The objective is to maximise the heat transferred per unit surface area of tube wall. The heat flow rate per unit area is given by

$$q = -\lambda \frac{dT}{dx} = \lambda \frac{\Delta T}{t} \tag{12.21}$$

The stress due to the internal pressure in a cylindrical thin-walled tube (Chapter 10, equation (10.7)) is

$$\sigma = p \frac{R}{t}$$

This stress must not exceed the yield strength $\sigma_y$ of the material of which it is made, so the minimum thickness $t$ needed to support the pressure $p$ is

$$t = p \frac{R}{\sigma_y} \tag{12.22}$$

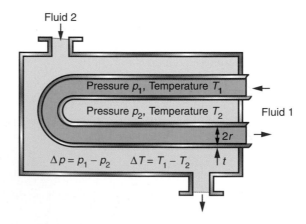

**Figure 12.16**   A heat exchanger. It must transmit heat yet withstand the pressure $p$.

Combining equations (12.21) and (12.22) to eliminate the free variable $t$ gives

$$q = \lambda \sigma_y \frac{\Delta T}{pR} \tag{12.23}$$

The best materials are therefore those with the highest values of the index $\lambda \sigma_y$. They are the ones at the top right of the $\lambda - \sigma_y$ chart of Figure 12.6—metals and certain ceramics. Ceramics are brittle and difficult to shape into thin-walled tubes, leaving metals as the only viable choice. The chart shows that the best of these are the alloys of copper and of aluminum. Both offer good corrosion resistance, but aluminum wins where cost and weight are to be kept low—in automotive radiators, for example. Commercial aluminum alloys used for this application are identified in Figure 12.11. They do not offer quite the highest values of $\lambda \sigma_y$ among the aluminum alloys—other factors are at work here. One is the continuous exposure to temperature during service, which can soften the precipitation-hardened alloys; another is that an efficient manufacturing route for thin-walled multi-channel tubing is extrusion, for which high ductility is needed.

Many aluminum alloys corrode in salt water. Marine heat exchangers use copper alloys even though they are heavier and more expensive because of their resistance to corrosion in seawater.

*Insulation: thermal walls*   Before 1940 few people had central heating or air-conditioning. Now most have both, and saunas and freezers and lots more. All consume power, and consuming power is both expensive and hard on the environment (Chapter 20). To save on both, it pays to insulate. Consider, as an example, the oven of Figure 12.17. The power lost per m² by conduction through its wall is

$$q = \lambda \frac{(T_i - T_o)}{t} \; \text{W/m}^2 \tag{12.24}$$

where $t$ is the thickness of the insulation, $T_o$ is the temperature of the outside and $T_i$ that of the inside. For a given wall thickness $t$ the power consumption is minimised by choosing materials

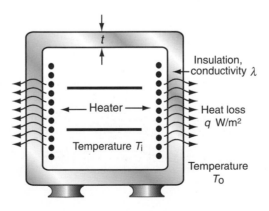

**Figure 12.17**   A heated chamber with heat loss by conduction through the insulation.

with the lowest possible λ. The chart of Figure 12.6 shows that these are foams, largely because they are about 95% gas and only 5% solid. Polymer foams are usable up to about 170 °C; above this temperature even the best of them lose strength and deteriorate. When insulating against heat, then, there is an additional constraint that must be met: that the material has a maximum service temperature $T_{max}$ that lies above the planned temperature of operation. Metal foams are usable to higher temperatures but they are expensive and—being metal—they insulate less well than polymers. For high temperatures ceramic foams are the answer. Low-density firebrick is relatively cheap, can operate up to 1000 °C and is a good thermal insulator. It is used for kiln and furnace linings.

*Using heat capacity: storage heaters*   The demand for electricity is greater during the day than during the night. It is not economic for electricity companies to reduce output, so they seek instead to smooth demand by charging less for off-peak electricity. Storage heaters exploit this, using cheap electricity to heat a large block of material from which heat is later extracted by blowing air over it.

Storage heaters also fill other more technical roles. When designing supersonic aircraft or re-entry vehicles, it is necessary to simulate the conditions they encounter—hypersonic air flow at temperatures up to 1000 °C. The simulation is done in wind tunnels in which the air stream is heated rapidly by passing it over a previously heated thermal mass before it hits the test vehicle, as in Figure 12.18. This requires a large mass of heat-storage material, and to keep the cost down the material must be cheap. The objective, both for the home heater and for the wind tunnel, is to store as much heat per unit cost as possible.

The energy stored in a mass $m$ of material with a heat capacity $C_p$ when heated through a temperature interval $\Delta T$ is

$$Q = mC_p\Delta T \tag{12.25}$$

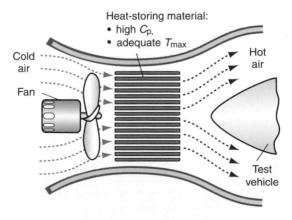

Figure 12.18   A wind tunnel with storage heating. The thermal mass is pre-heated, giving up its heat to the air blast when the fan is activated.

The cost of mass $m$ of material with a cost per kg of $C_m$ is

$$C = mC_m$$

Thus, the heat stored per unit material cost is

$$\frac{Q}{C} = \frac{C_p}{C_m} \qquad (12.26)$$

The best choice of material is simply that with the largest heat capacity, $C_p$, and lowest price, $C_m$. There is one other obvious constraint: the material must be able to tolerate the temperature to which it will be heated—say, $T_{max} > 150\,°C$ for the home heater, $T_{max} > 1000\,°C$ for the wind tunnel. It is hard to do this selection with the charts provided here, but easy to do it with the CES software: apply the limit on $T_{max}$ then use the index

$$M = \frac{C_p}{C_m}$$

to isolate the cheapest candidates. You will find that the best choice for the home heater is concrete (cast to a shape that provides some air channels). For the wind tunnel, in which temperatures are much higher, the best choice is a refractory brick with air channels for rapid heat extraction.

***Using phase change: thermal buffers***  When materials vaporise, or melt, or change their crystal structure while remaining solid (all of which are known as 'phase changes'), many things happen, and they can be useful. They change in volume. If solid, they may change in shape. And they absorb or release heat, the *latent heat L* (J/kg) of the phase change, without changing temperature.

Suppose you want to keep something at a fixed temperature without external power. Perfect insulation is not just impractical, it is impossible. But—if keeping things warm is the problem—the answer is to surround them by a liquid that solidifies at the temperature that you want to maintain. As it solidifies it releases its latent heat of fusion, holding the temperature precisely at its solidification temperature (i.e. its melting point) until all the liquid is solid—it is the principle on which some food warmers and 'back-woods' hand warmers work. The reverse is also practical: using latent heat of melting to keep things cool. For this, choose a solid that melts, absorbing latent heat, at the temperature you want to maintain—'cool bag' inserts for holiday travel work in this way. In both cases, change of phase is used as a *thermal buffer*.

***Shape-memory and super-elastic materials***  Almost all materials expand on heating and shrink on cooling. Some few have a different response, one that comes from a solid-state phase change. *Shape-memory materials* are alloys based on titanium or nickel that can, while still solid, exist in two different crystal configurations with different shapes, called *allotropes*. If you bend the material it progressively changes structure from the one allotrope to the other, allowing enormous distortion. Shape-memory alloys have a critical temperature. If below this critical temperature, the distorted material just stays in its new shape. Warm it up and, at its critical temperature, the structure reverts to the original allotrope and the sample springs back to its original shape. This has obvious application in fire alarms, sprinkler systems and other temperature-controlled safety systems, and they are used for all of these.

But what if room temperature is already above this critical temperature? Then the material does not sit waiting for heat, but springs back to its original shape as soon as it is released, just as if it were elastic. The difference is that the phase change allows distortions that are larger, by a factor of 100 or more, than straightforward elasticity without the phase change. The material is *super-elastic*. It is just what is needed for eyeglass frames that get sat on and otherwise mistreated, or springs that must allow very large displacements at constant restoring force—and these, indeed, are the main applications for these materials.

## 12.7 Summary and conclusions

All materials have:

- A specific heat, $C_p$: the amount of energy it takes to heat unit mass by unit temperature.
- A thermal expansion, $\alpha$: the change in its dimensions with change of temperature.
- A thermal conductivity $\lambda$ and thermal diffusivity $a$: the former characterising how fast heat is transmitted through the solid, the latter how long it takes for the temperature, once perturbed, to settle back to a steady pattern.
- Characteristic temperatures of its changes of phase or behaviour for a crystalline solid, its melting point $T_m$, and for non-crystalline solids, a glass transition temperature $T_g$; for design in all materials, an empirical temperature may be defined, the maximum service temperature $T_{max}$, above which, for reasons of creep or degradation, continuous use is impractical.
- Latent heats of melting and of vaporisation: the energy absorbed or released at constant temperature when the solid melts or evaporates.

These properties can be displayed as material property charts—examples of which are utilised in this chapter. The charts illustrate the great range of thermal properties and the ways they are related. To explain these we need an understanding of the underlying physics. The key concepts are that thermal energy is atomic-level vibration, that increasing the amplitude of this vibration causes expansion, and that the vibrations propagate as phonons, giving conduction. In metals the free electrons—the ones responsible for electrical conduction—also transport heat, and they do so more effectively than phonons, giving metals their high thermal conductivities. Thermal properties are manipulated by interfering with these processes: alloying to scatter phonons and electrons; foaming to dilute the solid with low-conductivity, low-specific-heat gases.

The thermal response of a component to temperature change can be a problem or an opportunity in design: thermal contraction can cause cracking or unwanted distortion; on the other hand, bi-material strips use controlled thermal distortion for actuation or sensing. It is the specific heat that makes an oven take 15 minutes to warm up, but it also stores heat in a way that can be recovered on demand. Good thermal conduction is not what you want in a coffee cup, but when it cools the engine of your car or the chip in your computer it is a big help. Thermal stresses and distortion are major issues in manufacturing too, from shaping processes to heat treatment and welding. And the transient cooling histories imposed, controlled by component geometry and the thermal diffusivity of the material, are central to understanding the evolution and control of the microstructure on which so many properties depend (Chapter 19).

This chapter has been about the thermal properties of materials. Heat, as we have said, also changes other properties, mechanical and electrical. We look more deeply into that in Chapter 13.

## **12.8** Further reading

Cottrell, A. H. (1964). *The Mechanical Properties of Matter*. London, UK: Wiley. Library of Congress Number 64−14262. (The best introduction to the physical understanding of the properties of materials, now out of print but worth searching for.)

Hollman, J. P. (1981). *Heat Transfer* (5th ed.). New York, USA: McGraw-Hill. ISBN 0-07-029618-9. (A good starting point: an elementary introduction to the principles of heat transfer.)

Tabor, D. (1969). *Gases, Liquids and Solids*. Harmondsworth, UK: Penguin Books. ISBN 14-080052-2. (A concise introduction to the links between inter-atomic forces and the bulk properties of materials.)

## **12.9** Exercises

| | |
|---|---|
| Exercise E12.1 | Define specific heat. What are its units? How would you calculate the specific heat per unit volume from the specific heat per unit mass? If you wanted to select a material for a compact heat-storing device, which of the two would you use as a criterion of choice? |
| Exercise E12.2 | What two metals would you choose from the $\lambda - a$ chart of Figure 12.5 to maximise the thermal displacement of a bi-metallic strip actuator? If the bi-metallic strip has a thickness 2 mm and an average thermal diffusivity $a$ of $5 \times 10^{-5}$ m$^2$/s, how long will it take to respond when the temperature suddenly changes? |
| Exercise E12.3 | A structural material is sought for a low-temperature device for use at $-20$ °C that requires high strength but low thermal conductivity. Use the $\lambda - \sigma_y$ chart of Figure 12.6 to suggest two promising candidates (reject ceramics on the grounds that they are too brittle for structural use in this application). |
| Exercise E12.4 | A new alloy has a density of 7380 kg/m$^3$. Its specific heat has not yet been measured. How would you make an approximate estimate of its value in the normal units of J/kg.K? What value would you then report? |
| Exercise E12.5 | The same new alloy of the last exercise has a melting point of 1540 K. Its thermal expansion coefficient has not yet been measured. How would you make an approximate estimate of its value? What value would you then report? |

Exercise E12.6    Interior wall insulation should insulate well, meaning low thermal conductivity, $\lambda$, but require as little heat as possible to warm up when the central heating system is turned on (if the wall absorbs heat the room stays cold). Use the $\lambda - a$ chart of Figure 12.5 to find the materials that do this best. (The contours will help.)

Exercise E12.7    An external wall of a house has an area of 15 m² and is of single-brick construction with a thickness 10 cm. The homeowner keeps the internal temperature at 22 °C. How much heat is lost through the wall per hour, if the outside temperature is a stable 0 °C? What percentage reduction in heat loss through the external wall would be achieved by reducing the internal temperature to 18 °C? The thermal conductivity of the brick is 0.8 W/m.K.

Exercise E12.8    The homeowner in the last exercise wishes to reduce heating costs. The wall could be thickened to a double-brick construction, with a cavity filled with polyurethane foam. The foam manufacturer quotes an effective thermal conductivity of 0.25 W/m.K for this wall. How much heat would be saved per hour, compared to the single brick, for internal temperatures of 22 °C and 18 °C?

Exercise E12.9    When a nuclear reactor is shut down quickly, the temperature at the surface of a thick stainless steel component falls from 600 °C to 300 °C in less than a second. Due to the relatively low thermal conductivity of the steel, the bulk of the component remains at the higher temperature for several seconds. The coefficient of thermal expansion of stainless steel is $1.2 \times 10^{-5}$ K$^{-1}$, Poisson's ratio $\nu = 0.33$, $E = 200$ GPa, and $\sigma_y = 585$ MPa.

(a) Find the thermal strain in the surface that is prevented from occurring by the underlying material.

(b) The induced thermal strain will be biaxial, as constraint applies in two perpendicular directions parallel to the surface. In this case the elastic stress induced is: $\varepsilon_{elastic} = (1 - \nu)\, \sigma\, /E$. Find the maximum possible elastic strain parallel to the surface, that is, if the induced stress were to reach the yield stress, and compare this with the thermal strain imposed. What do you conclude will happen to the surface layer when the shutdown occurs?

Exercise E12.10    Square porcelain tiles are to be manufactured with a uniform layer of glaze, which is thin compared to the thickness of the tile. They are fired at a temperature of 700 °C and then cooled slowly to room temperature, 20 °C. Two tile types are to be manufactured: the first is decorative, using tension in the surface to form a mosaic of cracks; the second uses compression in the surface to resist surface cracking, improving the strength. Use equation (12.11) in this chapter and the data below to select a suitable glaze (A, B or C) for each of these applications.

Young's modulus of glaze, $E = 90$ GPa, Poisson's ratio of glaze, $v = 0.2$, tensile failure stress of glaze $= 10$ MPa. Coefficients of thermal expansion:

|  | $\alpha \times 10^{-6}$ (K$^{-1}$) |
|---|---|
| Porcelain | 2.2 |
| Glaze A | 1.5 |
| Glaze B | 2.3 |
| Glaze C | 2.5 |

Exercise E12.11　You notice that the ceramic coffee mugs in the office get too hot to hold about 10 seconds after pouring in the hot coffee. The wall thickness of the cup is 2 mm.

(a) What, approximately, is the thermal diffusivity of the ceramic of the mug?

(b) Given that the volumetric specific heat of solids, $\rho C_p$, is more or less constant at $2 \times 10^6$ J/m$^3$.K, what approximately is the thermal conductivity of the cup material?

(c) If the cup were made of a metal with a thermal diffusivity of $2 \times 10^{-5}$ m$^2$/s, how long could you hold it?

Exercise E12.12　A flat heat shield is designed to protect against sudden temperature surges. Its maximum allowable thickness is 10 mm. What value of thermal diffusivity is needed to ensure that, when heat is applied to one of its surfaces, the other does not start to change significantly in temperature for at least 5 minutes? Use the $\lambda - a$ chart of Figure 12.5 to identify material classes that you might use for the heat shield.

Exercise E12.13　In the development of a new quenching facility for steel components, scale model tests were conducted on dummy samples containing a thermocouple located in the centre of the sample. The geometry of the samples was geometrically similar to the real components, but at one-third scale, and the quenching process was identical. The thermocouple readout showed that on quenching a sample from a furnace into cold water, the temperature fell by 50% of the temperature interval in a time of 15 seconds. How long do you estimate it would take to cool by the same amount in the real component? In a second trial, full-sized components were tested using aluminium dummy samples, with a suitably reduced furnace temperature. What time would you expect to find is required to cool the centre by 50% of the new temperature interval?

Exercise E12.14　A material is needed for a small, super-efficient pressurised heat exchanger. The text explained that the index for this application is $M = \lambda \sigma_y$. Plot contours of this index onto a copy of the $\lambda - \sigma_y$ chart of Figure 12.6 and hence find the two class of materials that maximises $M$.

## **12.10** Exploring design with CES

Use Level 2, Materials, throughout exercises.

Exercise E12.15    Use the 'Search' facility of CES to find materials for:
(a) Thermal insulation.
(b) Heat exchangers.

Exercise E12.16    The analysis of storage heaters formulated the design constraints for the heat-storage material, which can be in the form of a particle bed or a solid block with channels to pass the air to be heated. Use the selector to find the best materials. Here is a summary of the design requirements. List the top six candidates in ranked order.

| | |
|---|---|
| Function | • Storage heater |
| Constraints | • Maximum service temperature >150 °C |
| | • Non-flammable (a durability rating) |
| Objective | • Maximise specific heat/material cost, $C_p/C_m$ |
| Free variables | • Choice of material |

Exercise E12.17    The requirements for a material for an automobile radiator, described in the text, are summarised in the table. Use CES to find appropriate materials to make them. List the top four candidates in ranked order.

| | |
|---|---|
| Function | • Automobile heat exchanger |
| Constraints | • Elongation >10% |
| | • Maximum use temperature >180 °C |
| | • Price <5 $/kg |
| | • Durability in fresh water = very good |
| Objective | • Maximise thermal conductivity × yield strength, $\lambda\sigma_y$ |
| | • Wall thickness, $t$ |
| Free variables | • Choice of material |

Exercise E12.18    A structural material is sought for a low-temperature device for use at −20 °C that requires high tensile strength $\sigma_{ts}$ but low thermal conductivity $\lambda$. For reasons of damage tolerance the fracture toughness $K_{1c}$ must be greater than 15 MPa.m$^{1/2}$. Apply the constraint on $K_{1c}$ using a 'Limit' stage, then make a chart with $\sigma_{ts}$ on the $x$-axis and $\lambda$ on the $y$-axis, and observe which materials best meet the requirements. Which are they?

Exercise E12.19    Interior wall insulation should insulate well, meaning low thermal conductivity, $\lambda$, but require as little heat as possible to warm up to the desired room temperature when the central heating system is turned on—that means low specific heat. The thickness of the insulation is almost always limited by the cavity space between the inner and outer wall, so it is the specific heat per unit volume, not per unit mass, that is important here. To be viable the material must have enough stiffness and strength to support its own weight and be easy to install—take that to mean a modulus $E > 0.01$ GPa and a strength $\sigma_y > 0.1$ MPa. Make a selection based on this information. List the materials you find that best meet the design requirements.

**Exercise E12.20**  Here is the gist of an email one of the authors received as this chapter was being written. 'We manufacture wood-burning stoves and fireplaces that are distributed all over Europe. We want to select the best materials for our various products. The important characteristics for us are: specific heat, thermal conductivity, density ... and price, of course. What can your CES software suggest?'

Form your judgement about why these properties matter to them. Rank them, deciding which you would see as constraints and which as the objective. Consider whether there are perhaps other properties they have neglected. Then—given your starting assumptions of just how these stoves are used—use CES to make a selection. Justify your choice.

There is no 'right' answer to this question—it depends on the assumptions you make. The essence is in the last sentence of the last paragraph: justify your choice.

## **12.11** Exploring the science with CES Elements

**Exercise E12.21**  The text says that materials expand about 2% between 0 K and their melting point. Use CES Elements to explore the truth of this by making a bar chart of the expansion at the melting point, $\alpha T_m/10^6$ (the $10^6$ is to correct for the units of $\alpha$ used in the database).

**Exercise E12.22**  When a solid vaporises, the bonds between its atoms are broken. You might then expect that latent heat of vaporisation, $L_v$, should be nearly the same as the cohesive energy, $H_c$, since it is the basic measure of the strength of the bonding. Plot one against the other, using CES Elements. How close are they?

**Exercise E12.23**  When a solid melts, some of the bonds between its atoms are broken, but not all—liquids still have a bulk modulus, for example. You might then expect that the latent heat of melting, $L_m$, should be less than the cohesive energy, $H_c$, since it is the basic measure of the strength of the bonding. Plot one against the other, using CES Elements ($L_m$ is called the heat of fusion in the database). By what factor is $L_m$ less than $H_c$? What does this tell you about cohesion in the liquid?

Exercise E12.24   The latent heat of melting (heat of fusion), $L_m$, of a material is said to be about equal to the heat required to heat it from absolute zero to its melting point, $C_p T_m$, where $C_p$ is the specific heat and $T_m$ is the absolute melting point. Make a chart with $L_m$ on one axis and $C_p T_m$ on the other. To make the comparison right we have to change the units of $L_m$ in making the chart to J/kg instead of kJ/mol. To do this multiply $L_m$ by

$$\frac{10^6}{\text{Atomic weight in kg/kmol}}$$

using the 'Advanced' facility when defining axes in CES. Is the statement true?

Exercise E12.25   The claim was made in the text that the modulus $E$ is roughly proportional to the absolute melting point $T_m$. If you use CES Elements to explore this correlation you will find that it is not, in fact, very good (try it). That is because $T_m$ and $E$ are measured in different units and, from a physical point of view, the comparison is meaningless. To make a proper comparison, we use instead $k_B T_m$ and $E\Omega$, where $k_B$ is Boltzmann's ($1.38 \times 10^{-23}$ J/K) constant and $\Omega$ m$^3$/atom is the atomic volume. These two quantities are both energies, the first proportional to the thermal energy per atom at the melting point and the second proportional to the work to elastically stretch an atomic bond. It makes better sense, from a physical standpoint, to compare these.

Make a chart for the elements with $k_B T_m$ on the $x$-axis and $E\Omega$ on the $y$-axis to explore how good this correlation is. Correlations like these (if good) that apply right across the periodic table provide powerful tools for checking data, and for predicting one property (say, $E$) if the other (here, $T_m$) is known. Formulate an equation relating the two energies that could be used for these purposes.

Exercise E12.26   Above the Debye temperature, the specific heat is predicted to be $3R$, where $R$ is in units of kJ/kmol.K. Make a plot for the elements with Debye temperature on the $x$-axis and specific heat in these units on the $y$-axis to explore this. You need to insert a conversion factor because of the units. Here it is, expressed in the units contained in the database:

Specific heat in kJ/kmol $\cdot$ K = Specific heat in J/kg $\cdot$ K

$\times$ Atomic weight in kg/kmol/1000

Form this quantity, dividing the result by $R = 8.314$ kJ/kmol.K, and plot it against the Debye temperature. The result should be 3 except for materials with high Debye temperatures. Is it? Fit a curve by eye to the data. At roughly what temperature does the drop-off first begin?

Exercise E12.27    Explore the mean free path of phonons, $\ell_m$ in the elements using equation (12.10) of the text. Inverting it gives

$$\ell_m = 3\frac{\lambda}{\rho C_p c_o}$$

in which the speed of sound $c_o = \sqrt{E/\rho}$. Use the 'Advanced' facility when defining the axes in CES to make a bar chart of $\ell_m$ for the elements. Which materials have the longest values? Which have the shortest?

# Chapter 13
# Running hot: using materials at high temperatures

Shuttle flame. (Image courtesy of C. Michael Holoway, NASA Langley, USA)

## Chapter contents

# **13.1** Introduction and synopsis

Material properties change with temperature. Some do so in a simple linear way, easy to allow for in design: the density, the modulus and the electrical conductivity are examples. But others, particularly the yield strength and the rates of oxidation and corrosion, change in more rapid and complex ways that could lead to disaster, if not understood and allowed for.

This chapter explores the ways in which properties change with temperature, and discusses design methods to deal with the changes. To do this we must first understand *diffusion*, the movement of atoms through solids, and the processes of *creep* and *creep fracture*, the continuous deformation of materials under load at temperature. Diffusion theory lies directly behind the analysis and procedures for high-temperature design with metals and ceramics. Polymers, being molecular, are more complicated in their creep behaviour, but semi-empirical methods allow safe design with these too. Diffusion and hot deformation also relate closely to material processing—forging, rolling and heat treatment exploit the same physical mechanisms—so the chapter explores this too.

# **13.2** The temperature dependence of material properties

*Maximum and minimum service temperatures*   First, the simplest measures of tolerance to temperature are the *maximum* and *minimum service temperatures*, $T_{max}$ and $T_{min}$. The former tells us the highest temperature at which the material can reasonably be used without oxidation, chemical change or excessive deflection or 'creep' becoming a problem (the *continuous use temperature*, or *CUT*, is a similar measure). The latter is the temperature below which the material becomes brittle or otherwise unsafe to use. These are empirical, with no universally accepted definitions. The minimum service temperature for carbon steels is the ductile-to-brittle-transition temperature—a temperature below which the fracture toughness falls steeply. For elastomers it is about $0.8\ T_g$, where $T_g$ is the glass transition temperature; below $T_g$ they cease to be rubbery and become hard and brittle.

*Linear and non-linear temperature dependence*   Some properties depend on temperature $T$ in a linear way, meaning that

$$P \approx P_o\left(1 + \beta\frac{T}{T_m}\right) \tag{13.1}$$

where $P$ is the value of the property, $P_o$ its low-temperature value and $\beta$ is a constant. Figure 13.1 shows four examples: density, modulus, refractive index and—for metals—electrical resistivity. Thus, the density $\rho$ and refractive index $n$ decrease by about 6% on heating from cold to the melting point $T_m$ ($\beta \approx -0.06$), the modulus $E$ falls by a factor of 2 ($\beta \approx -0.5$), and the resistivity $R$ increases by a factor of about 7 ($\beta \approx +6$). These changes cannot be

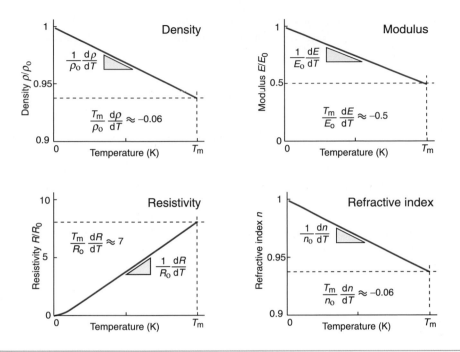

**Figure 13.1**   Linear dependence of properties on temperature: density, modulus, resistivity and refractive index.

neglected, but are easily accommodated by using the value of the property at the temperature of the design.

Other properties are less forgiving. Strength falls in a much more sudden way and the rate of *creep*—the main topic here—increases exponentially (Figure 13.2). This we need to explore in more detail.

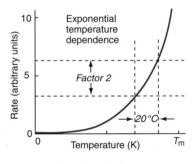

**Figure 13.2**   The exponential increase of rate of straining with temperature at constant load. The rate can double with a rise of temperature of only 20 °C.

*Viscous flow*   When a substance flows, its particles change neighbours; flow is a process of shear. Newton's law describes the flow rate in fluids under a shear stress $\tau$:

$$\dot{\gamma} = \frac{\tau}{\eta}$$

where $\dot{\gamma}$ is the shear rate and $\eta$ the viscosity (units Pa.s). This is a linear law, like Hooke's law, with the modulus replaced by the viscosity and the strain by strain rate (units $s^{-1}$). Viscous flow occurs at constant volume (Poisson's ratio = 0.5) and this means that problems of viscous flow can be solved by taking the solution for elastic deformation and replacing the strain $\varepsilon$ by the strain-rate $\dot{\varepsilon}$ and, using equation (4.10), Young's modulus $E$ by $3\eta$. Thus, the rate at which a rod of a very viscous fluid, like tar, extends when pulled in tension is

$$\dot{\varepsilon} = \frac{\sigma}{3\eta} \qquad (13.2)$$

The factor 1/3 appears because of the conversion from shear to normal stress and strain.

---

### Example 13.1

A tar has a viscosity of $\eta = 10^8$ Pa.s. A cube of tar with a cross-section of $A = 0.01$ mm$^2$ is placed under a steel plate with a mass m of 0.5 kg. By how much will the cube shorten in 1 day?

*Answer.* The stress $\sigma = mg/A = 490$ MPa. The compressive strain-rate is $\dot{\varepsilon} = \sigma/3\eta = 1.6 \times 10^{-6}$/s. The compressive strain in 1 day ($t = 8.64 \times 10^4$ s) is thus $\varepsilon = \dot{\varepsilon}t = 0.14$. The initial edge length of the cube is 100 mm, so it will shorten by 14 mm.

---

*Creep*   At room temperature, most metals and ceramics deform in a way that depends on stress but not on time. As the temperature is raised, loads that are too small to give permanent deformation at room temperature cause materials to *creep*: to undergo slow, continuous deformation with time, ending in fracture. It is usual to refer to the time-independent behaviour as the 'low-temperature' response, and the time-dependent flow as the 'high-temperature' response. But what, in this context, is 'low' and what is 'high'? The melting point of tungsten, used for lamp filaments, is over 3000 °C. Room temperature, for tungsten, is a very low temperature. The filament temperature in a tungsten lamp is about 2000 °C. This, for tungsten, is a high temperature: the filament slowly sags over time under its own weight until the turns of the coil touch and the lamp burns out.

Sometimes creep is desirable. Hot forging, rolling and extrusion are carried out at temperatures and stresses at which high strain-rate creep occurs, lowering the force and power required for the shaping operation. To design for controlled creep in service, or to design and operate a hot forming process, we need to know how the strain-rate $\dot{\varepsilon}$ depends on the stress $\sigma$ and the temperature $T$. That requires *creep testing*.

***Creep testing and creep curves***   Creep is measured in the way shown in Figure 13.3. A specimen is loaded in tension or compression, usually at constant load, inside a furnace that is maintained at a constant temperature, $T$. The extension is measured as a function of time. Metals, polymers and ceramics all have creep curves with the general shape shown in the figure.

The *initial elastic* and the *primary creep* strains occur quickly and can be treated in much the way that elastic deflection is allowed for in a structure. Thereafter, the strain increases steadily with time in what is called the *secondary creep* or the *steady-state creep* regime. Plotting the log of the steady-state creep rate, $\dot{\varepsilon}_{ss}$, against the log of the stress, $\sigma$, at constant $T$, as in Figure 13.4(a), shows that

$$\dot{\varepsilon}_{ss} = B\sigma^n \tag{13.3}$$

where $n$, the *creep exponent*, usually lies between 3 and 8, and for that reason this behaviour is called *power-law creep*. At low $\sigma$ there is a tail with slope $n \approx 1$ (the part of the curve labeled 'Diffusional flow' in Figure 13.4(a)). By plotting the natural logarithm (ln) of $\dot{\varepsilon}_{ss}$ against the reciprocal of the absolute temperature $(1/T)$ at constant stress, as in Figure 13.4(b), we find that:

$$\dot{\varepsilon}_{ss} = C \exp - \left( \frac{Q_c}{RT} \right) \tag{13.4}$$

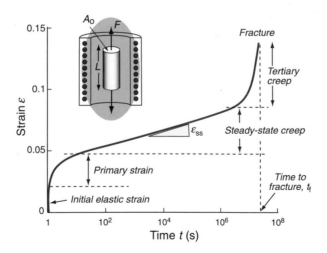

**Figure 13.3**   Creep testing and the creep curve, showing how strain $\varepsilon$ increases with time $t$ up to the fracture time $t_f$.

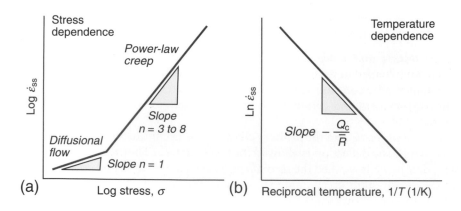

**Figure 13.4** The stress and temperature dependence of the creep rate.

Here $\overline{R}$ is the gas constant (8.314 J/mol.K) and $Q_c$ is called the *activation energy for creep*, with units of J/mol. The creep rate $\dot{\varepsilon}_{ss}$ increases exponentially in the way suggested by Figure 13.2; increasing the temperature by as little as 20 °C can double the creep rate. Combining these two findings gives

$$\dot{\varepsilon}_{ss} = C'\sigma^n \exp - \left(\frac{Q_c}{\overline{R}T}\right) \tag{13.5}$$

where $C'$ is a constant. Written in this way the constant $C'$ has weird units ($s^{-1} \cdot MPa^{-n}$) so it is more usual and sensible to write instead

$$\dot{\varepsilon}_{ss} = \dot{\varepsilon}_o \left(\frac{\sigma}{\sigma_o}\right)^n \exp - \left(\frac{Q_c}{\overline{R}T}\right) \tag{13.6}$$

The four constants $\dot{\varepsilon}_o$ (units: $s^{-1}$), $\sigma_o$ (units: MPa), $n$ and $Q_c$ characterise the steady state creep of a material; if you know these, you can calculate the strain-rate $\dot{\varepsilon}_{ss}$ at any temperature and stress using equation (13.6). They vary from material to material and have to be measured experimentally.

Equation (13.6) is an example of a *constitutive model*, in this case $\dot{\varepsilon} = f(\sigma, T)$. Mathematical models to describe material responses may be directly based on the underlying physics, or be purely empirical 'fits' to experimental data, or a combination of the two. The creep equation (13.6) is such a hybrid model: the power-law variation with stress, and the exponential variation with temperature, have a firm basis in diffusion and plasticity theory, providing indicative values for the material parameters $n$ and $Q_c$. Theoretical predictions for the other constants, $\dot{\varepsilon}_o$ and $\sigma_o$, have proved elusive, however; design values for a given material are therefore adjusted empirically by fitting to experimental data over suitable ranges of $\sigma$ and $T$.

Example 13.2

A stainless steel tie of length $L = 100$ mm is loaded in tension under a stress $\sigma = 150$ MPa at a temperature $T = 800\ °C$. The creep constants for stainless steel are: reference strain-rate $\dot{\varepsilon}_o = 10^6/s$, reference stress $\sigma_o = 100$ MPa, stress exponent $n = 7.5$ and activation energy for power-law creep $Q_c = 280$ kJ/mol. What is the creep rate of the tie? By how much will it extend in 100 hours?

*Answer.* The steady state creep rate is given by $\dot{\varepsilon}_{ss} = \dot{\varepsilon}_o \left(\dfrac{\sigma}{\sigma_o}\right)^n \exp - \left(\dfrac{Q_c}{RT}\right)$. Using the data listed above, converting the temperature to Kelvin, and using the gas constant $\overline{R} = 8.314$ J/mol.K gives $\dot{\varepsilon}_{ss} = 4.8 \times 10^{-7}/s$. After 100 hours ($3.6 \times 10^5$ seconds) the strain is $\varepsilon = 0.173$ and the 100 mm long tie has extended by $\Delta L = 17.3$ mm.

*Creep damage and creep fracture*   As creep continues, damage accumulates. It takes the form of voids or internal cracks that slowly expand and link, eating away the cross-section and causing the stress to rise. This makes the creep rate accelerate as shown in the tertiary stage of the creep curve of Figure 13.3. Since $\dot{\varepsilon} \propto \sigma^n$ with $n \approx 5$, the creep rate goes up even faster than the stress: an increase of stress of 10% gives an increase in creep rate of 60%.

Times to failure, $t_f$, are normally presented as *creep–rupture* diagrams (Figure 13.5). Their application is obvious: if you know the stress and temperature you can read off the life. If instead you wish to design for a certain life at a certain temperature, you can read off the design stress. Experiments show that

$$\dot{\varepsilon}_{ss} t_f = C \tag{13.7}$$

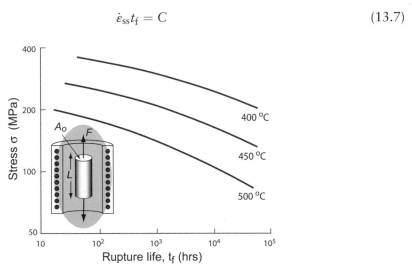

**Figure 13.5**   Creep rupture life as a function of stress and temperature. The scales are logarithmic, and the data are typical of carbon steel.

which is called the Monkman–Grant law. The Monkman–Grant constant, $C$, is typically 0.05–0.3. Knowing it, the creep life (meaning $t_f$) can be estimated from equation (13.6).

---

### Example 13.3

A component in a chemical engineering plant is made of the stainless steel of Example 13.2. It is loaded by an internal pressure that generates a hoop stress of 150 MPa.

(a) If the hoop creep rate $\dot{\varepsilon}_{ss} = 4.8 \times 10^{-7}/s$ at $T = 800\,°C$ under this stress, by what factor will it increase if the pressure is raised by 20%?

(b) If instead the pressure is held constant but the temperature is raised from $T_1 = 800$ °C to $T_2 = 850\,°C$, by what factor will the creep life $t_f$ be reduced?

*Answer.*

(a) The hoop stress is proportional to the pressure. Thus the factor by which the strain-rate increases when the pressure is raised is

$$\frac{\text{Creep rate at 180 MPa}}{\text{Creep rate at 150 MPa}} = \left(\frac{180}{150}\right)^{7.5} = 3.9; \quad \text{the creep rate increases to } \dot{\varepsilon}_{ss} = 1.9 \times 10^{-6}/s$$

(b) The factor by which the strain-rate is increased by the increase in temperature is

$$\frac{\text{Creep rate at } T_2 = 1123\text{K}}{\text{Creep rate at } T_1 = 1073\text{K}} = \exp-\frac{Q_c}{R}\left(\frac{1}{T_2} - \frac{1}{T_1}\right) = 4.0$$

The creep life $t_f = C/\dot{\varepsilon}_{ss}$ where $C$ is a constant. The creep life therefore is reduced by the same factor. The message here is that comparatively small changes in stress or temperature cause big changes in creep rate and creep life because of their strongly non-linear dependence on both $\sigma$ and $T$.

---

## 13.3 Charts for creep behaviour

*Melting point* Figure 13.6 shows melting points for metals, ceramics and polymers. Most metals and ceramics have high melting points and, because of this, they start to creep only at temperatures well above room temperature. *Lead*, however, has a melting point of 327 °C (600 K), so room temperature is almost half its absolute melting point and it creeps—a problem with lead roofs and cladding of old buildings. Crystalline polymers, most with melting points in the range 150–200 °C, creep slowly if loaded at room temperature; glassy polymers, with $T_g$ of typically 50–150 °C, do the same. The point, then, is that the temperature at which materials start to creep depends on their melting points. As a general rule, it is found that creep starts when $T \approx 0.35\,T_m$ for metals and $0.45\,T_m$ for ceramics, although alloying can raise this temperature significantly.

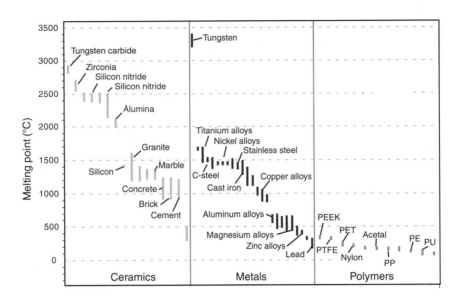

**Figure 13.6** Melting points of ceramics, metals and polymers.

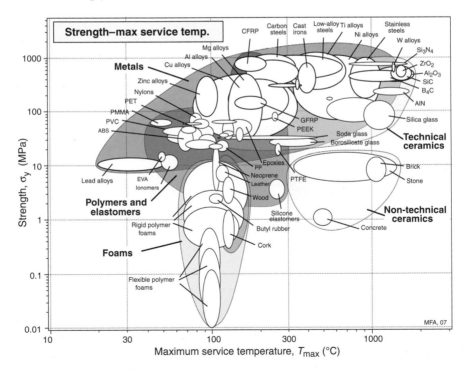

**Figure 13.7** The strength and the maximum service temperature of materials. The strengths decrease with temperature in ways described in the text.

***Maximum service temperature and strength*** Figure 13.7 charts the maximum service temperature $T_{max}$ and room temperature strength $\sigma_y$. It shows that polymers and low melting metals like the alloys of zinc, magnesium and aluminum offer useful strength at room temperature but by 300 °C they cease to be useful—indeed, few polymers have useful strength above 135 °C. Titanium alloys and low-alloy steels have useful strength up to 600 °C; above this temperature high-alloy stainless steels and more complex alloys based on nickel, iron and cobalt are needed. The highest temperatures require refractory metals like tungsten, or technical ceramics such as silicon carbide (SiC) or alumina ($Al_2O_3$).

***Creep strength at 950 °C and density*** Figure 13.8 is an example of a chart to guide selection of materials for load-bearing structures that will be exposed to high temperatures. It shows the creep strength at 950 °C, $\sigma_{950°C}$. Strength at high temperatures (and this is a very high temperature), as we have said, is rate-dependent, so to construct the chart we first have to choose an acceptable strain-rate, one we can live with. Here $10^{-6}$/s has been chosen. $\sigma_{950°C}$ is plotted against the density $\rho$. The chart is used in exactly the same way as the $\sigma_y$–$\rho$ chart of Figure 6.6, allowing indices like

$$M_t = \frac{\sigma_{950°C}}{\rho}$$

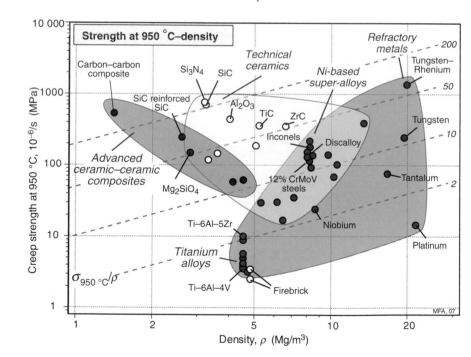

**Figure 13.8** A chart showing the strength of selected materials at a particularly high temperature—950 °C and a strain-rate of $10^{-6}$/s—plotted against density. Software allows charts like this to be constructed for any chosen temperature. (The chart was made with a specialised high-temperature materials database that runs in the CES system.)

to be plotted to identify materials for lightweight design at high temperature. Several such contours are shown.

The figure shows that titanium alloys, at this temperature, have strengths of only a few megapascals—about the same as lead at room temperature. Nickel- and iron-based super-alloys have useful strengths of 100 MPa or more. Only refractory metals like alloys of tungsten, technical ceramics like SiC, and advanced (and very expensive) ceramic—ceramic composites offer high strength.

At room temperature we need only one strength—density chart. For high-temperature design we need one that is constructed for the temperature and acceptable strain-rate or life required by the design. It is here that computer-aided methods become particularly valuable, because they allow charts like Figure 13.8 to be constructed and manipulated for any desired set of operating conditions quickly and efficiently.

## **13.4** The science: diffusion

Creep deformation (and several other responses of materials at temperature) require the relative motion of atoms. How does this happen, and what is its rate? To understand this, we go right down to the unit step—one atom changing its position relative to those around it, by *diffusion.*

***Diffusion*** Diffusion is the spontaneous intermixing of atoms over time. In gases and liquids this is familiar—the spreading of smoke in still air, the dispersion of an ink drop in water—and explained via the random motion and collisions of the atoms or molecules (Brownian motion). In crystalline solids the atoms are confined to lattice sites, but in practice they can still move and mix, if they are warm enough.

Heat, as we have said, is atoms in motion. In a solid at temperature $T$ they vibrate about their mean position with a frequency $\nu$ (about $10^{13}$ per second) with an average energy, kinetic plus potential, of $k_B T$ in each mode of vibration, where $k_B$ is Boltzmann's constant ($1.38 \times 10^{-23}$ J/atom.K). This is the average, but at any instant some atoms have less, some more. Statistical mechanics gives the distribution of energies. The Maxwell-Boltzmann equation[1] describes the probability $p$ that a given atom has an energy greater than a value $q$ Joules:

$$p = \exp - \left( \frac{q}{k_B T} \right) \tag{13.8}$$

Crystals, as we saw in Chapter 4, contain *vacancies*—occasional empty atom sites. These provide a way for diffusive jumps to take place. Figure 13.9 shows an atom jumping into a vacancy. To make such a jump, the atom marked in red (though it is the same sort of atom as the rest) must break away from its original comfortable site at A, its *ground state*, and squeeze between neighbours, passing through an *activated state*, to drop into the vacant site at B where

---

[1] James Clerk Maxwell (1831—1879), pre-eminent Scottish theoretical physicist, considered by many to be the equal of Newton and Einstein (the latter kept a photo of Maxwell in his study, alongside pictures of Newton and Faraday). Maxwell's input to kinetic theory of diffusion was a relative sideshow compared to his electromagnetic theory, which unified the understanding at the time of electricity, magnetism and optics.

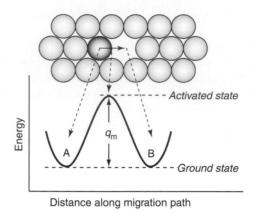

**Figure 13.9** A diffusive jump. The energy graph shows how the energy of the red atom (which may be chemically the same as the green ones, or may not) changes as it jumps from site A to site B.

it is once again comfortable. There is an energy barrier, $q_m$, between the ground state and the activated state to overcome if the atom is to move. The probability $p_m$ that a given atom has thermal energy this large or larger is just equation (13.8) with $q = q_m$.

So two things are needed for an atom to switch sites: enough thermal energy, and an adjacent vacancy. A vacancy has an energy $q_v$, so—not surprisingly—the probability $p_v$ that a given site be vacant is also given by equation (13.8), this time with $q = q_v$. Thus the overall probability of an atom changing sites is

$$p = p_v \, p_m = \exp - \left( \frac{q_v + q_m}{k_B T} \right) = \exp - \left( \frac{q_d}{k_B T} \right) \qquad (13.9)$$

where $q_d$ is called the *activation energy for self-diffusion*. Diffusion, and phenomena whose rates are governed by diffusion, are often said to be *thermally activated*. If instead the red atom were chemically different from the green ones (so it is a substitutional solid solution), the process is known as *inter-diffusion*. Its activation energy has the same origin. Interstitial solutes diffuse too: they have comfortable locations in their interstitial holes and need thermal activation to hop between them, but without the need for an adjoining vacancy (so in general their diffusion rates are much higher).

Figure 13.10 illustrates how mixing occurs by inter-diffusion. It shows a solid in which there is a *concentration gradient $dc/dx$* of red atoms: there are more in slice A immediately to the left of the central, shaded plane, than in slice B to its right. If atoms jump across this plane at random, there will be a net flux of red atoms to the right because there are more on the left to jump, and a net flux of white atoms in the opposite direction. The number in slice A, per unit area, is $n_A = 2\,r_o\,c_A$, and that in slice B is $n_B = r_o\,c_B$ where $2\,r_o$, the atom size, is the thickness of the slices and $c_A$ and $c_B$ are the concentration of red atoms on the two planes expressed as atom fractions. The difference is

$$n_A - n_B = 2\,r_o (c_A - c_B) = 4\,r_o^2\,\frac{dc}{dx}$$

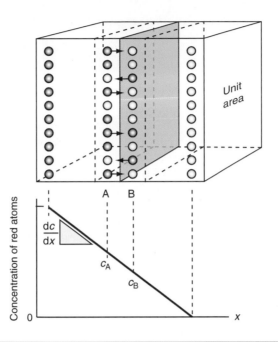

**Figure 13.10**  Diffusion in a concentration gradient.

(since $dc/dx = (c_A - c_B)/2\,r_o$). The number of times per second that an atom on slice A oscillates towards B, or one on B towards A, is $v/6$, since there are six possible directions in which an atom can oscillate in three dimensions, only one of which is in the right direction. Thus the net flux per second of red atoms across unit area from left to right is

$$J = -\frac{v}{6}\exp - \left(\frac{q_d}{k_B T}\right) 4\,r_o^2\,\frac{dc}{dx}$$

It is usual to work with the activation energy per mole, $Q_d$, rather than that per atom, $q_d$, so we write $Q_d = N_A\,q_d$ and $\overline{R} = N_A\,k_B$ (where $N_A$ is Avogadro's number, $6.022 \times 10^{23}$), and assemble the terms $4\,v\,r_o^2/6$ into a single *diffusion constant* $D_o$ to give

$$J = -D_o\exp - \left(\frac{Q_d}{\overline{R}T}\right)\frac{dc}{dx} \tag{13.10}$$

This is known as Fick's law:[2]

$$J = -D\,\frac{dc}{dx} \tag{13.11}$$

---

[2] Adolph Eugen Fick (1829–1901), German physicist, physiologist and inventor of the contact lens. He formulated the law that describes the passage of gas through a membrane, but which also describes the heat-activated mixing of atoms in solids.

where $D$ is called the *diffusion coefficient* given by

$$D = D_o \exp - \left( \frac{Q_d}{\overline{R}T} \right) \tag{13.12}$$

Values for $D_o$ and $Q_d$ have been established experimentally for self-diffusion and diffusion of solutes in all the major alloys and ceramics. Often they will be inferred by measuring the temperature dependence of a macroscopic phenomenon that scales directly with the diffusion coefficient. Note that the theory above considered atoms diffusing within the bulk of a crystal only containing vacancies. Most metals and ceramics are *polycrystalline* and are made up of grains, within which there are dislocations (Chapter 6). Grain boundaries and dislocations offer 'short-circuit' diffusion paths, as there is a bit more space between the atoms. Diffusion along grain boundaries and dislocation cores still requires an activation energy for atomic jumps, but different values of $D_o$ and $Q_d$ will apply.

There are two rules-of-thumb, useful in making estimates of diffusion rates when data are not available and to give a feel for the upper limit for diffusion rate. The first is that the activation energy for diffusion, normalised by $\overline{R}T_m$, is approximately constant for metals:

$$\frac{Q_d}{\overline{R}T_m} \approx 18 \tag{13.13a}$$

The second is that the diffusion coefficient of metals, evaluated at their melting point, is also approximately constant:

$$D_{T_m} = D_o \exp - \left( \frac{Q_d}{\overline{R}T_m} \right) \approx 10^{-12} \ \text{m}^2/\text{s} \tag{13.13b}$$

The CES Elements database contains data for $D_o$ and $Q_d$ for the elements—these equations are therefore investigated in the Exercises at the end of the chapter.

---

### Example 13.4

Alumina, $Al_2O_3$ is suggested for the membrane of a fuel cell. To assess the performance of the cell for oxygen ion transport, the diffusion coefficient of oxygen ions in the membrane at the maximum practical operating temperature $T = 1100°C$ is needed. If $D < 10^{-12} \ \text{m}^2/\text{s}$ the cell is impractical. The activation energy for oxygen diffusion in $Al_2O_3$ is $Q_d = 636$ kJ/mol and the pre-exponential diffusion constant $D_o = 0.19 \ \text{m}^2/\text{s}$. Will the cell work?

*Answer.* The oxygen diffusion coefficient at $T = 1100°C$ is

$$D = D_o \exp - \left( \frac{Q_d}{\overline{R}T} \right) = 0.19 \ \exp - \left( \frac{636,000}{8.314 \times 1373} \right) = 1.17 \times 10^{-25} \ \text{m}^2/\text{s}$$

This is very much less than $10^{-12} \ \text{m}^2/\text{s}$—the cell is not practical.

***Diffusion in liquids and non-crystalline solids***   The vacancies in a crystal can be thought of as *free volume*—free, because it is able to move, as it does when an atom jumps. The free volume in a crystal is thus in the form of discrete units, all of the same size—a size into which an atom with enough thermal energy can jump. Liquids and non-crystalline solids, too, contain free volume, but because there is no lattice or regular structure it is dispersed randomly between all the atoms or molecules. Experiments show that the energy barrier to atom movement in liquids is not the main obstacle to diffusion or flow. There is actually plenty of free volume among neighbours of any given particle, but it is of little use for particle jumps except when, by chance, it comes together to make an atom-sized hole—a temporary vacancy. This is a fluctuation problem just like that of the Maxwell-Boltzmann problem, but a fluctuation of free volume rather than thermal energy, and with a rather similar solution: diffusion in liquids is described by

$$D = \frac{2}{3} r_o^2 \, \nu \, \exp - \left( \frac{v_a}{v_f} \right) \tag{13.14}$$

where $v_f$ is the average free volume per particle, $v_a$ is the volume of the temporary vacancy, and $r_o$ and $\nu$ have the same meaning as before.

***Diffusion driven by other fields***   So far we have thought of diffusion driven by a concentration gradient. Concentration $c$ is a *field quantity*; it has discrete values at different points in space (the concentration field). The difference in $c$ between two nearby points divided by the distance between them defines the local concentration gradient, $dc/dx$.

   Diffusion can be driven by other field gradients. A stress gradient, as we shall see in a moment, drives diffusional flow and power law creep. An electric field gradient can drive diffusion in non-conducting materials. Even a temperature gradient can drive diffusion of matter as well as diffusion of heat.

***Transient diffusion problems***   Fick's law for diffusion of matter, equation (13.11), captures the rate at which atoms flow down a uniform concentration gradient, at steady-state (i.e. when the concentration at a given position doesn't change with time). But we have seen something similar in chapter 12: Fourier's Law for steady-state heat flow, equation (12.2). The analogy between mass flow and heat flow is even more apparent when we consider transient problems. Figure 13.11 shows a hypothetical problem in which two solid blocks of pure elements A and B are brought into perfect contact and heated up. We will assume that A and B form a solid solution in one another in any proportions—a few elements do this (Cu and Ni for instance). The concentration profile evolves steadily with time, heading (at very long times) to a uniform mixture of 50% A atoms to 50% B atoms. Looking back to the problem of quenching a hot steel plate into cold water (Figure 12.15), the shape of the curves is strikingly similar—in fact they are exactly the same. Let's explore why.

   Figure 13.11 shows that because the concentration profile is curved, the local concentration gradient changes continuously with position. So we can repeat the thought experiment of Chapter 12, considering the flux of atoms in and out of a small volume at a given value of $x$. As we have seen, the flux of diffusing atoms is proportional to the local concentration gradient, equation (13.11). So in a small time step $\delta t$, a slightly different number of atoms enter and leave the small volume, reflecting the small difference in the concentration gradient. The

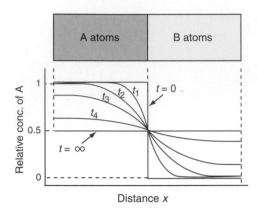

**Figure 13.11**  Intermixing by diffusion.

concentration of the element will therefore change by $\delta C$, just as the temperature changed a little in heat flow because of a difference in heat entering and leaving a small volume. So for the diffusion case, applying Fick's law to a small volume in a non-steady-state concentration profile, we can obtain the differential equation governing the concentration $C$ in terms of position $x$ and time $t$, when atoms are diffusing in one dimension (the $x$-direction)

$$\frac{\partial^2 C}{\partial x^2} = \frac{1}{D}\frac{\partial C}{\partial t} \qquad (13.15)$$

(where the curly '$\partial$' again signifies that this is a partial differential equation). In comparison with the heat flow equivalent in equation (12.15), we see that temperature and concentration are indeed directly analogous, with the diffusion coefficient $D$ replacing the thermal diffusivity, $a$. This means that mathematical solutions to transient diffusion problems can be lifted from heat flow analysis (and vice versa), provided the boundary conditions also match up. Look at the left half of Figure 13.11. Initially the relative concentration of A atoms is uniform and equal to 1, and the effect of applying the block of B at time $t = 0$ is to fix the concentration at the interface at a value of 0.5. The successive concentration profiles at times $t_1$, $t_2$, $t_3$, $t_4$ evolve between these limits—exactly as the temperature in Figure 12.15 started at $T_1$ and eventually settled to $T_0$. In the right half of Figure 13.11, they are all the same shape, just the other way up. For short times, the profile matches the error function solution, erf $(X)$, as in equation (12.17), and at long times we would adapt equation (12.20), with $l/2$ being the length of each block. Without going through the mathematics again, we can anticipate that there will be a characteristic atom diffusion distance for a time $t$, given by

$$x \approx \sqrt{Dt} \qquad (13.16)$$

For the problem of Figure 13.11, this would relate the time and depth at which the concentration has fallen halfway between its initial and final value (i.e. 0.75 in the left half, and 0.25 in the right half). These solutions apply directly to a number of problems in materials processing. We'll solve them now.

***Diffusion in surface hardening of steel: carburising***   Carbon steels are iron containing up to 1% carbon (by weight), giving hardening by solid solution and precipitation (as in Chapter 6). Many steel components are used in heavy-duty applications where the surfaces are in sliding and rolling contact, placing much greater demands on surface properties than the overall bulk of the component. We therefore use a lot of *surface engineering* techniques to enhance the material, just where we need it—these are revisited in Chapters 18 and 19. One method is *carburising*—diffusing more carbon into the surface of a hot steel component, to locally increase the concentration to that of a different, harder alloy. When combined with a surface heat treatment, this can increase the hardness of a steel by a factor of 5 or more, greatly improving the resistance to wear, for example (Chapter 11). Figure 13.12 shows the concentration profile for carburising a steel of carbon concentration $C_o$. The carburised layer is thin compared to the component dimensions—it is effectively a semi-infinite problem, with 1D diffusion. At a given temperature, there is a maximum possible concentration $C_s$ to which the carbon can be increased in solid solution—this sets the constant surface boundary condition. With a bit of thought, we can just write down what the concentration profile $C(x,t)$ must be:

$$C(x,t) = C_o + (C_s - C_o)\left(1 - \text{erf}\left(\frac{x}{2\sqrt{Dt}}\right)\right) \tag{13.17}$$

In this case the solution includes $(1 - \text{erf}(X))$, as the curves fall with increasing $x$, while the factor $(C_s - C_o)$ scales the function to match the boundary conditions. Re-arranging equation (13.17), we get a simple dimensionless solution:

$$\left(\frac{C(x,t) - C_o}{C_s - C_o}\right) = 1 - \text{erf}\left(\frac{x}{2\sqrt{Dt}}\right) \tag{13.18}$$

Figure 13.12 is plotted with these dimensionless axes—by scaling the depth $x$ with the characteristic diffusion distance $\sqrt{Dt}$ the solution for all times falls onto a single curve. The depth at which the concentration has increased to a value halfway between $C_o$ and $C_s$ is shown—it is simply $x = \sqrt{Dt}$. Let's use the solution to estimate the timescale of a typical carburising treatment.

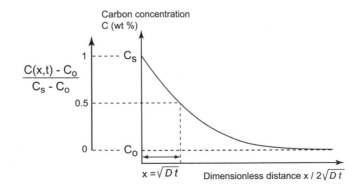

**Figure 13.12**   Concentration profile of carbon in iron during surface carburising. On dimensionless axes, *C(x,t)* falls onto a single curve.

### Example 13.5

It is proposed to harden a component of low carbon steel, of concentration 0.2 weight % C, by diffusing carbon into the surface from a carbon-rich environment at $T = 1000°C$. At this temperature, the maximum carbon concentration in solid solution is $C_s = 1.6$ weight % C. Iron at this temperature makes an FCC structure (called austenite), for which the diffusion constant for carbon $D_o = 8 \times 10^{-6}$ m$^2$/s, and the activation energy $Q_d = 131$ kJ/mol. The objective is to raise the carbon concentration to a minimum of 0.9 weight % C to a depth of 0.2 mm. For how long must the component be carburised to achieve the desired depth of hardening?

*Answer.* The diffusion coefficient at 1000 °C is

$$D = 8 \times 10^{-6} \exp - \left( \frac{131,000}{8.314 \times 1273} \right) = 3.35 \times 10^{-11} \text{ m}^2/\text{s}$$

The time to diffuse a distance $x = 0.2$ mm is approximately

$$t \approx \frac{x^2}{D} = \frac{4 \times 10^{-8}}{3.35 \times 10^{-11}} = 1.2 \times 10^3 \text{ s} \approx 20 \text{ minutes}$$

*Diffusion in doping of semiconductors*    In Chapter 14, it will be shown how semiconductors achieve their unusual electrical properties by *doping* high-purity silicon to achieve a delicate balance between very small amounts of atoms such as boron and phosphorus. This is achieved in extraordinarily clean processing facilities to allow only the right atoms to diffuse in—rogue impurities will ruin the device. A common doping process is to diffuse in a controlled quantity of boron atoms at the surface—'pre-deposition'—and then to remove the source of boron atoms and to 'drive-in' the dopant to a much greater depth. Figure 13.13 illustrates the

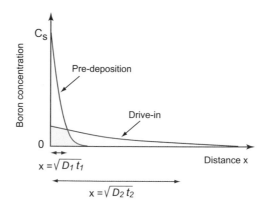

**Figure 13.13**    Concentration profile of boron in silicon during two-stage doping of semiconductors: pre-deposition, and drive-in.

concentration profiles after these two steps: the area under both curves is constant, being the total number of B atoms diffused in during pre-deposition. In this first stage, the profile will be exactly the same as for carburising. The mathematical solution for the drive-in stage takes a different mathematical form, but it too depends on $(x/\sqrt{Dt})$ and the extent of the redistribution of solute can be estimated using $x = \sqrt{Dt}$. You can plug in some numbers for this problem in the Exercises at the end of the chapter.

The carburising and semiconductor examples show just how simple and useful the characteristic diffusion distance formula is in materials processing. As with heat flow problems, if you want a quick estimate of how long something will take, or how far heat or atoms will have penetrated by diffusion, then reach no further than $x = \sqrt{at}$ or $x = \sqrt{Dt}$.

## **13.5** The science: creep

***Creep by diffusional flow***   Liquids and hot glasses are wobbly structures, a bit like a bag loosely filled with beans. If atoms (or beans) move inside the structure, the shape of the whole structure changes in response. Where there is enough free volume, a stress favours the movements that change the shape in the direction of the stress, and opposes those that do the opposite, giving viscous flow. The higher the temperature, the more is the free volume and the faster the flow.

But how do jumping atoms change the shape of a crystal? A crystal is not like a bag of loose beans—the atoms have well-defined sites on which they sit. In the cluster of atoms in Figure 13.9 an atom has jumped but the shape of the cluster has not changed. At first sight, then, diffusion will not change the shape of a crystal. But in fact it does, provided the material is *polycrystalline*—made up of many crystals meeting at grain boundaries. This is because the grain boundaries act as sources and sinks for vacancies. If a vacancy joins a boundary, an atom must leave it—repeat this many times and that face of the crystal is eaten away. If instead a vacancy leaves a boundary, an atom must join it and—when repeated—that face grows. Figure 13.14 shows the consequences: the slow extension of the polycrystal in the direction of stress. It is driven by a stress gradient: the difference between the tensile stress $\sigma$ on the horizontal boundaries from which vacancies flow, and that on the others, essentially zero, to which they go. If the grain size is $d$ the stress gradient is $\sigma/d$. The flux of atoms, and thus the rate at which each grain extends, $\delta d/\delta t$, is proportional to $D\sigma/d$. The strain-rate, $\dot{\varepsilon}$, is the extension rate divided by the original grain size, $d$, giving

$$\dot{\varepsilon} = C\,\frac{D\,\sigma}{d^2} = C\,\frac{\sigma}{d^2}\,D_o\,\exp-\left(\frac{Q_d}{RT}\right) \qquad (13.19)$$

where C is a constant. This is a sort of viscous flow, linear in stress. The smaller the grain size the faster it goes.

***Dislocation climb and power-law creep***   Plastic flow, as we saw in Chapter 5, is the result of the motion of dislocations. Their movement is resisted by dissolved solute atoms, precipitates particles, grain boundaries and other dislocations; the yield strength is the stress needed to force dislocations past or between them. Diffusion can unlock dislocations from obstacles in their path, making it easier for them to move. Figure 13.15 shows how it happens. Here a

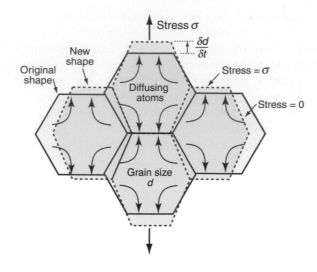

**Figure 13.14** Deformation by diffusion alone, giving *diffusional flow*.

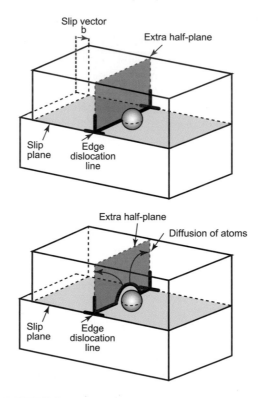

**Figure 13.15** Climb of a dislocation: the extra half-plane is eaten away by diffusion, allowing the dislocation to pass the obstacle. The result is power-law creep.

dislocation is obstructed by a particle. The glide force $\sigma_s b$ per unit length is balanced by the reaction $f_o$ from the precipitate. But if the atoms of the extra half-plane diffuse away, thus eating a slot into the half-plane, the dislocation can continue to glide even though it now has a step in it. The stress gradient this time is less obvious—it is the difference between the local stress where the dislocation is pressed against the particle and that on the dislocation remote from it. The process is called 'climb', and since it requires diffusion, it can occur at a measurable rate only when the temperature is above about $0.35T_m$.

Climb unlocks dislocations from the obstacles that pin them, allowing further slip. After a little slip, of course, the unlocked dislocations encounter the next obstacles, and the whole cycle repeats itself. This explains the *progressive* and *continuous* nature of creep. The dependence on diffusion explains the dependence of creep rate on *temperature*, with

$$\dot{\varepsilon}_{ss} = A\,\sigma^n \exp\,-\left(\frac{Q_c}{RT}\right) \tag{13.20}$$

with $Q_c \approx Q_d$. The power-law dependence on *stress* $\sigma$ is harder to explain. It arises partly because the stress-gradient driving diffusion increases with $\sigma$ and partly because the density of dislocations itself increases too. Note that equation (13.20) has exactly the same form as equation (13.6), the general constitutive model we introduced for creep. The constant A has just been sub-divided into a ratio of two parameters in equation (13.6), $\dot{\varepsilon}_o$ and $\sigma_o$ (to a power of $n$), giving these parameters a more physical meaning, being a strain-rate and a stress.

*Creep fracture* Diffusion, we have seen, gives creep. It also gives creep fracture. You might think that a creeping material would behave like toffee or chewing gum—that it would stretch a long way before breaking in two—but, for crystalline materials, this is very rare. Indeed, creep fracture (in tension) can happen at unexpectedly small strains, often only 2−5%, by the mechanism shown in Figure 13.16. Voids nucleate on grain boundaries that lie normal to the tensile stress. These are the boundaries to which atoms diffuse to give diffusional creep, coming from the boundaries that lie more nearly parallel to the stress. But if the tensile boundaries have voids on them, they act as sources of atoms too, and in doing so, they grow. The voids cannot support load, so the stress rises on the remaining intact bits of boundary, making the voids grow more and more quickly until finally (bang) they link.

Many engineering components (e.g. tie bars in furnaces, super-heater tubes, high temperature pressure vessels in chemical reaction plants) are expected to withstand moderate creep loads for long times (say 20 years) without failure. The loads or pressure they can safely carry are calculated by methods such as those we have just described. One would like to be able to test new materials for these applications without having to wait for 20 years to get the results. It is thus tempting to speed up the tests by increasing the load or the temperature to get observable creep in a short test time, and this is done. But there are risks. Tests carried out on the steep $n = 3 - 8$ branch of Figure 13.4(a), if extrapolated to lower stresses where $n \approx 1$, greatly underestimate the creep rate.

*Deformation mechanism diagrams* Let us now pull all of this together. Materials can deform by dislocation plasticity (Chapter 6), or, if the temperature is high enough, by diffusional flow or power-law creep. If the stress and temperature are too low for any of these, the deformation is elastic. This competition between mechanisms is summarised in *deformation mechanism*

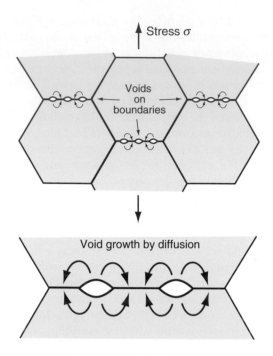

**Figure 13.16** Creep fracture caused by the diffusive growth of voids on grain boundaries.

*diagrams* of which Figure 13.17 is an example. It shows the range of stress and temperature in which we expect to find each sort of deformation and the strain-rate that any combination of them produces (the contours). Diagrams like these are available for many metals and ceramics, and are a useful summary of creep behaviour, helpful in selecting a material for high temperature applications—one appears in the examples at the end of this chapter. The figure also shows the typical hot working regime for the alloy, discussed next.

*Creep in hot forming processes* The ductility of most metals and alloys opens up a diverse range of shaping methods—rolling, extrusion and forging, for example (Figure 7.11). Since the power expended in shaping a component scales with the material strength, we would like to keep the yield stress down, but we then need high strength afterwards in service. One common solution to this conflict is hot forming—working the alloy at high temperature in a soft condition. Blacksmiths and weapon-makers have known about this for centuries—the expenditure of energy in metal shaping is self-evident when you are the person wielding the hammer. In heat-treatable alloys, such as many carbon steels and aluminium alloys, there is often the opportunity to add a secondary heat treatment to develop a high-strength microstructure. For really efficient processing, the hot forming step doubles up as the first stage of the heat treatment—extrusion and 'age hardening' of aluminium alloys are an example (see Chapter 19).

Hot forming temperatures are typically over 0.5 times the melting point of the alloy (in Kelvin), which means we are in the creep regime. But since speed is now of the essence, we are

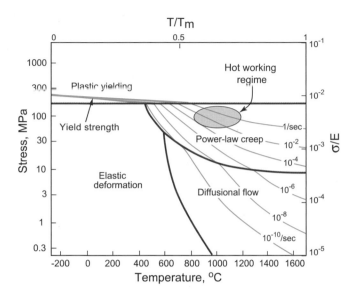

**Figure 13.17** Deformation mechanism map for a nickel alloy, showing the regime in which each mechanism operates, and the region typically used for hot working.

now at the high strain-rate end of things—limiting the softening that is achievable at a sensible production rate, as the stress rises strongly with strain-rate. Nonetheless, the material strength is well below the room temperature yield stress—let's see how much with an example.

---

### Example 13.6

A Ni alloy billet is to be hot forged between flat parallel dies at a temperature between 1100 and 1120 °C. The deformation mechanism map for the alloy is that shown in Figure 13.17. The initial billet is a cylinder of diameter 20 cm, and height 50 cm, with its axis vertical. The top platen descends at a speed of 5 cm/s and comes to rest when the billet has been shortened by 2.5 cm. Assume that the deceleration is constant with time, that is, the speed varies linearly.

(a) Estimate the initial and the average strain-rate during the process, and the time that the deformation takes.
(b) Locate the hot working regime for this process on the deformation mechanism map, and estimate the stress needed to perform the operation. By what factor is this below the room temperature yield stress?
(c) Estimate the load required for this process, assuming that the axial stress is approximately 1.5 times the uniaxial yield stress (a consequence of the friction acting between the workpiece and the platen, explained in Chapter 19).

*Answer.*

(a) On initial contact the strain-rate is given by the speed of the platen divided by the height of the cylinder $= 0.05/0.5 = 0.1s^{-1}$. For a linear deceleration in speed with time, the average speed is 2.5 cm/s, so the average strain-rate is approximately half of the initial value, and the deformation takes about 1 second.

(b) From Figure 13.17, the stress required for this strain-rate at 1100–1120 °C is approximately 70 MPa, compared to a room temperature yield stress around 200 MPa, so the hot strength is about three times lower.

(c) The axial stress required is $1.5 \times 70$ MPa $= 105$ MPa, giving a force of $105 \times 10^6 \times (\pi \times 0.2^2)/4 = 3.3$ MN. Forging on this scale needs a big, expensive machine.

The deformation mechanism map shows that power-law creep is the expected mechanism operating in hot forming, due to the need for high deformation rates. One special forming process exploits diffusional flow. The map suggests that this mechanism would normally be much too slow for a manufacturing process. But if the grain size is made very small, the strain-rate in diffusional flow can be increased to an acceptable level for a slow shaping process: equation (13.19) shows that a factor of 10 in grain size gives a factor of 100 in strain-rate. *Superplastic forming* uses very fine-grained alloys, effectively accelerating the mechanism of Figure 13.14. It is particularly useful for shaping lightweight Ti and Al alloys for aerospace applications, taking a closed hollow pre-form of partly bonded sheets and stretching them using internal pressure, to give complex lightweight stiffened panels.

*Creep in polymers*  Creep in crystalline materials, as we have seen, is closely related to diffusion. The same is true of polymers, but because most of them are partly or wholly amorphous, diffusion is controlled by free volume (equation (13.14)). Free volume increases with temperature (its fractional change per degree is just the volumetric thermal expansion, $3\alpha$), and it does so most rapidly at the glass transition temperature $T_g$. Thus polymers start to creep as the temperature approaches $T_g$, and that, for most, is a low temperature: 50–150 °C.

This means that the temperature range in which most polymers are used is near $T_g$ when they are neither simple elastic solids nor viscous liquids; they are *visco-elastic* solids. The physics of molecular sliding are more difficult to describe analytically than the atomic diffusion and dislocation behaviour in crystals, but we still need constitutive models for the relationship between stress, strain, strain-rate and temperature. So we often take a more pragmatic, empirical approach. If we represent the elastic behaviour by a spring, and the viscous behaviour by a dash-pot, then visco-elasticity (at its simplest) can be described by coupled sets of springs and dash-pots—Figure 13.18 shows three possible arrangements. For the springs we have the usual elastic response, $E = \sigma/\varepsilon$, while for the dash-pot, the stress depends on the strain-rate, via the viscosity $\eta = \sigma/\dot{\varepsilon}$. A good way to envisage the mechanical response is to apply a rapid rise to a uniform constant stress; Figure 13.18 shows the resulting strain histories. Applying the stress causes a combination of elastic strain and creep, but in different ways depending on the configuration.

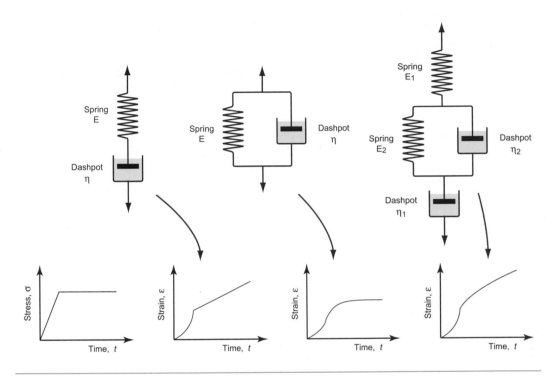

**Figure 13.18** Visco-elastic behaviour can be modelled using combinations of elastic springs and viscous dashpots. Different configurations produce different time-dependent responses to applied stress.

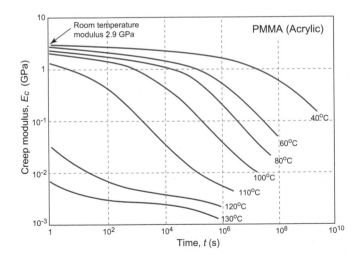

**Figure 13.19** The creep modulus of PMMA.

Real polymers can require elaborate systems of springs and dash-pots to describe them, and no one model will capture the response for all loading situations and temperatures. Experimental tests are conducted for conditions relevant to the design problem, and an appropriate spring and dash-pot model fitted to the data (by adjusting the moduli and viscosities of the springs and dash-pots). This approach of *polymer rheology* can be developed to provide criteria for design and numerical analysis. A simpler initial approach to mechanical design with polymers is to use graphical data for what is called the *creep modulus* $E_c$ to provide an estimate of the deformation during the life of the component. Figure 13.19 shows data for $E_c$ as a function of temperature $T$ and time $t$. Creep modulus data allow conventional elastic solutions to be used for design against creep. The service temperature and design life are chosen, the resulting creep modulus is read from the graph, and this is used instead of Young's modulus in any of the solutions to design problems listed in Chapter 5.

## 13.6 Materials to resist creep

***Metals and ceramics to resist creep*** Diffusional flow is important when grains are small (as they often are in ceramics) and when the component is subject to high temperatures at low loads. Equation (13.15) says that the way to avoid diffusional flow is to choose a material with a high melting temperature and arrange that it has a large grain size, $d$, so that diffusion distances are long. Single crystals are best of all; they have no grain boundaries to act as sinks and sources for vacancies, so diffusional creep is suppressed completely. This is the rationale behind the wide use of single-crystal turbine blades in jet engines.

That still leaves power-law creep. Materials that best resist power-law creep are those with high melting points, since diffusion and thus creep rates scale as $T/T_m$, and a microstructure that maximises obstruction to dislocation motion through alloying to give a solid solution and precipitate particles. Current creep-resistant materials, known as *super-alloys*, are remarkably successful in this. Most are based on iron, nickel or cobalt, heavily alloyed with aluminum, chromium and tungsten. The chart of Figure 13.8 shows both effects: the high melting point *refractory metals* and the heavily alloyed *super-alloys*.

Many ceramics have high melting points, meeting the first criterion. They do not need alloying because their covalent bonding gives them a large lattice resistance. Typical among them are the technical ceramics alumina ($Al_2O_3$), silicon carbide (SiC) and silicon nitride ($Si_3N_4$)—they too appear in Figure 13.8. The lattice resistance pins down dislocations but it also gives the materials low fracture toughness, even at high temperature, making design to use them difficult.

***Polymers to resist creep*** The creep resistance of polymers scales with their glass temperature. $T_g$ increases with the degree of cross-linking; heavily cross-linked polymers (e.g. epoxies) with high $T_g$ are therefore more creep resistant at room temperature than those that are not (like polyethylene). The viscosity of polymers above $T_g$ increases with molecular weight, so the rate of creep is reduced by having a high molecular weight. Finally, crystalline or partly crystalline polymers (e.g. high-density polyethylene) are more creep resistant than those that are entirely glassy (e.g. low-density polyethylene).

The creep rate of polymers is reduced by filling them with powdered glass, silica (sand) or talc, roughly in proportion to the amount of filler. PTFE on saucepans and polypropylene used for automobile components are both strengthened in this way. Much better creep resistance is obtained with composites containing chopped or continuous fibres (GFRP and CFRP), because much of the load is now carried by the fibres that, being very strong, do not creep at all.

| $T\,°C$ | Materials | Applications | $T\,K$ |
|---|---|---|---|
| 1200<br><br><br>1000 | Refractory metals: Mo, W, Ta<br>Alloys of Nb, Mo, W, Ta<br>Ceramics: Oxides<br>$Al_2O_3$, MgO, etc.<br>Nitrides, $Si_3N_4$,<br>Carbides, SiC | Rocket nozzles<br>Special furnaces<br>Experimental turbines | 1400 |
| <br>800 | Austenitic stainless steels<br>Nichromes, nimonics<br>Nickel based super-alloys<br>Cobalt based super-alloys<br>Iron based super-alloys | Gas turbines<br>Chemical engineering<br>Petrochemical reactors<br>Furnace components<br>Nuclear construction | 1200<br><br>1000 |
| 600 | Iron-based super-alloys<br>Ferritic stainless steels<br>Austenitic stainless steels<br>Inconels and nimonics | Steam turbines<br>Superheaters<br>Heat exchangers | |
| 400 | Low-alloy steels<br>Titanium alloys (up to 450 °C)<br>Inconels and nimonics | Heat exchangers<br>Steam turbines<br>Gas turbine compressors | 800<br><br>600 |
| 200 | Fibre-reinforced polymers<br>Copper alloys (up to 400 °C)<br>Nickel, monels and nickel-silvers<br>PEEK, PEK, PI, PPD, PTFE<br>and PES (up to 250 °C) | Food processing<br>Automotive (engine) | 400 |
| 0 | Most polymers (max temp: 60 to 150 °C)<br>Magnesium alloys (up to 150 °C)<br>Aluminum alloys (up to 150 °C)<br>Monels and steels | Civil construction<br>Household appliances<br>Automotive<br>Aerospace | 200 |
| −200 | Austenitic stainless steels<br>Aluminum alloys | Rocket casings, pipework, etc.<br>Liquid $O_2$ or $N_2$ equipment | |
| −273 | Copper alloys<br>Niobium alloys | Superconduction | 0 |

Figure 13.20 Materials for each regime of temperature.

*Selecting materials to resist creep*   Classes of industrial applications tend to be associated with certain characteristic temperature ranges. There is the cryogenic range, between −273 °C and roughly room temperature, associated with the use of liquid gases like hydrogen, oxygen or nitrogen. Here the issue is not creep, but the avoidance of brittle fracture. There is the regime at and near room temperature (−20 to +150 °C) associated with conventional mechanical and civil engineering: household appliances, sporting goods, aircraft structures and housing are examples. Above this is the range 150–400 °C, associated with automobile engines and with food and industrial processing. Higher still are the regimes of steam turbines and superheaters (typically 400–650 °C), and of gas turbines and chemical reactors (650–1000 °C). Special applications (lamp filaments, rocket nozzles) require materials that withstand even higher temperatures, extending as high as 2800 °C.

Materials have evolved to fill the needs of each of these temperature ranges (Figure 13.20). Certain polymers, and composites based on them, can be used in applications up to 250 °C, and now compete with magnesium and aluminum alloys and with the much heavier cast irons and steels, traditionally used in those ranges. Temperatures above 400 °C require special creep-resistant alloys: ferritic steels, titanium alloys (lighter, but more expensive) and certain stainless steels. Stainless steels and ferrous super-alloys really come into their own in the temperature range above this, where they are widely used in steam turbines and heat exchangers. Gas turbines require, in general, nickel-based or cobalt-based super-alloys. Above 1000 °C, the refractory metals and ceramics become the only candidates. Materials used at high temperatures will, generally, perform perfectly well at lower temperatures too, but are not used there because of cost.

The chart of Figure 13.8 showed the performance of materials able to support load at 950 °C. Charts like this are used in design in the same way as those for low-temperature properties.

# 13.7 Design to cope with creep

Creep problems are of four types:

- Those in which limited creep strain can be accepted but creep rupture must be avoided, as in the creep of pipework or of lead roofs and cladding on buildings.
- Those in which creep strain is design limiting, as it is for blades in steam and gas turbines where clearances are critical.
- Those involving more complex problems of creep strain, loss of stiffness and risk of buckling—a potential problem with space-frames of supersonic aircraft and space vehicles.
- Those involving stress relaxation—the loss of tension in a pre-tightened bolt, for instance.

*Predicting life of high-temperature pipework*   Chemical engineering and power generation plants have pipework that carries hot gases and liquids under high pressure. A little creep, expanding the pipe slightly, can be accepted; rupture, with violent release of hot, high-pressure fluid, cannot. We take, as an example, pipework in a steam-turbine power-generating station.

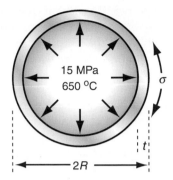

**Figure 13.21** A pressurised pipe.

Pipes in a typical 600 MW unit carry steam at 650 °C and a pressure $p$ of 15 MPa. The stress in the wall of a thin-walled pipe with a radius $R$ and a wall thickness $t$ carrying a pressure $p$, as in Figure 13.21, is

$$\sigma = \frac{pR}{t} \tag{13.21}$$

Suppose you have been asked to recommend a pipe required as a temporary fix while modifications are made to the plant. Space is constrained: the pipe cannot be more than 300 mm in diameter. The design life is six months. A little creep does not matter, but the pipe must not rupture. Type 304 stainless steel pipe with a diameter of 300 mm and a wall thickness of 10 mm is available. Will it function safely for the design life?

The stress in the pipe wall with these dimensions, from equation (13.21), is 225 MPa. Figure 13.22 shows stress–rupture data for Type 304 stainless steel. Enter 225 MPa at 650 °C and read off the rupture life: about seven hours. Not so good.

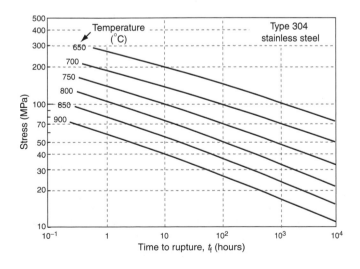

**Figure 13.22** Stress–rupture time data for Type 304 stainless steel.

So how thick should the pipe be? To find out, reverse the reasoning. The design life is six months—4380 hours. A safety-critical component such as this one needs a safety factor, so double it: 8760 hours. Enter this and the temperature on the stress–rupture plot and read off the acceptable stress: 80 MPa. Put this into equation (13.21) to calculate the wall thickness. To be safe the pipe wall must be at least 28 mm thick.

***Turbine blades*** Throughout the history of its development, the gas turbine has been limited in thrust and efficiency by the availability of materials that can withstand high stress at high temperatures. The origin of the stress is the centrifugal load carried by the rapidly spinning turbine disk and rotor blades. Where conditions are at their most extreme, in the first stage of the turbine, nickel- and cobalt-based super-alloys currently are used because of their unique combination of high-temperature strength, toughness and oxidation resistance. Typical of these is MAR-M200, an alloy based on nickel, strengthened by a solid solution of tungsten and cobalt and by precipitates of $Ni_3(Ti, Al)$, and containing chromium to improve its resistance to attack by gases.

When a turbine is running at a steady speed, centrifugal forces subject each rotor blade to an axial tension (equation (7.8)). If the blade has a constant cross-section, the tensile stress rises linearly from zero at its tip to a maximum at its root. As an example, a rotor of radius $r$ = 0.3 m rotating at an angular velocity $\omega$ of 10,000 rpm (1000 radians/s) induces an axial stress ($R\omega^2\rho x$) of order 150 MPa. (Here $\rho$ is the density of the alloy, about 8000 kg/m$^3$, and $x$ the distance from the tip; a typical blade is about 80 mm long.) Typical stress and temperature profiles for a blade in a medium-duty engine are shown in Figure 13.23. They are

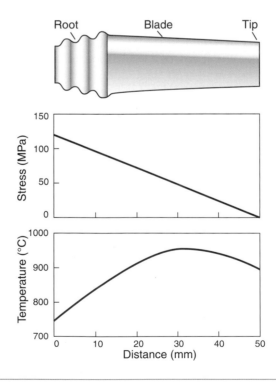

**Figure 13.23** A turbine blade, showing the stress and temperature profiles.

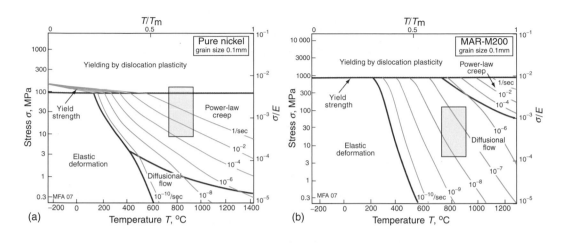

**Figure 13.24** (a) The stress–temperature profile of the blade plotted onto a deformation map for pure nickel. (b) The same profile plotted on a map for the alloy MAR-M200. The strain rates differ by a factor of almost $10^6$.

plotted as a shaded box onto two deformation mechanism maps in Figure 13.24. If made of pure nickel (Figure 13.24(a)) the blade would deform by *power-law creep*, at a totally unacceptable rate. The strengthening methods used in MAR-M200 with a typical as-cast grain size of 0.1 mm (Figure 13.24(b)) reduce the rate of power-law creep by a factor of $10^6$ and change the dominant mechanism of flow from power-law creep to diffusional flow. Further solution strengthening or precipitation hardening is now ineffective unless it slows this mechanism. A new strengthening method is needed: the obvious one is to increase the grain size or remove grain boundaries altogether by using a single crystal. This slows diffusional flow or stops it altogether, while leaving the other flow mechanisms unchanged. The power-law creep field expands and the rate of creep of the turbine blade falls to a negligible level.

The point to remember is that creep has contributions from several distinct mechanisms; the one that is dominant depends on the stress and temperature applied to the material. If one is suppressed, another will take its place. Strengthening methods are selective: a method that works well in one range of stress and temperature may be ineffective in another. A strengthening method should be regarded as a way of attacking a particular flow mechanism. Materials with good creep resistance combine strengthening mechanisms in order to attack them all; single-crystal MAR-M200 is a good example of this.

***Thermal barrier coatings*** The efficiency and power-to-weight ratio of advanced gas turbines, as already said, is limited by the burn temperature and this in turn is limited by the materials of which the rotor and stator blades are made. Heat enters the blade from the burning gas, as shown in Figure 13.25. Blades are cooled by pumping air through internal channels, leading to the temperature profile shown on the left. The surface temperature of

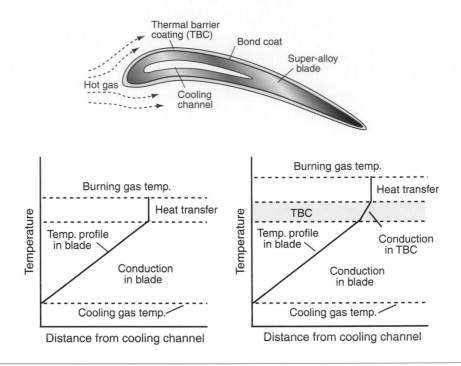

**Figure 13.25**   A cross-section of a turbine blade with a thermal barrier coating (TBC). The temperature profile for the uncoated blade is shown on the left, that for the coated blade on the right.

the blade is set by a balance between the heat transfer coefficient between gas and blade (determining the heat in) and conduction within the blade to the cooling channel (determining the heat out). The temperature step at the surface is increased by bleeding a trickle of the cooling air through holes in the blade surface. This technology is already quite remarkable—air-cooled blades operate in a gas stream at a temperature that is above the melting point of the alloy of which they are made. How could the burn temperature be increased yet further?

Many ceramics have higher melting points than any metal and some have low thermal conductivity. Ceramics are not tough enough to make the whole blade, but coating the metal blade with a ceramic to form a *thermal barrier coating* (TBC) allows an increase in gas temperature with no increase in that of the blade, as shown in Figure 13.25 on the right. How is the ceramic chosen? The first considerations are those of a low thermal conductivity, a maximum service temperature above that of the gas (providing some safety margin) and adequate strength. The $\alpha$–$\lambda$ chart (Figure 12.4) shows that the technical ceramic with by far the lowest thermal conductivity is zirconia ($ZrO_2$). The $T_{max}$–$\sigma_f$ chart (Figure 13.7) confirms that it is usable up to very high temperature and has considerable strength. Zirconia looks like a good bet—and indeed it is this ceramic that is used for TBCs.

As always, it is not quite so simple. Other problems must be overcome to make a good TBC. It must stick to the blade, and that is not easy. To achieve it the blade surface is first plated with a thin *bond coat* (a complex Ni−Cr−Al−Y alloy); it is the glue, so to speak, between blade and coating, shown in Figure 13.25. Most ceramics have a lower expansion coefficient than the super-alloy of the blade (compare $\alpha$ for $ZrO_2$ and Ni in Figure 12.4) so when the blade heats up it expands more than the coating and we know what that means: thermal stresses and cracking. The problem is solved (amazingly) by arranging that the coating is *already* cracked on a fine scale, with all the cracks running perpendicular to its surface, making an array of interlocking columns like a microscopic Giant's Causeway. When the blade expands the columns separate very slightly, but not enough for hot gas to penetrate to any significant extent; its protective thermal qualities remain. Next time you fly, reflect on all this—the plane you are in almost certainly has coated blades.

*Airframes* If you want to fly at speeds above Mach 1 (760 mph), aerodynamic heating becomes a problem. It is easiest to think of a static structure in a wind tunnel with a supersonic air stream flowing over it. In the boundary layer immediately adjacent to the structure, the velocity of the air is reduced to zero and its kinetic energy appears as heat, much of which is dumped into the skin of the structure. The surface temperature $T_s$ can be calculated; it is approximately

$$T_s \approx (1 + 0.2M_a^2)T_o$$

Here $M_a$ is the Mach number and $T_o$ the ambient temperature. Supersonic flight is usually at altitudes above 35,000 ft, where $T_o = -50\,°C$. Figure 13.26 shows what $T_s$ looks like.

This, not surprisingly, causes problems. First, creep causes a gradual change in dimension over time; wing deflection can increase, affecting aerodynamic performance. Second, thermal stress caused by expansion can lead to further creep damage because it adds to aerodynamic loads. Finally, the drop in modulus, $E$, brings greater elastic deflection and changes the buckling and flutter (meaning vibration) characteristics.

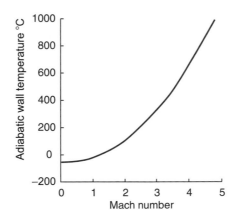

**Figure 13.26** Adiabatic heating as a function of speed.

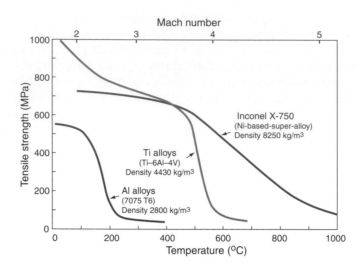

**Figure 13.27** The strength of alloys as a function of temperature.

In a large aircraft, the wing-tip displacement between ground and steady flight is about 0.5 m, corresponding to a strain in the wing spar (which is loaded in bending) of about 0.1%. To avoid loss of aerodynamic quality, the creep strain over the life of the aircraft must be kept below this. Figure 13.27 illustrates how the strength of three potential structural materials influences the choice. Where the strength is flat with temperature (on the left) the material is a practical choice. The temperature at which it drops off limits the Mach number, plotted across the top. Aluminum is light, titanium twice as dense, super-alloys four times so, so increasing the Mach number requires an increase in weight. A heavier structure needs more power and therefore fuel, and fuel too has weight—the structural weight of an aircraft designed to fly at Mach 3 is about three times that of one designed for Mach 1. Thus, it appears that there is a practical upper limit on speed. It has been estimated that sustained flight may be limited by the weight penalty to speeds below Mach 3.5.

*Creep relaxation* Creep causes pre-tensioned components to relax over time: bolts in hot turbine casings must be regularly tightened; pipe connectors carrying hot fluids must be readjusted. It takes only tiny creep strain $\varepsilon_{cr}$ to relax the stress—a fraction of the elastic strain $\varepsilon_{el}$ caused by pre-tensioning is enough, and $\varepsilon_{el}$ is seldom much greater than $10^{-3}$. Figure 13.28 shows a bolt that is tightened onto a rigid component so that the initial stress in its shank is $\sigma_i$. The elastic strain is then

$$\varepsilon_{el} = \frac{\sigma_i}{E}$$

If creep strain $\varepsilon_{cr}$ replaces part of $\varepsilon_{el}$ the stress relaxes. At any time $t$

$$\varepsilon_{tot} = \varepsilon_{el} + \varepsilon_{cr}$$

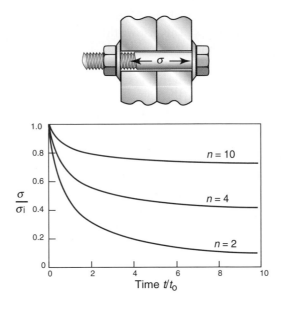

**Figure 13.28** Stress relaxation by creep.

This total strain $\varepsilon_{\text{tot}}$ is fixed because the component on which the bolt is clamped is rigid. Differentiating with respect to time, inserting equation (13.3) for $\dot{\varepsilon}_{\text{cr}}$ gives

$$\dot{\varepsilon}_{\text{tot}} = \dot{\varepsilon}_{\text{el}} + \dot{\varepsilon}_{\text{cr}} = \frac{\dot{\sigma}}{E} + B\sigma^n = 0$$

The time $t$ for the stress to relax from $\sigma_i$ to $\sigma$, for $n > 1$, is found by integrating over time:

$$t = \frac{1}{(n-1)BE}\left(\frac{1}{\sigma^{n-1}} - \frac{1}{\sigma_i^{n-1}}\right)$$

Solving for $\sigma/\sigma_i$ as a function of $t$ gives

$$\frac{\sigma}{\sigma_i} = \frac{1}{(t/t_o + 1)^{1/(n-1)}} \quad \text{with } t_o = (n-1)BE\sigma_i^{n-1}$$

The stress $\sigma/\sigma_i$ is plotted as a function of $t/t_o$ in Figure 13.28, for three values of $n$. The relaxation is most extensive when $n$ is small. As it rises, the drop in stress becomes less until, as $n$ approaches infinity, meaning rate-independent plasticity, it disappears. Since transient creep is neglected, these curves overestimate the relaxation time for the first tightening of the bolt, but improve for longer times.

## **13.8** Summary and conclusions

All material properties depend on temperature. Some, like the density and the modulus, change relatively little and in a predictable way, making compensation easy. *Transport properties,* meaning thermal conductivity and conductivity for matter flow (which we call diffusion), change in more complex ways. The last of these—diffusion—has a profound effect on mechanical properties when temperatures are high. To understand and use diffusion we need the idea of *thermal activation*—the ability of atoms to jump from one site to another, using thermal energy as the springboard. In crystals, atoms jump from vacancy to vacancy, but in glasses, with no fixed lattice sites, the jumps occur when enough free volume comes together to make an atom-sized hole. This atom motion allows the intermixing of atoms in a concentration gradient. External stress, too, drives a flux of atoms, causing the material to change shape in a way that allows the stress to do work.

Diffusion plays a fundamental role in the processing of materials, the subject of Chapters 18 and 19. It is also diffusion that causes creep and creep fracture. Their rates increase exponentially with temperature, introducing the first challenge for design: that of predicting the rates with useful accuracy. Exponential rates require precise data—small changes in activation energy give large changes in rates—and these data are hard to measure and are sensitive to slight changes of composition. And there is more than one mechanism of creep, compounding the problem. The mechanisms compete, the one giving the fastest rate winning. If you design using data and formulae for one, but under service conditions another is dominant, you are in trouble. Deformation mechanism maps help here, identifying both the mechanism and the approximate rate of creep.

A consequence of all this is that materials selection for high-temperature design, and the design itself, is based largely on empirical data rather than on modeling of the sort used for elastic and plastic design in Chapters 5, 7 and 10. Empirical data means plots, for individual alloys, of the creep strain-rate and life as functions of temperature and stress. For polymers a different approach is used. Because their melting points and glass temperatures are low, they are visco-elastic at room temperature. Then design can be based on plots of the *creep modulus*—the equivalent of Young's modulus for a creeping material. The creep modulus at a given temperature and time is read off the plot and used in standard solutions for elastic problems (Chapter 5) to assess stress or deflection.

The chapter ends with examples of the use of some of these methods.

## **13.9** Further reading

Cottrell, A. H. (1964). *The Mechanical Properties of Matter*. London, UK: Wiley. Library of Congress Number 64-14262. (*Now 40 years old, inevitably somewhat dated, and now out of print, but still the best introduction to the mechanical properties of matter that I have ever found. Worth hunting for.*)

Finnie, I., & Heller, W. R. (1959). *Creep of Engineering Materials*. New York, USA: McGraw-Hill. Library of Congress No. 58–11171. (*Old and somewhat dated, but still an excellent introduction to creep in an engineering context.*)

Frost, H. J., & Ashby, M. F. (1982). *Deformation Mechanism Maps*. Oxford, UK: Pergamon Press, ISBN 0-08-029337-9. (*A monograph compiling deformation mechanism maps for a wide range of materials, with explanation of their construction.*)

Penny, R. K., & Marriott, D. L. (1971). *Design for Creep*. New York, USA: McGraw-Hill. ISBN 07-094157-2. (*A monograph bringing together the mechanics and the materials aspects of creep and creep fracture.*)

Waterman, N. A., & Ashby, M. F. (1991). *The Elsevier Materials Selector*. Barking, UK: Elsevier Applied Science, ISBN 1-85166-605-2. (*A three-volume compilation of data and charts for material properties, including creep and creep fracture.*)

## **13.10** Exercises

Exercise E13.1    A high-temperature alloy undergoes power law creep when the applied stress is 70 MPa and the temperature is 1000 °C. The stress exponent $n$ is 5 and the activation energy $Q$ is 300 kJ/mol. Recent modification to the manufacturing process has improved the creep resistance. The stress can be raised to 80 MPa without changing the creep rate at 1000 °C. An aero-engine manufacturer is more interested in raising the operating temperature of their engine than increasing the working stresses. Estimate by how much the temperature could be raised by adopting the new manufacturing process.

Exercise E13.2    Pipework with a radius of 20 mm and a wall thickness of 4 mm made of 2 ¼ Cr Mo steel contains a hot fluid under pressure. The pressure is 10 MPa at a temperature of 600 °C. The table lists the creep constants for this steel. Calculate the creep rate of the pipe wall, assuming steady-state power-law creep.

| Material | Reference strain-rate $\dot{\varepsilon}_o$ (1/s) | Reference stress $\sigma_o$ (MPa) | Rupture exponent $m$ | Activation energy $Q_c$ (kJ/mol) |
|---|---|---|---|---|
| 2 ¼ Cr Mo steel | $3.48 \times 10^{10}$ | 169 | 7.5 | 280 |

Exercise E13.3    There is concern that the pipework described in the previous exercise might rupture in less than the design life of one year. If the Monkman-Grant constant for 2 ¼ Cr Mo steel is 0.06, how long will it last before it ruptures?

Exercise E13.4    If the creep rate of a component made of 2 ¼ Cr Mo steel must not exceed $10^{-8}$ / sec at 500 °C, what is the greatest stress that it can safely carry? Use the data listed in the previous two examples to find out.

Exercise E13.5    The designers of a chemical plant are concerned about the creep failure of a critical alloy tie bar. They have carried out creep tests on specimens of the alloy under the nominal service conditions of a stress $\sigma$ of 25 MPa at 620 °C, and found a steady-state creep rate $\dot{\varepsilon}$ of $3.1 \times 10^{-12}$ s$^{-1}$. In service it is expected that for 30% of the running time the stress and temperature may increase to 30 MPa and 650 °C respectively, while for the remaining time the stress and temperature will be at the nominal service values. Calculate the expected *average* creep rate in service.

It may be assumed that the alloy creeps according to the equation:

$$\dot{\varepsilon} = A\sigma^n \exp\left(-\frac{Q}{\bar{R}T}\right)$$

where $A$ and $Q$ are constants, $\bar{R}$ is the universal gas constant and $T$ is the absolute temperature. For this alloy, $Q = 160$ kJ mol$^{-1}$ and $n = 5$.

Exercise E13.6

(a) The steady-state creep strain-rate at constant temperature $\dot{\varepsilon}_{ss}$ depends on the applied stress according to

$$\dot{\varepsilon}_{ss} = B\sigma^n$$

where $B$ and $n$ are constants. State how $B$ varies with temperature and explain why this is the case.

(b) The figure below shows the steady-state creep response of a 12% Cr steel at 500 °C. Determine $B$ at this temperature, and the creep exponent $n$, for this material.

(c) A 12% Cr steel turbine blade of length 10 cm runs with an initial clearance of 100 μm between its tip and the housing in a generator operating at 515 °C. Determine the lifetime of the blade if it runs continuously, assuming a uniform tensile stress of 60 MPa is applied along its axis. (The creep activation energy $Q$ for 12% Cr steel is 103 kJ/mol.)

(d) Explain why in practice a more sophisticated calculation is required to determine the change in length of a rotating turbine blade.

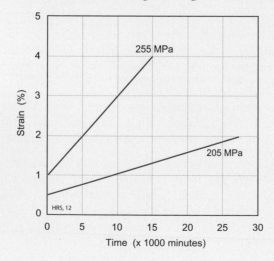

Exercise E13.7    Use the property chart in Figure 13.8 to identify materials that are suitable for carrying a tensile stress of 100 MPa at 950 °C, with a creep strain-rate below $10^{-6}$ s$^{-1}$. Use the contours of specific creep strength at this temperature, $\sigma_{950°C}/\rho$, to identify the lightest material. What other materials might be worth considering for operation at this temperature, according to Figure 13.7?

Exercise E13.8    The self-diffusion constants for aluminium are $D_o = 1.7 \times 10^{-4}$ m$^2$/s and $Q_d = 142$ kJ/mol. What is the diffusion rate in aluminium at 400 °C?

Exercise E13.9    Use your knowledge of diffusion to account for the following observations:

(a) The rate of diffusion of oxygen from the atmosphere into an oxide film is strongly dependent on the temperature and the concentration of oxygen in the atmosphere.

(b) Carbon diffuses far more rapidly than chromium through iron.

(c) Diffusion is more rapid in polycrystalline silver with a small grain size than in coarse-grained silver.

Exercise E13.10    Molten aluminium is cast into a copper mould having a wall thickness of 5 mm. The diffusion coefficient $D$ of aluminium in copper is 2.6 × $10^{-17}$ m$^2$/s at 500 °C and 1.6 ×$10^{-12}$ m$^2$/s at 1000 °C.

(i) Calculate the activation energy $Q$ for the diffusion of aluminium in copper. By considering the mechanisms of diffusion, list some factors that will affect the value of $Q$ in diffusion-controlled processes in crystalline solids.

(ii) Determine the coefficient of diffusion of aluminium in copper at 750 °C.

(iii) A useful 'rule-of-thumb' in diffusion problems states that the characteristic diffusion distance in time $t$ is given by $\sqrt{Dt}$, where $D$ is the diffusion coefficient for the temperature concerned. Make an order-of-magnitude estimate of the time taken for an aluminium atom to diffuse through the wall of the copper mould if the mould temperature is 750 °C.

Exercise E13.11    A steel component is nickel plated to give corrosion protection. To increase the strength of the bond between the steel and the nickel, the component is heated for 4 hours at 1000 °C. If the diffusion parameters for nickel in iron are $D_o = 1.9 \times 10^{-4}$ m$^2$/s and $Q_d = 284$ kJ/mol, how far would you expect the nickel to diffuse into the steel in this time?

Exercise E13.12    The diffusion coefficient at the melting point for materials is approximately constant, with the value $D = 10^{-12}$ m$^2$/s. What is the diffusion distance if a material is held for 12 hours at just below its melting temperature? This distance gives an idea of the maximum distance over which concentration gradients can be smoothed by diffusion.

Exercise E13.13    Find the diffusion coefficient $D$ for carbon in iron at 1000 °C, using the data below, and compare it with the thermal diffusivity of steel at 1000 °C, $a = 9 \times 10^{-6}$ m$^2$s$^{-1}$. Estimate how much faster heat 'diffuses' a distance $x$ in steel compared to the diffusion of carbon at 1000 °C.

For carbon diffusion in FCC iron: $D_o = 2.5 \times 10^{-5}$ m$^2$s$^{-1}$, and the activation energy $Q = 148$ kJ/mol.

Exercise E13.14  In the fabrication of a silicon semiconductor device, boron is diffused into silicon in two stages: a pre-deposition stage lasting 10 minutes at a temperature of 1000 °C, followed by a drive-in stage for one hour at 1100 °C. Compare the characteristic diffusion distances of these two stages. For the diffusion of boron in silicon, $D_o = 3.7 \times 10^{-6}$ m$^2$ s$^{-1}$, and the activation energy $Q = 333$ kJ/mol.

Exercise E13.15  A thick steel plate of thickness $2\,w = 0.2$ m is continuously cast into a long sheet. The casting process gives a non-uniform concentration of a key substitutional alloying element in the steel (due to the phenomenon of segregation, discussed in Guided Learning Unit 2). The initial distribution of solute is sinusoidal with the maximum concentration on the centre-line, falling to zero at the surface. A homogenisation treatment is proposed to try and redistribute the solute to a more uniform level. Solution of the differential equation for transient diffusion predicts that the concentration profile has the following form:

$$C(x,t) = C_o + C_o \left( \cos \frac{\pi x}{w} \right) \exp(-t/\tau), \text{ where } \tau = w^2/(\pi^2 D).$$

(a) Sketch the solute distribution for times $t = 0$, $\tau$ and $10\,\tau$.
(b) An attempt is made to homogenise the casting by holding at a high temperature for 12 hours. At the treatment temperature, the relevant diffusion constant $D = 4 \times 10^{-15}$ m$^2$s$^{-1}$. Will this treatment make any significant difference to the solute distribution?
(c) The casting process is modified to prevent the non-uniform solute distribution developing across the whole plate. A micro-probe analysis then identifies a similar sinusoidal variation of amplitude $\pm\, 0.2\, C_o$ across the grains in the casting, about a mean of $C_o$. If the grain size is 50 μm (i.e. in the solution for $C(x,t)$, $2w = 25$ μm), what is the maximum % deviation from the average concentration within the grain after the 12-hour heat treatment?

Exercise E13.16  A carbon steel component containing $C_o = 0.2$ weight % C is to be carburised at a temperature of 1000 °C, at which the maximum solubility of carbon is 1.6 weight %—this is the (constant) surface concentration $C_s$ throughout the treatment. The concentration as a function of time $t$ and depth below the surface $x$ is given by

$$C(x,t) = C_o + (C_s - C_o)\left\{ 1 - \text{erf}\left( \frac{x}{2\sqrt{Dt}} \right) \right\}$$

The diffusion coefficient $D$ for carbon diffusion in iron at $1000\,^{\circ}$C is $1.94 \times 10^{-11}$ m$^2$s$^{-1}$, and values for the error function are given in the table.

| X | 0 | 0.1 | 0.2 | 0.3 | 0.4 | 0.5 | 0.6 | 0.7 | 0.8 | 0.9 | 1.0 |
|---|---|-----|-----|-----|-----|-----|-----|-----|-----|-----|-----|
| erf(X) | 0 | 0.112 | 0.223 | 0.329 | 0.428 | 0.520 | 0.604 | 0.678 | 0.742 | 0.797 | 0.843 |

| X | 1.1 | 1.2 | 1.3 | 1.4 | 1.5 | $\infty$ |
|---|-----|-----|-----|-----|-----|----------|
| erf(X) | 0.880 | 0.910 | 0.934 | 0.952 | 0.966 | 1.0 |

Sketch the variation of carbon concentration C against distance $x$ below the surface, for treatment times of 15 minutes and 1 hour. By what factor does the diffusion depth increase between these two treatment times? Find the time required for carburisation to a depth of 0.3 mm, if the target concentration at this depth is $C = 0.5$ wt % carbon.

Exercise E13.17    (a) Explain briefly, with sketches as appropriate, the *differences* between the following pairs of mechanisms in metal deformation:
(i) room temperature yielding and power-law creep;
(ii) power-law creep and diffusional flow.
(b) Explain the following manufacturing characteristics:
(i) superplastic forming requires a fine grain size material;
(ii) creep resistant alloys are often cast as single crystals.

Exercise E13.18    (a) Briefly summarise the important microstructural characteristics of a creep-resistant material, distinguishing between features that influence diffusional flow and power-law creep.
(b) Polycrystalline copper has a steady-state creep rate of $10^{-4}$ s$^{-1}$ at $560\,^{\circ}$C when subjected to a given tensile stress. The activation energy $Q$ for self-diffusion in copper is 197 kJ/mol. Calculate the steady-state creep-rate of copper at $500\,^{\circ}$C at the same stress.

Exercise E13.19    Creep data for a stainless steel at two constant temperatures are shown in the figure below. The steel obeys the steady-state creep equation

$$\dot{\varepsilon} = A\,\sigma^n \exp\left(-Q/\overline{R}T\right)$$

where $\overline{R} = 8.314$ J/mol.K.

(a) A specimen of the steel of length 750 mm was subjected to a tensile stress $\sigma = 40$ MPa at a temperature $T = 538\,^{\circ}$C. Calculate the extension after 5000 hours.
(b) Use the creep data to evaluate the constants $n$ and $Q$.
(c) A solid cylinder of the same steel was designed for operation as a support for a turbine housing under a uniform tensile load $P = 6$kN and a temperature $T = 450\,^{\circ}$C. A routine check after 10,000 hours in service revealed that the extension of the cylinder was 50% higher than

expected, due to the applied load being higher than the design specification. The extension which had occurred was half of the maximum allowable before the component would need to be replaced.

 (i) Evaluate the load which had actually been applied, given that the operating temperature had been correctly maintained.
(ii) The loading on the component could not be corrected, but a drop in operating temperature of 20 °C was possible. Check whether this modification would be sufficient to ensure a further 40,000 hours of operation before replacement.

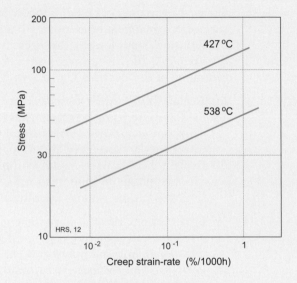

Exercise E13.20  The figure below shows a deformation-mechanism map for a nickel alloy. The contours are lines of equal strain-rate for steady-state creep in units of $s^{-1}$.

(a) What is the maximum allowable stress at 600 °C if the maximum allowable strain-rate is $10^{-10}\,s^{-1}$? What would be the creep rate if the stress were increased by a factor of 2, *or* the temperature raised by 100 °C?

(b) The strain-rate contours on the map for both diffusional flow and power-law creep are based on curve fits to the creep equation:

$$\dot{\varepsilon} = A\sigma^{n} \exp\left(-\frac{Q}{RT}\right)$$

Use the strain-rate contours on the map to estimate the value of the stress exponent $n$, for diffusional flow and for power-law creep.

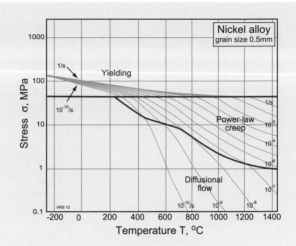

Exercise E13.21  A stainless steel suspension cable in a furnace is subjected to a stress of 100 MPa at 700 °C. Its creep rate is found to be unacceptably high. By what mechanism is creep occurring? What action would you suggest to tackle the problem? The figure shows the deformation mechanism map for the material.

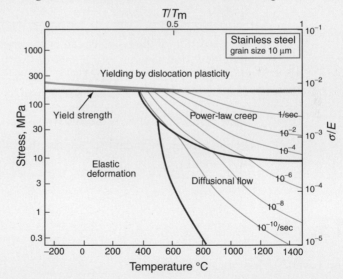

Exercise E13.22  The wall of a pipe of the same stainless steel as that of the previous exercise carries a stress of 3 MPa at the very high temperature of 1000 °C. In this application it, too, creeps at a rate that is unacceptably high. By what mechanism is it creeping? What action would you suggest to tackle the problem?

Exercise E13.23 It is proposed to make a shelf for a hot-air drying system from Acrylic sheet. The shelf is simply supported, as in the diagram, and has a width $w = 500$ mm, a thickness $t = 8$ mm and a depth $b = 200$ mm. It must carry a distributed load of 50 N at 60 C with a design life of 8000 hours (about a year) of continuous use. Use the creep modulus plotted in Figure 13.19 and the solution to the appropriate elastic problem (Chapter 5) to find out how much it will sag in that time.

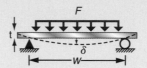

Exercise E13.24 In a series of stress relaxation experiments on PMMA conducted at 25 °C, a cylindrical bar of polymer is subjected to an instantaneous step-wise compressive strain of 1%, at time $t = 0$. The stress in the bar decays with time according to the following table.

| Stress (MPa) | 3.2 | 2.4 | 1.1 | 0.46 | 0.18 | 0.06 |
|---|---|---|---|---|---|---|
| Time (s) | 0 | 1 | 3 | 6 | 9 | 12 |

The relaxation of the polymer can be represented by a model of a spring of modulus $E$ (for which $E = \sigma/\varepsilon$) in series with a dashpot of viscosity $\eta$, (for which $\eta = \sigma/\dot{\varepsilon}$), as in Figure 13.18. The relaxation response is given by

$$\sigma = \sigma_0 \exp(-t/\tau)$$

where $\tau = \eta/E$, and $\sigma_0$ is the initial stress at $t = 0$.

Use the experimental data to estimate the values of $E$, $\eta$ and the relaxation time $\tau = \eta/E$.

## **13.11** Exploring design with CES

Use Level 2 unless otherwise stated.

Exercise E13.25 Use the Search facility in CES to find:
(a) Materials for turbine blades.
(b) Materials for thermal barrier coatings.

Exercise E13.26   Use the CES software to make a bar chart for the glass temperatures of polymers like that for melting point in the text (Figure 13.6). What is the range of glass temperatures for polymers?

Exercise E13.27   The analysis of thermal barrier coatings formulated the design constraints. Use the selector to find the best material. Here is a summary of the constraints and the objective.

| Function | • Thermal barrier coating |
|---|---|
| Constraints | • Maximum service temperature >1300 °C |
| | • Adequate strength: $\sigma_y > 400$ MPa |
| Objective | • Minimise thermal conductivity $\lambda$ |
| Free variable | • Choice of material |

Exercise E13.28   Find materials with $T_{max} > 500$ °C and the lowest possible thermal conductivity. Switch the database to Level 3 to get more detail. Report the three materials with the lowest thermal conductivity.

## **13.12** Exploring the science with CES Elements

Exercise E13.29   The claim is made in the text (equation (13.13a)) that the activation energy for self-diffusion, normalised by $\bar{R}T_m$, is approximately constant for metals:

$$\frac{Q_d}{\bar{R}T_m} \approx 18$$

Make a bar chart of this quantity and explore the degree to which it is true.

Exercise E13.30   The claim is made in the text (equation (13.13b)) that the diffusion coefficient of metals, evaluated at their melting point, is also approximately constant:

$$D_{T_m} = D_o \exp - \left(\frac{Q_d}{\bar{R}T_m}\right) \approx 10^{-12} \text{ m}^2/\text{s}$$

Make a bar chart of this quantity and explore the degree to which it is true.

Exercise E13.31   What is a 'typical' value for the pre-exponential, $D_o$? Is it roughly constant for the elements? Make a chart with the activation energy for diffusion, $Q_d$, on the x-axis and the pre-exponential $D_o$ on the y-axis to explore this.

Exercise E13.32 The diffusive jump shown in Figure 13.9 requires that the diffusing atom breaks its bonds in its starting position in order to jump into its final one. You might, then, expect that there would be at least an approximate proportionality between the activation energy for self-diffusion $Q_d$ and the cohesive energy $H_c$ of the material. Make a chart with $H_c$ on the x-axis and $Q_d$ on the y-axis. Report what you find.

# Chapter 14
# Conductors, insulators and dielectrics

Conductors (left) and insulators (right). Copper, aluminum and silver (if you can afford it) make good conductors. Pyrex, porcelain and ceramics, like alumina, make good insulators. (Images courtesy of the Copper Development Association)

# **14.1** Introduction and synopsis

York Minster, constructed between 1220 and 1400, is one of the great cathedrals of Europe. Many fine sermons have been preached from its pulpit, but when, in 1984, the Archbishop of York expressed his disbelief in the Virgin Birth, the Minster was struck by lightning and severely damaged. Some put this down to the wrath of God, but the reality is simpler: the cathedral was inadequately grounded (earthed). There is a message here: even when you are designing cathedrals, electrical properties matter.

This chapter is about the simplest of these: they relate to conduction, insulation and dielectric behaviour. *Electrical conduction* (as in lightning conductors) and *insulation* (as in electric plug casings) are familiar properties. Superconductivity, semiconduction and dielectric behaviour may be less so. *Superconductivity* is a special characteristic of many materials at very low temperatures, in which there is no resistance to current flow. *Semiconductors*, the materials of transistors and silicon chips, fall somewhere between conductor and insulator, as the name suggests. A *dielectric* is an insulator. It is usual to use the word 'insulator' when referring to its inability to conduct electricity, and to use 'dielectric' when referring to its behaviour in an electric field. Three properties are of importance here. The first, the *dielectric constant* (or *relative permittivity*), has to do with the way the material acquires a dipole moment (it *polarises*) in an electric field. The second, the *dielectric loss factor*, measures the energy dissipated when radio-frequency waves pass through a material, the energy appearing as heat (the principle of microwave cooking). The third is the *dielectric breakdown potential*, and this takes us back to York Minster. Lightning is dielectric breakdown, and it can be as damaging on a small scale—in a piece of electrical equipment, for example—as on a large one.

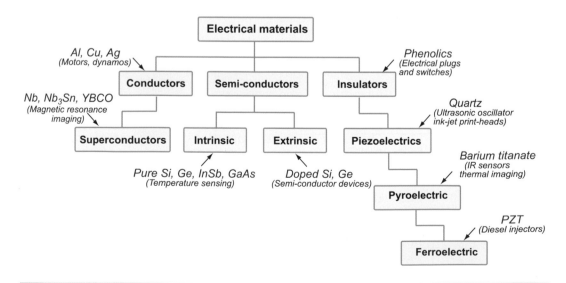

**Figure 14.1**   The hierarchy of electrical behaviour, with examples of materials and applications. Their nature and origins are described in this chapter.

There are many different kinds of electrical behaviour, all of them useful. Figure 14.1 gives an overview, with examples of materials and applications. This chapter introduces electrical properties and their physical origins, and gives examples of design for electrical performance.

## **14.2** Conductors, insulators and dielectrics

***Resistivity and conductivity*** The electrical resistance $R$ (units: ohms[1], symbol $\Omega$) of a rod of material is the potential drop $V$ (volts[2]) across it, divided by the current $i$ (A, or amps[3]) passing through it, as in Figure 14.2. This relationship is Ohm's Law:

$$R = \frac{V}{i} \tag{14.1}$$

Resistance also determines the power $P$ dissipated when a current passes through a conductor (measured in Watts[4]):

$$P = i^2 R \tag{14.2}$$

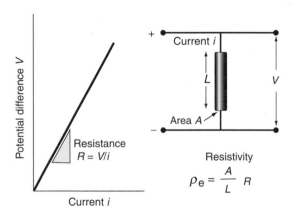

**Figure 14.2** Measuring electrical resistivity $\rho_e$. Its value ranges from 1 to $10^{24}$ $\mu\Omega$.cm.

---

[1] George Simon Ohm (1789–1854), dropped out of his studies at the University of Erlangen, preferring billiards to mathematics, but later became absorbed in the theory of electricity and magnetism.

[2] Alessandro Guiseppe Antonio Anastasio Volta (1745–1827), Italian physicist, inventor of the battery, or, as it was then known, the Voltaic Pile. His face used to appear on the 10,000 lire note.

[3] André Marie Ampère (1775–1836), French mathematician and physicist, largely self-taught and said to have mastered all known mathematics by the age of 14. His life was not a happy one: his father was guillotined during the French revolution, and he was miserably married and had difficult children—yet he remained extensively productive up to his death, contributing to the theories of electricity, light, magnetism and chemistry (he discovered fluorine).

[4] James Watt (1763–1819), instrument maker and inventor of the steam engine (the idea came to him whilst 'walking on a fine Sabbath afternoon'), which he doggedly developed. Unlike so many of the footnoted characters of this book, Watt, in his final years, was healthy, happy and famous.

The material property that determines resistance is the *electrical resistivity*, $\rho_e$. It is related to the resistance by

$$\rho_e = \frac{A}{L} R \tag{14.3}$$

where $A$ is the section and $L$ the length of a test rod of material—think of it as the resistance of a unit cube of the material. Its units in the metric system are $\Omega$.m, but it is commonly reported in units of $\mu\Omega$.cm. It has an immense range, from a little more than $10^{-8}$ in units of $\Omega$.m for good conductors (equivalent to 1 $\mu\Omega$.cm, which is why these units are still used) to more than $10^{16}$ $\Omega$.m ($10^{24}$ $\mu\Omega$.cm) for the best insulators. The electrical conductivity $\kappa_e$ is simply the reciprocal of the resistivity. Its units are Siemens per meter (S/m or $(\Omega.m)^{-1}$).

---

### Example 14.1

A potential difference of 0.05 volts is applied across a nickel wire 100 mm long and 0.2 mm in diameter. What current flows in the wire? How much power is dissipated in the wire? The resistivity $\rho_e$ of nickel is 9.5 $\mu\Omega$.cm.

*Answer.* The current is given by Ohm's law: $V = iR$ where $V$ is the potential difference, $i$ is the current and $R$ is the resistance. Converting the resistivity from $\mu\Omega$.cm to $\Omega$.m by multiplying by $10^{-8}$ gives $\rho_e = 9.5 \times 10^{-8}$ $\Omega$.m, so the resistance $R$ of the wire is

$$R = \frac{\rho_e L}{A} = 9.5 \times 10^{-8} \times \frac{0.1}{\pi(1 \times 10^{-4})^2} = 0.3 \ \Omega$$

The current in the wire is

$$i = \frac{V}{R} = \frac{0.05}{0.3} = 0.165 \text{ amps}$$

The power dissipated in the wire is $P = i^2 R = 8.33 \times 10^{-3}$ Watts.

---

*Temperature-dependence of resistivity* The resistivity $\rho_e$ of metals increases with temperature because thermal vibrations scatter electrons. The resistivity of semiconductors, by contrast, decreases as temperature increases because thermal energy allows more carriers to cross the band gap, entering the conduction band. The temperature dependence of resistivity $\rho_e$ is described by

$$\beta_e = \frac{1}{\rho_{e,o}} \frac{d\rho_e}{dT}$$

where $\rho_{e,o}$ is the value of the resistivity at room temperature, and $\beta_e$ has dimension $K^{-1}$. The dimensionless quantity

$$T_m \beta_e = \frac{T_m}{\rho_{e,o}} \frac{d\rho_e}{dT}$$

is the factor by which the resistivity increases between room temperature and the melting point $T_m$. For metals this factor lies between 2 and 10. Incandescent lamps (now rare) depend on this: as the filament heats, its resistance rises, and this stabilises the temperature at a value that does not cause burn-out.

*Superconductors*    When metals are cooled their resistivity falls. Many retain a finite resistivity at absolute zero, but some show a startling change, losing all resistance at a critical temperature, $T_c$, above 0 K (Figure 14.3 (a)). Below $T_c$ a current in a superconducting material flows without any resistive loss; above $T_c$ superconductivity is suppressed. Superconductivity is closely related to magnetism (Chapter 15). For now, we note that a material that is superconducting will expel a magnetic field completely (the Meissner effect). An applied magnetic field lowers the temperature at which superconduction starts, and a field greater than the zero-Kelvin critical field $H_{c,o}$ suppresses it completely (Figure 14.3 (b)). The critical field $H_c$ depends on temperature according to

$$H_c = H_{c,o}\left(1 - \left(\frac{T}{T_c}\right)^2\right) \tag{14.4}$$

Any electrical current induces a magnetic field. Thus the existence of a critical field implies an upper-limiting current density that can by carried by the superconductor before its own field quenches the superconduction.

The critical temperatures and fields of superconducting materials are plotted in Figure 14.4. Many pure metals become superconducting, though only at very low temperatures, and all have low critical fields, too low to be of much practical use. The intermetallic compounds $Nb_3Sn$, $NbTi$, $Nb_3Ge$ and $Nb_3Al$ have higher critical temperatures and fields—high enough to be practical for MRI scanners and superconducting energy storage. In 1986 a third class of superconductor was discovered with critical temperatures and fields that were higher still. They are (surprisingly) ceramics, with a high resistivity at room temperature, yet

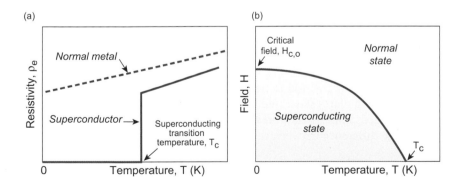

**Figure 14.3**    (a) Superconduction: a transition to zero resistivity at low temperatures; (b) the relationship between critical temperature and field.

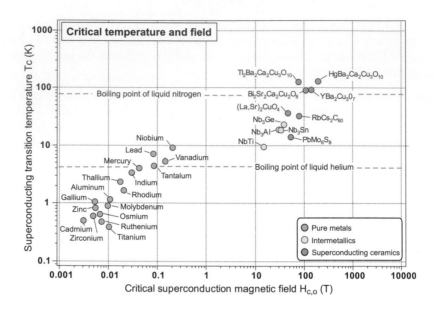

**Figure 14.4** The critical fields and temperatures for metallic and ceramic superconductors. Note the boiling points of liquid nitrogen and helium.

superconducting at temperatures as high as 130 K. This is significant because it is above the boiling point of liquid nitrogen, a cheap and widely available refrigerant, making super-conducting devices practical. The most widely used at present is the oxide ceramic $YBa_2Cu_3O_7$ ('YBCO'), used in high-field magnets and magnetic resonance imaging and mag-lev trains.

*Dielectric properties*  First, a reminder of what is meant by a *field*: it is a region of space in which objects experience forces if they have the right properties. Charge creates an *electric field*, *E*. The electric field strength between two oppositely charged plates separated by a distance *t* and with a potential difference *V* between them is

$$E = \frac{V}{t} \tag{14.5}$$

and is independent of position except near the edge of the plates.

Two conducting plates separated by a dielectric make a *capacitor* (Figure 14.5). Capacitors (sometimes called *condensers*) store charge. The charge *Q* (coulombs[5]) is directly proportional to the voltage difference between the plates, *V* (volts):

$$Q = CV \tag{14.6}$$

---

[5] Charles Augustin Coulomb (1736–1806), military physicist, laid the foundations of both the mathematical description of electrostatic forces and the laws of friction. Despite his appointment as Intendant des Eaux et Fontaines de Paris, he made no known contributions to hydrodynamics.

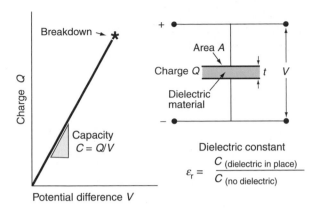

**Figure 14.5** Dielectric constant and dielectric breakdown. The capacitance of a condenser is proportional to the dielectric constant, and the maximum charge it can hold is limited by breakdown.

where $C$ (farads[6]) is the capacitance. The capacitance of a parallel plate capacitor of area A and spacing $t$, separated by empty space (or by air) is

$$C = \varepsilon_0 \frac{A}{t} \qquad (14.7)$$

where $\varepsilon_0$ is the *permittivity of free space* ($8.85 \times 10^{-12}$ F/m, where F is farads) . If the empty space is replaced by a dielectric, capacitance increases. This is because the dielectric *polarises*. The field created by the polarisation opposes the field $E$, reducing the voltage difference $V$ needed to support the charge. Thus the capacity of the condenser is increased to the new value

$$C = \varepsilon \frac{A}{t} \qquad (14.8)$$

where $\varepsilon$ is the *permittivity of the dielectric* with the same units as $\varepsilon_0$. It is usual to cite not this but the *relative permittivity* or *dielectric constant*, $\varepsilon_r$:

$$\varepsilon_r = \frac{C_{\text{with dielectric}}}{C_{\text{no dielectric}}} = \frac{\varepsilon}{\varepsilon_0} \qquad (14.9)$$

making the capacitance

$$C = \varepsilon_r \, \varepsilon_0 \frac{A}{t} \qquad (14.10)$$

---

[6] Michael Faraday (1791–1867), brilliant experimentalist both in physics and chemistry, discoverer of electromagnetic induction and inspiring lecturer at the Royal Institution, London.

Being a ratio, $\varepsilon_r$ is dimensionless. Its value for empty space and, for practical purposes, for most gases, is 1. Most dielectrics have values between 2 and 20, though low-density foams approach the value 1 because they are largely air. Ferroelectrics are special: they have values of $\varepsilon_r$ as high as 20,000. We return to ferroelectrics later in this chapter.

---

### Example 14.2

(a) A capacitor has a plate area of $A = 0.1$ m$^2$ separated by an air gap of $t = 100$ μm. What is its capacitance? (b) The air gap is replaced by a dielectric of thickness $t = 1$ μm with a dielectric constant $\varepsilon_r = 20$. What is its capacitance now?

*Answer.*

(a) The capacitance is $C_1 = \varepsilon_o \frac{A}{t} = 8.85 \times 10^{-12} \times \frac{0.1}{100 \times 10^{-6}} = 8.85 \times 10^{-9}\text{F} = 8.85$ pF

(b) The capacitance is $C_2 = \varepsilon_r \varepsilon_o \frac{A}{t} = 20 \times 8.85 \times 10^{-12} \times \frac{0.1}{10^{-6}} = 1.77 \times 10^{-5}\text{F} = 17.7$μF

---

Capacitance is one way to measure the dielectric constant of a material (Figure 14.5). The charge stored in the capacitor is measured by integrating the current that flows into it as the potential difference $V$ is increased. The ratio $Q/V$ is the capacitance. The dielectric constant $\varepsilon_r$ is calculated from equation (14.9).

Small capacitors, with capacitances measured in micro-farads (μF) or pico-farads (pF), are used in R-C circuits to tune oscillations and give controlled time delays. The time constant for charging or discharging a capacitor is

$$\tau = RC \tag{14.11}$$

where $R$ is the resistance of the circuit. When charged, the energy stored in a capacitor is

$$E_c = \frac{1}{2}\, QV = \frac{1}{2}\, CV^2 \tag{14.12}$$

'Super-capacitors', with capacitances measured in farads, store enough energy to power a hybrid car.

The *charge density*, $D$ (units Coulombs/m$^2$), on the surface of a dielectric like that of the condenser of Figure 14.5 is

$$D = \varepsilon_r\, \varepsilon_o\, E \tag{14.13}$$

where $E$ is the electric field (voltage across the dielectric divided by its thickness).

The *breakdown field* or *dielectric strength of a dielectric*, $E_b$ (units: V/m or, more usually MV/m), is the electrical field gradient at which an insulator breaks down and a damaging surge of current flows through it. It is measured by increasing, at a uniform rate, a 60 Hz alternating potential applied across the faces of a plate of the material in a configuration like that of Figure 14.5 until breakdown occurs, typically at a potential gradient of between 1 and 100

MV/m. The maximum charge density that a dielectric can carry is thus when the field is just below its breakdown field:

$$D_{max} = \varepsilon_r \varepsilon_o E_b \tag{14.14}$$

and the maximum energy density is

$$\frac{1}{2} \frac{CV^2}{At} = \frac{1}{2} \varepsilon_r \varepsilon_o E_b^2 \tag{14.15}$$

---

### Example 14.3

A supercapacitor is required that is able to store an energy of $E_c = 1$ kJ when a potential difference $V = 100$ Volts is applied to its plates. The dielectric constant of the dielectric is $\varepsilon_r = 10,000$ and its breakdown potential is $E_b = 20$ MV/m. What is the minimum area of plate required?

*Answer.* Breakdown occurs if $\frac{V}{t} \geq E_b$, requiring that the dielectric have a thickness $t \geq 5$ μm. The energy stored in a capacitor is $E_c = \frac{1}{2} CV^2$, so a capacitance of $C = \frac{2 \times 1000}{(100)^2} = 0.2$ F is needed. To achieve this with a dielectric of thickness $t \geq 5$ μm and dielectric constant 10,000 requires an area

$$A = \frac{Ct}{\varepsilon_r \varepsilon_o} = 11.3 \text{ m}^2.$$

---

The *loss tangent* and the *loss factor* take a little more explanation. We shall see in Section 14.4 that polarisation involves the small displacement of charge (either of electrons or of ions) or of molecules that carry a dipole moment when an electric field is applied to the material. An oscillating field drives the charge between two alternative configurations. This charge-motion is like an electric current that—if there were no losses—would be 90° out of phase with the voltage. In real dielectrics this current dissipates energy, just as a current in a resistor does, giving it a small phase shift, $\delta$ (Figure 14.6). The *loss tangent*, tan $\delta$, also called the *dissipation factor*, $D$, is the tangent of the loss angle (the phase shift between the voltage and current in Figure 14.6). The power factor, $P_f$, is the sine of the loss angle. When $\delta$ is small, as it is for the materials of interest here, all three are essentially equivalent:

$$\tan \delta \approx \sin \delta \approx P_f \approx D \tag{14.16}$$

More useful, for our purposes, is the *loss factor* $L$, which is the loss tangent times the dielectric constant:

$$L = \varepsilon_r \tan \delta \tag{14.17}$$

It measures the energy dissipated by a dielectric when in an oscillating field. If you want to select materials to minimise or maximise dielectric loss, then the measure you want is $L$.

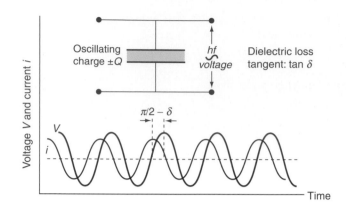

**Figure 14.6** Dielectric loss. The greater the loss factor, the greater is the microwave coupling and heating.

When a dielectric material is placed in a cyclic electric field of amplitude $E$ and frequency $f$, power $P$ is dissipated and the field is correspondingly attenuated. The power dissipated per unit volume, in $W/m^3$, is

$$P \approx f\, E^2\, \varepsilon \tan \delta = f\, E^2\, \varepsilon_0\, \varepsilon_r \tan \delta \qquad (14.18)$$

where, as before, $\varepsilon_r$ is the dielectric constant of the material and $\tan \delta$ is its loss tangent. This power appears as heat and is generated uniformly (if the field is uniform) through the volume of the material. Thus the higher the frequency or the field strength and the greater the loss factor $\varepsilon_r \tan \delta$, the greater is the heating and energy loss. Sometimes this dielectric loss is exploited in processing—for example, in radio frequency welding of polymers.

All dielectrics change shape in an electric field, a consequence of the small shift in charge that allows them to polarise; the effect is called *electrostriction*. Electrostriction is a one-sided relationship in that an electric field causes deformation, but deformation does not produce an electric field. *Piezoelectric* materials, by contrast, display a two-sided relationship between polarisation and deformation: a field induces deformation, and deformation induces charge differences between its surfaces, thus creating a field. The piezoelectric coefficient is the strain per unit of electric field, and although it is very small, it is a true linear effect, and this makes it useful: when you want to position or move a probe with nano-scale precision it is just what you need. *Pyroelectric* materials contain molecules with permanent dipole moments that, in a single crystal, are aligned, giving the crystal a permanent polarisation. When the temperature is changed, the polarisation changes, creating surface charges or, if the surfaces are connected electrically, a pyroelectric current—the principle of intruder-detection systems and of thermal imaging. *Ferroelectric* materials, too, have a natural dipole moment: they are polarised to start with, and the individual polarised molecules line up so that their dipole moments are parallel, like magnetic moments in a magnet. Their special feature is that the direction of polarisation can be changed by applying an electric field, and the change causes a change of shape. Applications of all of these are described in Section 14.5.

**Example 14.4**

(a) An oscillating voltage with an amplitude of 100 volts at a frequency $f = 60$ Hz appears across a nylon insulator thickness $x = 1$ mm. Nylon has a dielectric loss factor $L \approx 0.1$. How much power is dissipated per unit volume in the nylon? Assuming no heat loss, how fast will the temperature of the nylon rise? (Nylon has a volumetric specific heat $C_p = 3 \times 10^6$ J/m$^3$.K). (b) If the frequency was, instead, $f = 1$ GHz, what is the heating rate?

*Answer.*

(a) The amplitude of the field is $E = \frac{V}{x} = 10^5$ V/m. The power dissipated is

$$P = f\, E^2 \varepsilon_o L = 0.53 \ \text{W/m}^3$$

The rate of temperature rise is

$$\frac{dT}{dt} = \frac{P}{C_p} = 1.8 \times 10^{-7} \, ^\circ\text{C/s}$$

(b) If the frequency is raised to $f = 1$ GHz the heating rate becomes $\frac{dT}{dt} = 3 \, ^\circ\text{C/s}$

## **14.3** Charts for electrical properties

All materials have an electrical resistivity, and all dielectrics have a dielectric constant and loss factor. Property charts for these are useful because they allow comparison and selection from a large and varied population. Certain other properties are peculiar to only a handful of materials: useful piezoelectric and ferroelectric behaviour are examples. Selection is then a case of a choice among a small number of candidates, easily explored by direct comparison. Property charts for these are not necessary.

*Thermal conductivity and electrical resistivity*   The first chart, Figure 14.7, shows the thermal conductivity $\lambda$ and the electrical resistivity $\rho_e$. The first has a range of $10^5$, the second a range of $10^{28}$; no other material property has such a wide range. The chart is useful for selecting materials for applications in which these two properties are important. For metals, thermal and electrical conduction are linked because both depend on free electrons, giving the obvious correlation between the two at the upper left of the chart

$$\lambda \approx \frac{1400}{\rho_e} \tag{14.19}$$

($\lambda$ in W/m.K, $\rho_e$ in $\mu\Omega$.cm). This is known as the Wiedemann-Franz law.

*Strength and electrical resistivity*   The second chart, Figure 14.8, shows electrical resistivity $\rho_e$ and strength $\sigma_{el}$. Metals lie at the extreme left: copper (and silver, not shown) have the

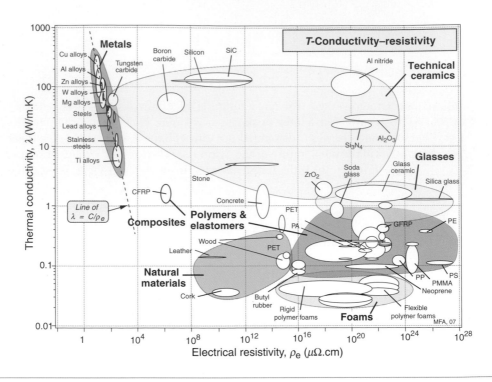

**Figure 14.7** Material property chart for electrical resistivity $\rho_e$ and thermal conductivity, $\lambda$.

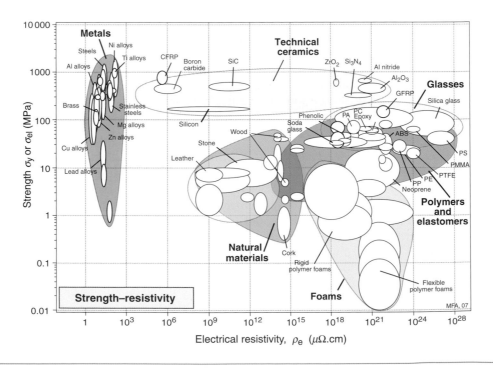

**Figure 14.8** Material property chart for electrical resistivity $\rho_e$ and strength (elastic limit) $\sigma_{el}$.

lowest resistivities, with aluminum only just above. The strongest alloys have resistivities that are 10 to 100 times larger. Ceramics such as alumina, aluminum nitride and silica, at the upper right, have very high resistivities and are strong; they are used for making insulators for power lines and substrates for electronic circuitry. Below them lie polymers that are less strong, but are flexible, easily moulded and excellent insulators; they are used for cable sheathing, insulation of switch gear and the like.

*Dielectric constant, dielectric breakdown and limiting electric energy density*   The chart of Figure 14.9 gives an overview of dielectric behaviour. Molecules containing polar groups, like nylon, polyurethane and ionic-bonded ceramics, react strongly to an electric field and polarise, giving them high dielectric constants. Those with purely covalent or Van der Waals bonding do not polarise easily and have low value. The dielectric strengths differ too—polymers have higher values than ceramics or glasses. The contours show values of the upper limiting energy density, which depends on both dielectric constant and dielectric strength (equation (14.15)).

*Strength and dielectric loss*   The final chart, Figure 14.10, shows dielectric loss factor $L$ and strength $\sigma_{el}$. Materials that are transparent to microwaves (radar uses microwave frequencies)

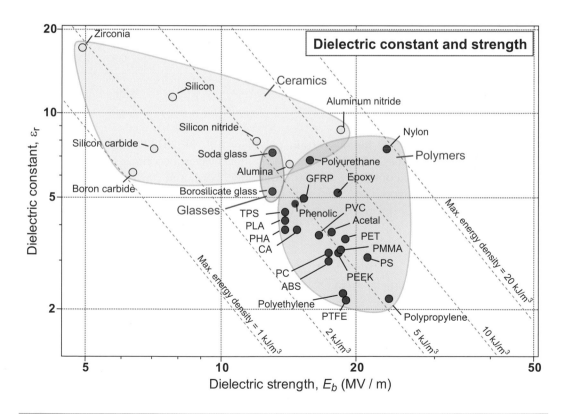

**Figure 14.9**   The dielectric properties of materials. The contours show the maximum energy density, $\frac{1}{2}\varepsilon_r\,\varepsilon_0\,E_b^2$, where $\varepsilon_0$ is the permittivity of vacuum, $8.85\times10^{-12}$ F/m.

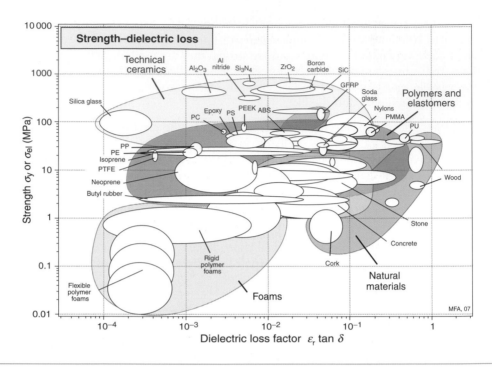

**Figure 14.10** Material property chart for dielectric loss, $\varepsilon_r \tan \delta$ and strength (elastic limit) $\sigma_{el}$.

lie at the left: polymer foams, certain polymers (PP, PE, PTFE) and some ceramics have particularly low values. Materials that absorb microwaves, becoming hot in the process, lie at the right: polymers containing a polar group, like nylon and polyurethane, and natural materials like woods (woods are dried by microwave heating). Water, which has a polar molecule, has a particularly large loss factor, which is why most foods can be cooked in a microwave oven.

## 14.4 Drilling down: the origins and manipulation of electrical properties

***Electrical conductivity*** An electric field, $E$ (volts/m), exerts a force $Ee$ on a charged particle, where $e$ is the charge it carries. Solids are made up of atoms containing electrons that carry a charge $-e$ and a nucleus containing protons, each with a positive charge $+e$. If charge-carriers can move, the force $Ee$ causes them to flow through the material—that is, it conducts. Metals are *electron-conductors*, meaning that the charge-carriers are the electrons. In ionic solids (which are composed of negatively and positively charged ions like $Na^+$ and $Cl^-$) the diffusive motion of ions allows *ionic conduction*, but this is only possible at temperatures at which diffusion is rapid. Many materials have no mobile electrons, and at room temperature they are too cold to be ionic conductors. The charged particles they contain still feel a force in an electric field and it is enough to displace the charges slightly, but they are unable to move more than a tiny fraction of the atom spacing. These are insulators; the small displacement of charge gives them dielectric properties.

How is it that some materials have mobile electrons and some do not? To explain this we need two of the stranger results of quantum mechanics. Briefly, the electrons of an atom occupy discrete energy states or orbits, arranged in shells (designated 1, 2, 3 etc. from the innermost to the outermost); each shell is made up of sub-shells (designated s, p, d and f), each of which contains 1, 3, 5 or 7 orbits respectively. The electrons fill the shells with the lowest energy, two electrons of opposite spin in each orbit; the *Pauli*[7] *exclusion principle* prohibits an energy state with more than two. When $n$ atoms (a large number) are brought together to form a solid, the inner electrons remain the property of the atom on which they started, but the outer ones interact (Figure 14.11). Each atom now sits in the field created by the charges of its neighbours. This has the effect of decreasing slightly the energy levels of electrons spinning in a direction favoured by the field of its neighbours and raising that of those with spins in the opposite direction, splitting each energy level. Thus the discrete levels of an isolated atom broaden, in the solid, into *bands* of very closely spaced levels. The number of electrons per atom that have to be accommodated depends only on the atomic number of the atoms. The electrons fill the bands from the bottom, lowest, energy level and progressively fill the levels above until all are on board, so to speak. The topmost filled energy level at zero Kelvin is called the Fermi[8] energy (more on this later). Above 0 K, electrons have a finite probability of occupying higher levels; the Fermi level is then defined as the energy level having the probability that it is exactly half filled with electrons. Levels of lower energy than the Fermi level tend to be entirely filled with electrons, whereas energy levels higher than

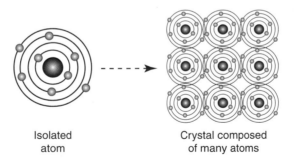

Isolated
atom

Crystal composed
of many atoms

Figure 14.11   When atoms are brought together to form crystals, the outermost electrons interact, and the discrete energy levels split into bands.

---

[7] Wolfgang Joseph Pauli (1900–1958), quantum theorist, conceiver of the neutrino and Nobel Prize winner. He was not, however, a happy man, requiring psychotherapy, which he received from none other than the great psychoanalyst Karl Jung.

[8] Enrico Fermi (1901–1954), devisor of the statistical laws known as Fermi statistics governing the behavior of electrons in solids. He was one of the leaders of the team of physicists on the Manhattan Project for the development of the atomic bomb.

the Fermi tend to be empty. If two materials with different Fermi levels are placed in contact, electrons flow from the material with the higher Fermi level into the one with the lower Fermi level, creating a charge difference between them. The electric field associated with this charge difference raises the lower Fermi level and lowers the higher Fermi level until the two are equal.

An electron at the Fermi level still has an energy that is lower than it would have if it were isolated in vacuum far from the atoms. This energy difference is called, for historical reasons, the *work function*, because it is the work that is required to remove an electron from the Fermi level to infinity. If you want to create an electron beam it can be done by heating the metal until some of its electrons have enough energy to exceed the work function, then accelerating them away with a field gradient.

Whether the material is a conductor or an insulator depends on how full the bands are, and whether or not they overlap. In Figure 14.12 the central column describes an isolated atom and the outer ones illustrate the possibilities created by bringing atoms together into an array, with the energies spread into energy bands. Conductors like copper, shown on the left, have an unfilled outer band; there are many very closely spaced levels just above the last full one, and—when accelerated by a field—electrons can use these levels to move freely through the material. In insulators, shown on the right, the outermost band with electrons in it is full, and the nearest empty band is separated from it in energy by a wide *band gap*. Semiconductors, too, have a band gap, but it is narrower—narrow enough that thermal energy can pop a few electrons into the empty band, where they conduct. Deliberate doping (adding trace levels of other elements) creates new levels in the band gap, reducing the energy barrier to entering the empty states and thus allowing more carriers to become mobile.

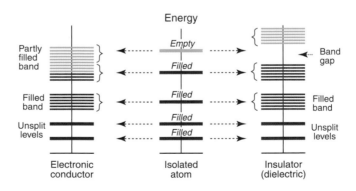

**Figure 14.12**   Conductors, on the left, have a partly filled outer band — electrons in it can move easily. Insulators, on the right, have an outer filled band, separated from the nearest unfilled band by a band gap.

***Electrical resistance***   If a field $E$ exerts a force $Ee$ on an electron, why does it not accelerate forever, giving a current that continuously increases with time? This is not what happens; instead, switching on a field causes a current that almost immediately reaches a steady value. Referring back to equations (14.1) and (14.3), the current density $i/A$ is proportional to the field $E$

$$\frac{i}{A} = \frac{E}{\rho_e} = \kappa_e \, E \tag{14.20}$$

where $\rho_e$ is the resistivity and $\kappa_e$, its reciprocal, is the electrical conductivity.

Broadly speaking, the picture is this. Conduction electrons are free to move through the solid. Their thermal energy $kT$ ($k$ = Boltzmann's constant, $T$ = absolute temperature) causes them to move like gas atoms in all directions. In doing this they collide with *scattering centres*, bouncing off in a new direction. Impurity or solute atoms are particularly effective scattering centres (which is why alloys always have a higher resistivity than pure metals), but electrons are scattered also by imperfections such as dislocations and by the thermal vibration of the atoms themselves. When there is no field, there is no *net* transfer of charge in any direction even though all the conduction electrons are moving freely. A field causes a drift velocity $v_d = \mu_e \, E$ on the electrons, where $\mu_e$ is the electron mobility and it is this that gives the current (Figure 14.13). The drift velocity is small compared with the thermal velocity; it is like a breeze in air—the thermal motion of the air molecules is far greater than the 'drift' that we feel as the breeze. The greater the number of scattering centres, the shorter is the mean-free path, $\lambda_{mfp}$, of the electrons between collisions, and the slower, on average, they move. Just as with thermal conductivity, the electrical conductivity depends on a mean-free path, on the density of carriers (the number $n_v$ of mobile electrons per unit volume) and the charge they carry. Thus the current density, $i/A$, is given by

$$\frac{i}{A} = n_v \, e \, v_d = n_v \, e \, \mu_e \, E$$

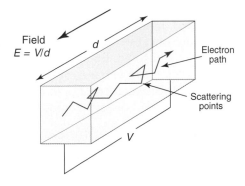

Figure 14.13   An electron, accelerated by the field E, is scattered by imperfections that create a resistance to its motion.

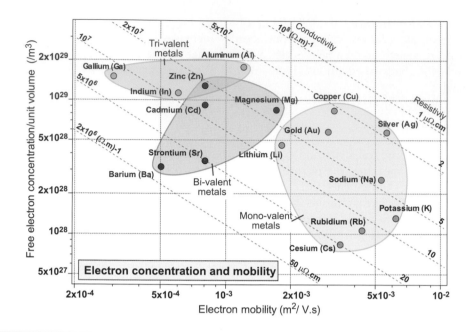

**Figure 14.14** The electron concentration and mobility of pure metals. Alloying changes the electron concentration a little but greatly reduces the mobility.

Comparing this with equation (14.20) gives the conductivity:

$$\kappa_e = n_v \, e \, \mu_e \qquad (14.21)$$

Thus the conductivity is proportional to the density of free electrons and to the drift velocity, and this is directly proportional to the mean-free path. The purer and more perfect the conductor, the higher is the conductivity and the lower its reciprocal, the resistivity.

Electron concentrations and mobilities for pure metals are plotted in Figure 14.14. Monovalent metals contribute one free electron per atom to the electron concentration, bivalent atoms contribute two, and so on, leading to the hierarchy evident in the figure. The typical mobilities, by contrast, decrease with increasing valence. The diagonal contours show the conductivity calculated from equation (14.21) with $e$, the electronic charge, set equal to $1.6 \times 10^{-19}$ coulombs. The other end of each contour is labelled with the resistivity in the convenience units $\mu\Omega.\text{cm}$.

---

### Example 14.5

The resistivity of pure copper is $\rho_e = 2.32 \ \mu\Omega.\text{cm}$. Each copper atom provides one free electron. The atomic volume of copper is $1.18 \times 10^{-29} \ \text{m}^3$. What is the electron mobility in copper?

*Answer.* If each atom contributes one electron the electron concentration in copper is

$$n_v = \frac{1}{\text{Atomic volume}} = 8.47 \times 10^{28} \ \text{per m}^3$$

The resistivity $\rho_e$ is related to this by

$$\rho_e = \frac{1}{\kappa_e} = \frac{1}{n_v \, e \, \mu_e} \quad \text{where } e = 1.6 \times 10^{-19} \text{ coulombs is the charge on an electron.}$$

Converting the resistivity from µΩ.cm to Ω.m by multiplying by $10^{-8}$ gives the electron mobility in copper as

$$\mu_e = \frac{1}{n_v \, e \, \rho_e} = 0.0032 \ \text{m}^2/\text{V.s}$$

Metals are strengthened (Chapter 7) by solid-solution hardening, work hardening or precipitation hardening. All of these change the resistivity too. By 'zooming in' on part of the strength-resistivity chart, we can see how both properties are affected. Figure 14.15 shows how the strength and resistivity of copper and aluminum are changed by alloying and deformation. Adding solute to either metal increases its strength, but the solute atoms also act as scattering centres, increasing the electrical resistivity too. Dislocations add strength (by what we called *work hardening*) and they too scatter electrons a little, though much less than solute.

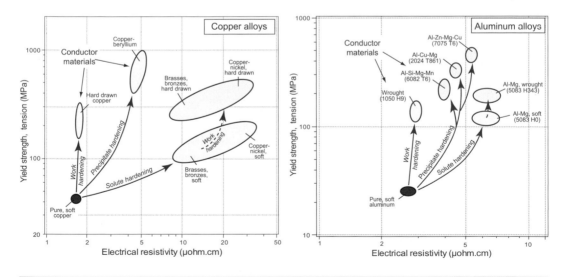

**Figure 14.15** The best choice of material for a cable is one with high strength and low resistivity, but strengthening mechanisms increase resistivity. Work hardening and precipitation hardening do so less than solute hardening.

Precipitates offer the greatest gain in strength; their effect on resistivity varies, depending on their size and spacing compared to the electron mean-free path and on the amount of residual solute left in the lattice. Precipitation hardening (with low residual solute) or work hardening are therefore the best ways to strengthen conductors. The two figures show that commercial conductor alloys have much greater strength, and only slightly greater resistivity than the pure metals.

The resistivity of metals increases with temperature because thermal vibration scatters electrons. Resistance decreases as temperature falls, which is why very high-powered electromagnets are pre-cooled in liquid nitrogen. As absolute zero is approached most metals retain some resistivity, but, as discussed earlier, a few suddenly lose all resistance and become superconducting between 0 and 10 K. The resistivity of semiconductors, by contrast, decreases as temperature increases, because in this case thermal energy allows more carriers to cross the band gap.

*Semiconductors* Semiconductors are based on elements with four valence electrons filling the valence band, with an energy gap separating them from the next available level (in the conduction band). They get their name from their electrical conductivities, much smaller than that of a metal but much larger than that of an insulator. Their conductivities lie in the range $10^{-4}$ to $10^{+4}$ $(\Omega.m)^{-1}$, corresponding to a range of resistivity of $10^4$ to $10^{12}$ $\mu\Omega.cm$. At 0 K pure semiconductors are perfect insulators because there are no electrons in their conduction band. The band gap, however, is narrow, allowing thermal energy (Chapter 12) to promote electrons from the valence to the conduction band at temperatures above 0 K, leaving holes behind in the valence band. The electrons in the conduction band are mobile and move in an electric field, carrying current. The holes in the valence band are also mobile (in reality it is the electrons in the valence band that move cooperatively, allowing the hole to change position), and this charge shift also carries current. Thus the conductivity of a semiconductor has an extra term

$$\kappa_e = n_v\, e\, \mu_e + n_h\, e\, \mu_h \qquad (14.22)$$

where $n_h$ is the number of holes per unit volume and $\mu_h$ is the hole mobility. In pure semiconductors each electron promoted to the conduction band leaves a hole in the valence bend, so $n_e = n_h$, the conductivity becomes

$$\kappa_e = n_v\, e\, (\mu_e + \mu_h) \qquad (14.23)$$

and the semiconduction is said to be *intrinsic* (a fundamental property of the pure material). The number of charge carriers can be controlled in another way, by doping with controlled, small additions of elements that have a different valence than that of the semiconductor itself. This creates free electrons and holes that simply depend on the concentration of dopant, and the semiconducting behaviour is said to be *extrinsic* (a characteristic of the doping, not of the base material).

Figure 14.16 compares the free electron densities and carrier mobilities (the sum of those of the electrons and those of the holes) with those of metals. The axes are the same of those of Figure 14.14, and the same data for metals appear on both. The scales have been expanded to include data for four intrinsic semiconductors at 300 K. The striking feature is that carrier mobility in semiconductors is much *larger* than in metals partly because the conduction band is practically empty. The low conductivity is due to the very small carrier densities.

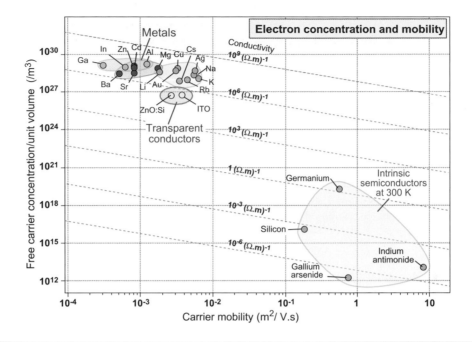

**Figure 14.16** The carrier concentration and mobility of intrinsic semiconductors at 300 K contrasted with those of the pure metals plotted in Figure 14.14. The carrier mobility in semiconductors is higher than in metals, but the carrier density is much lower. Transparent conductors indium tin oxide (ITO) and silicon-doped zinc oxide (ZnO:Si) are also shown.

---

### Example 14.6

The electron mobility $\mu_e$ in silicon at 300 K is $0.14\ \text{m}^2/\text{V.s}$ and that of holes $\mu_h$ is $0.04\ \text{m}^2/\text{V.s}$. The carrier density is $1.5 \times 10^{16}\ /\text{m}^3$. What is the conductivity of silicon at 300 K?

*Answer.* The conductivity is $\kappa_e = n_v\, e\, (\mu_e + \mu_h) = 1.5 \times 10^{16} \times 1.6 \times 10^{-19} \times (0.14 + 0.04) = 4.3 \times 10^{-4}\ (\Omega.\text{m})^{-1}$

---

*Dielectric behaviour* In the absence of an electric field, the electrons and protons in most dielectrics are symmetrically distributed and the material carries no net charge or dipole moment. In the presence of a field the positively charge particles are pushed in the direction of the field and negatively charged particles are pushed in the opposite direction. The effect is easiest to see in ionic crystals, since here neighbouring ions carry opposite charges, as on the left of Figure 14.17. Switch on the field and the positive ions (charge $+q$) are pulled in the field direction, the negative ones (charge $-q$) in the reverse, until the restoring force of the interatomic bonds just balances the force due to the field at a displacement of $\Delta x$, as on the

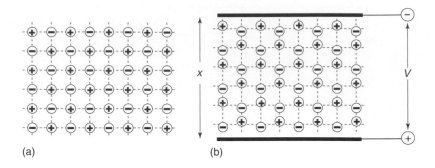

**Figure 14.17** An ionic crystal (a) in zero applied field, and (b) when a field $V/x$ is applied. The electric field displaces charge, causing the material to acquire a dipole moment.

right of the figure. Two charges $\pm q$ separated by a distance $\Delta x$ create a dipole with dipole moment, $d$, given by

$$d = q\Delta x \tag{14.24}$$

The polarisation of the material, $P$, is the volume-average of all the dipole moments it contains:

$$P = \frac{\sum d}{\text{Volume}} \tag{14.25}$$

Even in materials that are not ionic, like silicon, a field produces a dipole moment because the nucleus of each atom is displaced a tiny distance in the direction of the field and its surrounding electrons are displaced in the opposite direction. The resulting dipole moment depends on the magnitude of the displacement and the number of charges per unit volume, and it is this that determines the dielectric constant. The bigger the shift, the bigger the dielectric constant. Thus compounds with ionic bonds and polymers that contain polar groups like $-OH^-$ and $-NH^-$ (nylon, for example) have larger dielectric constants than those that do not.

*Dielectric loss*   Think now of polarisation in an alternating electric field. When the upper plate of Figure 14.17(b) is negative, the displacements are in the direction shown in the figure. When its polarity is reversed, it is the negative ions that are displaced upwards, the positive ions downwards; in an oscillating field, the ions oscillate. If their oscillations were exactly in phase with the field, no energy would be lost, but this is never exactly true, and often the phase shift is considerable. Materials with high dielectric loss usually contain awkwardly-shaped molecules that themselves have a dipole moment—the water molecule is an example (Figure 14.18). These respond to the oscillating field by rotating, but because of their shape they interfere with each other (you could think of it as molecular friction), and this dissipates energy that appears as heat—which is how microwave heating works. As equation (14.18) showed, the energy that is dissipated depends on the frequency of the electric field. Generally speaking, the higher the frequency, the greater the power dissipated but there are peaks at certain frequencies that are characteristic of the material structure.

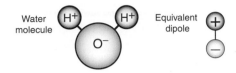

**Figure 14.18**   A water molecule. The asymmetry gives it a dipole moment.

### Example 14.7

An ionic crystal with the rock-salt structure is made up of equal numbers of positive and negative ions each carrying or lacking the charge carried by a single electron. The lattice parameter of the crystal is 0.45 nm. An electric field displaces the ions relative to each other by 1% of the lattice parameter. What is the resulting polarisation?

*Answer.* The dipole moment due to one pair of ions is

$$d = q\Delta x = 1.6 \times 10^{-19} \times 0.0045 \times 10^{-9}$$

$$= 0.72 \times 10^{-30} \ \text{C.m}$$

The volume of the unit cell is $V = (0.45 \times 10^{-9})^3 = 9.1 \times 10^{-29} \ \text{m}^3$, and each contains four pairs of ions.

Thus the polarisation is $P = \frac{4d}{V} = 0.032 \ \text{C/m}^2$

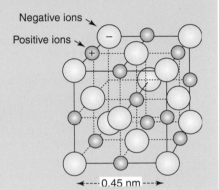

Negative ions

Positive ions

0.45 nm

trons to flow. In insulators they can't, because of the band gap. But if, at some weak spot, one electron is torn free from its parent atom, the force $Ee$ exerted by the field $E$ accelerates it, giving it kinetic energy; it continues to accelerate until it collides with another atom. A sufficiently large field can give the electron so much kinetic energy that, in the collision, it kicks one or more new electrons out of the atom it hits, creating electron-hole pairs, and they, in turn, are accelerated and gain energy. The result, sketched in Figure 14.19, is a cascade—an avalanche of charge. It can be sufficiently violent that the associated heating damages the material permanently. It can also be harnessed: avalanche-effect transistors use the effect to provide rapid switching of large currents.

The critical field strength to make this happen, called the *breakdown potential*, is hard to calculate: it is that at which the first electron breaks free at the weak spot, a defect in the material such as a tiny crack, void or inclusion that locally concentrates the field. The necessary fields are large, typically 1–15 MV/m. That sounds like a lot, but such fields are found in two very different circumstances: when the voltage is very high, or the distances are very small. In power transmission, the voltages are sufficiently high—20,000 volts or so—that breakdown can occur at the insulators that support the line; while in microcircuits and thin-film devices the

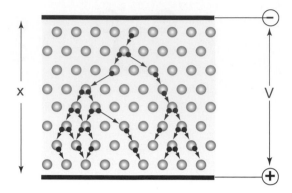

**Figure 14.19** Breakdown involves a cascade of electrons like a lightning strike.

distances between components are very small: a 1-volt difference across a distance of 1 micron gives a field of 1 MV/m.

***Piezoelectric materials*** The word 'Quartz' on the face of your watch carries the assurance that its time-steps are set by the oscillations of a piezoelectric quartz crystal. Piezoelectric behaviour is found in crystals in which the ions are not symmetrically distributed (the crystal structure lacks a centre of symmetry), so that each molecule carries a permanent dipole moment (Figure 14.20(a)). If you cut a little cube from such a material, with faces parallel to crystal planes, the faces would carry charge. This charge attracts ions and charged particles from the atmosphere just as a television screen does, giving an invisible surface layer that neutralises the charge. If you now squeeze the crystal, its ions move relative to each other, the

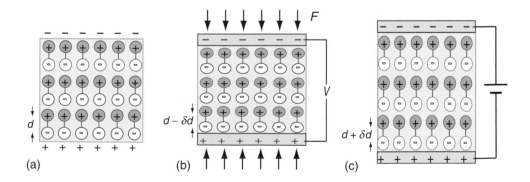

**Figure 14.20** A piezoelectric material has non-symmetrically distributed charge, giving it a natural dipole moment. The surface charge associated with this is neutralized by pick-up of ions, but if it is deformed, as at (b), the dipole moment changes, and the surfaces become charged. The inverse is also true: a field induces a change of shape, the basis of piezoelectric actuation, as at (c).

dipole moment changes, and the charge on the surface changes too (Figure 14.20(b)). Given time, the newly appeared charge would attract neutralising ions, but this does not happen immediately, giving a potential difference. This provides the basis of operation of electric microphones and pick-ups. The potential difference between the faces can be large—large enough to generate a spark across a narrow gap—and is the way that gas lighters work.

A strain, then, induces an electric field in a piezoelectric material. The inverse is also true: a field induces a strain. The field pulls the positive ions and pushes the negative ones, changing their spacing and so changing the shape of the crystal (Figure 14.20(c)). If a small strain produces a large field, then a large field will produce only a very small strain. But the strain is a linear function of field, allowing extremely precise, if small, displacements, used for positioning and actuation at the sub-micron scale.

Piezoelectric materials respond to a change in electric field faster than most materials respond to a stimulus of this or any other kind. Put them in a megahertz field and they respond with microsecond precision. That opens up many applications, some described later in this chapter. In particular, it opens up the world of ultrasonics—sound waves with frequencies starting at the upper limit of the human ear, 20 kHz, on up to 20,000 kHz and above.

*Pyroelectric materials*    The Greek philosopher Theophrastus[9] noted that certain stone, when warmed or cooled, acquired the ability to pick up straw and dry leaves. It was not until the twentieth century that this pyroelectric behaviour—polarisation caused by change of temperature—was understood and exploited.

Some materials have a permanent dipole moment because their positive and negative ions balance electrically but are slightly out of line with each other. If a thin disk of one of these is cut so that its faces are parallel to the plane in which the misalignment happens, the disc has a dipole moment of its own. The unit cells of pyroelectric materials are like this. The dipole moment per unit volume of the material is called the spontaneous polarisation $P_s$. This net dipole moment exists in the absence of an applied electric field and is equivalent to a layer of bound charge on each flat surface. Nearby free charges such as electrons or ions are attracted to the surfaces, neutralising the charge. Imagine that conductive electrodes are then attached to the surfaces and connected through an ammeter. If the temperature of the sample is constant, then so is $P_s$, and no current flows through the circuit. An increase in temperature, however, causes expansion and that changes the net dipole moment and the polarisation. Redistribution of free charges to compensate for the change in bound charge results in a current flow—the pyroelectric current—in the circuit. Cooling rather than heating reverses the sign of the current. Thus the pyroelectric current only flows while the temperature is changing; this is the way that intruder alarms, automatic doors and safety lights are activated.

In an open circuit the free charges remain on the electrodes, which has its uses too, described in the next section. Pyroelectric materials include minerals such as tourmaline (the one Theophrastus found), ceramics such as barium titanate, polymers such as polyvinylidene fluoride, and even biological materials, such as collagen.

---

[9] Theophrastus (372—287 BC), successor to Aristotle, a teacher of science and author of a ten-volume *History of Plants*, which remained the most important contribution to botanical science for the following 1200 years.

***Ferroelectric materials*** Ferroelectrics are a special case of piezoelectric behaviour. They, too, do not have a symmetric structure but have the special ability to switch asymmetry. Barium titanate ($BaTiO_3$), shown schematically in Figure 14.21, is one of these. Below a critical temperature, the Curie[10] temperature (about 120 °C for barium titanate), the titanium atom, instead of sitting at the centre of the unit cell, is displaced up, down, to the left or to the right, as in (a) and (b). Above the Curie temperature the asymmetry disappears and with it the dipole moment, as in (c). In ferroelectrics, these dipoles spontaneously align so that large volumes of the material are polarised even when there is no applied field.

In the absence of an external field a ferroelectric divides itself up into *domains*—regions in which all the dipoles are aligned in one direction—separated by *domain walls* at which the direction of polarisation changes (Figure 14.22). The domains orient themselves so that the dipole moment of one more or less cancels those of its neighbours. If a field is applied the domain walls move so that those polarised parallel to the field grow and those polarised across or against it shrink, until the entire sample is polarised (or 'poled') in just one direction. Figure 14.23 shows how the polarisation $P$ changes as the field $E$ is increased: $P$ increases, reaching a maximum at the *saturation polarisation* $P_s$. If the field is now removed, a large part of the polarisation remains (the *remanent polarisation*), which is only removed by reversing the field to the value $-E_c$, the *coercive field*. Figure 14.23 shows a complete cycle through full reverse polarisation, ending up again with full forward poling. The little inserts show the domain structures round the cycle.

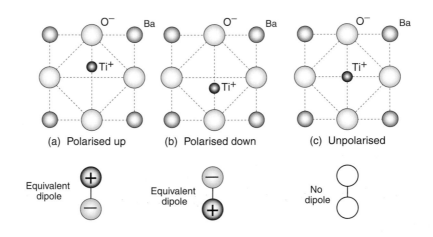

**Figure 14.21** Ferroelectric materials have a permanent dipole moment that can switch: here the $Ti^+$ ion can flip from the upper to the lower position. Above the Curie temperature, the asymmetry disappears.

---

[10] Pierre Curie (1859—1906), French physicist, discoverer of the piezoelectric effect and of magnetic transformations; and husband of the yet more famous Marie Curie. He was killed in a street accident in Paris, a city with dangerous traffic even in 1906.

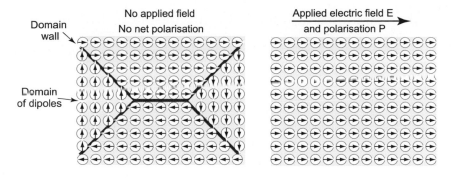

**Figure 14.22** Ferroelectric domains arrange themselves so that their dipole moments cancel. In an electric field $E$ they align, giving the crystal a net dipole moment, which can be large.

Ferroelectric materials have enormous dielectric constants. Those of normal materials with symmetric charge distributions lie in the range 2 to 20. Those for ferroelectrics can be as high as 20,000. It is this that allows their use to make super-capacitors that can store 1000 times more energy than conventional capacitors. Such is the energy density that super-capacitors now compete with batteries for energy storage.

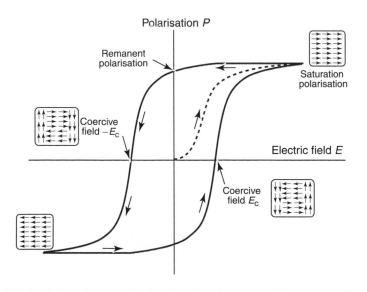

**Figure 14.23** A ferroelectric hysteresis curve.

## **14.5** Design: using the electrical properties of materials

***Transmitting electrical power***   The objective in power transmission is to minimise electrical loss, $P_L$ (equations (14.2) and (14.3), combined):

$$P_L = i^2 R = i^2 \rho_e \frac{L}{A} \qquad (14.26)$$

The best choice, for a given cross-section $A$ and length $L$, is that of a material with the lowest possible value of $\rho_e$; if the line is buried or supported along its entire length, the only other constraint is that of material cost. The chart of Figure 14.7 shows that, of the materials plotted on it, copper has the lowest resistivity, followed by aluminum. If instead the line is above ground and supported by widely spaced pylons, its strength and its density become important: high strength, to support self-weight and wind loads; low density to minimise self-weight.

Here we have a conflict: the materials with the lowest resistivity have low strength; those with high strength have high resistivities, for the reasons given in the last section. The aluminum and copper conductors used in overhead cables are those shown in Figure 14.15, precipitation hardened or work hardened to give strength with minimum loss of conductivity. Another solution is to create a hybrid cable in which two materials are combined, the first to give the conductivity, the second to give the strength. One such is sketched in Figure 14.24: a cable with a core of pure, high-conductivity copper or aluminum wrapped in a cage of high-strength carbon-steel wires. With a 50–50 mix the hybrid, when averaged, has roughly half the strength of the steel and twice the resistivity of the copper per unit length of cable, a better compromise than that offered by any single material. Many overhead cables use this approach, with a steel core surrounded by aluminum conductors; some use instead a combination of low- and high-strength aluminum alloys to reduce self-weight.

***Electrical insulation***   Electrical insulators keep electricity where it is supposed to be. The charts of Figures 14.7 and 14.8 show that most polymers are excellent electrical insulators, with resistivities above $10^{20}$ μΩ.cm. They are used for cable insulation, for casings of plugs, connectors and switchgear. Many ceramics, too, are good electrical insulators, notably alumina ($Al_2O_3$), silica ($SiO_2$) and glass (Pyrex); they are used in applications in which rigidity and resistance to heat are important, such as substrates for electronic circuits and high-voltage

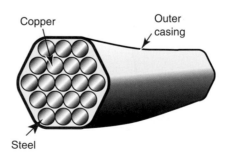

Copper

Outer casing

Steel

Figure 14.24   A hybrid cable made up of strands of copper reinforced with strands of steel.

**Figure 14.25**   An electric plug. The casing must be an insulator, be moldable, and have sufficient strength to tolerate working loads.

insulators. Where electrical isolation is vital, the objective is to maximise resistivity, but as with conduction there are usually additional constraints. Here is an example.

The electric plug, Figure 14.25, is perhaps the commonest of electrical products. It has a number of components, most performing more than one function. The most obvious are the casing and the pins, though there are many more (connectors, a cable clamp, fasteners, and, in some plugs, a fuse). Power plugs have two or three pins and are of robust construction. System plugs, like that of the parallel port of a computer, have 25 or more pins and are miniaturised and delicate, placing more importance on the mechanical as well as the electrical properties of the materials of the pins and casing. The task here is to select a material for the casing of the plug.

The shape is complicated, the part will be made in very large numbers, and pins have to be moulded-in. The only practical way to make it is by using a polymer. The designers set a required strength of at least 25 MPa, and, of course, the electrical resistivity should be as high a possible. The chart of Figure 14.10 suggests ABS, PMMA and PC as suitable candidates—they are strong enough and have high resistivity. But there is a problem: all three are thermoplastics; they soften if they get hot. If circuit overload or poor contact generates heat, the plug will soften, distort and even melt, presenting hazards. So the choice has to be a thermoset, which will char rather than melt. The chart suggests Phenolic—and that is what plugs are really made of. The CES software allows much more sophisticated suggestions: try the Exercises at the end of this chapter to refine the choice.

***Electrical insulation with thermal conduction: keeping microchips cool***   As microchips shrink in scale and clock-speeds rise, overheating becomes a problem. The Pentium chip and its

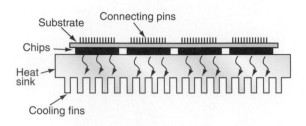

**Figure 14.26** A heat sink. It must conduct heat but not electricity.

successors in today's PCs already reaches 85 °C, requiring forced cooling. Multiple-chip modules (MCMs) pack over 100 chips onto a single substrate, and there the problem is more severe. Heating is kept under control by attaching the chip to a heat sink (Figure 14.26), taking care that they are in good thermal contact. The heat sink now becomes a critical component, limiting further development of the electronics.

To drain heat away from the chip as fast as possible, the heat sink must have the highest possible thermal conductivity, $\lambda$. Metals like copper and aluminum have high $\lambda$, but they are good electrical conductors and this is bad because it allows electrical coupling and stray capacitance between chip and heat sink, slowing the information transfer rate (remember the time delay in a circuit is $\tau = RC$). To prevent this, the heat sink must be an excellent electrical insulator.

Figure 14.7 is the chart with the properties we want: thermal conductivity $\lambda$ and electrical resistivity, $\rho_e$. The materials at the top right have high thermal conductivity and very high resistivity—they are electrical insulators. The best choice of material for the heat sink, then, is aluminum nitride, AlN, alumina, $Al_2O_3$ or silicon carbide SiC. It is these that are used for such heat sinks.

*Using dielectric loss: microwave heating* When you cook chicken in a microwave oven, you are using dielectric loss. Looking back at equation (14.18) we see that the dielectric loss, and thus the heating, is proportional to the loss factor $L = \varepsilon_r \tan \delta$, one axis of the chart of Figure 14.10. Chickens do not appear on the chart, but they *do* have a high loss factor, as do all foods that contain water. There are other ways of cooking chicken, but this one has the advantage that it is quick because microwaves absorb uniformly through the section, giving uniform heating. Microwave heating is not just useful in the kitchen; it is also widely used for industrial heating. Wood can be dried, adhesives cured, polymers moulded and ceramics sintered using microwave heating. As equation (14.18) implies, the higher the frequency $f$ of the electric field, the greater the heating, so it is common, in the industrial application of microwaves, to use gigahertz frequencies (0.3–3 GHz).

*Using dielectric loss: stealth technology* Radar, developed during the Second World War, played a major role in protecting Britain (and other countries) from air attack. It works by transmitting bursts of radio-frequency waves that are reflected back by objects in their path. The radar antenna measures the time it takes for the reflection to arrive back, from which the distance and speed of the object are computed.

The metal body of an aircraft is an excellent radar reflector, and its curved shape causes it to reflect in all directions—just what is needed for safe air-traffic control. But if your aim is to avoid detection, reflection is the last thing you want. Stealth technology combines three tricks to minimise it.

- A shape made up of flat planes that reflect the radar signal away from the detector unless it lies exactly normal to the surface.
- Nonmetallic structural materials—usually composites—that are semi-transparent to microwaves (see Chapter 16 for how to achieve transparency).
- Surface coatings of radar-absorbing materials (called RAMs) that absorb microwaves rather than reflecting them. RAMs use dielectric loss. A surface layer of a material with a large value of loss factor $\varepsilon_r \tan \delta$, a ferroelectric, for example attenuates the incoming wave, diminishing the reflected signal.

*Avoiding dielectric loss: materials for radomes*  The function of a *radome* is to shield a microwave antenna from the adverse effects of wind and weather while having as little effect as possible on the signal it sends and receives (Figure 14.27). When trying to detect incoming signals that are weak to begin with, even a small attenuation of the signal as it passes through the radome decreases the sensitivity of the system. Yet the radome must withstand structural loads, loads caused by pressure difference between the inside and outside of the dome, and—in the case of supersonic flight—high temperatures. So the problem here is the opposite of stealth coatings; it is how to design materials that do not attenuate electromagnetic waves. The answer is to seek those with exceptionally low values of $\varepsilon_r \tan \delta$. And since the radome carries structural loads, they should also be strong.

The chart of Figure 14.10 shows that some polymers have low loss factor, but they are not very strong. Reinforcing them with glass fibres, creating GFRPs, combines low loss with good strength. When, additionally, high temperatures are involved, the ceramics that appear on the upper left of Figure 14.10—silica ($SiO_2$), alumina ($Al_2O_3$) and silicon nitride ($Si_3N_4$)—are all employed.

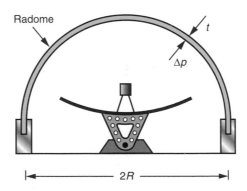

**Figure 14.27**  A radome. The shell must be transparent to microwaves, requiring a low dielectric loss factor.

***Piezoelectric actuation: ink-jet printers*** Ink-jet printing is now widely used to make black and white or colour prints. Many printers use thermal ('bubble-jet') technology in which a tiny heating wire boils a minute volume of ink in the print-head, creating a vapor bubble. The sudden pressure this creates ejects a droplet to ink. The bubble collapses as the wire cools, sucking more ink from a reservoir into the print-head to replace that in the droplet. Thermal technology is very simple, but it gives little control of the droplet size or shape.

The alternative is the piezo print-head, one unit of which is sketched in Figure 14.28. An electric field applied to the disk of piezoelectric material transmits a kick to the vibration plate, creating a pressure pulse that spits a droplet of ink from the nozzle. On relaxation, more ink is drawn in to replace it. Like bubble-jet printers, piezo print-heads have 300 to 600 nozzles. The great advantage of piezo technology is the control of the shape and duration of the pulse delivered by the piezo disk. The droplets of ink are significantly smaller, and the control of droplet size gives greater resolution and colour gradation.

***Pyroelectric thermal imaging*** Night vision, the location of survivors in disaster areas, the filming of wildlife—all these make use of thermal imaging. The technology is comparatively new: the first device for two-dimensional pyroelectric imaging was developed in the late 1970s. A typical set-up is sketched in Figure 14.29. A lens, made of a material transparent to infrared (such as germanium) focuses the IR radiation from an object on the left to form an image on the pyroelectric disk on the right. The focused IR rays warm points on the disk to different degrees, depending on the temperature and emissivity of the points on the object from which they came, and the change of temperature induces local polarisation. The disk is an insulator, so the image creates a pattern of charge on the surface of the disk that can be 'read' with an electron beam.

The problem with early devices was that lateral heat-diffusion in the disk during the time between successive electron-beam sweeps resulted in low resolution. The problem is overcome by reticulating the surface of the disk into 20 micron islands by ion-milling (a way of cutting very thin channels with a high-energy ion beam), thermally isolating each island from its

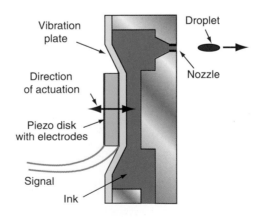

Figure 14.28 A piezo print-head.

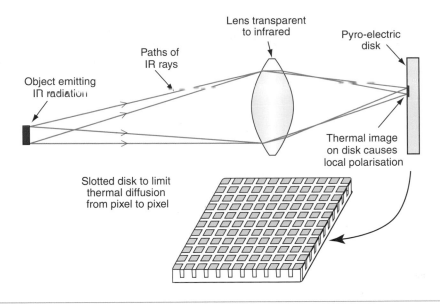

**Figure 14.29** Thermal imaging.

neighbours. Present-day imaging systems can detect a temperature change of about 0.2 °C on the target disk.

*Ferroelectric actuation: injectors for diesel engines* Fuel is injected into the combustion chamber of a diesel engine by micro-pumps, one for each cylinder. To meter the fuel accurately and adapt the quantity injected to the needs of the engine, the pump must allow precise proportional control. Until recently the standard actuator for this application was a solenoid (a coil with a moving ferromagnetic core), but to exert sufficient force to inject fuel into the high pressure in the cylinder, the solenoid had to be large and the inertia of the core limited actuation speed. The answer is to use ferroelectric actuators (Figure 14.30). Most actuators of this sort are based on the compound lead zirconium titanate (PZT). When a potential gradient $dV/dx$ is applied across the faces of a disk of PZT, it undergoes a strain $\varepsilon$ that is proportional to the electric potential gradient:

$$\varepsilon = k \frac{dV}{dx} \qquad (14.27)$$

where $k$ is the *ferroelectric coefficient* of the material. In practice $k$ is small, meaning that a high gradient is needed to get useful strain. The system voltage in cars or trucks is usually 20 volts, so the only way to get a large gradient is to use very thin slices of the PZT—typically 100 microns thick, giving a gradient $dV/dx$ of 200,000 volts per meter. The problem then is that the displacement (the strain $\varepsilon$ times the thickness of the disk, $x$) is small. The solution is to make a stack like that shown, expanded, on the right of Figure 14.30. The thin ($x = 100$ micron) slices of PZT are sandwiched between conducting electrodes, alternately charged, creating the gradient. A stack of 100 (still only 10 mm high) gives adequate displacements. These actuators

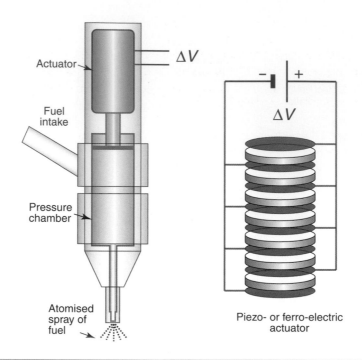

**Figure 14.30** A diesel injector. The actuator of the injector must be very precise to meter the fuel accurately, and must adapt its timing and stroke to match the conditions of the engine. Piezoelectric or ferroelectric stacks like that shown on the right, offer the best control.

are half the size and allow twice the pulse rate of solenoids, giving more precise control, performance and fuel economy.

## 14.6 Summary and conclusions

All materials contain charged particles: electrons, protons, ions, charged molecules and the like. An electric field $E$ exerts a force $qE$ on a charge $q$, pushing it in the direction of the field if $q$ is positive and in the opposite direction if $q$ is negative. If the particles can move freely, as electrons can in metals, an electric current flows through the material, which we refer to as conduction. Special types of conduction are also found—superconductivity at very low temperatures, and semiconduction in the materials of electronics such as silicon and germanium. If, instead, the particles cannot move freely, as is the case in insulators, they are still displaced slightly by the field, leading to dielectric polarisation.

Materials that polarise easily have a high dielectric constant. If used as the insulating film of a capacitor, the polarisation screens the electrodes from each other, increasing the capacitance. The slight displacement of charge has other consequences. One of these is electrostriction: the charge movement causes the material to change shape slightly. If the material is made up of ions and these are arranged in a way that lacks a centre of symmetry, a further effect appears: that of piezoelectric response. The lack of symmetry means that each molecule carries a dipole moment, and within a single crystal these are aligned. An electric field causes a change of shape, and a change of shape changes the molecular dipole moment, thus inducing a change in polarisation, generating a field. A change of temperature, too, changes the molecular dipole moment, so materials that are piezoelectric are also pyroelectric. A few materials have a further property: that of molecular asymmetry that can switch direction because ions can occupy one of several equivalent sites in the crystal. These materials are ferroelectric; the molecular dipoles spontaneously align to form polarised domains. In the relaxed state the domains take up orientations such that the field of one is cancelled by those of its neighbours, but when 'poled' by placing it in an electric field, the domains align, giving a large net polarisation that is retained when the field is removed.

All these effects have practical utility, examples of which are described in this chapter.

## 14.7 Further reading

Braithwaite, N., & Weaver, G. (1990). *Electronic Materials*. Oxford, UK: The Open University and Butterworth Heinemann. ISBN 0-408-02840-8. (One of the excellent Open University texts that form part of their Materials program.)

Callister, W. D., & Rethwisch, D. G. (2011). *Materials Science and Engineering: An Introduction* (8th ed.). New York, USA: John Wiley. ISBN 978-0-470-50586-1. (A well-established and comprehensive introduction to the science of materials)

Jiles, D. (2001). *Introduction to the Electronic Properties of Materials* (2nd ed.). Cheltenham, UK: Nelson Thornes. ISBN 0-7487-6042-3. (The author develops the electronic properties of materials, relating them to thermal and optical as well as electrical behaviour.)

Solymar, L., & Walsh, D. (2004). *Electrical Properties of Materials* (7th ed.). Oxford, UK: Oxford University Press. ISBN 0-19-926793-6. (A mature introduction to electrical materials.)

## 14.8 Exercises

| | |
|---|---|
| Exercise E14.1 | Why do metals conduct electricity? Why do insulators not conduct electricity, at least when cold? |
| Exercise E14.2 | A potential difference of 4 volts is applied across a tungsten wire 200 mm long and 0.1 mm in diameter. What current flows in the wire? How much power is dissipated in the wire? The resistivity $\rho_e$ of tungsten is 5.6 $\mu\Omega$.cm. |
| Exercise E14.3 | A gold interconnect 1 mm long and with a rectangular cross-section of 10 $\mu$m × 1 $\mu$m has a potential difference of 1 mV between its ends. What is the current in the interconnect and how much power is dissipated in it? The resistivity $\rho_e$ of gold is 2.5 $\mu\Omega$.cm. |

**Exercise E14.4**   The gold interconnect of the previous question (length 1 mm and rectangular cross-section of 10 μm × 1 μm with a potential difference of 1 mV across its length) carries power for 10 seconds. If no heat is lost by conduction, radiation or convection, how hot will the interconnect get? The resistivity $\rho_e$ of gold is 2.5 μΩ.cm, its specific heat is 130 J/kg.C and its density is 19300 kg/m$^3$.

**Exercise E14.5**   A 0.5 mm diameter wire must carry a current $i = 10$ amps. The system containing the wire will overheat if the power dissipation, $P = i^2 R$, exceeds 5 Watts/m. The table lists the resistivities $\rho_e$ of four possible candidates for the wire. Which ones meet the design requirement?

| Material | $\rho_e$, μΩ.cm |
|---|---|
| Aluminum | 2.7 |
| Copper | 1.8 |
| Nickel | 9.6 |
| Tungsten | 5.7 |

**Exercise E14.6**   Which metallic superconductors have a critical temperature above the boiling point of helium? Which ceramic superconductors have a critical temperature above the boiling point of liquid nitrogen?

**Exercise E14.7**   A superconducting magnet design envisages a superconducting coil cooled in liquid nitrogen, generating a magnetic field of 30 Tesla. Is it practical to use YBCO (YBa$_2$Cu$_3$O$_7$) as the conductor? The critical temperature $T_c$ of YBCO is 93 K and its zero-Kelvin critical field $H_{c,o}$ is 140 T. Liquid nitrogen boils at 77 K.

**Exercise E14.8**   What is a dielectric? What are its important dielectric properties? What is meant by polarisation? Why do some dielectrics absorb microwave radiation more than others?

**Exercise E14.9**   It is much easier to measure the electrical conductivity of a material than to measure its thermal conductivity. Use the Weidemann-Franz law to find the thermal conductivities of
(a) an alloy that has an electrical resistivity $\rho_e$ of 28 μΩ.cm.
(b) tungsten of the kind used for lamp filaments that has an electrical conductivity $\kappa_e$ of $9.9 \times 10^6$ S/m (as S/m is 1/Ω.m).

**Exercise E14.10**   The metal zinc has free electron concentration per unit volume, $n_v$ of $1.3 \times 10^{29}$/m$^3$ and an electron mobility $\mu_e$ of $8 \times 10^{-4}$ m$^2$/V.s. The charge carried by an electron, $e$, is $1.6 \times 10^{-19}$ Coulomb. Based on this information, what is the electrical conductivity of zinc? Handbooks list the measured resistivity of zinc as 5.9 μΩ.cm. Is this consistent with your calculation? (Watch the units.)

Exercise E14.11    The resistivity of pure silver is $\rho_e = 1.59\ \mu\Omega.\text{cm}$. Each silver atom provides one free electron. The atomic volume of silver is $1.71 \times 10^{-29}\ \text{m}^3$. What is the electron mobility in silver?

Exercise E14.12    Estimate the drift velocity of free electrons when a potential difference of 3 volts is applied between the ends of a copper wire 100 mm long. The electron mobility in copper is $\mu_e = 0.0032\ \text{m}^2/\text{V}$.

Exercise E14.13    The resistivity of brass with 60 atom% copper and 40 atom% of zinc is $\rho_e = 6.8\ \mu\Omega.\text{cm}$. Each copper atom contributes one free electron. Each zinc atom contributes two. The atomic volumes of copper and zinc are almost the same, and the electron concentration in copper is $n_v = 8.45 \times 10^{28}/\text{m}^3$. What is the electron mobility $\mu_e$ in brass? How does it compare with that of pure copper ($\mu_e = 0.0032\ \text{m}^2/\text{V.s}$)?

Exercise E14.14    Graphene is a semiconductor with remarkable electrical properties. It is reported that the free-electron mobility $\mu_e$ in graphene is $1.5\ \text{m}^2/\text{V.s}$ and that the effective free electron concentration $n_v$ per unit volume, $n_v$, is $5 \times 10^{25}/\text{m}^3$. If these values are accepted, what would you expect the resistivity of graphene to be?

Exercise E14.15    A power line is to carry 5 kA at 11 kV using pylons 460 m apart. The dip $d$, in a wire of weight $m_l$ per unit length strung between pylons $L$ apart at a tension $T$ is given by $d = L^2\, m_l\, /8T$. The maximum tension allowed is 0.8 of the yield stress, $\sigma_y$. If the maximum allowable dip is 6 m, which of the materials in the table could be used?

| Material | Electrical resistivity $\rho_e$, $\Omega$m | Yield stress $\sigma_y$, MPa | Density $\rho$, kg/m$^3$ |
|---|---|---|---|
| Aluminum | $1.7 \times 10^{-8}$ | 102 | 2700 |
| Copper | $1.5 \times 10^{-8}$ | 336 | 8900 |
| Steel | $55 \times 10^{-8}$ | 295 | 7800 |

Exercise E14.16    A material is required for a transmission line that gives the lowest full-life cost over a 20-year period. The total cost is the sum of the material cost and the cost of the power dissipated in Joule heating. The cost of electricity $C_E$ is $6 \times 10^{-3}$ \$/MJ. Material prices are listed in the table below. Derive an expression for the total cost per meter of cable in terms of the cross-sectional area $A$ (which is a free parameter), the material and electrical costs and the material parameters. Show that the minimum cost occurs when the two contributions to the cost are equal. Hence derive a performance index for the material and decide on the best of the materials in the table.

| Material | Electrical resistivity $\rho_e$, $\Omega$m | Density $\rho$, kg/m$^3$ | Price $C_m$, $/kg |
|----------|----------|----------|----------|
| Aluminum | $1.7 \times 10^{-8}$ | 2700 | 1.6 |
| Copper   | $1.5 \times 10^{-8}$ | 8900 | 5.2 |
| Steel    | $55 \times 10^{-8}$ | 7800 | 0.5 |

**Exercise E14.17**  In the discussion of conductors, a 50–50 mix of copper and steel strands was suggested for transmission cables. Using the values in the table for resistivity and strength, calculate the effective values for both for the cable, assuming a rule of mixtures. Plot this on copies of the $\sigma_y$–$\rho_e$ chart of Figures 14.8 and 14.15(a) to explore its performance. Assume both strength and resistivity of cables follow a rule of mixtures.

| Material | Strength $\sigma_y$, MPa | Resistivity $\rho_e$, $\mu\Omega$.cm |
|----------|----------|----------|
| High-strength steel, cold-drawn | 1700 | 22 |
| High-conductivity copper, cold-drawn | 300 | 1.7 |

Low density, we said, was important for long-span transmission lines because the self-weight becomes significant. Suppose you were asked to design a hybrid power-transmission cable with the lowest possible weight, what combination of materials would you choose? Use information from the charts for inspiration.

**Exercise E14.18**  Roughly 50% of all cork that is harvested in Portugal ends up as cork dust, a worthless by-product. As a materials expert, you are approached by an entrepreneur who has the idea of making useful products out of cork dust by compacting it and heating it, using microwave heating. The loss factor $L$ of cork is 0.21. The entrepreneur reckons he needs a power density $P$ of at least 2 kW per m$^3$ for the process to be economic. If the maximum field $E$ is limited to $10^2$ V/m, what frequency $f$ of microwaves will be needed?

**Exercise E14.19**  Derive the expression (equation (14.15))

$$\text{Max. energy density} = \frac{1}{2} \varepsilon_r \, \varepsilon_o \, E_b^2$$

for the maximum limiting electrical energy density that can be stored in a dielectric.

Exercise 14.20    You are asked to suggest a dielectric material for a capacitor with the highest possible energy density. What material would you suggest? Use Figure 14.9 to find out.

Exercise E14.21   The electron mobility $\mu_e$ in germanium at 300 K is 0.36 m²/V.s and that of holes $\mu_h$ is 0.19 m²/V.s. The carrier density is $2.3 \times 10^{19}$ /m³. What is the conductivity of germanium at 300 K?

Exercise E14.22   The atomic volume of silicon is $2.0 \times 10^{-29}$ m³ and the free electron density $n_e$ at 300 K is $1.5 \times 10^{16}$ /m³. What fraction of silicon atoms have provided a conduction electron?

## 14.9 Exploring design with CES

(Use Level 2 unless stated otherwise)

Exercise E14.23   Use the 'Search' facility of CES to search for
(a) Electrical conductors.
(b) Heat sinks.

Exercise E14.24   The analysis of heat sinks in Chapter 3 formulated the following design constraints and objective. They are summarised in the table. Use the selector to find the material that best meets them.

| | |
|---|---|
| Function | • Heat sink |
| Constraints | • Material must be good electrical insulator |
| | • Maximum operating temperature > 200 °C |
| Objective | • Maximise thermal conductivity |
| Free variable | • Choice of material |

Exercise E14.25   The sheathing of the cable shown in Figure 14.24 is moulded over the bundled wires. The material of the sheath must be electrically insulating, flexible, water-resistant and cheap. The table translates the requirements. Apply the constraints using a 'Tree' and a 'Limit' stage, then plot a bar chart of material price to find the three least expensive materials that meet all the constraints.

| | |
|---|---|
| Function | • Cable sheathing |
| Constraints | • Able to be moulded |
| | • Good electrical insulator |
| | • Durability in fresh water: very good |
| | • Young's modulus < 4 GPa |
| Objective | • Minimise material price |
| Free variable | • Choice of material |

Exercise E14.26    The pins of the plug of Figure 14.25 must conduct electricity well, be corrosion resistant so that contact remains good, and be hard enough to resist abrasion and wear. The table translates the requirements. Apply these and rank the promising candidates by material price.

| Function | • Connecting pins for power plug |
|---|---|
| Constraints | • Good electrical conductor |
| | • Durability in fresh water: very good |
| | • Hardness > 200 HV |
| Objective | • Minimise material price |
| Free variable | • Choice of material |

Exercise E14.27    The casing of the plug of Figure 14.25 performs both a mechanical and an electrical function. It must insulate, and it must also be rigid and resilient enough to tolerate service loads. (Resilient means it must be capable of some elastic distortion requiring a high-yield strain, $\sigma_y/E$).

| Function | • Plug casing |
|---|---|
| Constraints | • Able to be moulded |
| | • Good electrical insulator |
| | • Flammability: non-flammable or self-extinguishing |
| | • Young's modulus > 2 GPa |
| | • Yield strain $\sigma_y/E > 0.01$ |
| Objective | • Minimise material price |
| Free variable | • Choice of material |

## **14.10** Exploring the science with CES Elements

Exercise E14.28    Make a chart with atomic number on the $x$-axis (use a linear scale) and electrical conductivity $\kappa$ on the $y$-axis. Convert the values of resistivity $\rho_e$ (which are in units of $\mu\Omega.cm$ in the database) to those of conductivity in Siemens/m (S/m), use $\kappa = 10^8/\rho_e$. This can be made using the 'Advanced' facility in the axis-choice dialog box. Does conductivity vary in a periodic way across the periodic table? Which three elements have the highest conductivities?

Exercise E14.29    Explore the Wiedemann-Franz relation for the elements. To do so, make a chart of electrical conductivity, constructed as in the previous example, on the $x$-axis and thermal conductivity on the $y$-axis. Limit the selection to metals only by using a 'Limit' stage—State at 300 K. Fit an equation to the resulting plot. If you were now given a new element with an electrical conductivity of $4 \times 10^7$ S/m, what would you estimate its thermal conductivity to be? (Relations such as this one provide ways of building 'intelligent checking' of data into material property databases.)

Exercise E14.30  Make a chart with atomic number on the $x$-axis (use a linear scale) and dielectric constant on the $y$-axis. Do you see any general trend? If so, how would you explain it?

Exercise E14.31  It is proposed to use the temperature-dependence of electrical resistivity for sensing. Plot this quantity against relative cost to identify three cheap metals with high values of this temperature dependence.

Exercise E14.32  Equation (14.21) of the text derives the electrical conductivity as

$$\kappa_e = n_v \, e \, \mu_e$$

The CES Elements database contains data for the free electron concentration $n_v$ and the electron mobility $\mu_e$. You will find the electronic charge under 'Constants/Parameters' in the 'Advanced' facility. Make a chart using the 'Advanced' facility with $\kappa_e = n_v \, e \, \mu_e$ on the $y$-axis in the reciprocal of the resistivity on the $x$-axis. Make sure you have made all the units self-consistent. How well does this equation describe the measured conductivities of metals?

Exercise E14.33  Explore superconductivity across the periodic table. Use CES Elements to plot superconducting transition temperature, $T_c$ against the atomic number (use linear scales). Which element has the highest transition temperature? What is $T_c$ for this element?

# Chapter 15
# Magnetic materials

Magnet materials in action. The audio-frequency transformer on the left has a soft magnetic core. The DC motor on the right has a hard (permanent) magnetic stator and a soft magnetic rotor.

## **15.1** Introduction and synopsis

Migrating birds, some think, navigate using the earth's magnetic field. This may be debatable, but what is beyond question is that sailors, for centuries, have navigated in this way, using a natural magnet, *lodestone*, to track it. Lodestone is a mineral, magnetite ($Fe_3O_4$), which sometimes occurs naturally in a magnetised state. Today we know lodestone is a *ferri-magnetic* ceramic, or *ferrite*, examples of which can be found in every radio, television set and micro-wave oven. *Ferro-magnetic* materials, by contrast, are metals, typified by iron but including also nickel, cobalt and alloys of all three. Placed in a magnetic field, these materials become magnetised, a phenomenon called *magnetic induction*; on removal, some, called soft magnets, lose their magnetisation, while others, the hard magnets, retain it.

Magnetic fields are created by moving electric charge—electric current in electro-magnets, electron spin in atoms of magnetic materials. This chapter is about magnetic materials: how they are characterised, where their properties come from and how they are selected and used. It starts with definitions of magnetic properties and the way they are measured. As in other chapters, charts display them well, separating the materials that are good for one sort of application from those that are good for others. The chapter continues by drilling down to the origins of magnetic behaviour and concludes with a discussion of applications and the ma-terials that best fill them.

## **15.2** Magnetic properties: definition and measurement

***Magnetic fields in vacuum*** First, some definitions. When a current $i$ passes through a long, empty coil of $n$ turns and length $L$ as in Figure 15.1, a magnetic field is generated. The magnitude of the field, $H$, is given by Ampère's law as

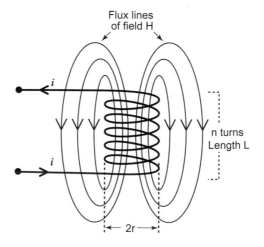

**Figure 15.1** A solenoid creates a magnetic field $H$; the flux lines indicate the field strength.

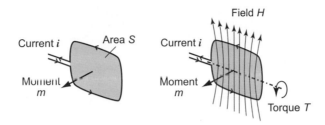

**Figure 15.2** Definition of magnetic moment and moment-field interaction

$$H = \frac{n\,i}{L} \qquad (15.1)$$

and thus has units of amps/meter (A/m). The field has both magnitude and direction—it is a vector field.

Magnetic fields exert forces on a wire carrying an electric current. A current $i$ flowing in a single loop of area $S$ generates a dipole moment $m$ where

$$m = i\,S \qquad (15.2)$$

with units A.m$^2$, and it too is a vector with a direction normal to the plane of S (Figure 15.2). If the loop is placed at right angles to the field $H$ it feels a torque $T$ (units: Newton.meter, or N.m) of

$$T = \mu_{\mathrm{o}}\,m\,H \qquad (15.3)$$

where $\mu_{\mathrm{o}}$ is called the *permeability of vacuum*, $\mu_{\mathrm{o}} = 4\,\pi \times 10^{-7}$ henry/meter (H/m). To link these we define a second measure of the magnetic field, one that relates directly to the torque it exerts on a unit magnetic moment. It is called the *magnetic induction* or *flux density*, B, and for vacuum or non-magnetic materials it is

$$B = \mu_{\mathrm{o}}\,H \qquad (15.4)$$

Its units are *tesla*, so a tesla[1] is 1 HA/m$^2$. A magnetic induction B of 1 tesla exerts a torque of 1 N.m on a unit dipole at right angles to the field $H$.

***Magnetic fields in materials***  If the space inside the coil of Figure 15.1 is filled with a material, as in Figure 15.3, the induction within it changes. This is because its atoms respond to the field by forming little magnetic dipoles in ways that are explained in Section 15.4. The material acquires a macroscopic dipole moment or *magnetisation*, M (its units are A/m, like $H$). The induction becomes

$$B = \mu_{\mathrm{o}}\,(H + M) \qquad (15.5)$$

---

[1] Nikola Tesla (1856–1943), Serbian-American inventor, discoverer of rotating magnetic fields, the basis of most alternating current machinery, inventor of the Tesla coil and of a rudimentary flying machine (never built).

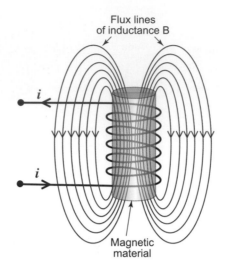

**Figure 15.3** A magnetic material exposed to a field *H* becomes magnetized, concentrating the flux lines.

The simplicity of this equation is misleading, since it suggests that $M$ and $H$ are independent; in reality $M$ is the response of the material to $H$, so the two are coupled. If the material of the core is ferro-magnetic, the response is a very strong one, and it is non-linear, as we shall see in a moment. It is usual to rewrite equation (15.5) in the form

$$B = \mu_R\, \mu_o\, H$$

where $\mu_R$ is called the *relative permeability*, and like the relative permittivity (the dielectric constant) of Chapter 14, it is dimensionless. The magnetisation, $M$, is thus

$$M = (\mu_R - 1)\, H = \chi\, H \tag{15.6}$$

where $\chi$ is the *magnetic susceptibility*. Neither $\mu_R$ or $\chi$ are constants—they depend not only on the material but also on the magnitude of the field, $H$, for the reason just given.

---

### Example 15.1

(a) A coil of length $L = 3$ cm and with $n = 100$ turns carries a current $i = 0.2$ A. What is the magnetic flux density $B$ inside the coil? (b) A core of silicon-iron with a relative permeability $\mu_R = 10\,000$ is placed inside the coil. By how much does this raise the magnetic flux density?

*Answer.*

(a) The field is $H = \dfrac{n\,i}{L} = \dfrac{100 \times 0.2}{3 \times 10^{-2}} = 666.7\ A/m$. The magnetic flux density is $B = \mu_o\, H = 8.38 \times 10^{-4}\ Tesla$

(b) The silicon-iron core raises the magnetic flux density to $B = \mu_R \mu_o H = 8.38\ Tesla$.

Example 15.2

A cylindrical coil with a length $L = 10$ mm with $n = 50$ turns carries a current $i = 0.01$ A. What is the field and flux density in the magnet? A ferrite core with a susceptibility $\chi = 950$ is placed in the coil. What is the flux density and what is the magnetisation of the ferrite?

*Answer.* The electro-magnet produces a field

$$H = \frac{n\,i}{L} = \frac{50 \times 0.01}{0.01} = 50 \; A\,/\,m$$

The flux density is

$$B_o = \mu_o \, H = 4\pi \times 10^{-7} \times 50 = 6.3 \times 10^{-5} \; Tesla$$

The relative permeability of the ferrite is $\mu_R = \chi + 1 = 951$. When it is inserted into the core the induction increases to

$$B = \mu_R \, \mu_o \, H = 951 \times 4\pi \times 10^{-7} \times 50 = 0.06 \; Tesla$$

The magnetisation is

$$M = \frac{B}{\mu_o} - H = \frac{0.06}{4\pi \times 10^{-7}} - 50 = 4.8 \times 10^4 \; A\,/\,m$$

(This is well below the saturation magnetisation of ferrites, see below.)

The magnetic pressure $P_{mag}$ (force per unit area) exerted by an electromagnet on a section of core material is:

$$P_{mag} = \frac{1}{2}\mu_o H^2 = \frac{1}{2}\frac{B^2}{\mu_o} \tag{15.7}$$

provided that the core does not saturate. Iron has a high saturation magnetisation $H_s$ of about $1.8 \times 10^6$ A/m, so the maximum pressure exerted by an electro-magnet with an iron core is

$$P_{mag} = \frac{1}{2} \times 4\,\pi \times 10^{-7} \times \left(1.8 \times 10^6\right)^2 = 2.9 \times 10^6 \; \text{N/m}^2 = 2 \; \text{MPa}$$

The energy density stored in a magnetic field per unit volume, $U/V$, has the same form as the magnetic pressure:

$$\frac{U}{V} = \frac{1}{2}\frac{B^2}{\mu_o} \; \text{joules/m}^3 \tag{15.8}$$

***Ferro-magnetic and ferri-magnetic materials*** Nearly all materials respond to a magnetic field by becoming magnetised, but most are paramagnetic with a response so faint that it is of no practical use. A few, however, contain atoms that have large dipole moments and the ability to spontaneously magnetise—to align their dipoles in parallel—as electric dipoles do in ferro-electric materials. These are called *ferro-magnetic* and *ferri-magnetic* materials (the second are called *ferrites* for short), and it is these that are of real practical use.

Magnetisation decreases with increasing temperature. Just as with ferro-electrics, there is a temperature, the Curie temperature $T_c$, above which it disappears, as in Figure 15.4. Its value for the materials we shall meet here is well above room temperature (typically 300–1000 °C), but making magnets for use at really high temperatures is a problem.

***Measuring magnetic properties*** Magnetic properties are measured by plotting an $M-H$ curve. It looks like Figure 15.5. If an increasing field $H$ is applied to a previously demagnetised sample, starting at A on the figure, its magnetisation increases, slowly at first and then faster, following the broken line, until it finally tails off to a maximum, the *saturation magnetisation* $M_s$ at the point B. If the field is now backed off, $M$ does not retrace its original path, but retains some of its magnetisation so that when $H$ has reached zero, at the point C, some magnetisation remains. The residual magnetisation is called the *remanent magnetisation* or *remanence*, $M_r$, and is usually only a little less than $M_s$. To decrease $M$ further, we must increase the field in the opposite direction until $M$ finally passes through zero at the point D when the field is $-H_c$, the *coercive field*, a measure of the resistance to demagnetisation. Some applications require $H_c$ to be as high as possible, others, as low a possible. Beyond point D the magnetisation $M$ starts to increase in the opposite direction, eventually reaching saturation again at the point E. If the field is now decreased again $M$ follows the curve through F and G back to full forward magnetic saturation again at B to form a closed $M-H$ circuit called the *hysteresis loop*.

Magnetic materials are characterised by the size and shape of their hysteresis loops. The initial segment AB is called the *initial magnetisation curve*, and its average slope (or sometime its steepest slope) is the magnetic susceptibility, $\chi$. The other key

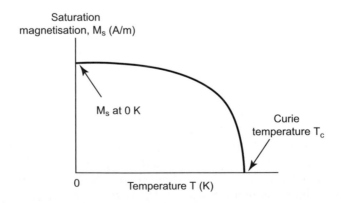

**Figure 15.4** Saturation magnetisation decreases with temperature, falling to zero at the Curie temperature $T_c$.

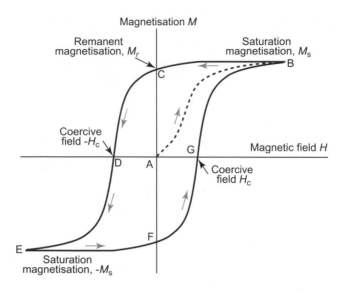

**Figure 15.5** A hysteresis curve, showing the important magnetic properties.

properties—the saturation magnetisation $M_s$, the remanence $M_r$ and the coercive field $H_c$—have already been defined. Each full cycle of the hysteresis loop dissipates an energy per unit volume equal to the area of the loop multiplied by $\mu_o$, the permeability of vacuum. This energy appears as heat—it is like magnetic friction. Many texts plot the curve of inductance $B$ against $H$, rather than the $M-H$ curve. Equation (15.5) says that $B$ is proportional to $(M + H)$. Since the value of $M$ for any magnetic materials worthy of the name is very much larger than the $H$ applied to it, $B \approx \mu_o M$ and the $B - H$ curve of a ferro-magnetic material looks very like its $M-H$ curve (it's just that the $M$ axis has been scaled by $\mu_o$).

There are several ways to measure the hysteresis curve, one of which is sketched in Figure 15.6. Here the material forms the core of what is in effect a transformer. The oscillating current through the primary coil creates a field $H$ that induces magnetisation $M$ in the material of the core, driving it round its hysteresis loop. The secondary coil picks up the inductance, from which the instantaneous state of the magnetisation can be calculated, mapping out the loop.

Magnetic materials differ greatly in the shape and area of their hysteresis loop, the greatest difference being that between *soft magnets*, which have thin loops, and *hard magnets*, which have fat ones, as sketched in Figure 15.7. In fact the differences are much greater than this figure suggests: the coercive field $H_c$ (which determines the width of the loop) of hard magnetic materials like Alnico is greater by a factor of about $10^5$ than that of soft magnetic materials like silicon-iron. There is more on this in the next two sections.

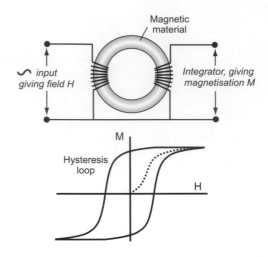

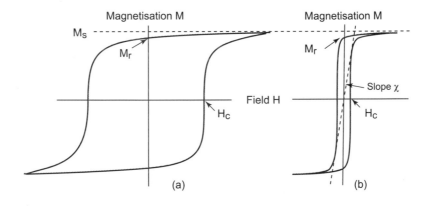

**Figure 15.6**  Measuring the hysteresis curve

**Figure 15.7**  Hysteresis loops. (a) a fat loop typical of hard magnets (b) a thin, square loop typical of soft magnets.

## 15.3 The big picture: charts for magnetic properties

*The Remanence–Coercive field chart*  The differences between families of soft and hard magnetic materials are brought out by Figure 15.8. The axes are remanent magnetisation $M_r$ and coercive field $H_c$. The saturation magnetisation $M_s$, more relevant for soft magnets, is only slightly larger than $M_r$, so an $M_s – H_c$ chart looks almost the same as this one. There are twelve distinct families of magnetic materials, each enclosed in a coloured envelope. The members of each family, shown as smaller ellipses, have unhelpful tradenames (such as 'H Ferrite YBM-1A')

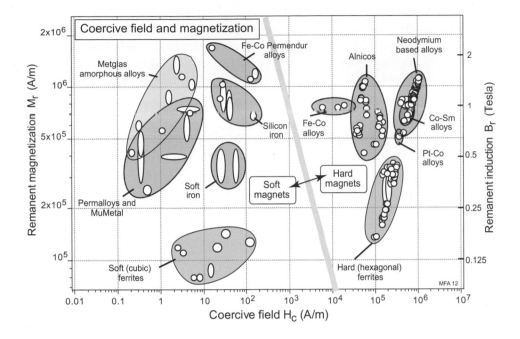

**Figure 15.8** Remanent magnetisation and coercive force. Soft magnetic materials lie on the left, hard magnetic materials on the right. The orange and blue envelopes enclose electrically conducting materials; the green ones enclose insulators.

so they are not individually labelled.[2] Soft magnets require high $M_s$ and low $H_c$; they are the ones on the left, with the best near the top. Hard magnets must hold their magnetism, requiring a high $H_c$, and to be powerful they need a large $M_r$; they are the ones on the right, with the best again at the top.

In many applications the electrical resistivity is also important because of eddy-current losses, which will be described later. Changing magnetic fields induce eddy currents in conductors but not in insulators. The orange and blue envelopes on the chart enclose metallic ferro-magnetic materials, which are good electrical conductors. The green ones describe ferrites, which are electrical insulators.

### Example 15.3

A material is required for a powerful permanent magnet that must be as small as possible and be resistant to demagnetisation by stray fields. Use the Chart of Figure 15.8 to identify your choice.

---

[2] This chart and the next two were made with Level 3 of the CES EduPack database, which contains data for 222 magnetic materials. Levels 1 and 2 do not include magnetic materials because of their specialized nature.

*Answer.* The requirement that the magnet be small and powerful requires a high remanent magnetisation $M_r$. The need to resist demagnetisation implies a high coercive field $H_c$. The choice, read from the chart, is the neodymium-based family of hard magnetic materials. (These are the magnets-of-choice for hybrid and electric car motors and for wind turbine generators.) These however are expensive, because they contain a rare earth element Nd. The Alnico family, based on aluminum, nickel and cobalt (hence the name) are cheaper.

*The Saturation magnetisation–Susceptibility chart* This is the chart for selecting soft magnetic materials (Figure 15.9). It shows the saturation magnetisation, $M_s$—the ultimate degree to which a material can concentrate magnetic flux—plotted against its susceptibility, $\chi$, which measures the ease with which it can be magnetised. Many texts use, instead, the saturation inductance $B_s$ and maximum relative permeability, $\mu_R$. They are shown on the other axes of the chart. As in the first chart, materials that are electrical conductors are enclosed in orange and blue envelopes, those that are insulators have green ones.

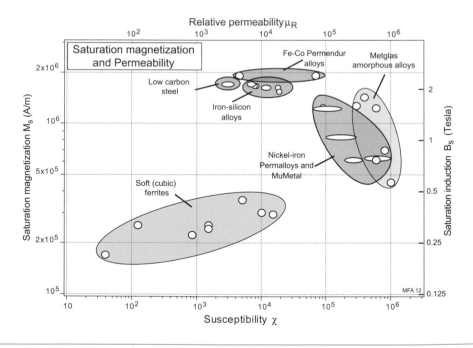

**Figure 15.9** Saturation magnetisation and susceptibility for soft magnetic materials. The orange and blue envelopes enclose electrically conducting materials; the green ones enclose insulators.

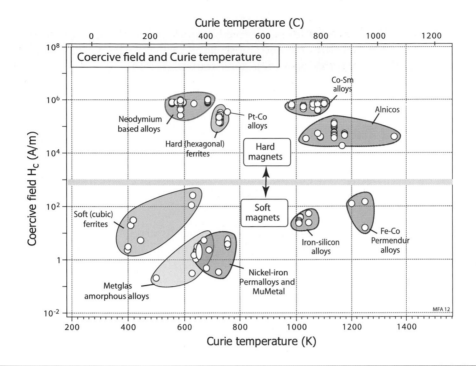

**Figure 15.10**  Coercive field and Curie temperature. The chart is useful for selecting materials for use above room temperature. The orange and blue envelopes enclose electrically conducting materials; the green ones enclose insulators.

*The Coercive field–Curie temperature chart*  Two things can disrupt ferro-magnetic behaviour: temperature and demagnetising fields. Ferro-magnetic response disappears above the Curie temperature, which, as the chart of Figure 15.10 shows, is as low as 100 °C for some materials and as high as 1000 °C for others. Remanent magnetisation is cancelled or reversed by fields exceeding the coercive field, a risk when permanent magnets must operate in environments with stray fields. The chart is useful for selecting magnetic materials for use at elevated temperatures—as actuators in control equipment of engines, for instance. As before, the colours distinguish conducting ferro-magnetic metals from insulating ferrites.

---

### Example 15.4

A permanent magnet is needed for an aerospace application. In use the magnet may be exposed to demagnetising fields as high as $3 \times 10^5$ A/m (0.38 Tesla) and temperatures of 600 °C. What material would you recommend for the magnet?

*Answer.* The coercive field must exceed the limit of $3 \times 10^5$ A/m. Figure 15.10 shows that this rules out Alnicos and most hard ferrites. The Curie temperature must comfortably exceed the operating temperature of 600 °C. The figure shows that neodymium-boron and hard ferrites cannot meet this requirement. This leaves the Cobalt-Samarium group of hard magnets, which comfortably meet both requirements.

## **15.4** Drilling down: the physics and manipulation of magnetic properties

The classical picture of an atom is that of a nucleus around which swing electrons, as in Figure 15.11. Moving charge implies an electric current, and an electric current flowing in a loop creates a magnetic moment, as in Figure 15.2. There is, therefore, a magnetic dipole associated with each orbiting electron. That is not all. Each electron has an additional moment of its own: its spin moment. A proper explanation of this requires quantum mechanics, but a way of envisaging its origin is to think of an electron not as a point charge but as slightly spread out and spinning on its own axis, again creating rotating charge and a dipole moment—and this turns out to be large. The total moment of the atom is the sum of the whole lot.

A simple atom like that of helium has two electrons per orbit, and they configure themselves such that the moment of one exactly cancels the moment of the other, as in Figure 15.11(a) and (c),

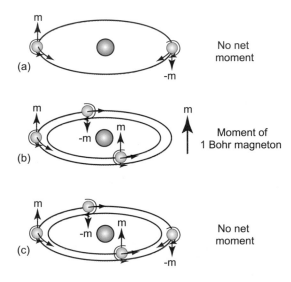

**Figure 15.11** Orbital and electron spins create a magnetic dipole. Even numbers of electrons filling energy levels in pairs have moments that cancel as at (a) and (c). An unpaired electron gives the atom a permanent magnetic moment, as at (b).

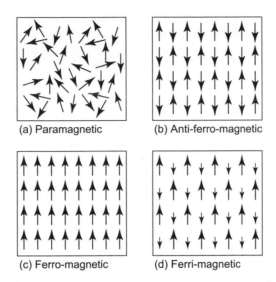

(a) Paramagnetic  (b) Anti-ferro-magnetic

(c) Ferro-magnetic  (d) Ferri-magnetic

**Figure 15.12**  Types of magnetic behaviour.

leaving no net moment. But now think of an atom with three, not two, electrons as in Figure 15.11(b). The moments of two may cancel, but there remains the third, leaving the atom with a net moment represented by the red arrow at the right of Figure 15.11(b). Thus atoms with electron-moments that cancel are non-magnetic; those with electron moments that don't cancel carry a magnetic dipole. Simplifying a little, one unpaired electron gives a magnetic moment of $9.27 \times 10^{-24}$ A.m$^2$, called a Bohr[3] magneton; two unpaired electrons gives two Bohr magnetons, three gives three and so on.

Think now of the magnetic atoms assembled into a crystal. In most materials the atomic moments interact so weakly that thermal motion is enough to randomise their directions, as in Figure 15.12(a). Despite their magnetic atoms, the structure as a whole has no magnetic moment; these materials are *paramagnetic*. In a few materials, though, something quite different happens. The fields of neighbouring atoms interact such that their energy is reduced if their magnetic moments line up. This drop in energy is called the *exchange energy*, and it is strong enough that it beats the randomising effect of thermal energy so long as the temperature is not too high. The shape of the Curie curve of Figure 15.4 shows how thermal energy overwhelms the exchange energy as the Curie temperature is approached. They may line up anti-parallel, head to tail, so to speak, as in Figure 15.12(b), and we still have no net magnetic moment; such materials are called *anti-ferro-magnets*. But in a few elements, notably iron, cobalt and nickel, exactly the opposite happens: the moments spontaneously align so that—if all are parallel—the structure has a net moment that is the sum of those of all the atoms it contains. These materials are *ferro-magnetic*. Iron has three unpaired electrons per atom,

---

[3] Niels Henrik David Bohr (1885–1962), Danish theoretical physicist, elucidator of the structure of the atom, contributor to the Manhattan Project and campaigner for peace.

cobalt has two and nickel just one, so the net moment if all the spins are aligned (the saturation magnetisation $M_s$) is greatest for iron, less for cobalt and still less for nickel.

Compounds give a fourth possibility. The materials we have referred to as ferrites are oxides; one class of them has the formula $MFe_2O_4$ where M is also a magnetic atom, such as cobalt, Co. Both the Fe and the Co atoms have dipoles but they differ in strength. They line up in the anti-parallel configuration, but because the moments differ, the cancellation is incomplete, leaving a net moment M; these are *ferri-magnets*, ferrites for short. The partial cancellation and the smaller number of magnetic atoms per unit volume means they have lower saturation magnetisation than, say, iron. But as they are oxides they have other advantages, notably that they are electrical insulators.

*Domains*   If atomic moments line up, shouldn't every piece of iron, nickel or cobalt be a permanent magnet? Although they are magnetic materials they do not necessarily have a magnetic moment. Why not?

A uniformly magnetised rod creates a magnetic field, $H$, like that of a solenoid. The field has an energy

$$U = \frac{1}{2}\mu_o \int_V H^2 dV \tag{15.9}$$

where the integral is carried out over the volume V within which the field exists. Working out this integral is not simple, but we don't need to do it. All we need to note is that the smaller is $H$ and the smaller the volume V that it invades, the smaller is the energy. If the structure can arrange its moments to minimise its overall $H$ or get rid of it entirely (remembering that the exchange energy wants neighbouring atom-moments to stay parallel) it will try to do so.

---

### Example 15.5

An electro-magnet generates a field of $H = 10,000\ A/m$. What is the magnetic energy density?

*Answer.*
The magnetic energy density is $U_o = \frac{1}{2}\mu_o H^2 = \frac{1}{2}(4\pi \times 10^{-7} \times 10^8) = 62.8\ J/m^3$.

---

Figure 15.13 illustrates how this can be done. The material splits up into *domains* within which the atomic moments are parallel, but with a switch of direction between mating domains to minimise the external field. The domains meet at *domain walls*, regions a few atoms thick in which the moments swing from the orientation of one domain to that of the other. Splitting into parallel domains of opposite magnetisation, as at Figure 15.13(b) and (c), reduces the field substantially; adding end regions magnetised perpendicular to both as at (d) kills it almost completely. The result is that most magnetic materials, unless manipulated in some way, adopt a domain structure with minimal external field—which is the same as saying that, while magnetic, they are not magnets.

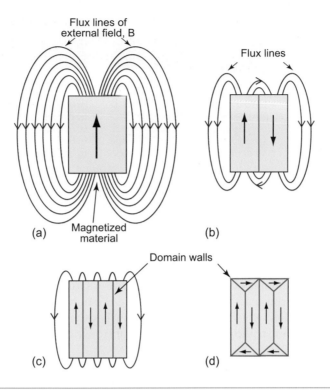

**Figure 15.13** Domains allow a compromise: the cancellation of the external field while retaining magnetisation of the material itself. The arrows show the direction of magnetisation.

How can they be magnetised? Figure 15.14, left, shows the starting domain-wall structure. Placed in a magnetic field, created, say, with a solenoid, the domains already aligned with the field have lower energy than those aligned against it. The domain wall separating them feels a force, the *Lorentz*[4] *force*, pushing it in a direction to make the favourably oriented domains grow at the expense of the others. As they grow, the magnetisation of the material increases, moving up the $M-H$ curve, finally saturating at $M_s$ when the whole sample is one domain, oriented parallel to the field, as in the right of Figure 15.14.

The saturation magnetisation, then, is just the sum of all the atomic moments contained in a unit volume of material when they are all aligned in the same direction. If the magnetic dipole per atom is $n_m\, m_B$ (where $n_m$ is the number of unpaired electrons per atom and $m_B$ is the magnetic moment of a Bohr magneton) then the saturation magnetisation is

$$M_s = \frac{n_m\, m_B}{\Omega} \qquad (15.10)$$

[4] Hendrik Antoon Lorentz (1853–1928), Dutch mathematical physicist, friend and collaborator with Raleigh and Einstein, originator of key concepts in optics, electricity, relativity and hydraulics (he modelled and predicted the movement of water caused by the building of the Zuyderzee).

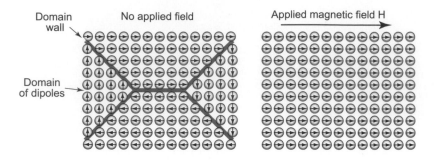

Domain wall

Domain of dipoles

No applied field

Applied magnetic field H

**Figure 15.14** An applied field causes domain boundaries to move. At saturation the sample has become a single domain, as on the right. Domain walls move easily in soft magnetic materials. Hard magnetic materials are alloyed to prevent their motion.

where $\Omega$ is the atomic volume. Iron has the highest because its atoms have the most unpaired electrons and they are packed close together. Anything that reduces either one reduces the saturation magnetisation. Nickel and cobalt have lower saturation because their atomic moments are less; alloys of iron tend to be lower because the non-magnetic alloying atoms dilute the magnetic iron. Ferrites, which are oxides, have much lower saturation, both because of dilution by oxygen and because of the partial cancellation of atomic moments sketched in Figure 15.12(d).

---

### Example 15.6

Nickel has a saturation magnetisation of $M_s = 5.2 \times 10^5$ A/m. Its atomic volume is $\Omega = 1.09 \times 10^{-29}$ m$^3$. What is the effective number of unpaired electrons $n_m$ in its conduction band?

*Answer.* The saturation magnetisation is given by $M_s = \dfrac{n_m m_B}{\Omega}$. The value of the Bohr magneton is $m_B = 9.27 \times 10^{-24}$ A.m$^2$. The resulting value of $n_m$ is $n_m = 0.61$.

---

### Example 15.7

Iron has a density $\rho$ of 7870 kg/m$^3$, and an atomic weight $A_{wt}$ of 55.85 kg/kmol. The net magnetic moment of an iron atom, in Bohr magnetons per atom, is $n_m = 2.2$. What is the saturation magnetisation $M_s$ of iron?

*Answer.* The number of atoms per unit volume in iron is

$$n = \frac{1}{\Omega} = \frac{N_A \, \rho}{A_{wt}} = 8.49 \times 10^{28}/\text{m}^3$$

(here $N_A$ is Avogadro's number, $6.022 \times 10^{26}$ atoms/kmol).

The saturation magnetisation $M_s$ is the net magnetic moment per atom $n_m\, m_B$ times the number of atoms per unit volume, $n$:

$$M_s = n\ n_m\ m_B$$

where $m_B - 9.27 \times 10^{-24}$ A.m$^2$ is the magnetic moment of one Bohr magneton. Thus the saturation magnetisation of iron is

$$M_s = 8.49 \times 10^{28} \times 2.2 \times 9.27 \times 10^{-24} = 1.73 \times 10^6 \text{ A/m}$$

There are, nonetheless, good reasons for alloying and making and using ferrites. One relates to the coercive field, $H_c$. A good permanent magnet retains its magnetisation even when placed in a reverse field, and this requires a high $H_c$. If domain walls move easily, the material is a soft magnet with a low $H_c$; if the opposite, it is a hard one. To make a hard magnet it is necessary to pin the domain walls to stop them from moving. Impurities, foreign inclusions, precipitates and porosity all interact with domain walls, tending to pin them in place. They act as obstacles to dislocation motion too (Chapter 6), so magnetically hard materials are mechanically hard as well. And, there are subtler barriers to domain-wall motion. One is *magnetic anisotropy*, arising because certain directions of magnetisation in the crystal are more favourable than others. Another relates to the shape and size of the sample itself: small, single-domain magnetic particles are particularly resistant to having their magnetism reversed, which is why they can be used for information storage.

The performance of a hard magnet is measured by its *energy product*, roughly proportional to the product $B_r H_c$, both shown on the chart of Figure 15.8. The higher the energy product the more difficult it is to demagnetise it.

## **15.5** Materials selection for magnetic design

Selection of materials for their magnetic properties follows the same process of translating design requirements followed by screening and ranking. In this case, however, the first screening step—'must be magnetic'—immediately isolates the special set of materials shown in the property charts of this chapter. The key selection issues are then

- whether a soft or hard magnetic response is required;
- whether a fat or a thin $M-H$ hysteresis loop is wanted;
- what frequency of alternating field the material must carry;
- whether the material must operate above room temperature in a large external field.

*Soft magnetic devices* Electromagnets, transformers, electron lenses and the like have magnetic cores that must magnetise easily when the field switches on, be capable of creating a high flux density, yet lose their magnetism when the field is switched off. They do this by using soft, low $H_c$, magnetic materials, shown on the left of the chart of Figure 15.8. Soft magnetic

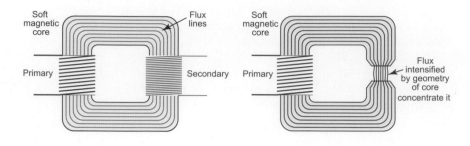

**Figure 15.15** Soft magnets 'conduct' magnetic flux and concentrate it if the cross-section of the magnet is reduced.

materials are used to 'conduct' and focus magnetic flux. Figure 15.15 illustrates this. A current-carrying coil induces a magnetic field in the core. The field, instead of spreading in the way shown in Figure 15.3, is trapped in the extended ferro-magnetic circuit, as shown here. The higher the susceptibility $\chi$, the greater is the magnetisation and flux density induced by a given current loop, up to a limit set by its saturation magnetisation $M_s$. The magnetic circuit conducts the flux from one coil to another, or from a coil to an air-gap where it is further enhanced by reducing the section of the core, as shown on the right of Figure 15.15. The field in the gap is used to actuate or to focus an electron beam. Thus the first requirement of a good soft magnet is a high $\chi$ and a high $M_s$ with a thin hysteresis loop like that sketched in Figure 15.16. The chart of Figure 15.9 shows these two properties. Permendur (Fe-Co-V alloys) and Metglas amorphous alloys are particularly good, but being expensive they are used only in small

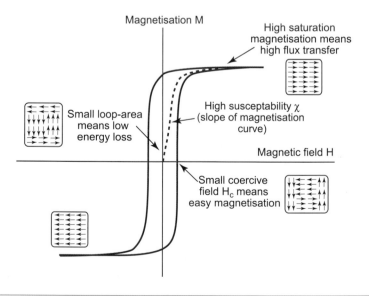

**Figure 15.16** A characteristic hysteresis loop for a soft magnet.

Table 15.1    Materials for soft magnetic applications

| Application | Frequency $f$ | Material requirements | Material choice |
|---|---|---|---|
| Electro-magnets | <1 Hz | High $M_s$, high $\chi$ | Silicon iron<br>Fe–Co alloys (Permendur) |
| Motors, low-frequency transformers | <100 Hz | High $M_s$, high $\chi$ | Silicon-iron<br>AMA (amorphous Ni–Fe–P alloys) |
| Audio amplifiers, loudspeakers, microphones | <100 kHz | High $M_s$, very low $H_c$, and high $\chi$ | Ni–Fe (Permalloy, Mumetal) |
| Microwave and UHF applications | <MHz | High $M_s$, very low $H_c$, electrical insulator | Cubic ferrites: $MFe_2O_4$ with M = Cu / Mn / Ni |
| Giga-hertz devices | >100 MHz | Ultra-low hysteresis, excellent insulator | Garnets |

devices. Silicon-iron (Fe 1–4% Si) is much cheaper and easier to make in large sections; it is the staple material for large transformers.

Most soft magnets are used in AC devices, and then energy loss, proportional to the area of the hysteresis loop, becomes a consideration. Energy loss is of two types: the hysteresis loss already mentioned, proportional to switching frequency $f$, and eddy current losses caused by the currents induced in the core itself by the AC field, proportional to $f^2$. In cores of transformers operating at low frequencies (meaning $f$ up to 100 Hz or so) the hysteresis loss dominates, so for these we seek high $M_s$ and a thin loop with a small area. Eddy-current losses are present but are suppressed when necessary by *laminating*: making the core from a stack of thin sheets interleaved with thin insulating layers that interrupt the eddy currents. At audio frequencies ($f < 50kHz$) losses of both types become greater because they occur on every cycle and now there are more of them, requiring an even narrower loop and thinner laminates; here the more expensive Permalloys and Permendurs are used. At higher frequencies still ($f$ in the MHz range) eddy current loss, rising with $f^2$, becomes very large and the only way to stop it is to use magnetic materials that are electrical insulators. The best choice is Ferrites, shown in green envelopes on all three charts, even though they have a lower $M_s$. Above this ($f > 100MHz$) only the most exotic ceramic magnets will work. Figure 15.16 summarises the hysteresis loop characteristics for soft magnets; Table 15.1 lists the choices, guided by the charts.

## Example 15.8

A magnetic material is needed to act as the driver for an acoustic frequency fatigue-testing machine. Use the information in Table 15.1 to identify the classes of magnetic materials that would be good choices.

*Answer.* The application requires high saturation magnetisation, $M_s$, to provide the loads required for fatigue testing. It also requires low hysteresis loss at acoustic frequencies to minimize power consumption and avoid heating. The table indicates that the Ni-Fe Permalloys or Mumetals would be good choices.

*Hard magnetic devices* Many devices use permanent magnets: fridge door seals, magnetic clutches, loudspeakers and earphones, DC motors (as in hybrid and electric cars), and power generation (as in wind turbines). For these, the key property is the remanence, $M_r$, since this is the maximum magnetisation the material can offer without an imposed external field to keep it magnetised. High $M_r$, however, is not all. When a material is magnetised and the magnetising field is removed, the field of the magnet itself tries to demagnetise it. It is this demagnetising field that makes the material take up a domain structure that reduces or removes the net magnetisation of the sample. A permanent magnet must be able to resist being demagnetised by its own or other fields. A measure of its resistance is its coercive field, $H_c$, and this too must be large. A high $M_r$ and a high $H_c$ means a fat hysteresis loop like that of Figure 15.17. Ordinary iron and steel can become weakly magnetised with the irritating result that your screwdriver sticks to screws and your scissors pick up paper clips when you wish they wouldn't, but neither are 'hard' in a magnetic sense. Here 'hard' means 'hard to demagnetise', and that is where high $H_c$ comes in; it protects the magnet from itself.

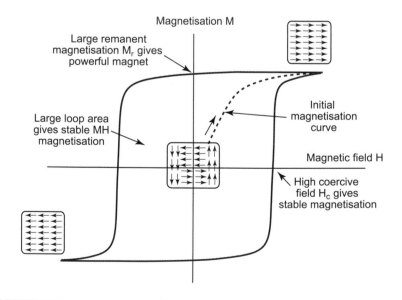

**Figure 15.17** A characteristic hysteresis loop for a hard magnet.

Table 15.2  Materials for hard magnetic applications

| Application | Material requirements | Material choice |
|---|---|---|
| Large permanent magnets | High $M_r$, high $H_c$, and high-energy product $(MH)_{max}$; low cost | Alnicos<br>Hexagonal ferrites |
| Small high-performance magnets | Acoustic, electronic and miniature mechanical applications, electric propulsion, wind turbine generators | Cobalt-samarium<br>Neodymium-iron-boron<br>(Platinum-cobalt) |
| Information storage | Thin, 'square' hysteresis loop | Elongated particles of $Fe_2O_3$, $Cr_2O_3$ or hexagonal ferrite |

Hard magnetic materials lie on the right of the $M_r$–$H_c$ chart of Figure 15.8 and the upper half of Figure 15.8. The workhorses of the permanent magnet business are the Alnicos and the hexagonal ferrites. An Alnico (there are many) is an alloy of aluminum, nickel, cobalt and iron—as mentioned earlier—with values of $M_r$ as high as $10^6$ A/m and a huge coercive field: $5 \times 10^4$ A/m. The hexagonal ferrites, too, have high $M_r$ and $H_c$ and, like the cubic soft-magnet ferrites, they are insulators. Being metallic, the Alnicos can be shaped by casting or by compacting and sintering powders; the ferrites can only be shaped by powder methods. Powders, of course, don't have to be sintered—you can bond them together in other ways, and these offer opportunities. Magnetic particles mixed with a resin can be moulded to complex shapes. Bonding them with an elastomer creates magnets that are flexible, as discussed below.

The charts of Figures 15.8 and 15.10 show three families of hard magnets with coercive fields that surpass those of Alnicos and hexagonal ferrites. These, the products of research over the last 30 years, provide the intense permanent-magnet fields that have allowed the miniaturisation of earphones, microphones and motors. The oldest of these is the precious-metal Pt-Co family but they are so expensive that they have now all but disappeared. They are out-performed by the cobalt-samarium family based on the inter-metallic compound $SmCo_5$, and they, in turn, are out-performed by the more recent development of the neodymium-iron-boron family. None of these are cheap, but in the applications they best suit, the quantity of material is sufficiency small or the product-value sufficiently high that this is not an obstacle. Table 15.2 lists the choices, guided by the charts.

*Flexible magnets*  Flexible magnets are made by mixing ferrite powders into an elastomeric resin that is then extruded, moulded or calendered (rolled). Typical of these is the range of magnets with strontium ferrous oxide particles embedded in styrene butadiene rubber (SBR) or vinyl (flexible PVC). Table 15.3 lists typical magnetic properties.

Flexible magnets can be bent, punched or cut without loss of magnetic properties. They are relatively cheap and because of this they are used in signs, promotional magnets and door seals. Like all ferrites, the remanent induction of flexible magnets decreases with increasing temperature, but the limiting factor for flexible magnets is not the ferrite but the binder, which degrades or flows at temperatures above about 470 K (200 °C).

Table 15.3  Properties of flexible elastomers at room temperature

| Property | Value range |
|---|---|
| Remanent induction $B_r$ | 0.15–0.25 T |
| Coercive force $H_c$ | $1 \times 10^5$–$2 \times 10^5$ A/m |
| Maximum energy product | 4.5–15 kJ/m$^3$ |

### Example 15.9

A flexible magnet with a remanent induction of 0.2 T is used as a seal on a refrigerator door. If the area of contact $A$ of the magnet with the steel frame of the fridge is 0.005 m$^2$ what clamping force does it exert?

*Answer.* From equation (15.7) the force exerted by the flexible magnet is

$$F = P_{mag}\, A \;\; = \frac{1}{2}\, A \frac{B^2}{\mu_o} = \frac{1}{2}\, 0.005 \frac{0.2^2}{4\pi \times 10^{-7}} = 80 \text{ N}$$

*Magnetic storage of information*   Magnetic information storage requires hard magnets, but those with an unusual loop shape: they are rectangular (called 'square'). The squareness means that a unit of the material can flip in magnetisation from $+M_r$ to $-M_r$ and back when exposed to fields above $+H_c$ or $-H_c$. The unit thus stores one binary bit of information. The information is read by moving the magnetised unit past a read head where it induces an electric pulse. The choice of the word 'unit' was deliberately vague because it can take several forms: discrete particles embedded in a polymer tape or disk, or as a thin magnetic film that has the ability to carry small stable domains called 'bubbles' that are magnetised in one direction, embedded in a background that is magnetised in another direction.

Figure 15.18 shows one sort of information storage system. A signal creates a magnetic flux in a soft magnetic core, which is often an amorphous metal alloy (AMA, the blue envelopes on the charts) because of its large susceptibility. The direction of the flux depends on the direction of the signal-current; switching this reverses the direction of the flux. An air gap in the core allows field to escape, penetrating a tape or disk in which sub-micron particles of a hard magnetic material are embedded. When the signal is in one direction, all particles in the band of tape or disk passing under the write-head are magnetised in one direction; when the signal is reversed, the direction of magnetisation is reversed. The same head is used for reading. When the tape or disk is swept under the air gap, the stray field of the magnetised bands induce a flux in the soft iron core and a signal in the coil, which is amplified and processed to read it.

Why use fine particles? It is because domain walls are not stable in very small particles: each particle is a single domain. If the particle is elongated, it prefers to be magnetised parallel to its long axis. Each particle then behaves like a little bar magnet with a north (N) and a south (S)

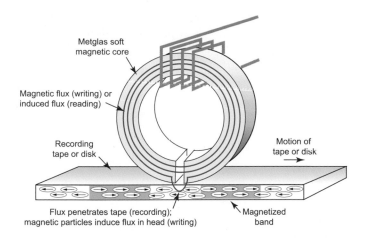

**Figure 15.18** A magnetic read-write head. The gap in the soft magnetic head allows flux to escape, p3enetrating the tape and realigning the particles.

pole at either end. The field from the write-head is enough to flip the direction, turning a N-S magnetisation into one that is S-N, a binary response that is well suited to information storage. Re-writable tapes and disks use particles of $Fe_2O_3$ or $CrO_2$, typically 0.1 microns long and with an aspect ratio of 10:1. Particles of hexagonal ferrites $(MO)(Fe_2O_3)_6$ have a higher co-ercive field, so a more powerful field is required to change their magnetisation—but having done so, it is hard to erase it. This makes them the best choice for read-only applications like the identification strip of credit and swipe-cards.

***Superconducting magnets***   Superconducting magnets allow high-field solenoids with no steady-state power consumption. For many years the obstacle to their widespread use was the low critical temperature, $T_c$, requiring that the magnet be cooled with liquid helium (boiling point 4.2 K). The development, about 25 years ago, of superconducting oxides with critical temperatures above the boiling point of liquid nitrogen (77.4 K) makes superconducting magnets a more practical, though still difficult, proposition (see Figure 14.4). The high-$T_c$ superconductors also have a high critical field—the magnetic field strength that quenches superconduction—that sets an upper limit to the performance of a superconductor. The critical field depends on temperature in the way shown in Figure 15.19 for four of them. Ordinary electromagnets get hot because of the resistance of the core, limiting them to flux densities of about 2 T. Superconducting magnets dissipate no energy in the windings (though energy is expended cooling them to below $T_c$), allowing them to reach 80 T. They are used in MRI scanners, NMR equipment, mass spectrometers, magnetic separation processes and particle accelerators such as those at CERN that recently detected the Higgs Boson.

***The Meissner effect and magnetic suspension***   When a superconductor is placed in a magnetic field it behaves like any other conductor in a moving field—a current is induced in it. But because it has no resistance, the current continues to flow, preventing the flux lines of the field from entering. The result is that the field exerts a repulsive pressure (given by equation (15.7))

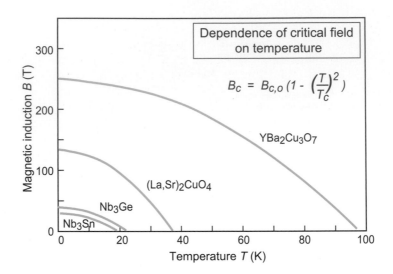

**Figure 15.19** The dependence of the critical field on temperature, falling from a peak of $B_{c,o}$ at 0 K to zero at the superconducting critical temperature $T_c$. (Don't confuse this with the decrease in magnetisation with temperature, Figure 15.4, which has a similar shape.)

on the superconductor in the same way that two bar magnets repel when two north or south poles are face-to-face. It sounds as if this so-called Meissner effect might be used to make magnetic bearings and levitation suspensions, but in practice most magnetic bearings rely on magnetic attraction rather than repulsion. The shaft or track is surrounded by a symmetrical array of electro-magnets that pull it towards them. This is intrinsically unstable because the closer the shaft or track gets to one of the magnets the greater is the attractive force it experiences. The instability is overcome by gap-sensors that monitor the position of the shaft (or of the vehicle in magnetic levitation systems), adjusting the power in the individual electromagnets continuously to correct the position.

*Superconducting magnetic energy storage (SMES) systems*   These systems store energy as a magnetic field, set up by passing current through a superconducting coil. Because there is no resistance, the current and the field, once established, continue without consuming further power. Energy is recovered by discharging the current through an external load.

Superconducting magnets require refrigeration, which does consume power, thus reducing efficiency, especially over long periods. High temperature superconductors (HTS) such as YBCO (yttrium barium copper oxide), which become superconducting at ~85 K, require less power for refrigeration than low temperature ~4 K superconductors (LTS), but they still need some. For this reason superconductors are used for short term storage (e.g. for frequency regulation of the national grid).

The energy-density stored in a magnetic field is:

$$\frac{U}{V} = \frac{1}{2}\frac{B^2}{\mu_o} \text{ Joules/m}^3$$

where $B$ is the magnetic flux density and $\mu_o$ is the permittivity of free space. The energy density is maximised by maximising $B$, but this cannot exceed the upper critical field $B_c$ of the magnet windings. The upper critical field of YBCO at a temperature of zero Kelvin is between 100 and 250 T, depending on the orientation of the superconducting current with respect to the crystal structure of the YBCO. A flux density of 100 T corresponds to an energy density of 4,000 MJ/m$^3$. This is three times as much as a lithium ion battery (1,200 MJ/m$^3$) making it interesting for energy storage, but it is still a lot less than diesel or petroleum (35,000 MJ/m$^3$).

## 15.6 Summary and conclusions

The classical picture of an atom is that of a nucleus around which swing electrons in discrete orbits, each electron spinning at the same time on its own axis. Both spins create magnetic moments that, if parallel, add up, but if opposed, cancel to a greater or lesser degree. Most materials achieve near-perfect cancellation either within the atomic orbits or, if not, by stacking the atomic moments head to tail or randomising them so that, when added, they cancel. A very few, most of them based on iron, cobalt or nickel, have atoms with residual moments and an inter-atomic interaction that causes them to line up to give a net magnetic moment or magnetisation. Even these materials can find a way to screen their magnetisation by segmenting themselves into domains: a ghetto-like arrangement in which atomic moments segregate into colonies or domains, each with a magnetisation that is oriented such that it tends to cancel that of its neighbours. A strong magnetic field can override the segregation, creating a single unified domain in which all the atomic moments are parallel, and if the coercive field is large enough, they remain parallel even when the driving field is removed, giving a 'permanent' magnetisation.

There are two sorts of characters in the world of magnetic materials. There are those that magnetise readily, requiring only slight urging from an applied field to do so. They transmit magnetic flux, and require only a small reversal of the applied field to realign themselves with it. And there are those that, once magnetised, resist realignment; they give us permanent magnets. The charts of this chapter introduced the two, displaying the properties that most directly determine their choice for a given application.

## 15.7 Further reading

Braithwaite, N., & Weaver, G. (1990). *Electronic Materials*. Oxford, UK: The Open University and Butterworth Heinemann. ISBN 0-408-02840-8. (One of the excellent Open University texts that form part of their Materials program.)

Callister, W. D., & Rethwisch, D. G. (2011). *Materials Science and Engineering: An Introduction* (8th ed.). New York, USA: John Wiley. ISBN 978-0-470-50586-1. (A well-established and comprehensive introduction to the science of materials.)

Campbell, P. (1994). *Permanent Magnetic Materials and Their Applications*. Cambridge, UK: Cambridge University Press.

Douglas, W. D. (1995). Magnetically Soft Materials. In *Properties and Selection of Non-Ferrous Alloys and Special Purpose Materials: Vol. 2.* (9th ed.). *ASM Metals Handbook* (pp. 761–781) Metals Park. Ohio, USA: ASM.

Fiepke, J. W. (1995). Permanent magnet materials. In *Properties and Selection of Non-Ferrous Alloys and Special Purpose Materials: Vol. 2.* (9th ed.). *ASM Metals Handbook* (pp. 782–803) Ohio, USA: ASM: Metals Park.

Jakubovics, J. P. (1994). In *Magnetism and Magnetic Materials* (2nd ed.). London, UK: The Institute of Materials. ISBN 0-901716-54-5. (A simple introduction to magnetic materials, short and readable.)

## **15.8** Exercises

| | |
|---|---|
| Exercise E15.1 | A cylindrical coil with a length $L = 30$ mm with $n = 75$ turns carries a current $i = 0.1$ A. What is the field in the magnet? A silicon-iron core with a relative permeability $\mu_R = 3000$ is placed in the core of the coil. What is the magnetisation of the silicon-iron? |
| Exercise E15.2 | Sketch a $M–H$ curve for a ferro-magnetic material. Identify the important magnetic properties. |
| Exercise E15.3 | A neodymium-boron magnet has a coercive field $H_c$ of $1.03 \times 10^6$ A/m, a saturation magnetisation $M_s = 1.25 \times 10^6$ A/m, a remanent magnetisation $M_r = 1.1 \times 10^6$ A/m, and a large energy product. Sketch approximately what its $M–H$ hysteresis curve looks like. |
| Exercise E15.4 | Why are some elements ferro-magnetic when others are not? |
| Exercise E15.5 | What, when describing magnetic materials, is a ferrite? What are its characteristics? |
| Exercise E15.6 | What is a Bohr magneton? A magnetic element has two unpaired electrons and an exchange interaction that causes them to align such that their magnetic fields are parallel. Its atomic volume, $\Omega$, is $3.7 \times 10^{-29}$ m$^3$. What would you expect its saturation magnetisation, $M_s$, to be? |
| Exercise E15.7 | Cobalt has a density $\rho$ of 8900 kg/m$^3$, and atomic weight $A_{wt}$ of 58.93 kg/kmol. The net magnetic moment of a cobalt atom, in Bohr magnetons per atom, is $n_m = 1.8$. What is the saturation magnetisation $M_s$ of cobalt? |
| Exercise E15.8 | The element with the largest saturation magnetisation is Holmium, with $M_s = 3.0 \times 10^6$ A/m (although it has a miserably low Curie temperature of 20 K). The density $\rho$ of Holmium is 8800 kg/m$^3$ and its atomic weight $A_{wt}$ is 165 kg/kmol. What is the atomic magnetic moment $n_m$, in Bohr magnetons per atom, of Holmium? |
| Exercise E15.9 | The soft iron laminations of transformers are made of 3wt% (6at%) silicon-iron. Silicon-iron has a density $\rho$ of 7650 kg/m$^3$, and a mean atomic weight $A_{wt}$ of 54.2 kg/kmol. The net magnetic moment of an iron atom, in |

Bohr magnetons per atom, is $n_m = 2.2$; that of silicon is zero. What is the saturation magnetisation $M_s$ of 3wt% silicon-iron?

| | |
|---|---|
| Exercise E15.10 | Nickel has a density $\rho$ of 8900 kg/m$^3$, and atomic weight $A_{wt}$ of 58.7 kg/kmol. and a saturation magnetisation $M_s$ of $5.2 \times 10^5$ A/m. What is the atomic magnetic moment $n_m$, in Bohr magnetons per atom, of nickel? |
| Exercise E15.11 | A coil of 50 turns and length 10 mm carries a current of 0.01 A. The core of the coil is made of a material with a susceptibility $\chi = 10^4$. What is the magnetisation, $M$, and the induction $B$? |
| Exercise E15.12 | An inductor core is required for a low frequency harmonic filter. The requirement is for low loss and high saturation magnetisation. Using the charts of Figures 15.8 and 15.9 as data sources, which class of magnetic material would you choose? |
| Exercise E15.13 | A magnetic material is required for the core of a transformer that forms part of a radar speed camera. It operates at a microwave frequency of 500 kHz. Which class of material would you choose? |
| Exercise E15.14 | Which soft magnetic material has the greatest saturation magnetisation, $M_s$, allowing it to concentrate magnetic field most effectively? Which has the greatest relative permeability, $\mu_R$, allowing the highest inductance $B$ for a given applied field $H$? The Saturation magnetisation–Permeability chart (Figure 15.9) can help. |
| Exercise E15.15 | The product $B_r H_c$ is a crude measure of the 'power' of a permanent magnet (if the $B-H$ curve were a perfect rectangle, it becomes equal to what is called the energy product, a more realistic measure of the power). Draw contours of $B_r H_c$ onto a copy of the Remanent induction–Coercive field chart (Figure 15.8) and use these to identify the most powerful permanent magnet materials. |
| Exercise E15.16 | A soft magnetic material is needed for the core of a small high-frequency power transformer. The transformer gets hot in use; forced air cooling limits the rise in temperature to 200 °C. Eddy current losses are a problem if the core is electrically conducting. What material would you recommend for the core? The Coercive field–Curie temperature chart (Figure 15.10) can help. |
| Exercise E15.17 | A material is required for a flexible magnetic seal. It must be in the form of an elastomeric sheet and must be electrically insulating. How would you propose to make such a material? |
| Exercise E15.18 | What are the characteristics required of materials for magnetic information storage? |

Exercise E15.19    An electro-magnet is designed to pick up and move car parts—the car part becomes a temporary part of the core, completing the magnetic circuit. The field $H$ is created by a coil 100 mm long with a cross-section of 0.1 $m^2$ with 500 turns of conductor carrying a current of 25 A. What force can the electro-magnet exert?

Exercise E15.20    What, approximately, is the maximum induction $B$ that is possible in a superconducting coil made of YBCO (the oxide superconductor $YBa_2Cu_3O_7$) at the temperature of liquid nitrogen (77 K)? Use Figure 15.19 to find out. What is the maximum energy density that can be stored in such a coil? How does this compare with crude oil with an energy density of 35 MJ/litre? (This is not a fair comparison because the superconductor requires ancillary equipment to support it, but make the comparison anyway.)

## 15.9 Exploring design with CES

Open CES at Level 3—it opens in the 'Browse' mode. At the head of the Browse list are two pull-down menus reading 'Table' and 'Subset'. Open the 'Subset' menu and choose 'Magnetic'. Records for 233 magnetic materials are displayed, listing their magnetic properties and, where available, mechanical, thermal, electrical and other properties too.

To select, click on 'Select' in the main toolbar and choose MaterialsUniverse > Magnetic materials in the dialog box. Now we can begin.

Exercise E15.21  Use the CES 'Search' facility to find materials for:
(a) Transformer cores.
(b) Electric motors.

Exercise E15.22  Find by browsing the records for:
(a) Cast Alnico 3. What is the value of its coercive force $H_c$?
(b) The amorphous alloy Metglas 2605-Co. What is the value of its coercive force $H_c$?
What do these values tell you about the potential applications of these two materials?

Exercise E15.23  Find by browsing the records for:
(a) Ferrite G (Ni-Zn ferrite). What are the values of its coercive force $H_c$ and resistivity $\rho_e$?
(b) 2.5 Si-Fe soft magnetic alloy. What are the values of its coercive force $H_c$ and resistivity $\rho_e$?
If you were asked to choose one of these for a transformer core, what would be your first question?

Exercise E15.24 Make a bar chart of saturation induction $B_s$ (the saturation magnetisation $M_s = B_s/\mu_o$, so the two are proportional). Report the three materials with the highest values.

Exercise E15.25 Make a bar chart of coercive force, $H_c$. Report the material with the lowest value.

Exercise E15.26 A soft magnetic material is required for the laminated rotor of AC induction electric motor. The material is to be rolled and further shaped by stamping, requiring an elongation of at least 40%. To keep hysteresis losses to a minimum it should have the lowest possible coercive force. Find the two materials that best meet these requirements, summarised below. Report them and their trade names.

| | |
|---|---|
| Function | • Motor laminations |
| Constraints | • Ferro- or ferri-magnetic material |
| | • Ductile (elongation > 40%) |
| Objective | • Minimise coercive force |
| Free variable | • Choice of material |

Exercise E15.27 A magnetic material is required for the read/write head of a VCR. It must have high hardness to resist wear and the lowest possible coercive force to give accurate read/write response.

(a) Use CES to identify possible candidates that meet these requirements, summarised in the table.

(b) Now add the further constraint that, for lower loss and high frequency damping, the magnetic material must have an electrical resistivity above $10^8$ $\mu\Omega$.cm.

| | |
|---|---|
| Function | • Magnetic read-head |
| Constraints | • Ferro- or ferri-magnetic material |
| | • High hardness for wear resistance |
| | • (Vickers hardness > 500 HV) |
| Objective | • Minimise coercive force |
| Free variable | • Choice of material |

## **15.10** Exploring the science with CES Elements

Exercise E15.28 Make a chart with Atomic number on the $x$-axis and Magneton moment per atom (Bohr magneton) on the $y$-axis (use linear scales) to identify the ferromagnetic elements. How many are there? Which have the highest magneton moment per atom?

Exercise E15.29  According to equation (15.10) of the text, the saturation magnetisation is

$$M_s = \frac{n_m \, m_B}{\Omega}$$

where $n_m$ is the magnetic dipole per atom in units of Bohr magnetons, $\Omega$ is the atomic volume, and $m_B$ is the value of a Bohr magneton ($9.27 \times 10^{-24}$ A/m$^2$). The saturation induction (in Tesla) $B_s = \mu_o \, M_s$. Make a chart showing $B_s$ on the $y$-axis and on the $x$-axis plot the CES parameter called 'saturation magnetisation' (which is actually $B_s$, in Tesla, for the relevant elements). How accurately is this equation obeyed?

# Chapter 16
# Materials for optical devices

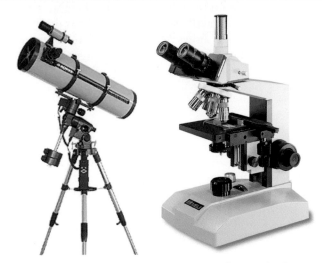

Instruments using the ability of materials to reflect and refract.
(Image of microscope courtesy of A−Z Microscope Corporation,
California, USA)

## **16.1** Introduction and synopsis

It was at one time thought that the fact that light could travel through space—from the sun to earth, for instance—must mean that space was not really empty but filled with 'luminiferous ether'. It was not until the experiments of Michelson[1] and Morley in 1857 that it was realised that light did not need a 'material' for its propagation but could propagate through totally empty space at what is now seen as the ultimate velocity: $3 \times 10^8$ m/s.

Light is electromagnetic radiation. When radiation strikes materials, things can happen. Materials interact with radiation by reflecting it, absorbing it, transmitting it and refracting it. This chapter is about these interactions, the materials that do them best and the ways we use them. The chapter opening page shows two: a reflecting telescope and a refracting microscope, each of which depend on the optical properties of materials.

## **16.2** The interaction of materials and radiation

Electromagnetic (e-m) radiation permeates the entire universe. Observe the sky with your eye and you see the visible spectrum, the range of wavelengths we call 'light' (0.40–0.77 μm). Observe it with a detector of X-rays or γ-rays and you see radiation with far shorter wavelengths (as short as $10^{-4}$ nm, one thousandth the size of an atom). Observe it instead with a radio telescope and you pick up radiation with wavelengths measured in millimeters, meters or even kilometers, known as radio and microwaves. The range of wavelengths of radiation is vast, spanning eighteen orders of magnitude (Figure 16.1). The visible spectrum is only a tiny part of it—but even that has entrancing variety, giving us colours ranging from deep purple through blue, green and yellow to deep red.

The intensity $I$ of an e-m wave, proportional to the square of its amplitude, is a measure of the energy it carries. When radiation with intensity $I_0$ strikes a material, a part $I_R$ of it is reflected, a part $I_A$ absorbed and a part $I_T$ may be transmitted. Conservation of energy requires that

$$\frac{I_R}{I_0} + \frac{I_A}{I_0} + \frac{I_T}{I_0} = 1 \qquad (16.1)$$

The first term is called the *reflectivity* of the material, the second the *absorptivity* and the last the *transmittability* (all dimensionless). Each depends on the wavelength of the radiation, on the nature of the material and on the state of its surfaces. They can be thought of as properties of the material in a given state of surface polish, smoothness or roughness.

In optics we are concerned with wavelengths in the visible spectrum. Materials that reflect or absorb all visible light, transmitting none, are called *opaque*, even though they may transmit in the near visible (infrared or ultraviolet). Those that transmit a little diffuse light are called *translucent*. Those that transmit light sufficiently well that they you can see through them are

---

[1] Albert A. Michelson (1852–1937), Prussian-American experimental physicist, who, with E. W. Morley, first demonstrated that the speed of light is independent of the earth's motion, a finding central to the establishment of the theory of relativity.

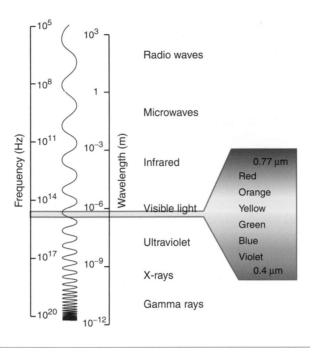

**Figure 16.1** The spectrum of electro-magnetic (e-m) waves. The visible spectrum lies between the wavelengths 0.4 and 0.77 microns.

called *transparent*; and a subset of these that transmit almost perfectly, making them suitable for lenses, light-guides and optical fibres are given the additional title of *optical quality*. To be transparent a material must be a dielectric. Metals, by contrast, are opaque; they reflect most of the light that strikes them.

*Specular and diffuse reflection* *Specular* surfaces are microscopically smooth and flat, meaning that any irregularities are much smaller than the wavelength of light. A beam of light striking a specular surface at an incident angle $\theta_1$ suffers specular reflection, which means that it is reflected as a beam at an angle $\theta_2$ (Figure 16.2, left) such that

$$\theta_1 = \theta_2. \tag{16.2}$$

*Diffuse* surfaces are irregular; the law of reflection (equation (16.2)) still holds locally but the incident beam is reflected in many different directions because of the irregularities, as on the right of Figure 16.2.

*Absorption* If radiation can penetrate a material, some is absorbed. The greater the thickness $x$ through which the radiation passes, the greater the absorption. The intensity $I$, starting with the initial value $I_o$, decreases such that

$$I = I_o \exp - \beta x \tag{16.3}$$

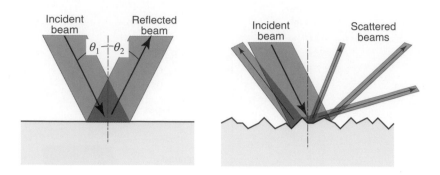

**Figure 16.2** Perfectly flat, reflective surfaces give specular reflection, such that $\theta_1 = \theta_2$. The angles of incidence and reflection are always equal, but the rough surface gives diffuse reflection even though the angles of incidence and reflection are still, locally, equal.

where $\beta$ is the absorption coefficient, with dimensions of $m^{-1}$ (or, more conveniently, $mm^{-1}$). The absorption coefficient depends on wavelength with the result that white light passing through a material may emerge with a colour corresponding to the wavelength that is least absorbed—that is why a thick slab of ice looks blue.

*Transmission*   By the time a beam of light has passed completely through a slab of material, it has lost some intensity through reflection at the surface at which it entered, some in reflection at the surface at which it leaves and some by absorption in between. Its intensity is

$$I = I_o \left(1 - \frac{I_R}{I_o}\right)^2 \exp - \beta x \tag{16.4}$$

The term $(1 - I_R/I_o)$ occurs to the second power because intensity is lost through reflection at both surfaces.

---

### Example 16.1

The polymer EVA has an absorption coefficient for light in the optical range of $\beta = 5/m$. Its absorption coefficient for ultraviolet light is $\beta = 1000/m$. If a window is covered with a 2 mm thick sheet of EVA, what fraction of optical and UV light will be absorbed in it?

*Answer.* The fraction of light intensity that is absorbed in passing through a thickness $x$ of a material is

$$\frac{I_{absorbed}}{I_o} = 1 - \exp - \beta x$$

Inserting the data for $\beta$ gives the fraction of optical frequencies absorbed to be 1%; the fraction of UV is 86%.

Example 16.2

Water has an absorption coefficient for blue light of $\beta = 0.01/m$ and for red light of $\beta = 0.3/m$. How much are the red and blue components of white light absorbed when a beam of white light passes through 10 meters of water? What is the consequent colour of the light perceived by an observer 10 meters from the source?

*Answer.* The transmitted fraction of the blue component is $\frac{I}{I_o} = \exp - \beta x = 0.9$; thus 10% is absorbed. The transmitted fraction of the red component is $\frac{I}{I_o} = 0.05$; thus 95% is absorbed. The water, in consequence, appears blue.

*Refraction*   The velocity of light in vacuum, $c_o = 3 \times 10^8$ m/s, is as fast as it ever goes. When it (or any other electromagnetic radiation) enters a material, it slows down. The *index of refraction*, $n$ is the ratio of its velocity in vacuum, $c_o$, to that in the material, $c$:

$$n = \frac{c_o}{c} \tag{16.5}$$

This retardation makes a beam of light bend or *refract* when it enters a material of different refractive index. When a beam with an angle of incidence $\theta_1$ passes from a material 1 of refractive index $n_1$ into a material 2 of index $n_2$, it deflects to an angle $\theta_2$, such that

$$\frac{\sin \theta_1}{\sin \theta_2} = \frac{n_2}{n_1} \tag{16.6}$$

as in Figure 16.3(a); the equation is known as Snell's law[2]. The refractive index depends on wavelength, so each of the colours that make up white light is diffracted through a slightly different angle, producing a spectrum when light passes through a prism. When material 1 is vacuum or air, for which $n_1 = 1$, the equation reduces to

$$\frac{\sin \theta_1}{\sin \theta_2} = n_2 \tag{16.7}$$

Equation (16.6) says that when light passes from a material with index $n_1 = n$ into air with $n_2 = 1$, it is bent away from the normal to the surface, like that in Figure 16.3(b). If the incident angle, here $\theta_1$, is slowly increased, the emerging beam tips down until it becomes parallel with the surface when $\theta_2 = 90°$ and $\sin \theta_2 = 1$. This occurs at an incident angle, from equation (16.6), of

$$\sin \theta_1 = \frac{1}{n} \tag{16.8}$$

---

[2] Willebrord Snell (1591–1626), Dutch astronomer, also known as Snell van Royen, or Snellius, who first derived the relationship between the different angles of light as it passes from one transparent medium to another. The lunar crater Snellius is named after him.

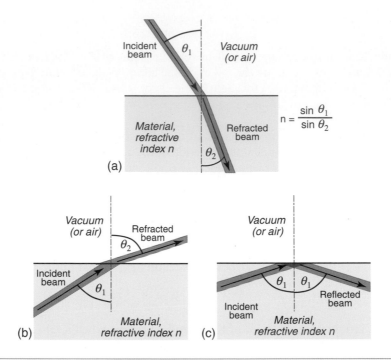

**Figure 16.3** Refraction and total internal reflection.

For values of $\theta_1$ greater than this, the equation predicts values of $\sin \theta_2$ that are greater than 1, and that can't be. Something strange has to happen, and it does: the ray is totally reflected back into the material as in Figure 16.3(c). This *total internal reflection* has many uses, one being to bend the path of light in prismatic binoculars and reflex cameras. Another is to trap light within an optical fibre, an application that has changed the way we communicate.

---

Example 16.3

A material has a refractive index of $n = 1.3$. At what internal incident angle is a beam of light first totally reflected?

*Answer.* The angle of internal reflection is $\theta_{\text{crit}} = \sin^{-1}\left(\frac{1}{n}\right) = 50°$

---

Reflection is related to refraction. When light travelling in a material of refractive index $n_1$ is incident normal to the surface of a second material with a refractive index $n_2$, the reflectivity is

$$R = \frac{I_R}{I_o} = \left(\frac{n_2 - n_1}{n_2 + n_1}\right)^2 \tag{16.9a}$$

If the incident beam is in air, with a refractive index very close to 1, this becomes

$$R = \frac{I_R}{I_o} = \left(\frac{n-1}{n+1}\right)^2. \tag{16.9b}$$

Thus materials with a high refractive index have high reflectivity. When a transparent material with a high reflectivity is coated with a second transparent material with a lower reflectivity, rays are reflected from both the front and the back surfaces of the coating. The quadratic nature of equations (16.9a,b) causes the overall reflectivity—the sum of the intensities from the front surface and from the back—to decrease. If the refractive index of the starting material is $n_2$, that of the coating is $n_1$ and that of the surrounding medium (usually air) is $n_o$, then the optimum value for the index $n_1$ of the coating is

$$n_1^{opt} = \sqrt{n_o \, n_2} \tag{16.10}$$

Thus the optimum refractive index for a coating to minimise reflectivity (and thereby maximise transmission) for a flint glass window with a refractive index of 1.61 is $n_1^{opt} = \sqrt{1.61} = 1.27$.

---

### Example 16.4

Single crystal sapphire is a form of alumina ($Al_2O_3$) used for precision optics and electronics. If its (mean) refractive index is 1.66, what is its reflectance in air?

*Answer.* From equation (16.9b)

$$R = \left(\frac{n-1}{n+1}\right)^2 = 0.06$$

---

Before leaving Snell and his law, ponder for a moment the odd fact that the beam is bent at all. Light entering a dielectric, as we have said, slows down. If that is so, why does it not continue in a straight line, just more slowly? Why should it bend? To understand this we need Fresnel's[3] construction. We think of light as advancing via a series of wave-fronts. Every point on a wave-front acts as a source, so that, in a time $\Delta t$ the front advances by $v\Delta t$, where $v$ is the velocity of light in the medium through which it is passing ( $v = c$ in vacuum or air)—as shown in Figure 16.4 with *Wave-front 1* advancing $c\Delta t$. When the wave enters a medium of higher refractive index it slows down so that the advance of the wave-front within the medium is less than that outside, as shown with *Wave-front 2* advancing $v\Delta t$ in the figure. If the angle of incidence $\theta_1$ is not zero, the wave-front enters the second medium progressively, causing it to bend, so that when it is fully in the material it is travelling in a new direction, characterised by the angle of refraction, $\theta_2$. Simple geometry then gives equation (16.6).

---

[3] Augustin Jean Fresnel (1788–1827), French physicist and engineer, known for his research on the wave theory of light and of diffraction and polarisation.

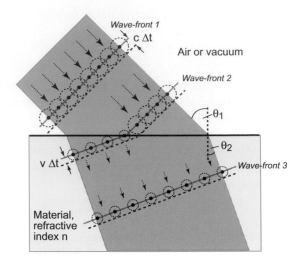

**Figure 16.4** The Fresnel construction, explaining why a light beam is bent on entering a material of higher refractive index.

*Interference*   The rainbow sheen of a compact disc when viewed obliquely, the iridescence of soap bubbles and oil films, even the colours of peacock feathers and butterfly wings are caused by interference. When the scale of the film or surface structure is comparable with the wavelength of light, then interference occurs.

When a transparent layer coats a surface, the impinging light is both reflected from it and transmitted through. If the transmitted portion is reflected at the back surface of the film it returns to the front surface and interferes with the incoming light (Figure 16.5). The nature of

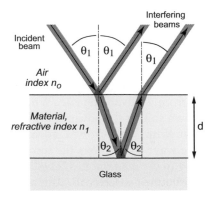

**Figure 16.5** Light waves reflected from the front and back surface of thin films interfere, reinforcing each other in some directions and cancelling in others.

the interference depends on the difference in the *optical path length* between the incoming and the reflected light. Optical path length $L_{opt}$ is defined by

$$L_{opt} = \sum_i n_i d_i \qquad (16.11)$$

where $n_i$ is the refractive index of the *i*th material along the path, and $d_i$ is the length of the segment of path. The difference in optical path length between two rays determines whether the interference is constructive or destructive. If this difference is $\Delta L_{opt}$, the phase difference between the two rays is $\Delta L_{opt}$, the phase difference between the two rays is

$$\Delta\varphi = \frac{2\pi \Delta L_{\text{opt}}}{\lambda}$$

The two waves are in phase with

$$\frac{\Delta L_{\text{opt}}}{\lambda} = 0, \pm 1, \pm 2, ..... \qquad (16.12a)$$

And they are half a cycle out of phase when

$$\frac{\Delta L_{\text{opt}}}{\lambda} = \pm\frac{1}{2}, \pm\frac{3}{2}, \pm\frac{5}{2}, ..... \qquad (16.12b)$$

Interference is constructive when the waves are in phase, destructive interference when they are half a cycle out of phase (Figure 16.6(a) and (b)). If the two interfering rays have the same amplitude, constructive interference doubles it, destructive interference kills it completely. All this only works if the differences in optical path length are small. If they are more than a few wavelengths the phase relationship between the rays is lost.

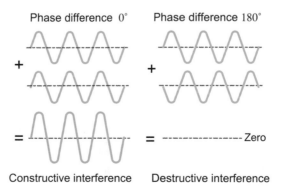

Phase difference $0°$      Phase difference $180°$

Constructive interference      Destructive interference

**Figure 16.6** Interference depends on the phase difference between two waves from the same source.

### Example 16.5

The track separation on a DVD is $d = 0.74$ microns. Suppose white light falls on a DVD at right angles to its surface. The DVD is viewed from an angle $\theta = 45°$. Which wavelength of the white light will be reinforced most strongly in the direction of viewing?

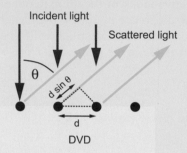

*Answer.* The illumination is a plane wave normal to the surface so that the incoming rays are in phase. Each line of dots on the DVD scatters light—think of it as light source. Then the difference in optical path for rays scattered from adjacent lines is

$$\Delta L_{opt} = n_o\, d \sin \theta$$

where $n_o$ is the refractive index of air, which we take to be 1. Reinforcement (constructive interference) occurs when this is an integral number of wavelengths, that is when

$$d \sin \theta = i\lambda$$

where $i$ is an integer. Inserting $d = 0.74$ microns, $\theta = 45°$ and $i = 1$ gives

$$\lambda = 0.52 \text{ microns}.$$

This corresponds to a vivid green colour. Any larger value of $i$ gives wavelengths that are outside the visible spectrum.

## 16.3 Charts for optical properties

*Refractive index and dielectric constant* Figure 16.7 shows the refractive index, $n$, of dielectrics, plotted against the dielectric constant $\varepsilon_R$ (defined in Chapter 14). Those shown as white bubbles are transparent to visible light; those with red bubbles are not but transmit radiation of longer wavelengths (germanium, for instance, is used for lenses for infrared imaging). Note the wide and continuous range of the refractive index of glasses, determined by their chemistry. This ability to control $n$ by manipulating composition is central to the selection

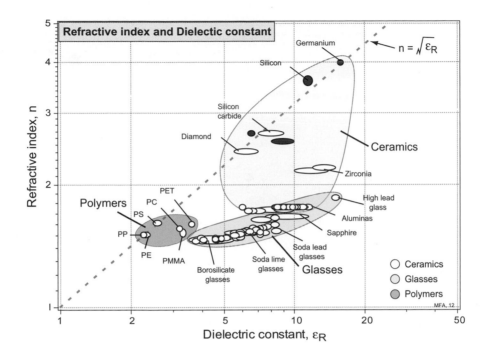

**Figure 16.7**    A chart of refractive index and dielectric constant, showing the approximate relationship $n = \sqrt{\varepsilon_R}$. (This chart and the next were made using Level 3 of the CES database.) The relationship is not well followed because the two properties are often measured at different frequencies, $n$ at optical frequencies, $\varepsilon_R$ at radio frequencies or below.

of materials for lenses and optical fibres. The relationship between $n$ and $\varepsilon_R$ is explored in Section 16.4.

*Reflectivity and refractive index*    The spectacular refraction and sparkle of diamond comes from its high refractive index, giving reflectance and extensive total internal reflection. Figure 16.8 plots these two properties. As in the previous chart, transparent materials are shown as white circles, those that are opaque in the visible range as red. As the reflectance depends on $n$, the materials fall on a characteristic curve. High refractive index gives high reflectivity, and it is here that diamond ($n = 2.42$) excels; this, combined with its unsurpassed hardness and durability ('diamonds are forever') and its scarcity, create its unique desirability as jewelry. Costume jewelry uses cheap alternatives that mimic diamond in having high refractive index: zircon, $ZrSiO_4$ ($n = 1.98$) and cubic zirconia, $ZrO_2$ ($n = 2.17$). Plastic jewelry can't come close; as the chart shows, polystyrene (PS), with $n = 1.58$, has the highest refractive index among polymers (one reason it is used for the so-called 'jewel' cases in which CDs are packaged) but it is much lower than those of real jewels.

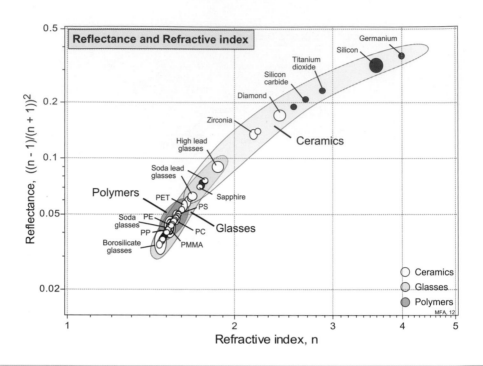

**Figure 16.8** A chart of refractive index and reflectance. The materials with open circles are transparent to visible light; those with red circles are transparent to infra-red and below, but not to visible radiation. Diamond is special because of its high refractive index, reflectance and transparency in the visible spectrum.

## 16.4 Drilling down: the physics and manipulation of optical properties

Light, like all radiation, is an electromagnetic (e-m) wave. The coupled fields are sketched in Figure 16.9. The electric part fluctuates with a frequency $v$ that determines where it lies in the spectrum of Figure 16.1. A fluctuating electric field induces a fluctuating magnetic field that is exactly $\pi/2$ out of phase with the electric one, because the induction is at its maximum when the electric field is changing most rapidly. A plane-polarised beam looks like this one: the electric and magnetic fields lie in fixed planes. Natural light is not polarised; the wave also rotates so that the plane containing each wave continuously changes.

Light slows down when it enters a dielectric because the electric field polarises the ions or molecules in the way that was illustrated in Figure 14.17. The field displaces charge to create electric dipoles, the strength of which depends on the charge and the displacement, so larger ions tend to have a larger dipole moment, giving a larger dielectric constant and a larger refractive index. Ordinary soda-lime glass (roughly 70% $SiO_2$, 20% $Na_2O$ and 10% CaO) has a refractive index of 1.52. If most of these oxides are replaced by the oxide of lead (PbO), the large Pb ion increases the polarisability and the refractive index increases to 2.0. Dielectrics

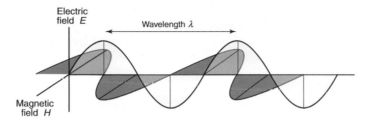

**Figure 16.9** An electro-magnetic wave. The electric component (yellow) is $\pi/2$ out of phase with the magnetic component (red).

with an amorphous or cubic structure are optically isotropic, meaning that their refractive index is independent of the direction that the light enters the material. Non-cubic crystals are anisotropic, meaning that their refractive index depends on direction, which is greatest in directions with the greatest ion-density.

Many aspects of radiation are most easily understood by thinking of it as a wave. Others need a different picture—that of radiation as discrete packets of energy, or *photons*. The idea of a wave that is also discrete energy units is not intuitive—it is another of the results of quantum theory that do not correspond to ordinary experience. The energy $E_{\mathrm{ph}}$ of a photon of radiation of frequency $\nu$ or wavelength $\lambda$ is

$$E_{\mathrm{ph}} = h\,\nu = \frac{h\,c}{\lambda} \qquad (16.13)$$

where $h$ is Planck's[4] constant ($6.626 \times 10^{-34}$ J.s) and $c$ is the speed of the radiation. Thus radiation of a given frequency has photons of fixed energy, regardless of its intensity—an intense beam simply has more of them. This is the key to understanding reflection, absorption and transmission.

---

### Example 16.6

What is the photon energy of light of wavelength $\lambda = 0.5$ $\mu$m in vacuum?

*Answer.* The table on the inside front cover of this book lists Planck's constant as $h = 6.6 \times 10^{-34}$ J.s and the speed of light in vacuum as $c = 3 \times 10^{8}$ $m/s$. The photon energy is $E_{\mathrm{ph}} = h\,\nu = \frac{h\,c}{\lambda} = 4.0 \times 10^{-19}$ J. It is more usual to give photon energies in electron volts (eV). An electron volt (eV) is the energy of an electron after it has fallen through a potential difference of 1 volt: thus 1 eV = $1.6 \times 10^{-19}$ J (it is listed in the Energy table on the inside back cover). Thus the energy of a photon of light of wavelength 0.5 $\mu$m is $E_{\mathrm{ph}} = 2.5$ eV.

---

[4] Max Karl Ernst Ludwig Planck (1858–1947), a central figure in the development of quantum theory; it was he who formulated equation (16.9).

***Why aren't metals transparent?*** Recall from Chapter 14 that the electrons in materials circle their parent atom in orbits with discrete energy levels, and only two can occupy the same level. Metals have an enormous number of very closely spaced levels in their conduction band; the electrons in the metal only fill part of this number. Filling the levels in a metal is like pouring water into a container until it is part full: its surface is the Fermi level, and levels above it are empty. If you 'excite' the water—say by shaking the container—some of it can slosh to a higher level. If you stop sloshing, it will return to its Fermi level.

Radiation excites electrons, and in metals there are plenty of empty levels in the conduction band into which they can be excited. But here quantum effects cut in. A photon with energy $h\nu$ can excite an electron only if there is an energy level that is exactly $h\nu$ above the Fermi level—and in metals there is. So all the photons of a light beam are captured by electrons of a metal, regardless of their wavelength. Figure 16.10 shows, on the left, what happens to just one.

What next? Shaken water settles back, and the electron does the same. In doing so it releases a photon with exactly the same energy that excited it in the first place, but in a random direction. Any photons moving into the material are immediately recaptured, so none make it more than about 0.01 μm (about 30 atom diameters) below the surface. All, ultimately, re-emerge from the metal surface—that is, they are reflected. Many metals (silver, aluminum and stainless steel, for example) reflect all wavelengths almost equally well, so if exposed to white light they appear silver. Others (copper, brass, bronze) reflect some colours better than others and appear coloured because, in penetrating this tiny distance into the surface some wavelengths are slightly absorbed.

Reflection by metals, then, has to do with electrons at the top of the conduction band. These same electrons provide electrical conduction. For this reason, the best metallic reflectors are the metals with the highest electrical conductivities—the reason that high-quality mirrors use silver, and cheaper ones use aluminum.

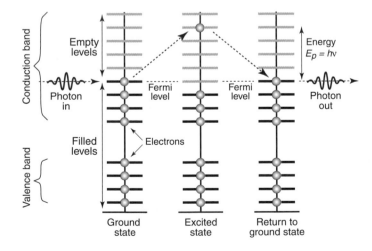

**Figure 16.10** Metals absorb photons, capturing their energy by promoting an electron from the filled part of the conduction band into a higher, empty level. When the electron falls back, a photon is re-emitted.

***How does light get through dielectrics?*** If electrons snatch up photons, how is it that some materials are transparent—light goes straight through them? Non-metals interact with radiation in a different way. Part may be reflected but much enters the material, inducing both dielectric and magnetic responses. Not surprisingly, then, the velocity of an e-m wave depends on the dielectric and magnetic properties of the material through which it travels. In vacuum the velocity is

$$c_0 = \frac{1}{\sqrt{\varepsilon_0 \, \mu_0}} \qquad (16.14)$$

where $\varepsilon_0$ is the electric permittivity of vacuum and $\mu_0$ its magnetic permeability, as defined in Chapters 14 and 15. Within a material its velocity is

$$c = \frac{1}{\sqrt{\varepsilon \, \mu}} \qquad (16.15)$$

where $\varepsilon = \varepsilon_R \, \varepsilon_0$ and $\mu = \mu_R \, \mu_0$ are the permittivity and permeability of the material. The refractive index, therefore, is

$$n = \frac{c_0}{c} = \sqrt{\varepsilon_R \, \mu_R} \qquad (16.16)$$

The relative permeability $\mu_R$ of most dielectrics is very close to unity—only the magnetic materials of Chapter 15 have larger values. Thus

$$n = \sqrt{\varepsilon_R} \qquad (16.17)$$

The chart of Figure 16.7 has $n$ as one axis and $\varepsilon_R$ as the other. The diagonal line is a plot of this equation (the log scales make it a straight line). Polymers and very pure materials like diamond, silicon and germanium lie close to the line but the agreement with the rest is not so good. This is because refractive index and dielectric constant depend on frequency. Refractive index is usually measured optically, and that means optical frequencies, around $10^{15}$ Hz. Dielectric constants are more usually measured at radio frequencies or below—$10^6$ Hz or less. If the two are measured at the same frequency, equation (16.17) holds.

---

### Example 16.7

A dielectric material has a high-frequency dielectric constant of $\varepsilon_R = 3.5$. What would you estimate its refractive index to be?

*Answer.* The refractive index is approximately $n \approx \sqrt{\varepsilon_R} = 1.87$.

---

The reason that radiation of certain wavelengths can enter a dielectric is that its Fermi level lies at the top of the valence band, just below a band gap (Chapter 14 and Figure 16.11). The conduction band, with its vast number of empty levels, lies above it. To excite an electron

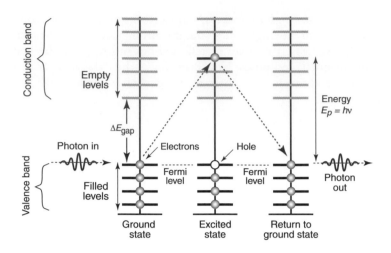

**Figure 16.11** Dielectrics have a full band, separated from the empty conduction band by an energy gap. The material cannot capture photons with energy less than $\Delta E_{gap}$, meaning that, for those frequencies, the material is transparent. Photons with energy greater than $\Delta E_{gap}$ are absorbed, as illustrated here.

across the gap requires a photon with an energy at least as great as the width of the gap, $\Delta E_{gap}$. Thus radiation with photon energy less than $\Delta E_{gap}$ cannot excite electrons—there are no energy states within the gap for the electron to be excited into. The radiation sees the material as transparent, offering no interaction of any sort, so it goes straight through.

Electrons are, however, excited by radiation with photons with energies greater than $\Delta E_{gap}$ (i.e. higher frequency, shorter wavelength). These have enough energy to pop electrons into the conduction band, leaving a 'hole' in the valence band from which they came. When they jump back, filling the hole, they emit radiation of the same wavelength that first excited them, and for these wavelengths the material is not transparent (Figure 16.11). The critical frequency $\nu_{crit}$, above which interaction starts, is given by

$$h\nu_{crit} = \Delta E_{gap} \tag{16.18}$$

The material is opaque to frequencies higher than this. Thus bakelite is transparent to infrared light because its frequency is too low and its photons too feeble to kick electrons across the band gap. The visible spectrum has higher frequencies with more energetic photons, exceeding the band gap energy; they are captured and reflected.

Although dielectrics can't absorb radiation with photons of energy less than that of the band gap, they are not all transparent. Most are polycrystalline and have a refractive index that depends on direction; then light is *scattered* as it passes from one crystal to another. Imperfections, particularly porosity, do the same. Scattering, sketched in Figure 16.12, makes the material appear translucent or even opaque, even though, when made as a perfect single crystal, it is completely transparent. Thus sapphire (alumina, $Al_2O_3$), used for watch crystals and cockpit windows of aircraft, is transparent, but the polycrystalline, slightly porous form of

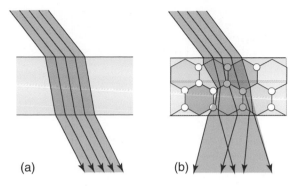

**Figure 16.12** (a) A pure glass with no internal structure and a wide band-gap is completely transparent. (b) Light entering a polycrystalline ceramic or a partly-crystalline polymer suffers multiple refraction and is also scattered by porosity, making it translucent.

the same material that is used for electronic substrates is translucent or opaque. Scattering explains why some polymers are translucent or opaque: their microstructure is a mix of crystalline and amorphous regions with different refractive indices. It explains, too, why some go white when you bend them: it is because light is scattered from internal micro-cracks—the crazes described in Chapter 6.

---

### Example 16.8

An insulator has a band gap of $\Delta E_{gap} = 1$ eV. For which wavelengths of radiation will it be transparent?

*Answer.* The material will be transparent to frequencies below $v_{crit} \leq \frac{\Delta E_{gap}}{h}$ where Planck's constant is $h = 6.6 \times 10^{-34}$ J.s. Thus $v_{crit} \leq 2.4 \times 10^{14}$ Hz. This corresponds to a range of wavelengths $\lambda_{crit} = \frac{c}{v_{crit}} \geq 1.24$ μm.

---

*Colour* If a material has a band gap with an energy $\Delta E_{gap}$ that lies within the visible spectrum, the wavelengths with energy greater than this are absorbed and those with energy that is less are not. The absorbed radiation is re-emitted when the excited electron drops back into a lower energy state, but this may not be the one it started from, so the photon it emits has a different wavelength than the one that was originally absorbed. The light emitted from the material is a mix of the transmitted and the re-emitted wavelengths, and it is this that gives it a characteristic colour.

More specific control of colour is possible by doping—the deliberate introduction of impurities that create a new energy level in the band gap, as in Figure 16.13. Radiation is absorbed as before but it is now re-emitted in two discrete steps as the electrons drop first into

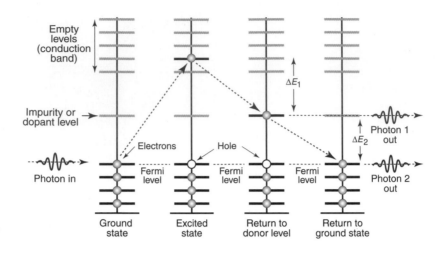

**Figure 16.13**  Impurities or dopants create energy levels in the band-gap. Electron transitions to and from the dopant level emits a photons of specific frequency and colour.

the dopant level, emitting a photon of frequency $\nu_1 = \Delta E_1/h$ and from there back into the valence band, emitting a second photon of energy $\nu_2 = \Delta E_2/h$. Particularly pure colours are created when glasses are doped with metal ions: copper gives blue; cobalt, violet; chromium, green; manganese, yellow.

*Photochromic* materials change colour when subjected to sunlight. Photochromic glass absorbs visible light and darkens, absorbing visible light, when exposed to ultraviolet light. The glass contains nano-crystalline silver halide particles that ionise, causing the colour change; additional copper salts sensitise the silver halide to ultraviolet light.

Photochromics respond passively and cannot be electronically or manually controlled. *Thermochromic* materials passively change colour when their temperature environments changed. Thermochromics depend on a thermally induced chemical reaction or phase transformation that in turn affects colour.

*Electrochromic* materials change colour in an electric field. Electrochromic glass dims to a darker colour while still remaining transparent. Electrochromic polymers switch from transparent to opaque in an electric field; they are used for electrochromic ink. Electrochromic windows allow dynamic control of daylight and solar heat gain in buildings, vehicles, aircraft and ships.

*Dichroic* (literally, 'two-colour') surfaces exhibit colour changes to the viewer as a function of either the angle of incident light or the angle of the viewer with respect to the orientation of the plane on which the light strikes. Dichroic materials are built up of thin films with different light transmission, absorption and reflection qualities. The colour derives either from interference between rays reflected from different surfaces of the thin films that coat the surface of the dichroic object, or from differential absorption of differing wavelengths within the layers. Dichroic glasses are known for their striking visual effects, but they are expensive. Polymeric *radiant colour films* exhibit similar angular-dependent colour effects but are less expensive.

---

### Example 16.9

An insulator has a dopant-created energy level that lies $\Delta E = 2.1$ eV below the conduction band. If electrons are excited into the conduction band and fall back into this level, what is the wavelength of the radiation that is emitted? Is it in the optical spectrum? If so, what is its colour?

*Answer.* The frequency of the emitted radiation is

$$v = \frac{\Delta E}{h} = \frac{2.1 \times 1.6 \times 10^{-19}}{6.6 \times 10^{-34}} = 5.1 \times 10^{14} \ /s.$$

Its wavelength is $\lambda = \frac{c}{v} = 0.59$ µm. Referring to Figure 16.1, this corresponds to a colour in the green part of the spectrum.

---

*Fluorescence, phosphorescence and electroluminescence*   Electrons can be excited into higher energy levels by incident photons, provided they have sufficient energy. Energetic electrons, like those of the electron beam of a cathode ray tube, do the same. In most materials the time delay before the electrons drop back into lower levels, re-emitting the energy they captured, is extremely short, but in some there is a delay. If, on dropping back, the photon they emit is in the visible spectrum, the effect is that of *luminescence*—the material continues to glow even when the incident beam is removed. When the time delay is fractions of seconds, it is called *fluorescence*, used in fluorescent lighting where it is excited by ultraviolet from a gas discharge, and in CRT TV tubes where it is excited by the scanning electron beam. When the time delay is longer it is called *phosphorescence*; it is no longer useful for creating moving images but is used instead for static displays like that of watch faces, where it is excited by electrons ($\beta$-particles) released by a mildly radioactive ingredient in the paint. In *chemoluminescence* the excitation derives instead from a chemical reaction, as in fireflies and the luminous fish of the deep ocean.

*Photoconductivity*   Dielectrics are true insulators only if there are no electrons in the conduction band, since if there are any, a field will accelerate them giving an electric current. Dielectrics with a band gap that is sufficiently narrow that the photons of visible light excite electrons across it, become conducting (though with high resistance) when exposed to light. The greater the intensity of light, the greater the conductivity. Photo light meters use this effect; the meter is simply a bridge circuit measuring the resistance of a photo-conducting element such as cadmium sulfide.

## 16.5 Optical design

Table 16.1 lists the refractive indices of common optical materials. There are many methods for producing thin glass or polymer films or coatings with controlled optical properties. Many of these methods rely on processing by physical vapor deposition (PVD) or chemical vapor deposition (CVD).

Table 16.1 The refractive index of optical materials and coatings

| Material class | Material | Refractive index |
|---|---|---|
| **Polymers** | Acrylic glass | 1.49 |
| | Polycarbonate | 1.59 |
| | PMMA | 1.49 |
| | PETg | 1.57 |
| **Glasses** | Crown glass (pure) | 1.52 |
| | Flint glass (pure) | 1.61 |
| | Crown glass (doped) | 1.48–1.76 |
| | Flint glass (doped) | 1.52–1.93 |
| | Borosilicate glass | 1.47 |
| **Coatings** | Magnesium fluoride | 1.38 |
| | Titanium dioxide | 2.71 |
| | Silica | 1.46 |

***Using reflection and refraction*** Telescopes, microscopes, cameras and car headlights all rely on the focusing of light. Reflecting telescopes and car headlights use metallised glass or plastic surfaces, ground or moulded to a concave shape; the metal, commonly, is silver because of its high reflectivity across the entire optical spectrum.

Refracting telescopes, microscopes and cameras use lenses. Here the important property is the refractive index and its dependence on wavelength (since if this is large, different wavelengths of light are brought to focus in different planes), and the reflectivity (since reflected light is lost and does not contribute to the image). As the chart of Figure 16.7 shows, most elements and compounds have a fixed refractive index. Glasses are different: their refractive index can be tuned to any value between 1.5 and more than 2. Adding components with light elements like sodium (as in soda-lime glass) gives a low refractive index; adding heavy elements like lead (as in lead glass or 'crystal') gives a high one.

***Using total internal reflection*** Optical fibres have revolutionised the digital transmission of information: almost all landlines now use these rather than copper wires as the 'conductor'. A single fibre consists of a core of pure glass contained in a cladding of another glass with a larger refractive index, as in Figure 16.14. A digitised signal is converted into optical pulses by a light-emitting diode. The stream of pulses is fed into the fibre where it is contained within the core, even when the fibre is bent, because any ray striking the core-cladding interface at a low angle

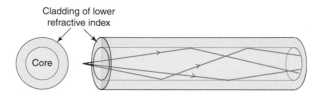

Figure 16.14 An optical fibre. The graded refractive index traps light by total internal reflection, making it follow the fibre even when it is curved.

suffers total internal reflection. The purity of the core—a silica glass—is so high that absorption is very small; occasional repeater stations are needed to receive, amplify and retransmit the signal to cover long distances.

*Using interference*    The total amount of light impinging on a surface must in sum be reflected, transmitted or absorbed (equation 16.1). If the amount of light reflected from a surface is reduced, more must be transmitted. Anti-reflective coatings, common on computer screens, eyeglasses and camera lenses, are designed to reduce reflection (cutting down glare), to increase light transmission through a surface and to increase contrast. They consist of single or multiple layers of transparent thin films with different refractive indices, chosen to produce destructive interferences for reflected light and constructive interferences for light that is transmitted. The effectiveness varies with the wavelength of the light and its incident angle. In current practice, antireflective coatings are designed for optimal performance for specified wavelengths and angles, commonly for infrared (IR), visible, and ultraviolet (UV) spectra, depending on the application.

Fundamental effects of layering are shown in Figures 16.5 and 16.15. Consider a design intended to optimise light transmission in a system consisting of a simple one-layer coating on glass. The light reflects twice, once from the surface to the air and once from that between the layer and the glass. The reflectivity of a material depends on its refractive index in the way described by equation (16.9). If $R_f$ and $R_g$ are the reflectivities at the surface-to-air and surface-to-glass interfaces, the transmission at each interface is $T_f = (1 - R_f)$ and $T_g = (1 - R_g)$ and, neglecting absorption, the total transmission is $T_f T_g$.

Suppose now that the layer is very thin, and a plane wave falls on its surface. If the angle of incidence is $0°$, meaning that the wave is normal to the layer surface, then waves reflected from the back surface travel along a longer optical path than those reflected from the front. The optical path difference for a wave reflected at the air surface and one reflected from the back of the layer, for normal illumination, is

$$\Delta L_{opt} = 2\,n_1\,d$$

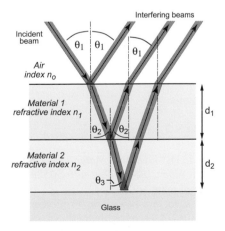

**Figure 16.15**   Multi-layers allow fine tuning of the reflectance and transmittance of glass.

When, instead, the incident wave is not normal to the surface, the optical path difference can be longer, as illustrated in the figure. Destructive interference occurs when this is equal to $\frac{1}{2}\lambda$, $\frac{3}{2}\lambda$, etc. Thus a single layer can be tuned for destructive interference, cancelling reflection, for a single wavelength and for angle of viewing, though wavelengths and angles near these will show partial cancellation. Cancellation for a wider range of wavelengths and viewing angles is possible with multi-layers like that shown in Figure 16.15, often achieved with alternating layers of silica ($SiO_2$, which has a low refractive index), and titanium dioxide ($TiO_2$, which has an exceptionally large one).

## 16.6 Summary and conclusions

Materials interact with electromagnetic radiation in several ways: reflecting it, refracting it and absorbing it. This is because radiation behaves both as a wave of frequency $v$ and as a stream of discrete photons with energy $E_{ph} = hv$. Electrons in materials capture photons, grabbing their energy, provided there are energy levels exactly $E_{ph}$ above their ground state for them to occupy. After a time delay that is usually very short, the electrons drop back down to their ground state, re-emitting their energy as new photons. In metals there are always usable energy levels so every photon hitting a metal is captured and thrown back out, giving reflection. In a dielectric, the atoms ignore low frequency, low energy photons and capture only those with energy above a critical level set by the band gap. Those that are not caught pass straight through; for these frequencies, the material is transparent. Doping introduces extra slots within the band gap for electrons to fall into as they release their energy, and in doing so they emit photons of specific frequency and colour.

When radiation enters a dielectric, it slows down. A consequence of this is that a beam entering at an angle is bent—the phenomenon of refraction. It is this that allows light to be focussed by lenses, reflected by prisms and trapped in fine, transparent fibres to transmit information.

Of all the transparent materials at our disposal, glasses offer the greatest range of refractive index and colour. Glasses are based on amorphous silica ($SiO_2$). Silica is an extremely good solvent, allowing a wide range of other oxides to be dissolved in it over a wide range of concentrations. It is this that allows the optical properties of glasses to be adjusted and fine-tuned to match design needs.

## 16.7 Further reading

Callister, W. D., & Rethwisch, D. G. (2011). *Materials Science and Engineering: An Introduction* (8th ed.). New York, USA: John Wiley. ISBN 978-0-470-50586-1. (A well-established and comprehensive introduction to the science of materials.)

Jiles, D. (2001). *Introduction to the Electronic Properties of Materials*. Cheltenham, UK: Nelson Thompson. ISBN 0-7487-6042-3. (A refreshingly direct approach to the thermal, electrical, magnetic and optical properties of materials and their electronic origins.)

# **16.8** Exercises

Exercise E16.1    The absorption coefficient of polyethylene for optical frequencies is $86.6 \text{ cm}^{-1}$. How thick a slab of polyethylene is required to reduce the transmitted light intensity by absorption to one-half of its initial value?

Exercise E16.2    The fraction of light that is transmitted through a 100 mm panel of PMMA is 0.97. Neglecting reflection at both surfaces, what is the absorption coefficient of PMMA?

Exercise E16.3    Derive the expression for the fraction of light that passes through a transparent panel, allowing for both reflection and absorption:

$$I = I_{\text{o}} \left( 1 - \frac{I_{\text{R}}}{I_{\text{o}}} \right)^2 \exp - \beta x$$

where $\beta$ is the absorption coefficient and $x$ is the length of the optical path in the material.

Exercise E16.4    Define refractive index. Give examples of devices that make use of refraction.

Exercise E16.5    The prism for a reflex camera is made from a silica glass with a coefficient of refraction of 1.46. To present a 'through the lens' image to the user of the camera, light rays entering the prism must be totally reflected. At what range of angles can they strike the internal surfaces of the prism ?

Exercise E16.6    A unique design for an underwater camera calls for a prism viewfinder that will be immersed in water. If the prism, as in the last exercise, is made from a silica glass with a coefficient of refraction of 1.46, what range of incident angles will lead to total internal reflection? The refractive index of water is 1.33.

Exercise E16.7    A 300 mm thick block of glass of refractive index 1.66 is placed over a line on a sheet of paper so that only part of the line is covered (see figure below). The line is viewed from above at an angle of 45° to the glass and normal to the line. Refraction in the glass displaces the image of the part of the line it covers. By how much is the part of the line viewed through the glass displaced relative to the part that is not covered?

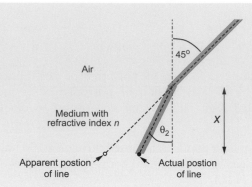

**Exercise E16.8**  Define reflectance. Give examples of devices that make use of reflection.

**Exercise E16.9**  The reflectivity of window glass is 4% per surface. What is its refractive index? Neglecting absorption, what fraction of the intensity of the light striking the window passes through?

**Exercise E16.10**  Compare the reflectance in air of polystyrene (refractive index 1.55) with that of PTFE (refractive index 1.36).

**Exercise E16.11**  The refractive index of PMMA is 1.52. What would you estimate its dielectric constant $\varepsilon_R$ to be?

**Exercise E16.12**  A crown glass sheet is partly coated by an acrylic film with thickness $t_1 = 4$ microns and refractive index $n_1 = 1.49$. A beam of red light strikes the plate at right angles to its surface. What is the difference in optical path length between a ray passing through the coated and the uncoated part of the glass? If the wavelength of red light is $\lambda = 0.65$ microns, is there potential for constructive or destructive interference between the two rays when they emerge from the glass?

**Exercise E16.13**  A filter is required to screen out UV light (wavelengths $\lambda$ around 0.35 microns) from crown glass windows. It is suggested that magnesium fluoride, $MgF_2$, (refractive index $n_1 = 1.38$) be used as an interference coating because it is hard wearing and easily deposited by physical vapor deposition. What thickness of $MgF_2$ coating is required to give maximum cancellation?

**Exercise E16.14**  A textured metal plate has a grid of fine parallel scratches with a regular spacing of $d = 1.6$ microns. Suppose white light falls normal to the surface of the plate. At what angle will the red component of the light, frequency $\nu = 5 \times 10^{14}$ per second, be diffracted most strongly when viewed normal to the scratches?

**Exercise E16.15**  Why are metals good reflectors of radiation?

Exercise E16.16   It is proposed to replace soda-glass windows of a green house with poly-carbonate (PC). Will the PC windows reflect more? The refractive index of the glass is 1.5 and that of PC is 1.6.

Exercise E16.17   Waterford glass and Steuben glass, used for expensive ornamental and cut-glass objects, are high-lead glasses, meaning that they contain oxides of lead. Using the information shown on the chart of Figure 16.8, can you explain the choice of composition?

Exercise E16.18   An X-ray system has a beryllium window to transmit the beam. The absorption coefficient of beryllium for the wavelength of X-rays of interest here is $3.02 \times 10^2$ m$^{-1}$. If the window is 2 mm thick, what fraction of the incident beam intensity will pass through the window? The rest of the equipment is shielded with 4 mm of lead, with absorption coefficient for X-rays of $3.35 \times 10^6$ m$^{-1}$. What fraction of the intensity of the incident beam will escape through the casing?

Exercise E16.19   What principle and material would you choose to make a light-sensing switch for a greenhouse?

Exercise E16.20   What principle and material would you choose to make a heat-sensing switch to turn the lights off in the garage when no one is moving around in it?

Exercise E16.21   An optical fibre has a glass core with a refractive index of $n_1 = 1.48$, clad with a glass with refractive index $n_2 = 1.45$. What is the maximum angle that the incoming optical signal can deviate from the axis of the core while still remaining trapped in it?

## **16.9** Exploring design using CES

Use Level 2 unless otherwise stated.

Exercise E16.22   Use the 'Search' facility to find materials for:
(a) Lenses.
(b) Mirrors.

Exercise E16.23   You are asked to suggest the best choice of cheap polymer for a new line of injection-moulded costume jewelry. Its transparency must be of optical quality, and it should have the highest possible refractive index. The table summarises the requirements.

| Function | • Plastic costume jewelry |
| Constraints | • Transparency of optical quality |
| | • Injection mouldable |
| | • Price < $5/kg |
| Objective | • Maximise refractive index |
| Free variable | • Choice of material |

**Exercise E16.24** A material is required for the mirror backing of a precision reflecting telescope. It must have a modulus of at least 50 GPa so that it does not deflect under its own weight; it must be hard so that the surface (which will be silvered), once ground, does not distort; it must have the lowest possible thermal expansion to minimise thermal distortion; and it must be able to be moulded or cast to its initial shape before grinding. Use CES to find suitable candidates.

| Function | • Mirror backing |
| Constraints | • Young's modulus > 50 GPa |
| | • Processing: mouldablity 5 |
| Objective | • Highest hardness with lowest thermal expansion coefficient |
| Free variable | • Choice of material |

**Exercise E16.25** Find materials that are optically clear, have a mouldability rating of 5 and have excellent resistance to fresh and to salt water for use as contact lenses.

## 16.10 Exploring the science with CES Elements

**Exercise E16.26** The Hagen-Rubens law—an empirical law—says that the reflectivity of metals, $R$, is given by

$$R \approx 1 - 0.02\sqrt{\rho_e}$$

where the electrical resistivity $\rho_e$ is in $\mu\Omega.cm$. Make a chart with $R$ on the $y$-axis and electrical conductivity $1/\rho_e$ on the $x$-axis. Use it to find the three elements with the highest reflectivities.

**Exercise E16.27** The text explained why the refractive index $n$ was related to the dielectric constant $\varepsilon_R$ by the equation

$$n = \sqrt{\varepsilon_R}$$

Make a chart with $\sqrt{\varepsilon_R}$ on the $x$-axis and $n$ on the $y$-axis to see how accurate this relationship is. Which elements best obey it?

# Chapter 17

# Durability: oxidation, corrosion, degradation

Corrosion (image courtesy of Norbert Wodhnl
© Norbert Wodhnl).

## **17.1** Introduction and synopsis

The Gospel according to St. Matthew reminds us that we live in a world in which 'moth and rust doth corrupt'—a world of corrosion, degradation and decay (see chapter opening picture). Not a happy thought with which to start a chapter. But by understanding these forces we can, to a degree, control them. This chapter describes ways in which materials degrade or corrode, the underlying mechanisms and what can be done to slow them down.

Durability is a key material attribute, one central to the safety and economy of products. It is one of the more difficult attributes to characterise, quantify and use for selection for the following reasons:

- It is a function not just of the material but also of the environment in which it operates.
- There are many mechanisms, some general, some peculiar to particular materials and environments.
- Material combinations (as in galvanic corrosion) and configuration (as in crevice corrosion) play a role.

Figure 17.1 shows some of the considerations involved. The central players are *materials* and *environments*. But the fact that a given material is resistant to a given environment is not enough—there are many other considerations, some of them listed in the figure. First there is

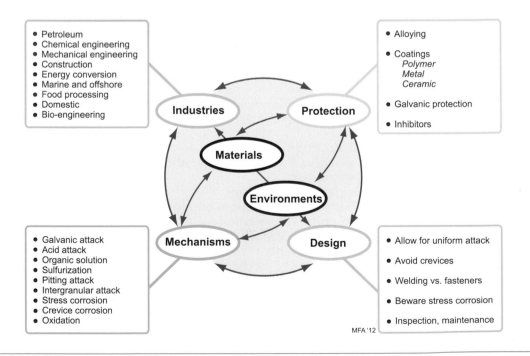

**Figure 17.1** Factors affecting durability.

the *industrial sector* in which the material is to be used: some are limited to lightweight materials, some to non-flammable materials, some to bio-compatible materials. In some applications a certain amount of corrosion can be tolerated (construction, shipping); in others the slightest attack is unacceptable (bio-medical equipment, for instance). Second, there are many *mechanisms* of attack, some general, some appearing only under special conditions. Third, there are *protection* methods, some generally applicable (like painting), some specific to particular combinations of material and environment (such as inhibitors). And finally there are issues of *design*, often specific to a given industry. Thus there are *preferred material choices* for use in a given environment—those that, through experience, best meet both the primary constraint of resisting attack and the secondary constraints of stiffness, strength, cost, and the like. We return to these in Section 17.7.

Much of our current understanding of degradation is empirical, enlightened by an understanding of many underlying mechanisms. We start by examining these. There are the chemical reactions triggered by radiation and heat, changing the properties in irreversible ways. There are electro-chemical mechanisms that appear in attack by aqueous solutions, acids and alkalis. Organic solvents attack polymers by a number of mechanisms. The mechanisms are well understood, but this alone is not sufficient to guide material choice without also drawing on experience. For this reason this chapter has an unusually large number of unusually large tables: they store the empirical knowledge of corrosion resistance, preferred material and preferred corrosion-protection methods. Don't try to 'read' the tables; simply inform yourself of their structure and use them as look-up tables when making a material selection. The Exercises at the end of the chapter provide practice.

Because many different properties are involved here, the exploration of the science is split into two: one delves into oxidation, flammability and photo-degradation, the other into corrosion by liquids. Each section ends with a description of ways in which degradation is inhibited.

# **17.2** Oxidation, flammability and photo-degradation

The most stable state of most elements is as an oxide. For this reason the earth's crust is almost entirely made of simple or complex oxides: silicates like granite, aluminates like basalt, carbonates like limestone. Techniques of thermo-chemistry, electro-chemistry and synthesis allow these to be refined into the materials we use in engineering, but these are not, in general, oxides. From the moment materials are made they start to re-oxidise, some extremely slowly, others more quickly; and the hotter they are, the faster it happens. For safe high-temperature design we need to understand rates of oxidation.

***Definition and measurement***   Oxidation rates of metals are measured in the way sketched in Figure 17.2: a thin sheet of the material (to give a lot of surface) is held at temperature $T$ for an increasing time $t$, and the gain or loss in weight, $\Delta m$, is measured. If the oxide adheres to the material, the sample gains weight in a way that is either linear ($\Delta m \propto t$) or parabolic ($\Delta m \propto t^{1/2}$) in time $t$; if instead the oxide is volatile, the sample loses weight linearly with time ($\Delta m \propto - t$), as in Figure 17.3.

Polymers, too, oxidise, but in a more spectacular way. That brings us to *flammability*.

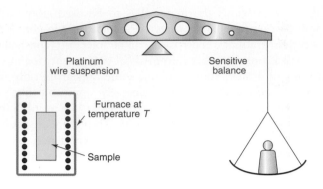

Figure 17.2  Measuring oxidation rates.

*Flammability*   Ceramics and glasses do not burn. Metals, if dispersed as a fine powder, are potentially combustible, but in bulk flammability is not normally an issue, even for magnesium. Most polymers, on the other hand, are inherently flammable, although to differing degrees: some burn spontaneously if ignited; others are self-extinguishing, burning only when directly exposed to flame.

There are several ways to characterise flammability. The most logical is *the Limiting oxygen index (LOI)*: it is the oxygen concentration, in %, required to maintain steady burning. Fresh air has about 21% oxygen in it. A polymer with an oxygen index lower than 21% will burn freely in air; one with an oxygen index that is larger will extinguish itself unless a flame is played onto it—then it burns. Thus a high oxygen index means resistance to self-sustained burning. Less logical, but much used, is the Underwriters Laboratory (UL) rating in which a strip of polymer 1.6 mm thick is held horizontally (H) or vertically (V) and ignited. Its response is recorded as a code such as HB, meaning 'horizontal burn'. The scales are compared in Tables 17.1.

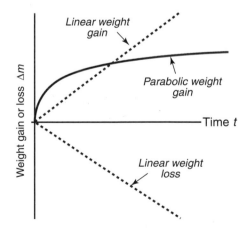

Figure 17.3  Oxidation rates: linear weight gain, parabolic gain and linear loss.

Table 17.1   Flammability ratings compared

| Oxygen index (LOI) | UL 94 (1.6mm) rating | Flammability |
|---|---|---|
| Up to 16 | Unrated | Highly flammable |
| 16–20 | HB | |
| 21–24 | HB | Slow burning |
| 25–29 | V-2 | |
| 30–43 | V-0 | Self-extinguishing |
| Over 44 | V-0 or better (e.g. 5V) | |

*Photo-degradation*   You don't have to set fire to a polymer for it to oxidise. Polymers and elastomers age when exposed to light (particularly UV) and oxygen, causing loss of strength, stiffness and toughness, discoloration and loss of gloss. This is countered by additives: *anti-oxidants*, *light stabilisers* and *fluorescent whitening agents*. Some of these are so universal that the 'standard' grade of the polymer already contains enough of one or another of them to give it acceptable resistance; PP, ABS, PS, PET, PMMA, PC, nylons and PU all need UV protection.

# 17.3 Oxidation mechanisms

*Mechanisms of oxidation*   The rate of oxidation of most metals in dry air at room temperature is too slow to be an engineering problem; indeed, slight oxidation can be beneficial in protecting metals from corrosion of other sorts. But heat them up and the rate of oxidation increases, bringing problems. The driving force for a metal to oxidise is its *free energy of oxidation*—the energy released when it reacts with oxygen—but a big driving force does not necessarily mean rapid oxidation. The rate of oxidation is determined by the *kinetics* (rate) of the oxidation reaction, and that has to do with the nature of the oxide. When any metal (with the exception of gold, platinum and a few others that are even more expensive) is exposed to air, an ultra-thin surface film of oxide forms on it immediately, following the oxidation reaction

$$M(metal) + O(oxygen) = MO(oxide) + energy \qquad (17.1)$$

The film now coats the surface, separating the metal beneath from the oxygen. If the reaction is to go further, oxygen must get through the film, or the metal must get through the film.

The weight gain shown in Figure 17.3 reveals two different types of behaviour. For some metals the weight gain per unit area, $\Delta m$, is linear, and this implies that the oxidation is progressing at a constant rate:

$$\frac{d\Delta m}{dt} = k_\ell \quad \text{giving} \quad \Delta m = k_\ell\, t \qquad (17.2)$$

where $k_\ell\,(\text{kg/m}^2.\text{s})$ is the *linear kinetic constant*. This is because the oxide film is porous or cracks (and, when thick, spalls off) and does not protect the underlying metal. Some metals behave better than this. The film that develops on their surfaces is compact, coherent and strongly bonded to the metal. For these the weight gain per unit area of surface is parabolic, slowing up with time, and this implies an oxidation rate with the form

$$\frac{d(\Delta m)}{dt} = \frac{k_p}{\Delta m} \quad \text{giving} \quad \Delta m^2 = k_p t \tag{17.3}$$

where $k_p$ ($kg^2/m^4.s$) is the *parabolic kinetic constant*.

---

### Example 17.1

A metal is found, by testing, to oxidise at 500 °C with a parabolic rate constant $k_p = 3 \times 10^{-10}$ $kg^2/m^4.s$. What is the weight gain per unit area after 1000 hours?

*Answer.* The weight gain is $\Delta m = (k_p t)^{1/2} = 0.033$ kg per $m^2$.

---

In these equations $\Delta m$ is the weight gain per unit area—that means the weight of the oxygen that has been incorporated into the oxide. The mass of the oxide itself is greater than this because it contains metal atoms (M) as well as oxygen (O). If the formula of the oxide is MO, the mass per unit area of the oxide is

$$m_{ox} = \left(\frac{A_m + A_o}{A_o}\right) \Delta m$$

where $A_m$ is the atomic weight of the metal and $A_o$ is that of oxygen. The thickness of the oxide is simply $x = m_{ox}/\rho_{ox}$ where $\rho_{ox}$ is the oxide density. Using equation (17.3) we find

$$x^2 = \left(\frac{A_m + A_o}{A_o}\right)^2 \frac{k_p}{\rho_{ox}^2} t \tag{17.4}$$

The oxide film, once formed, separates the metal from the oxygen. To react further either oxygen atoms must diffuse inward through the film to reach the metal, or metal atoms must diffuse outward through the film to reach the oxygen. The driving force is the free energy of oxidation, but the rate of oxidation is limited by the rate of diffusion, and the thicker the film, the longer this takes.

---

### Example 17.2

The oxide of the last example is MgO. The atomic weight of magnesium is 24.3 kg/kmol and that of oxygen is 16 kg/kmol and the density of MgO is 3550 $kg/m^3$. How thick will the oxide be after 1000 hours at 500 °C?

*Answer.* The mass $m$ of the oxide is $m = \frac{40.3}{16} \Delta m = 0.083 kg/m^2$.
Thickness $x = \frac{\text{mass per unit area}}{\text{density}} = 2.3 \times 10^{-5} m = 23$ microns.

---

Figure 17.4 shows a growing oxide film. The reaction creating the oxide $MO$

$$M + O = MO$$

goes in two steps. The metal first forms an ion, say $M^{2+}$, releasing electrons:

$$M = M^{2+} + 2e^-$$

The electrons are then absorbed by oxygen to give an oxygen ion:

$$O + 2e^- = O^{2-}$$

The problem is that the first of these reactions occurs at the metal side of the oxide film whereas the oxygen reaction is on the other side. Either the metal ions and the electrons must diffuse out to meet the oxygen, or the oxygen ions and electron holes (described in Chapter 14) must diffuse in to find the metal. If the film is an electrical insulator, as many oxides are, electrons cannot move through it, so it is the oxygen that must diffuse in. The concentration gradient of oxygen is that in the gas $C_o$ divided by the film thickness, $x$. The rate of growth of the film is proportional to the flux of atoms diffusing through the film, giving

$$\frac{dx}{dt} \propto D\frac{C_o}{x} \approx D_o \left( \exp - \frac{Q_d}{RT} \right) \frac{C_o}{x} \tag{17.5}$$

where $D$ is the diffusion coefficient, $D_o$ and $Q_d$ are the pre-exponential and the activation energy for oxygen diffusion. Integrating gives

$$x^2 \propto D_o \left( \exp - \frac{Q_d}{RT} \right) C_o t$$

This has the same form as equation (17.4) with

$$k_p \propto D_o \, C_o \exp - \frac{Q_d}{RT} \tag{17.6}$$

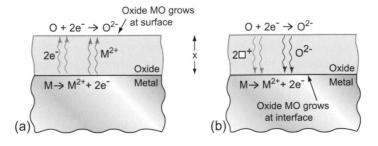

**Figure 17.4**  Oxidation mechanisms (a) growth by metal ion diffusion and electron conduction, (b) growth by diffusion of oxygen ions and holes.

This explains why oxidation rates rise steeply with rising temperature, and why the growth is parabolic. The most protective films are those with low diffusion coefficients, and this means that they have to have high melting points. This is why the $Al_2O_3$ oxide film on aluminum, the $Cr_2O_3$ film on stainless steel and chrome plate, and the $SiO_2$ film on high-silicon cast iron are so protective.

Not all oxides grow with parabolic kinetics, as we have seen. Those that show linear weight gain do so because the oxide that forms is not compact, but has cracks or spalls off because of excessive volume change, leaving the fresh surface continually exposed to oxygen. Linear weight loss, the other behaviour, also occurs when the oxide is volatile, and simply evaporates as it forms.

---

### Example 17.3

The activation energy for oxygen diffusion in MgO is 370 kJ/mol. If the test in Example 17.2 had been carried out at the temperature $T_1 = 450\,°C$ rather than at $T_o = 500\,°C$ how thick would the oxide be after 1000 hours? The gas constant $R = 8.314\times10^{-3}$ kJ/mol.K.

*Answer.* The thickness scales with temperature as $x \propto \left(\exp - \frac{Q_d}{RT}\right)^{1/2}$, so the thickness $x_1$ after 1000 hours at 450 °C is less than that, $x_o$, at 500 °C by the factor

$$\frac{x_1}{x_o} = \left(\exp - \frac{Q_d}{R}\left(\frac{1}{T_1} - \frac{1}{T_o}\right)\right)^{1/2} = \left(\exp - 4.45 \times 10^4\left(\frac{1}{723} - \frac{1}{773}\right)\right)^{1/2} = 0.137$$

Thus the thickness after the test at 450 °C is 3.1 microns.

---

## 17.4 Resistance to oxidation, burning and photo-degradation

Elements for heaters, furnace components, power generation and chemical engineering plants all require materials that can be used in air at high temperatures. If it is a pure metal that you want, it has to be platinum (and indeed some special furnaces have platinum windings) because the oxide of platinum is less stable than the metal itself. If you want something more afford-able, it has to be an alloy.

Oxides, of course, are stable in air at high temperature—they are already oxidised. One way to provide high-temperature protection is to coat metals like cast irons, steels or nickel alloys with an oxide coating. Stoves are protected by enamelling (a glass coating, largely $SiO_2$); turbine blades are coated with plasma-sprayed TBCs based on the oxide zirconia ($ZrO_2$) (see Chapter 13, Section 13.6). But coatings of this sort are expensive, and, if damaged, they cease to protect.

There is another way to give oxidation resistance to metals, and it is one that repairs itself if damaged. The oxides chromium ($Cr_2O_3$), of aluminum ($Al_2O_3$), of titanium ($TiO_2$) and of silicon ($SiO_2$) have very high melting points, so the diffusion of either the metal or oxygen through them is very slow. They are also electrical insulators so electrons cannot move through them either. The mechanism of oxidation requires both diffusion and conduction, so the rate constant $k_p$ is very small: the oxide stops growing when it is a few molecules thick. These oxides adhere well and are very protective; it is they that make these otherwise-reactive metals so passive. The films can be artificially thickened by *anodising*—an electro-chemical process for increasing their protective power. Anodised films accept a wide range of coloured dyes, giving them a decorative as well as a protective function.

If enough chromium, aluminum or silicon can be dissolved in a metal like iron or nickel— 'enough' means 18 to 20%—a similar protective oxide grows on the alloy. And if the oxide is damaged, more chromium, aluminum or silicon immediately oxidises, repairing the damage. Stainless steels (typical composition Fe—18% Cr—8% Ni), widely used for high-temperature equipment, and nichromes (nickel with 10 to 30% chromium), used for heating elements, derive their oxidation resistance in this way; so too do the aluminum bronzes (copper with 10% aluminum), and high-silicon cast irons (iron with 16—18% silicon). To see just how important this way of suppressing oxidation has become, look back for a moment at Figure 13.20. It is the table guiding material choice at low and high temperature. All the metals above the 400° C mark rely on alloy methods for their protection.

***Flammability: how do polymers burn and how do you stop them?***   The combustion of a polymer is an exothermic reaction in which hydrocarbons are oxidised to $CO_2$ and $H_2O$. That sounds simple, but it isn't. Combustion is a *gas phase reaction*; the polymer or its decomposition products must become gaseous for a fire to begin. When you light a candle you are melting the wax and raising it to the temperature at which it *pyrolyses* (400—800 °C) forming gaseous hydrocarbon decomposition products. These gases react in the flame to produce heat, which melts and pyrolyses more wax, keeping the reaction going.

A fully developed fire results when an ignition source like a spark or cigarette ignites combustible material such as paper. The heat of the fire radiates out causing other combustible materials (particularly polymers and fabrics) to decompose into a flammable gas mix. Flash-over (Figure 17.5) occurs when these gases ignite, instantly spreading the fire over the entire area and producing temperatures of over 1000 °C.

Combustion involves the reaction of *free-radicals*. At high temperatures hydrogen, together with one electron from its covalent bond (symbol **H·**) is freed from a carbon atom of the polymer molecule, leaving a highly reactive free radical **R·** in the gas. The hydrogen radical reacts with oxygen and the hydrocarbon radical to give $CO_2$ and heat, releasing the **H·** again to propagate the reaction further.

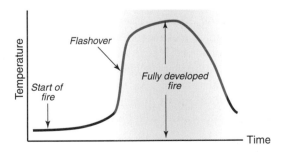

**Figure 17.5**   The temperature rise during a fire.

*Flame retardants* work in one of two ways. Some scavenge the free radicals, tying them up harmlessly and so retarding or snuffing out the combustion reaction. These are usually compounds containing chlorine or bromine; they, too, pyrolyse to give free radicals **Cl·** and **Br·** that attach themselves to the hydrocarbon radical, removing it from the reaction. Others work by creating a protective layer of water vapor between the solid polymer and the gaseous decomposition products, limiting heat transfer, cooling it and reducing pyrolisation. Typical among these is the addition of $Mg(OH)_2$ that decomposes at about 300 °C, releasing $H_2O$ and leaving inert $MgO$. Polymers containing flame retardants are identified by adding 'FR' to their names.

*What causes photo-degradation?*   As we have seen, commercial polymers are long-chain, high molecular weight, hydrocarbons. When exposed to radiation chemical reactions are triggered that change their chemical composition and molecular weight. These reactions, called *photo-oxidation* or *photo-degradation*, also create free radicals. They trigger a change in the physical and optical properties of the polymer, making them brittle and, if transparent, turning them white or gray. Once this starts it sets off a chain reaction that accelerates degradation unless stabilisers are used to interrupt the oxidation cycle. Heat, too, can trigger degradation in an oxygen-containing atmosphere.

*Ultraviolet absorbers* such as benzophenone or benzotriazole work by preferentially absorbing UV radiation, following the grandly named Beer-Lambert law, which states the obvious: that the amount of UV radiation absorbed increases with an increase in the sample thickness and stabiliser concentration. In practice, high concentrations of absorbers and sufficient thickness of the polymer are required before enough absorption takes place to effectively retard photo-degradation. *Hindered amine stabilisers* (HALS) are a more efficient way to stabilise against light-induced degradation of most polymers. They do not absorb UV radiation but act to inhibit degradation of the polymer; their efficiency and longevity are due to a cyclic process wherein the HALS are regenerated rather than consumed. Low concentrations are enough to give good stabilisation.

# 17.5 Corrosion: acids, alkalis, water and organic solvents

Table 17.2 lists some of the more common liquids with which products have contact, with an indication of where they are found. The same environments appear in the tables that follow.

*Aqueous solutions, acids and alkalis*   Even when pure, water causes corrosion if oxygen is available. The rusting of steel in fresh water alone presents enormous economic and reliability problems. Add dissolved ions (like NaCl) and the corrosion rate increases dramatically.

Strong acids like $H_2SO_4$ (sulfuric acid), $HNO_3$ (nitric acid) and HCl (hydrochloric acid) are widely used in the chemical industry. So too are strong alkalis like NaOH (caustic soda). Household products contain some of these in dilute form, but in the home it is organic acids like vinegar (acetic acid) and milder alkalis like washing soda (sodium carbonate) that are more commonly found.

Dunk a metal into an acid and you can expect trouble. The sulfates, nitrates, chlorides and acetates of most metals are more stable than the metal itself. The question is not whether the metal will be attacked, but how fast. As with oxidation, this depends on the nature of the corrosion products and on any surface coating the metal may have. The oxide film on aluminum, titanium and chromium protect them from some acids provided the oxide itself is not damaged. Alkalis, too, attack some metals. Zinc, tin, lead and aluminum react with NaOH, for instance, to give zincates, stannates, plumbates and aluminates, as anyone who has heated washing soda in an aluminum pan will have discovered.

Ways of combating corrosion by water, acids and alkalis are described in Section 17.7. Corrosion, however, can sometimes be beautiful: the turquoise patina of copper on roofs and the rich browns of bronze statues are created by corrosion.

*Oils and organic solvents*   Organic liquids are ubiquitous. We depend on them as fuels and lubricants, for cooking, for removing stains, as face cream, as nail varnish and much more—any material in service will encounter them. Metals, ceramics and glasses are largely immune to them, but not all polymers can tolerate organic liquids without problems. The identification of solvents that are incompatible with a given polymer is largely based on experience, trial and error and intuition guided by such rules as 'like dissolves like', meaning that non-polar solvents like benzene ($C_6H_6$) or carbon tetrachloride ($CCl_4$) tend to attack non-polar polymers like polypropylene or polyethylene, while polar solvents such as acetone ($CH_3COCH_3$) tend to attack polar polymers like PMMA and nylon.

# **17.6** Drilling down: mechanisms of corrosion

*Ions in solution and pH*   Pure water, $H_2O$, dissociates a little to give a hydrogen ion, $H^+$, and a hydroxyl ion, $OH^-$:

$$H_2O \leftrightarrow H^+ + OH^- \tag{17.7}$$

The product of the concentrations of the two ions is constant: increase one and the other falls. This is known as Law of Mass Action:

$$[H^+] \cdot [OH^-] = \text{constant} \tag{17.8}$$

where the square brackets mean 'molar concentration', the number of moles of an ion per liter of water.

---

### Example 17.4 Molar concentrations

Seawater contains 35 grams of salt, NaCl, per liter. What is the molar concentration of $Cl^-$ ions in seawater?

*Answer.* One molecule of NaCl, when dissolved, creates one $Cl^-$ ion, so the molecular concentration of $Cl^-$ is the same as that of NaCl. The atomic weight of sodium is 22.99 kg/kmol and that of chlorine is 35.45 kg/kmol, so the molecular weight of NaCl is 58.44 kg/kmol. A one mol solution of NaCl therefore contains 58.44 grams per liter of NaCl. Thus the molar concentration of salt and thus of $Cl^-$ ions in seawater is $M = \frac{35}{58.44} = 0.599$ mol/liter.

Table 17.2 Commonly encountered liquid environments*

| Environment | Where encountered |
| --- | --- |
| *Water and aqueous solutions* | |
| Water (fresh) | Fresh water is ubiquitous: for any object, exposure to high humidity, rain or washing acquires a film of water containing (unless distilled) dissolved oxygen and, usually, other impurities. |
| Water (salt) | Materials in marine environments are exposed to salt water and wind-carried spray. Seawater varies in composition depending on the location. It is typically 3.5%. |
| Soils | Soils differ greatly in composition, moisture content and pH. The single most important property of a soil that determines its corrosive behaviour is its electrical resistivity—a low resistivity means that the water in the soil has high concentration of dissolved ions. A resistivity below $10^9$ $\mu\Omega.cm$ is very corrosive; one with a resistivity above $2 \times 10^{10}$ $\mu\Omega.cm$ is only slightly corrosive. The choice of material for use in soil depends on this and on the pH: acidic (peaty) soils have a low pH; alkali soils (those containing clay or chalk) have a high one. |
| Body fluids | Body fluids include blood, urine, saliva, sweat and gastric fluids. All are water-based with high ion content, some acidic, stimulating electro-chemical and acid attack. |
| *Acids and alkalis* | |
| Acetic acid, $CH_3COOH$ | Acetic acid is an organic acid made by the oxidation of ethanol. It is used in the production of plastics, dyes, insecticides and other chemicals. Dilute acetic acid (vinegar) is used in cooking. |
| Hydrochloric acid, HCl | Hydrochloric acid is used as a chemical intermediate, for ore reduction, for pickling steel, in acidising oil wells, and in other industrial processes. In dilute form it is a component of household cleaners. |
| Hydrofluoric acid, HF | Hydrofluoric acid is used for the etching of glass, synthesis of fluorocarbon polymers, aluminum refining, as an etchant for silicon-based semiconductors, and to make $UF_6$ for uranium isotope separation. |
| Nitric acid, $HNO_3$ | Nitric acid is used in the production of fertilisers, dyes, drugs and explosives. |
| Sulfuric acid, $H_2SO_4$ | Sulfuric acid, of central importance in chemical engineering, is used in making fertilisers, chemicals, paints and in petrol refining. It is a component of acid rain. |
| Sodium hydroxide, NaOH | Sodium hydroxide of this concentration is found in some household cleaners, the making of soap, and in the cleaning of food processing. |
| *Fuels, oils and solvents* | |
| Benzene, $C_6H_6$ | Benzene is used as an industrial solvent, as well as in the synthesis of plastics, rubbers, dyes and certain drugs. It is also found in tobacco smoke. |
| Carbon tetrachloride, $CCl_4$ | Carbon tetrachloride is used principally to manufacture refrigerants, as a dry-cleaning solvent and as a pesticide (now banned in the United States). |

Table 17.2 Commonly encountered liquid environments* *continued*

| Environment | Where encountered |
|---|---|
| Crude oil | Refined petroleum is not corrosive to metals, but crude oil usually contains saline water, sulfur compounds and other impurities, some acidic. |
| Diesel oil | Diesel oil is a specific fractional distillate of crude oil. It is the primary fuel for truck, shipping and nonelectric and diesel-electric rail transport. Its use for car propulsion is increasing. Diesel oil acts both as fuel and as lubricant in the engine. |
| Kerosene (paraffin oil) | Paraffin is used as aviation fuel, as well as being commonly used for heating and lighting on a domestic level. It is used to store highly reactive metals to isolate them from oxygen. |
| Lubricating oil | Oil is used as a lubricant in most metal systems with moving parts. Typically, these are mineral oils and frequently contain sulfur. Low-sulfur synthetic oils can also be produced. Lubricating oil is less corrosive than petroleum and diesel oil because it is based on hydrocarbons with higher molecular weight. |
| Petroleum (gasoline) | Petroleum is a volatile distillate of crude oil. It is used mainly to power engines, for cars, light aircraft and agricultural equipment. It often contains additives such as lead, ethanol or dyes. |
| Silicone fluid, $((CH_3)_2SiO)_n$ | Silicone oils are synthetic silicon-based polymers. They are exceptionally stable and inert. They are used as brake and hydraulic fluids, vacuum pump oils, and as lubricants for metals and for textile threads during sewing and weaving of fabrics. |
| Vegetable oil | Vegetable oils are derived from olive, peanut, maize, sunflower, rape and other seed and nut crops. They are widely used in the preparation of foods. They are the basis of bio-fuels. |
| *Alcohols, aldehydes, ketones* | |
| Acetone, $CH_3COCH_3$ | Acetone is the simplest of ketones. It is widely used as a degreasing agent and a solvent, and as a thinning agent for polyester resins and other synthetic paints (commonly nail varnish), as a cleaning agent and as an additive in automobile fuels. It is used in the manufacture of plastics, drugs, fibres and other chemicals. |
| Ethyl alcohol, $C_2H_5OH$ | Ethanol is made by fermentation and is thus in alcoholic beverages. It is used medically as a solvent for disinfectants and for cleaning wounds before dressing them. It is used industrially as a solvent, a dehydrating agent and as a 'green' fuel for cars. |
| Methyl alcohol, $CH_3OH$ | Methanol is used in glass cleaners, stains, dyes, inks, antifreeze, solvents, fuel additives and as an extractant for oil. It is also used as a high-energy fuel for cars, aircraft and rockets, and is a possible fuel for fuel cells. |
| Formaldehyde, $CH_2O$ | Formaldehyde is used as a disinfectant in medical applications. It is used industrially to make many resins (including melamine resin and phenol formaldehyde resin) and glues, including those used in plywood. It is found in car exhausts and tobacco smoke. It is the basis of embalming fluids. |

* The CES software contains information of this sort for a larger number of environments.

In pure water there are equal numbers of the two types of ion, $[H^+] = [OH^-]$, and the value of the constant, when measured, is $10^{-14}$. Thus the molar concentration of both ion types is $10^{-7}$; one $H_2O$ molecule in $10^7$ is ionised. The pH of the ionised water is defined as the negative of the log of the hydrogen ion concentration:

$$pH = -\log_{10}[H^+] \tag{17.9}$$

so, for pure water, pH = 7.

Acids dissociate in water to give $H^+$ ions. Sulfuric acid, for instance, dissociates as:

$$H_2SO_4 \leftrightarrow 2H^+ + SO_4^{2-} \tag{17.10}$$

This pushes up the concentration of $[H^+]$ and, because of equation (17.8), it pulls down the concentration of $[OH^-]$; weak acids have a pH of 4–6; strong ones a pH down to 0. Alkalis do the opposite. Sodium hydroxide, for example, dissociates when dissolved in water, to give $OH^-$ ions:

$$NaOH \leftrightarrow Na^+ + OH^- \tag{17.11}$$

Weak alkalis have a pH of 8–10, strong ones a pH up to 13 (Figure 17.6).

Corrosion by acids and alkalis is an electro-chemical reaction. One half of this is the dissociation reaction of a metal M into a metal ion, $M^{z+}$, releasing electrons $e^-$:

$$M \rightarrow M^{z+} + ze^-$$

where z, an integer of value 1, 2 or 3, is the valence of the metal. Acidic environments, with high $[H^+]$ (and thus low pH) stimulate this reaction; thus a metal such as copper, in sulfuric acid solution, reacts rapidly:

$$\left. \begin{array}{l} Cu \rightarrow Cu^{2+} + 2e^- \\[2mm] H_2SO_4 \leftrightarrow 2H^+ + SO_4^{2-} \end{array} \right\} \quad Cu^{2+} + SO_4^{2-} + 2H^+ + 2e \rightarrow CuSO_4 + H_2 \tag{17.12}$$

Some metals are resistant to attack by some acids because the reaction product, here $CuSO_4$, forms a protective surface layer; thus lead-lined containers are used to process sulfuric acid because lead sulfate is protective.

Alkalis, too, cause corrosion via an electro-chemical reaction. Zinc, for instance, is attacked by caustic soda via the steps

$$\left. \begin{array}{l} Zn \rightarrow Zn^{2+} + 2e^- \\[2mm] 2NaOH \leftrightarrow 2Na^+ + 2OH^- \end{array} \right\} \quad Zn^{2+} + 2Na^+ + 2OH^- + 2e^- \rightarrow Na_2ZnO_2 + H_2$$

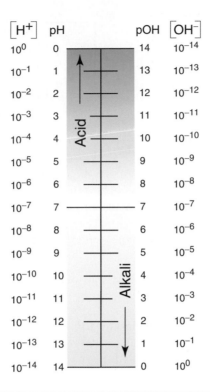

Figure 17.6  The scales of pH and pOH.

*Fresh and impure water*  Corrosion, as we have just seen, is the degradation of a metal by an electro-chemical reaction with its environment. Figure 17.7 illustrates in more detail the idea of an electro-chemical reaction. If a metal is placed in a conducting solution like salt water, it dissociates into ions, releasing electrons, as the iron is shown doing in the figure, via the anodic reaction

$$Fe \leftrightarrow Fe^{2+} + 2e^{-} \tag{17.13}$$

The electrons accumulate on the iron giving it a negative charge that grows until the electrostatic attraction starts to pull the $Fe^{2+}$ ions back onto the metal surface, stifling further dissociation. At this point the iron has a potential (relative to a standard, the *hydrogen standard*) of $-0.44$ volts. Each metal has its own characteristic potential (called the *standard reduction potential*), shown on the left of Figure 17.8. The extra electrons enter the band structure of the metal just above the Fermi level, so the energy associated with the Fermi level determines how strongly they are held.

If two metals are connected together in a cell, like the iron and copper samples in Figure 17.9, a potential difference equal to their difference in the reduction potential

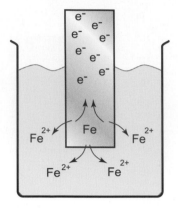

**Figure 17.7** Ionization.

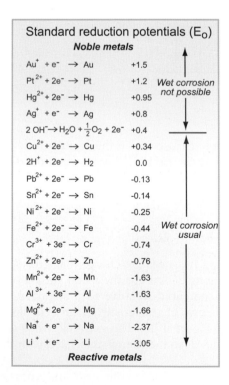
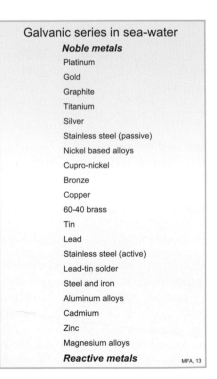

## Standard reduction potentials ($E_o$)

### *Noble metals*

| | | |
|---|---|---|
| $Au^+ + e^- \rightarrow Au$ | +1.5 | |
| $Pt^{2+} + 2e^- \rightarrow Pt$ | +1.2 | *Wet corrosion* |
| $Hg^{2+} + 2e^- \rightarrow Hg$ | +0.95 | *not possible* |
| $Ag^+ + e^- \rightarrow Ag$ | +0.8 | |
| $2\,OH^- \rightarrow H_2O + \frac{1}{2}O_2 + 2e^-$ | +0.4 | |
| $Cu^{2+} + 2e^- \rightarrow Cu$ | +0.34 | |
| $2H^+ + 2e^- \rightarrow H_2$ | 0.0 | |
| $Pb^{2+} + 2e^- \rightarrow Pb$ | -0.13 | |
| $Sn^{2+} + 2e^- \rightarrow Sn$ | -0.14 | |
| $Ni^{2+} + 2e^- \rightarrow Ni$ | -0.25 | |
| $Fe^{2+} + 2e^- \rightarrow Fe$ | -0.44 | *Wet corrosion* |
| $Cr^{3+} + 3e^- \rightarrow Cr$ | -0.74 | *usual* |
| $Zn^{2+} + 2e^- \rightarrow Zn$ | -0.76 | |
| $Mn^{2+} + 2e^- \rightarrow Mn$ | -1.63 | |
| $Al^{3+} + 3e^- \rightarrow Al$ | -1.63 | |
| $Mg^{2+} + 2e^- \rightarrow Mg$ | -1.66 | |
| $Na^+ + e^- \rightarrow Na$ | -2.37 | |
| $Li^+ + e^- \rightarrow Li$ | -3.05 | |

### *Reactive metals*

## Galvanic series in sea-water

### *Noble metals*

Platinum

Gold

Graphite

Titanium

Silver

Stainless steel (passive)

Nickel based alloys

Cupro-nickel

Bronze

Copper

60-40 brass

Tin

Lead

Stainless steel (active)

Lead-tin solder

Steel and iron

Aluminum alloys

Cadmium

Zinc

Magnesium alloys

### *Reactive metals*

MFA, 13

**Figure 17.8** Standard reduction potentials of metals (left) and galvanic series in sea water (right) for engineering alloys.

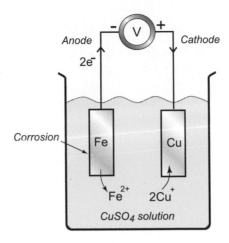

**Figure 17.9** A bi-metal corrosion cell. The corrosion potential is the potential to which the metal falls relative to a hydrogen standard.

(Figure 17.8, left hand side) appears between them. The corrosion potential of iron, $-0.44$, differs from that of copper, $+0.34$, by 0.78 volts, so if no current flows in the connection the voltmeter will register this difference. If a current is now allowed, electrons flow from the iron (the *anode*) to the copper (the *cathode*); the iron ionises (that is, it corrodes), following the anodic reaction of equation (17.13) and—if the solution were one containing copper sulphate—copper ions, $Cu^{2+}$, plate out onto the copper following the cathodic reaction

$$Cu^{2+} + 2e^- \leftrightarrow Cu \tag{17.14}$$

Reduction potentials tell you which metal will corrode if you make a bi-metal cell like that of Figure 17.9. Electroplating, you might say, is 'un-corrosion', and that is a helpful analogy. If you want to reverse the corrosion path, returning ions to the surface they came from, apply a potential exceeding the corrosion potential—indeed, the term 'reduction potential' means the voltage that must be exceeded to cause electroplating.

A corrosion cell is a kind of battery. Simple batteries have electrodes separated by an electrolyte. Choosing electrode materials with a big difference in their standard electrode potentials, $E_o$, is a start, because this difference is the open-circuit voltage of the battery if the electrolyte contains a standard (meaning 1 molar) solution of metal ions and the metals are pure. If the solution is non-standard the reduction potential changes in a way described by the *Nernst equation*. The reduction potential $E$ when the molar concentration is $M = C_{ion}$ instead of $M = 1$ is

$$E = E_o + \frac{0.059}{z} \log_{10}(C_{ion}) \tag{17.15}$$

where $z$ is the valence of the ion.

---

### Example 17.5 Using the Nernst equation

A zinc electrode is put in an electrolyte that is a solution of 2 grams of zinc as $Zn^{2+}$ ions in a litre of water. What is its corrected reduction potential at 25 °C? The atomic weight of zinc is 65.4 kg/kmol.

*Answer.* The zinc ions have a valence $z = 2$, a standard reduction potential $E_o$ of $-0.76$ and a molar concentration

$$C_{ion} = \frac{2}{65.4} = 0.031 \text{ M}$$

Thus the corrected reduction potential is

$$E = -0.76 + \frac{0.059}{2} \log_{10}(0.031) = -0.76 - 0.045 = -0.0805 \text{ Volts}$$

---

There are other non-standard influences that are not so easily fixed. Engineering metals are almost all alloys. Some, like aluminum, titanium and stainless steel, form protective surface layers, making them passive (more corrosion-resistant). But in some solutions the passive layer breaks down and the behaviour becomes 'active', corroding more quickly. This has led engineers to formulate a more practical, empirical ordering called a *galvanic series*. They rank engineering alloys by their propensity to corrode in common environments such as sea water when joined to another metal (Figure 17.8, right side). The ranking is such that any metal will become the anode (and corrode) if joined to any metal above it in the list, and it will become the cathode (and be protected) if joined to one below, when immersed in sea water. Thus if you make a mild-steel tank with copper rivets and fill it with sea water, the steel will corrode. But if instead you made the tank out of aluminum with mild steel rivets, it is the aluminum that corrodes. The corrosion is most intense just around the rivet, where you least want it to happen.

The rate at which corrosion or electro-plating takes place is determined by the corrosion current, $I$ (amps), according to *Faraday's equation*. It describes the weight of metal corroded away, $w$ kg, as a function of time $t$ (seconds)

$$w = \frac{I\,t\,M}{z\,F} \text{ kg} \tag{17.16}$$

Here $M$ is the atomic mass of the metal, $z$ (as before) is the valence of its ions and $F$ is Faraday's constant $= 96,500$ Coulombs (the units of electrical charge, and the same units as the quantity $I\,t$).

---

### Example 17.6 Using Faraday's equation

What product of current density $(A/m^2)$ and time are needed to plate a 10 micron thick layer of nickel onto an object? The density of nickel is 8910 kg/m$^3$, its atomic weight is $M = 58.7$ kg/kmol and nickel ions have a valence of $z = 2$.

---

*Answer.* The volume of nickel required per unit area is $10^{-5}$ m$^3$/m$^2$. This corresponds to a mass of nickel per unit area of $w_A = 0.089$ kg. Inverting Faraday's equation, expressed per unit area (subscript $A$), gives

$$I_A\, t = \frac{w_A\, z\, F}{M}$$

where $I_A$ is the current density and $t$ is the time. Inserting values for the parameters gives

$$I_A\, t = 293 \text{ A.s (or Coulombs) per m}^2.$$

Thus the plating requires 0.1 amp/m$^2$ for 2930 (49 minutes) or 1 amp/m$^2$ for 293 seconds (4.9 minutes).

Suppose now that the liquid is not a copper sulfate solution but just water (Figure 17.10). Water dissolves oxygen, so unless it is specially de-gassed and protected from air, there is oxygen in solution. The iron and the copper still dissociate until their corrosion potential difference is established but now, if the current is allowed to flow, there is no reservoir of copper ions to plate out. The iron still corrodes but the cathodic reaction has changed; it is now the *hydrolysis reaction*

$$H_2O + O + 2e^- \leftrightarrow 2OH^- \tag{17.17}$$

While oxygen can reach the copper, the corrosion reaction continues, creating Fe$^{2+}$ ions at the anode and OH$^-$ ions at the cathode. They react to form insoluble Fe(OH)$_2$ which ultimately oxidises further to Fe$_2$O$_3$.H$_2$O, which is what we call rust.

Thus connecting dissimilar metals in either pure water or water with dissolved salts is a bad thing to do: corrosion cells appear that eat up the metal with the lower (more negative) corrosion potential. Worse news is to come: it is not necessary to have two metals—both anodic and cathodic reactions can take place on the *same* surface. Figure 17.11 shows how this happens.

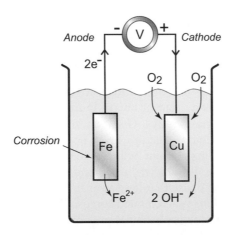

Figure 17.10  A bi-metal cell containing pure water in which oxygen can dissolve.

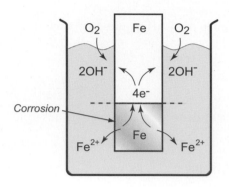

**Figure 17.11**   A corrosion cell created by differential access to oxygen.

Here an iron sample is immersed in water with access to air. The part of the sample nearest the water surface has an easy supply of oxygen; that further away does not. On the remoter part the ionisation reaction of equation (17.13) takes place, corroding the iron and releasing electrons that flow up the sample to the near-surface part where oxygen is plentiful, where they enable the hydrolysis reaction of equation (17.17). The hydrolysis reaction has a corrosion potential of +0.81 volts—it is shown on Figure 17.8—and the difference between this and that of iron, −0.44 volts, drives the corrosion. If the sample could be cut in two along the broken line in Figure 17.11 and a tiny voltmeter inserted, it would register the difference: 1.23 volts.

Differential oxidation corrosion is one of the most common and most difficult forms of corrosion to prevent: where there is water and a region with access to oxygen and one that is starved of it, a cell is set up. Only metals above the hydrolysis reaction potential of +0.81 volts in Figure 17.8 are immune.

*Selective corrosion*   Often, wet corrosion, instead of being uniform, occurs selectively, and when it does, it can lead to failure more rapidly than the uniform rate would suggest. Stress and corrosion acting together ('stress-corrosion' and 'corrosion fatigue') frequently lead to local-ised attack, as do local changes in microstructure such as those at a welded joint. Briefly, the localised mechanisms are these.

*Intergranular corrosion* occurs because grain boundaries have chemical properties that differ from those of the grain, because the lower packing density there gives these atoms higher energy. Grain boundaries can also be electrochemically different due to the presence of pre-cipitates on the boundaries that do not form within the bulk of the grains. *Pitting corrosion* is preferential attack that can occur at breaks in the natural oxide film on metals, or at precip-itated compounds in certain alloys. *Galvanic attack* at the microstructural level appears in alloys with a two-phase microstructure, made up of grains alternately of one composition and of another. The two compositions will, in general, lie at slightly different points on the reduction potential scale. If immersed in a conducting solution, thousands of tiny corrosion cells appear, causing the phase that lies lower in reduction potential to be eaten away.

*Stress corrosion cracking* is accelerated corrosion, localised at cracks in loaded components. As we saw in Chapter 8, fracture becomes possible when the energy released by the growth of a crack exceeds the energy required to make the two new surfaces plus that absorbed in the plastic zone at the crack tip. Brittle materials have little or no crack-tip plasticity so the energy needed to drive the crack is small. Tough materials have a large crack-tip plastic zone that absorbs a great deal of energy, resisting crack advance. A corrosive environment provides a new source of energy—the chemical energy released when the material corrodes. Crack advance exposes new, unprotected surfaces and it is also at the crack tip that the elastic strain energy is highest, so the corrosive reaction is most intense there. The result is that cracks grow under a stress intensity $K_{scc}$ that can be far below $K_{1c}$. Examples are brass in ammonia, mild steel in caustic soda, and some aluminum and titanium alloys in salt water. Gases—particularly atomic hydrogen—can diffuse into steels, embrittling the surrounding material. The rate of diffusion depends exponentially on temperature (Chapter 13, equation (13.12)) with an activation energy $Q_d$ that is characteristic of the diffusing atom. It is then found that the rate of crack advance follows the same exponential dependence on temperature with the same characterising activation energy.

Polymers, too, can be embrittled by certain environments: polycarbonate in alkalis, PMMA (Plexiglas) in acetone, some nylons and polyesters in acids are examples. The stress intensity driving the crack may come from working loads or from internal stresses left by moulding. To avoid the second, components made of materials that are sensitive to stress corrosion cracking are heat-treated to remove internal stress before going into service.

*Corrosion fatigue* refers to the accelerated rate at which fatigue cracks grow in a corrosive environment. The endurance limit of some steels in salt water is reduced by a factor as large as 4. The rate of fatigue crack growth, too, increases to a level that is larger than the sum of the rates of corrosion and of fatigue acting together.

## **17.7** Fighting corrosion

In fighting corrosion, there are four broad strategies (Figure 17.12):

* *Judicious* design, meaning informed material choice and choice of geometry and configuration.
* *Protective coatings*, either passive (meaning that the coating simply excludes the corrosive medium) or active (meaning that the coating protects even when incomplete).

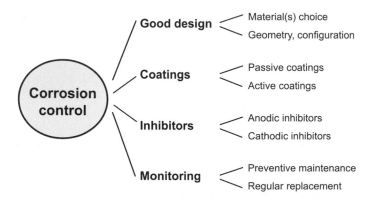

Figure 17.12  Strategies for controlling corrosion.

Table 17.3 Corrosion-resistance ranking of materials in common environments

| Environment | | Cast iron | High carbon steel | Medium carbon steel | Low carbon steel | Low alloy steel | Stainless steel | Aluminum alloys | Copper alloys | Lead alloys | Magnesium alloys | Nickel alloys | Titanium alloys | Zinc alloys | Borosilicate glass | Glass ceramic | Silica glass | Soda-lime glass | Brick | Concrete | Stone | Alumina | Aluminum nitride | Boron carbide | Silicon | Silicon carbide | Silicon nitride | Tungsten carbides | Zirconia |
|---|---|---|---|---|---|---|---|---|---|---|---|---|---|---|---|---|---|---|---|---|---|---|---|---|---|---|---|---|---|
| Alcohols and aldehydes | Methyl alcohol (methanol) | B | B | B | B | B | A | C | A | A | A | A | A | A | A | A | A | A | A | A | A | A | A | A | A | A | A | A | A |
| | Formaldehyde (40%) | A | A | A | A | A | A | A | A | A | A | A | A | A | A | A | A | A | A | A | A | A | A | A | A | A | A | A | A |
| | Ethyl alcohol (ethanol) | B | B | B | B | B | B | A | A | A | A | A | A | A | A | A | A | A | A | A | A | A | A | A | A | A | A | A | A |
| | Acetone | A | A | A | A | A | A | A | A | A | A | A | A | A | A | A | A | A | A | A | A | A | A | A | A | A | A | A | A |
| Fuels, oils and solvents | Vegetable oils | A | A | A | A | A | A | A | A | A | A | A | A | A | A | A | A | A | A | A | A | A | A | A | A | A | A | A | A |
| | Silicone fluids | A | A | A | A | A | B | A | A | A | A | A | A | A | A | A | A | A | A | A | A | A | A | A | A | A | A | A | A |
| | Petroleum (gasoline) | A | A | A | A | A | A | A | A | A | A | A | A | A | A | A | A | A | A | A | A | A | A | A | A | A | A | A | A |
| | Paraffin oil (kerosene) | A | A | A | A | A | A | A | A | A | A | A | A | A | A | A | A | A | A | A | A | A | A | A | A | A | A | A | A |
| | Lubricating oil | A | A | A | A | A | A | A | A | A | A | A | A | A | A | A | A | A | A | A | A | A | A | A | A | A | A | A | A |
| | Diesel oil | A | A | A | A | A | A | A | A | A | A | A | A | A | A | A | A | A | A | A | A | A | A | A | A | A | A | A | A |
| | Carbon tetrachloride | A | A | A | A | A | A | A | A | A | A | A | A | A | A | A | A | A | A | A | A | A | A | A | A | A | A | A | A |
| | Benzene | A | A | A | A | A | B | A | A | A | A | A | A | A | A | A | A | A | A | A | A | A | A | A | A | A | A | A | A |
| Acids and alkalis | Sodium hydroxide (60%) | B | B | B | B | B | A | D | A | D | B | A | C | B | A | A | B | A | A | A | A | A | A | A | A | A | A | A | A |
| | Sodium hydroxide (10%) | A | B | B | B | B | A | A | B | B | B | A | C | C | A | A | A | A | A | A | A | A | A | A | A | A | A | A | A |
| | Sulfuric acid (10%) | B | D | D | D | D | B | D | C | B | D | B | C | U | A | A | A | A | B | C | A | A | A | A | A | A | A | A | A |
| | Nitric acid (10%) | D | D | D | D | D | A | C | B | B | B | A | D | A | C | B | C | U | A | A | A | A | B | C | A | A | A | A | A |
| | Hydrofluoric acid (40%) | D | D | D | D | D | D | C | U | D | D | D | D | D | D | D | D | D | D | D | D | D | D | D | D | D | D | D | D |
| | Hydrochloric acid (10%) | D | D | D | D | D | A | C | U | C | U | D | B | A | D | A | A | A | A | B | A | A | A | A | A | A | A | A | A |
| | Acetic acid (10%) | C | C | C | C | C | A | C | U | B | U | A | A | C | A | A | A | B | C | U | A | A | A | A | A | A | A | A | A |
| Aqueous environments | Soils, alkaline (clay) | B | B | B | B | B | A | A | A | A | C | U | A | A | A | A | A | A | A | A | A | A | A | A | A | A | A | A | A |
| | Soils, acidic (peat) | B | B | B | B | B | A | D | A | A | C | U | A | A | B | A | A | A | A | A | A | A | A | A | A | A | A | A | A |
| | Water (salt) | C | C | C | C | A | B | A | A | C | U | A | A | A | A | B | A | A | A | A | A | A | A | A | A | A | A | A | A |
| | Water (fresh) | B | B | B | B | A | A | A | A | A | A | A | A | A | A | A | A | A | A | A | A | A | A | A | A | A | A | A | A |

A: excellent;  B: acceptable;  C: limited use;  D: unacceptable

| Environment | | | | Aqueous environments | | | | | Acids and alkalis | | | | | | | Fuels, oils and solvents | | | | | | | | | Alcohols and aldehydes | | | |
|---|---|---|---|---|---|---|---|---|---|---|---|---|---|---|---|---|---|---|---|---|---|---|---|---|---|---|---|---|
| Material | | | | Water (fresh) | Water (salt) | Soils, acidic (peat) | Soils, alkaline (clay) | Acetic acid (10%) | Hydrochloric acid (10%) | Hydrofluoric acid (40%) | Nitric acid (10%) | Sulfuric acid (10%) | Sodium hydroxide (10%) | Sodium hydroxide (60%) | Benzene | Carbon tetrachloride | Diesel oil | Lubricating oil | Paraffin oil (kerosene) | Petroleum (gasoline) | Silicone fluids | Vegetable oils | Acetone | Ethyl alcohol (ethanol) | Formaldehyde (40%) | Methyl alcohol (methanol) |
| **Composites** | | Al/SiC composite | | A | A | D | A | C | B | D | C | D | D | D | A | A | A | A | A | A | A | A | A | A | A | A |
| | | CFRP | | A | B | U | U | U | A | D | U | U | U | A | B | A | A | A | A | A | A | A | D | A | A | D |
| | | GFRP | | B | A | C | U | U | A | D | B | A | C | A | A | A | A | A | A | A | B | A | D | B | C | D |
| **Natural materials** | Woods and paper | Bamboo | | B | A | B | U | B | B | D | B | B | D | A | B | A | A | A | B | B | B | D | B | B | B | B |
| | | Cork | | A | B | U | U | U | B | D | B | B | D | B | B | B | B | B | B | B | B | B | U | B | B | B |
| | | Leather | | B | C | D | D | U | C | D | U | C | D | U | C | U | C | U | B | B | B | U | C | B | B | B |
| | | Paper and cardboard | | D | D | D | D | D | D | D | D | D | D | D | D | B | D | B | B | B | B | B | B | B | B | B |
| | | Wood | | B | B | D | U | B | U | U | B | B | U | B | B | B | B | B | U | B | D | B | B | B | B | B |
| **Polymers** | Elastomers | Butyl rubber | | A | B | B | A | D | A | U | U | D | U | A | D | D | D | D | U | A | D | D | B | D | B | D |
| | | EVA | | A | A | D | A | A | C | A | D | B | A | U | D | D | A | D | D | A | U | A | A | D | A | A |
| | | Isoprene rubber (IR) | | A | A | A | A | A | U | A | U | D | A | A | D | U | A | D | U | A | D | A | A | A | A | A |
| | | Natural rubber (NR) | | A | A | A | A | A | A | A | U | A | A | D | D | A | A | A | D | A | U | D | A | A | D | D |
| | | Neoprene (CR) | | A | A | A | A | A | A | U | U | A | A | A | D | A | A | A | D | U | A | A | A | A | A | A |
| | | Polyurethane (elPU) | | A | A | A | A | D | U | U | U | U | U | D | D | D | C | A | B | C | A | D | A | D | A | A |
| | | Silicone elastomers (SI) | | A | A | D | A | A | A | U | C | C | A | A | D | U | D | U | U | U | U | U | D | U | D | D |
| | Thermoplastics | ABS | | B | A | A | A | B | U | C | A | C | A | U | A | A | A | A | A | A | A | D | A | D | C | A |
| | | Cellulose polymers (CA) | | B | A | A | A | A | C | U | D | A | A | U | D | A | A | A | B | A | A | A | A | U | A | A |
| | | Ionomer (I) | | A | A | A | A | A | A | C | A | C | A | C | U | A | A | A | A | A | A | D | A | D | A | A |
| | | Nylons (PA) | | A | A | A | A | B | U | U | D | U | C | A | D | A | C | A | A | C | C | A | A | A | A | A |
| | | Polycarbonate (PC) | | A | A | C | A | A | A | A | A | A | A | A | A | A | A | A | A | A | A | A | D | U | D | A |
| | | Polyetheretherketone (PEEK) | | A | A | A | A | A | A | D | A | A | A | A | D | A | A | A | A | A | A | D | A | D | D | D |
| | | Polyethylene (PE) | | A | A | A | A | A | A | A | A | A | A | A | A | B | B | B | B | A | A | A | A | A | A | A |
| | | PET | | A | A | A | A | A | C | D | B | D | A | A | C | A | A | A | A | A | B | A | D | A | A | A |
| | | Acrylic (PMMA) | | A | A | C | A | A | A | A | A | A | A | A | U | A | A | A | A | A | A | D | U | A | D | A |
| | | Acetal (POM) | | A | B | U | U | U | C | A | D | D | A | A | D | B | U | U | U | U | B | A | C | U | U | D |
| | | Polypropylene (PP) | | A | A | A | A | A | A | A | A | A | A | A | A | C | A | A | A | A | A | A | A | A | A | A |
| | | Polystyrene (PS) | | A | A | C | C | U | A | D | D | A | A | C | A | U | U | U | B | C | B | D | D | A | D | D |
| | | Teflon (PTFE) | | A | A | A | A | C | A | D | U | A | A | A | A | A | A | A | A | A | A | D | A | U | A | A |
| | | Polyurethane (tpPUR) | | A | B | U | U | U | A | B | A | A | C | D | C | U | B | B | B | B | B | U | D | A | D | A |
| | | Polyvinylchloride (tpPVC) | | A | A | C | D | C | A | D | D | A | D | A | B | D | D | D | A | D | A | C | A | A | C | A |
| **Thermosets** | | Epoxies | | A | A | A | C | U | A | D | U | A | U | U | A | A | A | A | A | A | A | B | D | D | D | A |
| | | Phenolics | | A | C | C | D | C | A | D | A | A | D | D | B | A | A | A | A | A | A | D | U | C | C | C |
| | | Polyester | | A | A | C | D | C | A | D | D | A | D | D | D | A | A | A | A | A | B | D | D | D | D | A |

A: excellent;  B: acceptable;  C: limited use;  D: unacceptable

- *Corrosion inhibitors*, chemicals added to the corrosive medium that retard the rate of the corrosion reaction.
- *Monitoring*, with protective maintenance or regular replacement.

We start with the first: design.

*Design: material choice*    Materials differ greatly in their vulnerability to attack by corrosive media, some surviving unscathed while others, in the same environment, are severely attacked. There are no simple rules or indices for predicting susceptibility. The best that can be done is a ranking, based on experience. Table 17.3 lists the resistance of common engineering materials to 23 commonly encountered fluids, on a four-point scale ranging from A (excellent resistance to attack; life: tens of years) to D (unacceptably poor resistance; life: days or weeks). Tables like these (or their computer-based equivalent[1]) allow screening to eliminate materials that might present severe problems. But, as explained in Section 17.1, there is more to it than that. The best choice of material depends not just on the environment in which it must operate, but—for economic reasons—on the application: it may be cheaper to live with some corrosion (providing regular replacement) than to use expensive materials with better intrinsic resistance. And there is the potential for protection of vulnerable materials by coatings, by inhibitors or by electro-chemical means. Experience in balancing these influences has led to a set of 'preferred' materials for use in any given environment. Table 17.4 presents these for the 23 environments that appear in the tables of this chapter.

---

### Example 17.7

Use Tables 17.3 and 17.4 to identify materials that you might use for pipework to handle 10% sodium hydroxide. Consider material combinations as well as single materials.

*Answer*. From Table 17.3, stainless steels, copper alloys and titanium alloys have an A rating in NaOH. So, too, do polyethylene and polypropylene, so they could be used to line a simple steel pipe.
From Table 17.4 stainless steel, nickel alloys (ranked B) and magnesium alloys (also B) are used.

---

*Design: geometry and configuration*    Here are the Dos and Don'ts.

- *Allow for uniform attack*. Nothing lasts forever. Some corrosion can be tolerated, provided it is uniform, simply by allowing in the initial design for the loss of section over life.
- *Avoid fluid trapping* (Figure 17.13). Simple design changes permit drainage and minimise the trapping of water and potentially corrosive dirt.

---

[1] The CES Edu software, for example, has rankings for 55 environments; other such sources are listed under Further Reading.

- *Suppress galvanic attack* (Figure 17.14). Two different metals, connected electrically and immersed in water with almost anything dissolved in it, create a corrosion cell like that of Figures 17.9 or 17.10. The metal that lies lower on the scale of reduction potential is the one that is attacked. The rate of attack can be large and localised at or near the contact. Attaching an aluminum body shell to a steel auto chassis risks the same fate, as incautious car-makers have found. Even the weld metal of a weld can differ from the parent plate in reduction potential enough to set up a cell. The answer is to avoid bi-metal couples if water is around or, when this is impossible, to insulate them electrically from each other and reduce the exposed area of the nobler metal to minimise the rate of the cathodic reaction.
- *Avoid crevices* (Figure 17.15). Crevices trap moisture and create the conditions for differential-aeration attack. The figure shows riveted lap joints (the rivet, of course, made from the same metal as the plate). If water can get under the plate edge, it will. The water-air surface has free access to oxygen, but the metal between the plates does not. The result is corrosion of the joint surface, just where it is least wanted. The answer is to join by welding or soldering, or to put sealant in the joint before riveting. This is not just a joint problem— differential aeration is very hard to avoid. It is the reason that the legs of jetties and piers rust just below the water line, and why some edges of the tank are more severely rusted than others, as shown in the picture at the opening of this chapter.
- *Consider cathodic protection* (Figure 17.16). If attaching a metal to one lower in reduction potential causes the lower one to corrode, it follows that the upper one is protected. So connecting buried steel pipework to zinc plates sets up a cell that eats the zinc but spares the pipes. The expensive bronze propellers of large ships carry no surface protection (the conditions are too violent for it to stay in place). Copper and bronze lie above steel in the reduction-potential table and electrical isolation is impossible. Bronze connected to steel in salt water is a recipe for disaster. The solution is to shift the disaster elsewhere by attaching zinc plates to the hull around the propeller shaft. The zinc, in both examples, acts as a *sacrificial anode* protecting both the bronze and the steel. The term 'sacrificial' is accurate—the zinc is consumed—so the protection only lasts as long as the zinc, requiring that it be replaced regularly.
- *Beware of stress corrosion and corrosion-fatigue.* Stresses arising from design loads can be calculated; residual stresses from cold work or from thermal contraction in the heat-affected zone of welds are not so easy to predict. Their interaction with a corrosive environment can be dangerous, giving localised stress-corrosion cracking, or adding to cyclic loads to give corrosion-fatigue. Internal stresses cannot exceed the yield strength of the material (if they did the material would yield) so, as a rule of thumb, select materials that have as low a yield strength (or hardness) as is consistent with the other constraints of the design.
- *Design for inspection and maintenance.* If you can't reach it, you can't fix it. Design for durability ensures that parts liable to corrosion are easy to inspect, clean and, if necessary, replace.

*Coatings*    Metals, as we have said, are reactive. Only gold and a couple of other precious metals are stable as metals rather than as oxides. Coatings allow reactive metals—and particularly steels, the backbone of mechanical engineering—to be used in environments in which, uncoated, they would corrode rapidly. They work in more than one way:

Table 17.4 Preferred materials and coatings for given environments

| Environment | Metals | Polymers and composites | Ceramics and glasses |
|---|---|---|---|
| **Water and aqueous solutions** | | | |
| Water (fresh) | Aluminum alloys, stainless steels, galvanised steel, copper alloys | PET, HDPE, GFRP. All polymers are corrosion free in fresh water, though some absorb up to 5%, causing swelling. | Glass, concrete, brick, porcelain |
| Water (salt) | Copper, bronze, stainless steels, galvanised steels, lead, platinum, gold, silver | PET, HDPE, GFRP. All polymers are corrosion-free in salt water, though some absorb up to 5%, causing swelling. | Glass, concrete, brick |
| soil | Steel, bare in high-resistivity soil, coated or with galvanic protection in those with low resistivity. | HDPE, PP, PVC, most polymers (except PHB, PLA and those that are bio-degradable) corrode only slowly in soil. | Brick, pottery, glass, concrete |
| Body fluids | Cobalt-chromium alloys, nickel-titanium alloys (nitinols), nickel-chromium alloys, precious metals, implants, silver amalgam, stainless steel, titanium, gold, platinum | Acrylic, silicone, ultra high mol. wt. polyethylene (UHMWPE) | Hydroxyapatite, alumina bio-ceramic, bioglass ceramic, calcium phosphate bio-ceramic, glass-ionomer, vitreous carbon, zirconia bio-ceramic |
| **Acids and alkalis** | | | |
| Acetic acid, $CH_3COOH$ (10%) | Aluminum, stainless steel, nickel and nickel alloys, titanium, monel | HDPE, PTFE | Glass, porcelain, graphite |
| Hydrochloric acid, HCl (10%) | Copper, nickel and nickel alloys, titanium, monel, molybdenum, tantalum, zirconium, platinum, gold, silver | HDPE, PP, GFRP, rubber | Glass |
| Hydrofluoric acid, HF (40%) | Lead, copper, stainless steel, carbon steels, monel, hastelloy C, platinum, gold, silver | PTFE, fluorocarbon polymers, rubber | Graphite |
| Nitric acid $HNO_3$ (10%) | Stainless steel, titanium, 14.5% silicon cast iron, alloy 20 (40Fe, 35Ni, 20Cr, 4Cu) | PTFE, PVC, phenolics, PE-CTFE | Glass, graphite |

Table 17.4   Preferred materials and coatings for given environments *continued*

| Environment | Metals | Polymers and composites | Ceramics and glasses |
|---|---|---|---|
| Sulfuric acid, $H_2SO_4$ (10%) | Stainless steel, copper, nickel-based alloys, lead, tungsten, 14.5% silicon, cast iron, zirconium, tantalum, alloy 20 (40Fe, 35Ni, 20Cr, 4Cu), platinum, gold, silver | PET, PTFE, PE-CTFE, phenolics | Glass, graphite |
| Sodium hydroxide, NaOH (10%) | Sodium hydroxide, NaOH (10%) | PVC, LDPE, HDPE, PTFE, PE-CTFE | Glass, graphite |
| **Fuels, oils and solvents** | | | |
| Benzene, $C_6H_6$ | Carbon steel, stainless steel, aluminum, brass | PTFE | Glass, graphite |
| Carbon tetrachloride, $CCl_4$ | Carbon steel, stainless steel, aluminum, hastelloy C, monel, platinum, gold, silver | POM, PTFE, rubber | Glass, graphite |
| Crude oil | Carbon steel, aluminum, stainless steels | HDPE, PTFE, epoxies, polyamides | Glass, porcelain, enamelled metal |
| Diesel oil | Carbon steel, stainless steel, brass, copper, aluminum, monel | HDPE, PP, PTFE, buna (nitrile) rubber, viton, GFRP | Glass |
| Kerosene (paraffin oil) | Carbon steel, stainless steel, aluminum, monel | HDPE, PP, PTFE, GFRP | Glass |
| Lubricating oil | Stainless steel, carbon steel, aluminum | HDPE, PP, PTFE, rubber | Glass |
| Petroleum (gasoline) | Carbon steel, stainless steel, brass, aluminum, hastelloy C, alloy 20 (40Fe, 35Ni, 20Cr, 4Cu) | PTFE, HDPE, PP, buna (nitrile) rubber, viton, GFRP | Glass |
| Silicone fluid, $((CH_3)2SiO)_n$ | Carbon steel, aluminum | HDPE, PP, PET | Glass |
| Vegetable oil | Aluminum, carbon steel, stainless steel | HDPE, PP, PET | Glass |
| **Alcohols, aldehydes, ketones** | | | |
| Acetone, $CH_3COCH_3$ | Aluminum, carbon steel, stainless steel | PTFE, HDPE, PP copolymer (PPCO) | Glass, graphite |
| Ethyl alcohol, $C_2H_5OH$ | Steel, stainless steel, copper | PTFE, HDPE, PP, GFRP | Glass |
| Methyl alcohol, $CH_3OH$ | Steel, stainless steel, lead, monel | PTFE, HDPE, PP | Glass |
| Formaldehyde, $CH_2O$ | Stainless steel, monel, hastelloy C | PTFE, HDPE, PET, POM, nitrile or butyl rubber | Glass, GFRP |

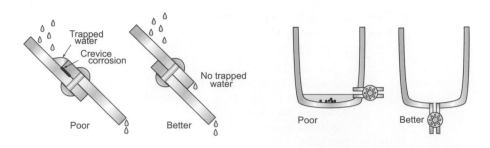

**Figure 17.13** Design changes to prevent trapped fluids reduce the rates of corrosion.

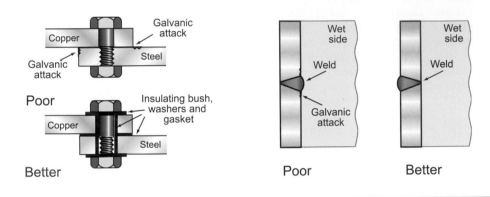

**Figure 17.14** Design changes to avoid or minimize galvanic attack.

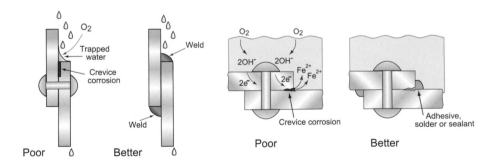

**Figure 17.15** Design changes to avoid crevice corrosion.

- *Passive coatings* separate the material from the corrosive environment. Some, like chrome, nickel or gold plate, are inherently corrosion resistant. Polymeric coatings—paints and powder-coats—provide an electrically-insulating skin that stifles electro-chemical reactions by blocking the passage of electrons. Ceramic and glass coatings rely on the extreme chemical stability of oxides, carbides and nitrides to encase the metal in a refractory shell.

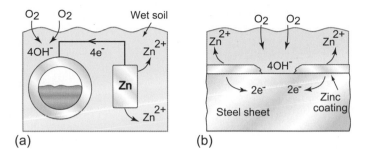

Figure 17.16 (a) Protection of steel pipes by a zinc sacrificial anode. (b) Galvanized steel plate; the zinc protects the steel even when scratched.

Passive coatings only work if they are perfect (and perfection, in this world, is unusual). If scratched, cracked or detached, corrosion starts.

- *Active coatings* work even when damaged. Some—like zinc—are metals that lie low on the reduction-potential scale of Figure 17.8. Zinc on steel acts as a sacrificial coat, becoming the anode of the corrosion cell, leaving the underlying steel unscathed (Figure 17.16(b)). Polymeric coatings can be made to work in this way by dispersing zinc powder in them. Others contain corrosion inhibitors (described below) that leach into the corrosive medium, weakening its potency. Cement coatings inhibit attack by providing an alkaline environment.
- *Self-generated coatings* rely on alloying, almost always with chromium, aluminum or silicon, in sufficient concentrations that a film of $Cr_2O_3$, $Al_2O_3$ or $SiO_2$ forms spontaneously on the metal surface. Stainless steels, aluminum bronzes and silicon iron rely on protection of this sort. When scratched, the film immediately re-grows, providing on-going protection.

With this background, glance at Table 17.5. It lists coatings all of which you will have encountered even if you didn't recognise them at the time. Modern engineering is totally dependent on them. They are applied in a number of ways, some illustrated in Figure 17.17.

- *Metallic coatings* rely on the intrinsic stability of the coat (gold), on its ability to form a self-generated, protective, oxide (chromium, aluminum, chromides and aluminides) or on its ability to provide cathodic protection (zinc, cadmium). *Galvanised iron* is steel sheet with a thin coating of zinc. The zinc acts as a sacrificial anode: if scratched, the exposed steel does not rust because the zinc protects it even when it does not cover it. Other platings look good but do not protect as well. A plating of copper sets up a cell that works in the opposite direction, eating the iron at a scratch rather than the copper. Chromium plating looks as if it should protect since Cr is below Fe in the table, but the chromium protects itself so well with its oxide film that it becomes passive and no cell is set up.
- *Polymeric coatings* rely either on a solvent that, after application, evaporates (acrylic, alkyds) or on a polymerisation reaction triggered by oxygen (linseed-oil based paints) or by an added catalyst (epoxies). Painting with *solvent-based paints* was, until recently, the most widely used method, but the evaporating solvent is generally toxic; it is a VOC (a volatile organic

Table 17.5  Coatings to enhance durability

| Type and process | Materials | Typical application |
|---|---|---|
| *Metallic coatings* | | |
| Electro-plating | Copper, nickel, chromium, silver, gold | Passive protection of steel |
| Electroless deposition | Nickel | Passive protection of steel |
| Hot dip | Zinc | Galvanised steel sheet |
| Metal spraying | Aluminum, zinc | Protection of steel components |
| Physical vapor deposition | Aluminum | Decorative protection |
| Chemical vapor deposition | Chromides, aluminides | High-temperature protection |
| Roll-cladding | Aluminum, nickel, copper | Cladding, coinage |
| *Polymer coatings* | | |
| Painting and spraying | Linseed oil-based paints | Protection of wood and steel |
| | Acrylics, epoxies | Automobile, aircraft bodies |
| | Alkyds | Cheap metallic components |
| | Cellulose acetate/butyrate | Cheap metallic components |
| Hot dip and | Nylons | Household goods |
| Thermal spray | PTFE, PE, PP | External cladding, nonstick coats |
| | Neoprene, vinyls | Tank linings |
| *Ceramic coatings* | | |
| Chemical reaction | Anodising ($Al_2O_3$), chromating ($Cr_2O_3$) | Protection of aluminium components |
| Plasma spray | Alumina ($Al_2O_3$), zirconia ($ZrO_2$) | Protection from oxidation—turbine and chemical engineering plant |
| Vapor deposit | Titanium nitride (TiN), zirconium nitride (ZrN) | Wear and corrosion resistance |
| Enamelling | Glass | Household goods |
| Cement coat | Portland cement, potassium silicate, calcium aluminate | Corrosion protection of steel, particularly from acid attack |

compound) and, for heath reasons, their use is discouraged. *Water-based paints* overcome the problem but do not yet produce quite as good a paint film. *Polymer powder coating* works by flame-spraying the polymer, or by dipping the component when hot into a fluidised bed of polymer powder, causing a thick film of polymer to melt onto the surface. While intact, paint films and polymer powder coats work well, but as soon as the coating is damaged the protection ceases unless the underlying metal has a subcoat of something—like zinc—that protects it.

- *Ceramic and glass (enamel) coatings* have to be applied hot, sufficiently so to sinter the particles of ceramic or to melt the glass onto the surface of the metal. Only cement coatings are applied cold.

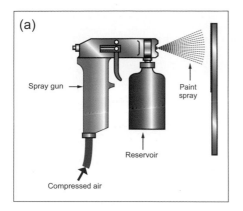

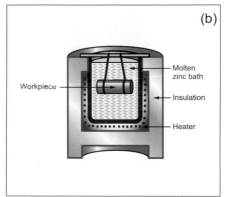

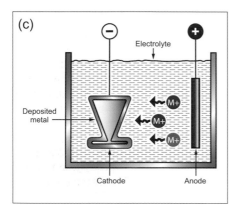

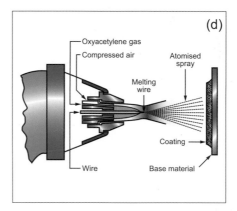

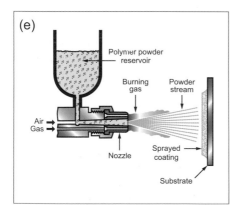

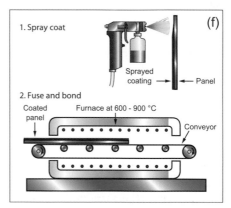

**Figure 17.17** Ways of applying coatings. (a) Paint spraying. (b) Hot dip galvanizing. (c) Electroplating. (d) Metal flame spraying. (e) Polymer powder spraying. (f) Enameling.

Example 17.8

A steel component is nickel-plated. Would you expect the nickel to act as a passive or an active barrier to aqueous corrosion? Why?

*Answer.* Nickel lies above steel in the reduction-potential chart of Figure 17.8. It is more noble than steel, and will become the cathode in the corrosion cell. That means it acts only as a passive barrier—the steel will corrode at holes or cracks in the nickel coating.

*Corrosion inhibitors*   Corrosion inhibitors are chemicals that, when dissolved or dispersed in a corrosive medium, reduce ('inhibit') the rate of attack. Some work (like indigestion tablets) by lowering the pH or by coating the part with a gooey film. Inhibitors suppress either the anodic or the cathodic step of the corrosion reaction (see Section 17.6). The choice of inhibitor depends both on the environment and on the material to be protected. Table 17.6 lists the more common ones and the metals and environments for which they work best. Many have alarmingly polysyllabic names that only specialised corrosion consultants try to remember. The things that are useful to know are itemised here.

- *PTFE $((CF_2—CF_2)_n)$* is used as a wide-spectrum inhibitor in hydrocarbon fuels, solvents and lubricants, most commonly in engine oil. It works by forming a protective layer over component surfaces, shielding them from full contact with the medium. Corrosion rates are reduced by up to 90%. Inadequate additions of PTFE lead to incomplete coverage and pitting corrosion, worsening the situation and leading to premature failure.
- *Calcium bicarbonate $(Ca(HCO_3)_2)$* is used in central heating, tap water and water cooling systems, where it lowers the pH and thus inhibits acidic attack. It makes the water hard, causing calcium deposits (lime scale) when water boils.
- *Thiourea $(C(NH_2)_2S)$* is an organic compound with a composition like that of urea but with the oxygen atom replaced by sulfur. It is the basic component of a range of organic anodic inhibitors that work by raising the activation energy of the corrosion reaction, thus decreasing the corrosion rates. Small additions can reduce the corrosion rate in hydrochloric acid by as much as 97%. Thiourea, however, is potentially carcinogenic, limiting its use to systems that do not involve human contact.
- *Arsenic (III) oxide $(As_2O_3)$* inhibits acidic attack of metals and, because it is an anodic inhibitor, it works well in a wide variety of acids. Even at low concentrations (0.5%) it can provide inhibition. It is, of course, toxic.
- *Polyphosphates* is the name of a family of polymers containing phosphate $(—PO_4)$ groups. They are found in living organisms where they perform biological functions (DNA includes polyphosphates). They inhibit corrosion of most metals in water. Polyphosphates are not toxic, allowing their use in drinking water.
- *Chromates* are mostly used to inhibit corrosion of metals in non-acidic environments (a few—the alkali chromates—can be used to inhibit corrosion in acids; they are used to protect aluminum). They work by forming an adsorbed layer of chromate ions, passivating the surface—that is, rendering it inactive. Chromates, like so many of these scary compounds, can be highly toxic, limiting their use.

Table 17.6 Corrosion inhibitors and the materials for which they work

| Environment | Inhibitors (materials) |
|---|---|
| *Water and aqueous solutions* | |
| Water (fresh) | Calcium bicarbonate (steel, cast iron), polyphosphate (Cu, Zn, Al, Fe), calcium hydroxide (Cu, Zn, Fe), sodium silicate (Cu, Zn, Fe), sodium chromate (Cu, Zn, Pb, Fe), potassium dichromate (Mg), sodium nitrite (Monel), benzoic acid (Fe), calcium and zinc metaphosphates (Zn). |
| Water (salt) | Sodium nitrite (Fe), sodium silicate (Zn), calcium bicarbonate (all metals), amyl stearate (Al), methyl-substituted dithiocarbamates (Fe) |
| Soils | Calcium nitrate is added to concrete to inhibit corrosion of steel reinforcement in soils. |
| *Acids and alkalis* | |
| Acetic acid, $CH_3COOH$ (10%) | Thiourea, arsenic oxide, sodium arsenate (all for Fe). |
| Hydrochloric acid, HCl (10%) | Ethylaniline, mercaptobenzotriazole, pyridine and phenylhydrazine, ethylene oxide (all used for Fe), phenylacridine (Al), napthoquinone (Al), thiourea (Al), chromic acid (Ti), copper sulphate (Ti). |
| Hydrofluoric acid, HF (40%) | Thiourea, arsenic oxide, sodium arsenate (all for Fe) |
| Nitric acid $HNO_3$, (10%) | Thiourea, arsenic oxide, sodium arsenate (all for Fe), hexamethylene tetramine (Al), alkali chromate (Al) |
| Sulfuric acid, $H_2SO_4$ (10%) | Phenylacridine (Fe), sodium chromate (Al), benzyl thiocyanate (Cu and Brass), hydrated calcium sulphate (Fe), aromatic amines (Fe), chromic acid (Ti), copper sulphate (Ti) |
| Sodium hydroxide, NaOH (10%) | Alkali silicates, potassium permanganate, glucose (all for Al) |
| *Fuels, oils and solvents* | |
| Benzene, $C_6H_6$ | Anthraquinone (Cu and Brass) |
| Carbon tetrachloride, $CCl_4$ | Formamide (Al), aniline (Fe, Sn, brass, monel and Pb), mesityl oxide (Sn) |
| Diesel oil | PTFE suspension (all metals), chlorinated hydrocarbons (all metals), poly-hydroxybenzophenone (Cu) |
| Kerosene (paraffin oil) | PTFE suspension (all metals), chlorinated hydrocarbons (all metals), poly-hydroxybenzophenone (Cu) |
| Lubricating oil | PTFE suspension (all metals), chlorinated hydrocarbons organozinc compound selected such as zinc dithiophosphate and zinc dithiocarbamate (all metals), poly-hydroxybenzophenone (Cu) |
| Petroleum (gasoline) | PTFE suspension (all metals), chlorinated hydrocarbons (all metals) |
| Vegetable oil | Anthraquinone (Cu and brass) |
| *Alcohols, aldehydes, ketones* | |
| Ethyl alcohol, $C_2H_5OH$ | Potassium dichromate, alkali carbonates or lactates (Al), benzoic acid (Cu and Brass), alkaline metal sulphides (Mg), ethylamine (Fe), ammonium carbonate with ammonium hydroxide (Fe) |
| Methyl alcohol, $CH_3OH$ | Sodium chlorate with sodium nitrate (Al), alkaline metal sulphides (Mg), neutralised stearic acid (Mg), Polyvinylimidazole (Cu). |

*Monitoring, maintenance and replacement*   Regular inspection allows early indications of corrosion to be detected. Maintenance—painting, recoating or repair—can then be carried out, minimising down-time and risk of failure. There is a design issue here: the system must allow inspection of all potentially vulnerable surfaces and permit access for restorative work. For safety-critical components a different strategy is adopted: that of replacement at prescribed, regular intervals too short for corrosion to have caused any serious damage.

## 17.8 Summary and conclusions

Corrosion is like cancer: it will kill you unless something else (like wear) kills you first. Almost as much money must have been invested in corrosion research as in research on cancer, such is its damaging effect on the economy. This research has revealed what goes on at an atomic scale when materials react with oxygen, acids, alkalis, aqueous solutions, aerated water and organic solvents.

All involve electrochemical reactions or reactions involving free radicals. In oxidation of metals, the metal ionises, releasing electrons; the electrons combine with oxygen molecules to give oxygen ions, that in turn combine with metal ions to form the oxide. As the oxide film grows, ions and electrons must diffuse through it to enable further growth; its resistance to this diffusion, and its ability to adhere to the metal surface determine how protective it is. Polymers oxidise through the action of free radicals. Radiation, including UV light, can generate the free radicals, as does heat. Polymers are protected by doping them with UV filters and with additions that absorb free radicals.

In corrosion, the electro-chemical reactions occur in the corrosive fluid rather than in the solid. A metal ionises when placed in water. The extent of ionisation determines its reduction potential. Those with the lowest (most negative) reduction potential are the most vulnerable to attack, and will corrode spontaneously even in pure water if oxygen is present. If a metal with a low reduction potential is in electrical contact with one of higher potential, the first corrodes and the second is protected.

In designing materials to resist corrosion we rely on the ability of a few of them—chromium, aluminum, silicon and titanium—to form a protective oxide film that adheres strongly to the surface. By using these as alloying elements, other metals can be given the same protection. Alternatively, metals can be protected by coating their surfaces with a corrosion-resistant film of polymer or ceramic.

## 17.9 Further reading and software

Bradford, S. A. (1993). *Corrosion Control*. New York, USA: Van Nostrand Reinhold. ISBN 0-442-01088-5. (A readable text emphasising practical ways of preventing or dealing with corrosion.)

DeZurik Materials Selection Guide. <http://www.dezurik.com/>. (Dezurik MSG allows the selection of materials to operate in a given environment.)

Fontanta, M. G. (1986). *Corrosion Engineering* (3rd ed.). New York, USA: McGraw Hill. ISBN 0-07-021463-8. (A text focusing on the practicalities of corrosion engineering rather than the science, with numerous examples and data.)

Ford, F. P., & Andresen, P. L. (1997). Design for Corrosion Resistance. In Vol. 20. *Materials Selection and Design, 545–72 Metals Park*, *ASM Handbook*. Ohio, USA: ASM International. ISBN 0-97170-386-6. (A recent guide to design for corrosion resistance, emphasising fundamentals, by two experts in the field.)

National Physical Laboratory (2000). *Stress Corrosion Cracking*. <http://www.npl.co.uk/upload/pdf/stress.pdf.> (A concise summary of stress corrosion cracking and how to deal with it.)

ProFlow Dynamics Chemical Resistance Selector (2008). <http://www.proflowdynamics.com/viewcorrosion.aspx.> (A free online database for finding a material suitable for use in specified environments.)

Schweitzer, P. A. (1995). *Corrosion Resistance Tables* (Vols. 1–3) (4th ed.). New York, USA: Marcel Dekker. (The ultimate compilation of corrosion data in numerous environments. Not bed-time reading.)

Schweitzer, P. A. (1998). *Encyclopedia of Corrosion Technology*. New York, USA: Marcel Dekker. ISBN 0-8247-0137-2. (A curious compilation, organised alphabetically, that mixes definitions and terminology with tables of data.)

Tretheway, K. R., & Chamberlain, J. (1995). *Corrosion for Science and Engineering* (2nd ed.). Harlow, UK: Longman Scientific and Technical. ISBN 0-582-238692. (An unusually readable introduction to corrosion science, filled with little bench-top experiments to illustrate principles.)

Waterman, N. A., & Ashby, M. F. (1991). *Elsevier Materials Selector* (3 Vols.). Oxford, UK: Elsevier. ISBN 1-85-166-605-2 and, in the CRC edition, ISBN 0-8493-7790-0. (A three-volume compilation of materials data for design, with extensive tables and guidelines for the oxidation and corrosion characteristics of metals and polymers.)

# **17.10** Exercises

| | |
|---|---|
| Exercise E17.1 | What plastic would you choose to contain hydrofluoric acid? Use Table 17.3 to find out. |
| Exercise E17.2 | Use Table 17.3 to find a flexible polymer (that is, an elastomer) suitable to make a tube through which diesel oil will flow. |
| Exercise E17.3 | A 600 litre (200 gallon) tank for storing rain water is made of aluminum. It is intended for use in the garden, partly buried in top soil. The tank was extensively tested and found to survive without noticeable corrosion. The soil in which it was buried was peaty, and thus acidic. When marketed, customers complained of corrosion problems. On investigation it was found that the soil in the affected regions was clay, and strongly alkaline. Could this have been anticipated? Use Table 17.3 to find out. |
| Exercise E17.4 | You rent a furnished apartment and discover, from the labels on the curtains, that those in the bedroom are made of a material with an LOI flammability rating of 17 and that those in the kitchen are made of a material with a UL94 rating of V–0. Would you be happy with these ratings? |

| Exercise E17.5 | By what mechanisms do metals oxidise? What determines the rate of oxidation? |
|---|---|
| Exercise E17.6 | A 10 mm square sheet of a metal, exposed on both front and back surfaces, gains weight by 4.2 mg when heated for 20 hours in air at 300 °C. If the kinetic of oxidation are linear, what is the linear rate constant $k_l$ for this metal at 300 °C? |
| Exercise E17.7 | A 10 mm square sheet of a metal, exposed on both front and back surfaces, gains weight by 3.6 mg when heated for 10 hours in air at 700 °C. If the kinetic of oxidation are parabolic, what is the parabolic rate constant $k_p$ for this metal at 700 °C? |
| Exercise E17.8 | Copper oxidises in air at 1000 °C with parabolic kinetics, forming a surface film of $Cu_2O$. The parabolic rate constant at this temperature, $k_p$, is $2.1 \times 10^{-6}$ kg$^2$ / m$^4$.s. What is the weight gain per unit area of copper surface after 1 hour? How thick will the oxide be after this time? (The atomic weight of copper is 63.5 kg/kmol, that of oxygen is 16 kg/kmol and the density of $Cu_2O$ is 6000 kg/m$^3$). |
| Exercise E17.9 | The oxidation kinetics of titanium to $TiO_2$ is limited by oxygen diffusion, with an activation energy $Q_d$ of 275 kJ/mol. If the oxide film grows to a thickness of 0.08 microns after 1 hour at 800 °C, how thick a film would you expect if it had been grown at 1000 °C for 30 minutes? |
| Exercise E17.10 | A soil is described as 'peaty'. What does this mean in terms of pH? The soil has an electrical resistivity of $2 \times 10^8$ µohm.cm. Would you expect it to be particularly corrosive? Use Table 17.2 to find out. |
| Exercise E17.11 | Household vinegar has a pH of 2.8. What is the molar concentration of $[H^+]$ ions? |
| Exercise E17.12 | A solution of NaOH (caustic soda) has a pH of 10. What is the molar concentration of $[OH^-]$ ions? |
| Exercise E17.13 | A solution of copper sulfate ($CuSO_4$) contains 100 grams of $CuSO_4$ per litre of water. The atomic weight of copper is 63.5, that of sulfur is 32.1 and that of oxygen is 16.0 kg/kmol. What is the molecular concentration of copper ions, $Cu^{2+}$, in the solution? |
| Exercise E17.14 | What is meant by the standard reduction potential? A copper and a platinum electrode are immersed in a bath of dilute copper sulfate. What potential difference would you expect to measure between them? If they are connected so that a current can flow, which one will corrode? |
| Exercise E17.15 | In a study of slowly propagating cracks in high-strength steel plates under constant stress, it was found that in a moist air environment the crack growth rate increased with temperature as follows: |

| Growth rate ($\mu$m s$^{-1}$) | 0.70 | 2.20 | 8.70 | 29.1 |
|---|---|---|---|---|
| Temperature (°C) | 5 | 25 | 55 | 87 |

Show, using an appropriate plot, that for these conditions crack propagation is a thermally activated process, and determine the activation energy. It is believed that diffusion of one of the elements listed below is the rate-controlling mechanism. Decide which of these elements is likely to be involved.

| Diffusing element in $\alpha$ iron | Activation energy for diffusion (kJ/mol) |
|---|---|
| Hydrogen | 38 |
| Nitrogen | 72 |
| Carbon | 84 |
| Iron | 285 |

**Exercise E17.16**  A polymer coating is required to protect components of a microchip processing unit from attack by hydrogen fluoride (HF). Use Tables 17.3 and 17.4 to identify possible choices.

**Exercise E17.17**  Metal pipework on an oil rig must carry hydrochloric acid solution to acidify the well. Use Tables 17.3, 17.4 and 17.6 to explore ways of providing and protecting the pipe.

**Exercise E17.18**  A food processing plant uses dilute acetic acid for pickling onions. The acid is piped to and from holding tanks. Select a suitable material for the pipes and tanks, given that, to have sufficient strength and toughness to tolerate external abuse they must be made of a metal.

**Exercise E17.19**  An automaker is concerned about the consequences of the introduction of bio-methanol ($CH_3OH$) or bio-ethanol ($C_2H_5OH$) into auto fuels. The particular concern is the corrosion of aluminum components, particularly the engine block, by methanol or ethanol. What steps could be taken to avoid this?

**Exercise E17.20**  The automaker of Exercise E17.19 is also concerned that spillage of methanol or ethanol bio-fuel might damage the GFRP bodywork of some models. Is the concern justified?

**Exercise E17.21**  The waste stream of a fertiliser plant includes dilute sulfuric acid. The dilute acid is stored in surface tanks some distance from the plant. It is suggested that the ducting to carry the acid to the tanks could, most economically, be made of wood. Is this a totally crazy suggestion?

**Exercise E17.22**  *Using the Nernst equation.* A copper electrode is put in an electrolyte that is a solution of 3 grams of copper as $Cu^{2+}$ ions in a litre of water. What is its corrected reduction potential at 25 °C? The atomic weight of copper is 63.5 kg/kmol.

Exercise E17.23    *Using the Faraday equation.* What current and time are needed to plate a 1 mm thick layer of copper onto an object with a surface area of 0.1 m$^2$? The density of copper is 8940 kg/m$^3$, its atomic weight is $M = 63.5$ kg/kmol and copper ions have a valence of $z = 2$.

Exercise E17.24    Mild sheet-steel guttering is copper-plated to protect it from corroding. The guttering acts as a drain for sea water. If the coating is damaged, exposing the steel, will the guttering corrode in a damaging way? If instead the guttering is zinc-plated, will it be better or less well protected? Use the Galvanic Series in sea water to find out.

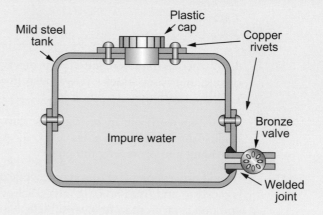

Exercise E17.25    The diagram shows a proposed design for an outdoor water tank. What aspects of the design might cause concern for corrosion?

## **17.11** Exploring design with CES

Use Level 2 unless otherwise stated.

Exercise 17.26    Find, by using 'Browse' or 'Search', the record for the nickel-chromium alloys called Nichromes. What are their main applications?

Exercise 17.27    Find, by using 'Browse' or 'Search', the record for Polymer powder-coating (remember to search in the ProcessUniverse, not the MaterialsUniverse). What are the three ways of applying a polymer powder coating?

Exercise 17.28   Use a 'Limit' stage, applied to the Level 2 Surface Treatment data table to find surface treatment processes that impart resistance to gaseous corrosion.

Exercise 17.29   Use a 'Limit' stage, applied to the Level 2 Surface Treatment data table to find surface treatment processes that impart resistance to aqueous corrosion.

Exercise 17.30   Use the CES 'Search' facility to find materials for food-processing equipment.

Exercise 17.31   Plastic cases for electrical plugs and switch-gear should not be made of flammable materials. Use the 'Select' facility in CES to find polymers that are non-flammable or self-extinguishing.

Exercise 17.32   A vat is required to hold hot caustic soda (NaOH), a strong alkali. Use the 'Select' facility in CES to find metals that resist strong alkalis very well.

Exercise 17.33   Pipework is required for a gherkin-pickling plant to carry vinegar (a dilute acid) at 100 °C from one vat to another. The liquid is under pressure, requiring a material with a strength of at least 100 MPa, and for ease of installation it must be able to be bent, requiring a ductility of at least 10%. Find the four cheapest materials that meet the constraints summarised in the table.

| Function | Pipework for hot acetic acid |
|---|---|
| Constraints | Durability in dilute acid = very good |
| | Maximum operating temperature > 100 °C |
| | Yield strength > 100 MPa |
| | Elongation > 10% |
| Objective | Minimise cost |
| Free variable | Choice of material |

Exercise 17.34   Find materials that have excellent resistance to corrosion in 10% NaOH (sodium hydroxide) and have high strength-to-weight ratio, $\sigma_y/\rho$.

- Open CES Edu Level 2 and select the subset 'Edu Level 2 with durability properties'.
- Open a LIMIT stage. Open the relevant 'Durability' folder. Find the environment and choose 'Excellent', then 'Apply'.
- Open a GRAPH stage. Make a chart with 'Density' on the $x$-axis and 'Yield strength' on the $y$-axis.
- Put on a selection line of slope 1, click above the line and then move it upwards to leave a small subset of materials. What do you find?

## 17.12 Exploring the science with CES Elements

Exercise 17.35    The Work function (Chapter 14) is the energy required to pluck an electron from the top of the Fermi level of a crystal and drag it away until it is isolated in vacuum. The Standard reduction potential of Figure 17.8 involves electrons dropping into energy levels just above the Fermi level. You might suspect that the two were in some way related. Make a graph with Standard reduction potential on the $x$-axis and Work function on the $y$-axis to find out if they are. What is your conclusion?

# Chapter 18
# Heat, beat, stick and polish: manufacturing processes

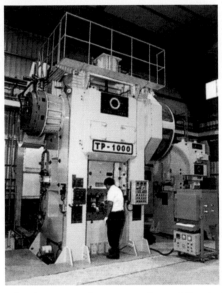

A 1000 tonne forging press. (Image courtesy of the Fu Sheng Group, Sporting Goods Division)

## **18.1** Introduction and synopsis

Materials and processing are inseparable. Ores, minerals and oil are processed to create the stuff of engineering in clean, usable form. These are shaped into components, finished and joined to make products. Chapter 2 defined a hierarchy for these processes; in this chapter we explore their technical capabilities and methods for choosing which to use. Increasingly, legislation and public opinion require that materials are recycled at the end of product life. New processes are emerging to extract usable material from what used to be regarded as waste.

The strategy for choosing a process parallels that for materials: *screen out* those that cannot do the job, *rank* those that can (usually seeking those with the lowest cost) and explore the *documented knowledge* of the top-ranked candidates. Many technical characteristics of a process can be described by numeric and non-numeric data: the size of part it can produce, the surface finish of which it is capable and, critically, the material families or classes it can handle and the shapes it can make. Constraints on these provide the inputs for screening, giving a short list of processes that meet the requirements of the design.

There is another aspect to this. It relates to the influence of processing on the properties of the materials themselves. Materials, of course, have properties before they are shaped, joined or finished, and processing is used to enhance properties such as strength. But the act of processing can also damage properties by introducing porosity, cracks and defects or by changing the microstructure in a detrimental way. These interactions, central to the choice and control of processes, are explored further in Chapter 19. For now, it is sufficient to recognise that success in manufacturing often relies on accumulated know-how and expertise, resident, in the past, in the heads of long-established employees but now captured as design guidelines, best practice guides and software providing the information needed for the documentation stage.

## **18.2** Process selection in design

*The selection strategy*    The taxonomy of manufacturing processes was presented in Chapter 2. There, it was established that processes fall into three broad families: *shaping*, *joining* and *finishing* (Figures 2.4 and 2.5). Each is characterised by a set of attributes listed, in part, in Figures 2.6 and 2.7.

The strategy for materials selection was outlined in Chapter 3. The procedure, whether dealing with original design or refining an existing one, was to translate the design requirements into a set of constraints and objectives. The constraints are used to screen and the objectives to rank, delivering a preferred short-list of materials that best meet both, finally seeking documentation to guide the final choice. The strategy for selecting processes follows a parallel path, shown in Figure 18.1. It uses constraints on process attributes to screen, together with an appropriate objective to rank, culminating as before with a search for information documenting the details of the top-ranked candidates.

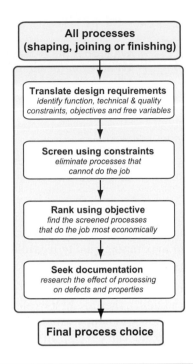

**Figure 18.1**   The selection strategy applied to processes. As for materials, the four steps are translation, screening, ranking and supporting documentation. All can be implemented in software.

***Translation***   As we saw in earlier chapters, the *function* of a component dictates the initial choice of material and shape. This choice exerts constraints on the choice of processes. It is helpful to think of two types of constraint: *technical*—can the process do the job at all? And *quality*—can it do so sufficiently well? One technical constraint applies across all process families: it is the compatibility of material and process. Shape, too, is a general constraint, although its influence on the choice of shaping, joining and finishing processes differs. Quality includes precision and surface finish, together with avoidance of defects and achieving the target for material properties. The usual *objective* in processing is to minimise cost. The *free variables* are largely limited to choosing the process itself and its operating parameters (such as temperatures, flow rates and so on). Table 18.1 summarises the outcome of the translation stage. The case studies of Section 18.9 illustrate its use for each process family.

***Screening***   Translation, as with materials, leads to constraints for process attributes. The screening step applies these, eliminating processes that cannot meet the constraints. Some process attributes are simple numeric ranges—the size or mass of component the process can handle, the precision or the surface smoothness it can achieve. Others are non-numeric—lists of materials to which the process can be applied, for example. Requirements such as 'made of

Table 18.1   Function, constraints, objective and free variables

| | |
|---|---|
| Function | • What does the process do? |
| Constraints | • What technical and quality limits must be met? |
| Objective | • Minimise cost |
| Free variables | • Choice of process and process operating conditions |

magnesium and weighing about 3 kg' are easily compared with the process attributes to eliminate those that cannot shape magnesium or cannot handle a component as large (or as small) as 3 kg.

*Ranking*   Ranking, it will be remembered, is based on *objectives*. The most obvious objective in selecting a process is that of minimising cost. In certain demanding applications it may be replaced by the objective of maximising quality regardless of cost, though more usually it is a trade-off between the two that is sought.

*Documentation*   Screening and ranking do not cope adequately with the less tractable issues (such as corrosion behaviour); they are best explored through a documentation search. This is just as true for processes. Old hands who have managed a process for decades carry this information in their heads and can develop an almost mystical ability to detect problems that defy immediate scientific explanation. But jobs for life are rare today; experts move, taking their heads with them. Expertise is stored more accessibly in compilations of design guidelines, best practice guides, case studies and failure analyses. The most important technical expertise relates to productivity and quality. Machines all have optimum ranges of operating conditions under which they work best and produce products with uncompromised quality. Failure to operate within this window can lead to manufacturing defects, such as excessive porosity, cracking or residual stress. This in turn leads to scrap and lost productivity, and, if passed on to the user, may cause premature failure. At the same time material properties depend, to a greater or lesser extent, on process history through its influence on microstructure. Processes are chosen not just to make shapes but to fine-tune properties.

So, documentation and knowledge of how processes work (or don't work) are critical in design. There is only so far that screening and ranking can take us in finding the right process.

## 18.3 Process attributes: material compatibility

The main goal in defining process attributes is to identify the characteristics that discriminate between processes, to enable their selection by screening. Each of the three process families—shaping, joining and surface treatment—has its own set of characterising attributes, described here and in the next three sections. It is also of interest to explore the physical limits to these for a given process, and to examine how the properties of the material being processed limit the rate of the process itself.

One process attribute applies to all three families—compatibility with material—so we examine this first.

*Material—process compatibility* Figure 18.2 shows a material—process compatibility matrix. Shaping processes are at the top, with compatible combinations marked by dots the colour of which identifies the material family. Its use for screening is straightforward—specify the material and some processes are immediately eliminated (or the reverse, as a screening step in material selection). The diagonal spread of the dots in the matrix reveals that each material class—metals, polymers, etc.—has its own set of process routes. This largely reflects the underlying process physics. Shaping is usually eased by melting or softening the material, and the melting temperatures of the material classes are quite distinct (Figure 13.6). There are some overlaps—powder methods are compatible with both metals and ceramics, moulding with

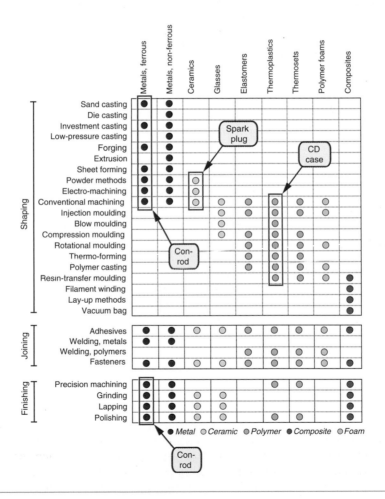

**Figure 18.2** The material—process compatibility matrix for shaping, joining and finishing processes colour-coded by material. Ignore the boxes on first reading; they refer to case studies to come.

both polymers and glasses. Machining (when used for shaping) is compatible with almost all families. Joining processes using adhesives and fasteners are very versatile and can be used with most materials, whereas welding methods are material specific. Finishing processes are used primarily for the harder materials, particularly metals—polymers are moulded to shape and rarely treated further except for decorative purposes. We will see why later.

*Coupling of material selection and process selection*   The choices of material and of process are clearly linked. The compatibility matrix shows that the choice of material isolates a sub-domain of processes. The process choice feeds back to material selection. If, say, *aluminum alloys* were chosen in the initial material selection stage and *casting* in the initial process selection stage, then only the subset of aluminum alloys that are *castable* can be used. If the component is then to be *age hardened* (a heat-treatment process), material choice is further limited to *heat-treatable aluminum casting alloys*. So we must continually iterate between process and material, checking that inconsistent combinations are avoided. The key point is that neither material nor process choice can be carried too far without giving thought to the other. Once past initial screening, it is essentially a co-selection procedure.

## 18.4 Shaping processes: attributes and origins

*Mass and section thickness*   There are limits to the size of component that a process can make. Figure 18.3 shows the limits. The colour coding for material compatibility has been retained, using more than one colour when the process can treat more than one material family. Size can be measured by volume or by mass, but since the range of either one covers many orders of magnitude, whereas densities span only a factor of about 10, it doesn't make much difference which we use—big things are heavy, whatever they are made of. Most processes span a mass range of about a factor of 1000 or so. Note that this attribute is most discriminating at the extremes; the vast majority of components are in the 0.1 to 10 kg range, for which virtually any process will work.

It is worth noting what the ranges represent. Each bar spans the size range of which the process is capable without undue technical difficulty. All can be stretched to smaller or larger extremes but at the penalty of extra cost because the equipment is no longer standard. During screening, therefore, it is important to recognise 'near misses'—processes that narrowly failed, but that could, if needed, be reconsidered and used.

Figure 18.4 shows a second bar chart: that for the ranges of section thickness of which each shaping process is capable. It is the lower end of the ranges—the minimum section thickness—where the physics of the process imposes limits. Their origins are the subject of the next subsection.

*Physical limits to size and section thickness*   Casting and moulding both rely on material flow in the liquid or semi-liquid state. Lower limits on section thickness are imposed by the physics of flow. Viscosity and surface tension oppose flow through narrow channels, and heat loss from the large surface area of thin sections cools the flowing material, raising the viscosity before the channel is filled (Figure 18.5). Polymer and glass viscosities increase steadily as temperature drops. Pure metals solidify at a fixed temperature, with a step increase in viscosity, but for alloys this happens over a range of temperature, known as the 'mushy zone', in which

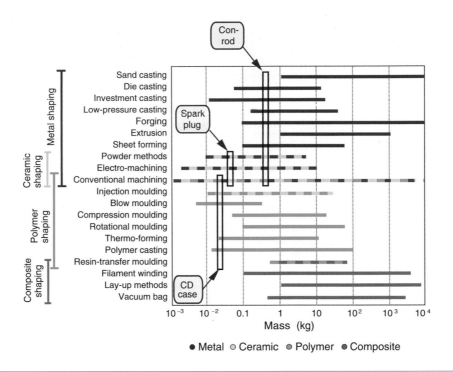

**Figure 18.3** The process—mass range bar chart for shaping processes. Ignore, on first reading, the boxes; they refer to case studies to come. The colours refer to the material, as in Figure 18.2. Multiple colours indicate compatibility with more than one family.

the alloy is part liquid, part solid. The width of this zone can vary from a few degrees centigrade to several hundred—so metal flow in castings depends on alloy composition. In general, higher pressure die-casting and moulding methods enable thinner sections to be made, but the equipment costs more and the faster, more turbulent flow can entrap more porosity and cause damage to the moulds.

Upper limits to size and section in casting and moulding are set by problems of shrinkage. The outer layer of a casting or moulding cools and solidifies first, giving it a rigid skin. When the interior subsequently solidifies, the change in volume can distort the product or crack the skin, or cause internal cavitation. Problems of this sort are most severe where there are changes of section, since the constraint introduces tensile stresses that may cause *hot tearing*—cracking caused by constrained thermal contraction. Different compositions have different susceptibilities to hot tearing—another example of coupling between material, process and design detail.

Much of the documentation for casting and moulding processes concerns guidance on designing both the component shape and the mould geometry to achieve the desired cross-sections while avoiding defects. Even when the component shape is fixed, there is freedom to choose where the material inlets are built into the mould (the 'runners') and where the air and excess material will escape (the 'risers'). Figure 18.6 shows an example of good practice in designing the cross-sectional shape of a polymer moulding.

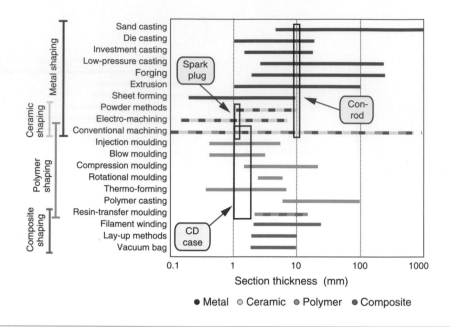

**Figure 18.4**  The process—section thickness bar chart for shaping processes colour coded by material, annotated with the target values for the case studies.

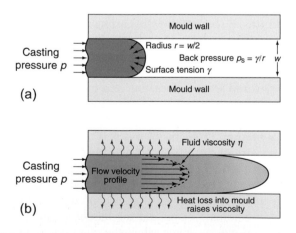

**Figure 18.5**  Flow of liquid metal or polymer into thin sections is opposed by surface tension (a) and by viscous forces (b). Loss of heat into the mould increases viscosity and may cause premature solidification.

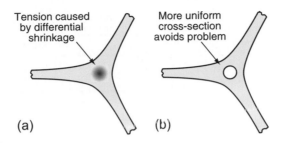

**Figure 18.6** (a) Differential shrinkage causes internal stress, distortion and cavitation in thick sections or at changes of section. (b) Good design keeps section thickness as uniform as possible, avoiding the problem.

Powder methods for metals and ceramics, too, depend on flow. Filling a mould with powder uses free flow under gravity plus vibration to bed the powder down and achieve uniform filling. This may be followed by compression ('cold compaction'). Once full of powder, the mould is heated to allow densification by sintering or, if the pressure is maintained, by hot isostatic pressing (HIPing).

Metal shaping by deformation—hot or cold rolling, forging or extrusion—also involves flow. Solid metals flow by plastic deformation or by creep—Chapter 13 described how the flow rate depends on stress and temperature. Much forming is done hot because the hot yield stress is lower than that at room temperature and work hardening, which drives the yield strength up during cold deformation, is absent. The thinness that can be rolled, forged or extruded is limited by plastic flow in much the same way that the thinness in casting is limited by viscosity: the thinner the section, the greater the required roll-pressure or forging force.

Figure 18.7 illustrates the problems involved in forging or rolling very thin sections. Friction changes the pressure distribution on the die and under the rolls. When they are well lubricated, as in (a), the loading is almost uniaxial and the material flows at its yield stress $\sigma_y$. With friction, as in (b), the metal shears at the die interface and the pressure ramps up because the friction resists the lateral spreading, giving a 'friction hill'. The area under the pressure distribution is the total forming load, so friction increases the load. The greater the aspect ratio of the section (width/thickness), the higher the maximum pressure needed to cause yielding, as in (c). This illustrates the fundamental limit of friction on section thickness—very thin sections simply stick to the tools and will not yield, even with very large pressures. Friction not only increases the load and limits the aspect ratio that can be formed, it also produces distortion in shape—'barrelling'—shown in both (b) and (c). A positive aspect of the high contact pressures in forging is the repeatability with which the metal can be forced to take up small features machined on the face of the tooling, exploited in making coins.

***Very small or very thin objects—pushing the limits*** So how do you make very small things? The obvious way to do it is to cut them out of bigger ones. That means machining: turning, shaping, drilling and milling. The upper limit on size is set by the budget available to buy the machine and the electricity bill for its use. Lower limits—now we are thinking of milligrams and fractions of a millimeter—are set by the stiffness of the machine and the material itself.

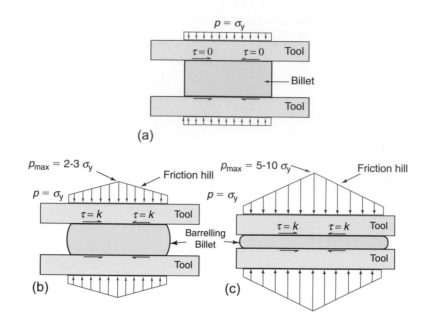

**Figure 18.7** The influence of friction and aspect ratio on open die forging. (a) Uniaxial compression with very low friction. (b) With sticking friction the contact pressure rises in a 'friction hill' causing barrelling. (c) The greater the aspect ratio, the greater the pressure rise and the barrelling.

One continuous shaping process, rolling, is able to produce sections as thin as 10 μm—the thickness of aluminum kitchen foil. The foil only yields and thins just as it enters and leaves the roll bite—in between, the huge aspect ratio and frictional constraint mean that the high pressure actually flattens the steel rolls elastically. The process is helped by 'pack rolling'—feeding two sheets through together, with lubricant in between so that they can be peeled apart as they exit the rolls. This is why kitchen foil is shiny on one side (where contact was with the roll) and matte on the other.

Suppose you want to go still smaller. Flow won't work, and the sections are so thin that they simply bend if you try to machine them. Now you need chemistry. Conventional chemistry—etching, chemical milling, electro-discharge milling—allows features on the same scale as that of precision machining. To get really small, we need the screening, printing and etching techniques used to fabricate electronic devices. These are used to make MEMS (micro-electro-mechanical systems)—tiny sensors, actuators, gears and even engines. At present the range of materials is limited by this silicon-based technology, but some commercial devices, such as the inertial triggers for airbags, are made in this way.

*Shape*   A key attribute of a process is the families of shapes it can make. It is also one of the most difficult to characterise. A classification scheme is illustrated in Figure 18.8, with three generic classes of shape, each subdivided in two. The merit of this approach is that processes map onto specific shapes. Figure 18.9 shows the matrix of viable combinations.

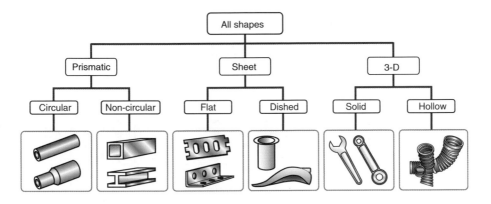

**Figure 18.8** The classification of shape. Subdivision into more than six classes is difficult, and does not give useful further discrimination.

The prismatic shapes shown on the left of Figure 18.8, made by rolling, extrusion or drawing, have a special feature: they can be made in continuous lengths. The other shapes cannot—they are discrete, and the processes that make them are called batch processes. The distinction between continuous and discrete (or batch) processes is a useful one, bearing on the cost of manufacture. Batch processes require repeated cycles of setting-up, which can be time-consuming and costly. On the other hand, they are often net-shape or nearly so, meaning that little further machining is needed to finish the component. Casting and moulding processes are near net-shape. Forging, on the other hand, is limited by the strains and shape changes by solid-state plasticity, starting from standard stock (bar, rod or sheet) to create a 3-D shape in several stages requiring a sequence of dies.

Continuous processes are well suited to long, prismatic products such as railway track or standard stock material such as tubes, plate and sheet. Cylindrical rolls produce sheets. Shaped rolls make more complex profiles—rail track is one of these. Extrusion is a particularly versatile continuous process, since complex prismatic profiles that include internal channels and longitudinal features such as ribs and stiffeners can be manufactured in one step.

***Tolerance and roughness*** We think of the precision and surface finish of a component as aspects of its *quality*. They are measured by the *tolerance*, $T$, and the *surface roughness*, $R$. When the dimensions of a component are specified the surface quality is specified as well, though not necessarily over the entire surface. Surface quality is critical in contacting surfaces such as the faces of flanges that must mate to form a seal or sliders running in grooves. It is also important for resistance to fatigue crack initiation and for aesthetic reasons. The tolerance $T$ on a dimension $y$ is specified as $y = 100 \pm 0.1$ mm, or as $y = 50^{+0.01}_{-0.001}$ mm, indicating that there is more freedom to oversize than to undersize. Surface roughness, $R$, is specified as an upper limit, for example, $R < 100$ μm. The typical surface finish required in various products is shown in Table 18.2. The table also indicates typical processes that can achieve these levels of finish.

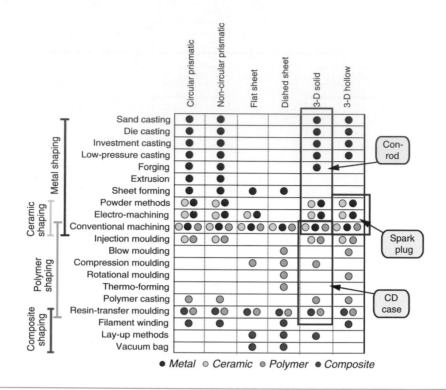

**Figure 18.9** The shape–process compatibility matrix. A dot indicates that the pair forms a viable combination colour-coded as before.

Surface roughness is a measure of the irregularities of the surface (Figure 18.10). It is defined as the root-mean-square (RMS) amplitude of the surface profile:

$$R^2 = \frac{1}{L} \int_0^L y^2(x)\mathrm{d}x \tag{18.1}$$

It used to be measured by dragging a light, sharp stylus over the surface in the $x$-direction while recording the vertical profile $y(x)$, like playing a gramophone record. *Optical*

**Table 18.2** Typical levels of finish required in different applications, and suitable processes

| Roughness (μm) | Typical application | Process |
|---|---|---|
| $R = 0.01$ | Mirrors | Lapping |
| $R = 0.1$ | High-quality bearings | Precision grind or lap |
| $R = 0.2\text{–}0.5$ | Cylinders, pistons, cams, bearings | Precision grinding |
| $R = 0.5\text{–}2$ | Gears, ordinary machine parts | Precision machining |
| $R = 2\text{–}10$ | Light-loaded bearings, non-critical | Machining components |
| $R = 3\text{–}100$ | Non-bearing surfaces | Unfinished castings |

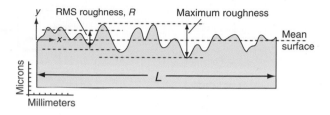

**Figure 18.10** A section through a surface, showing its irregular surface (artistically exaggerated in the vertical direction). The irregularity is measured by the RMS roughness, *R*.

*profilometry*, which is faster and more accurate, has now replaced the stylus method. It scans a laser over the surface using interferometry to map surface irregularity. The tolerance *T* is obviously greater than $2R$; indeed, since *R* is the root-mean-square roughness, the peak roughness, and hence absolute lower limit for tolerance, is more like $5R$. Real processes give tolerances that range from $10R$ to $1000R$.

Figures 18.11 and 18.12 show the characteristic ranges of tolerance and roughness of which processes are capable, retaining the colour coding for material family. Data for finishing processes are added below the shaping processes. Sand casting gives rough surfaces; casting into

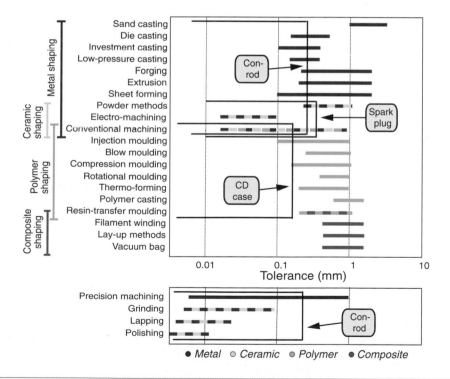

**Figure 18.11** The process—tolerance bar chart for shaping processes and some finishing processes, enabling process chains to be selected. The chart is colour-coded by material and annotated with the target values for the case studies.

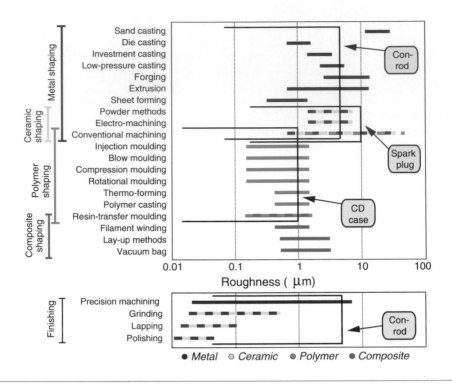

**Figure 18.12** The process—roughness bar chart for shaping processes and some finishing processes, enabling process chains to be selected.

metal dies gives a smoother one. No shaping processes for metals, however, do better than $T = 0.1$ mm and $R = 0.5$ μm. Machining, capable of high dimensional accuracy and surface finish, is used commonly after casting or deformation processing to bring the tolerance or finish up to the desired level, creating a *process chain*. Metals and ceramics can be surface-ground and lapped to a high precision and smoothness: a large telescope reflector has a tolerance approaching 5 μm and a roughness of about 1/100 of this over a dimension of a metre or more. But precision and finish carry a cost: processing costs increase exponentially as the requirements for both are made more severe. It is an expensive mistake to overspecify precision and finish.

Moulded polymers inherit the finish of the moulds and thus can be very smooth (Figure 18.12); machining to improve the finish is rarely necessary. Tolerances better than ±0.2 mm are seldom possible (Figure 18.11) because internal stresses left by moulding cause distortion and because polymers creep in service.

## 18.5 Joining processes: attributes and origins

The design requirements to be met in choosing a joining process differ from those for choosing a shaping process. Even material compatibility is more involved than for shaping, in that joints are often to be made between dissimilar materials.

*Material compatibility* Processes for joining metals, polymers, ceramics and glasses differ. Adhesives will bond to some materials but not to others; methods for welding polymers differ from those for welding metals; and specific metals often require specific types of welds. The material-process matrix (Figure 18.2) includes four classes of joining process.

When the joint is between dissimilar materials, further considerations arise, both in manufacture and in service. The process must obviously be compatible with both materials. Adhesives and fasteners generally allow joints between different materials; many welding processes do not. Practicality in service is a different issue. If two materials are joined in such a way that they are in electrical contact, a corrosion couple appears if close to water or a conducting solution. This can be avoided by inserting an insulating interlayer between the surfaces. Thermal expansion mismatch gives internal stresses in the joint if the temperature changes, with risk of damage. Identifying good practice in joining dissimilar materials is part of the documentation step.

*Joint geometry and mode of loading* Important joining-specific considerations are the geometry of the joint, the thickness of material they can handle and the way the joint will be loaded. Figures 18.13 and 18.14 illustrate standard joint geometries and modes of loading.

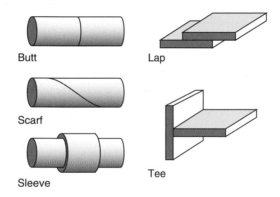

Figure 18.13  Joint geometries.

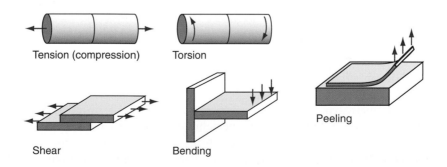

Figure 18.14  Standard modes of loading for joints.

The choice of geometry depends on the shapes and thicknesses of the parts to be joined, and the mode of loading—it is a coupled problem. Adhesive joints support shear but are poor in peeling—think of peeling off sticky tape. Adhesives need a good working area—lap joints are fine, butt joints are not. Rivets and staples, too, are well adapted for shear loading of lap joints but are less good in tension. Welds and threaded fasteners are more adaptable, but here, too, matching choice of process to geometry and loading is important. Screening on each in isolation correctly keeps options open, but the details of the coupling must be researched at the documentation stage.

When making joints, sheet thickness is an issue. Joining processes like stapling, riveting and sewing can be used with thin sections but not with thick. The thickness that can be welded is restricted, to varying degrees, by heat transfer considerations. Some processes are better adapted than others to joining greatly unequal sections: adhesives can bond a very thin layer to a very thick one, whereas riveting and sewing work best for sections that are nearly equal. In practice the thickness, too, is coupled to geometry.

*Secondary functionalities and manufacturing constraints*    Most joints carry load, but that is not all. Joints may need to conduct or insulate against the conduction of heat or electricity, or they may be required to operate at an elevated operating temperature. A joint may also serve as a seal, and be required to exclude gases or liquids.

Manufacturing conditions also impose constraints on joint design. Certain joining processes impose limits on the size of the assembly; electron beam welding, for instance, usually requires that the assembly fit within a vacuum chamber. On-site joining operations conducted out of doors require portable equipment, possibly with portable power supplies; an example is the hot-plate welding of PVC and PE pipelines. Portability, on the other hand, effectively means there is no limit on the size of the assembly—bridges, shipbuilding and aircraft make this self-evident.

An increasingly important consideration is the requirement for economic disassembly at the end of product life to facilitate recycling and reuse. Threaded fasteners can be disassembled and adhesives can be loosened with solvents or heat.

*Physical limits in joining*    Thermal joining processes—arc, laser and friction welding, for example—operate by imposing a local thermal cycle to the joint area to produce a bond in the liquid or solid state. Heat transfer and conduction set the physical limit to the size of joint. Here the intensity of heating is important. For translating heat sources of power $Q$ moving at speed $v$ along the joint line, the heat input per unit length, $Q/v$ (J/m), is commonly used as a characteristic welding parameter. Peak temperatures and cooling rates scale with $Q/v$. Heat input is thus important in determining the extent of the 'heat-affected zone'—the region round the weld affected by the thermal cycle imposed. The microstructural changes induced depend on the peak temperature and the cooling rate—especially in steels, which are commonly joined by welding. And local thermal cycles induce internal residual stresses. Because of their importance, microstructural changes during welding have been researched in depth and are now well understood; handbooks on the subject provide extensive documentation.

Process economics are dictated by the set-up time and the actual time for making the joint. For thermal processes, conduction of heat governs the operating speed with a given level of power, both in making the joint and in cooling of the component before it can be handled and moved. Some adhesives require prolonged curing times before they can take load, again slowing the throughput. And many processes take much longer to set up than to make the joint itself, in order to avoid misalignment or lack of cleanliness in the joint region. Inventive workstation design can help to improve the economics by arranging that multiple set-up and product cooling areas run in parallel with a high-speed joining machine, a strategy used to enhance the economic performance of capital-intensive processes like laser and electron beam welding.

## **18.6** Surface treatment (finishing) processes: attributes and origins

*Material compatibility*   Material–process compatibility for surface treatments is shown at the bottom of the matrix in Figure 18.2. As noted previously, surface finishing is more important for metals than for polymers.

*The purpose of the treatment*   The most discriminating finishing-specific attribute proves to be the purpose of applying the treatment. Table 18.3 illustrates the diversity of functions that surface treatments can provide. Some protect, some enhance performance, still others are primarily aesthetic. All surface treatments add cost but the added value can be large. Protecting a component surface extends product life and increases the interval between maintenance cycles. Coatings on cutting tools enable faster cutting speeds and greater productivity. And surface hardening processes may enable the substrate alloy to be replaced with a cheaper material—for example, using a plain carbon steel with a hard carburised surface or a coating of hard titanium nitride (TiN), instead of using a more expensive alloy steel.

*Secondary compatibilities*   Material compatibility and design function are the first considerations in selecting a finishing process. But there are others. Some surface treatments leave the dimensions, precision and roughness of the surface unchanged. Deposited coatings obviously change the dimensions a little, but may still leave a perfectly smooth surface. Others build up a relatively thick layer with a rough surface, requiring re-finishing. Component geometry also influences the choice. 'Line-of-sight' deposition processes coat only the surface at which they

Table 18.3   The design functions provided by surface treatments

- Corrosion protection (aqueous)
- Corrosion protection (gases)
- Wear resistance
- Friction control
- Fatigue resistance
- Thermal conduction
- Thermal insulation
- Electrical insulation
- Magnetic properties
- Decoration
- Color
- Reflectivity

are directed, leaving inaccessible areas uncoated; others, with what is called 'throwing power', coat flat, curved and re-entrant surfaces equally well. Many surface-treatment processes require heat. These can only be used on materials that can tolerate the rise in temperature. Some paints are applied cold, but many require a bake at up to 150 °C. Heat treatments like carburising or nitriding to give a hard surface layer require prolonged heating at temperatures up to 800 °C. This thermal exposure of the substrate can change the underlying microstructure, not necessarily for the better. There are some innovative developments to match the thermal treatments required in both substrate and surface, achieving two treatments in one go. An example is the paint-bake cycle for heat-treatable aluminum alloy automotive panels, which cures and hardens the paint while simultaneously age hardening the sheet itself.

*Physical limits to surface treatments*   For thermal processes, the surface heating intensity and absorptivity, and the thermal properties of the substrate, dictate the rise time in temperature. For a stationary heat source of given power density, the temperature rises with the square root of time $\sqrt{t}$. This in turn determines the depth of surface layer that reaches the temperature at which the desired microstructural change takes place. Since diffusion of matter and heat follows the same mathematical equations, the depth of treatment in diffusion processes such as carburising has the same dependence on time. For a spray-coating process delivering a given volumetric flow rate, the coating thickness increases linearly with time.

## 18.7 Estimating cost for shaping processes

Estimating process costs accurately—at the precision needed for competitive contract bidding, for example—is a specialised job. It is commonly based on interpolation or extrapolation of known costs for similar, previous jobs, appropriately scaled. Our interest here is not in estimating cost for this purpose; it is to compare approximate costs of alternative process routes. This can be useful even when imprecise. We illustrate the method with a cost model for one of the most common choices: that of a batch shaping process. It requires certain *user-specified inputs*, such as local labor costs, as well as values for *cost-related attributes* of each process.

*The cost model*   The manufacture of a component consumes resources (Figure 18.15), each of which has an associated cost. The final cost is the sum of those of the resources it consumes. They are defined in Table 18.4. Thus, the cost of producing a component of mass $m$ entails the cost $C_m$ ($/kg) of the materials and consumable feed-stocks from which it is made. It involves the cost of dedicated *tooling*, $C_t$ ($), and that of the capital *equipment*, $C_c$ ($), in which the tooling will be used. It requires *time*, chargeable at an overhead rate $\dot{C}_{oh}$ (thus with units of $/h), in which we include the cost of labor, administration and general plant costs. It requires energy, which is sometimes charged against a process step if it is very energy intensive, but more usually is treated as part of the overhead and lumped into $\dot{C}_{oh}$, as we shall do here. Finally, there is the cost of information, meaning that of research and development, royalty or license fees; this, too, we view as a cost per unit time and lump it into the overhead.

Consider now the manufacture of a component (the 'unit of output') weighing $m$ kg, and made of a material costing $C_m$ $/kg. The first contribution to the unit cost is that of the material $mC_m$. Processes rarely use exactly the right amount of material but generate a scrap fraction $f$, which ends up in runners, risers, machining swarf and rejects. Some is recycled, but the

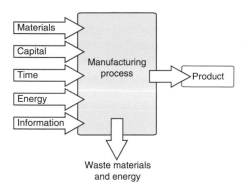

Figure 18.15 The inputs to a manufacturing process.

Table 18.4 Symbols, definitions and units in the cost model

| Resource | | Symbol | Unit |
|---|---|---|---|
| *Materials:* | Including consumables | $C_m$ | $/kg |
| *Capital:* | Cost of tooling | $C_t$ | $ |
| | Cost of equipment | $C_c$ | $ |
| *Time:* | Overhead rate, including labor, administration, rent, etc. | $\dot{C}_{oh}$ | $/h |
| *Energy:* | Cost of energy | $C_e$ | $/h |
| *Information:* | R & D or royalty payments | $C_i$ | $/year |

material cost per unit needs to be magnified by the factor $1/(1 - f)$ to account for the fraction that is lost. Hence the material contribution per unit is:

$$C_1 = \frac{mC_m}{(1 - f)} \tag{18.2}$$

The cost $C_t$ of a set of tooling—dies, moulds, fixtures and jigs—is what is called a *dedicated cost*: one that must be wholly assigned to the production run of this single component. It is written off against the numerical size $n$ of the production run. However, tooling wears out at a rate that depends on the number of items processed. We define the tool life $n_t$ as the number of units that a set of tooling can make before it has to be replaced. We have to use one tooling set even if the production run is only one unit. But each time the tooling is replaced there is a step up in the total cost to be spread over the whole batch. To capture this with a smooth function, we multiply the cost of a tooling set $C_t$ by $(1 + n/n_t)$. Thus, the tooling cost per unit takes the form:

$$C_2 = \frac{C_t}{n}\left(1 + \frac{n}{n_t}\right) \tag{18.3}$$

The *capital cost of equipment*, $C_c$, by contrast, is rarely dedicated. A given piece of equipment—a powder press, for example—can be used to make many different components by installing different die-sets or tooling. It is usual to convert the capital cost of *nondedicated* equipment and the cost of borrowing the capital itself into an overhead by dividing it by a *capital write-off time*, $t_{wo}$, (5 years, say) over which it is to be recovered. The quantity $C_c/t_{wo}$ is then a cost per hour—provided the equipment is used continuously. That is rarely the case, so the term is modified by dividing it by a *load factor*, $L$—the fraction of time for which the equipment is productive. This gives an effective hourly cost of the equipment, like a rental charge, even though it is your own piece of kit. The capital contribution to the cost per unit is then this hourly cost divided by the production rate/hour $\dot{n}$ at which units are produced:

$$C_3 = \frac{1}{\dot{n}} \left( \frac{C_c}{L t_{wo}} \right) \tag{18.4}$$

Finally, there is the general background hourly *overhead rate* $\dot{C}_{oh}$ for labor, energy and so on. This is again converted to a cost per unit by dividing by the production rate $\dot{n}$ units per hour:

$$C_4 = \frac{\dot{C}_{oh}}{\dot{n}} \tag{18.5}$$

The total shaping cost per part, $C_s$, is the sum of these four terms, $C_1$–$C_4$, taking the form:

$$C_s = \frac{m C_m}{(1-f)} + \frac{C_t}{n} \left( 1 + \frac{n}{n_t} \right) + \frac{1}{\dot{n}} \left( \frac{C_c}{L t_{wo}} + \dot{C}_{oh} \right) \tag{18.6}$$

To emphasise the simple form of this equation, it can be written:

$$C_s = C_{material} + \frac{C_{dedicated}}{n} + \frac{\dot{C}_{capital} + \dot{C}_{overhead}}{\dot{n}} \tag{18.7}$$

This equation shows that the cost has three essential contributions—a material cost per unit of production that is independent of batch size and rate, a dedicated cost per unit of production that varies as the reciprocal of the production volume ($1/n$), and a gross overhead per unit of production that varies as the reciprocal of the production rate ($1/\dot{n}$). The dedicated cost, the effective hourly rate of capital write-off and the production rate can all be defined by a representative range for each process; target batch size $n$, the overhead rate $\dot{C}_{oh}$, the load factor $L$ and the capital write-off time $t_{wo}$ must be defined by the user.

Figure 18.16 uses equation (18.6). It is a plot of cost, $C_s$, against batch size, $n$, comparing the cost of casting a small aluminum component by three alternative processes: sand casting, die casting and low-pressure casting. At small batch sizes the unit cost is dominated by the 'fixed' costs of tooling (the second term on the right of equation (18.6)). As the batch size $n$ increases, the contribution of this to the unit cost falls (provided, of course, that the tooling has a life that is greater than $n$) until it flattens out at a value that is dominated by the 'variable' costs of material, labor and other overheads. Competing processes differ in tooling cost $C_t$, equipment cost $C_c$ and production rate $\dot{n}$. Sand-casting equipment is cheap but slow. Die-casting equipment costs much more but is also much faster. Mould costs for low-pressure die casting are

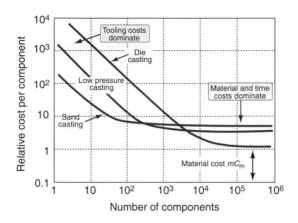

**Figure 18.16** The cost of casting a small aluminum component by three competing processes.

greater than for sand casting; those for high pressure die casting are higher still. The combination of all these factors for each process causes the $C_s-n$ curves to cross, as shown in Figure 18.16.

The crossover means that the process that is cheapest depends on the batch size. This suggests the idea of an *economic batch size*—a range of batches for which each process is likely to be the most competitive. Equation (18.6) allows the cost of competing processes to be compared if data for the parameters of the model are known. If they are not, the simpler economic batch size provides an alternative way of ranking. Figure 18.17 is a bar chart of this attribute, for the same processes and using the same colour coding as the earlier charts. Processes such as investment casting of metals and lay-up methods for composites have low tooling costs but are slow; they are economic when you want to make a small number of components but not when you want a large one. The reverse is true of the die casting of metals and the injection moulding of polymers: they are fast, but the tooling is expensive.

## 18.8 Computer-aided process selection

*Computer-aided selection* As with the material charts, the process charts of this chapter give an overview, but the number of processes that can be shown on any one of them is limited. Selection using them is practical when there are few constraints, but when there are many—as there usually are—checking that a given process meets them all is cumbersome, and the cost model cannot be presented in a useful way because of the need for user-defined parameters. All these problems are overcome in the CES implementation of the method.

Its database contains records for processes, organised into the families, classes and members shown in Figures 2.5–2.7 of Chapter 2. Each record contains attribute data for a process, with each attribute stored as a range spanning its usual values, and is linked to the records for the materials it can process. It also contains limited documentation in the form of text, images and references to sources of information about the process. The data are interrogated by the same

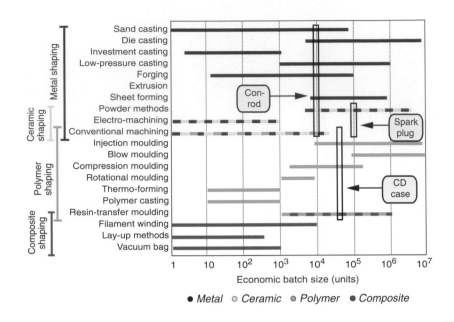

**Figure 18.17** The process–economic batch size bar chart for shaping processes colour coded by material.

search engine that is used for material selection, allowing superimposed stages like those suggested by Figure 18.18. On the left is a stage that limits the process choice to those able to handle a chosen material family, class or member (here, thermoplastics). In the centre is a simple query interface for screening on one or more process attributes; the desired upper or lower limits for constrained attributes are entered and the search engine rejects all processes with attribute values that do not lie within the limits. On the right is a bar chart, constructed by the software, for any numeric attribute (here, economic batch size). For screening, a selection box is superimposed on the bar chart with upper and lower edges that lie at the constrained values of the attribute. This retains the processes with values that penetrate the box, rejecting those that lie outside it.

The software includes the batch-process cost model described in the text. A dialog box allows the user to edit default values of the user-defined parameters $L$, $t_{wo}$, $\dot{C}_{oh}$ and so forth. The software then retrieves approximate values for the economic process attributes $C_c$, $C_t$, $\dot{n}$ from the database where they are stored as ranges. It allows the data to be presented in a number of ways, two of which are shown in Figures 18.19 and 18.20. The first is a plot of cost against batch size for a single process (here, injection moulding), in the manner of the earlier Figure 18.16. The user-defined parameters are listed on it. The bandwidth derives from the ranges of the economic attributes: a simple shape, requiring only simple dies, lies near the lower edge; a more complex one, requiring multi-part dies, lies near the upper edge. Figure 18.20 shows an alternative presentation. Here the range of cost for making a chosen batch size (here, 10,000) of a component by a number of alternative processes is

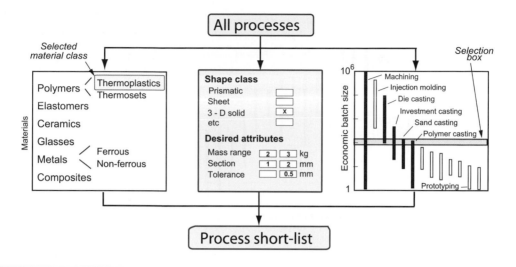

**Figure 18.18** Schematic of the screening steps in selecting process (here for shaping). First specify classes such as material and shape; then screen on numerical attributes, plotted graphically as bar charts (as on the right). In the CES software, target design requirements can also be entered through dialogue boxes, shown in the centre.

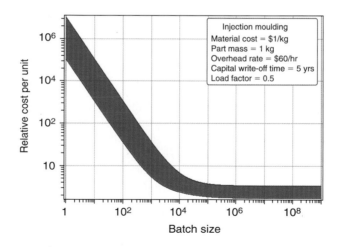

**Figure 18.19** The output of the cost model as implemented in the CES software, showing how unit cost changes with batch size. The range reflects the ranges in the input parameters to the model.

plotted as a bar chart. The user-defined parameters are again listed. Other selection stages can be applied in parallel with this one (as in Figure 18.18) applying constraints on material, shape and so on, causing some of the bars to drop out. The effect is to rank the surviving processes by cost.

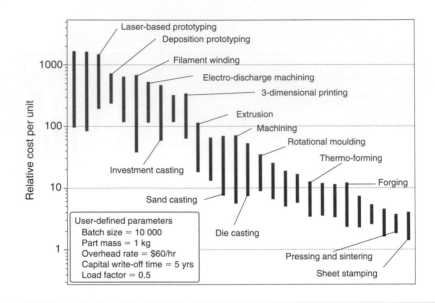

**Figure 18.20**    A bar chart made with the CES software showing unit cost of a set of processes for a specified component at a given batch size. The size of each bar reflects the range in the input parameters of the model.

## 18.9 Case studies

Here we have examples of the selection methodology in use. In each case, translation of the design requirements leads to a target list of process attributes for screening. Three shaping problems are approached using the charts of this chapter, with the cost model applied to the last of the three. These are followed by examples of selecting joining and surface treatment processes. For these we use the CES software; the charts show only the broad classes of joining and finishing processes, whereas the level of detail in the software is much greater. In each case, aspects of processing beyond the scope of simple screening are noted. These relate to defects, or material properties, or economic limitations—all of which can be addressed by seeking documentation.

*Shaping a ceramic spark plug insulator*    The anatomy of a spark plug is shown schematically in Figure 18.21. It is an assembly of components, one of which is the insulator. This is to be made of a ceramic, alumina, in an axisymmetric, hollow 3-D shape. The insulator is part of an assembly, fixing its dimensions. Given the material and dimensions, the expected mass can be estimated and the minimum section thickness identified. The insulator must seal within its casing, setting limits on precision and surface finish. Spark plugs are made in large numbers—the projected batch size is 100,000. Table 18.5 lists the constraints. The lower part of the table flags up key process-sensitive outcomes that cannot be checked by screening, but should be investigated by a documentation search for the processes that successfully pass the screening steps.

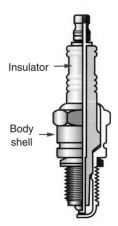

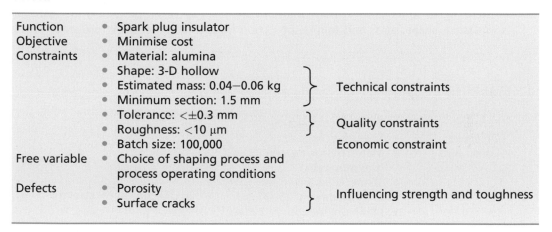

**Figure 18.21** A section through a spark plug. The insulator is cylindrical but not prismatic, making it 3-D hollow in the shape classification.

**Table 18.5** Translation for shaping a spark plug insulator and potential process-sensitive defects

| | | |
|---|---|---|
| Function | • Spark plug insulator | |
| Objective | • Minimise cost | |
| Constraints | • Material: alumina | |
| | • Shape: 3-D hollow | |
| | • Estimated mass: 0.04–0.06 kg | Technical constraints |
| | • Minimum section: 1.5 mm | |
| | • Tolerance: <±0.3 mm | Quality constraints |
| | • Roughness: <10 μm | |
| | • Batch size: 100,000 | Economic constraint |
| Free variable | • Choice of shaping process and process operating conditions | |
| Defects | • Porosity | Influencing strength and toughness |
| | • Surface cracks | |

The constraints are plotted on the compatibility matrices and bar charts of Figures 18.2–18.4, 18.11 and 18.12. The material compatibility chart shows only three process groups for ceramics, and all three can handle the required shape. All three can cope with the mass and the section thickness. All three can also meet the quality constraints on roughness and tolerance, although powder methods are near their lower limit. The constraint on economic batch size (Figure 18.17) identifies powder methods as the preferred choice. Applying

**Table 18.6**    Short-list of processes for shaping a spark plug insulator

| |
| --- |
| Powder pressing and sintering |
| Powder injection moulding |

the same steps using the CES software identifies two specific variants of powder methods that meet the constraints, listed in Table 18.6. Powder injection moulding is indeed the process used to make spark plug insulators.

A documentation search details good practice in die design, powder quality, die filling and the cycle of pressure and temperature important for controlling porosity, finish and production rate.

*Shaping a thermoplastic CD case*    The unsatisfactory qualities of the 'jewel' cases of CDs were noted in Chapter 3 and Figure 3.7. The current material choice is polystyrene, but before considering a change of material we might wish to check our processing options for thermoplastics in this geometry. Table 18.7 shows the translation for the CD case. The dimensions and mass are set by the standardised size of CDs. The shape is specified as 'solid, 3-D' as it is not a flat panel but has features at the edges to make the hinges and tabs to hold the sleeve in place. These features also require a good tolerance. Low surface roughness is needed to preserve transparency. We assume a target batch size of 50,000.

Refer again to the compatibility matrices and bar charts. Screening on the material (thermoplastic), shape, mass and minimum section leads to three possible processes: machining,

**Table 18.7**    Translation for shaping a CD case and potential process-sensitive defects

| | | |
| --- | --- | --- |
| Function | • Case for compact disc | |
| Objective | • Minimise cost | |
| Constraints | • Material: polystyrene (thermoplastic) | |
| | • Shape: solid 3-D | ⎫ |
| | • Estimated mass: 25–35 g | ⎬ Technical constraints |
| | • Minimum section: 1–2 mm | ⎭ |
| | • Tolerance: 0.2 mm | ⎫ |
| | • Roughness: 1 μm | ⎬ Quality constraints |
| | • Batch size: 50 000 | Economic constraint |
| Free variable | • Choice of shaping process and process operating conditions | |
| Defects | • Cracks | |
| | • Trapped bubbles | ⎫ |
| | • Sink marks | ⎬ Influencing strength and visual quality |
| | • Poor surface finish | ⎭ |

Table 18.8   Short-list of processes for shaping a polymer CD case

| |
|---|
| Injection moulding |
| Compression moulding |

injection moulding and compression moulding. Compression moulding is just beyond its normal mass range. All three processes meet the targets for tolerance and roughness. The final bar chart shows that machining is not economic for the given batch size, but the other two are acceptable. Table 18.8 summarises the surviving processes.

Good die design to include the protruding ridges and tabs on a CD case requires expert input. These re-entrant features exclude compression moulding using uniaxial loading and open dies. Following up the avoidance of defects, a documentation search is principally a matter of good die design with correct location of the runners and risers.

*Shaping a steel connecting rod*   Figure 18.22 shows a schematic connecting rod ('con-rod') to be made of a medium carbon steel. The minimum section is around 8 mm, and an estimate of the volume gives a mass around 0.35 kg. Dimensional precision is important to ensure clearances at both the little and the big end. The con-rod carries cyclic loads with the conse-quent risk of fatigue crack initiation, so a low surface roughness overall is necessary (the bores will be finished by a subsequent machining operation). A modest batch size of 10,000 is required. Control of properties and of defects is dominated by the need to avoid fatigue failure. Table 18.9 summarises.

These constraints are plotted on the matrices and charts. Screening on material and shape eliminates low-pressure casting, die casting, extrusion and sheet forming, but plenty of options survive. Sand casting is disqualified by the tolerance constraint. Investment casting, conven-tional machining, forging and powder methods remain. All achieve the target roughness, and

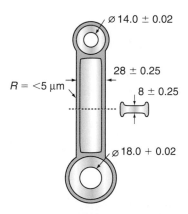

Figure 18.22   A steel connecting rod. The precision required for the bores and bore facing is much higher than for the rest of the body, requiring subsequent machining. Dimensions in mm (unless otherwise stated).

**Table 18.9** Translation for shaping a steel connecting rod and potential process-sensitive defects

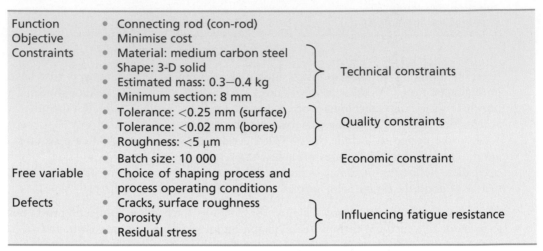

| | |
|---|---|
| Function | • Connecting rod (con-rod) |
| Objective | • Minimise cost |
| Constraints | • Material: medium carbon steel |
| | • Shape: 3-D solid |
| | • Estimated mass: 0.3–0.4 kg ⎬ Technical constraints |
| | • Minimum section: 8 mm |
| | • Tolerance: <0.25 mm (surface) |
| | • Tolerance: <0.02 mm (bores) ⎬ Quality constraints |
| | • Roughness: <5 μm |
| | • Batch size: 10 000  Economic constraint |
| Free variable | • Choice of shaping process and process operating conditions |
| Defects | • Cracks, surface roughness |
| | • Porosity ⎬ Influencing fatigue resistance |
| | • Residual stress |

surface tolerance, though not that required for the bores. However, by following the shaping process with a finishing process, the tolerance requirement can be reached. All the finishing processes at the bottom of Figures 18.11 and 18.12 give better tolerance and roughness than specified—precision machining is most appropriate for finishing the bores and the faces surrounding them. Finally, the economic batch size chart suggests that forging and powder methods are suitable—machining (from solid) and investment casting are not economic for a batch size of 10,000. Table 18.10 summarises.

This case study reveals that there is close competition for the manufacture of carbon steel components in the standard mid-range size and shape of a connecting rod. Surface finish and accuracy present some problems, but finishing by machining is not difficult for carbon steel. The final choice will be determined by detailed cost considerations and achieving the required fatigue strength. The properties will be sensitive to the composition of the steel and its heat treatment—something illustrated further in the next chapter.

Cost is considered further by applying the model of Section 18.7. Table 18.11 lists values for the parameters in the cost model for three processes: die casting, forging and powder methods. Assuming the same amount of material is used in each process, the costs are all normalised to the material cost. Figure 18.23 shows the computed relative cost per part against batch size for the three processes.

**Table 18.10** Short-list of processes for shaping a steel connecting rod

| |
|---|
| Die casting |
| Forging + machining |
| Powder methods + machining |

Table 18.11   Process cost model input data for three shaping methods to make the con-rod

| Parameters | Die casting | Forging | Powder methods |
|---|---|---|---|
| *Material, $mC_m/(1-f)$ | 1 | 1 | 1 |
| *Basic overhead, $\dot{C}_{oh}$ (per hour) | 100 | 100 | 100 |
| Capital write-off time, $t_{wo}$ (years) | 5 | 5 | 5 |
| Load factor | 0.5 | 0.5 | 0.5 |
| *Dedicated tooling cost, $C_t$ | 17 500 | 125 | 1000 |
| *Capital cost, $C_c$ | 50 000 | 110 000 | 1 000 000 |
| Production rate, $\dot{n}$ (per hour) | 50 | 50 | 2 |

* Costs normalised to $mC_m/(1-f)$.

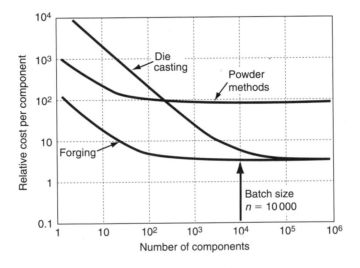

**Figure 18.23**   The cost model for three possible processes to manufacture the steel con-rod, using representative constant values for the model parameters. For the target batch size of 10,000, the cheapest option indicated is forging.

For the batch size of 10,000 forging comes out the cheapest, followed closely by die casting. Powder is more expensive, principally because the slower production rate and high capital cost give a large overhead contribution per part. Since there is a spread of estimated cost (due to uncertainty in the inputs), and forging is likely to be followed by machining, the cost difference between forging and die casting is not significant—both remain on the list for investigation of the achievable properties.

*Joining a steel radiator*   Figure 18.24 shows a section through a domestic radiator made from corrugated pressed sheet steel. The task is to choose a joining process for the seams between the sheets. As always, the process must be compatible with the material (here, low-carbon steel sheet of thickness 1.5 mm). The lap joints carry only low loads in service but

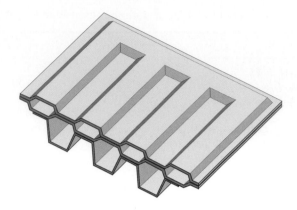

**Figure 18.24** A section through a domestic radiator. The three pressed steel sections are joined by lap joints.

handling during installation may impose tension and shear. They must conduct heat, be watertight and be able to tolerate temperatures up to 100 °C. There is no need for the joints to be disassembled for recycling at the end of life, since the whole thing is steel. Quality constraints are not severe, but distortion and residual stress will make fit-up of adjacent joints difficult. Low carbon steels are readily weldable, so it is unlikely that there is a risk of any welding process causing cracking or embrittlement problems—but this can be checked by seeking documentation. Table 18.12 summarises the translation.

To tackle this selection we use the Cambridge Engineering Selector. It stores records and attributes for 52 joining processes. Applying the material compatibility check first, 32

**Table 18.12** Translation for joining a steel radiator and outcome (potential process-sensitive defects and properties)

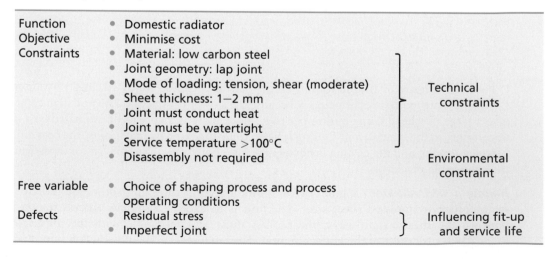

| Function | • Domestic radiator | |
|---|---|---|
| Objective | • Minimise cost | |
| Constraints | • Material: low carbon steel | |
| | • Joint geometry: lap joint | |
| | • Mode of loading: tension, shear (moderate) | Technical |
| | • Sheet thickness: 1–2 mm | constraints |
| | • Joint must conduct heat | |
| | • Joint must be watertight | |
| | • Service temperature >100°C | |
| | • Disassembly not required | Environmental constraint |
| Free variable | • Choice of shaping process and process operating conditions | |
| Defects | • Residual stress | Influencing fit-up |
| | • Imperfect joint | and service life |

Table 18.13   Short-list of processes for joining a steel radiator

Brazing
Electron beam welding
Explosive welding
Metal inert gas arc welding (MIG)
Tungsten inert gas arc welding (TIG)
Laser beam welding
Manual metal arc welding (MMA)
Oxyacetylene welding
Riveting
Soldering

processes are found to be suitable for joining low carbon steel—those eliminated are principally polymer-specific processes. Further screening on joint geometry, mode of loading and section thickness reduces the list to 20 processes. The requirement to conduct heat is then the most discriminating—only 10 processes now pass. Water resistance and operating temperature do not change the short-list further. The processes passing the screening stage are listed in Table 18.13.

Quality and economic criteria are difficult to apply as screening steps in selecting joining processes. At this point we seek documentation for the processes. This reveals that explosive welding requires special facilities and permits (hardly a surprise). Electron beam and laser welding require expensive equipment, so use of a shared facility would be necessary to make them economic. Resistance spot welding is screened out because it failed the requirement to be watertight. This is only necessary for the edge seams, so internal joint lines could be spot welded. This highlights the need for judgement in applying automated screening steps. Always ask: has an obvious (or existing) solution been eliminated? Why was this? Has something narrowly failed, so that a low-cost option could be accommodated with modest redesign? Could a second process step, like sealing the spot-welded joint with a mastic, correct a deficiency? The best way forward is to turn constraints on and off, and adjust numerical limits, exploring the options that appear.

*Surface hardening a ball bearing race*   The balls of ball bearing races run in grooved tracks (Figure 18.25). As discussed in Chapter 11, the life of a ball race is limited by wear and by fatigue. Both are limited by using hard materials. Hard materials, however, are not tough, incurring the risk that shock loading or mishandling might cause the race to fracture. The solution is to use an alloy steel, which has excellent bulk properties, and to apply a separate surface treatment to increase the hardness where it matters. We therefore seek processes to surface harden alloy steels for wear and fatigue resistance. The precision of both balls and race is critical, so the process must not compromise the dimensions or the surface smoothness. Table 18.14 summarises the translation.

The CES system contains records for 44 surface treatment processes. Many are compatible with alloy steels. More discriminating is the purpose of the treatment—to impart fatigue and wear resistance—reducing the list to eight. Imposing the requirement for a very smooth surface

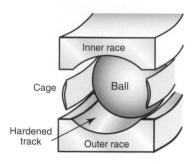

**Figure 18.25**   A section through a ball bearing race. The surface of the race is to be hardened to resist wear and fatigue crack initiation.

**Table 18.14**   Translation for surface hardening a ball bearing race and potential process-sensitive defects

| Function | • Ball bearing race | |
|---|---|---|
| Objective | • Minimise cost | |
| Constraints | • Material: alloy steel | |
| | • Function of treatment: hardening for fatigue and wear resistance | Technical constraints |
| | • Curved surface coverage: good | |
| | • Precision and surface finish not compromised | |
| Free variable | • Choice of surface treatment process and process operating conditions | |
| Defects | • Cracks | Influencing surface toughness and fatigue resistance |
| | • Residual stress | |

knocks out those that coat or deform the surface because these compromise the finish. The short-list of processes that survive the screening is given in Table 18.15. Adding the further constraint that the curved surface coverage must be *very good* leaves just the first two: carburising and carbonitriding, and nitriding.

To get further we turn to documentation for these processes. The hardness of the surface and the depth of the hardened layer depend on process variables: the time and temperature of the treatment and the composition of the steel. And economics, of course, enters. Ball races are made in enormous batches and although their sizes vary, their geometry does not. This is where dedicated equipment, even if very expensive, is viable.

Table 18.15   Short-list of processes for hardening an alloy
steel ball race

| |
| --- |
| Carburising and carbonitriding |
| Nitriding |
| Induction and flame hardening |
| Laser surface hardening and melting |
| Vapor metallising |

# 18.10 Summary and conclusions

Selecting a process follows a strategy much like that for materials: *translation* of requirements, *screening* of the options, *ranking* of the processes that survive screening, ending with a search of *documentation* for the most promising candidates. For processes, translation is straight-forward—the requirements convert directly into the attributes of the process to be used in the screening step. Process attributes capture their technical capabilities and aspects of product quality. All process families require screening for compatibility with material to which they are to be applied, but after this the attributes and constraints are specific to the family. For shaping, component geometry, precision and finish are most discriminating. For joining it is joint shape and mode of loading. For surface treatments it is the purpose for which the treatment is applied (wear, corrosion, aesthetics, etc.). Ranking is based on cost if this can be modeled, as it can for batch processes. Cost models combine characteristic process attributes such as tooling and capital costs and the production rate, with user-specified parameters such as overhead rates and required batch size. The economic batch size provides a crude alternative to the cost model—a preliminary indication of the batch size for which the process is commonly found to be economic.

Translation of the design requirements frequently points to important aspects of processing that fall beyond a simple screening process. The avoidance of defects, achieving the required modification of material properties, and determining the process speed (and thus economics), all involve coupling between the process operating conditions, the material composition and design detail. This highlights the importance of the documentation stage in making the final choice of processes, and the need to refine the material and process specifications in parallel. The next chapter explores aspects of process-property interactions in more depth.

# 18.11 Further reading

Ashby, M. F. (2005). *Materials Selection in Mechanical Design* (3rd ed.). Oxford, UK: Butterworth-Heinemann. Chapter 4. ISBN 0-7506-6168-2. (The source of the process selection methodology and cost model, with further case studies, giving a summary description of the key processes in each class.)

ASM Handbook Series (1971–2004). Volume 4, *Heat Treatment*; Volume 5, *Surface Engineering*; Volume 6, *Welding, Brazing and Soldering*; Volume 7, *Powder Metal Technologies*; Volume 14, *Forming and Forging*; Volume 15, *Casting*; and Volume 16, *Machining*; ASM International, Metals

Park, OH, USA. (A comprehensive set of handbooks on processing, occasionally updated, and now available online at <www.asminternational.org/hbk/index.jsp>.)

Bralla, J. G. (1998). *Design for Manufacturability Handbook* (2nd ed.). New York, USA: McGraw-Hill. ISBN 0-07-007139-X. (Turgid reading, but a rich mine of information about manufacturing processes.)

Campbell, J. (1991). *Casting*. Oxford, UK: Butterworth-Heinemann. ISBN 0-7506-1696-2. (The fundamental science and technology of casting processes.)

Houldcroft, P. (1990). *Which Process? Abington*. Cambridge, UK: Abington Publishing. ISBN 1-85573-008-1. (The title of this useful book is misleading—it deals only with a subset of joining process: the welding of steels. But here it is good, matching the process to the design requirements.)

Kalpakjian, S., & Schmid, S. R. (2003). *Manufacturing Processes for Engineering Materials* (4th ed.). New Jersey, USA: Prentice-Hall, Pearson Education. ISBN 0-13-040871-9. (A comprehensive and widely used text on material processing.)

Lascoe, O. D. (1988). *Handbook of Fabrication Processes*. Columbus, OH, USA: ASM International, Metals Park. ISBN 0-87170-302-5. (A reference source for fabrication processes.)

Swift, K. G., & Booker, J. D. (1997). *Process Selection, from Design to Manufacture*. London, UK: Arnold. ISBN 0-340-69249-9. (Details of 48 processes in a standard format, structured to guide process selection.)

## **18.12** Exercises

Exercise E18.1    (a) Explain briefly why databases for initial process selection need to be subdivided into generic process classes, whereas material selection can be conducted on a single database for all materials.

            (b) After initial screening of unsuitable processes, further refinement of the process selection usually requires detailed information about the materials being processed and features of the design. Explain why this is so, giving examples from the domains of shaping, joining and surface treatment.

Exercise E18.2    A manufacturing process is to be selected for an aluminum alloy piston. Its weight is between 0.8 and 1 kg, and its minimum thickness is 4—6 mm. The design specifies a precision of 0.5 mm and a surface finish in the range 2—5 μm. It is expected that the batch size will be around 1000 pistons. Use the process attribute charts earlier in the chapter to identify a subset of possible manufacturing routes, taking account of these requirements. Would the selection change if the batch size increased to 10,000?

Exercise E18.3    The choice of process for shaping a CD case was discussed in Section 18.5, but what about the CDs themselves? These are made of polycarbonate (another thermoplastic), but the principal difference from the case is precision. Reading a CD involves tracking the reflections of a laser from microscopic pits at the interface between two layers in the disc. These pits are typically 0.5 μm in size and spacing, and 1.5 μm apart; the tolerance must be better than 0.1 μm for the disc to work.

First estimate (or measure) the mass and thickness of a CD. Use the charts to find a short-list of processes that can meet these requirements, and will be compatible with the material class. Can these processes routinely achieve the tolerance and finish needed? See if you can find out how they are made in practice.

Exercise E18.4   A process is required for the production of 10,000 medium-carbon steel engine crankshafts. The crankshaft is a complex three-dimensional shape about 0.3 m in length with a minimum diameter of 0.06 m. At the position of the bearings, the surface must be very hard, with a surface finish better than 2 μm and a dimensional accuracy of $\pm$ 0.1mm. Elsewhere the surface finish can be 100 μm, and the tolerance is 0.5 mm. Use the process attribute charts and other information in the chapter to identify a suitable fabrication route for the crankshaft.

Exercise E18.5   A heat-treatable aluminium alloy automotive steering arm has a mass of 200 g, a minimum thickness of 6 mm and a required tolerance of 0.25 mm. Production runs over 50,000 are expected. Discuss the implications for the manufacture of the component by forging, using the process attribute charts. Suggest an alternative processing route, commenting on how this might influence the choice of Al alloy and final properties.

Exercise E18.6   A small polyethylene bucket is to be manufactured by injection moulding or rotational moulding. The designer wishes to estimate the manufacturing cost for various batch sizes, using the cost equation

$$\text{Cost per part } C = C_{\text{material}} + \frac{C_{\text{dedicated}}}{n} + \frac{\dot{C}_{\text{capital}} + \dot{C}_{\text{overhead}}}{\dot{n}}$$

Define the meaning of all of the parameters in the equation. Using the data in the table, determine the cheapest process for batch sizes of 1000 and 5000.

| Process | $C_{\text{material}}$ (£) | $C_{\text{dedicated}}$ (£) | $\dot{C}_{\text{capital}} + \dot{C}_{\text{overhead}}$ (£/hr) | $\dot{n}$ (parts/hr) |
|---|---|---|---|---|
| Injection moulding | 0.25 | 5000 | 20 | 120 |
| Rotational moulding | 0.20 | 1000 | 20 | 40 |

Exercise E18.7   In a cost analysis for casting a small aluminium alloy component, costs were assigned to tooling and overheads (including capital) in the way shown in the following table. The costs are in units of the material cost of one component. Use the simple cost model presented in equation (18.7) to identify the cheapest process for a batch size of (i) 100 units and (ii) $10^6$ units.

| Process | Sand casting | Investment casting | Pressure die | Gravity die |
|---|---|---|---|---|
| Material, $C_{material}$ | 1 | 1 | 1 | 1 |
| Capital + Overhead, $\dot{C}_{capital} + \dot{C}_{overhead}$ (hr$^{-1}$) | 500 | 500 | 500 | 500 |
| Dedicated, $C_{dedicated}$ | 50 | 11,500 | 25,000 | 7,500 |
| Rate $\dot{n}$ (hr$^{-1}$) | 20 | 10 | 100 | 40 |

Exercise E18.8    Look at the products around your house (or your garage or garden shed), identifying as many different types of *joints* as you can, and the process used. Select a few examples and draw up a list of design requirements that the joining process needed to satisfy.

Exercise E18.9    Look at the products around your house (or your garage or garden shed), identifying as many different types of *surface treatments* as you can. Select a few examples and draw up a list of design requirements that the surface treatment process needed to satisfy, in particular the function of the surface treatment.

## 18.13 Exploring design with CES

Exercise E18.10    Use the 'Browse' facility in CES to find:
  (a) The record for the shaping process injection moulding, thermoplastics. What is its economic batch size? What does this term mean?
  (b) The machining process water-jet cutting (records for machining processes are contained in the Shaping data table). What are its typical uses?
  (c) The joining process friction-stir welding. Can it be used to join dissimilar materials?
  (d) The surface treatment process laser hardening. What are the three variants of this process?

Exercise E18.11    Use the 'Search' facility in CES to find:
  (a) Processes used for boat building.
  (b) Processes to make bottles.
  (c) Processes to make tail-light assemblies.
  (d) Processes for decoration.

Exercise E18.12    Use CES Level 3 to explore the selection of casting process for the products listed. First check compatibility with material and shape, and make reasonable estimates for the product dimensions (to assess mass and section thickness). Then include appropriate values for tolerance, roughness, and economic batch size.

(a) Large cast iron water pipes.

(b) 10,000 Zn alloy toy cars (60 mm long).

(c) Small Ni—Co super-alloy (MAR-M432) gas turbine blades (best possible tolerance and finish).

(d) Large brass ship propeller.

Exercise E18.13 A small nylon fan is to be manufactured for a vacuum cleaner. The design requirements are summarised in the following table. Use CES Level 3 to identify the possible processes, making allowance if necessary for including a secondary finishing process. Suggest some aspects of the design that may merit investigation of supporting information on the selected processes.

| Function | Vacuum cleaner fan | |
|---|---|---|
| Objective | Minimise cost | |
| Constraints | Material: nylon | |
| | Shape: 3-D solid | |
| | Mass: 0.1—0.2 kg | Technical constraints |
| | Minimum section: 4 mm | |
| | Tolerance: <0.5 mm | Quality constraints |
| | Roughness: <1 μm | |
| | Batch size: 10,000 | Economic constraint |
| Free variable | Choice of shaping process | |

Exercise E18.14 The selection of process for a connecting rod was discussed earlier in the chapter. Conduct the selection using CES Level 3, making reasonable estimates for any unspecified requirements. Explore the effect of changing to a 3 m con-rod for a ship, of approximate cross-section $10 \times 10$ cm, in a batch size of 10.

Exercise E18.15 Process selection for an aluminum piston was investigated in Exercise E18.2. Further investigation of the economics of gravity die casting and ceramic mould casting is suggested. Plot the cost against batch size for these processes, assuming a material cost of $2/kg and a piston mass of 1 kg. The overhead rate is $70/hour, the capital write-off time is 5 years and the load factor is 0.5. Which process is cheaper for a batch size of 1000? Assume that as the piston is simple in shape, it will fall near the bottom of each cost band.

Exercise E18.16 Two examples of selection of secondary processes were discussed in this chapter. The design requirements were summarised for joining processes for a radiator in Table 18.12 and for hardening a steel bearing race in Table 18.14. Use CES Level 3 to check the results obtained. For the radiator problem, use the 'Pass—Fail table' feature in CES to see if other processes could become options if the design requirements were modified.

## 18.14 Exploring the science with CES Elements

Exercise E18.17  Casting processes require that the metal be melted. Vapour methods like vapour metallising require that the metal be vapourised. Casting requires energy: the latent heat of melting is an absolute lower limit (in fact it requires more than four times this). Vapourisation requires the latent heat of vapourisation, again as an absolute lower limit. Values for both are contained in the Elements database. Make a plot of one against the other. Using these lower limits find, approximately, how much more energy-intensive vapour methods are compared with those that simply melt.

# Chapter 19

# Follow the recipe: processing and properties

The Khafji rig disaster. (Image courtesy of Thomas Brinsko with Bic Alliance Magazine)

## **19.1** Introduction and synopsis

Some people are better at multi-tasking than others. The good ones should feel at home in manufacturing. Shaping, finishing and assembling components into products to meet the technical, economic and aesthetic expectations of consumers today requires the balancing of many priorities. Earlier chapters have introduced intrinsic properties—strength, resistivity and so on—that depend intimately on microstructure, and microstructure depends on processing. Microstructure is not accessible to observation during processing, so controlling it requires the ability to predict how a given process step will cause it to evolve or change. So manufacturers have their work cut out to turn materials reliably into good-quality products, while making themselves a decent profit. Manufacturing involves more than making materials into the right shapes and sticking them together; it is also responsible for producing properties on target.

There is a parallel with cooking. The recipe lists the ingredients and cooking instructions: how to mix, beat, heat, finish and present the dish. A good cook draws on experience (and creativity) to produce new dishes with pleasing flavour, consistency and appearance; you might think of these as the attributes of the dish. But if the dish is not a success the first thing the cook might do is to cut right through it and examine what went wrong with its *microstructure*. Suppose, for instance, the dish is a fruit cake, then the distribution of porosity and fruit constitutes aspects of its microstructure.

Microstructure is key to engineering properties too. Some of its components are very small—precipitate particles can consist of clusters of a few atoms. Others are larger—grains range in size from microns to millimeters. The goal of this chapter is to explore how microstructures evolve during processing, with the focus on *processing for properties*. The ability to tune microstructure and properties is central to materials processing and design, so it brings with it the need for good process understanding and control. Failure to 'follow the recipe' leads, at best, to scrap and lost revenue and, at worst, to engineering failures. The essence of the chapter is therefore captured by the statement:

$$\text{Composition} + \text{Processing} \; \rightarrow \; \text{Microstructure} + \text{Properties}$$

The chapter opens with an illustrative example—making an aluminum bicycle frame—to highlight some general features of processing for properties. Microstructural features were introduced in many previous chapters, in 'drilling down' into the underlying science, so these are drawn together in an overview across the length scales, for reference. In materials processing, the evolution of the microstructure is governed primarily by the thermal (and sometimes deformation) history. When drilling down in this context, particularly for metals, the central topics are *phase diagrams* and the thermodynamics and kinetics of *phase transformations*.

The main text of the chapter provides a concise overview of these topics. A detailed coverage of how to read and interpret phase diagrams, and the microstructural evolution in common processes, is provided in Guided Learning Unit 2, *Phase Diagrams and Phase Transformations*, for those seeking more depth. A wide range of manufacturing processes for metals and non-metals are then considered. In each case process schematics explain how the process works, highlighting the key role of thermal history and cooling rate in determining the microstructure evolution. As in earlier chapters, property charts are used to illustrate the property changes imparted by composition and processing, with examples from each of the material classes.

## 19.2 Processing for properties

We noted earlier that the mantra for this chapter is: *Composition + Processing → Microstructure + Properties*. So here is an example to highlight some general issues about this interaction.

*Aluminum bike frames*    The material chosen for our illustrative bike frame example is a heat-treatable aluminum alloy, for its good stiffness and fatigue resistance at low weight (Chapters 5 and 10). Figure 19.1 illustrates the main steps in the process history. It makes a number of points.

Point 1: Materials processing involves more than one step. This bike frame is an example of a *wrought product* (i.e. one that undergoes some deformation processing) involving a chain of processes. In contrast, shaped metal casting, powder processes and polymer moulding are *near-net-shape processes*—the raw material is turned into the shape of the component in a single step, leaving only finishing operations (including heat treatment), so the number of steps is lower. Wrought alloys start life as a casting, but this ingot casting is most economic on a large scale, as in Figure 19.1(a). Standard compositions are cast in large plant, and transported to different factories for processing into different products. For the bike frame, the ingot is sliced into billets and extruded into tube, as in Figure 19.1(b). It is shaped directly from a solid circular billet, around 200 mm in diameter, to a hollow tube, 30 mm in diameter and 3 mm thick—a huge shape change in one step. The tube is then cut to length before the final stages of age hardening for optimum strength and assembly by arc welding (Figure 19.1(c) and (d)). In passing, we note that the composition of the chosen alloy must be adjusted at the outset, via processing in the liquid state. It cannot be changed after casting (other than at the surface). Metallic elements dissolve freely in one another in the liquid state, but it is very difficult to add them in the solid state; diffusion is much too slow.

Point 2: Each process step has a characteristic *thermal history*. Each temperature cycle is determined by the way the process step works, and what the target outcome is. Casting by definition involves melting, solidification and cooling. Extrusion is done hot to reduce the strength and to increase the ductility. This enables the material to undergo the large plastic strain, allowing fast throughput at relatively low extrusion force. The quality of surface finish is also determined at this stage. At the same time, extrusion enhances the microstructure. Hot deformation refines the grain structure in the as-cast billet, and helps to homogenise the alloy. Then as the tube emerges from the extrusion press, forced cooling quenches it to trap solute elements in solid solution, thus avoiding the need for a subsequent solution heat treatment (the first step of age hardening a heat-treatable aluminum alloy). In a non-heat-treatable aluminum alloy, strength comes from solid solution and work hardening, not precipitation, so the extrusion is often done cold to maximise the

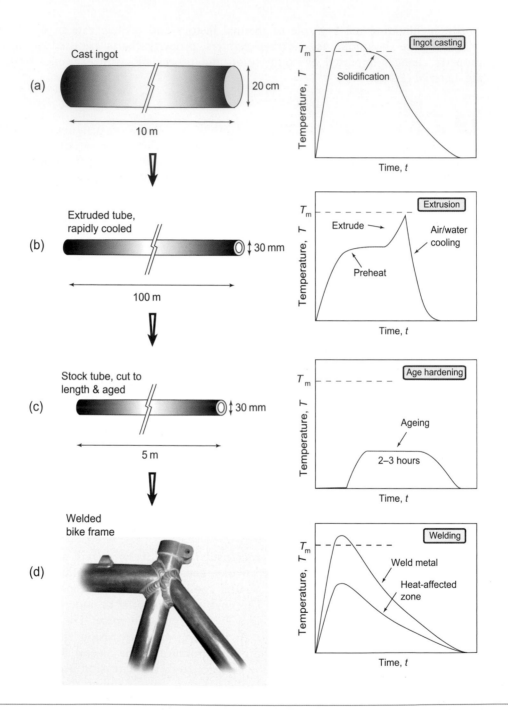

**Figure 19.1** A schematic process history for the manufacture of a bike frame in heat-treatable aluminum alloy: (a) casting of large ingot, cut into billets for extrusion; (b) hot extrusion into tube, incorporating the solution heat treatment and quench; (c) sections cut to length and age hardened; (d) assembly into the frame by arc welding.

work hardening, without a further heat treatment. Note therefore the interplay between deformation and thermal histories, and how this is critical for controlling the shape and final properties simultaneously, while the detail depends on the particular alloy variant being processed.

Point 3: Designers should watch out for unintended side-effects in the joining stage. Bike frame assembly here is by arc welding, with its own temperature cycle back to the melting point. In the 'heat-affected zone' around the welds the alloy doesn't melt, but is hot enough for the microstructure to evolve—the precipitates carefully produced by prior age hardening can disappear back into solution. As the whole age-hardening cycle is difficult to repeat on the completed frame, the heat-affected zone ends up with different, inferior properties to the rest of the frame (discussed further in Section 19.5).

Point 4: Design focuses on the properties of finished products, but some of these properties (strength, ductility, thermal conductivity, etc.) are also critical *during processing*. And there are other properties that are only really relevant for processing—for example, the viscosity of the melt in metal casting or polymer moulding. Design and processing characteristics can often be in conflict—strong alloys are more difficult to process and are thus more expensive. So in choosing materials for a component it is important to examine their suitability for processing as well as for performance in service. Clever processing tricks are exploited to resolve the conflicts—the bike frame is hot-formed (when it is soft) and then heat-treated (after shaping) to produce the high strength for service.

This introductory example shows how combinations of composition and processing are used to manipulate microstructure and properties. Further examples are given later for all material classes. The goal is to understand the interactions between processing and the target 'ideal' properties, as stored in a property database. But the bicycle frame case study shows that we need to keep our eyes open in this game, and to be aware that we often have to compromise between design objectives and manufacturing realities, and that things can go badly wrong. The importance of the 'Seek Documentation' stage in material and process selection should not be underestimated!

## **19.3** Microstructure of materials

Before exploring what goes on inside materials in a range of processes, it is helpful to see the 'big picture' in relation to microstructure. Properties reflect microstructure from atoms to grains to cracks, spanning length scales over many orders of magnitude. Here we draw together all of the microstructural features introduced earlier in the book, by way of an overview.

*Metals* Figure 19.2 summarises the main microstructural features in metals. Starting at the bottom with atoms, we have crystalline packing (with the exception of the unusual amorphous metals)—responsible for elastic moduli and density. Atom-scale defects—vacancies and solute atoms—were introduced in Chapter 6. Thermal, electrical, optical and magnetic behaviour are most directly influenced by this atomic scale of microstructure. Vacancies are responsible for diffusion so this, and the service and processing phenomena it causes (creep, sintering, heat treatments), also depend on atomic-scale structure. Strength, toughness and fatigue depend on

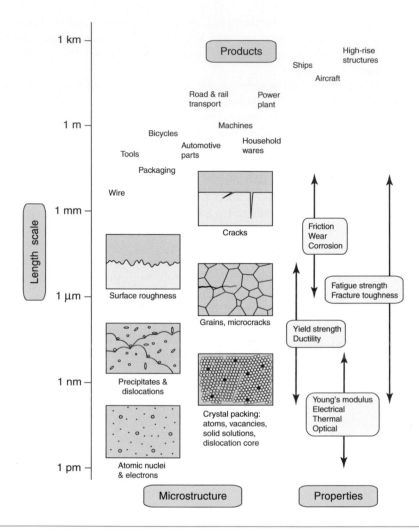

**Figure 19.2** Microstructural features in metals, showing their length scale and the properties that they determine. Each interval on the length scale is a factor of 1000.

structure at a slightly larger scale, that of dislocations and the obstacles to their motion sketched in Figure 19.2: precipitates and grain boundaries. Grains themselves usually fall into the 1–50 μm scale—a similar scale to the roughness on metal surfaces and of the porosity caused by the gases trapped in a metal when it is cast. Fatigue cracks start at grain scale, but grow to the dimensions of the component itself at fracture. Metal products themselves span a huge range: kitchen foil is around 10 μm thick, automotive panels a millimetre or so, ship propellers are several metres in diameter, and bridges and buildings reach the kilometre scale.

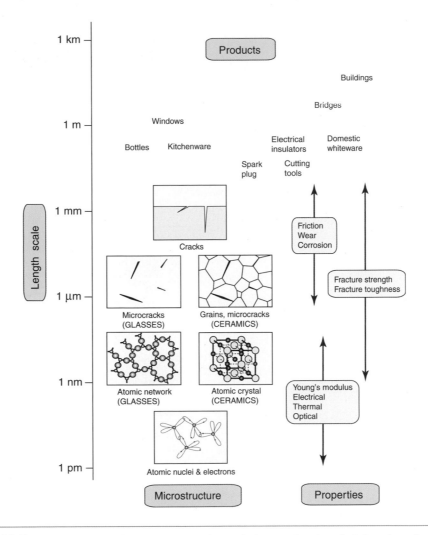

**Figure 19.3** Microstructural features in ceramics and glasses, showing their length scale and the properties that they determine. Each interval on the length scale is a factor of 1000.

***Ceramics and glasses*** Ceramics are crystalline, glasses are amorphous. Figure 19.3 shows both near the bottom of our materials length scale. Most of the properties directly reflect the atomic layout and the intrinsically strong nature of the covalent or ionic bonding—from elastic modulus to electrical insulation. Atomic scales also govern viscosity in molten glass and diffusion in compacted ceramic powders. The exceptions are strength and toughness. Cracks dominate failure in ceramics and glasses, so these are the key features—closely related to grain size (in ceramics) and to the surface finish.

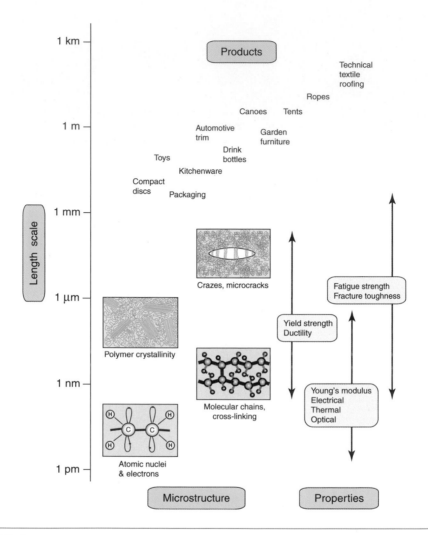

**Figure 19.4** Microstructural features in polymers and elastomers, showing their length scale and the properties that they determine. Each interval on the length scale is a factor of 1000.

***Polymers and elastomers*** Polymers and elastomers are inherently molecular rather than atomic. Figure 19.4 shows that there is wide diversity at this fine scale: polymer molecules can be amorphous, crystalline, cross-linked or aligned by drawing. Most properties again reflect behaviour at this scale, with strength and toughness (and optical properties) bringing in larger features such as crazes and cracks. Polymer foams and fibre-reinforced polymer composites have additional length scales relating to their architecture—for example, pore size and fibre diameter. These and other 'hybrid' materials are discussed further in Section 19.7.

# **19.4** Microstructure evolution in processing

The role of shaping processes is to produce the right shape with the right final properties. Achieving the first requires control of viscous flow or plasticity. Achieving the second means control of the nature and rate of microstructural evolution. Secondary processing (joining and surface treatment) may cause the microstructure to evolve further. Examples of microstructural evolution are discussed in Sections 19.5 and 19.6, but first some general principles on why and how it happens. The emphasis is on metals, reflecting their great diversity in both alloy chemistry and process variants.

*Phase diagrams, phase transformations and other structural change* A *phase* is defined as a region of a material with a specified atomic arrangement. This means that a phase has characteristic physical properties, such as density. Steam, water and ice are the vapour, liquid and solid phases of $H_2O$. Engineering alloys such as steel also melt and then vapourise if we heat them enough. More interestingly, they can contain different phases simultaneously in the solid state, depending on the composition and temperature. It is this that gives us precipitation-hardened steels and non-ferrous alloys, made up of one phase, the precipitate, in a matrix of the other. The precipitates may contain only a handful of atoms, but they are nonetheless a separate single phase with a defined crystallographic structure.

*Phase diagrams* are essential tools of the trade for processing of metals and ceramics: they are maps showing the phases expected as a function of composition and temperature, if the material is in a state of thermodynamic equilibrium. By this we mean the state of lowest *free energy* (defined in more detail later), such that there is no tendency to change.

A *phase transformation* occurs when the phases present change—for example, melting of a solid, solidification of a liquid, or a solid solution forming a dispersion of fine precipitates within a crystal. Phase diagrams show us when changes such as these are expected as the temperature rises and falls. Phase transformations are thus central to microstructural control in metal processing.

A change in the phases present requires a *driving force* and a *mechanism*. The first is determined by thermodynamics, the second by kinetic theory. The term driving force, somewhat confusingly, actually means a change in free energy between the starting point (the initial phase or phases) and the end point (the final state). A ball released at the top of a ramp will roll down it because its potential energy at the bottom is lower than that at the top. Here the driving force is the potential energy. Phase changes are driven by differences in free energy—solids melt on heating because being a liquid is the state of lowest free energy above the melting point. And chemical compounds can have lower free energy than solid solutions, so an alloy may decompose into a mixture of phases (both solutions and compounds) if, in so doing, the total free energy is reduced.

Microstructures can therefore only evolve from one state to another if it is energetically favourable to do so—thermodynamics points the way for change. But having energy available is not sufficient in itself to make the change happen. Changes in microstructure also need a *mechanism* for the atoms to rearrange from one structure to another. For example, when a liquid solidifies, there is an interface between the liquid and the growing solid. This is where kinetic theory comes in—atoms transfer from liquid to solid by diffusion at the interface.

Virtually all solid-state phase changes also involve diffusion to move the atoms around to enable new phases to form. Diffusion occurs at a rate that is strongly dependent on temperature—we calculated it in Chapter 13. The overall rate of structural evolution depends on both the magnitude of the driving force (no driving force, no evolution) and the kinetics of the diffusion mechanism by which it takes place. The rates of structural evolution are of great significance in processing. First they govern the length scale of the important microstructural features—grains, precipitates and so on—and thus the properties. But they also affect process economics, via the time it takes for the desired structural change to occur.

Solidification and precipitation involve changes in the phases present. There are other important types of structure evolution in processing that do not involve a phase change, but the principles are the same: there must be a reduction in energy, and a kinetic mechanism, for the change to occur. Concentrated solid solutions have higher free energy than dilute ones, so gradients in concentration tend to smooth out, if heated to give the solute atoms enough diffusive mobility. Recovery and recrystallisation are mechanisms that change the grain structure of a crystal, even though the phases making up the grains do not change. The driving force comes from the stored elastic energy of the dislocations in a deformed crystal, released as the density of dislocations falls; the mechanism of recrystallisation is diffusive transfer of atoms across the interface between the deformed and new grain structures. We revisit these examples later in the chapter, in relation to manufacturing processes in which they occur.

***Thermodynamics of phases*** Materials processing principally deals with liquids and solids—only a few special coating processes deal with gases (for example, vapour deposition onto a substrate). Drilling down into the origins of material properties in earlier chapters, we have seen that solid phases can form as pure crystals of a single element, as solid solutions of two or more elements, or as compounds with well-defined crystal structure and stoichiometry (by which we mean the relative proportions of the different elements are simple integers: alumina $Al_2O_3$, iron carbide $Fe_3C$, and so on).

So if we take an alloy of a given composition and hold it at a fixed temperature, what determines whether it will be solid, liquid or a mixture of the two? Or a solid solution, a compound or a mixture of solid phases? The answer lies in thermodynamic equilibrium and the Gibbs[1] free energy, $G$, defined as

$$G = U + pV - TS = H - TS \qquad (19.1)$$

The thermodynamic variables in this equation require some explanation. The *internal energy U* is the intrinsic energy of the material associated with the chemical bonding between the atoms and their thermal vibration. We have encountered it before in the cohesive energy responsible for recoverable elastic strain and stored energy under load (Chapter 4), or thermal expansion (Chapter 12). The pressure $p$, volume $V$ and temperature $T$ are the thermodynamic state variables familiar from the universal gas law. The combination $U + pV$ is so common

---

[1] Josiah Willard Gibbs (1839–1903), a seventh generation American academic, spent almost his entire career at Yale, receiving its first Ph.D. in Engineering (on the form of gear teeth). A formative visit to Europe, working with scientists such as Kirchhoff and Helmholtz, set him on the path towards his great work *On the Equilibrium of Heterogeneous Substances*. This established him as a founding father of physical chemistry and chemical thermodynamics.

that it is frequently combined as the enthalpy, *H*. In materials processing, we are concerned mostly with changes in free energy rather than absolute values. Pressure is usually at or close to atmospheric, and volume changes in metals are modest in the liquid or solid state, so changes in enthalpy are usually dominated by the internal energy *U*.

Entropy *S* is a complex concept in thermodynamics associated with how exchanges of heat at a given temperature change the enthalpy of a system, and whether the exchange is reversible. This is central to determining the efficiencies of gas-based systems that generate and exchange heat to do mechanical work, such as internal combustion engines and gas turbines. For material phase changes, a common interpretation of entropy is as a measure of the disorder in the system. For a solid crystal to melt, heat must be supplied (at constant temperature) to overcome the intrinsic 'latent heat' of melting. We will see later that this is an enthalpy change, with a corresponding increase in the entropy. The rise in entropy is associated with the greater disorder in the liquid state, compared to the regular packed state of the solid. Changes in entropy of a crystalline solid are relatively small (due to the regular atomic packing). Entropy is much more important in polymers, since the molecules have more freedom to adopt different configurations.

So, for any composition and temperature we could (at least in principle) calculate the free energy of alternative material states—liquid solution, solid solution, mixtures of phases and so on. Out of the unlimited number of possible states, that with the lowest free energy is the state at *thermodynamic equilibrium*. Higher energy states may form, but there is then the possibility of energy being released by a change to equilibrium. Only at equilibrium is no change physically possible. In practical alloy processing it is often easy enough to reach equilibrium or near-equilibrium states—hence the importance of phase diagrams, which map the equilibrium phases expected for a given alloy.

*Introduction to Guided Learning Unit 2: phase diagrams and phase transformations (Parts 1–4)* Phase diagrams and phase transformations are big topics in process metallurgy. In the following subsections, we give a concise summary of phase diagrams and the underlying science of phase transformations. Guided Learning Unit 2 goes into depth on how to interpret phase diagrams and relate them to microstructural evolution in common processes. The rest of this chapter will be intelligible without diverting into the unit, but a more complete grasp is available by working through the Guided Learning text and exercises. It is recommended that Guided Learning Unit 2 is visited in two installments. Parts 1–4 cover phase diagrams and how to read them: work through this now. Parts 5–8 cover common phase transformations—you will be prompted in the text when it is best to go back to these sections.

*Phase diagrams: illustrative examples* Real alloys can contain up to 10 or more different elements, though many contain only two. The starting point for understanding all alloys is to consider the principal element mixed with one other element, making a binary alloy (or system). Carbon steels are based on iron-carbon, stainless steels on iron-chromium and so on. Phase diagrams for binary alloys are plots of temperature against composition, showing the fields in which the equilibrium phase or phases are fixed. Figure 19.5 shows a simple example: the phase diagram for the binary lead-tin (Pb-Sn) system. This and more complex diagrams are built up and explained in more detail in Guided Learning Unit 2.

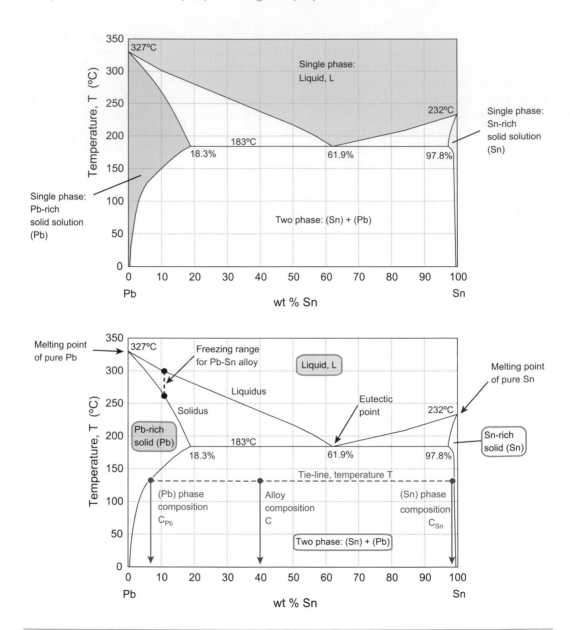

**Figure 19.5** Phase diagram for the binary Pb-Sn system, showing a number of key features. In the upper figure, the single-phase fields are: liquid—peach; Pb-rich solid solution (Pb)—green; Sn-rich solid solution (Sn)—yellow. These are separated by two phase fields, the relevant phases being identified by a horizontal tie-line, as shown in the (Pb)–(Sn) field in the lower figure. Also shown are the melting points of the pure elements, the freezing range between liquidus and solidus boundaries for alloy compositions and the eutectic point at the bottom of the liquid field.

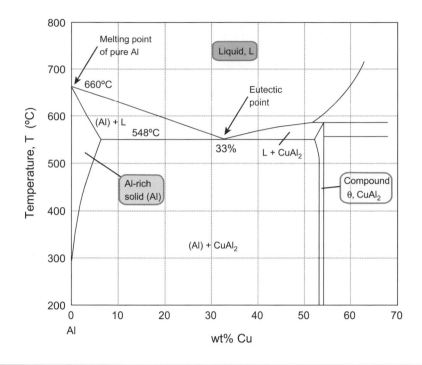

**Figure 19.6** Phase diagram for part of the binary Al-Cu system, showing the compound CuAl₂ at 53 wt% Cu.

Note the following features of Figure 19.5:

- The diagram divides up into single- and two-phase regions, separated by phase boundaries.
- At any point in the two-phase regions, the phases present are those found at the phase boundaries at either end of a horizontal 'tie-line' through the point defining the composition and temperature concerned.
- Both Pb and Sn will dissolve one another to some extent (Sn is more soluble in Pb than the reverse), with the maximum solubility in both cases being at the same temperature.
- The pure elements (Pb and Sn) have a unique melting point (at which the material can change from 100% solid to 100% liquid).
- Alloys show a 'freezing range' between the boundaries known as the liquidus and solidus, so there is no longer a single melting point.
- At one special point, the *eutectic*, the alloy can change from 100% liquid to 100% two-phase solid at a fixed temperature (it is the only place at which three phases can co-exist); this temperature is the lowest temperature at which 100% liquid is possible.

The Pb-Sn system only contains solid solutions, and mixtures of the two. The common appearance of compounds on phase diagrams is illustrated in Figure 19.6, part of the aluminum-copper (Al-Cu) system. The compound CuAl₂ contains one Cu atom for every two Al atoms: 33 atom % (at%) Cu. Converted to percentage by weight (see Guided Learning Unit 2), the

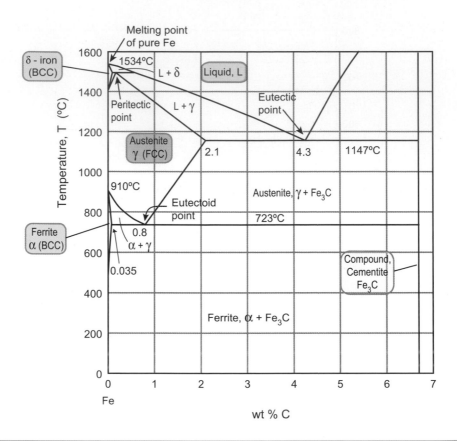

**Figure 19.7** Phase diagram for part of the binary Fe-C system, up to the compound iron carbide (cementite) Fe₃C at 6.7 wt% C. The single phases are: liquid, and 3 solid solutions of carbon in iron: α-iron (ferrite), γ-iron (austenite) and δ-iron. The Fe-C system contains a eutectic point, and the austenite field closes at the bottom with a eutectoid point, and at the top with a peritectic point.

compound is Al-53 wt% Cu. The phase diagram shows a thin vertical single-phase field for CuAl₂. The small spread of composition shows that the stoichiometry need not be exactly in the 1:2 atomic ratio of Cu:Al, but some excess Al can be accommodated in the CuAl₂ crystal structure. Most compounds show some spread in their composition, but some are strictly stoichiometric—the single-phase field appears as a vertical line on the diagram. Note that between pure Al and CuAl₂, the form of the phase diagram is similar to the binary system in Figure 19.5, with a eutectic point at 33 wt% Cu.

Figure 19.7 shows the important Fe-C phase diagram. It too has a compound (iron carbide, Fe₃C) and a eutectic point (at 4.3 wt% C and 1147 °C) but in other respects it is a bit more complicated, with a number of additional features (explained more fully in Guided Learning Unit 2):

- Pure iron shows *three* phases in the solid state: as the temperature increases iron changes from α-iron (BCC lattice) to γ-iron (FCC lattice) before reverting to BCC δ-iron and then melting.

- The single-phase *austenite* field (solid solution of carbon in FCC γ) has a *eutectoid point* at its base—at this point the single-phase austenite at a composition of 0.8 wt% C can transform on cooling to a mixture of *ferrite* (a solid solution of carbon in BCC α) and *cementite* (the compound $Fe_3C$).
- The austenite field closes at the top in a *peritectic point*, at which single-phase austenite with this composition transforms on heating to a mixture of liquid and δ-iron.

As discussed in Guided Learning Unit 2 (Part 8), an important practical consequence of this diagram relates to heat treatments that exploit the difference in solubility of carbon in FCC and BCC iron.

*Thermodynamics and kinetics of phase transformations*   We noted earlier that the equilibrium state was that with the lowest free energy at a given composition and temperature. The boundaries on phase diagrams show where two different states can both be in equilibrium. So what happens if we start at equilibrium in one field of the diagram and change the temperature until we cross a boundary? The free energy of any state varies with temperature. Figure 19.8 shows schematically the variation in free energy with temperature for the liquid and solid state of a pure metal. The two curves cross at the melting point—liquid and solid can co-exist here. Above the melting point, liquid is the state of lower free energy; below it is solid. That is the equilibrium state. If we cool a liquid beyond its melting point, its free energy as a liquid is greater than that as a solid—the system can release energy if it solidifies. This change in free energy (per unit volume) $\Delta G$ is shown in the figure. It is commonly called the *driving force*, even though it is not actually a force at all. Note from the figure that on heating a solid above its melting point, there is then a driving force to melt to

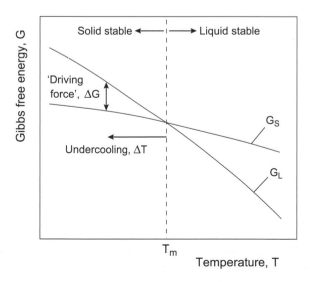

**Figure 19.8**   Schematic variation of free energy G with temperature for pure liquid ($G_L$) and solid ($G_S$). Above the melting point $T_m$, $G_L < G_S$ and liquid is the equilibrium state; below $T_m$, solid is stable, and $G_S < G_L$. The driving force $\Delta G$ is the difference between the two curves when, for example, liquid is undercooled by an amount $\Delta T$ below its melting point, as shown.

a liquid. And in both cases the driving force increases with distance away from the equilibrium temperature $T_E$. The difference $\Delta T$ above or below $T_E$ is known as the superheating or the undercooling, respectively.

It is straightforward to show how the driving force varies with the undercooling $\Delta T$. Consider liquid and solid at the melting point $T_m$, where the free energies are equal:

$$G_S = G_L \ \text{ so } \ (H - T_m S)_S = (H - T_m S)_L \tag{19.2}$$

Hence the changes in enthalpy and entropy between solid and liquid are related by

$$\Delta S = \frac{\Delta H}{T_m} \tag{19.3}$$

The enthalpy change $\Delta H \ (= H_L - H_S)$ may be familiar already as the 'latent heat of melting'—the heat that must be applied to change solid to liquid at its melting point. The entropy increases in proportion, as in equation (19.3), giving no net change in free energy $G$.

Now if we undercool the liquid to a temperature $T$ below the melting point, the driving force is given from equation (19.2) as

$$\Delta G = \Delta H - T \Delta S \tag{19.4}$$

For modest undercooling, neither $\Delta H$ nor $\Delta S$ change much with temperature, so we can substitute from equation (19.3), giving

$$\Delta G = \Delta H - T \left( \frac{\Delta H}{T_m} \right) = \frac{\Delta H}{T_m} (T_m - T) = \frac{\Delta H}{T_m} \Delta T \tag{19.5}$$

So to a first approximation the driving force for a phase change on cooling increases in proportion to the undercooling (or conversely the superheating on heating). Solid-state phase changes follow exactly the same principles, but with driving forces that turn out to be of the order of around one-third those for solidification. There is still 'latent heat' released at constant temperature when a solid crystal changes into another solid phase (or phases), but it is smaller. Similarly the entropy change associated with a change between two crystals is lower, as both still have a high degree of order.

As noted earlier, microstructures also evolve *without* the phases changing. An example is *precipitate coarsening*, in which a population of precipitates of various sizes evolves such that the average size increases. Small ones dissolve back into the surrounding matrix, and their atoms diffuse over and attach to the larger ones. We can still approach this using thermodynamics, considering free energy before and after the change. But in this case it is not a change of phase driving the process, but the energy associated with the *interface* between the precipitates and the surrounding matrix. This surface energy between two different crystals is associated with the atomic misfit between the phases across the boundary between them. It is analogous to the 'surface tension' that tries to pull water droplets to be spherical. The surface energy of the system is the surface energy per unit area, $\gamma$, times the total area of interface. Hence coarsening occurs because the total surface area falls if the precipitates form a smaller number of larger particles.

Other important examples in metals processing are recrystallisation and grain growth. Deformed metals are packed full of dislocations, with an associated energy per unit length (introduced as equation (6.15)). Recrystallisation wipes out the deformed grains, replacing them with new grains of much reduced dislocation density (see later)—the driving force is the change in stored dislocation energy. The grains may then grow, meaning boundaries migrate such that small grains disappear and the average grain size increases. Grain boundaries also have surface energy associated with the misfit in the lattice—larger grains means less area of boundary, and a reduction in system energy.

Returning now to the solidification example, equation (19.5) shows that the more we cool down, the more energy is available to drive solidification. We might therefore expect solidification (or any other phase change) to go faster and faster the more we cool a liquid below the equilibrium temperature. But this is not the case. To borrow a mathematical phrase, we might say that having thermodynamics on our side is a necessary condition, but it is not sufficient. It is also essential to have a *kinetic mechanism* by which the microstructural change takes place. Atoms are being asked to relocate, which in the vast majority of cases means they must diffuse (even if they only need to hop a fraction of the atomic spacing). This brings in a further dependence on temperature, and also (crucially) a dependence on the time available—both things that are strongly influenced by the manufacturing process (witness the bike frame example in Section 19.2).

The kinetics of diffusion were discussed in Chapter 13, principally in relation to creep of materials when used at elevated temperatures. Diffusion is at the heart of material processing too. Recall from Section 13.4 that there is a strong exponential temperature-dependence of the rate of diffusion, known as Arrhenius' law:

$$\text{Diffusion rate} \propto \exp(-Q/RT) \tag{19.6}$$

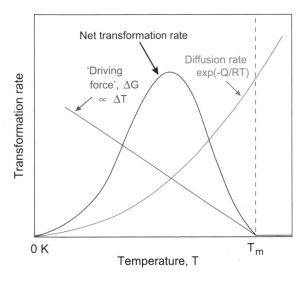

**Figure 19.9** Schematic variation of transformation rate with temperature, showing the competition between driving force and diffusion rate, leading to a maximum rate at some intermediate undercooling.

The rate of a phase transformation therefore depends on both the driving force and the diffusion rate. Their temperature-dependencies are in opposition—the more we cool, the higher the driving force but the lower the diffusion rate. How do they trade off? Figure 19.9 shows what happens. At low undercooling, diffusion is rapid but the driving force is small, falling to zero at the equilibrium temperature—the net transformation rate is low, limited by the driving force. At the other extreme (approaching zero Kelvin) the driving force is maximised, but diffusion is now so slow that nothing can happen. Somewhere in between, a maximum transformation rate is found where both thermodynamics and kinetics are favourable. This is an important and counter-intuitive outcome—diffusion-controlled changes go fastest at some particular undercooling, not when they are hottest.

*Nucleation and growth*    There is a further refinement we should know about in materials processing. When a liquid solidifies, solid first has to appear from somewhere, after which the interface between solid and liquid can migrate to enable atoms to switch from one phase to the other at the boundary. We call these two stages *nucleation* and *growth* (see Figure 19.10). Nucleation presents another subtlety of thermodynamics. At first glance, there is a driving force, so surely off it goes? But the first nuclei to form generate an interface that wasn't there before (between the liquid and solid).

It was noted earlier that interfaces between phases have an associated surface energy (due to the atomic mismatch). So some energy has to be provided to create this interface, and the source of this energy is the free energy of the transformation. But the energy released depends on the *volume* of the new phase formed, whereas the energy penalty at the interface depends on the *area* of its surface. For small particles of the new phase the surface area energy represents a significant barrier to transformation; for larger particles the volumetric energy release dominates. As a consequence of this trade-off there is a *critical radius* for nucleation. Only above this size do nuclei become stable and grow—smaller nuclei are unstable and melt back into the liquid.

The size of nuclei formed spontaneously is a statistical phenomenon, but it depends on the same factors as the growth rate (the driving force and the diffusion rate). The nucleation rate therefore follows the same form of temperature-dependence as shown in Figure 19.9 (although

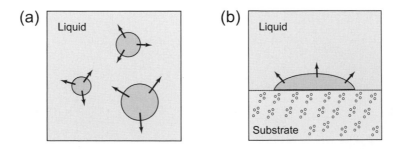

**Figure 19.10** Nucleation and growth, illustrated for solidification. In (a) the solid nucleus forms spontaneously as a sphere within the liquid (homogeneous nucleation); in (b) the solid nucleus forms as a spherical cap on a solid substrate, such as the mould wall (heterogeneous nucleation).

the temperatures at which maximum nucleation and growth rates occur do not coincide). But as both tend to zero at the equilibrium temperature and at zero Kelvin, the net transformation rate must still peak somewhere in between. This is why Figure 19.9 may be regarded as the temperature-dependence of the overall transformation rate.

The spontaneous formation of new phases, such as solid crystals within the bulk of a liquid, is strictly known as *homogeneous nucleation*. Figure 19.10 also shows an alternative way to start a transformation, with the new phase attaching itself to some other interface in the system, such as the surface of the mould holding the liquid. This is called *heterogeneous nucleation*. It turns out that the amount of undercooling needed for this route is significantly smaller than for homogeneous nucleation, so in practice heterogeneous behaviour dominates. Qualitatively we can see why. Figure 19.10 shows the same volume of new solid formed by both nucleation mechanisms. Three different surface energies are involved in the heterogeneous case (between liquid, solid and the additional material such as the mould), with various interfaces being created and eliminated. But the net effect is that the solid-liquid interface achieves a large radius (bigger than the critical radius) with far fewer atoms. The solid-liquid interface can then grow stably as if it is part of a much bigger nucleus, but with a much reduced surface energy penalty. Metals processing often exploits heterogeneous nucleation to influence the length scale of the new microstructure being formed (particularly grains). Examples are provided later.

***Time-temperature-transformation (TTT) diagrams and the critical cooling rate***   Foundry engineers worry about how long it takes for a casting to solidify, rather than the rate at which the solid-liquid interface is moving. But it is clear that these are directly related—inversely in fact, the faster the transformation rate, the sooner it will both start and finish. First we flip Figure 19.9 on its side, so that it shows temperature against rate of transformation (Figure 19.11); then we can infer the figure shown to the right, in which the time taken to reach a given fraction transformed is a minimum where the rate is a maximum, and the time tends to

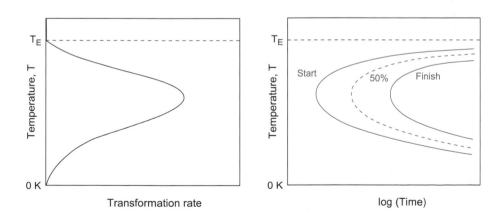

**Figure 19.11**   Temperature against the rate of transformation, and the corresponding form of temperature-time-transformation (TTT) diagram—maximum rate leads to minimum transformation time. 'C-curves' are shown for the start, 50% completion and finish of the transformation.

infinity for the temperatures at which the rate falls to zero. The characteristic shape of the curve leads to the name of 'C-curve' for these plots of diffusion-controlled phase transformations. A set of C-curves, as in the figure, depicts the onset and progress of a transformation to its completion. The net picture is called a *time-temperature-transformation (TTT) diagram*. For some heat treatments (notably steels) these are as central to processing as the phase diagram.

TTT diagrams are widely used to study diffusion-controlled transformation rates. The C-curve shape reflects the transformation rate as a function of temperature, assuming that we somehow cool rapidly from above the transformation temperature and hold isothermally at a fixed undercooling. This is fine in the laboratory with small samples, but real processing involves continuous cooling from above the transformation temperature to room temperature. In this scenario we reach for a variant set of curves called *CCT diagrams* (i.e. 'continuous-cooling transformation'). Without going into details, the essential C-shape is retained in continuous cooling (though the curves are shifted to longer times and lower temperatures).

Figure 19.12 shows the temperature histories that are valid for each type of diagram. From this we should extract one core idea: in continuous cooling there is a *critical cooling rate (CCR)*, that just avoids the onset of the diffusional transformation. If the cooling rate exceeds this value then (in general) the initial state at high temperature is retained, at a temperature when it is no longer the equilibrium state, but is unable to transform because the kinetics are much too sluggish. In thermodynamics terms this is a *metastable* state. Forcing components to cool at an accelerated rate is a common trick in heat treatment, to deliberately bypass equilibrium and achieve an alternative microstructure as a route to better properties than slow cooling.

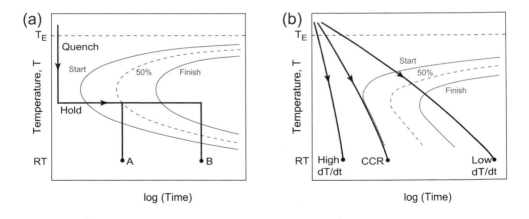

**Figure 19.12** (a) A TTT diagram, interpreted for quench-and-hold isothermal treatments to room temperature (RT); at B the transformation is 100% complete, while at A a 'split transformation' has occurred, with only 50% transformed; (b) a CCT diagram, broadly the same shape, but for continuous cooling histories. The *critical cooling rate (CCR)* is the minimum rate that avoids the start of any diffusion-controlled transformations.

*Introduction to Guided Learning Unit 2: phase diagrams and phase transformations (Parts 5—8)*    At this point, we recommend that you finish working through the Guided Learning Unit 2 (Parts 5—8). These sections illustrate the evolution of microstructure for a number of important processes (solidification and heat treatment), in relation to the relevant phase diagrams.

The rest of this chapter illustrates a range of manufacturing processes, and the microstructural evolution that takes place during processing. These case studies illustrate how control of composition and cooling history determine the microstructure in industrial processes, and thus how we can achieve the desired properties, if we follow the recipe correctly.

## **19.5** Metals processing

*Metals—overview*    It was clear in Chapter 18 that a favourable characteristic of metals is the diversity of manufacturing processes available to make and assemble components of complex shape. We now explore how, during processing, the underlying microstructure and thus properties are evolving. This includes both properties relevant to the process itself (for example, ductility for deep drawing a beer can) and the properties needed in service (for example, strength and toughness for automotive and aerospace alloys).

To illustrate the breadth of metals processing and property manipulation, we draw on examples from casting, deformation, heat treatment, joining and surface engineering. Steels (and other ferrous alloys) are the dominant engineering alloys, followed by aluminum alloys, so these feature in most of the examples. But similar principles apply to all the alloy systems, such as those based on Cu, Ti, Ni, Zn, Mg and so on. Most alloy systems offer both cast and wrought variants, and many are also processed as powder. Usually either cast or wrought alloys dominate an alloy class. This reflects subtle differences in the ease of deformation of the underlying crystal structures in different elements. For example, Zn, Mg and Ti have the 'hexagonal close-packed (HCP)' crystal structure (see Chapter 4). This is inherently less ductile than the FCC or BCC crystal structures found in Al, Fe and Cu—so Zn, Mg and Ti are mostly cast, with only limited deformation processing being conducted, and usually then done hot (when the ductility is better). Ductile Al, Fe and Cu are mostly wrought, but all come as casting alloys too.

Casting and wrought alloys in a given metal system tend to have quite different compositions. Good castability requires much higher levels of alloying additions than the relatively dilute wrought alloys (to lower the melting point). Casting leads to coarser microstructures and poorer strength and toughness than in wrought alloys (as illustrated for aluminum alloys in Chapter 8).

*Solidification: metal casting*    Casting is a relatively cheap shaping route, and is well suited to making complex 3-D shapes. It is used over a wide range of size, from large ship propellers to engine parts, machinery and sculptures, down to toys, household fittings and medical implants. Tweaking the chemistry of castings to improve their properties has been a bit of a black art for centuries. Nowadays it is possible to use sophisticated software tools to model everything from the choice of composition to the way to pour the metal into the mould in order to control the grain size and to minimise porosity and residual stress.

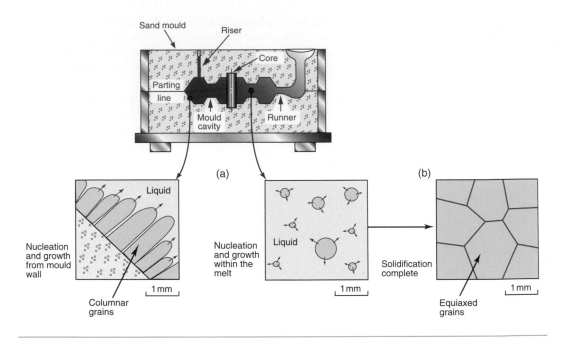

**Figure 19.13** Solidification in metal casting: (a) nucleation and growth of solid crystals in the melt; (b) impingement of growing solid crystals forms the grains and grain boundaries.

In casting, a liquid above its melting point is poured into a mould, where it cools by thermal conduction. New solid forms by *nucleation*, with crystals forming spontaneously in the melt, or on the walls of the mould, or on foreign particles in the melt itself (Figure 19.13(a)). The nuclei grow by diffusive attachment of atoms at the liquid–solid interface. Solidification is complete when crystals growing in opposing directions impinge on one another, forming grain boundaries—each original nucleus is the origin of a grain in the solid (Figure 19.13(b)). The grain size of the casting thus depends on the number of nuclei, since each one grows into a grain. Achieving a fine grain size, often important in achieving optimal properties, is assisted by controlling nucleation through the addition of *inoculants*—fine-scale solid particles that stimulate heterogeneous nucleation. An example is the addition of $TiB_2$ to aluminum castings. At the other extreme, we sometimes wish to grow a single crystal, with no grain boundaries at all. An example is the casting of Ni superalloy turbine blades for jet engines. In this case it is a question of retaining just one nucleus, something that can be done by cooling the liquid slowly from one end so that the first nucleus to appear is made to grow along the entire length of the component. Examples of cast microstructures for different alloy compositions are illustrated in Guided Learning Unit 2.

The rates of nucleation and growth (and thus grain size) also depend on the imposed cooling history and this is governed by heat flow. This is a multi-parameter problem: it depends on the thermal properties of the metal and the mould, the contact between the two, the initial liquid temperature, the release of latent heat on solidification, and the size and shape of the casting. Here is an example of the design–process–material coupling discussed in Chapter 18: the geometry of the design combines with the type of casting process (mould material, metal temperature, etc.) and the choice of alloy to dictate the cooling rate; and the microstructural response to that cooling rate (and thus properties) depends on the alloy chemistry.

A further complication in casting is that the liquid has a uniform concentration of solute but the solid does not. The solubility of impurities and solutes in the solid is less than that in the liquid, so they are rejected ahead of a growing crystal, enriching the remaining liquid. Concentration gradients build up in the solid grains with higher solute and impurity levels in the last part to solidify (i.e. the grain boundaries). This *segregation* is explained further in Guided Learning Unit 2. It can be a source of problems such as embrittlement or corrosion sensitivity at the boundaries. Castings therefore are often held at high temperature for prolonged periods (*homogenisation*) to enable some levelling out of the solute concentration gradients. In the case of impurity segregation, it is impractical and expensive to try and remove them from the melt. The trick is to render them harmless by giving them something else to react with, to form a solid compound distributed throughout the casting. For example, manganese is added to carbon steels in order to wrap up the impurity sulfur as harmless $MnS$, rather than letting it concentrate at the grain boundaries as brittle $FeS$.

The difference in solubility between liquid and solid is also a problem with respect to gaseous impurities dissolved in the melt. On solidification, most of the dissolved gas is rejected into the remaining liquid, but must eventually be released as a separate phase—in this case as trapped bubbles of gas. As a result, castings inevitably contain some *porosity*. Good casting practice traps these bubbles as microporosity throughout, rather than sweeping all the gas into a big cavity in the middle of the component. Better still, but more expensive, the liquid is 'outgassed' under vacuum just before it is poured. In the case of steel-making, high levels of dissolved gas are a direct consequence of previous processing of the melt—the carbon content is reduced to the required level by injecting oxygen into high carbon molten iron, but this leaves oxygen and carbon monoxide/dioxide in solution. To deal with this, reactive elements such as aluminum may be added to the melt prior to casting. These react with oxygen in solution to form solid oxides, which are again trapped as harmless particles in the casting—a trick known as 'killing' the steel. It is another example of how a small alloying addition is used to fix a problem due to the presence of a trace impurity.

*Deformation processing of metals* Solid-state deformation processes (rolling, forging, extrusion, drawing) exploit the plastic response of metals, in particular their *ductility*, that is, their ability to remain intact without damage when subjected to large strains and shape changes. Most processes are compressive rather than tensile, to avoid the problem of necking in tension. And in many cases forming is conducted hot, to exploit the reduced yield stress—alloys at temperature can effectively be forced to creep at high strain-rates. Wrought metal products (i.e. shaped by deformation) are as ubiquitous as castings: beams, columns and tubes for buildings and offshore structures; bodies, panels and engine parts for every type of vehicle; machinery, tools, pipework, wire, food and drink containers—the list goes on. Deformation not only makes the required shape, it also drives the internal changes in microstructure. The component temperature, strain-rate and strain control both the flow stress during shaping, but also the material internal state during and after forming.

First, the temperature determines the phases present (see Guided Learning Unit 2, and Figures 19.6 and 19.7). This affects the material strength during forming, but has implications for what happens during cooling afterwards. For example, hot rolling of carbon steels takes place in the *austenite* field, at temperatures of 900 °C or more (Figure 19.7). On cooling, rolled carbon steels will typically follow equilibrium, leading to a mixture of grains of *ferrite* and *pearlite* (the latter being a two-phase mixture of iron carbide and ferrite). The mechanism of this transformation is described in Guided Learning Unit 2. But the grain size and proportions of ferrite and pearlite (and hence properties) vary in fine detail, depending on the steel composition and the cooling rate. And the thermal history will depend on the initial temperature, the size and shape of the component and how the product is handled after shaping (e.g. thin strip is coiled and as a result remains hot for longer).

The second generic microstructural issue in deformation processing is the development of grain size. We have seen how casting leads to the initial solid grain size. Perhaps surprisingly, the grain size can be manipulated further in the solid-state. In the first instance, when the component changes shape, the grains inside follow suit to accommodate the overall strain. Figure 19.14(a) shows schematically how rolling 'pancakes' the grains. This can give

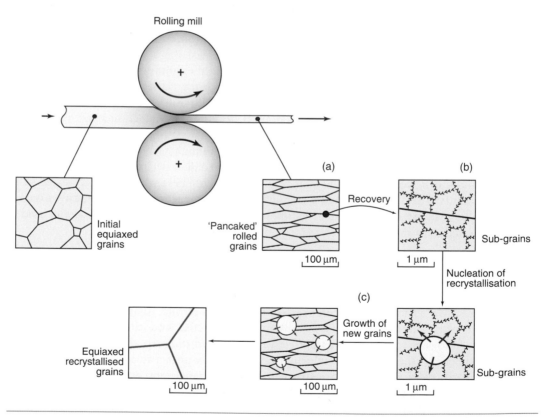

**Figure 19.14** Grain structure evolution in the solid state by deformation and annealing: (a) grains follow the shape change in the component; (b) the recovery mechanism: dislocations re-arrange as subgrains (note the magnified scale); (c) the recrystallisation mechanism: new grains form by migration of boundaries from a few subgrain nuclei, replacing the deformed microstructure.

a bit of useful longitudinal strengthening, rather like a fibre composite. More significantly the microstructure *within* the grains is modified: work hardening occurs, increasing the dislocation density. If this is the final shaping step, this is exploited to strengthen the final product, for example, deep drawing the walls of an aluminum beverage can, or drawing a copper wire. For earlier forming stages, however, work hardening is a problem as it pushes up the forming loads, and limits the ductility. It is therefore necessary to *anneal* the metal (soften with heat), and this heat treatment offers the opportunity to control the grain structure too.

Annealing involves *recovery* and *recrystallisation*, which are solid-state changes in the dislocation and grain structure at elevated temperature. This may require a separate heat treatment after forming, but they can also take place concurrently during hot deformation. In *recovery*, dislocations interact and rearrange into organised patterns forming *subgrains*, which are like tiny grains within grains (see Figure 19.14(b)). More prolonged heating leads to *recrystallisation*, shown in Figure 19.14(c). Grain boundaries migrate by atoms hopping from old to new grains at the boundary between the two. This sweeps up and eliminates most of the stored dislocations, removing the work hardening completely. This lowers the yield stress, making it easier to work the material to shape—something blacksmiths have known for centuries, though they had no idea of the microstructural changes.

As noted earlier, recrystallisation has a driving force (the stored dislocation energy) and a kinetic mechanism (boundary migration), though it is not a phase transformation. In common with solidification, however, the resulting grain size depends on the number of nuclei, here the number of sub-grains that exceed a critical size and migrate during recrystallisation. The metallurgy of recrystallisation is a complex matter—it depends on the initial grain size, the deformation conditions (temperature, strain and strain-rate), and the annealing temperature. Furthermore, all the nucleation is heterogeneous—either on grain boundaries (as shown in Figure 19.14(c)), or from small hard particles of intermetallic phases within the grains (so-called 'particle-stimulated nucleation'). This population of particles is itself a consequence of the alloy chemistry, and the upstream processing stages of casting and homogenisation when the particles are formed.

The scope for manipulating the grain structure during forming is therefore a major distinction between the cast and wrought alloys. The grain sizes in wrought alloys are much smaller (typically 10–100 µm). But it is one of the most coupled, multi-variable problems to control, dependent on alloy composition, deformation rate, process temperatures and component geometry and strain. It is therefore difficult to get a uniform grain size in a formed part, as the strain, strain-rate and temperature vary with location.

As if this wasn't enough, there is another important side-effect of deformation and annealing: the concept of crystal *texture*. The crystal planes in a casting are random—each nucleus forms in isolation within a melt. Not only does deformation 'pancake' the grains, it also tends to align the crystallographic planes with respect to the axes of the deformation. Perhaps unexpectedly, recrystallisation does not restore randomness—there remains a statistical distribution of orientations. This doesn't sound very important, but it is significant in working with sheet metal, since textured metals have yield properties that are *anisotropic* (i.e. different in different directions). This can be a problem if you want to make something smooth and circular out of a sheet, like a beverage container.

***Heat treatment of metals*** Wrought metal products, and some castings and powder processed parts, are often subjected to heat treatment. This serves various purposes. Annealing to soften an alloy before further forming is one. Normalising is another—slow cooling from high temperature as the final step in making a component, the object being to minimise the final residual stress, and to produce a microstructure of lower strength but high toughness. Many alloys also undergo final heat treatments to enhance strength by *precipitation hardening*. Maximising strength without loss of toughness is essential for many heavy duty applications of steels (e.g. gears, cranks, cutting tools). And it is essential in combination with low density in the aluminum alloys used widely for aircraft structures.

In a typical heat treatment, the shaped component is heated to high temperature, cooled at a controlled rate and (usually) reheated to an intermediate temperature. This exploits the solid-state phase changes that occur with temperature, the use of a quench to avoid transformations that are ineffective for hardening and finally low temperature precipitation for maximum strength. Figure 19.15 illustrates a common sequence:

- Solutionize at elevated temperature, to form a solid solution.
- Quench to room temperature at a cooling rate $dT/dt$ above the critical cooling rate (defined in Figure 19.12); this retains the high-temperature solid solution state at room temperature, avoiding the diffusion-controlled transformations to equilibrium phases.
- Precipitate fine-scale phases in the subsequent reheat.

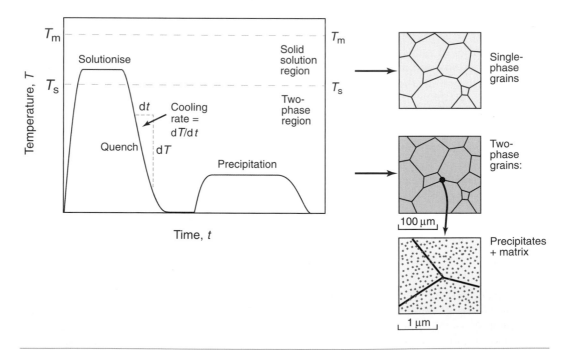

**Figure 19.15** Schematic thermal profile for heat treatments that produce precipitation hardening, illustrating microstructural changes being induced: a solid solution at high temperature, and precipitation of hardening particles (shown at an enlarged scale) at a lower temperature.

Precipitation hardening is a generic mechanism in many alloy systems, but two heat treatments of this type are particularly prevalent: the 'quench and temper' of steels, and 'age hardening' of aluminum alloys. These are discussed in more detail in Guided Learning Unit 2. Although both follow the general picture illustrated in Figure 19.15, there are subtle microstructural details in each case.

First consider age hardenable Al alloys. On quenching the supersaturated solid solution is found to have modest strength. Then during ageing the yield strength rises to a peak, significantly harder than that produced by coarse equilibrium precipitates on slow cooling. But this is not only because of the fine scale of the precipitates formed—age hardening exploits the formation of metastable phases rather than the equilibrium phases. The effectiveness of age hardening for enhancing strength in aluminum alloys was illustrated in Chapter 8 on a fracture toughness–strength property chart (Figure 8.18). Below we trace the responses of carbon steels on the same property chart.

Consider now the quench and temper treatment for carbon steels, which proceeds as follows. The first stage is to solutionise in the FCC austenite field (austenitisation), dispersing the carbon into solid solution. Quenching to room temperature traps the carbon in supersaturated solution. But there is also the FCC to BCC phase change in iron to be accommodated. In spite of the quench this still takes place, by an unusual diffusionless mechanism (see Guided Learning Unit 2). The result is a metastable phase called *martensite*. The supersaturation of carbon causes severe distortion of the BCC iron lattice, giving martensite a high yield strength but very low fracture toughness. In this state the component is too brittle to place in service, and is susceptible to cracking during the quench due to induced thermal stresses. It is essential to restore the toughness, and this is achieved by tempering at an intermediate temperature. This leads to precipitation of the equilibrium phase, iron carbide. The yield strength falls as toughness is restored, but the final yield strength is substantially higher than that of the ferrite-pearlite microstructure formed on slow cooling.

The effectiveness of quenching and tempering may be illustrated via a fracture toughness–strength property chart, shown in Figure 19.16. The figure is annotated with selected micrographs. First the effect of carbon content on the properties in the normalised condition can be seen: from 0 to 0.8 wt% carbon the strength increases steadily as the proportion of pearlite to ferrite increases. Then taking a medium carbon steel (0.4 wt% C) as an example, we see how quenching produces the unusual microstructure and properties of martensite, en route to the final tempered condition (which can be fine tuned by choice of temper temperature, as shown in the figure). Note that the precipitation during tempering is too fine-scale to be resolved in an optical micrograph—the appearance remains similar to martensite. Quenching and tempering are further enhanced by making *alloy steels*—carbon steels with modest alloying additions—as discussed in the case studies below.

*Joining processes*  Thermal welding of metals involves heating and cooling. This may cause phase transformations: (1) in the 'weld metal', which is like a localised miniature casting, melting and resolidification occur; (2) in all the heated regions of the weld, solid-solid phase changes occur on cooling, as in a heat treatment process. The solid region surrounding the melt pool where microstructural changes occur is referred to as the 'heat-affected zone'. The evolution of microstructure and the consequent property changes in the HAZ depend on the thermal cycles imposed. Once again the thermal history couples the material, the way the

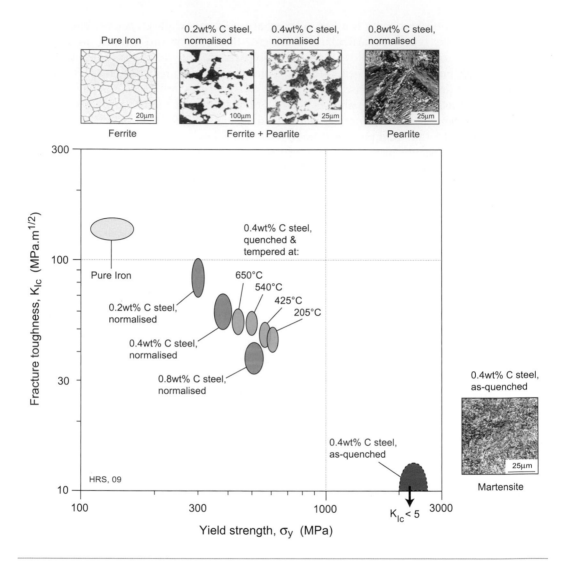

**Figure 19.16** Fracture toughness—yield strength property chart for plain carbon steels, showing the effect of carbon content in the normalised condition, and the effect of quenching and tempering for medium carbon steel. Selected micrographs show the corresponding microstructures. (Images courtesy: ASM Micrograph Center, ASM International, 2005)

process is operated (e.g. power input) and design details such as joint thickness. Thermal welding can have major consequences for joint performance—depending on the alloy and the welding process, the material may be softened, or embrittled or lose its corrosion resistance.

Figure 19.17 illustrates the contrasting responses to welding that are commonly observed in low-carbon and low-alloy steels, and heat-treatable aluminum alloys. Hardness profiles are the

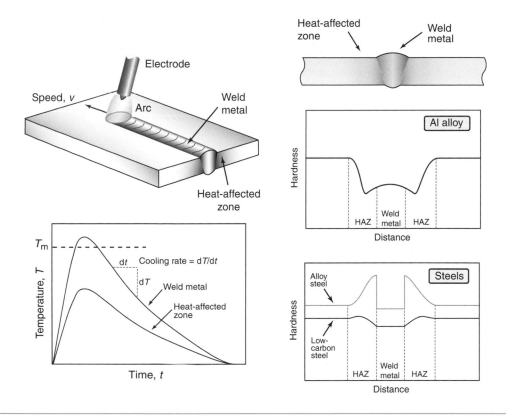

**Figure 19.17** A weld cross-section with corresponding thermal histories in the weld metal and heat-affected zone. To the right are typical hardness profiles induced across welds in heat-treatable aluminum alloys, low carbon steel and low alloy steel.

simplest first indicator of property changes across a welded joint. Low carbon steels are considered to be weldable, as on cooling the material reverts to ferrite and pearlite—the weld and HAZ have similar hardness to the original material. Low alloy steels form martensite more easily, as the critical cooling rate is lower, increasing the risk of weld embrittlement (the high hardness of martensite being a characteristic indicator). Oil rigs and bridges have collapsed without warning as a result of this behaviour, usually because the welding process was conducted incorrectly, leading to cooling rates that are too high. Heat-treatable aluminum alloys are age hardened to their peak strength—on welding there is only one way for the hardness to go, and that is down. Over-ageing due to the weld thermal cycle can lead to a permanent loss of 50% or more of the strength. In more weldable Al alloys, the metastable precipitates dissolve instead, enabling some subsequent recovery of strength by room temperature ('natural') ageing.

Welding metallurgy is therefore a big field of study—welds are often the 'Achilles heel' in design, being the critical locations that determine failure and allowable stresses. This may be

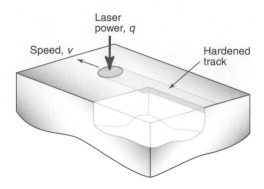

**Figure 19.18**   Laser hardening: a surface treatment process that modifies microstructure. The traversing laser beam induces a rapid thermal cycle, causing phase changes on both heating and cooling. The track below the path of the laser has a different final microstructure of high hardness.

because the material properties are damaged in some way at the joint (as shown earlier), or it may also be due to other side-effects of the process (residual stresses, or stress concentrations or crevices, potentially causing fatigue and corrosion problems). Some mechanical joining processes (e.g. friction welding) also lead to local changes in microstructure and properties, by deformation as well as heat. Adhesive technologies have the advantage that they don't impose, deform or heat the components being joined, leaving the microstructure and properties intact. But they are not trouble free—they still concentrate stress and contain defects, requiring good design and process control.

*Surface engineering*   Surface treatments exploit many different mechanisms and processes to change the surface microstructure and properties. Some simply add a new coating material, with its own microstructure and properties, leaving the substrate unchanged—the only problem is then making sure they stick. Others induce near-surface phase transformations by local heating and cooling—*transformation hardening*. Figure 19.18 shows an example—laser hardening of steels. The laser acts as an intense heat source traversing the surface, inducing a rapid thermal cycle in a thin surface layer. Heating depths of order 1 mm into the austenite field is readily achieved, with conduction into the cold material underneath providing the quench needed to form martensite. The high hardness of martensite gives excellent wear resistance, and the brittleness is not a problem when it is confined to a thin layer on a tough substrate.

Direct diffusion of atoms into the surface is also feasible, but only for interstitial solute atoms at high temperature for diffusion to occur over a distance of any significance. Chapter 13 revealed that, even near the melting point, there is a physical limit to diffusion distances ($\sqrt{Dt}$) of around 0.2 mm for a 1 hour hold, which is sufficient for surface treatments (and also for bulk processing on a very small scale, as in semiconductor devices). This mechanism is used to increase the carbon content (and thus hardness) of steels—the surface treatment known as *carburising*. Even better is a hybrid surface treatment: first carburise, and then transformation harden too, giving very hard high carbon martensite.

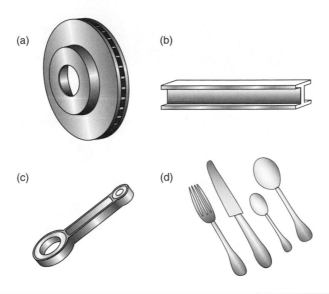

Figure 19.19   A selection of products made from alloys based on iron: (a) cast iron: brake disk; (b) low carbon steel: I-beam; (c) low alloy steel: connecting rod; (d) stainless steel: cutlery.

*Case studies: ferrous alloys*   For structural and mechanical applications, steels and other alloys based on iron dominate. They are intrinsically stiff, strong and tough and mostly low cost. High density is a drawback for transport applications, allowing competition from light alloys, wood and composites. Figure 19.19 illustrates the diversity of applications for ferrous alloys. These reflect the many classes of alloy—different chemistries combined with different process histories. Examining each in turn highlights the key material property, composition and processing factors at work.

*Cast iron: brake disk*   Brake disks (Figure 19.19(a)) use sliding friction on the brake pads to decelerate a moving vehicle, generating a lot of heat in the process (Chapter 11). The key material properties are therefore hardness (strength) for wear resistance, good toughness and a high maximum service temperature. Low weight would be nice for a rotating part in a vehicle, but this is secondary. All ferrous alloys fit the bill—so why cast iron? This is because casting is the cheapest way to make the moderately complicated shape of the disk, and the as-cast microstructure usually does not require further heat treatment to provide the hardness required. The carbon content of cast irons is typically 2–4% (by weight) giving a lot of iron carbide, the hard compound in ferrous alloys giving excellent precipitation hardening. Cast irons can also be transformation hardened at the surface to enhance their wear resistance (e.g. by laser hardening).

Cast irons illustrate another way in which minor alloying additions are used to make significant microstructural modifications, with consequent property changes. Figure 19.20(a) shows the microstructure of gray cast iron, containing thin flakes of graphite as one of the phases present. The surrounding matrix is a mixture of ferrite and iron carbide. In this form cast iron has excellent *machinability*, that is, it can be machined quickly with minimal or no

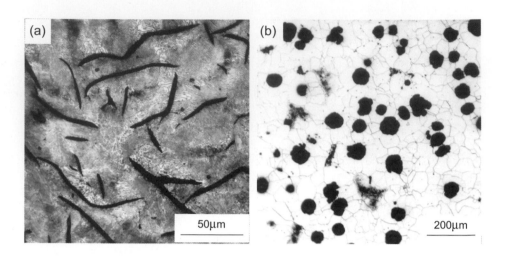

**Figure 19.20** Alternative microstructures in cast iron containing 3.5 wt% carbon: (a) gray cast iron, showing flake graphite; (b) nodular cast iron (at a lower magnification), showing spherodised graphite. (Images courtesy: ASM Micrograph Center, ASM International, 2005)

lubricant, an example of a property that is important during processing, rather than in service. The graphite flakes are brittle and crack-like, making it easy for chips to break off during machining. But in service this brittleness may be a problem. Cast iron microstructures can be toughened using an alloying trick called *poisoning*. The addition of a small amount of cerium or magnesium, sometimes with a prolonged annealing treatment, has the effect of radically changing the morphology of graphite in iron, so that the graphite spherodises to minimise the surface area of interface. By avoiding the crack-like flake form of graphite, the fracture toughness and failure strength of the cast iron are enhanced (illustrated later). The surrounding matrix may be ferritic or pearlitic, depending on the alloy and the cooling history. Figure 19.20(b) shows such a 'nodular' cast iron containing spheroidal graphite, with in this case a surrounding matrix of ferrite. Note that the carbon content in both figures is the same. This poisoning technique is also used to enhance toughness in Al-Si casting alloys—small sodium additions lead to the brittle Si forming fine-scale rounded clumps, rather than the plates and needles characteristic of solidification of eutectic compositions (see Guided Learning Unit 2).

***Plain carbon steel: I-beams, cars and cans*** If we had to single out one universally dominant material, we might well choose mild steel—iron containing 0.1–0.2% carbon. Almost all structural sections (Figure 19.19(b)), automotive alloys and steel packaging (beer and food cans) are made of mild steel (or a variant enhanced with a few other alloying additions). All of these applications are wrought—the alloy is deformed extensively to shape. The microstructure consists of ferrite and a small amount of pearlite. This provides the excellent ductility needed both for processing and in service. The fracture toughness is high, and combined with the inherent strength of iron, which may be increased by deformation processing (work hardening). Ductility, toughness and decent strength are vital in a structural material—we would rather a bridge sagged a little rather than broke in two, and in a car crash our lives depend on the energy absorption of the front of the vehicle.

*Alloy steels and heat treatment: cranks, tools and gears*   The moving and contacting parts of machinery are subjected to very demanding conditions—high bulk stresses to transmit loads in bending and torsion (as in a crank or a drive shaft), high contact stresses where they slide or roll over one another (as in gears) and often reciprocating loads promoting fatigue failure (as in connecting rods, Figure 19.19(c)). Strength with good toughness is everything. Plain carbon steels up to 0.8 wt% C give effective precipitation hardening, particularly when quenched and tempered (Figure 19.16). But the density of iron is a problem. Make the steel stronger, and use less of it, and we can save weight. This is particularly true for fast-moving parts in engines, since the support structure can also be made lighter if the inertial loading is reduced (so-called 'secondary weight savings'). By now the solution should not be a surprise: yet more alloying, and more processing.

A bewildering list of additions to carbon steel can be used to improve the strength—Mn, Ni, Cr, V, Mo and W are just the most important! Some contribute directly to the strength, giving a solid solution contribution (e.g. high-alloy tool steels, with up to 20% tungsten). More subtle though is the way quite modest additions (<5% in low-alloy steels) affect the alloy's response to the quench and temper heat treatment (discussed earlier for plain carbon steels). The problem with quenching plain carbon steels is that the critical cooling rate is high. As a result, only thin sections can be quenched to form martensite— the cooling rate within the component being physically limited by heat flow, governed by the component thickness and the inherently low thermal conductivity of steel (compared to other metals).

Alloying solves the problem by slowing down the diffusional transformations to ferrite and iron carbide, shifting the C-curves of Figure 19.12 to much longer times. This enables slower quench rates to achieve the target martensitic microstructure. Hence bigger components can be quenched to this supersaturated state, and thus tempered. The technical term for the ability of a steel to form martensite is *hardenability*—the higher the hardenability, the lower the critical cooling rate, or the larger the component that will form 100% martensite. In high alloy steels, such as those used for cutting tools, the effect is so pronounced that cooling in air is sufficient to form martensite. An additional benefit of using alloy steels is that a bit of additional precipitation strength comes from the formation of alloy carbides (rather than iron carbide). Compounds such as tungsten carbide survive to higher temperatures than iron carbide without dissolving back into the surrounding iron—a useful source of high-temperature precipitation strength (e.g. for cutting tools).

So in its full complexity, heat treatment of steels presents another example in which the resulting properties depend on coupling between material, process and design detail—alloy composition determines the hardenability, and the quenching process and component size determine the cooling rate, and hence the microstructural outcome.

*Stainless steel: cutlery*   The city of Sheffield in England made its name on stainless steel cutlery (Figure 19.19(d)), and still manufactures steel (of all types) on a major scale. The addition of substantial amounts of chromium (up to 20% by weight) imparts excellent corrosion resistance to iron, avoiding one of its more obvious failings: rust. It works because the chromium reacts more strongly with the surrounding oxygen, protecting the iron from attack (Chapter 17). Nickel is also usually added for other reasons—one being the preservation of the material's toughness at the very low cryogenic temperatures needed for the stainless steel pressure vessels used to store liquefied gasses. Another is that both chromium and nickel provide solid solution hardening—this and work hardening (during rolling and forging) being the usual routes to strength in stainless steels, though some Fe-Cr-C alloys are also heat treated for precipitation strength.

*Ferrous alloys: summary*   We have seen that in all of these applications, strength and toughness dominate the property profile. The variation of these properties for plain carbon steels was illustrated in Figure 19.16. In summary, we can map all classes of ferrous alloys on this chart. Figure 19.21 assembles the data for a selection of alloys from the classes discussed in the case studies discussed earlier. This paints a remarkable picture. Starting with soft, tough pure iron (top left), we can manipulate it, by changing composition and process history, to produce more or less any combination of strength and toughness we like—increasing the strength by a factor of 20 or more. Every final material (with the exception of as-quenched martensite) is above the typical fracture toughness threshold ($<15$ MPa.m$^{1/2}$) for structural

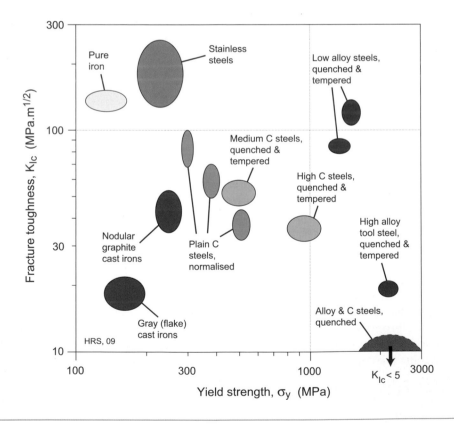

**Figure 19.21**   Fracture toughness—yield strength property chart for ferrous alloys. Composition and heat treatment enable a wide range of combinations of these properties.

or mechanical application. Cast irons primarily offer low cost and lower pouring temperatures than pure iron, for a modest increase in strength, but note the effectiveness of poisoning in restoring toughness by forming nodular rather than flake graphite (Figure 19.20). The normalised condition is standard for structural use, unless the corrosion resistance of stainless steels is required. The chart shows the benefit of quench-and-temper treatments in raising the strength without damaging the toughness, particularly in alloy steels. This figure explains why ferrous alloys are so important—no other material is so versatile, with literally hundreds of different steels and ferrous alloys available commercially.

## 19.6 Non-metals processing

*Ceramics and glasses* Ceramics and glasses are somewhat niche materials. Ceramic applications exploit their high hardness (cutting and grinding tools), high-temperature resistance (furnace liners) and high electrical resistivity (spark plugs, electronics, overhead power line insulators). Glasses are exploited mainly for optical properties, both in sheet and fibre form, and for containers such as bottles. But as brittle materials they have a problem. No matter what other advantages they have, they are susceptible to fracture, which may be traced back in many cases to the inherent defects built in during processing. Defect control is therefore a key factor in ceramic and glass processing for mechanical applications. The microcracks in ceramics largely reflect the size of powders used in the first place, but good-quality processing is then needed to avoid porosity. This is surprisingly difficult to achieve, even in simple shapes. Loose powders contract by a factor of 2 or more when they are pressed and sintered together. Achieving uniform contraction of this magnitude is not easy, since different regions of the product densify first and then constrain the further compaction of the remainder.

Glasses are easier to handle as they are melt processed. Nonetheless, great care is taken over achieving a smooth finish—plate glass is produced mostly by flotation on a bath of molten tin. Ceramics and glasses are also very susceptible to thermal stresses during cooling. Being processed at high temperature, but inherently stiff and strong, stresses are not easily relieved by creep during cooling. In one instance, however, thermal stresses are exploited to good effect. Glass can be made more fracture resistant by deliberately generating a residual stress profile through-thickness, by rapid cooling of the surfaces. This leads to beneficial compression at the surfaces and tension within—the surface being the likely location of the scratches and microcracks that cause fracture under tensile load.

Ceramic compositions tend to be largely discrete, each being a specific compound (alumina, silicon carbide, etc.). Some ceramic-metal composites are produced, such as tungsten carbide and cobalt, to make a hard, tough material for dies and cutting tools. In contrast, glass compositions can be tuned continuously over a wide range to achieve particular combinations of color and other optical properties (e.g. the graded compositions used in optical fibres). The adaptable spread of optical properties was evident in the property charts of Chapter 15. This is largely a matter of tweaking the chemistry rather than the process. More recent developments with glass processing are associated with coating technology, for scratch and chip resistance (in windscreens), or for self-cleaning and high reflectivity in high-rise buildings.

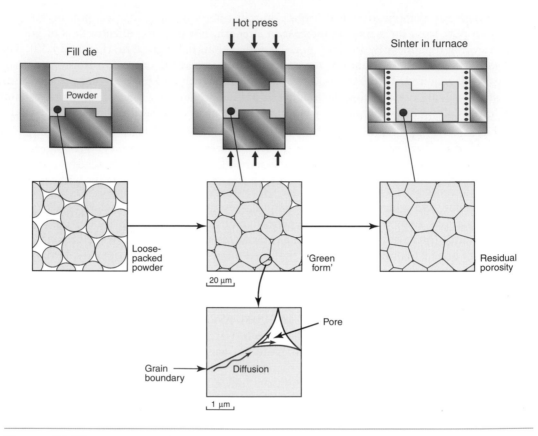

Figure 19.22   Mechanism of powder compaction in hot pressing and sintering.

***Powder processing***   Ceramics, and also many metals, can be shaped by filling a mould with loose powder and compacting it. The main microstructural evolution is the shrinkage of porosity during compaction, with each particle becoming a final grain (Figure 19.22). Powder compaction mechanisms are closely related to diffusional flow and creep (Chapter 13). In purely thermal compaction (sintering), atoms diffuse along the particle boundaries to fill in the pores. Compaction is accelerated by imposing external pressure as well as temperature—'hot isostatic pressing' (or HIPing)—giving particle deformation by creep.

***Polymers and elastomers***   Polymers are versatile materials in many respects—easy to mould into complex shapes at low cost, easy to join together by snap fitting and easy to colour. Bulk polymers provide a spread of Young's modulus, strength and fracture toughness, all relatively low. Polymer chemistry, and the use of additives and fillers, gives freedom to adapt the bulk properties, though the ability to make large changes in strength is more limited than for metals. But in contrast, polymers do offer scope for changing elastic properties, which metals do not. Polymers find widest application in packaging, and for diverse household and workplace applications (computer and phone casings, toys, pens, kitchen appliances and containers and so on). Design of these products mostly leads to only modest mechanical performance, but low cost and aesthetics dominate.

Chapter 4 introduced the structure of polymers, and how chemistry is used to produce the thousands of variants available today. Long-chain molecules are produced by polymerisation of a basic monomer unit of carbon, bonded with hydrogen, chlorine, oxygen and so on. Each monomer has a discrete chemistry—$C_2H_4$ makes polyethylene, for example. Unlike metal (and non-metal) atoms, monomers do not readily combine with one another. Only selected combinations will polymerise together, to make copolymers (with two monomers), terpolymers (with three) and so on. For example, ABS is a terpolymer, combining monomers of acrylonitrile, butadiene and styrene (hence the name). Blending polymers extends the range further, giving a mixture of more than one type of chain molecule—for example, ABS—PVC. And the molecular architecture itself can be manipulated. As described in Chapter 4, thermosets and elastomers are cross-linked, and some thermoplastics will partially crystallise. Furthermore, bulk polymers usually contain additives and fillers, for all sorts of reasons—for example, glass or ceramic particles or fibres (for stiffness, strength or colour), flame retardants, plasticisers or UV protection.

*Polymer moulding*   Most of the microstructural variety in polymers has only a modest impact on processing—most processes work with most polymers. Thermoplastics are moulded as viscous liquids. Additives change the viscosity, so the ease of moulding will be influenced by the volume fraction of filler. Injection moulding and extrusion dominate. Polymers mostly prefer to form an amorphous structure, but in some thermoplastics partial crystallinity may develop, the extent depending on the cooling rate. All polymers shrink as the mould cools from the moulding temperature to room temperature, because of both thermal contraction and the loss of free volume caused by crystallisation. Allowance must be made for this when the mould is designed. Fusion welding works only for non-cross-linked thermoplastics, which can be remelted.

Processing can play a significant role in determining the final molecular arrangement, and thus properties, if the molecules are first aligned mechanically by the flow induced during the shaping process, and cooling is then rapid enough to freeze in the alignment. Figure 19.23 shows the blow moulding of a drinks bottle, from a preform closed tube called a *parison*. In some variants the parison is stretched longitudinally with an internal rod prior to blow moulding. The stretching process of blow moulding aligns the molecules in the bottle wall, enhancing stiffness and strength. The degree of stretch is even customised to give greater enhancement in the circumferential direction, which carries the greater tensile stress under pressure.

*Polymer fibres*   The most dramatic impact of processing on polymer properties is in making fibres. These are drawn from the melt, or for even better results, cold drawn. Now the molecular alignment is much more marked, so that the fibre properties exploit the covalent bonding along the chain. We can illustrate the resulting strength and stiffness on a property chart. To emphasise the lightweight performance with respect to other materials, Figure 19.24 shows the specific strength $\sigma_y/\rho$ and specific stiffness $E/\rho$. Bulk polymers, in spite of all the chemical trickery, span only about a decade in strength and stiffness, and are outperformed by metals, particularly on specific stiffness. Standard polymer fibres of nylon or polyester are stiffer and over 10 times stronger than the bulk polymer. They quickly overtake most natural fibres, such as cotton and hemp, but not silk, which is a particularly good natural fibre.

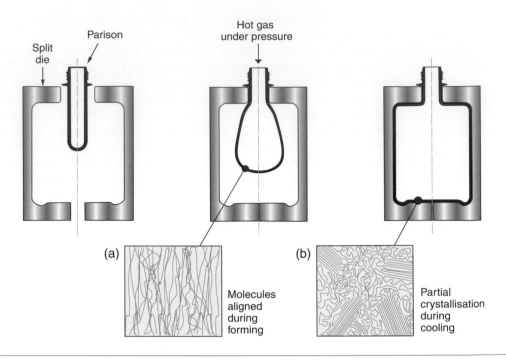

**Figure 19.23** Evolution of molecular architecture in polymer moulding: (a) alignment of molecules during viscous flow in shaping; (b) partial crystallisation during cooling.

However, aramid fibres (such as Kevlar), and special grades of polyethylene fibres, are up to 100 times stiffer and stronger than bulk polymer, with specific properties close or superior to carbon and glass fibres.

Fibre properties are all very impressive—the problem though is getting the fibres into a form that is useful to the engineer. One option is composites—aramid fibres are used in the same way as carbon and glass fibres, in a matrix of epoxy resin. As discussed in earlier chapters, the resulting composites shown in Figure 19.24 lie between those of the matrix and fibre, where they compete strongly with the best of the metals, such as low-alloy steels. Alternatively, we use polymer fibres in exactly the way that natural fibres have been used for millennia: twisted, tangled or woven to make rope, cables, fabrics and textiles. Fibre-based materials are a whole discipline of materials engineering in themselves, with many important applications. The geometric structure of ropes and textiles is essential to the way they work—it is this that gives them flexibility in bending, combined with enormous stiffness and strength in tension. The drape of clothing exploits the differences in stiffness parallel to and at 45 degree to the weave (the 'bias'). Ropes can be stretched, coiled and knotted elastically due to their twisted and woven architecture. The loss of alignment of the fibres in making a rope or weave does mean that their load-carrying capacity is never quite as good as that of the fibres themselves in their pristine form. Nonetheless it is clear from the property chart that cables based on the best polyethylene fibres, for example, can potentially compete with conventional steel cabling.

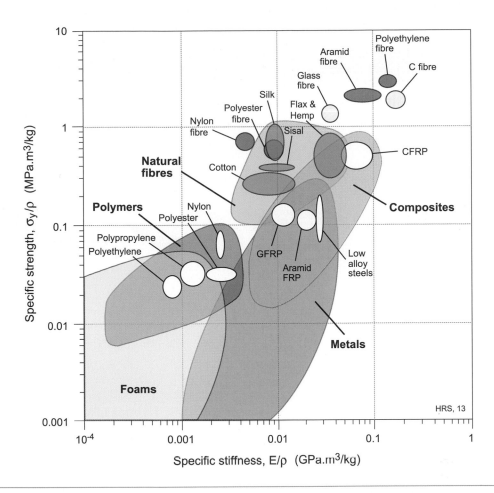

**Figure 19.24**   Specific strength-specific stiffness property chart. Bulk polymers are stiffened and strengthened by drawing into fibres, competing with natural, carbon and glass fibres. Most products use the fibres in a composite, such as CFRP, or woven into a fabric.

# **19.7** Making hybrid materials

Hybrids were defined in Chapter 2 as a combination of materials from different classes, one of which could just be air (as in foams). Making a hybrid is often a processing challenge, so we conclude the discussion of processing for properties with some examples. First we might ask: why bother with hybrids at all? The answer is simple: hybrids can occupy spaces on property charts not occupied by monolithic materials. Sometimes design requirements can be in direct conflict, with no single material able to provide the property profile needed—for example, high strength (for which alloys are best) and high electrical conduction (for which pure metals are best). Hybrids allow innovative design solutions and improvements in performance, exploiting

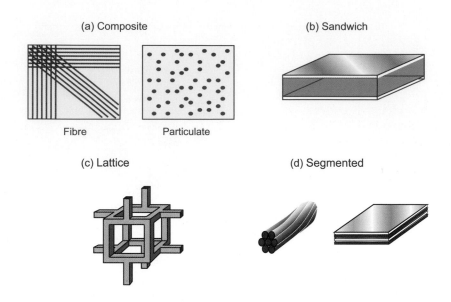

**Figure 19.25**  Classes of hybrid materials: (a) fibre and particulate composites; (b) sandwich panels; (c) lattice structures, such as foams; (d) segmented materials, such as ropes and laminates.

the individual properties of the component materials. Hybrids take a variety of forms—a simple classification is illustrated in Figure 19.25. Composites are the most straightforward—particles or fibres of one material embedded in another. Next are sandwich structures—skins of high-performance material on a lower grade core. Lattices (random or regular) include foams and other highly porous materials, and segmented materials are made of interlocking strands or layers of materials—ropes and cables are often hybrids of several types of material.

This classification is not 100% robust, since the boundaries get a bit fuzzy at the edges. For example, modern 'nanocomposites' are made by mixing nanometer-scale particulates into a molten metal and casting it—by Figure 19.25 this is a hybrid. However, heat treatment of steels produces an iron matrix with a fine dispersion of carbides—essentially the same microstructure, also produced for strength, yet we classify this as a metallic alloy. And at the other end of the scale, designing with a sandwich 'material' merges into designing a sandwich 'structure' (a ski, an aircraft floor panel or an industrial door). Essentially, we might define hybrids as structured materials whose component materials are routinely available in bulk and used in their own right, with the bulk materials and the hybrid having definable material properties. In spite of some non-uniqueness in this definition, hybrid materials can all be thought of as some combination of component (bulk) materials plus configuration and scale. Configuration defines the proportions, geometric integration and shape of the components; scale defines the length scale on which they are mixed. The length scales of hybrids are mostly relatively coarse—fibres of order $1-10$ μm, and the layers of laminates and sandwich panels of order 0.1–10 mm. Bulk material microstructures span everything from the atomic scale (nanometers) up to the micron scale (see Figures 19.2–19.4). However, since hybrids combine the properties of the component materials, their properties reflect microstructure all the way from atomic to product scale.

*Making foams*    Foams extend the range of stiffness, strength and density axes of the property charts downward by several orders of magnitude. Some foams are made by trapping bubbles in a melt and solidifying the froth as a slab. Others are made by mixing in chemicals that react to form a gas, injecting it into a cavity and allowing it to inflate in situ as the gas bubbles are formed internally. The mechanical properties of an ideal foam depend systematically on the relative density (the fraction of solid), as shown in Chapter 4 for Young's modulus. In practice it is found that the properties are also sensitive to the variability in pore size. It is important in processing to control the size of the largest pores, which initiate failure. Metal foams and rigid polymer foams can be made to shape and then cut and assembled into other structured materials (often as the cores of sandwich materials themselves).

*Making composites*    Most polymers are in fact particulate composites—bulk polymer melts to which filler powders have been added before moulding to shape. This is a hybrid that we opt to classify as a polymer variant—largely because the scale of the additives is very fine, almost at the scale of the polymer's own molecular microstructure. Metal–matrix composites, on the other hand, are regarded as distinct hybrids—SiC particles in aluminum alloys, for example—but they too are processed in exactly the same way as the matrix aluminum alloy. Long fibre-reinforced polymer composites—CFRP, GFRP and Kevlar (aramid FRP)—require dedicated manufacturing techniques. Their potential performance in lightweight design has been clearly illustrated in earlier chapters, but can the ideal properties be achieved routinely in practice? The answer depends on the complexity of the component geometry and the impact of joining. Composites are straightforward to produce in flat panels. 'Prepreg' layers of partly cured resin containing unidirectional fibres are stacked, usually aligned in several directions (e.g. 0, 90 degree, ±45 degree) to give reasonably uniform isotropy in the properties and cured under pressure in an autoclave.

Circular tubes are also easy to make by filament winding. More complex shapes are more difficult to laminate, particularly if the section changes in thickness. Joints are particularly troublesome. Fibre composites cannot be welded, and drilling holes for bolts and rivets damages the fibres and provides stress concentrations that can initiate cracks and delaminations around the hole. The best technique is adhesive joining, and the increasing use of composites has stimulated the development of improved adhesives. Careful joint design and manufacture are necessary to avoid premature failure in or around the joint.

*Making laminates and sandwich panels*    Most interest in laminates and sandwich panels reflects the potential to exploit fibre-reinforced composites in this form—a hybrid within a hybrid. Sandwich panels provide bending stiffness (and strength) at low weight. The lightweight core separates the skins, to maximise the second moment of area, without contributing directly to the bending resistance. Good-quality adhesive joints between skin and core are again critical for success. Comparison of sandwich panels with monolithic-shaped materials (such as I-beams) would normally be made on characteristics such as 'flexural rigidity' *EI*, as opposed to strict material properties. Laminates offer another hybrid solution. Various GFRP and aluminum laminates, such as GLARE and ARALL, have been developed for aerospace structures, as they offer a competitive alternative to purely aluminum or composite fabrication. And woods have had something of a renaissance in large-scale civil engineering in the form of laminates such as Glulam. The laminating process distributes the defects inherent in natural wood, reducing property variability, and enables much larger beam structures to be built. Once again, the glue between the layers in all laminates is a key part of the hybrid.

## 19.8 Summary and conclusions

Manufacturing processes are central to achieving the target properties in a material. Shaping processes such as casting, forming and moulding always impose a thermal history and often a deformation history. These do more than make the shape—they also govern the evolution of the internal microstructure on which properties depend. Further processing at the surface can locally change the microstructure and properties there, for example, for wear resistance. And joining components together can change things further—joints often being the design-limiting location, or the source of disaster. So processing offers many opportunities for innovation, together with responsibility for careful control. Metals show the greatest versatility—not surprising with the many different elements and alloy compositions available to work with, the diversity of phases that form in the solid state and their inherent castability and/or formability. One class alone—ferrous alloys—covers a wide domain of the key structural properties of strength and toughness. Other alloy systems use the same principles of modifying chemistry, forming and heat treatment to develop different micro-structures and thus property profiles.

  Ceramics, glasses and polymers all have some scope for variation of composition and process. Blending, cross-linking and crystallinity provide modest ranges in the properties of polymers, including the elastic modulus. Most effective is the production of fibres, taking bulk polymers into completely new territory, well beyond the performance of natural fibres. Making use of these fibre properties in practical products needs imaginative hybrid construction, as composites, ropes, textiles and fabrics. Hybrids also extend to other forms—sandwich panels, foams and laminates. Processing again plays its part in the practi-calities of using hybrids, with the development of high-performance adhesives receiving a boost from the drive to exploit their potential in design.

## 19.9 Further reading

Ashby, M. F., & Jones, D. R. H. (2005). *Engineering Materials II* (3rd ed.). ISBN 0-7506-6381-2. (Popular treatment of material classes, and how processing affects microstructure and properties.)
ASM Handbook Series (1971–2004). Volume 4, *Heat Treatment*; Volume 5, *Surface Engineering*; Volume 6, *Welding, Brazing and Soldering*; Volume 7, *Powder Metal Technologies*; Volume 14, *Forming and Forging*; Volume 15, *Casting*; and Volume 16, *Machining*; ASM International, Metals

Park, OH, USA. (A comprehensive set of handbooks on processing, occasionally updated, and now available online at <*www.products.asminternational.org/hbk/index.jsp*>)

Bralla, J. G. (1998). *Design for Manufacturability Handbook* (2nd ed.). New York, USA: McGraw-Hill. ISBN 0-07-007139-X. (Turgid reading, but a rich mine of information about manufacturing processes.)

Kalpakjian, S., & Schmid, S. R. (2003). *Manufacturing Processes for Engineering Materials* (4th ed.). New Jersey, USA: Prentice-Hall, Pearson Education. ISBN 0-13-040871-9. (A comprehensive and widely used text on material processing.)

Lascoe, O. D. (1988). *Handbook of Fabrication Processes*. Columbus, OH, USA: ASM International, Metals Park. ISBN 0-87170-302-5. (A reference source for fabrication processes.)

# **19.10** Exercises

| | |
|---|---|
| Exercise E19.1 | Briefly explain the meaning of the following terms in relation to casting processes: homogeneous nucleation, heterogeneous nucleation, inoculants, segregation, aluminium-killed steel, poisoning. |
| Exercise E19.2 | Briefly explain the meaning of the following terms in relation to deformation processing: annealing, recovery, recrystallisation, texture. |
| Exercise E19.3 | Briefly explain the meaning of the following terms in relation to bulk and surface heat treatment of metals: age-hardening aluminium alloys, martensite, tempering, hardenability, transformation hardening, carburising. |
| Exercise E19.4 | Exercises E19.1–E19.3 cover many of the reasons that alloying additions are made in metals processing. See if you can identify even more reasons for alloying, with examples. |
| Exercise E19.5 | Briefly explain the meaning of the following terms in relation to polymer processing: copolymer, polymer blend, crystallisation, fibres. Why is it possible to weld and recycle thermoplastics but not a thermoset or an elastomer? |
| Exercise E19.6 | The following table shows typical data for strength and fracture toughness of a selection of copper alloys, both cast and wrought. Sketch a property chart (on log scales) and plot the data. Use the chart to answer the following: |

(a) How do the cast and wrought alloys compare on fracture toughness, at comparable strength?

(b) Rank the strengthening mechanisms (as indicated in the table) in order of effectiveness.

(c) Do the trends observed in (a, b) follow a similar pattern to that seen in aluminum alloys (Figure 8.18)?

| Alloy | Process route | Main strengthening mechanisms | Yield strength (MPa) | Fracture toughness (MPa.m$^{1/2}$) |
|---|---|---|---|---|
| Pure Cu | Cast | None | 35 | 105 |
| Pure Cu | Hot-rolled | Work | 80 | 82 |
| Bronze (10% Sn) | Cast | Solid solution | 200 | 55 |
| Brass (30% Zn) | Cast | Solid solution | 90 | 80 |
| Brass (30% Zn) | Wrought + annealed | Solid solution | 100 | 75 |
| Brass (30% Zn) | Wrought | Solid solution + work | 400 | 35 |
| Cu−2% Be | Wrought + heat treated | Precipitation | 1000 | 17 |

Exercise E19.7    The following table shows the yield strength for a number of pure metals in the annealed condition, together with typical data for the strongest wrought alloys available based on these metals. By what factor is the strength increased in each alloy system, relative to that of the pure metal on which it is based? Which shows the greatest absolute increase in strength?

| Base element | Yield strength (MPa) (pure metal) | Strongest wrought alloy | Yield strength (MPa) (alloy) |
|---|---|---|---|
| Iron | 120 | Fe-Mo-Co-Cr-W-C quenched and tempered tool steel | 2750 |
| Aluminium | 25 | Al-Zn-Mg-Cu age-hardened | 625 |
| Copper | 35 | Cu-Be-Co-Ni age-hardened | 1250 |
| Magnesium | 65 | Mg-Al-Zn-Nd age-hardened | 435 |
| Nickel | 70 | Ni-Be cold-worked | 1590 |

Exercise E19.8    The ceramic alumina $Al_2O_3$ is hot or cold pressed and sintered with a variety of binders to facilitate densification. These binders have a much lower Young's modulus than the alumina, and there is usually some residual porosity, giving nominal compositions from 85 to 100% alumina. The table shows the resulting density, Young's modulus and compressive strength for a range of aluminas. Plot a graph showing the variation of these properties with alumina composition, accounting for the trends in each.

| Composition (%Al$_2$O$_3$) | Density (Mg/m$^3$) | Young's modulus (GPa) | Compressive strength (GPa) |
|---|---|---|---|
| 85.0 | 3.40 | 220 | 1.8 |
| 88.0 | 3.47 | 250 | 2.0 |
| 90.0 | 3.52 | 276 | 2.5 |
| 94.0 | 3.67 | 314 | 2.3 |
| 97.0 | 3.80 | 355 | 2.3 |
| 99.5 (fine grain) | 3.85 | 375 | 6.3 |
| 100.0 | 3.96 | 400 | 2.6 |

Exercise E19.9 The excellent specific properties of natural and artificial fibres were highlighted in Figure 19.24. It is also of interest to explore other property combinations that measure performance—for example, the maximum elastic stored energy. The following table summarises typical data for various fibres, together with some bulk polymers, and steel, for comparison.

Calculate the following performance indices for the materials given: (a) maximum elastic stored energy (per unit volume), $\sigma_f^2/E$; (b) maximum elastic stored energy (per unit mass), $\sigma_f^2/E\rho$.

Which material appears best on each criterion? Which criterion would be more important for climbing ropes? Why is it not practical for the properties of fibres to be exploited to their maximum potential?

| | Hemp | Spider web silk | Bulk nylon | Aramid fibre | Polyester fibre | PE fibre | Nylon fibre | Alloy steel wire |
|---|---|---|---|---|---|---|---|---|
| Young's Modulus (GPa) | 8 | 11.0 | 2.5 | 124 | 13 | 2.85 | 3.9 | 210 |
| Strength (MPa) | 300 | 500 | 63 | 3930 | 784 | 1150 | 616 | 1330 |
| Density (kg/m$^3$) | 1490 | 1310 | 1090 | 1450 | 1390 | 950 | 1140 | 7800 |

# 19.11 Exploring design with CES

Exercise E19.10 Use CES Level 3 data to explore the properties of Cu-Ni alloys. Extract data for Young's modulus, yield stress, fracture toughness, and electrical resistivity for annealed pure Cu and Ni, and a selection of alloys in between (e.g. 10%,

30%, 70% Ni). Sketch how the properties vary with Ni composition (between 0 and 100%). (Note that in CES you can plot the composition of the dominant element on a chart axis—try plotting each property against the %Cu, selecting both Cu-Ni and Ni-Cu alloys to cover the full range, together with pure Cu and Ni.)

(a) Which properties follow an approximate linear rule of mixtures between the values for pure Cu and Ni?
(b) The alloys in CES are not binary Cu-Ni alloys. Which properties appear most strongly influenced by the other alloying additions?

Exercise E19.11 Use CES Level 3 to plot the strength and fracture toughness of Mg and its alloys. How do cast and wrought alloy variants compare?

Exercise E19.12 The records for polymers in CES include the % filler as a parameter. Plot each of the following properties against the % filler for PA, PP and epoxy, and explore how effective fillers are at enhancing the Young's modulus and strength of these polymers. How does the addition of filler affect the fracture toughness and the price/kg?

Exercise E19.13 Exercise E19.8 investigated three properties of alumina $Al_2O_3$, as a function of the alumina composition. Use CES Level 3 to plot density, Young's modulus and compressive strength against the composition parameter 'Al2O3 (alumina) (%)', for all the ceramics listed under 'Alumina'. Do these plots confirm the trends found in Exercise E19.8? Explain any outliers.

# Chapter 20
# Materials, processes and the environment

Power from the wind. (Image courtesy of Leica Geosystems, Switzerland)

## **20.1** Introduction and synopsis

The practice of engineering consumes vast quantities of materials and is dependent on a continuous supply of them. We start by surveying this consumption, emphasising the materials used in the greatest quantities. Increasing population and improving living standards cause this consumption rate to grow—something it cannot do forever. Finding ways to use materials more efficiently is a prerequisite for a sustainable future.

There is also a more immediate problem: present-day material usage already imposes stress on the environment in which we live. The environment has some capacity to cope with this, so that a certain level of impact can be absorbed without lasting damage. But it is clear that current human activities exceed this threshold with increasing frequency, diminishing the quality of the world in which we now live and threatening the well-being of future generations. *Design for the environment* is generally interpreted as the effort to adjust our present product design efforts to correct known, measurable environmental degradation; the time-scale of this thinking is ten years or so, an average product's expected life. *Design for sustainability* is the longer-term view: that of adaptation to a lifestyle that meets present needs without compromising the needs of future generations. The time-scale here is less clear—it is measured in decades or centuries—and the adaptation required is much greater.

## **20.2** Material consumption and its growth

*Material consumption*   Speaking globally, we consume roughly 10 billion ($10^{10}$) tonnes of engineering materials per year. Figure 20.1 gives a perspective: it is a bar chart of the consumption of the materials used in the greatest quantities. It has some interesting messages. On the extreme left, for calibration, are hydrocarbon fuels—oil and coal—of which we currently consume a colossal 9 billion tonnes per year. Next, moving to the right, are metals. The scale is logarithmic, making it appear that the consumption of steel (the first metal) is only a little greater than that of aluminum (the next); in reality, the consumption of steel exceeds, by a factor of ten, that of all other metals combined. Steel may lack the high-tech image that attaches to materials like titanium, carbon-fibre-enforced composites and (most recently) nano-materials, but make no mistake, its versatility, strength, toughness, low cost and wide availability are unmatched.

Polymers come next: 50 years ago their consumption was tiny; today the combined consumption of commodity polymers polyethylene (PE), polyvinyl chloride (PVC), polypropylene (PP) and polyethylene-terephthalate (PET) begins to approach that of steel.

The really big ones, though, are the materials of the construction industry. Steel is one of these, but the consumption of wood for construction purposes exceeds that of steel even when measured in tonnes per year (as in the diagram), and since it is a factor of 10 lighter, if measured in m³/year, wood totally eclipses steel. Bigger still is the consumption of concrete, which exceeds that of all other materials combined. The other big ones are asphalt (roads), glass, and man-made and natural fibres.

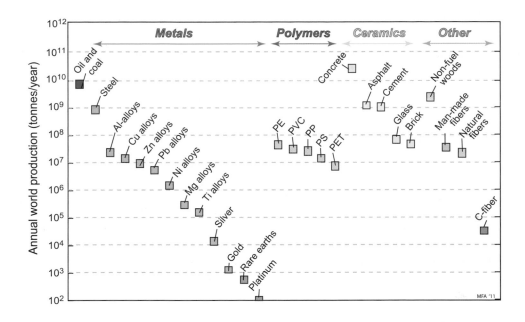

**Figure 20.1**    The consumption of hydrocarbons (left hand column) and of engineering materials (the other columns).

The last column of all illustrates things to come: it shows today's consumption of carbon fibre. Just 20 years ago this material would not have even crept onto the bottom of this chart. Today its consumption is approaching that of titanium and is growing fast.

The columns on this figure describe broad classes of materials, so—out of the thousands of materials now available—they probably exceed 99% of all material consumption when measured in tonnes. This is important when we come to consider the impact of materials on the environment, since impact scales with consumption.

***The growth of consumption***    Most materials are being consumed at a rate that is growing exponentially with time (Figure 20.2), simply because both population and living standards grow exponentially. One consequence of this is dramatised by the following statement: at a global growth rate of just 3% per year we will mine, process and dispose of more 'stuff' in the next 25 years than in the entire history of human engineering. If the current rate of consumption in tonnes per year is C then exponential growth means that

$$\frac{dC}{dt} = \frac{r}{100}C \qquad (20.1)$$

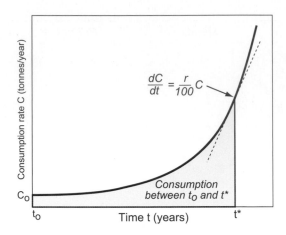

**Figure 20.2** Exponential growth. Consumption C doubles in a time $t_d \approx 70/r$ where $r$ is the annual growth rate.

where, for the generally small growth rates we deal with here (1–5% per year), $r$ can be thought of as the percentage fractional rate of growth per year. Integrating over time gives

$$C = C_0 \exp\left\{\frac{r(t - t_0)}{100}\right\} \tag{20.2}$$

where $C_0$ is the consumption rate at time $t = t_0$. The *doubling-time* $t_D$ of consumption rate is given by setting $C/C_0$ to give

$$t_D = \frac{100}{r}\log_e(2) \approx \frac{70}{r} \tag{20.3}$$

After a period of stagnation in the 1990s, steel consumption is growing again, driven by growth in China; at 4% per year it doubles about every 18 years. Polymer consumption is rising at about 5% per year—it doubles every 14 years. During times of boom—the 1960s and 1970s, for instance—polymer production increased much faster than this, peaking at 18% per year (it doubled every 4 years).

---

### Example 20.1 Calculating growth rates

World production $P$ of silver in 1950 was 4000 tonnes/year. By 2010 it had grown to 21,000 tonnes/year. Assuming exponential growth, what is the annual growth rate of production of silver?

*Answer.* Exponential growth of production $P$ tonnes per year is described by

$$P = P_0 \exp\left\{\frac{r(t - t_0)}{100}\right\}$$

where $P_0$ is the rate of production at time $t = t_0$. Setting $P = 21,000$ tonnes per year, $P_0 = 4000$ tonnes per year and $(t - t_0) = 60$ years, then solving for the growth rate $r$ gives

$$r = \frac{100}{(t - t_0)} \ln\left(\frac{P}{P_0}\right) = 2.8\% \text{ per year.}$$

### Example 20.2 Cumulative growth

A total of 5 million cars were sold in China in 2007; in 2008 the sale was 6.6 million. What is the annual growth rate of car sales, expressed as % per year? If there were 15 million cars already on Chinese roads by the end of 2007 and this growth rate continues, how many cars will there be in 2020, assuming that the number that are removed from the roads in this time interval can be neglected?

*Answer.* Starting with equation (20.2)

$$C = C_0 \exp\left\{\frac{r(t - t_0)}{100}\right\}$$

We enter $C = 6.6 \times 10^6$ and $C_0 = 5 \times 10^6$ and the time interval $(t - t_0) = 1$ year, and solve for $r$. The result is $r = 27.8\%$ per year.

The cumulative number of cars entering use in the subsequent 13 years is found from the integral of this equation over time

$$Q_{t^*} = \int_{t_0}^{t^*} C\, dt = \frac{100 C_0}{r}\left(\exp\left\{\frac{r(t^* - t_0)}{100}\right\} - 1\right)$$

Entering $C_0 = 5 \times 10^6$ (the number in 2007), $r = 27.8\%$ per year and $(t^* - t_0) = 13$, gives the additional number of cars by 2020 at $Q_{t^*} = 650 \times 10^6$. To this (if we are picky) we must add the number already there in 2007, giving a final total of $655 \times 10^6$—half a billion—an increase on the number in 2007 by a factor of 100. (This colossal number is still only equivalent to one car per three-person family, less than the current car ownership per family in the United States.)

*Strategic (critical) materials*  The mineral resource bases from which many materials are drawn are so large and widely distributed that the health of the supply chain is not a concern. The resource bases supporting the steel and aluminum industry are examples—both are vital to the economy, but it is the resource of energy rather than that of material that could limit their production. But there are others that are a cause of concern. They are the materials for which the known reserves are limited in size or are localised in countries from which supply cannot be guaranteed, or both. Governments classify materials as 'strategic' or 'critical' if their supply is concentrated in one country or could be restricted by a few corporate interests, and because they are used in products that are important economically or for national security.

Table 20.1 gives some examples of materials classified as strategic. It is interesting to ask 'Why?' The answer lies partly in their specialised uses and partly in the global location of the main ore bodies from which they are drawn. The members of the first group have applications in general engineering. The high electrical conductivity of copper gives it exceptional economic importance—replacing it by any affordable substitute results in increased resistive losses, lower electrical efficiencies and higher energy costs. Manganese has no satisfactory substitute as an alloying addition in carbon and stainless steels. Columbium (niobium), tantalum and vanadium, through their ability to form very stable carbides, have become essential ingredients of high-strength steels. The alloys of titanium have a unique position as light, high-strength, corrosion-resistant alloys.

The elements in the other three groups have more specialised applications. Electronic engineering, with its dependence on semiconducting devices, has created demand for a much wider spectrum of elements (it is said that a mobile phone contains most of the periodic table). Chemical and petroleum engineering rely heavily on catalysts drawn from the platinum group. Laser, magnet and battery technologies increasingly rely on rare earths for their performance. Many of these elements are drawn from small, highly localised resource bases: 80% of the world production of platinum group elements comes from South Africa; 90% of all rare earth supply comes from China. For these materials it is the unique properties, the rarity and the extreme localisation of the sources that makes them 'strategic'.

The picture, then, is one of a global economy ever more dependent on a supply of materials, almost all drawn from non-renewable resources, with many whose future availability is uncertain. To manage these in a sustainable way requires an understanding of the material life cycle. We turn to this next.

## 20.3 The material life cycle and criteria for assessment

*Life-cycle assessment and energy*   The materials life cycle is sketched in Figure 20.3. Ore and feedstock, drawn from the earth's resources, are processed to give materials; these are manufactured into products that are used, and, at the end of their lives, discarded, a fraction perhaps entering a recycling loop, the rest committed to incineration or landfill. Energy and materials are consumed at each point in this cycle (we shall call them 'phases'), with an associated penalty of $CO_2$ , $SO_x$, $NO_x$ and other emissions—heat, and gaseous, liquid and solid waste, collectively called environmental '*stressors*'. These are assessed by the technique of *life-cycle analysis* (LCA). A rigorous LCA examines the life cycle of a product and assesses in detail the eco-impact created by one or more of its phases of life, cataloguing and quantifying the stressors. This requires information for the life history of the product at a level of precision

Table 20.1 Elements that, by reason of critical use or extreme localisation, are deemed 'strategic'

| Element | Critical applications | Principal global sources* |
|---|---|---|
| *General engineering* | | |
| Copper | Electrical conduction in all electro-mechanical things | Canada, Chile, Mexico |
| Manganese | Essential alloying element in steels | South Africa, Russia, Australia |
| Columbium (Niobium) | Micro-alloyed steels, superalloys, superconductors | Brazil, Canada, Russia |
| Tantalum | Ultra-compact capacitors for mobile electronic equipment, alloying element in steels | Australia, China, Thailand |
| Vanadium | High-speed tool steels, microalloyed steels | South Africa, China |
| Cobalt | Cobalt-based superalloys, alloying element in steels | Zambia, Canada, Norway |
| Titanium | Light, high-strength, corrosion-resistant alloys | China, Russia, Japan |
| Rhenium | High-performance turbines | Chile |
| *Electronics* | | |
| Lithium | Lithium-ion batteries, Al-Li alloys for aircraft | Russia, Kazakhstan, Canada |
| Gallium | Gallium arsenide PV devices, semiconductors | Canada, Russia, China |
| Indium | Transparent conductors, InSb semiconductors, LEDs | Canada, Russia, China |
| Germanium | Solar cells | China |
| *Platinum group* | | |
| Platinum | Catalyst in chemical engineering and auto exhausts | South Africa, Russia |
| Palladium | Catalyst in chemical engineering and auto exhausts | South Africa, Russia |
| Rhodium | Catalyst in chemical engineering and auto exhausts | South Africa |
| *Rare earths* | | |
| Lanthanum | High-refractive-index glass, hydrogen storage, battery-electrodes, particularly for hybrid cars | China, Japan, France |

Table 20.1 Elements that, by reason of critical use or extreme localisation, are deemed 'strategic'—*continued*

| Element | Critical applications | Principal global sources* |
|---|---|---|
| Cerium | Catalysis, alloying element in aluminum alloys | China, Japan, France |
| Praseodymium | Rare-earth magnets, materials for lasers | China, Japan, France |
| Neodymium | Rare-earth permanent magnets, lasers | China, Japan, France |
| Promethium | Nuclear batteries (beta-emissions to electric power) | China, Japan, France |
| Samarium | Rare-earth magnets, lasers, neutron capture, lasers | China, Japan, France |
| Europium | Red and blue phosphors, lasers, mercury-vapor lamp | China, Japan, France |
| Gadolinium | Rare-earth magnets, high-refractive-index glass or garnets, lasers, X-ray tubes, computer memory, neutron capture | China, Japan, France |
| Terbium | Green phosphors, lasers, fluorescent lamp | China, Japan, France |
| Dysprosium | Rare-earth magnets, lasers | China, Japan, France |
| Holmium | Lasers | China, Japan, France |
| Erbium | Lasers, vanadium steel | China, Japan, France |
| Ytterbium | Infrared lasers, high-temperature superconductors (YBCO) | China, Japan, France |
| Lutetium | Catalysts in petroleum industry | China, Japan, France |
| Lanthanum | High-refractive-index glass, hydrogen storage, battery-electrode, camera lens | China, Japan, France |

* Sources: USGS (2002); US Department of Energy (2010); Jaffee and Price (2010).

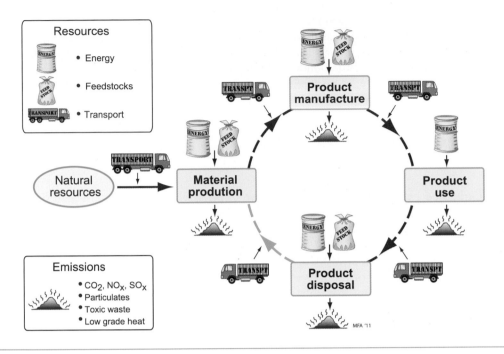

**Figure 20.3** The material life-cycle. Ore and feedstock are mined and processed to yield a material. This is manufactured into a product that is used and at the end of its life, discarded or recycled. Energy and materials are consumed in each phase, generating waste heat and solid, liquid and gaseous emissions.

that is only available after the product has been manufactured and used. It is a tool for the evaluation and comparison of existing products, rather than one that guides the design of those that are new. A full LCA is time-consuming and expensive, and it cannot cope with the problem that 80% of the environmental burden of a product is determined in the early stages of design when many decisions are still fluid. This has led to the development of more approximate 'streamline' LCA methods that seek to combine acceptable cost with sufficient accuracy to guide decision-making, the choice of materials being one of these decisions. But even then there is a problem: a designer, seeking to cope with many interdependent decisions that any design involves, inevitably finds it hard to know how best to use data of this type. How are $CO_2$ and $SO_x$ emissions to be balanced against resource depletion, toxicity or ease of recycling?

This perception has led to efforts to condense the eco-information about a material production into a single measure or *indicator*, normalising and weighting each source of stress to give the designer a simple, numeric ranking. The use of a single-valued eco-indicator is criticised by some. The grounds for criticism are that there is no agreement on normalisation or weighting factors, and that the method is opaque since the indicator value has no simple physical significance. But on one point there is international agreement: the Kyoto Protocol of 1997 committed the developed nations that signed it to progressively reduce carbon emissions, meaning $CO_2$. At the national level the focus is more on reducing energy consumption, but since this and $CO_2$ production are closely related, they are nearly equivalent. Thus there is a certain logic in basing design decisions on energy consumption or $CO_2$ generation; they carry more conviction and public understanding than the use of a more obscure eco-indicator. We shall follow this route, using energy as our measure. Before doing this, however, we need some definitions.

## 20.4 Definitions and measurement: embodied energy, process energy and recycling energy

*Embodied energy* $\mathbf{H_m}$ *and* $\mathbf{CO_2}$ *footprint*   The *embodied energy* of a material is the energy that must be committed to create 1 kg of usable material—1 kg of steel stock, or of PET pellets, or of cement powder, for example—measured in MJ/kg. The $CO_2$ *footprint* is the associated release of $CO_2$, in kg/kg. It is tempting to try to estimate embodied energy via the thermodynamics of the processes involved—extracting aluminum from its oxide, for instance, requires the provision of the free energy of oxidation to liberate it. This much energy must be provided, it is true, but it is only the beginning. The thermodymamic efficiencies of processes are low, seldom reaching 50%. Only part of the output is usable—the scrap fraction ranges from a few per cent to more than 10%. The feedstocks used in the extraction or production themselves carry embodied energy. Transport is involved. The production plant itself has to be lit, heated and serviced. And if it is a dedicated plant, one that is built for the sole purpose of making the material or product, there is an 'energy mortgage'—the energy consumed in building the plant in the first place.

Embodied energies are more properly assessed by *input–output analysis*. For example, for a material such as ingot iron, cement powder or PET granules, the embodied energy/kg is found by monitoring over a fixed period of time the total energy input to the production plant (including that smuggled in, so to speak, as embodied energy of feedstock) and dividing this by the quantity of usable material shipped out of the plant. The upper part of Figure 20.4 shows, much simplified, the inputs to a PET production facility: oil derivatives such as naptha and other feedstock, direct power (which, if electric, is generated with a production efficiency of about 34%), and the energy of transporting the feedstock to the facility. The plant has an hourly output of usable PET granules. The embodied energy of the PET, $(H_m)_{PET}$, with usual units of MJ/kg, is then given by

$$(H_m)_{PET} = \frac{\sum \text{Energies entering plant per hour}}{\text{Mass of PET granules produced per hour}}$$

*The processing energy*   $H_p$ associated with a material is the energy, in MJ, used to shape, join and finish 1 kg of the material to create a component or product. Thus polymers, typically, are moulded or extruded; metals are cast, forged or machined; ceramics are shaped by powder methods. A characteristic energy per kg is associated with each of these. Continuing with the PET example, the granules now become the input (after transportation) to a facility for blow-moulding PET bottles, as shown in the lower part of Figure 20.4. There is no need to list the

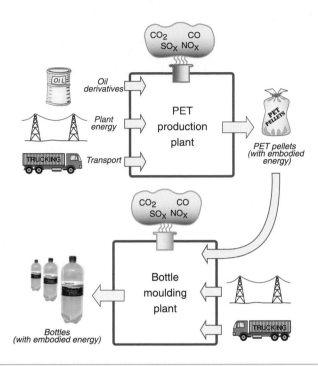

Figure 20.4   An input-output diagram for PET production (giving the embodied energy/kg of PET) and for bottle production (giving the embodied energy/bottle).

inputs again—they are broadly the same, the PET itself bringing with it its embodied energy $(H_m)_{PET}$. The output of the analysis is the energy committed per bottle produced.

There are many more steps before the bottle reaches a consumer and the contents are drunk: collection, filtration and monitoring of the water (plus other ingredients, if any), transportation of drink and bottles to a bottling plant, labelling, delivery to a central warehouse, distribution to retailers and possibly refrigeration prior to sale. All have energy inputs, which, when totalled, give the energy cost of as simple a thing as a plastic bottle of cold drink.

The *end-of-life potential* summarises the possible utility of the material at life's end: the ability to be recycled back into the product from which it came; the lesser ability to be down-cycled into a lower-grade application; the ability to be biodegraded into usable compost; the ability to yield energy by controlled combustion; and, failing all of these, the ability to be buried as landfill without contaminating the surrounding land then or in the future.

***Recycling: ideals and realities***   We buy, use and discard paper, packaging, cans, bottles, television sets, computers, furniture, tires, cars, even buildings. Why not retrieve the materials they contain and use them again? What could be simpler?

If you think that, think again. First, some facts. There are (simplifying again) two sorts of 'scrap', by which we mean material with recycling potential. In-house scrap is the off-cuts, ends and below-specification material left in a production facility when the usable material is shipped out. Here ideals are realised: almost 100% is recycled, meaning that it goes back into the primary production loop. But once a material is released into the outside world the picture changes. It is processed to make parts that may be small, very numerous and widely dispersed; it is assembled into products that contain many other materials; it may be painted, printed or plated; and its subsequent use contaminates it further. To reuse it, it must be collected (not always easy), separated from other materials, identified, decontaminated, chopped and processed. Collection is time-intensive, and this makes it expensive. Imperfect separation causes problems: even a little copper or tin damages the properties of steel; residual iron embrittles aluminum; heavy metals (lead, cadmium, mercury) are unacceptable in many alloys; PVC contamination renders PET unusable, and dyes, water and almost any alien plastic renders a polymer unacceptable for its original demanding purpose, meaning that it can only be used in less demanding applications (a fate known as 'down-cycling').

Despite these difficulties, recycling can be economic, both in cash and energy terms. This is particularly so for metals: the energy commitment per kg for recycled aluminum is about one-tenth of that for virgin material; that for steel is about one-third. Some inevitable contamination is countered by addition of virgin material to dilute it. Metal recycling is both economic and makes important contributions to the saving of energy and the efficient use of materials.

The picture for plastics is less rosy. The upper part of Figure 20.5 illustrates this for PET. Bottles are collected and delivered to the recycling plant as mixed plastic—predominately PET,

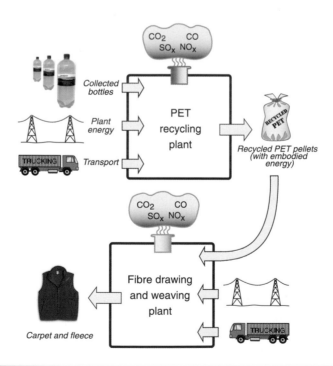

Figure 20.5   A simplified input-output diagram for the recycling of plastics to recover PET, and its use to make lower-grade products such as fleece.

Table 20.2   The energy-absorbing steps in recycling PET

| | |
|---|---|
| 1. Collection | 7. Melting |
| 2. Inspection | 8. Filtration |
| 3. Chopping | 9. Pelletising |
| 4. Washing | 10. Packaging |
| 5. Flotation—separation | 11. Plant heating, lighting |
| 6. Drying | 12. Transport |

but with PE and PP bottles too. Table 20.2 lists the steps required to recycle the PET, each one consuming energy, with the results listed in Table 20.3. Some energy is saved, but not a lot—typically 50%.

Recycling of PET, then, can offer an energy saving. But is it economic? Time, in manufacture, is money. Collection, inspection, separation and drying are slow processes, and every minute adds dollars to the cost. Add to this the fact that the quality of recycled material is less good than the original, limiting its use to less demanding products, as suggested by the lower part of Figure 20.5—recycled PET cannot be used for bottles. Table 20.3 lists the current market price of granules of five commodity polymers in the virgin and the recycled state. If the recycled stuff were as good as new it would command the same price; in reality it commands little more than half. Thus, using today's technology, the cost of recycling plastics is high and the price they command is low—not a happy combination.

The consequences of this are brought out by Figure 20.6. It shows the current recycle fraction of commodity metals and plastics and other materials. The recycle fraction is the fraction of current supply that derives from recycling. For metals it is high: most of the lead and almost half the steel and one-third of the aluminum we use today has been used at least once before. For plastics the only small success is PET with a recycle fraction of about 20%, with similar fractions for glass and brick. For the rest, the contribution is 15% or less, and for many it is zero. Oil price inflation and restrictive legislation could change all this, but for the moment, that is how it is.

Table 20.3   Embodied energy and market price of virgin and recycled plastics

| Polymer | Embodied energy[+], (MJ/kg) | | Price* ($/kg) | |
|---|---|---|---|---|
| | Virgin | Recycled | Virgin | Recycled |
| HDPE | 82 | 40 | 1.9 | 0.9 |
| PP | 82 | 40 | 1.8 | 1.0 |
| PET | 85 | 55 | 2.0 | 1.1 |
| PS | 101 | 45 | 1.5 | 0.8 |
| PVC | 66 | 37 | 1.4 | 0.9 |

[+] Approximate values; see CES EduPack for details.
* Spot prices, December 2005.

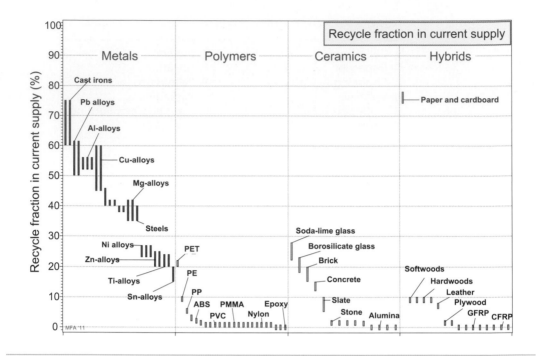

**Figure 20.6**   The fractional contribution of recycled material to current consumption. For metals, the contribution is large; for polymers, small (2005 data).

*The energy demands of products*   With this background, we can proceed to look at the way products consume energy in each of the four life phases of Figure 20.3. The procedure is to tabulate the main components of the product together with their material and weight. The embodied energy and energy of processing associated with the product are estimated by multiplying the mass of the different parts (the 'bill of materials') by the appropriate energies $H_m$ and $H_p$, and summing. The use-energy of energy-using products may be estimated from information for the power, the duty cycle and the source from which the power is drawn. To this should be added the energy associated with maintenance and service over the useful life of the product. The energy of disposal is more difficult: some energy may be recovered by incineration, some saved by recycling, but, as already mentioned, there is also an energy cost associated with collection and disassembly. Transport costs can be estimated from the distance of transport and the energy/km.kg of the transport mode used. You will find data for transport energies at the end of this chapter.

Despite the uncertainty in some of the data, the outcome of this analysis is revealing. Figure 20.7 presents the evidence for a range of product groups. It has two significant features, with important implications. The product groups in the top row all consume energy as an unavoidable consequence of their use, and for these the use phase overwhelmingly dominates the life energy. The products in the bottom row depend less heavily on energy but are material-intensive; for these it is the embodied energy of the material that dominates. If large changes

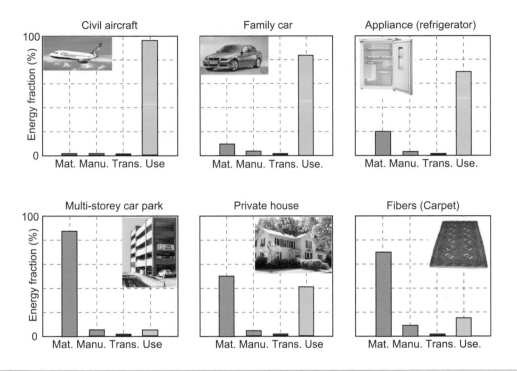

**Figure 20.7** Approximate values for the energy consumed at each phase of Figure 20.3 for a range of products. The columns show the approximate embodied energy ('Mat.'), energy to manufacture ('Manu.'), transport energy ('Trans.'), and use-energy over design life ('Use').

are to be achieved, it is the dominant phase that must be the first target; when the differences are as great as those shown here, a reduction in the others makes little impact on the total, and the precision of the data for $H_m$ and $H_p$ is not the issue—an error of a factor of 2 changes the outcome very little.

## 20.5 Charts for embodied energy

***Bar charts for embodied energy*** Figures 20.8 and 20.9 show the embodied energy per kg and per m$^3$ for materials. When compared per unit mass, metals, particularly steels, appear as attractive choices, demanding much less energy than polymers. But when compared on a volume basis, the ranking changes, and polymers lie lower than metals. The light alloys based on aluminum, magnesium and titanium are particularly demanding, with energies that are high by either measure. This prompts the question: what measure should we choose to make meaningful comparisons if we wish to minimise the embodied energy of a product? The answer is the same as the one we used with the objectives of minimising mass or cost: it is to minimise embodied energy *per unit of function*. To do that we need the next two charts.

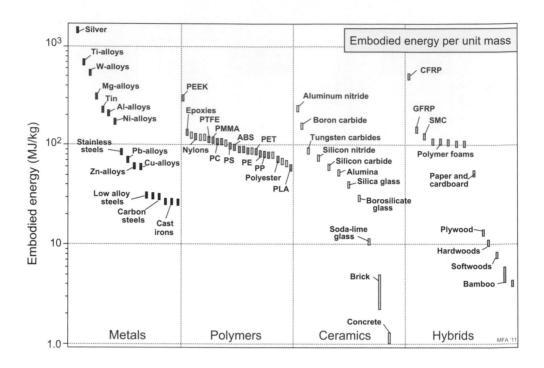

**Figure 20.8** Bar chart of embodied energy of basic materials by weight. By this measure polymers are more energy intensive than many metals.

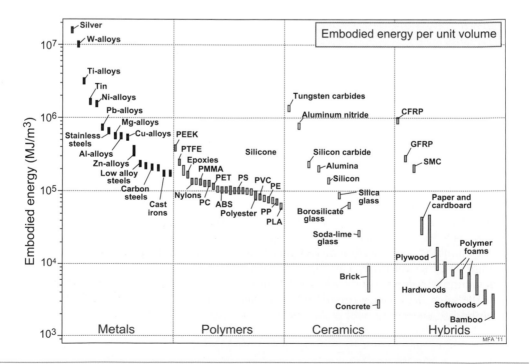

**Figure 20.9** Bar chart of embodied energy of basic materials by volume. By this measure polymers are less energy intensive than any metal.

*Property charts for embodied energy in structural design* Earlier chapters discussed the property trade-offs in problems of structural design. The function of the design might be, for example, to support a load without too much deflection, or without failure, while minimising the mass. For this, modulus–density or strength–density were used. If the objective becomes minimising the energy embodied in the material of the product, while providing structural functionality, we need equivalent charts for these.

Figures 20.10 and 20.11 are a pair of materials selection charts for minimising energy $H_m$ per unit stiffness and strength. The first shows modulus $E$ plotted against $H_m\rho$; the guidelines give the slopes for three of the commonest performance indices. The second shows strength $\sigma_y$ plotted against $H_m\rho$; again, guidelines give the slopes. The two charts give data suitable for minimum energy design limited by stiffness or strength. They are used in exactly the same way as the $E-\rho$ and $\sigma_y-\rho$ charts for minimum mass design.

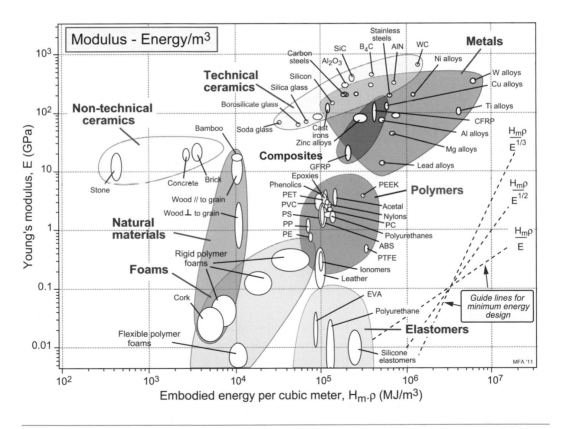

**Figure 20.10** The Modulus — Embodied energy chart. It is the equivalent of the $E-\rho$ chart of Figure 4.8 and is used in the same way.

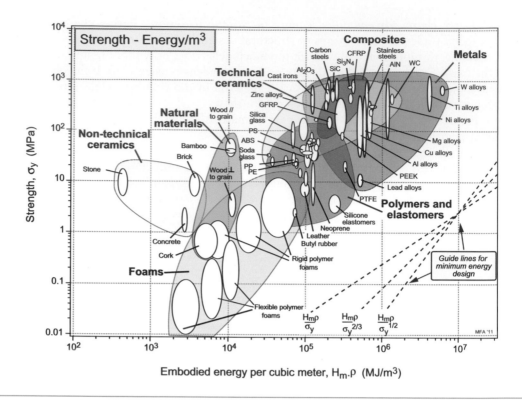

**Figure 20.11**    The Strength — Embodied energy chart. It is the equivalent of the $\sigma_y - \rho$ chart of Figure 6.7 and is used in the same way.

## 20.6 Design: selecting materials for eco-design

For selection of materials in environmentally responsible design we must first ask: which phase of the life cycle of the product under consideration makes the largest impact on the environment? The answer guides the effective use of the data in the way shown in Figure 20.12.

*The material production phase*    If material production consumes more energy than the other phases of life, it becomes the first target. Drink containers provide an example: they consume materials and energy during material extraction and container production, but, apart from transport and possible refrigeration, not thereafter. Here, selecting materials with low embodied energy and using less of them are the ways forward. Figure 20.7 made the point that large civil structures—buildings, bridges, roads—are material intensive. For these the embodied energy of the materials is the largest commitment. For this reason architects and civil engineers concern themselves with embodied energy as well as the thermal efficiency of their structures.

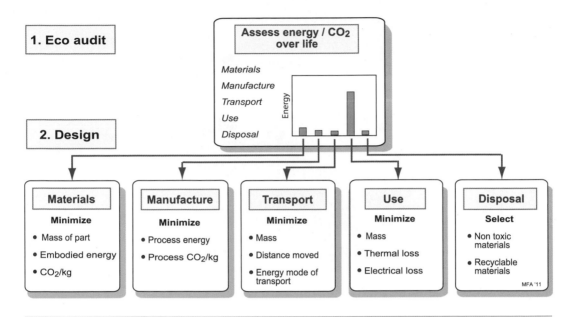

**Figure 20.12**  Rational use of the database starts with an analysis of the phase of life to be targeted. The decision then guides the method of selection to minimise the impact of the phase on the environment.

*The product manufacture phase*  The energy required to shape a material is usually much less than that to create it in the first place. Certainly it is important to save energy in production. But higher priority often attaches to the local impact of emissions and toxic waste during manufacture, and this depends crucially on local circumstances. Clean manufacture is the answer here.

*The product use phase*  The eco-impact of the use phase of energy-using products has nothing to do with the embodied energy of the materials themselves—indeed, minimising this may frequently have the opposite effect on use energy. Use energy depends on mechanical, thermal and electrical efficiencies; it is minimised by maximising these.

Fuel efficiency in transport systems (measured, say, by MJ/km) correlates closely with the mass of the vehicle itself; the objective then becomes that of minimising mass. The evidence for this can be seen in Figure 20.13, showing the fuel consumption of some 4000 European models of car against their unladen mass, segregated by engine type (super-sport and luxury cars, shown as red symbols, are separated out—for these, fuel economy is not a design priority). The lines show linear fits through the data, the lowest, through the green symbols, for diesel-powered cars, the one above, through the blue symbols, for those with petrol engines. One hybrid model is included (yellow symbol). The correlation between fuel consumption and weight is clear. Here the solution is *minimum mass design*, discussed extensively in earlier chapters; it is just as relevant to eco-design as to performance-driven design.

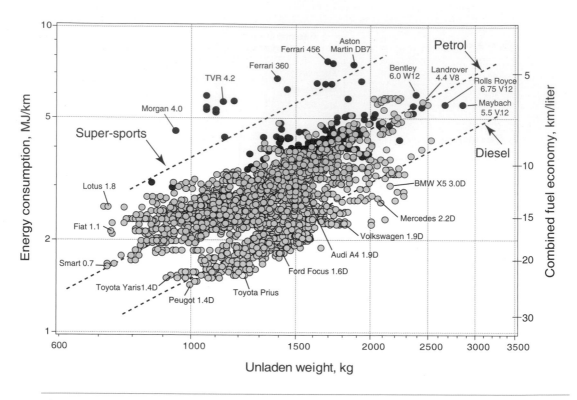

**Figure 20.13** Energy consumption and fuel economy of 2005 model European cars, plotted against the unladen weight. The open red symbols are diesels, the open black are petrol driven and the full red symbols are for cars designed with performance, above all else, in mind. The broken lines are best fits to data for each type. Note the near-linear dependence of energy consumption on weight.

Energy efficiency in refrigeration or heating systems is achieved by minimising the heat flux into or out of the system; the objective is then that of minimising thermal conductivity or thermal inertia. Energy efficiency in electrical generation, transmission and conversion is maximised by minimising the ohmic losses in the conductor; here the objective is to minimise electrical resistance while meeting necessary constraints on strength, cost and so on. Material selection to meet these objectives is well documented in other chapters and the texts listed under Further reading at the end of this chapter.

***The product disposal phase***   The environmental consequences of the final phase of product life have many aspects. The ideal is summarised in the following guidelines:

- Avoid toxic materials such as heavy metals and organometallic compounds that, in landfill, cause long-term contamination of soil and groundwater.
- Examine the use of materials that cannot be recycled, since recycling can save both material and energy; but keep in mind the influence they have on the other phases of life.

- Seek to maximise recycling of materials for which this is possible, even though recycling may be difficult to achieve for the reasons already discussed.
- When recycling is impractical seek to recover energy by controlled combustion.
- Consider the use of materials that are biodegradable or photo-degradable, although these are ineffectual in landfill because the anaerobic conditions within them inhibit rather than promote degradation.

Implementing these guidelines requires information for toxicity, potential for recycling, controlled combustion and biodegradability. The CES software provides simple checks of each of these.

***Case study: crash barriers***   Barriers to protect driver and passengers of road vehicles are of two types: those that are static (the central divider of a freeway, for instance) and those that move (the bumper of the vehicle itself) (Figure 20.14). The static type lines tens of thousands of miles of road. Once in place they consume no energy, create no $CO_2$ and last a long time. The dominant phases of their life in the sense of Figure 20.7 are those of material production and manufacture. The bumper, by contrast, is part of the vehicle; it adds to its weight and thus to its fuel consumption. The dominant phase here is that of use. This means that, if eco-design is the objective, the criteria for selecting materials for the two sorts of barrier will differ.

The function of a barrier is to transfer load from the point of impact to the support structure. Either this structure will collapse, absorbing energy, or the barrier and supports react against the vehicle, and the energy is absorbed in the crush elements designed into the vehicle. To do this the material of the barrier must have adequate strength, $\sigma_y$, and the ability to be shaped and joined cheaply and (thinking of the disposal phase of life) recyclable. The material for the car bumper must meet these constraints with minimum mass, since this will reduce the use energy. As we know from Chapter 7, for strength when loaded in bending, this means materials with high values of the index (equation (7.18))

$$M_1 = \frac{\sigma_y^{2/3}}{\rho}$$

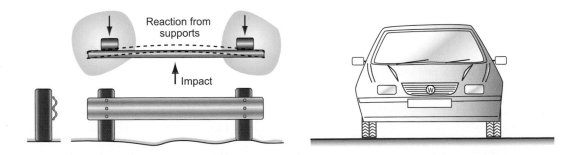

**Figure 20.14**  Two crash barriers, one static, the other — the bumper — attached to something that moves. Both are loaded in bending. Different eco-criteria are needed for each.

where $\sigma_y$ is the tensile strength and $\rho$ is its density. For the static barrier embodied energy, not weight, is the problem. If we change the objective to that of *minimum embodied energy*, we require materials with large values of

$$M_2 = \frac{\sigma_y^{2/3}}{H_m\rho}$$

where $H_m$ is the embodied energy per kg of material.

The chart of Figure 7.8 guides selection for the mobile barrier, where we seek strength at low weight. CFRPs excel by this criterion, but they are not recyclable. Heavier, but recyclable, are alloys of magnesium, titanium and aluminum. Ceramics are excluded both by their brittleness and the difficulty of shaping and joining them.

The $\sigma_y - H_m\rho$ chart of Figure 20.11 guides the selection for static barriers, where we seek strength at low embodied energy. The index $M_2$ is plotted on it in Figure 20.15. The chart

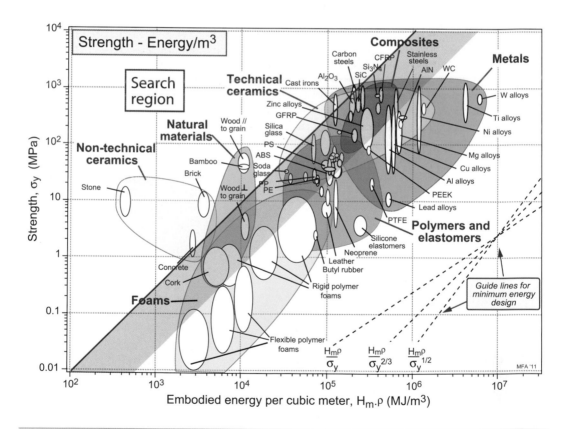

**Figure 20.15** The selection of materials for strength at minimum embodied energy. The best choices (rejecting ceramics because they are brittle) are cast iron, steel and wood.

shows that embodied energy per unit strength (leaving ceramics aside because of brittleness) is minimised by making the barrier from carbon steel, cast iron or wood; nothing else comes close.

Stiffness-limited design is treated in a similar way. Achieving it at minimum mass was the subject of Chapters 4 and 5. To do so at minimum embodied energy just requires that $\rho$ is replaced by $H_m\rho$.

## **20.7** Materials and sustainability

Here is a much-quoted definition of sustainable development: 'Sustainable development is development that meets the needs of the present without compromising the ability of future generations to meet their own needs' (*The Brundtland Report of the World Council on Economic Development*, WCED, 1987).

It sounds right. But how is it to be achieved? And where do materials fit in? The definition gives no concrete guidance.

So let's try another view of sustainability, one expressed in the language of accountancy: the Triple Bottom Line or 3BL (Figure 20.16). The idea is that a corporation's ultimate success and health should be measured not just by the traditional financial bottom line, but also by its social—ethical and environmental performance. Instead of just reporting the standard bottom line of the income and outgoings ('Prosperity') the balance sheet should also include the bottom lines of two further accounts: one tracking impact on the environmental ('Planet') and one tracking the social ('People') balance. In this view, sustainable business practice requires that the bottom lines of all three columns show positive balances, represented by the 'Sustainable' sweet-spot on Figure 20.16. Many businesses now claim to implement 3BL reporting; indeed the Dow Jones Sustainability Index[1] of leading industries is based on it. But is it really possible for all three bottom lines to be

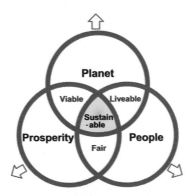

**Figure 20.16** Triple Bottom Line, or 'PPP' thinking.

[1] Dow Jones Sustainability Index (2012). <http://www.sustainability-index.com/>

positive at the same time? And again: where do materials fit in? The 3BL concept does not give clear guidance.

***The three capitals*** We make better progress if we separate the circles of Figure 20.16 and unpack their content, so to speak. Here a view of sustainability seen through the lens of economics can help. Global or national 'wealth' can be seen as the sum of three components: the *net manufactured capital*, the *net human capital* and the *net natural capital*. They are defined like this:

- *Manufactured capital*—Industrial capacity, institutions, roads, built environment, financial wealth (GDP).
- *Human capital*—Health, education, skills, technical expertise, accumulated knowledge, happiness.
- *Natural capital*—Clean atmosphere, fresh water, fertile land, productive oceans, accessible minerals and fossil energy.

All can (with some difficulty) be quantified in a common measure, dollars, say. The sum of all three, the *net comprehensive capital*, is a measure of national or global wealth (Figure 20.17).

Assigning precise values to the growth or decline of each capital has obvious difficulties. But it is possible to assign a *sign* and *order of magnitude* to the change in each capital, reporting whether it is positive or negative, large or small. 'Strong sustainability', in this picture, is development that delivers positive growth in all three capitals. 'Weak sustainability' is development that delivers positive comprehensive capital, ensuring that the sum of the capitals passed on to future generations is positive, even if one of them is diminished.

The main force, today, that drives change in the three capitals is the pressure for economic growth. An economy that grows is seen as healthy; one that does not is stagnant, and one that is in recession is sick. Positive economic growth is seen as so essential to the welfare of a nation that its influence on natural and human capitals is sometimes treated as secondary. Economic

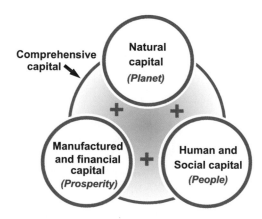

**Figure 20.17**  The three capitals and their sum.

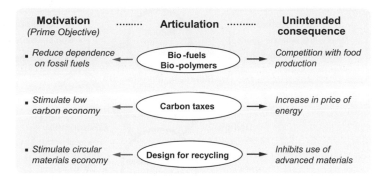

**Figure 20.18**  Competing articulations of sustainable technology conflict. The Motivation provides the Prime Objective.

growth may contribute to human capital by enabling greater education and health care, for instance, or it may diminish it by encouraging unfair labour practices and social inequality. And unfettered economic growth must, in the long run, diminish natural capital by consuming irreplaceable natural resources.

***Competing articulations of sustainable development***  Perceptions such as these have stimulated activities to diminish the undesired impacts of economic growth—particularly to diminish resource consumption and emission release. These activities, of which there are many examples, are presented by their proponents as contributions to sustainable development. Each has a particular motivation—to reclaim scarce elements from mobile phones, for example, or to reduce the carbon emissions from cars. We will refer to them as 'articulations' of sustainable technology. The difficulty with almost all of them is that they conflict: an articulation that addresses one facet of the problem may aggravate another.

Figure 20.18 gives some examples. Advocates of *bio-fuels and bio-polymers* do so because they diminish dependence on fossil hydrocarbons, but the land and water required to grow them is no longer available for the cultivation of food. *Carbon taxes* are designed to stimulate a low-carbon economy, but they increase the price of energy, and hence of materials and products. *Design for recycling* is intended to meet the demand for materials with less drain on natural resources, but it constrains the use of lightweight composites because most cannot be recycled. The motivation for *ethical sourcing of raw materials* (sourcing them only from nations with acceptable records of human rights) is that of social responsibility but a side-effect is to increase the prices paid for resources. The many different articulations of sustainable technology aim to support one or another of the three capitals of Figure 20.17, but they generally address a single facet of a multifaceted challenge and very few support all three—nor do they often support each other.

Thus sustainable technology is not one thing, it is many. While the starting point may be an environmental one, sustainability extends far beyond eco-design to embrace economics, legislation, social and ethical issues. Examination of a longer list of articulations drawn from

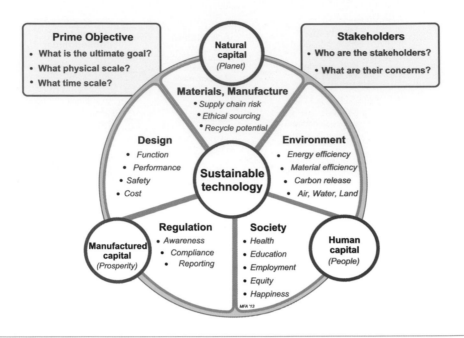

**Figure 20.19** Necessary components of sustainable technology.

journals of sustainable technology[2] suggests that the central issues might be grouped under the five broad headings shown in the central circle of Figure 20.19. Each heads a checklist for what might be called a 'sustainability analysis' of a design, scheme, project or product:

- *Materials and Manufacture*: ethical sourcing, supply-chain risk, recycle potential.
- *Design*: product function, performance, material selection, safety and cost.
- *Environment*: material efficiency, energy efficiency, bio-efficiency, preserving clean air, water and land.
- *Regulation*: awareness of, and compliance with, national and international agreements, legislation, directives, restrictions and agreements.
- *Society*: health, education, shelter, employment, equity, happiness.

This reasoning suggests a five-step strategy for assessing a design or project that claims to contribute to sustainable development. It is illustrated in Figure 20.20 and summarised as follows.

Any articulation of sustainability has an underlying motive—a *Prime Objective*—with both a physical scale and a time scale. If the articulation is going to make a significant difference the

---
[2] *World Journal of Science, Technology and Sustainable Development*, ISSN: 2042-5945. *World Journal of Science, Technology and Sustainable Development*, ISSN: 2042-5945. *Journal of Technology Management & Innovation*, ISSN: 0718-2724. *Journal of Clean Technology and Environmental Sciences*, ISSN: 1052-1062. *Clean Technologies and Environmental Policy*, ISSN: 1618-9558.

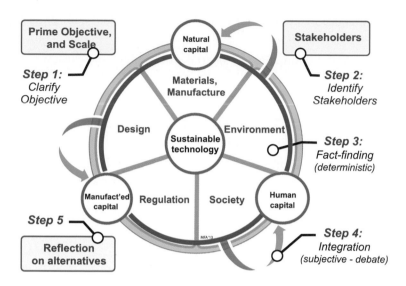

**Figure 20.20** The five-step sustainability analysis of technology.

physical scale is likely to be large and may demand significant natural resources. Thus the first step is *to clarify the Prime Objective and its scale, physical and temporal.*

Stakeholders are involved. If their concerns are not addressed the articulation will face opposition and may fail to gain acceptance. The second step, then, is to *identify the stakeholders and their concerns*—they set the context in which the assessment is carried out.

The third step is one of *fact-finding.* The questions posed by the five headings in the cartwheel of Figure 20.19 are straightforward, factual. Each can be researched; relevant information can be assembled from generally available books, databases and the Internet, guided by checklists. Conclusions about each can be drawn from this information in an objective way.

The fourth step, *integration,* is one of informed debate, drawing together the five blocks of information from Step 3 to form a balanced judgment about their impacts on the three capitals. Each capital has a base value. The technology, if implemented, will change these values. If the changes increase human and manufactured capital and reduce the drain on natural capital, and do so on a nontrivial scale, the technology makes a contribution to a more sustainable society. If positive change in one capital comes at the expense of increased drain on another, the relative merits require further debate.

The fifth and last step is that of *reflection on alternatives.* Is the Prime Objective achieved? Does it do so on a scale that makes a significant difference? Do the negative impacts on the three capitals outweigh the benefits? Have the stakeholders' concerns been met? Can the analysis suggest a new, less damaging way of achieving the Prime Objective? There is no completely 'right' answer to questions of sustainability—instead, there is a thoughtful, well-researched response that recognises the many conflicting facets and seeks the most productive compromise.

The varied concerns are best illustrated by an example: issues of sustainability raised by proposals for wind farms.

*Wind farms*   Many nations have undertaken to reduce the carbon emissions arising from electric power generation and seek at the same time to diversify their sources of electrical power. One strategy is to encourage the building of wind farms that feed electricity into the national grid. At the start of 2012 there were about 200,000 wind turbines worldwide, averaging 2 MW in power. The number, globally, is increasing at 25% per year, meaning that roughly 50,000 new turbines are installed each year.

**Step 1: Prime Objectives and scale**   These are defined in the project statement. There are two: to reduce national carbon emissions, and to provide a more diverse portfolio of electric power sources. To make any real difference to emissions, power from wind farms must have a lower carbon footprint than that of conventional power, and it must make a significant (20%, say) contribution to the total, setting an approximate target on the desired scale, at present seen as 50,000 new units per year.

**Step 2: Stakeholders and their concerns**   The national press reports initiatives to promote wind farms and the reactions these provoke. The reports identify most of the stakeholders and their known concerns. Here are two examples.

- 'Government and industry slam "spurious" anti-wind farm headlines' (*The Times*, 16 April 2012). The British government defends its policy of encouraging wind farms.
- 'Strike a blow against wind-farm bullies' (*The Times*, 35 February 2013). A columnist calls for protests against the siting of wind farms in Cornwall, the Lake District and other landscapes he loves.

Stakeholders need to be heard, reassured, persuaded or compensated if large-scale wind power is to be sustainable. Among them are:

- *National and Local Government*: Many nations have made commitments to reduce carbon emissions over a defined time period and see wind farms as able to contribute. To encourage their construction, nations impose taxes on carbon emissions and subsidise renewable energy projects. The erection of wind farms also creates jobs, attractive to government.
- *Energy providers*: Electricity-generation from fossil fuels releases carbon to atmosphere. Carbon taxes or carbon-trading schemes and carbon penalties create financial incentives for energy providers to reduce the use of fossil fuels.
- *Wind turbine makers*: Developing a manufacturing base for wind turbines requires considerable investment. Turbine makers want assurance that government policy on renewable energy is consistent and transparent, that incentives will not suddenly be withdrawn and that the supply chain for essential materials is secure.
- *Local communities*: There is opposition to land-based wind turbines from communities from which the turbines are audible or visible. Even off-shore wind farms are found objectionable by some. Feed-in tariffs for small-scale generation and compensation for acoustic intrusion aim to make turbines more acceptable.
- *The public at large*: To some, wind turbines are both necessary and beautiful, but others object to them and their associated power distribution systems because the power they

generate is intermittent and expensive, because they are visually and acoustically intrusive and because they are harmful to wildlife. They point out that the scale of deployment of wind farms has to be very large if they are to generate a significant fraction (say, 20%) of the nation's electrical power, and that energy-storage systems to deal with intermittent power generation add cost and require space.

**Step 3: Fact-finding**

*Materials and Manufacture*    Most of the materials of a wind turbine are conventional: carbon and stainless steels, concrete, copper, aluminum and polymer matrix composites. One is exceptional. The generators of wind turbines uses Neodymium-Boron rare-earth permanent magnets—their composition is shown in Table 20.4. Annual construction of 50,000 new turbines per year, each requiring 25 kg of Nd, creates a demand for 1,250 tonnes of Neodymium per year.

Table 20.4

| Nd-B magnets | Weight % |
|---|---|
| *Neodymium (Nd) | 30 |
| Iron (Fe) | 66 |
| Boron (B) | 1 |
| Aluminum (Al) | 0.3 |
| *Niobium (Cb) (Nb) | 0.7 |
| *Dysprosium (Dy) | 2 |

Starred (*) elements are on the critical list.

From which nations is Neodymium sourced? What proportion is this of global supply? Neodymium is co-produced with other rare-earth metals, of which it forms 15% on average. The US Geological Survey reports the global production of rare earths as 133,600 tonnes per year in 2011, corresponding to an annual production of Nd of 20,000 tonnes per year. Table 20.5 shows that over 95% derives from a single nation, China. The current rate of building wind turbines, as given in the question, carries a requirement of 1,250 tonnes of Neodymium per year. This is 6% of current global production.

Table 20.5

| Rare-earth producing nation | Tonnes/year 2011 |
|---|---|
| China | 130,000 |
| India | 3,000 |
| Brazil | 550 |
| Malaysia | 30 |
| World | 133,580 |

Source: Minerals.usgs.gov/minerals/pubs/commodity

*Design*   Permanent magnets for electric turbines require high remanent induction with high coercive force. Figure 15.8 of Chapter 15 plots these two properties. Neodymium-based magnets have by far the largest values of this pair of properties. If a substitute were to be sought, the next best choice would be the AlNiCo group of magnets, but all have a smaller remanent induction and a much smaller coercive force. There is no obvious substitute for Neodymium.

*The Environment*   The Prime Objective of a wind farm is to generate electrical power with low carbon emissions. It meets this objective only if the carbon emissions associated with its construction are quickly offset by the low carbon emissions during life to give net emissions per kW.hr that are lower than a conventional fossil-fuel power station. Various studies allow comparison of the carbon emission per kW.hr of delivered power for alternative systems—one is reproduced here as Figure 20.21. The data are approximate but sufficiently precise to establish that wind power has the ability to generate electrical power with significantly lower carbon emissions than gas- or coal-fired power stations when averaged over life. This, however, neglects power distribution: wind farms need windy places, often far from where the power will be used, and they may need energy storage systems to smooth intermittent generation. Present studies of wind power tend to ignore these two factors.

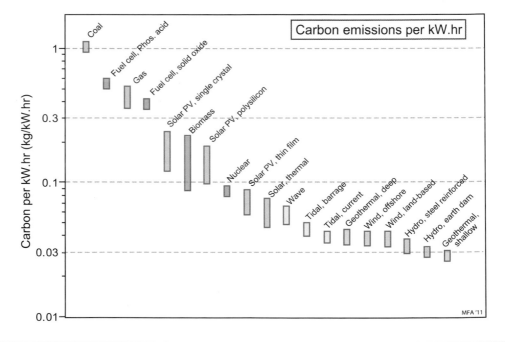

**Figure 20.21**   The carbon footprint per kW.hr of electrical power sources.

*Regulation*   A Web search for legislation relating to renewable energy and carbon emissions reveals a number that are relevant to the building of wind farms. They are listed in Table 20.6. From these we learn that making and installing wind farms is made financially attractive by 'green' subsidies and feed-in tariffs, but these have also been changed (usually down-graded) at short notice, making the market unpredictable.

Table 20.6

| Relevant Regulation and Legislation |
| --- |
| Carbon off-setting |
| Carbon tax |
| Carbon trading |
| Energy-Efficient Buildings Directive |
| Feed-in Tariffs |
| US Business Investment Tax Credit |
| US Recovery Act 1603 Program |

*Society*   The manufacture of wind turbines creates jobs. Against this must be set the cost. Without subsidies, electricity from wind farms is more expensive than that from gas-fired power stations. This cost is passed on to industrial and domestic consumers and, via subsidies, to the taxpayer. Per unit of generating power, wind farms require a land area that is almost 1000 times greater than a gas-fired power station, and although this land can still be used for agriculture, the scale of the visual intrusion is considerable. To put this in perspective, if the electric power requirement of the state of New York (average 33 kW.hr per day or 1.4 kW per person, population 19.5 million) were to be met by wind power alone, the necessary wind farms would occupy 21% of the area of the entire state (area 131,255 km$^2$).

The findings of the sustainability assessment thus far are summarised in Figure 20.22.

**Step 4: Integration**   This is the moment to reflect on and debate the relative importance of the information unearthed in the fact-finding step, using the effect on the three capitals as a framework. It will, inevitably, require an element of personal judgment and advocacy. The function of the previous three steps is to help inform the debate. Here is one view to set it off.

*Natural Capital*   The Prime Objective in building wind farms was to reduce greenhouse gas emissions and to diversify the sources of electrical power. The studies cited above suggested that they can do both. Their dependence on critical elements, particularly Neodymium, might give concern, but the placement of wind turbines is fixed and known, and large groups of them are managed by a single organisation, making the recovery, reconditioning or recycling of critical elements at end of life straightforward.

The beauty of the countryside is a component of natural capital. All power-generating plants occupy space and are visually intrusive. The problem with wind farms is the scale of this intrusion if they are to contribute significantly to national needs for power. The long-term impact of acoustic intrusion is not known.

*Human Capital*   Large-scale deployment of wind farms creates employment. If these jobs and wealth they generate are in Nations that are well governed, have fair distribution of wealth and equality of job opportunity, a contribution is made to Human Capital.

Some argue that the visual and acoustic intrusion of wind turbines represents a significant loss of quality of life. Against this must be set the reduction in emissions and in atmospheric pollution that can significantly damage human health.

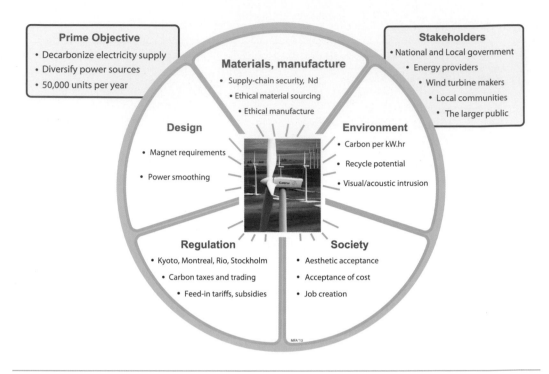

**Figure 20.22** Fact-finding for wind farms.

There is another aspect: that of independence and national security. A mix of energy sources increases independence and a distributed rather than a centralised power system is more robust, harder to disrupt and less vulnerable to a single catastrophic event.

*Manufactured Capital*  The typical design-life of a wind turbine is 25 years. Building 50,000 turbines per year is a significant investment in energy infrastructure. Is it a good investment? Some argue that it is not because, without a subsidy, the electricity they produce is more expensive than that from gas-fired power stations. Governments have been inconsistent in dealing with subsidies, encouraging investment at one moment and then cutting the subsidy with little warning the next. Much will depend on the price and predictability of hydrocarbon fuels over the next 25 years and the (at-present externalised) cost of carbon-induced climate change.

*Have the Stakeholders concerns been addressed?*  Wind farms contribute to governments' target of power from renewable energy. The concerns of energy providers and wind turbine makers for long-term commitment by government is not, at the present time, met, probably constraining investment. The concerns of local residents could be addressed by a design-focus on reduced acoustic signature, and the dislike of a neighbouring wind farm could be alleviated by compensation or reduced energy tariffs. The broader aesthetic concerns have no easy solution.

**Step 5: Reflection** This is the point for relaxed thinking.

Energy is one of humankind's most basic needs and electrical energy is the most versatile and valuable. We are in transition from a carbon-powered economy to one powered in other ways but the detailed shape of the future is not yet clear. A distributed energy mix in the economy is desirable. Wind farms together with other low-carbon power systems (hydro, geothermal, photo-voltaic and thermal solar, and nuclear) can all make some contribution, but for now the dominant source power continues to be fossil fuels. Perhaps we just have to live with wind farms as one, perhaps temporary, contribution while striving for cleaner ways to derive power from gas and coal.

## 20.8 Summary and conclusions

Rational selection of materials to meet environmental objectives starts by identifying the phase of product-life that causes greatest concern: production, manufacture, use or disposal. Dealing with all of these requires data not only for the obvious eco-attributes (energy, $CO_2$ and other emissions, toxicity, ability to be recycled and the like) but also data for mechanical, thermal, electrical and chemical properties. Thus if material production is the phase of concern, selection is based on minimising the embodied energy or the associated emissions ($CO_2$ production, for example). But if it is the use phase that is of concern, selection is based instead on low weight, or excellence as a thermal insulator, or as an electrical conductor while meeting other constraints on stiffness, strength, cost, and so on. The charts of this book give guidance in meeting these constraints and objectives. The CES databases provide data and tools that allow more sophisticated selection.

Eco-design is one aspect of sustainable development. But it is not the only one. Sustainable development requires clean energy and responsibly sourced materials that, as far as it is possible, are recovered and reused at the end of product life. It requires an economy that generates sufficient wealth to provide for daily needs and for investment in education, health and industrial infrastructure. And, if it is to really be sustainable, it must be equitable, not merely catering to the wishes of a sector of society but benefitting society as a whole. Rational judgment of the sustainability of a given project or plan ('articulation') requires facts and the willingness to debate the implications of the facts on prosperity, people and the planet.

## 20.9 Appendix: some useful quantities

| Energy contents of fuels | |
| --- | --- |
| Coal, lignite | 15–19 MJ/kg |
| Coal, anthracite | 31–34 MJ/kg |
| Oil | 11.69 kWh / litre = 47.3 MJ/kg |
| Petrol (gasoline) | 35 MJ/litre = 45 MJ/kg |
| Gas | 10.42 kWh / cubic meter |
| LPG | 13.7 kWh / litre = 46.5–49.6 MJ/kg |

| Approximate energy requirements of transport systems in MJ per tonne-km. | |
|---|---|
| Sea freight | 0.16 |
| Barge (river freight) | 0.36 |
| Rail freight | 0.25 |
| 2-axle (14 ton) truck | 1.5 |
| 3-axle (24 ton) truck | 1.1 |
| 4-axle (32 ton) truck | 0.94 |
| Air freight | 6.5—15, depending on type and size of plane |

**Conversion factors**
1 BthU = 1.06 kJ
1 kWh/kg (sometimes written kW/kg/hr) = 3.6 MJ/kg
A barrel of oil = 42 US gallons = 159 litres =138 kg = 6210 MJ
At \$100 per barrel, a dollar buys 62 MJ

## 20.10 Further reading

Ashby, M. F. (2012). *Materials and the Environment: Eco-Informed Material Choice*. Oxford, UK: Butterworth Heinemann. ISBN 978-0-12-385971-6. (A more advanced text developing the ideas presented in this chapter.)

Brundlandt, D. (1987). *Report of the World Commission on the Environment and Development*. Oxford, UK: Oxford University Press. ISBN 0-19-282080-X. (A much-quoted report that introduced the new and potential difficulties of ensuring a sustainable future.)

Dasgupta, P. (2010). Nature's Role in Sustaining Economic Development. *Phil Trans Roy Soc B, 365*, 5—11. (A concise exposition of the ideas of economic, human and natural capital.)

Dieter, G. E. (1991). *Engineering Design: A Materials and Processing Approach* (2nd ed.). New York, USA: McGraw-Hill. ISBN 0-07-100829-2. (A well-balanced and respected text focussing on the place of materials and processing in technical design.)

Elkington, J. (1998). *Cannibals with Forks: The Triple Bottom Line of 21st-Century Business*. Stony Creek, CT, USA: New Society Publishers.

Goedkoop, M., Effting, S., & Collignon, M. (2000). The Eco-Indicator 99: A Damage-Oriented Method for Life-Cycle Impact Assessment, Manual for Designers (14 April 2000) <http://www.pre.nl>. (An introduction to eco-indicators, a technique for rolling all the damaging aspects of material production into a single number.)

Graedel, T. E. (1998). *Streamlined Life-Cycle Assessment*. New Jersey, USA: Prentice Hall. ISBN 0-13-607425-1. (An introduction to LCA methods and ways of streamlining them.)

Jaffe, R., & Price, J. (2010). *Critical Elements for New Energy Technologies*. American Physical Society, USA: American Physical Society Panel on Public Affairs (POPA) study.

Kyoto Protocol. (1997). *United Nations, Framework Convention on Climate Change*. Document FCCC/CP1997/7/ADD.1 (http://cop5.unfccc.de) O 14040/14041/14042/14043, 'Environmental Management—Life-Cycle Assessment' and subsections. Geneva, Switzerland. (The international consensus on combating climate change.)

Mulder, K., Ferrer, D., & Van Lente, H. (2011). *What Is Sustainable Technology?* Sheffield, UK: Greenleaf Publishing. ISBN 978-1-906093-50-1. (A set of invited essays on aspects of sustainability, with a perceptive introduction and summing up.)

Porritt, J. (2012). http://www.forumforthefuture.org/project/five-capitals/overview.

US Department of Energy. (2010). *Critical Materials Strategy for Clean Technology.* www.energy.gov/
news/documents/criticalmaterialsstrategy.pdf. (A report on the role of rare earth elements and other
materials in low-carbon energy technology.)

US Department of Energy. (2010). *Critical Materials Strategy.* Office of Policy and International Affairs,
materialstrategy@hq.doe.gov, www.energy.gov. (A broader study than MRS 2010, above, but
addressing many of the same issues of material critical to the energy, communication and defence
industries, and the priorities for securing adequate supply.)

Wagner, L. W. (2002). *Materials in the Economy: Material Flows, Scarcity and the Environment.* USG.S.
Circular 2112, US Department of the Interior (www.usgs.gov). (A readable and perceptive summary of
the operation of the material supply chain, the risks to which it is exposed, and the environmental
consequences of material production.)

WCED. (1987). *Report of the World Commission on the Environment and Development.* Oxford, UK:
Oxford University Press.

## 20.11 Exercises

| | |
|---|---|
| Exercise E20.1 | What is meant by embodied energy per kilogram of a metal? Why does it differ from the thermodynamic energy of formation of the oxide, sulphide or silicate from which it was extracted? |
| Exercise E20.2 | What is meant by the process energy per kilogram for casting a metal? Why does it differ from the latent heat of melting the metal? |
| Exercise E20.3 | Why is the recycling of metals more successful than that of polymers? |
| Exercise E20.4 | The world consumption of CFRP is rising at 8% per year. How long does it take to double? |
| Exercise E20.5 | The global production of platinum in 2011 stands at about 178 tonnes per year, most from South Africa. The catalytic converter of a car requires about 1 gram of precious metal catalyst, most commonly platinum. Car manufacture in 2011 was approximately 52 million vehicles. If all have platinum catalysts, what fraction of the world production is absorbed by the auto industry? If the production rate of cars is growing at 4% per year and that of platinum is constant, how long will it be before the demand exceeds the total global supply? |
| Exercise E20.6 | Global water consumption has tripled in the last 50 years. What is the growth rate, $r\%$, in consumption $C$ assuming exponential growth? By what factor will water consumption increase between now (2013) and 2050? |
| Exercise E20.7 | Use the Internet to research rare earth elements. What are they? Why are they important? Why is there concern about their availability? |

Exercise E20.8    The price of cobalt, copper and nickel have fluctuated wildly in the past decade. Those of aluminum, magnesium and iron have remained much more stable. Why? Research this by examining uses (which metals are used in high value-added products?) and the localisation of the producing mines. The USGS website listed under Further reading is a good starting point.

Exercise E20.9    Which phase of life would you expect to be the most energy intensive (in the sense of consuming fossil fuel) for the following products:

- A toaster.
- A two-car garage.
- A bicycle.
- A motorbike.
- A wind turbine.
- A ski lift.

Indicate, in each case, your reasoning in one sentence.

Exercise E20.10    Car tyres create a major waste problem. Use the Internet to research ways in which the materials contained in car tyres can be used, either in the form of the tyre or in some decomposition of it.

Exercise E20.11    What are the US CAFE rules relating to car fuel economy? Use the Internet to find out and report your findings in ten sentences or less.

Exercise E20.12    The European Union's ELV (End of Life Vehicles) Directive on the disposal of vehicles at the end of life dictates what fraction of the weight of the vehicle must be recycled and what fraction is permitted to go to landfill. What are the current fractions? Use the Internet to find out.

Exercise E20.13    *Embodied energies.* Window frames are made from extruded aluminum. It is argued that making them instead from extruded PVC would be more environmentally friendly (meaning that less embodied energy is involved). If the section shape and thickness of the aluminum and the PVC windows are the same, and both are made from virgin material, is the claim justified? The table lists the data you will need.

| Material | Density, $kg/m^3$ | Embodied energy, MJ/kg |
|---|---|---|
| Aluminum | 2700 | 210 |
| PVC | 1440 | 82 |

Exercise E20.14    *Recycling energies.* The aluminum window frame of Exercise E20.13 is, in reality, made not of virgin aluminum but of 100% recycled aluminum with an embodied energy of 26 MJ/kg. Recycled PVC is not available, so the

PVC window continues to use virgin material. Which frame now has the lower embodied energy?

Exercise E20.15  *Recycling energies.* It is found that the quality of the window frame of Exercise E20.14, made from 100% recycled aluminum, is poor because of the pick-up of impurities. It is decided to use aluminum with a 'typical' recycled content of 44% instead. The PVC window is still made from virgin material. Which frame now has the lower embodied energy? The data you will need is given in the previous two exercises.

Exercise E20.16  *Precious metals.* A chemical engineering reactor consists of a stainless steel chamber and associated pipework weighing 3.5 tonnes, supported on a mild steel frame weighing 800 kg. The chamber contains 20 kg of loosely packed alumina spheres coated with 200 grams of palladium, the catalyst for the reaction. Compare the embodied energies of the components of the reactor, using data in the table.

| Material | Embodied energy (MJ/kg) |
|---|---|
| Stainless steel | 81 |
| Mild steel | 33 |
| Alumina | 53 |
| Palladium | 41,500 |

Exercise E20.17  The embodied energy of a mobile phone is about 400 MJ. The charger, when charging, consumes about 1 Watt. The phone is used for two years, during which it is charged overnight (8 hours) every night. Is the use energy larger or smaller than the energy to make the phone in the first place? What conclusions can you draw from your result? (In making the comparison, remember that electrical power is generated from primary fuel with an efficiency of about 38%.)

Exercise E20.18  The embodied energy of a mid-sized car is about 70 GJ. Assume the car is driven 250,000 km (150,000 miles) over its life, during which it averages 10 km/litre (23 miles per US gallon) of petrol (gasoline). Compare the energy involved in making the car with the energy consumed over its life. The energy content of petrol is 35 MJ/litre. What conclusions can you draw?

Exercise E20.19  *Transport energies.* Cast-iron scrap is collected in Europe and shipped 19,000 km to China where it is recycled. The energy to recycle cast iron is 5.2 MJ/kg. How much does the transport stage add to the total energy for recycling by this route? Is it a significant increase?

Exercise E20.20  *Transport energies.* Bicycles, weighing 15 kg, are manufactured in South Korea and shipped to the West Coast of the United States, a distance of 10,000 km. On unloading they are transported by 32-tonne truck to the point of sale, Chicago, a distance of 2900 km. What is the transport energy per bicycle?

To meet Christmas demand, a batch of the bicycles is air-freighted from South Korea directly to Chicago, a distance by air of 10,500 km. What is the transport energy then? Transport energies are listed in Section 20.9.

The bikes are made almost entirely out of aluminum. How do these transport energies compare with the total embodied energy of the bike? Take the embodied energy of aluminum to be 210 MJ/kg.

**Exercise E20.21**  Show that the index for selecting materials for a strong panel, loaded in bending, with the minimum embodied energy content is

$$M = \frac{\sigma_y^{1/2}}{H_m \rho}$$

**Exercise E20.22**  Use the $E - H_m \rho$ chart of Figure 20.10 to find the metal with a modulus $E$ greater than 100 GPa and the lowest embodied energy per unit volume.

**Exercise E20.23**  Use the $\sigma_y - H_m \rho$ chart of Figure 20.11 to find materials for strong panels with minimum embodied energy content.

**Exercise E20.24**  Car bumpers used to be made of steel. Most cars now have extruded aluminium or glass-reinforced polymer bumpers. Both materials have a much higher embodied energy than steel. Take the weight of a steel bumper to be 20 kg, and that of an aluminum one to be 14 kg; a bumper set (two bumpers) weighs twice as much. Find an equation for the energy consumption in MJ/km as a function of weight for petrol engine cars using the data plotted in Figure 20.13 of the text.
(a) Work out how much energy is saved by changing the bumper set of a 1500 kg car from steel to aluminum.
(b) Calculate whether, over an assumed life of 200,000 km, the switch from steel to aluminum has saved energy. You will find the embodied energies of steel and aluminum in the CES Level 2 database. Ignore the differences in energy in manufacturing the two bumpers—it is small. (The energy content of gasoline is 44 MJ/litre.)
(c) The switch from steel to aluminum increases the price of the car by $60. Using current pump prices for gasoline, work out whether, over the assumed life, it is cheaper to have the aluminum bumper or the steel one.

**Exercise E20.25**  A total of 16 million cars were sold in China in 2010; in 2008 the sale was 6.6 million. What is the annual growth rate of car sales, expressed as per cent per year? If there were 16 million cars already on Chinese roads by the end of 2010 and this growth rate continues, how many cars will there be in

2020, assuming that the number that are removed from the roads in this time interval can be neglected?

Exercise E20.26 | Prove the statement made in the text that, 'at a global growth rate of just 3% per year we will mine, process and dispose of more "stuff" in the next 25 years than in the entire history of human engineering'. For the purpose of your proof, assume consumption started with the dawn of the industrial revolution, 1750.

# 20.12 Exploring design with CES

Exercise E20.27 | Rank the three common commodity materials *low-carbon steel, age-hardening aluminum alloy* and *Polyethylene* by embodied energy/kg and embodied energy/m$^3$, using data drawn from Level 2 of the CES Edu database (use the means of the ranges given in the databases). Materials in products perform a primary function—providing stiffness, strength, heat transfer and the like. What is the appropriate measure of embodied energy for a given function?

Exercise E20.28 | Plot a bar chart for the embodied energies of metals and compare it with one for polymers, on a 'per unit yield strength' basis, using CES. You will need to use the 'Advanced' facility in the axis-selection window to make the function

$$\text{Energy per unit strength} = \frac{\text{Embodied energy} \times \text{Density}}{\text{Yield strength}} = \frac{H_m \rho}{\sigma_y}$$

Which materials are attractive by this measure?

Exercise E20.29 | Drink containers co-exist that are made from a number of different materials. The masses of five competing container types, the material of which they are made, and the specific energy content of each are listed in the table. Which container type carries the lowest overall embodied energy per unit of fluid contained?

| Container type | Material | Mass, g | Embodied energy, MJ/kg |
|---|---|---|---|
| PET 400 ml bottle | PET | 25 | 84 |
| PE 1 litre milk bottle | High-density PE | 38 | 80 |
| Glass 750 ml bottle | Soda glass | 325 | 14 |
| Al 440 ml can | 5000 series Al alloy | 20 | 200 |
| Steel 440 ml can | Plain carbon steel | 45 | 23 |

Exercise E20.30      Iron is made by the reduction of iron oxide ($Fe_2O_3$) with carbon, aluminum by the electro-chemical reduction of Bauxite, basically $Al_2O_3$. The enthalpy of oxidation of iron to its oxide is 5.5 MJ/kg, that of aluminum to its oxide is 20.5 MJ/kg. Compare these with the embodied energies of cast iron and of carbon steel, and of aluminum, retrieved from the CES 06 database (use mean values of the ranges). What conclusions do you draw?

Exercise E20.31      Calculate the energy to mould PET by assuming it to be equal to the energy required to heat PET from room temperature to its melting temperature, $T_m$. Compare this with the actual moulding energy. You will find the moulding energy, the specific heat and the melting temperature in the Level 2 record for PET in CES (use mean values of the ranges). Assume that the latent heat of melting is equal to that to raise the temperature from room temperature to the melting point. What conclusions do you draw?

Exercise E20.32      Use CES to plot the *Annual world production* of materials against their *Price*. What trend is visible?

Exercise E20.33      Make a bar chart of $CO_2$ footprint divided by embodied energy using the 'Advanced' facility in CES EduPack Level 2 software. Which material has the highest ratio? Why?

Exercise E20.34      Figure 20.9 of the text shows a plot of the embodied energy of materials per $m^3$. Use CES to make a similar plot for the carbon footprint per $m^3$ of material. Use the 'Advanced' facility in the axis selection window to plot kg $CO_2/m^3$ by multiplying kg $CO_2$/kg by the density in kg/$m^3$.

Exercise E20.35      Compare the eco-indicator values of materials with their embodied energy. To do so, make a chart with 'Embodied energy × Density' on the x-axis and 'Eco-indicator value' on the y-axis. (Ignore the data for foams since these have an artificially inflated volume). Is there a correlation between the two? Is it linear? Given that the precision of both could be in error by 10% are they significantly different measures? Does this give a way of estimating, approximately, eco-indicator values where none are available?

# 20.13 Exercises on sustainability

These exercises require a looser way of thinking. There is no 'right' answer. Instead there is a thoughtful, well-researched response that recognises the many conflicting facets of sustainable technology and that seeks compromises that offer the greatest common good with the least harm.

Exercise E20.36 In discussions of sustainable development the term *comprehensive global capital* appears. What does it mean and what are its three main components? In your judgement, does it seem possible to adopt a life-style that results in growth, or no diminution, of all three?

Exercise E20.37 Globally, affluence is increasing. Would you expect that Global Natural Capital would be influenced by this increase, and in which direction? The Gross Domestic Product (GDP) per capita is a measure of affluence per person per year. The Ecological Footprint is a measure of the human impact on Natural Capital, per person per year. A value greater than 1 means that Natural Capital is being drained faster than it can be replaced. You may find the following URLs useful in forming a judgment.

The Ecological Footprint: http://www.footprintnetwork.org/en/index.php/GFN/page/basics_introduction/

GDP of countries: http://en.wikipedia.org/wiki/List_of_countries_by_GDP_(PPP)

Exercise E20.38 Use research and judgment to form an opinion about the viability of extraction of critical elements from electronic waste. Critical elements are those that are rare or highly localised (leading to supply-chain constraints) or with no known effective substitutes or of strategic importance. The table lists five of these with the current grade of ore from which they are mined. Use the Internet to find the approximate concentration of these in mobile phones. Does it equal or exceed that of the ores from which they are currently extracted?

| Critical elements | Typically mined ore grade, wt % |
| --- | --- |
| Platinum | 0.00025 |
| Gold | 0.0015 |
| Silver | 0.055 |
| Cobalt | 0.5 |
| Copper | 2.0 |

# Guided Learning Unit 1: Simple Ideas of Crystallography

## Contents

## Introduction and synopsis

Though they lacked the means to prove it, the ancient Greeks suspected that solids were made of discrete atoms that were packed in a regular, orderly lattice to give crystals. Today, with the 21st century techniques of X-ray and electron diffraction and lattice-resolution microscopy available to us, we know that all solids are indeed made of atoms, and that most (but not all) are crystalline. The common engineering metals and ceramics are made of many small crystals, or *grains*, stuck together at *grain boundaries* to make *polycrystalline aggregates*. The properties of the material—its strength, stiffness, toughness, conductivity and so forth—are strongly influenced by the underlying crystallinity. So it is important to be able to describe it.

Crystallography is a language for describing the 3-dimensional arrangement of atoms or molecules in crystals. Simple crystalline and non-crystalline structures were introduced earlier in the book. Here we go a little further, exploring a wider range of structures, interstitial space (important in understanding the hardness of steel), and ways of describing planes and directions in crystals (crucial in the production of turbine blades and in the manufacture of microchips).

Exercises are provided in Parts 1—5, so do these as you go along; solutions are provided at the end of the unit.

## PART 1: Crystal structures

Three-dimensional crystal lattices may be characterised by the repeating geometry of the atomic arrangement they contain, particularly their degree of symmetry. Most atomic bonding gives a well-characterised equilibrium spacing between neighbouring atoms, so for the purposes of understanding crystal packing, the atoms may be treated as hard spheres. Spheres can be packed to fill space in various different ways—in fact it can be shown that there are 14 distinguishable three-dimensional lattices. If you are a crystallographer or mineral scientist you need to know about all of them. But engineering materials, for the most part, have simple structures. We shall concentrate on these.

Each lattice is characterised by a unit containing a small number of atoms, which repeats itself in 3D—the *unit cell*.

> **DEF.** The *unit cell* of a crystal structure is the unit of the structure, chosen so that it packs to fill space, and which, translated regularly, builds up the entire structure. The *primitive unit cell* is the smallest such cell.

Figure GL1.1 shows the most important unit cells. The first lattice is the *triclinic* unit cell, which is the most general—the edge lengths (or *lattice constants*) a, b and c are all different and none of the angles is 90°. The other 13 space (or *Bravais*[1]) lattices are

---

[1] Auguste Bravais (1811—1863), French crystallographer, botanist and physicist, sought to explain the shapes of mineral crystals by analysing the figures formed by points distributed regularly in space. Others had tried this and had concluded that the number of such 'lattices' distinguishable by their symmetry was finite and small, but it took Bravais to get it right, demonstrating that there are exactly 14.

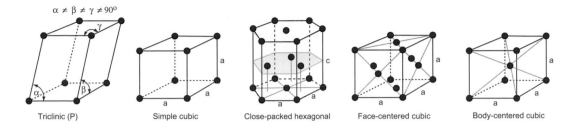

**Figure GL1.1** The general triclinic unit cell, and the four most important lattices for engineering materials: simple cubic (SC), close-packed hexagonal (CHP), face-centred cubic (FCC), and body-centred cubic (BCC).

all special cases of this one—for example, for all the cubic unit cells, a = b = c and $\alpha$ = $\beta$ = $\gamma$ = 90°.

The great majority of the 92 stable elements are metallic, and of these, the majority (68 in all) have one of the last three simple structures in Figure GL1.1: close-packed hexagonal (CPH, or sometimes HCP), face-centred cubic (FCC) and body-centred cubic (BCC). Both CPH and FCC lattices are close-packed—the spheres fill as much space as possible—and can be constructed by stacking close-packed planes of spheres. Let's investigate this as an Exercise (solutions are provided at the end of the Guided Learning Unit).

## Exercises

E.1 Figure GL1.2 shows a close-packed layer of spheres, with a partial second layer stacked on top. Identify two alternative ways to stack the third layer, and draw them on the two views below. Look at the alignment between the atoms in the first and third layers. What do you observe?

**Figure GL1.2** Two layers of close-packed planes.

*The close-packed hexagonal (CPH) structure* The CPH structure is usually described by a hexagonal unit cell, with an atom at each corner, one at the centre of the hexagonal faces and three in the middle layer. In this packing, the close-packed layers are clearly seen, and alternate close-packed layers are aligned (as you should have observed for one case in Exercise E.1), giving an ABAB ... sequence (Figure GL1.3). Unit cells are often drawn with the atoms reduced in size, to show clearly where there centres lie with respect to one another, but in reality atoms will touch in certain *close-packed directions*. Of the metallic elements, 30 have this structure. They include the following:

| Material | Typical uses |
|---|---|
| Zinc | Die-castings, plating |
| Magnesium | Lightweight structures |
| Titanium, its alloys | Light, strong components for airframes and engines, biomedical and chemical engineering |
| Cobalt | High temperature superalloys, bone-replacement implants |
| Beryllium | The lightest of the light metals; its use is limited by expense and potential toxicity |

HCP metals have the following characteristics.

• They are ductile, allowing them to be forged, rolled and drawn, but in a more limited way than FCC metals.
• Their structure makes them more anisotropic than FCC or BCC metals.

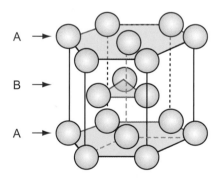

Figure GL1.3 A CPH structure of packed spheres showing the ABA stacking.

*The face-centred cubic (FCC) structure* The FCC structure is described by a cubic unit cell with one atom at each corner and one at the centre of each face. Atoms touch along the diagonals of the cube faces, and the centre atoms in any pair of adjoining faces also touch one another. The close-packed layers are not so obvious in FCC, but can be seen by looking down any diagonal of the cube. Figure GL1.4 shows that the close-packed layers are now stacked in an ABCABC ... sequence (the other case in Exercise E.1).

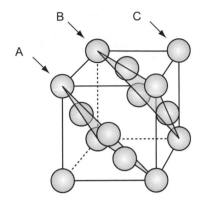

Figure GL1.4   An FCC structure of packed spheres showing the ABC stacking.

Among the metallic elements, 17 have the FCC structure. Engineering materials with this structure include the following.

| Material | Typical uses |
|---|---|
| Aluminum and its alloys | Airframes, space frames and bodies of trains, trucks, cars, drink cans |
| Nickel and its alloys | Turbine blades and disks |
| Copper and $\alpha$-brass | Conductors, bearings |
| Lead | Batteries, roofing, cladding of buildings |
| Austenitic stainless steels | Stainless cookware, chemical and nuclear engineering, cryogenic engineering |
| Silver, gold, platinum | Jewelry, coinage, electrical contacts |

FCC metals have the following characteristics:

- They are very ductile when pure, work hardening rapidly but softening again when annealed, allowing them to be rolled, forged, drawn or otherwise shaped by deformation processing.
- They are generally tough, resistant to crack propagation (as measured by their fracture toughness, $K_{1c}$).
- They retain their ductility and toughness to absolute zero, something very few other structures allow.

*The body-centred cubic (BCC) structure*   The BCC structure is described by a cubic unit cell with one atom at each corner and one in the middle of the cube (Figure GL1.5). The atoms touch along the internal diagonals of the cube, and this structure is not made up of close-packed planes.

Of the metallic elements, 21 have this structure (most are rare earths). They include the following.

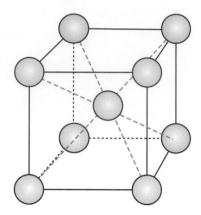

Figure GL1.5   A BCC structure of non-close-packed spheres.

| Material | Typical uses |
|---|---|
| Iron, mild steel | The most important metal of engineering: construction, cars, cans and more |
| Carbon steels, low alloy steels | Engine parts, tools, pipelines, power generation |
| Tungsten | Lamp filaments |
| Chromium | Electroplated coatings |

BCC metals have the following characteristics:

- They are ductile, particularly when hot, allowing them to be rolled, forged, drawn or otherwise shaped by deformation processing.
- They are generally tough, resistant to crack propagation (as measured by their fracture toughness, $K_{1c}$) at and above room temperature.
- They become brittle at low temperatures. The change happens at the 'ductile-brittle transition temperature', limiting their use below this.
- Their strength depends on temperature, even at low temperatures.
- They can generally be hardened with interstitial solutes.

## Exercises

E.2   (a) For the CPH structure, show that the ratio of the lattice constants, $c{:}a$ is equal to 1.633.

(b) The atomic radius of CPH magnesium is 0.1605 nm. Find the lattice constants, $c$ and $a$, for Mg.

E.3 Find the ratio of the lattice constant (the size of the cube) to the atomic radius for: (a) FCC; (b) BCC.

*Atomic packing fraction and theoretical density* The density of a crystalline material depends on the mass of the atoms and the way in which they are packed. We can find the theoretical density of the crystal directly from the mass and volume of the unit cell. An important first step is to be clear how many atoms are fully contained within a given unit cell (since atoms on the boundaries of the unit cell are shared with neighbouring cells). This also allows us to find the *atomic packing fraction*, that is, the proportion of space that is occupied by the atoms, represented as solid spheres. Let's do a worked example, for the FCC lattice.

Example GL1.1

Find the atomic packing fraction for the FCC lattice.

*Answer.* The volume of the FCC unit cell, with a lattice constant $a$, is equal to $a^3$. This can be expressed in terms of the atomic radius, using the result from Exercise E.3(a): $a = 2\sqrt{2}\,R$. Hence the unit cell volume is $16\sqrt{2}\,R^3$. Now consider the FCC unit cell in Figure GL1.4. Atoms on the corners are shared between 8 unit cells, while those on the faces are shared between 2. The number of atoms within a single unit cell is thus:

$$8 \times 1/8 \text{ (corners)} + 6 \times 1/2 \text{ (faces)} = 4 \text{ atoms}$$

Hence the atomic packing fraction for FCC =

$$\frac{4 \times \frac{4}{3}\,\pi R^3}{16\sqrt{2}\,R^3} = 0.74\ (74\%).$$

## Exercises

E.4 Use the same method to find the atomic packing fractions for: (a) CPH; (b) BCC. Compare these with the value for FCC, and comment on the differences.

The theoretical density is found as follows. First the mass of 1 atom is given by $A/N_A$, where $A$ is the atomic mass (in kg) of 1 mol of the element, and $N_A = 6.022 \times 10^{23}$ is Avogadro's number[2] (the number of atoms per mol). So if the number of atoms per unit cell is $n$, and its volume is $V_c$, the theoretical density is:

$$\rho = \frac{n\,A}{V_c\,N_A}$$

Since the atomic mass and the crystal packing are both physically well-defined, this explains why the densities of metals (and ceramics) have narrow ranges, and there is no scope to modify the density of a solid (e.g. by processing a metal differently).

---

### Example GL1.2

The diameter of an atom of nickel is 0.2492 nm. Calculate the theoretical density of FCC nickel (the atomic mass of Ni is 58.71 kg/kmol).

*Answer.*    From Exercise E.3(a), $a = D\sqrt{2}$, where the atomic diameter is $D$. Hence the lattice constant for Ni is $\sqrt{2} \times 0.2492$ nm $= 0.3524$ nm. Hence the volume of the unit cell $V_c = (0.3524 \times 10^{-9})^3$ m.

The mass of a nickel atom $= A/N_A = 58.71 \times 10^{-3}/(6.022 \times 10^{23}) = 9.749 \times 10^{-26}$ kg. As the number of atoms per unit cell $n = 4$, the theoretical density of nickel is:

$$\rho_{Ni} = 4 \times 9.749 \times 10^{-26} / \left(0.3524 \times 10^{-9}\right)^3 = 8911 \text{ kg/m}^3.$$

---

### Exercises

E.5    Gold has an FCC structure, a density of 19.3 Mg/m$^3$ and an atomic mass of 196.967 kg/kmol. Estimate the dimensions of the unit cell and the atomic diameter of gold.

E.6    (a) Determine the theoretical density for BCC iron, Fe (for which the atomic mass is 55.847 kg/kmol and the lattice constant is 0.2866 nm (at room temperature)).

---

[2] Amedeo Avogadro (1776–1856), Italian professor at the University of Turin, though originally a church lawyer and schoolteacher. The hypothesis and associated constant named after him stem from his essay on the molecular theory of gases, postulating that there are a fixed number of molecules in a given volume of an ideal gas, at constant temperature and pressure. Sacked from his chair for revolutionary activities against the king, the university's official stance was that he was on leave to concentrate on research, but this particular sabbatical lasted for 10 years before he was reinstated.

(b) A sample of pure iron is heated from room temperature, 20 °C, to 910 °C. The sample expands uniformly, such that the strain $\varepsilon$ in all directions is $\varepsilon = \alpha \, \Delta T$, where $\alpha$ is the thermal expansion coefficient, and $\Delta T$ is the temperature change. Find the expected value for the lattice constant at 910 °C, and thus the % decrease in the density compared to the room temperature value.

(c) At 910 °C, pure iron undergoes an important phase transformation to FCC (more on this in Guided Learning Unit 2). Use the atomic packing fractions of BCC and FCC to determine the % increase in density during the transformation of the pure Fe sample, and thus find the % reduction in the linear dimensions of the sample when this occurs.

E.7 $\alpha$-titanium has a CPH structure, a density of 4.54 Mg/m$^3$ and an atomic mass of 47.9 kg/kmol. Calculate the dimensions of the unit cell and the atomic diameter of titanium.

## PART 2: Interstitial space

*Interstitial space* is the unit of space between the atoms of molecules. The FCC and CPH structures both contain interstitial space of two sorts: *tetrahedral* and *octahedral*. They are shown for the FCC structure in Figure GL1.6(a) and (b).

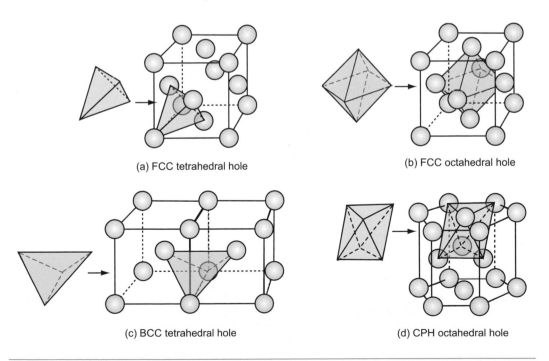

(a) FCC tetrahedral hole

(b) FCC octahedral hole

(c) BCC tetrahedral hole

(d) CPH octahedral hole

**Figure GL1.6** (a) and (b) The two types of interstitial hole in the FCC structure, (c) the tetrahedral hole of the BCC structure and (d) the octahedral hole in the CPH structure.

The holes are important because foreign atoms, if small enough, can fit into them. For both structures, the tetrahedral hole can accommodate, without strain, a sphere with a radius of 0.22 of that of the host. The octahedral holes are larger: they can hold a sphere that is … but wait, that is Exercise E.8. The atoms of which crystals are made, in reality, are slightly squashy, so that foreign atoms that are larger than the holes can be squeezed into the interstitial space.

This is particularly important for carbon steel. Carbon steel is iron with carbon in some of the interstitial holes. Iron is BCC, and, like the FCC structure, it contains holes (Figure GL1.6(c)). They are tetrahedral. They can hold a sphere with a radius 0.29 times that of the host without distortion. Carbon goes into these holes, but, because it is too big, it distorts the structure. It is this distortion that gives carbon steels much of their strength.

The CPH structure, like the FCC, has both octahedral and tetrahedral interstitial holes. The larger of the two, the octahedral hole, is shown in Figure GL1.6(d). The holes have the same sizes as those in the FCC structure.

We shall encounter these interstitial holes in another context later: they provide a way of understanding the structures of many oxides, carbides and nitrides.

Now more Exercises, in the interstitial space before the next subsection.

## Exercises

E.8   Calculate the diameter of the largest sphere that will fit into the octahedral hole in the FCC structure. Take the diameter of the host-spheres to be unity.

E.9   The CPH structure contains tetrahedral interstitial holes as well as octahedral ones. Identify a tetrahedral hole on the CPH lattice of Figure GL1.7.

E.10  From Figure GL1.6(c), there also appears to be an octahedral hole in BCC, in the middle of the shared face. Do the atoms around this hole form a regular octahedron? Can you

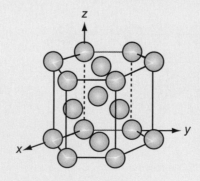

**Figure GL1.7**   The CPH structure.

identify another identical octahedron in the BCC lattice and locate the position of the hole on the unit cell? Find the size of the largest sphere that could fit into this space, taking the diameter of the host spheres to be unity. Hence explain why carbon sits in the tetrahedral holes in BCC iron.

E.11   Derive the result given in the text, that the largest sphere that fits in a tetrahedral hole in FCC or CPH lattices has a diameter equal to 0.22 times that of the host spheres.

## PART 3: Describing planes

The properties of crystals depend on *direction*. The elastic modulus of hexagonal titanium (CPH), for example, is greater along the hexagonal axis than normal to it. This difference can be used in engineering design. In making titanium turbine blades, for example, it is helpful to align the hexagonal axis along a turbine blade to use the extra stiffness. Silicon (which also has a cubic structure) oxidises in a more uniform way on the cube faces than on other planes, so silicon 'wafers' are cut with their faces parallel to a cube face. For these and many other reasons, a way of describing planes and directions in crystals is needed. *Miller*[3] *indices* provide it.

First, let's describe planes. The *Miller indices of a plane* are the *reciprocals* of the intercepts the plane makes with the three axes that define the edges of the unit cell, reduced to the smallest integers.

Figure GL1.8 illustrates how the Miller indices of a plane are found. Draw the plane in the unit cell so that it does *not* contain the origin—if it docs, displace it along one axis until it doesn't. Extend the plane until it intersects the axes. Measure the intercepts, in units of the cell edge-length, and take the reciprocals. (If the plane is parallel to an axis, its intercept is at infinity and the reciprocal is zero.) Reduce the result to the smallest set of integers by multiplying through to get rid of fractions or dividing to remove common factors. The six little sketches show, in order, the (100), the (110), the (111), the (211), the $(11\bar{1})$ and the (112) planes; the $\bar{1}$ means that the plane intercepts the $y$-axis at the point $-1$. Check that you get the same indices.

The Miller indices of a plane are always written in *round* brackets: (111). But there are many planes of the 111-type; two are shown in Figure GL1.8. The *complete family* is described by putting the indices in curly brackets, thus:

$$\{111\} = (111), (11\bar{1}), (1\bar{1}1), (\bar{1}11)$$

---

[3] William Hallowes Miller (1801—1880), British mineralogist, devised his index system (the 'Millerian system') in 1839. He also discovered a mineral which he named—wait for it—Millerite.

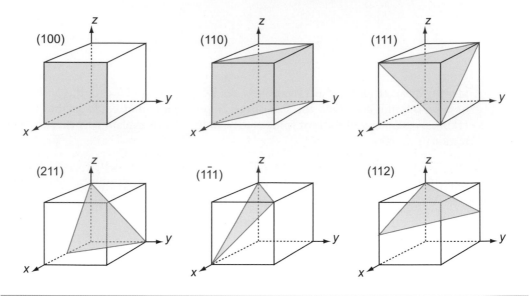

Figure GL1.8   The procedure for finding the Miller indices of a plane. The indices of six common planes are shown.

## Exercises

E.12   What are the Miller indices of the planes shown in the six sketches of Figure GL1.9(a), (b) and (c)? (Remember that, to get the indices, the plane must not pass through the origin, and remember too to get rid of fractions or common factors.)

E.13   When iron is cold it cleaves (fractures in a brittle way) on the (100) planes. What will be the angle between cleavage facets?

E.14   Mark, on the cells of Figure GL1.9(d), (e) and (f), the following planes: $(\bar{1}11), (120), (\bar{1}\bar{1}2)$.

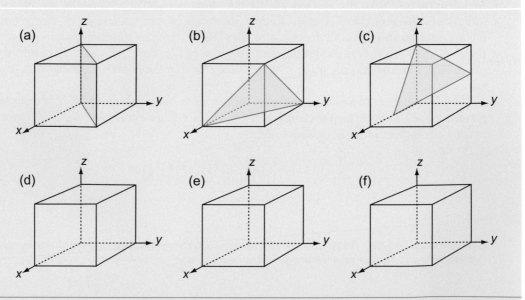

Figure GL1.9   Identify the planes shown in the upper three figures. Draw those required in the question in the lower three.

## PART 4: Describing directions

The *Miller indices of a direction* are the components of a vector (*not* reciprocals) that starts from the origin, along the direction, reduced to the smallest integer set.

Figure GL1.10 shows how to find the indices of a direction. Draw a line from the origin, parallel to the direction, extending it until it hits a cell edge or face. Read off the coordinates of the point of intersection. Get rid of any fraction or common factor by multiplying all the components by the same constant. The six sketches show the [010], the [011], the [111], the [021], the [$\bar{2}$12] and the [$\bar{1}$10] directions. The $\bar{1}$ means that the *y*-coordinate of the intersection point is −1.

The Miller indices of a direction are always written in *square* brackets: [100]. As with planes, there are several directions of the 100-type. The *complete family* is described by putting the indices in angle-brackets, thus:

$$<100> = [100], [010], [001]$$

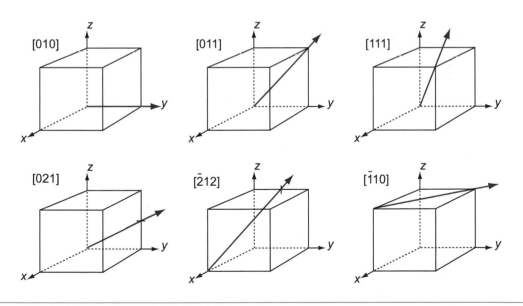

**Figure GL1.10** Miller indices of directions. Always translate the direction so that it starts at the origin.

## Exercises

E.15   Identify the Miller indices of the three directions shown in the sketches of Figure GL1.11(a), (b) and (c).

E.16   Mark, on the cells of Figure GL1.11(d), (e) and (f), the following directions: $[1\bar{1}1]$, $[210]$ and $[22\bar{1}]$.

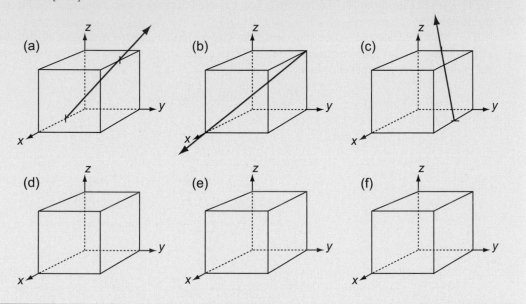

**Figure GL1.11**   Identify the directions shown in the upper three figures. Draw those required in the question in the lower three.

## PART 5: Ceramic crystals

Technical ceramics give us the hardest, most refractory materials of engineering. Those of greatest economic importance include:

- Alumina, $Al_2O_3$ (spark plug insulators, substrates for microelectronic devices).
- Magnesia, $MgO$ (refractories).
- Zirconia, $ZrO_2$ (thermal barrier coatings, ceramic cutting tools).
- Uranium dioxide, $UO_2$ (nuclear fuels).
- Silicon carbide, $SiC$ (abrasives, cutting tools).
- Diamond, $C$ (abrasives, cutting tools, dies, bearings).

The ceramic family also gives us many of the functional materials—those that are semi-conductors, are ferromagnetic, show piezo-electric behaviour and so on. Their structures often look complicated, but when deconstructed, so to speak, a large number turn out to be comprehensible (and for good reasons) as atoms of one type arranged on a simple FCC, CPH or BCC lattice with the atoms of the second type (and sometimes a third) inserted into the interstices of the structure of the first.

***The diamond cubic (DC) structure***  We start with the hardest ceramic of the lot—diamond—of major engineering importance for cutting tools, abrasives, polishes and scratch-resistant coatings, and at the same time as a valued gemstone for jewelry. Silicon and germanium, the foundation of semiconductor technology, have the same structure.

Figure GL1.12 shows the unit cell. Think of it as an FCC lattice with an additional atom in four of its eight tetrahedral interstices, labeled 1, 2, 3 and 4 in the figure. The tetrahedral hole is far too small to accommodate a full-sized atom, so the others are pushed farther apart, lowering the density. The cause is the 4-valent nature of the carbon, silicon and germanium atoms—they are happy only when each has 4 nearest neighbors, symmetrically placed around them. That is what this structure does.

Silicon carbide, like diamond, is very hard, and it too is widely used for abrasives and cutting tools. The structures of the two materials are closely related. Carbon lies directly above silicon in the periodic table. Both have the same crystal structure and are chemically similar. So it comes as no surprise that a compound of the two with the formula SiC has a structure like that of diamond, with half the carbon atoms replaced by silicon, as in Figure GL1.13 (it is the atoms marked 1, 2, 3 and 4 of Figure GL1.12 that are replaced).

| Materials with the diamond-cubic structure | Comment |
| --- | --- |
| Carbon, as diamond | Cutting and grinding tools, jewelry |
| Silicon, germanium | Semiconductors |
| Silicon carbide | Abrasives, cutting tools |

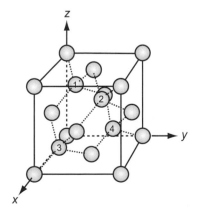

Figure GL1.12   The diamond-cubic (DC) structure.

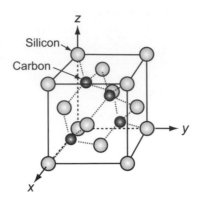

**Figure GL1.13**   The structure of silicon carbide.

## Exercises

E.17   How many atoms are there in the unit cell of the DC structure, shown in Figure GL1.12? Show that the lattice constant for DC is equal to $2.309\ D$, where $D$ is the atomic diameter. Hence find the atomic packing fraction for the DC structure, and comment on the result compared to a close-packed structure.

E.18   The lattice constant of DC silicon carbide is $a = 0.436$ nm. The atomic mass of silicon is 28.09 kg/kmol, and that of carbon is 12.01 kg/kmol. What is the theoretical density of silicon carbide?

*Ceramics with the Rocksalt (Halite) structure*   Several important oxides have the formula MO, where M is a metal ion. Oxygen ions are large, usually bigger than those of the metal. When this is so the oxygen packs in an FCC structure, the metal atoms occupy the octahedral holes in this lattice. The resulting structure is sketched in Figure GL1.14. This is known as the Rocksalt (or Halite) structure because it is that of sodium chloride, NaCl, with chlorine where the oxygens are and sodium where the magnesiums are in the figure.

Sodium chloride (common salt) itself is a material of engineering importance. It has been used for road beds and buildings, there are plans to bury nuclear waste in rocksalt deposits; and large single crystals of rocksalt are used for the windows of high-powered lasers.

| Materials with the Rocksalt structure | Comment |
| --- | --- |
| Magnesia, MgO | A refractory, and ceramic with useful strength |
| Ferrous oxide, FeO | One of several oxides of iron |
| Nickel oxide, NiO | Ceramic superconductors |
| All alkali halides, including salt, NaCl | Feedstock for chemical industry, nuclear waste storage |

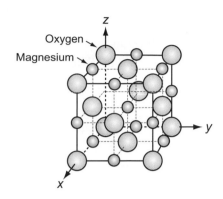

**Figure GL1.14**  The halite structure of MgO, typical of many simple oxides.

## Exercises

E.19  (a) Consider the halite structure shown in Figure GL1.14. If the oxygen ions form a true FCC lattice, then the oxygen ions should touch on the face diagonals. Determine the diameter of the largest metal ion that can fit into the locations shown in the figure, without distorting the lattice, relative to an oxygen diameter of unity.

(b) The diameters of oxygen and magnesium ions are 0.252 and 0.172 nm, respectively. Are the oxygen ions actually close-packed in the MgO lattice?

E.20  (a) Sodium chloride (NaCl) has a lattice constant $a = 0.564$ nm. Determine the density of NaCl, given that the atomic masses of sodium and chlorine are 22.989 kg/kmol and 35.453 kg/kmol, respectively.

(b) The ratio of the diameters of sodium and chlorine ions is 0.69. Show that the chlorine and sodium atoms are touching along the edges of the unit cell, and find the size of the gaps between the chlorine atoms on the diagonal of the face of the unit cell.

*Oxides with the Corundum structure*  A number of oxides have the formula $M_2O_3$, among them alumina ($Al_2O_3$). The oxygen ions, the larger of the two, are arranged in a CPH stacking. The M ions occupy two-thirds of the octahedral holes in this lattice, one of which is shown for alumina, filled by an Al ion, in Figure GL1.15. The hole is not big enough to accommodate the aluminium ion, so the oxygen lattice is pushed slightly apart, and it is not actually close-packed (but the atom centres fit the CPH structure). It is also the case that in this instance the lattice of Figure GL1.15 is not a unit cell. A larger and more complicated unit cell is needed, so that when these are stacked to form the lattice, the metal ions also fall into a regular pattern of hexagons, occupying the correct proportion of holes. This need not concern us further here.

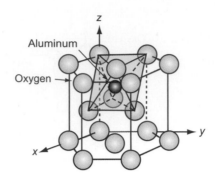

Figure GL1.15   The M atoms (here aluminum) of the Corundum structure lie in the octahedral holes of a CPH oxygen lattice.

| Materials with the Corundum structure | Comment |
|---|---|
| Alumina, $Al_2O_3$ | The most widely used technical ceramic |
| Iron oxide, $Fe_2O_3$ | The oxide from which iron is extracted |
| Chromium oxide, $Cr_2O_3$ | The oxide that gives chromium its protective coating |
| Titanium oxide, $Ti_2O_3$ | The oxide that gives titanium its protective coating |

*Oxides with the Fluorite structure*   The metal atoms in the important technical ceramic zirconium dioxide, $ZrO_2$, and the nuclear fuel uranium dioxide, $UO_2$, being far down in the periodic table, are large in size. Unlike the oxides described earlier, the M of the $MO_2$ in these is larger than the O. In these it is the M atoms that form a close-packed FCC structure and it is the oxygens that fit into the tetrahedral interstices in it, as shown in Figure GL1.16.

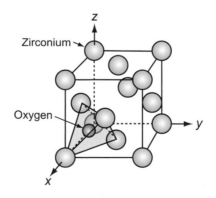

Figure GL1.16   The M atoms of fluorite-structured oxides pack in an FCC array, with oxygen in the tetrahedral interstices.

| Materials with the Fluorite structure | Comment |
|---|---|
| Zirconia, $ZrO_2$ (slightly distorted fluorite) | The toughest high-temperature ceramic |
| Urania, $UO_2$ | Nuclear fuel |
| Thoria, $ThO_2$ | Nuclear fuel |
| Plutonia, $Pu_2O_3$ | A product of nuclear fuel reprocessing, a fuel in itself |

## Exercises

E.21 Uranium dioxide $UO_2$ has the fluorite structure shown in Figure GL1.16. The oxygen ions are too large to fit into the locations shown in the figure, pushing the metal lattice further apart (so it is not strictly close-packed).

(a) Assuming the oxygen ions touch all the surrounding uranium ions, find the lattice constant for $UO_2$, given that the ionic diameters of uranium and oxygen are 0.222 and 0.252 nm, respectively.

(b) Determine the theoretical density of $UO_2$, given that the atomic masses of uranium and oxygen are 238.05 kg/kmol and 16.00 kg/kmol, respectively.

## PART 6: Polymer crystals

Many polymers crystallise. The long chains line up and pack to give an ordered, repeating structure, just like any other crystal. The low symmetry of the individual molecules means that the choice of lattice is limited. Figure GL1.17 is a typical example; it is polyethylene.

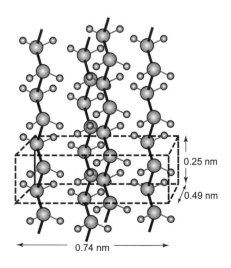

Figure GL1.17   The structure of a crystalline polyethylene.

Few engineering polymers are completely crystalline, but many have as much as 90% crystallinity. Among those of engineering importance are:

| Material | Typical uses |
|---|---|
| Polyethylene, PE, 65–90% | Bags, tubes, bottles |
| Nylon, PA, 65% | High-quality parts, gears, catches |
| Polypropylene, PP, 75% | Mouldings, rope |

## Answers to exercises

E.1

Figure GL1.18 shows the two configurations for the third layer. In the RH case, the third layer is aligned directly over the first layer, so the stacking goes ABA. In the LH case, the third layer occupies the remaining possible location, offset from the first layer, so the stacking goes ABC.

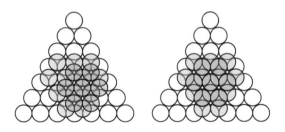

**Figure GL1.18**   Two alternative ways to stack close-packed planes.

E.2

(a) Figure GL1.19 shows a regular tetrahedron between 4 atoms in the CPH unit cell: 3 atoms in the base-plane of the unit cell, and the atom from the middle layer that sits on top of them. The height of the unit cell $c$ is twice the height $h$ of the tetrahedron, and the side-length of the hexagonal base, $a = 2R$, where $R$ is the atomic radius.

From the extracted tetrahedron to the left, and using Pythagoras' theorem, the perpendicular height $d$ of the equilateral base triangle is given by $d^2 = a^2 - (a/2)^2$, so $d = a\sqrt{3}/2$. Then from the vertical triangle, the height $h$ of the tetrahedron is given by $h^2 = d^2 - (d/3)^2 = 8d^2/9 = 2a^2/3$, so $h = a\sqrt{3}/\sqrt{2}$. Hence $c{:}a = 2h{:}a = 2 \times \sqrt{3}/\sqrt{2} = 1.633$.

(b) For Mg, $R = 0.1605$ nm, so $a = 2R = 0.321$ nm and $c = 1.633 \times 0.321 = 0.524$ nm.

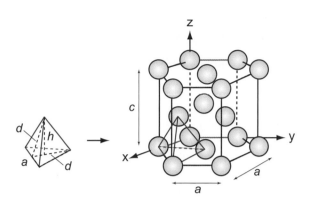

**Figure GL1.19** Dimensions of the CPH unit cell and internal tetrahedron.

E.3 (a) In the FCC unit cell, atoms touch along the diagonal of the face, which therefore has a length $= 4R$. Hence for a cubic unit cell of side length $a$: $a^2 + a^2 = (4R)^2$ and so the ratio of lattice constant to atomic radius for FCC $= a/R = 2\sqrt{2}$.
(b) In the BCC unit cell, atoms touch along the diagonal of the cube, which therefore has a length $= 4R$. For a cubic unit cell of side length $a$, first find the diagonal of a face is $a\sqrt{2}$. Hence $(a\sqrt{2})^2 + a^2 = (4R)^2$, so the ratio of lattice constant to atomic radius for BCC $= a/R = 4R\sqrt{3}$.

E.4 (a) In the CPH unit cell, the corner atoms are shared between 6 cells, the hexagonal face atoms between 2, and there are 3 complete internal atoms. Hence the number of atoms per unit cell $n = 12 \times 1/6$ (corners) $+ 2 \times 1/2$ (faces) $+ 3 \times 1$ (internal) $= 6$. The base area of the hexagonal unit cell $= a^2 3\sqrt{3}/2$ and the height $c = a\, 2\sqrt{2}/\sqrt{3}$ (from Exercise E.2). Hence the volume of the CPH unit cell $V_c =$ base $\times$ height $= a^2\, 3\sqrt{3}/2 \times a\, 2\sqrt{2}/\sqrt{3} = a^3\, 3\sqrt{2}$. Since $a = 2R$, $V_c = R^3\, 24\sqrt{2}$. Hence the atomic packing fraction $= \dfrac{6 \times \frac{4}{3} \pi R^3}{R^3\, 24\sqrt{2}} = 0.74$ (74%), i.e. the same as FCC (as both are close-packed).
(b) The number of atoms per BCC unit cell $n = 8 \times 1/8$ (corners) $+ 1 \times 1$ (internal) $= 2$. From Exercise E.3, $a = 4\,R/\sqrt{3}$, so the volume of the BCC unit cell $V_c = (4\,R/\sqrt{3})^3 = 64R^3/3\sqrt{3}$. Hence the atomic packing fraction $= \dfrac{2 \times \frac{4}{3} \pi R^3}{64\,R^3 /3\sqrt{3}} = 0.68$ (68%). BCC is not close-packed, for which the packing fraction is 74%.

E.5 The theoretical density is given by $\rho = n\,A/ (V_c\, N_A)$. For FCC gold, the density is 19.3 Mg/m³, the number of atoms in the unit cell $n$ is 4, the atomic mass is 196.967 kg/kmol. Hence the volume of the Au unit cell $V_c = 4 \times 196.967 \times 10^{-3} / (19{,}300 \times 6.022 \times 10^{23}) = 1.0168 \times 10^{-28}$ m³. Hence a hard sphere model predicts the Au lattice constant $a = (1.0168 \times 10^{-28}\,\text{m})^{1/3} = 0.4193$ nm, and an atomic diameter $= a/\sqrt{2} = 0.297$ nm.

**E.6**

(a) For BCC iron, the number of atoms in the unit cell $n$ is 2, the atomic mass is 55.847 kg/kmol, and the volume of the Fe unit cell $V_c = 2.3549 \times 10^{-29}\,\text{m}^3$. Hence the theoretical density $= \rho = n\,A/(V_c\,N_A) = 2 \times 55.847 \times 10^{-3} / (2.3549 \times 10^{-29} \times 6.022 \times 10^{23}) \approx 7.88$ Mg/m$^3$.

(b) At 910 °C, the fractional increase in the unit cell size $= \Delta T \times \alpha = 890 \times 12 \times 10^{-6} = 10.7 \times 10^{-3}$ (1.07%). New volume $V' =$ old volume $\times (1.0107)^3 \approx 1.032\,V$. The mass is unchanged, so the density *decreases* by this factor to 7.876/1.032 $\approx 7.63$ Mg /m$^3$ (BCC iron at 910 °C).

(c) From Exercise E.4, the ratio of the atomic packing factors for FCC and BCC unit cells $= 0.74/0.68 = 1.088$. Hence the BCC to FCC transformation gives an 8.8% *increase* in density. Hence the density of FCC Fe at 910 °C $\approx 7.63 \times 1.088 \approx 8.30$ Mg /m$^3$. The dimensions will contract by a strain $\varepsilon$ given by $(1 + \varepsilon)^3 = 1/1.088$, hence the contraction strain $= -0.0277$, i.e. $\approx 2.8\%$ reduction in linear dimensions.

**E.7**

The theoretical density is given by $\rho = n\,A/(V_c\,N_A)$. For CPH $\alpha$-titanium, the density is 4.54 Mg/m$^3$, the number of atoms in the unit cell $n$ is 6, the atomic mass is 47.9 kg/kmol. Hence the volume of the CPH Ti unit cell $V_c = 6 \times 47.9 \times 10^{-3} / (4540 \times 6.022 \times 10^{23}) = 1.015 \times 10^{-28}\,\text{m}^3$. Hence as $V_c = a^3\,3\sqrt{2}$ and $c = a\,2\sqrt{2}/\sqrt{3}$ (Exercise E.4), a hard sphere model predicts the Ti lattice constants are: $a = (1.015 \times 10^{-28}/3\sqrt{2})^{1/3} = 0.288$ nm; and $c = a\,2\sqrt{2}/\sqrt{3} = 0.471$ nm. The atomic diameter is the same as the lattice constant $a$, 0.288 nm.

**E.8**

Figure GL1.20 shows how to calculate the octahedral hole size in the FCC structure. The face diagonals are close-packed directions, so the atom spacing along the diagonal is 1 unit. The cell edge thus has length $\sqrt{2} = 1.414$ units. That is also the separation of the centres of the atoms at opposite corners of

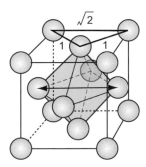

**Figure GL1.20**   The calculation of the interstitial hole size in the FCC lattice.

the octahedral hole. The atoms occupy 1 unit of this, leaving a hole that will just contain a sphere of diameter 0.414 unit without distortion.

E.9      A tetrahedral hole of the CPH structure is shown in Figure GL1.21.

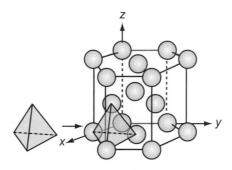

Figure GL1.21    One of the tetrahedral holes of the CPH structure.

E.10      Figure GL1.22 shows an apparent location for an octahedral hole in BCC. It is not a regular octahedron—the sides are not of equal length. The spacing between the two atoms at the centre of the cubes is the same as the side length of the cube, so locations midway along each edge of the cubes also make octahedral holes of the same shape.

     The side length of the cube is $a = 4\,R/\sqrt{3}$ (Exercise E.3) so the gap between the atoms at the cube corners is $a = (4R/\sqrt{3}) - 2R = 0.3094R$. So the largest sphere that would sit in this octahedral hole is 0.155 times the diameter of the host spheres. The tetrahedral holes in BCC are much larger: they fit an atom of diameter 0.29 times the host diameter, and as they are immediately adjacent (above and below), interstitial atoms will sit in the tetrahedral holes.

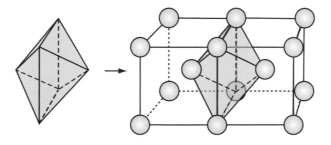

Figure GL1.22    An apparent octahedral hole in BCC.

E.11    Figure GL1.23 shows the geometry of the tetrahedral hole in FCC. The centre of the interstitial hole is marked by a small circle in the middle figure. This point sits at the apex of two identical, isosceles triangles (shaded), lying in vertical planes at right angles to one another. The left hand figure shows the dimensions of one of these triangles. Atoms touch along the bottom edge, length $2R$, where $R$ is the host radius. If an interstitial atom of radius $r$ just fits in the hole, then the inclined edges are of length $R + r$. And finally the height of the triangle is $a/4$, for a lattice constant $a$, so the vertical dimension of the triangle is also $= (2\sqrt{2})R\,/\,4 = R/\sqrt{2}$. Hence from Pythagoras' theorem: $(R + r)^2 = R^2 + (R/\sqrt{2})^2$, from which $r/R = 0.22$.

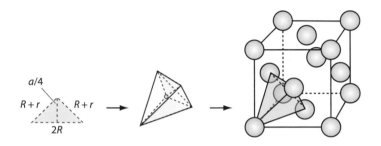

**Figure GL1.23**   The geometry of the tetrahedral hole in FCC.

E.12    The indices of the planes are (a) $(\bar{1}10)$, (b) $(11\bar{1})$, (c) $(412)$.

E.13    The planes are parallel to the cube faces—the cleavage facets will meet at 90 degrees.

E.14    Figure GL1.24 shows the planes.

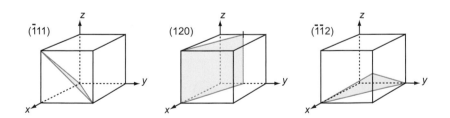

**Figure GL1.24**   The planes of Exercise E.14.

E.15    The directions are (a) $[011]$, (b) $[1\bar{1}\bar{1}]$, (c) $[\bar{1}\bar{1}2]$.

E.16    Figure GL1.25 shows the directions.

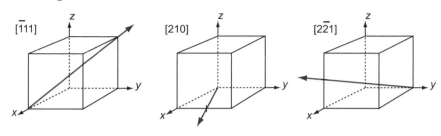

Figure GL1.25   The directions of Exercise E.16.

E.17    In diamond cubic there are 8 atoms per unit cell: 8 × 1/8 (corners) + 6 × 1/2 (faces) + 4 (internal). Referring to the DC unit cell in Figure GL1.12, atom number 3 sits in the middle of a regular tetrahedron, similar to that shown in Figure GL1.23. The geometry of a DC tetrahedron is shown in Figure GL1.26. Now all the spheres are of equal size, touching on the inclined sides. The atom spacing along the horisontal edges of the tetrahedron is equal to half the length of the diagonal of one face of the cube, i.e. $a/\sqrt{2}$, where $a$ is the lattice constant.

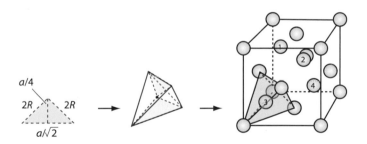

Figure GL1.26   The geometry of the tetrahedral packing in DC.

From Pythagoras' theorem: $(2R)^2 = (a/4)^2 + (a/2\sqrt{2})^2$, from which $R = \sqrt{3}\,a/8$, or $a = 8R/\sqrt{3} = 4D/\sqrt{3} = 2.309D$. Hence the atomic packing fraction =
$$\frac{8 \times \frac{4}{3}\pi R^3}{8^3\,R^3\,/3\sqrt{3}} = 0.34 \ (34\%).$$ This is remarkably low compared to close-packing (74%).

E.18    For DC silicon carbide, there are 4 Si atoms and 4 C atoms in the unit cell. The mass of the unit cell is therefore $[(4 \times 28.09) + (4 \times 12.01)] \times 10^{-3} / 6.022 \times 10^{23} = 26.636 \times 10^{-26}$ kg. The volume of the SiC unit cell $= (0.436 \times 10^{-9})^3 = 8.288 \times 10^{-29}$ m$^3$. Hence the theoretical density $= 26.636 \times 10^{-26}$ kg / $8.288 \times 10^{-29}$ m$^3 \approx 3.21$ Mg/m$^3$.

E.19

(a) For FCC oxygen, the size of the unit cell $= D\sqrt{2}$, for an oxygen diameter of $D$. The gap between the atoms along an edge of the cell is therefore $D\sqrt{2} - D = 0.414\,D$. This is the diameter of the largest sphere that will just fit in the gap, hence the ratio of the diameters is 0.414.

(b) In MgO, the diameter ratio of Mg to O is $0.172/0.252 = 0.68$. Since this is greater than 0.414, the oxygen ions are pushed apart by the Mg and are not truly close-packed (i.e. no longer touching on the diagonals of the unit cell).

E.20

(a) For the halite unit cell of sodium chloride, there are 4 Cl atoms (as these form an FCC packing) and hence 4 Na atoms (as the compound is NaCl). The mass of the unit cell is therefore $[(4 \times 35.453) + (4 \times 22.989)] \times 10^{-3} / 6.022 \times 10^{23} = 38.819 \times 10^{-26}$ kg. The volume of the NaCl unit cell $= (0.564 \times 10^{-9})^3 = 17.94 \times 10^{-29}$ m$^3$. Hence the theoretical density $= 38.819 \times 10^{-26}$ kg $/ 17.94 \times 10^{-29}$ m$^3 \approx 2.16$ Mg/m$^3$.

(b) The length of the edge of the NaCl cell $=$ diameter of Cl $+$ diameter of Na. For a Cl diameter of unity, the unit cell size is thus 1.69, and the diagonal of one face $= \sqrt{2} \times 1.69 = 2.39$. Ions only touch along this diagonal if its length is equal to 2 (i.e. twice the Cl diameter). There are thus two equal gaps between Cl ions on the diagonal, equal to $0.39/2 = 0.195$ times the diameter of the Cl.

E.21

(a) The fluorite structure of UO$_2$ is FCC packing of U with O in every tetrahedral hole. The geometry of the filled hole, which is similar to those in earlier exercises, is shown in Figure GL1.27: U and O ions touch (radii $R$ and $r$, respectively), while the other dimensions are given in terms of the lattice constant $a$. Note that the U ions are not touching along the cube face diagonals, which will be longer than $4R$ (checked below).

From Pythagoras' theorem: $(R + r)^2 = (a/4)^2 + (a/2\sqrt{2})^2$, from which $a = 4(R + r)/\sqrt{3}$, or $2(D + d)/\sqrt{3}$ (for U and O diameters of $D$ and $d$ respectively). Hence the lattice constant for UO$_2 = 2\,(0.222 + 0.252)/\sqrt{3} = 0.547$ nm. (Note that the length of the diagonal of the face is therefore 0.774 nm, which is

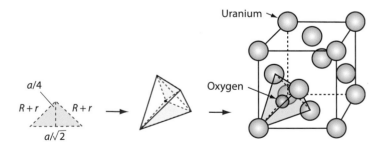

**Figure GL1.27**   The geometry of the tetrahedral packing in UO$_2$ with the Fluorite structure.

greater than $2D$ for U, equal to 0.444 nm, that is the U ions are not touching, as expected).

(b) The unit cell contains 4 U and 8 O atoms, so the mass of the unit cell is $[(4 \times 238.05) + (8 \times 16.00)] \times 10^{-3}/6.022 \times 10^{23} = 179.4 \times 10^{-26}$ kg. The volume of the $UO_2$ unit cell $= (0.547 \times 10^{-9})^3 = 16.37 \times 10^{-29}$ m$^3$. Hence the theoretical density $= 179.4 \times 10^{-26}$ kg / $16.37 \times 10^{-29}$ m$^3 \approx 11.0$ Mg/m$^3$.

# Guided Learning Unit 2:
# Phase diagrams and phase transformations

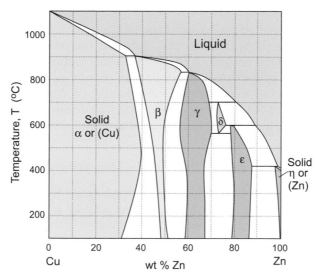

## Contents

## Introduction and synopsis

Phases and phase transformations are more familiar in everyday life than you may realise. Rain and snow signify the products of phase transformations in the sky succumbing to gravity—the word 'precipitation' is used in phase transformation theory, and this is no coincidence. We make roads safer in the ice and snow by spreading salt—lowering the melting point of the water by changing its composition, and causing a phase change. Bubbles rising in a glass of beer signify gases dissolved in the beer forming a separate phase, while in a boiling pan of water the bubbles of steam are formed by the water itself changing phase from liquid to vapour.

Phase diagrams and phase transformations are central to understanding microstructure evolution (and hence properties) in relation to processing. Manufacturing involves shaping and assembling engineering products and devices, while simultaneously providing the material properties required by the design. Most material processing operations involve a *thermal history* (e.g. cooling from a high-temperature shaping or deposition process, or producing a controlled diffusional change in a solid product). The details of the thermal history govern the way phase transformations take place, generating the microstructure. Our mantra for this topic is thus:

$$\text{Composition} + \text{Processing} \rightarrow \text{Microstructure} + \text{Properties}$$

*Phase diagrams* provide some fundamental knowledge of what the *equilibrium* structure of a metallic (or ceramic) alloy is, as a function of temperature and composition. The real structure may not be the equilibrium one, but equilibrium gives a starting point from which other (non-equilibrium) structures can often be inferred.

This Guided Learning Unit aims to provide a working knowledge of:

- What a phase diagram is.
- How to read it.
- How phases change on heating and cooling.
- The resulting microstructures.

When you have worked through this unit you should be able to do the following:

1. Interpret the equilibrium phases at any point on binary phase diagrams.
2. Predict the microstructures that may result from simple processing histories relating to solidification and heat treatment of important engineering alloys.
3. Understand how equilibrium and non-equilibrium microstructure evolution in alloy heat treatment relate to the form of time-temperature-transformation (TTT) diagrams.

Key definitions are marked 'DEF' as they appear, and exercises are provided throughout for each topic. Do these as you go along to build up your knowledge systematically. Full solutions are provided at the end of each section—use these to check your answers, not as a short-cut! Further exercises (without solutions) are provided for practice at the end of the unit.

**Part 1** contains some essential terminology and definitions.

**Parts 2—4** show you how to read and interpret simple phase diagrams, describe the important iron-carbon phase diagram, and give examples of some more complex phase diagrams.

**Parts 5—7** introduce phase transformations and show how phase diagrams can be used to predict microstructure evolution during slow cooling (for example in solidification, and in the solid state during cooling to room temperature).

**Part 8** extends the theory of phase transformations to examples of non-equilibrium cooling in heat treatment of steels and other alloys, relating this to the TTT (time-temperature-transformation) diagram.

The unit fits best with Chapter 19 in two installments: Parts 1—4, dealing with phase diagrams; Parts 5—8, covering microstructure evolution in relation to phase diagrams.

## PART 1: Key terminology

*Alloys and components*

**DEF.** A *metallic alloy* is a mixture of a metal with other metals or non-metals. Ceramics too can be mixed to form *ceramic alloys*.

Examples are:

- *Brass*: a mixture of copper (Cu) and zinc (Zn).
- *Carbon steel*: based on iron (Fe) and carbon (C).
- *Spinel*: a ceramic alloy made of magnesia (MgO) and alumina ($Al_2O_3$).

**DEF.** The *components* are the chemical elements that make up alloys.

Components are given capital letters: A, B, C or the element symbols Cu, Zn, C. In *brass* the main components are Cu and Zn. In *carbon steel* the main components are Fe and C.

**DEF.** A *binary alloy* contains two components. A *ternary alloy* contains three; a *quaternary alloy*, four; and so on.

*Concentration* Alloys are defined by stating the components and their concentrations, in weight or atom %.

DEF. The *weight* % of component A:

$$W_A = \frac{\text{Weight of component A}}{\sum \text{Weights of components}} \times 100$$

The *atom* (or *mol*) % of component A:

$$X_A = \frac{\text{Number of atoms (or mols) of component A}}{\sum \text{Number of atoms (or mols) of all components}} \times 100$$

To convert between weight and mols:

(Weight in grams)/(Atomic or molecular wt in grams/mol)
= Number of mols

(Number of mols) × (Atomic or molecular wt in grams/mol)
= Weight in grams (g)

*Phases*    For pure substances, the idea of a phase is familiar: ice, water and steam are the solid, liquid and gaseous states of pure $H_2O$—each is a distinct phase. Processing of metallic alloys leads to microstructures in which the component elements are distributed in a number of ways. In the liquid state for metals, more or less everything dissolves completely. But in the solid state, things are more complex—for example, in a binary alloy the solid microstructure usually takes one of three forms (examples later):

- A single solid solution.
- Two separated solid solutions.
- A chemical compound, with a separated solid solution.

Recall that a *solid solution* is a solid in which one (or more) element is 'dissolved' in another so that they are homogeneously dispersed, at an atomic scale. Some solid solutions may be so dilute that they are effectively the pure component.

A region of a material that has a homogeneous atomic structure is called a *phase*. A phase can be identified as a cluster of as few as 10 or so atoms, but it is usually much more. In the three types of solid microstructures listed above, each solid solution or chemical compound would be an identifiable phase.

DEF. All parts of an alloy microstructure with the same atomic structure are a single *phase*.

## Exercises (reminder: answers at the end of each section)

E.1   A 1.5 kg sample of $\alpha$-brass contains 0.45 kg of Zn, and the rest is Cu. The atomic weight of copper is 63.5 and zinc 65.4. Write down the concentration of copper in $\alpha$-brass, in wt%, $W_{Cu}$. Find the concentrations of copper and zinc in the $\alpha$-brass, in at%, $X_{Cu}$ and $X_{Zn}$.

E.2   An alloy consists of $X_A$ at% of A with an atomic weight $a_A$, and $X_B$ at% of B with an atomic weight $a_B$. Derive an equation for the concentration of A in wt%. By symmetry, write down the equation for the concentration of B in wt%.

### Constitution, equilibrium and thermodynamics

**DEF.** The *constitution* of an alloy is described by:

(a) The phases present.
(b) The weight fraction of each phase.
(c) The composition of each phase.

At *thermodynamic equilibrium*, the constitution is stable: there is no further tendency for it to change. The independent *state variables* determining the constitution are temperature, pressure and composition. Hence, the *equilibrium constitution* is defined at constant temperature and pressure for a given alloy composition.

Thermodynamics controls the phases in which mixtures of elements can exist as a function of the state variables—this is discussed in Thermodynamics of Phases in Section 19.4. The key parameter for a given composition, temperature $T$ and pressure $p$, is the *Gibbs free energy*, $G$, defined as (equation (19.1)):

$$G = U + pV - TS = H - TS$$

where $U$ is the *internal energy*, $V$ is the volume, $H$ is the enthalpy $(U + pV)$ and $S$ is the entropy. The internal energy $U$ is the sum of the atomic vibration and the bond energies between the atoms. For the liquid and solid states most relevant to materials processing, $U$ dominates the enthalpy ($pV$ is small), so $H \approx U$. Entropy $S$ is a measure of the disorder in the system—when an alloy solidifies there is a decrease in entropy because the solid has a regular lattice, whereas the liquid does not.

Each possible state—liquid solution, solid solution, mixtures of phases and so on—has an associated free energy, and that with the lowest free energy is the state at thermodynamic equilibrium.

> **DEF.** The *equilibrium constitution* is the state of lowest Gibbs free energy $G$, for a given composition, temperature and pressure. An alloy in this state shows no tendency to change—it is thermodynamically stable.

*Phase diagrams*   As noted earlier, pressure has a limited influence on material processing, as this primarily involves the liquid and solid states. From now on, we will therefore consider these material states to be controlled by the remaining two state variables: temperature and composition. Two-dimensional maps with these state variables as axes are called *phase diagrams*.

> **DEF.** A *phase diagram* (or equilibrium diagram) is a diagram with $T$ and composition as axes, showing the equilibrium constitution.

The phase diagram of an alloy made of components A and B, for all combinations of $T$ and $X_B$, defines the *A–B system*. Binary systems have two components, ternary systems have three, and so on. Commercial alloys may contain 10 or more elements, but in all cases there is one principal element (copper alloys, aluminium alloys and so on) to which other elements are added. The starting point for understanding the essential behaviour of any alloy is therefore to consider the binary systems for the principal component element and one or more alloying elements in turn.

## Answers to exercises, Part 1

| | |
|---|---|
| E.1 | $W_{Cu} = (1.5 - 0.45)/1.5 = 70\%$. <br> No. mols Cu $= 1050/63.5 = 16.535$; No. mols Zn $= 450/65.4 = 6.881$. <br> Atomic fraction of Cu: $X_{Cu} = 16.535/(16.535 + 6.881) = 71\%$; hence $X_{Zn} = (100 - 71)\% = 29\%$. |
| E.2 | Consider 100 atoms: mass of A atoms $= a_A x_A$ and mass of B atoms $= a_B x_B$. <br> Hence $W_A = \dfrac{a_A X_A}{a_A X_A + a_B X_B}$ and by inspection, $W_B = \dfrac{a_B X_B}{a_A X_A + a_B X_B}$ |

## PART 2: Simple phase diagrams, and how to read them

***Melting point, liquidus and solidus***   Consider first a pure material A heated from the solid state. The *melting temperature* is the unique temperature at which the phase change to the liquid state occurs, and solid and liquid can co-exist—see Figure GL2.1(a). Similarly, liquid changes to vapour at a unique (higher) temperature, the boiling point. In practical phase diagrams boiling is of little interest, so the diagram usually is limited to liquid and solid states.

Now if we consider the binary A–B system and add a second axis for composition, the behaviour illustrated in Figure GL2.1(b) is commonly observed. This figure shows that the upper limit of 100% solid and lower limit of 100% liquid separate, and there is not a unique melting point—this is known as *partition*. In the region between the two *phase boundaries* on the figure, liquid (L) and solid (S) are stable together, in proportions that depend on the temperature (see later). These boundaries have special names, defined next.

> **DEF.** The phase boundary that limits the bottom of the *liquid* field is called the *liquidus line*; the line giving the upper limit of the single-phase *solid* field is called the *solidus line*.

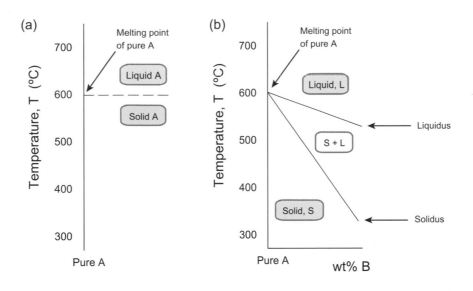

Figure GL2.1   (a) One-dimensional phase diagram for a pure substance: a temperature scale showing the phase boundary between solid and liquid—the melting point; (b) the A-rich end of a binary A–B phase diagram, illustrating partition of the melting point between solidus and liquidus boundaries.

Had we considered pure B first, it too would have a unique melting point and shown partition as we added some A. So what happens as we cover all possible compositions between pure A and pure B? One possible outcome is illustrated in Figure GL2.2, the phase diagram for the Cu-Ni system (the basis of several alloys used for coinage, explaining the terminology 'coppers' in the UK, and 'nickels' in the USA). This *isomorphous phase diagram* is the simplest possible example: 'isomorphous' meaning 'single-structured'. Here the solid state is a solid solution for all compositions, all the way from pure A to pure B. Since the atomic structure of this solid solution is the same at all compositions (with only the proportions of Cu and Ni atoms varying), it is a single phase.

It turns out that in the solid state this behaviour is very unusual—in virtually every other atomic mixture, there is a limit to the amount of an element that can be dissolved in another. We will explore this *solubility limit* with an everyday example: a cup of tea.

*Solubility limits*   For the purposes of illustration, we will think of tea as hot water, and sugar as a component (albeit molecular). Add a spoonful of sugar to hot tea, and it dissolves—the sugar disperses into solution. Those with a very sweet tooth may keep spooning until there is solid sugar sitting on the bottom of the cup. The tea has reached its *solubility limit* at this temperature and has become *saturated*: it will not dissolve any more sugar. This saturated tea now co-exists with a second phase, solid sugar. This too is a saturated solution, as the sugar absorbs as much tea as it can. This is characteristic of mixtures of two equilibrium phases, both of which are solutions—both phases are saturated (i.e. both are as impure as possible).

If we add sugar to cold tea, we find less sugar will dissolve—the saturation limit (in wt% sugar) rises with temperature. Harder to observe, but equally true, the sugar too will absorb

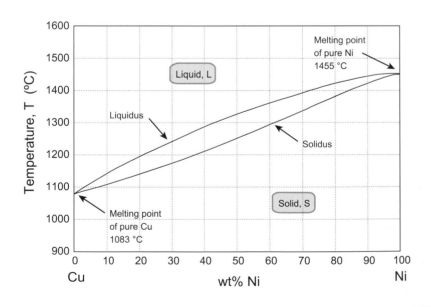

Figure GL2.2   Isomorphous phase diagram for the Cu-Ni system.

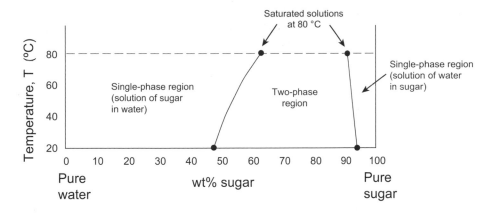

Figure GL2.3  Schematic phase diagram for sugar and tea (water)—the saturation level of both solutions increases with temperature.

more water as it is heated. Conducting this experiment quantitatively over a range of temperatures would lead to the partial phase diagram shown in Figure GL2.3. The boundaries between the single- and two-phase regions are known as *solvus boundaries*.

What happens at a given temperature and composition is determined by the thermodynamics of mixing sugar and water. Minimising free energy dictates whether there are one or two phases, and in the two-phase regions fixes the proportions and compositions of the phases.

*Building a simple binary diagram, and the eutectic point*  The picture in Figure GL2.3 is also found in many metallic mixtures. As an exemplar binary alloy we will consider the Pb-Sn system, the basis for many years of solders used for electronic joints (but rapidly falling out of favour due to environmental health concerns). Figure GL2.4 shows a partially complete Pb-Sn phase diagram. The lower part replicates the sugar-tea behaviour: the solubility of Pb in Sn (and of Sn in Pb) increases with temperature. Note that the solubilities of these elements in one another is low, especially Pb in Sn on the right of the diagram. The upper part of the diagram shows the partition behaviour from the melting points of the pure elements, as in Figure GL2.1(b).

To complete the diagram, first consider where the falling solidus boundaries meet the rising solvus boundaries. Figure GL2.4 suggests that the points where they intersect will be the points of maximum solubility (highest saturation) in the single-phase solids. This closes the regions or *fields* representing single-phase solid solutions. For good thermodynamic reasons, which will not be elaborated here, peak saturation of Sn in Pb occurs at the same temperature as that for Pb in Sn. It is essentially a corollary of the fact that in a two-phase region, both phases are as impure as possible. Below this temperature we have a mixture of solid solutions, with a horizontal boundary linking the two points of maximum solubility, as shown in Figure GL2.5(a), closing the two-phase field, (Pb) + (Sn). Note the nomenclature adopted: (Pb) for a Pb-rich single solid phase, and (Sn) for Sn-rich single solid phase.

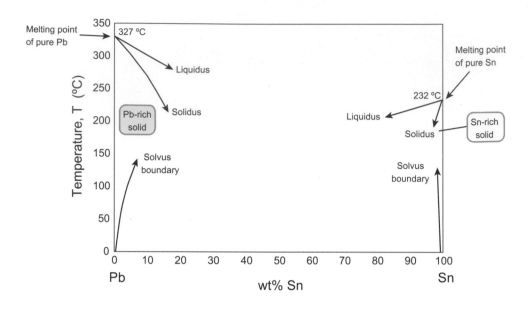

Figure GL2.4   Partial phase diagram for the Pb-Sn system, showing the limiting behaviour at high and low temperatures.

Finally: what happens to the two liquidus boundaries? Again, thermodynamics dictates that these meet the horizontal line at a single point, and the liquid field closes in a shallow 'V'—see Figure GL2.5(a). This point on the diagram is very important, and is known as a *eutectic point*. At this special temperature and composition, two solid phases and liquid of that composition can co-exist. This also closes the two (liquid + solid) regions above the horizontal line through the eutectic. The single-phase fields are highlighted on the phase diagram in Figure GL2.5(b).

**DEF.** The lower limit of the single-phase liquid field formed by the intersection of two liquidus lines is called the *eutectic point*.

For the Pb-Sn system the eutectic point is at the composition $W_{Sn} = 61.9$ wt%, and temperature T = 183 °C. Eutectics will be discussed again later in relation to the microstructures that form when a eutectic composition solidifies. For the time being we simply note that the eutectic composition gives the lowest temperature for which 100% liquid is stable. For this reason, casting, brazing and soldering alloys are often eutectic or near-eutectic in composition.

***Reading a binary phase diagram: phase compositions***   The state variables (temperature and composition) define a point on the phase diagram: the *constitution point*. The first thing to

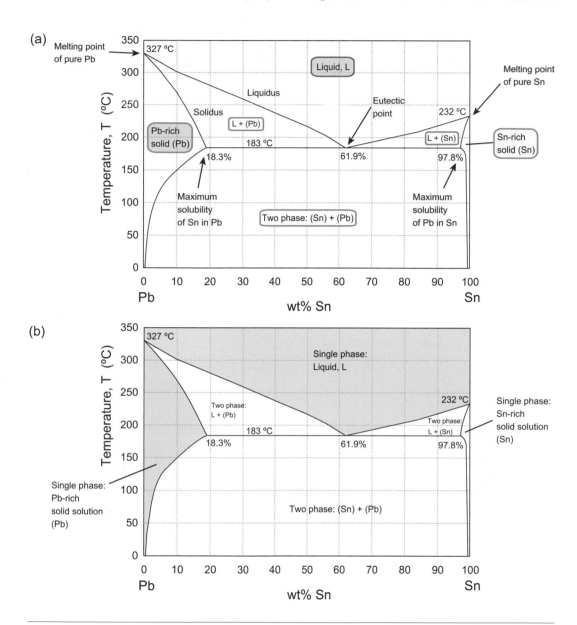

**Figure GL2.5**  The completed phase diagram for the Pb-Sn system, showing (a) the solubility limits for the single phases, and the eutectic point closing the liquid field; (b) the single-phase fields (shaded), separated by two-phase fields.

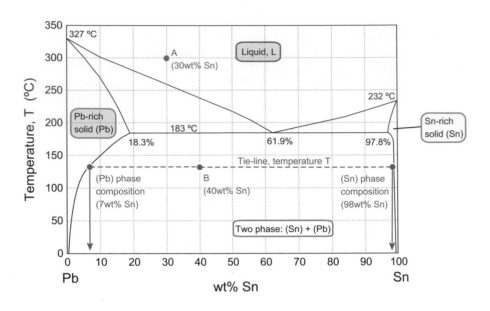

**Figure GL2.6**   Phase diagram for the Pb-Sn system, illustrating constitution points in single- and two-phase fields, and the tie-line defining the phases and compositions in the two-phase field.

establish at a constitution point is the number of phases present, one or two (in a binary system). Single-phase regions are always labelled, either with the notation in Figure GL2.5, with (Pb) for Pb-rich solid and so on, or with a Greek character ($\alpha$, $\beta$, $\gamma$, etc). When a phase diagram is traversed at a given temperature, crossing phase boundaries takes us from a single-phase field to a two-phase field (or vice versa)—see Figure GL2.5(b). In the two-phase regions, a horizontal line through the constitution point ending at the adjacent phase boundaries identifies the phases present: they are the single phases beyond those boundaries. This line is called a *tie-line* (Figure GL2.6).

At a constitution point in a single-phase region, the *phase composition* is simply the composition of the alloy itself (e.g. point A in Figure GL2.6). In two-phase regions, the phase compositions are given by the values on the phase boundaries at the ends of the tie-line through the constitution point (e.g. point B in Figure GL2.6). Recall that these are the saturation limits of the single-phase fields on the other sides of the boundaries.

> **DEF.** In a single-phase region, phase and alloy compositions coincide. In a two-phase region the phase compositions lie on the phase boundaries at either end of a horizontal tie-line through the constitution point.

Consider points A and B on the Pb-Sn phase diagram in Figure GL2.6. Constitution point A (temperature 300 °C, alloy composition Pb-30 wt% Sn) lies in the single-phase liquid field; the phase composition is also Pb-30 wt% Sn. Constitution point B (temperature 130 °C, alloy composition Pb-40 wt% Sn) lies in a two-phase field with two solid phases identified from the ends of the tie-line: (Pb) and (Sn); the phase compositions are Pb-7 wt% Sn and Pb-98 wt% Sn, respectively.

## Exercises

E.3     Use the Pb-Sn diagram in Figure GL2.7 to answer the following questions.
   (a) What are the values of the state variables (composition and temperature) at constitution points 1 and 2, and what phases are present?
   (b) Mark the constitution points for Pb-70 wt% Sn and Pb-95 wt% Sn alloys at 210 °C. What phases are present in each case?
   (c) The alloy at constitution point 1 is cooled very slowly to room temperature, maintaining equilibrium. Identify the temperature of the phase boundary at which a change in the phases occurs. What phase(s) is present below the phase boundary?
   (d) The alloy at constitution point 2 is cooled slowly to room temperature. Identify the temperatures at which phase changes occur, and the phase(s) before and after each change.

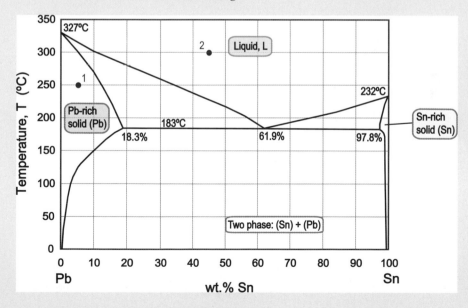

**Figure GL2.7** Pb-Sn phase diagram.

E.4    Use the Pb-Sn diagram in Figure GL2.7 to answer the following questions.

(a) The constitution point for a Pb-25 wt% Sn alloy at 250 °C lies in a two-phase field. Construct a tie-line on the figure and read off the two phases and their compositions.

(b) The alloy is slowly cooled. Identify the phases and their compositions: (i) at 200 °C; (ii) at 150 °C.

(c) Indicate with arrows on the figure the lines along which the compositions of the two phases move during slow cooling from 250 °C to 200 °C. The overall composition of the alloy stays the same, of course. How can this be maintained as the compositions of the phases change?

*Reading a binary phase diagram: proportions of phases*    In a two-phase field at constant temperature, the compositions of the phases are fixed at the saturation limits—the values on the boundaries at the ends of the tie-line. So different compositions at this temperature will contain different *proportions* of each phase, in such a way as to conserve the overall fractions of the two elements. The proportions of each phase (by weight) in a two-phase region can be found from the phase diagram using the *lever rule*. These weight fractions are fixed by the requirement that matter is conserved—the derivation is given as an exercise later. Consider Pb-20 wt% Sn at 250 °C in Figure GL2.8, a constitution point in the two-phase field: liquid

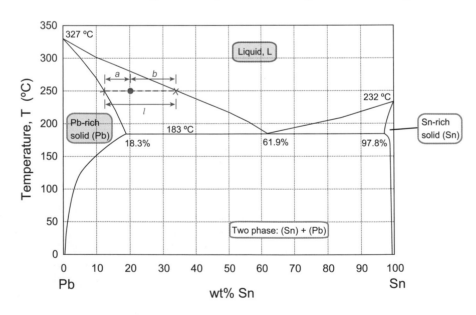

**Figure GL2.8**    Phase diagram for the Pb-Sn system, illustrating the lever rule for finding the weight fractions of the phases in a two-phase field.

plus Pb-rich solid. To find the proportions of each phase, first construct a tie-line through the constitution point and read off the compositions of the phases:

$$Pb\text{-}rich\ solid\ \text{with}\ W^{SOL}_{Sn} = 12\%; \quad Liquid\ \text{with}\ W^{LIQ}_{Sn} = 34\%$$

The tie-line is of length $\ell$, and the lengths of the segments to either side of the constitution point are $a$ and $b$, respectively. For the example alloy of composition $W_{Sn} = 20\%$:

$$\ell = W^{LIQ}_{Sn} - W^{SOL}_{Sn} = 34 - 12 = 22\%$$

$$a = W_{Sn} - W^{SOL}_{Sn} = 20 - 12 = 8\%$$

$$b = W^{LIQ}_{Sn} - W_{Sn} = 34 - 20 = 14\%$$

The weight fractions of liquid and solid in the alloy are $W^{LIQ} = a/\ell$ and $W^{SOL} = b/\ell$. Hence:

$$W^{LIQ} = 8/22 = 36\%$$

$$W^{SOL} = 14/22 = 64\%$$

This illustrates why the name is the lever rule—it is analogous to balancing two weights on either side of a pivot, with the shorter distance being that to the greater weight (for moment equilibrium).

Note the following:

- $W^{SOL} + W^{LIQ} = a/\ell + b/\ell = (a + b)/\ell = 1$ (as expected, the two fractions sum to unity).
- At the left-hand end of the tie-line, $W^{SOL} = 1$ ($a = 0$, $b = \ell$).
- At the right-hand end of the tie-line, $W^{LIQ} = 1$ ($a = \ell$, $b = 0$).

To summarise: to find the weight fractions of the phases in *any* two-phase region (liquid-solid, or two solid phases):

- Construct the tie-line through the constitution point.
- Read off the three compositions (for the alloy, and the two ends of the tie-line).
- Apply the lever rule.

Alternatively, and more approximately, the lengths can be measured directly from the phase diagram. Note that the proportions of the phases only vary linearly with composition along the tie-line if the diagram has a *linear weight % scale*.

Some phase diagrams have linear *atom %* scales, though they may also show a non-linear weight % scale along the top of the diagram (examples later). In this case, the lever rule for weight fraction cannot be applied by direct measurement, but the equations for weight fractions can still be applied by finding the three compositions (in wt%) and evaluating $a$, $b$ and $\ell$ as in the previous example. Note that the concept of the atom fraction of a phase is not particularly useful, so the lever rule is not generally applied to linear atom % scales.

## Exercises

E.5    Derive the lever rule for a general mixture of two phases, $\alpha$ and $\beta$. Let the composition of the alloy be C (wt% of alloying element), the compositions of the phases be $C_\alpha$ and $C_\beta$, and the weight fractions of the phases be $W_\alpha$ and $W_\beta$. [Hints: first find an expression conserving the mass of the alloying element between the alloy and the two phases, then define $a$, $b$ and $\ell$ in this notation and use the overall conservation of mass expressed in $W_\alpha + W_\beta = 1$.]

E.6    Using the Pb-Sn phase diagram in Figure GL2.7, consider the Pb-Sn alloy with composition $W_{Sn} = 25\%$. What are the approximate proportions by weight of the phases identified in Exercise E.4, at 250 °C, 200 °C and 150 °C?

*Intermediate phases*   Many systems show *intermediate phases*: compounds that form between components. Examples are $CuAl_2$, $Al_3Ni$ and $Fe_3C$. If the components are both metallic, they are called intermetallic compounds. Thermodynamically, compounds form because the particular combination of components is able to form as a single phase with a specific lattice of lower free energy than, say, a mixture of two phases. Example lattices are given in Guided Learning Unit I: Crystallography.

The atomic % of components in a compound is called its *stoichiometry*. Compounds are written in the form noted earlier, $A_xB_y$ where $x$ and $y$ are integers. The at% of the components in an intermediate compound can easily be stated by inspection, $x/(x + y)$ and $y/(x + y)$, for example $Fe_3C$ contains 25 at% C. In general the integer values $x$ and $y$ are small, since the number of atoms that define the repeating unit of the crystal lattice is also small. Compounds therefore usually appear on phase diagrams with at% scales at simple integer ratios: 25%, 33%, 50% and so on. In principle they therefore plot as a vertical line representing the single phase (an example follows).

As a single phase of fixed composition, intermediate phases have unique melting points (like pure components). The higher degree of thermodynamic stability means that compounds often have higher melting points. The liquid field for compositions on either side often shows a falling liquidus line, with eutectics forming between the compound and a solid solution, or between two compounds if the system shows more than one. Figure GL2.9(a) shows the unusual (and very untypical) silver-strontium phase diagram (notice the *at% scale*). This is not exactly a well-known engineering alloy, but illustrates the 'ideal' behaviour of compounds on phase diagrams. It has four intermetallic compounds (the vertical lines), and looks like five separate phase diagrams back-to-back: the Ag-$Ag_5Sr$ diagram, the $Ag_5Sr$-$Ag_5Sr_3$ diagram and so on. The liquidus boundary falls from each single phase melting point, forming five eutectics. Note that the solidus lines are vertical—they coincide with the lines representing the single-phase compound. In all the adjoining two-phase solid fields the compositions of the phases are fixed, and do not vary with temperature.

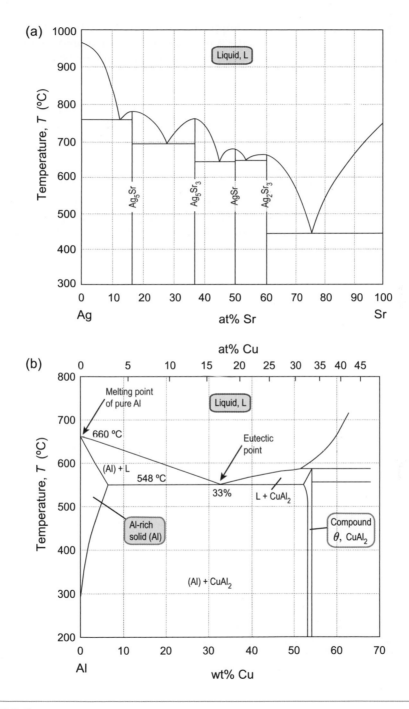

Figure GL2.9 Phase diagrams showing intermediate compounds: (a) the silver-strontium Ag-Sr system; (b) part of the aluminium-copper Al-Cu system.

Compounds of this type give the impression that there are two two-phase fields meeting at a vertical boundary, which violates the fundamental thermodynamics. Traversing a diagram at constant temperature must show one-phase, two-phase, one-phase and so on as boundaries are crossed, with tie-lines in the two-phase fields ending at single-phase boundaries. This remains the case here—there *is* a single-phase field in between the two-phase fields, it is one of the compounds—but the field has essentially collapsed to a single line.

A more typical example with engineering significance is illustrated in Figure GL2.9(b), showing part of the Al-Cu diagram, on a wt% scale (with the corresponding at% scale across the top). This diagram is the basis of the important 'age-hardening' Al-Cu alloys, used widely in aerospace. A compound forms at 33 at% Cu: it is therefore $CuAl_2$ (given the name θ-phase, to signify that it is a single phase). The liquidus boundaries fall to a eutectic point at 33 wt% Cu and 548 °C. Now in contrast to the silver-strontium diagram, the θ field is not a single vertical line, but a tall thin region with a small spread in composition. In other words $CuAl_2$ can tolerate a small amount of excess Al while remaining a single phase—some of the Cu atoms are replaced by Al, and the stoichiometry may not be exactly 1:2. We can think of it as a solid solution of Al in $CuAl_2$. Most practical compounds show some tendency to form a solid solution over a small range of composition close to stoichiometric, giving a thin single-phase field rather than a vertical line. In consequence they are less easily overlooked or misinterpreted. In some cases the spread of composition is so great that it ceases to be meaningful to distinguish it as a compound at all, and simply to consider it as a solid solution. But there is no rigorous definition as to how much spread in composition is allowed before it is not considered to be a compound any more.

## Exercises

E.7          Use Figure GL2.9 to answer the following:

(a) For an Ag-90 at% Sr alloy at 600 °C:
(i) Plot the constitution point on the phase diagram.
(ii) Identify the phases present, and find their compositions in at%.
(iii) The temperature is slowly reduced to 500 °C. Will the phase compositions and proportions change?

(b) For an Ag-30 at% Sr alloy at 600 °C:
(i) Plot the constitution point on the phase diagram.
(ii) Identify the phases present, and find their compositions in at%.
(iii) Will the proportions change if the temperature is reduced to 500 °C? Why is this?

(c) The atomic weight of Ag is 107.9 and that of Sr is 87.6. Calculate the compositions of the four intermetallic compounds in the Ag-Sr system in *weight%*.

(d) For an Al-4 wt% Cu alloy:
(i) Calculate the composition in at% Cu (atomic masses of Al and Cu: 26.98 and 63.54, respectively).

(ii) At 550 °C, identify the phase(s) present, and find its composition (in wt%) and proportion by weight.
(iii) Repeat for 250 °C.

## Answers to exercises, Part 2

E.3

(a) Constitution point 1: composition: Pb-5 wt% Sn; temperature: 250 °C. Pb-rich solid (Pb).
Constitution point 2: composition: Pb-45 wt% Sn; temperature: 300 °C. Liquid.

(b) See Figure GL2.10: Pb-70 wt% Sn: Liquid; Pb-95 wt% Sn: Liquid plus Sn-rich solid (Sn).

(c) At approximately 125 °C: (Pb) changes to two-phase (Pb-rich solid + Sn-rich solid).

(d) At approximately 225 °C: Liquid changes to two-phase (Liquid + Pb-rich solid). At 183 °C: (Liquid + Pb-rich solid) changes to (Pb-rich solid + Sn-rich solid).

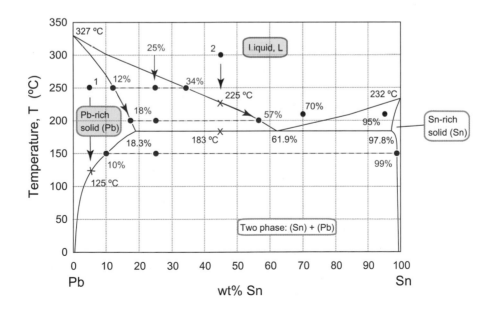

Figure GL2.10   Pb-Sn diagram.

E.4          (a) 250 °C: Pb-rich solid (12 wt% Sn) and Liquid (34 wt% Sn).
                  (b) (i) 200 °C: Pb-rich solid (18 wt% Sn) and Liquid (57 wt% Sn).
                      (ii) 150 °C: Pb-rich solid (10 wt% Sn) and Sn-rich solid (99 wt% Sn).
                  (c) See Figure GL2.10. As the phase compositions change with temperature, the overall composition is maintained by changes in the *proportions* of the phases.

E.5          Consider unit mass of the alloy, so that the mass of the alloying element $= C$, the mass of phase $\alpha = W_\alpha$ and the mass of phase $\beta = W_\beta$.
The mass of the alloying element in phase $\alpha = W_\alpha C_\alpha$ and in phase $\beta = W_\beta C_\beta$. Hence to conserve mass of the alloying element:

$$C = W_\alpha C_\alpha + W_\beta C_\beta \qquad\qquad \text{(GL2.1)}$$

From Figure GL2.11:

$$\ell = C_\beta - C_\alpha \text{ and } a = C - C_\alpha \text{ and } b = C_\beta - C \qquad \text{(GL2.2)}$$

Since $W_\alpha + W_\beta = 1$, $W_\alpha = 1 - W_\beta$.
Substituting into equation (GL2.1):
$$C = (1 - W_\beta)C_\alpha + W_\beta C_\beta = W_\beta(C_\beta - C_\alpha) + C_\alpha.$$
Rearranging, and comparing with equations (GL2.2):

$$W_\beta = (C - C_\alpha)/(C_\beta - C_\alpha) = a/\ell.$$

Following the same procedure for $W_\alpha$ the result will be:
$$W_\alpha = (C_\beta - C)/(C_\beta - C_\alpha) = b/\ell.$$

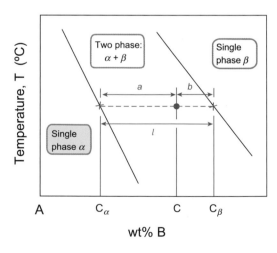

**Figure GL2.11**   Derivation of the lever rule.

E.6      Using the compositions identified in Exercise E.4 (or measuring from the figure):

(a) 250 °C: proportion of Pb-rich solid = $(34 - 25)/(34 - 12) = 41\%$, and 59% liquid.

(b) 200 °C: proportion of Pb-rich solid = $(57 - 25)/(57 - 18) = 82\%$, and 18% liquid.

(c) 150 °C: proportion of Pb-rich solid $(99 - 25)/(99 - 10) = 83\%$ and 17% Sn-rich solid.

E.7     
(a) See Figure GL2.12(a): at 600 °C: liquid, Ag-83 at% Sr and pure Sr (solid). At 500 °C, the liquid concentration has changed to Ag-78 at% Sr, while the second phase remains pure Sr (solid). The proportion of solid increases.

(b) See Figure GL2.12(a): at 600 °C: solid compounds $Ag_5Sr$ (16.7 at% Sr) and $Ag_5Sr_3$ (37.5 at% Sr). At 500 °C, the phases remain the same in both composition (as perfect compounds) and in proportions (as the phase boundaries are vertical).

(c) Atomic weights: Ag 107.9 and Sr 87.6. By inspection, compound atomic fractions are:

$$Ag_5Sr : 16.7 \text{ at\% Sr}; \; Ag_5Sr_3 : 37.5 \text{ at\% Sr};$$
$$AgSr : 50 \text{ at\% Sr}; \; Ag_2Sr_3 : 60 \text{ at\% Sr}.$$

Using the wt% formula from Exercise E.2 (with A = Sr, B = Ag):

$$Ag_5Sr : 14.0 \text{ wt\% Sr}; \; Ag_5Sr_3 : 32.8 \text{ wt\% Sr};$$
$$AgSr : 44.8 \text{ wt\% Sr}; \; Ag_2Sr_3 : 54.9 \text{ wt\% Sr}.$$

(d) (i) Consider 1 kg of Al-4 wt% Cu: so mass of Cu = 40 g, mass of Al = 960 g.

No. mols Cu = $40/63.54 = 0.63$; No. mols Al = $960/26.98 = 35.6$.

Atomic fraction of Cu: $X_{Cu} = 0.63/(0.63 + 35.6) = 1.74$ at% Cu.

(ii) See Figure GL2.12(b). At 550 °C, single-phase Al-rich solid. Composition = alloy composition Al-4 wt% Cu (proportion 100%).

(iii) See Figure GL2.12(b). At 250 °C, two solid phases: (essentially) pure Al + compound $CuAl_2$ (composition 53 wt% Cu). Proportions from lever rule: $(53 - 4)/(53 - 0) = 92\%$ (Al) and 8% $CuAl_2$.

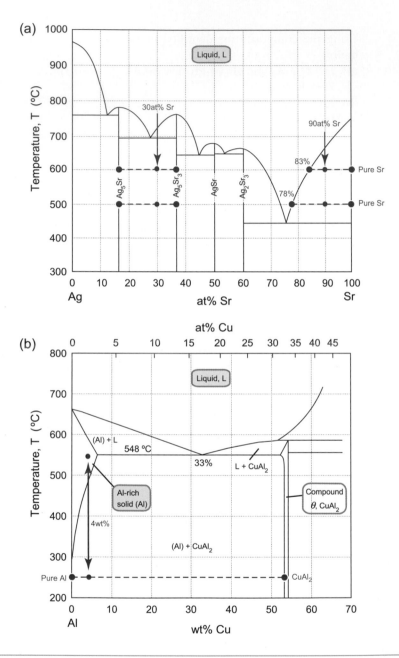

Figure GL2.12 (a) Ag-Sr phase diagram; (b) Al-Cu phase diagram.

## PART 3: The iron-carbon diagram

The *iron-carbon* phase diagram is important in engineering as it provides the basis for understanding all *cast irons* and *carbon steels* and their heat treatment. First, we consider pure iron. The low temperature form of iron is called *ferrite* (or $\alpha$-iron), with a BCC lattice (body-centred cubic, defined in Guided Learning Unit 1). On heating pure iron changes to *austenite* (or $\gamma$-iron) at 910 °C, and switches to a face-centred cubic (FCC) lattice. Pure austenite is stable up to 1391 °C, when it changes back to BCC $\delta$-iron, before melting at 1534 °C.

A key characteristic of the iron-carbon system is the extent to which iron dissolves carbon in interstitial solid solution, forming single phases. This is where the changes between BCC and FCC are significant. In Guided Learning Unit 1, it was shown that the *interstitial holes* are larger in FCC than in BCC. This leads to low solubility of carbon in BCC ferrite and $\delta$-iron, and much higher solubility in FCC austenite. Note that the same names (ferrite, austenite and $\delta$) are applied equally to the different states of pure iron and to the solid solutions they form with carbon. The nomenclature 'iron-rich solid (Fe)' is not much help to us, as there are three distinct variants.

Figure GL2.13 shows the completed iron-carbon diagram, up to 6.7 wt% carbon. This upper limit corresponds to 25 at% C, at which composition the compound *iron carbide* $Fe_3C$ is formed (also given the name *cementite*). This part of the Fe-C system covers all the main cast irons and carbon steels. This diagram is more complicated than those shown previously, and needs a bit of breaking down. Figure GL2.14 shows expanded extracts to clarify the diagram.

First consider the picture below 1000 °C, up to 2.0 wt% carbon (Figure GL2.14(c)). This shows the low solubility of carbon in ferrite, with a maximum of 0.035 wt% at 723 °C. Below the transformation temperature of ferrite to austenite (910 °C) the picture resembles the partition behaviour seen below the melting point of a pure element, with two-phase boundaries falling from this temperature and a two-phase region in between. But in this case the upper phase is a solid solution (austenite), rather than a liquid. But at the temperature of maximum C solubility in ferrite (723 °C), the lower limit of the austenite field also forms a 'V', giving the minimum temperature at which austenite forms as a single phase, at a composition of 0.8 wt% C.

This feature on a phase diagram is called a *eutectoid point*, and is particularly important in the context of carbon steels (as illustrated later when we consider their microstructures and heat treatments). Note the similarity in shape to a eutectic, with the key difference that the phase above the 'V' is a single solid phase (as opposed to a single liquid phase in the case of a eutectic).

**DEF.** The lower limit of a single-phase solid field formed by two falling phase boundaries intersecting in a 'V' is called a *eutectoid point*.

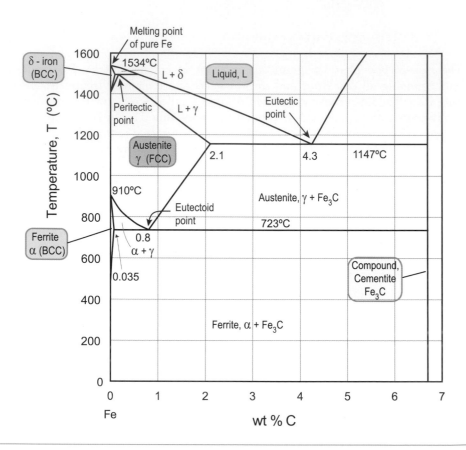

Figure GL2.13   The iron-carbon phase diagram up to 6.7 wt% carbon, the region covering cast irons and carbon steels.

Following the rising boundary of the austenite field to the right and above the eutectoid point, we reach a point of maximum solubility. Figure GL2.14(b) shows the top-right region of the phase diagram, including this point. This shows exactly the eutectic structure seen earlier. The eutectic temperature coincides with the temperature of maximum solubility of C in austenite, with falling solidus and liquidus lines enclosing the two-phase liquid + austenite region. Tie-lines in the two-phase regions to the right of and below the eutectic point end at the compound, iron carbide.

Completing the austenite field introduces another new feature, a *peritectic point*. Figure GL2.14(a) shows an expanded view of the diagram at the top of the austenite field, including the δ-iron field. The austenite field closes in an inverted 'V' at the peritectic point; that is, the maximum temperature at which this single phase forms. This temperature coincides with the temperature at which δ-iron has its maximum solubility, giving a horizontal line through the peritectic point. Above the line is a two-phase field, of which one is a liquid—here

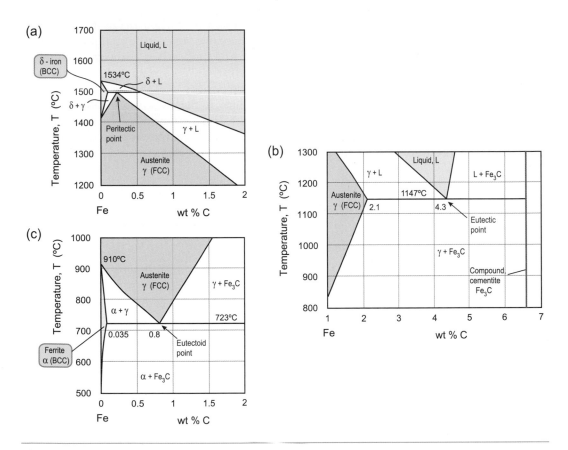

**Figure GL2.14** Expanded views of parts of the iron-carbon phase diagram: (a) high temperature and low wt% C: the peritectic point; (b) high temperature and high wt% C: the eutectic point; (c) low temperature and low wt% C: the eutectoid point.

liquid + $\delta$-iron. The phases in the other two-phase fields are readily identified from tie-lines, $\delta$ + austenite ($\gamma$) to the left of the peritectic, and $\gamma$ + liquid to the right.

To summarise the key nomenclature of the iron-carbon system, the single phases are:

*Ferrite*: $\alpha$-iron (BCC) with up to 0.035 wt% C dissolved in solid solution.
*Austenite*: $\gamma$-iron (FCC) with up to 2.1 wt% C dissolved in solid solution.
*$\delta$-iron*: (BCC) with up to 0.08 wt% C dissolved in solid solution.
*Cementite*: $Fe_3C$, a compound, at the right-hand edge of the diagram.

The system has a eutectic point at 4.3 wt% C, a eutectoid point at 0.8 wt% C and a peritectic point at 0.2 wt% C.

In passing, it should perhaps be noted that this iron-carbon diagram is not strictly an equilibrium diagram. Iron carbide is in fact a metastable state—true equilibrium is reached in thermodynamic terms in two-phase mixtures of iron and carbon. However, in most circumstances iron carbide forms readily in preference to carbon as a separate phase—it is an example of the *mechanism* of phase transformations taking priority over the thermodynamics and free energy. The phase diagram including iron carbide is therefore most commonly used as a pseudo-equilibrium phase diagram. The difference does become apparent in the solidification of cast irons, alloys containing 1–4 wt% C (i.e. approaching the eutectic composition). In this case carbon can form as an equilibrium phase in the microstructure.

## Exercises

E.8    Use Figures GL2.13 and GL2.14 to answer the following:
(a) For a Fe-0.4 wt% C alloy at 900 °C and 600 °C:
    (i) Plot the constitution points on the phase diagram.
    (ii) Identify the phases present, and find their compositions in wt%.
    (iii) If the temperature is slowly reduced from 900 °C to 600 °C, at what temperatures are phase boundaries crossed? Identify the phases present after each boundary is crossed.
(b) How does slow cooling from 900 °C to 600 °C differ for a Fe-0.8 wt% C alloy?

## Answers to Exercises, Part 3

E.8    (a) (i) See Figure GL2.15.
    (ii) 900 °C: single phase austenite ($\gamma$), composition 0.4 wt% C; 600 °C: two-phase, ferrite ($\alpha$), composition effectively 0 wt% C + cementite (iron carbide, $Fe_3C$), composition 6.7 wt% C.
    (iii) At 800 °C, austenite transforms to austenite + ferrite; at 723 °C, the remaining austenite transforms to more ferrite, + cementite.
(b) For 0.8 wt% C, austenite transforms directly to ferrite + cementite at the eutectoid point (at 723 °C), rather than first forming some ferrite over a temperature interval above 723 °C.

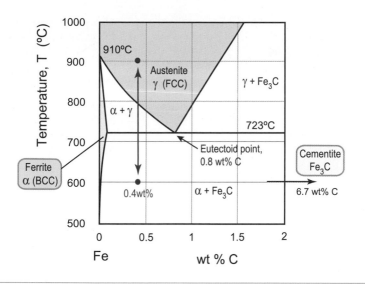

**Figure GL2.15**   Fe-C phase diagram.

## PART 4: Interpreting more complex phase diagrams

The phase diagrams covered so far illustrate almost all the features found in binary systems. This section completes the story, and shows how even very complicated diagrams can be interpreted if the rules are applied carefully.

*Eutectics, eutectoids, peritectics and peritectoids*   Three of these features were seen on the iron-carbon diagram. Here we define the fourth—a *peritectoid point*—and show all four together for clarity. The nomenclature is a little confusing on first encounter.

A peritectoid point is similar in appearance to a peritectic, being an inverted 'V' corresponding to an upper limit of formation of a single solid phase. But the difference is that the two-phase field above is formed of two solid phases (whereas in a peritectic, one is liquid). So, to help remember which is which:

- *eutec*— means a normal V meeting a horizontal line, whereas *peritec*— means an inverted 'V' meeting a horizontal line.
- *—tic* means a liquid phase is involved, whereas *—toid* means all phases are solid.

Figure GL2.16 shows all four for comparison. Single solid phases are denoted by Greek letters, liquid by L. Note that each point involves *three* phases—a single phase inside the 'V', and two different phases across the horizontal line.

Compared to eutectics and eutectoids, peritectics and peritectoids are of much less engineering significance. An unusual exception is the growth of single crystals of the new high

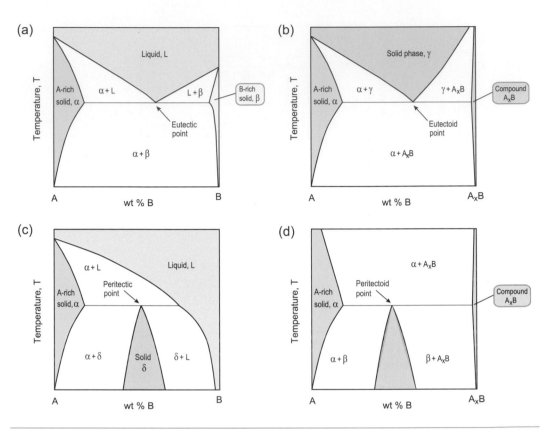

**Figure GL2.16**    Schematic views of: (a) eutectic point; (b) eutectoid point; (c) peritectic point; (d) peritectoid point.

temperature superconductors based on yttrium-barium-copper oxide. This is conducted by very slow cooling through a peritectic transformation.

***Ceramic phase diagrams***    Ceramics are mostly compounds of a metal with one of the elements O, C or N. They form with specific stoichiometry to satisfy the electronic balance between the elements; for example, alumina $Al_2O_3$. In some cases, we are interested in mixtures of ceramics (or ceramic alloys). An example was mentioned earlier: *spinel* is made of magnesia (MgO) and alumina ($Al_2O_3$). Earth scientists in particular need phase diagrams for ceramics to help interpret natural minerals and microstructures.

Phase diagrams for ceramics work in exactly the same way as for metal systems, with the elements replaced by the pure compounds. Figure GL2.17 shows the silica-alumina

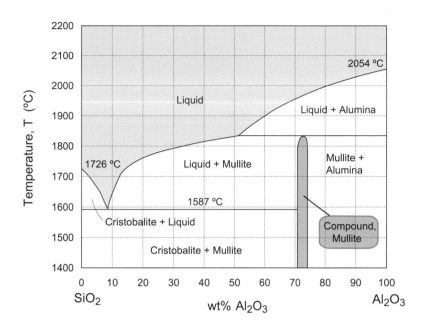

Figure GL2.17   Phase diagram for the binary ceramic silica-alumina ($SiO_2$-$Al_2O_3$) system.

($SiO_2$-$Al_2O_3$) system. It forms an intermediate single phase, known as *mullite*. Note that the top of this single-phase field closes in a peritectic point—above this point, the two-phase field is liquid + $Al_2O_3$.

## Exercises

E.9

The two-phase diagrams (or parts of diagrams) in Figure GL2.18 both have a eutectic point.

Mark the eutectic point on each figure, and find the eutectic temperature and composition in wt% in each system. (Note: The atomic masses of Cr and Ni are 52.00 and 58.71, respectively.)

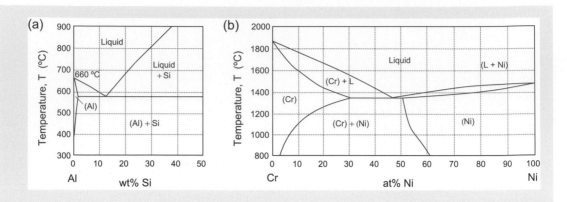

Figure GL2.18    (a) Aluminium-silicon Al-Si, and (b) chromium-nickel Cr-Ni phase diagrams.

E.10    The phase diagram for the copper-zinc system (which includes *brasses*) is shown in Figure GL2.19. Use the diagram to answer the following questions.

(a)  (i) Shade the single-phase regions.

(ii) Highlight the eutectoid point and five peritectic points in the copper-zinc system, and write down their compositions and temperatures.

(b)  The two common commercial brasses are: 70/30 brass: $W_{Cu} = 70\%$, and 60/40 brass: $W_{Cu} = 60\%$. Locate their constitution points on the diagram at 200 °C.

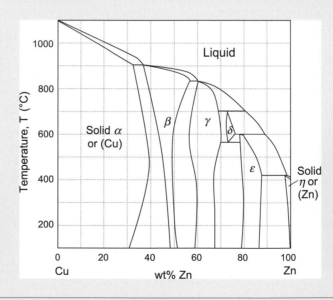

Figure GL2.19    Copper-zinc Cu-Zn phase diagram.

(i) What distinguishes the two alloys?
(ii) What roughly is the melting point of 70/30 brass?
(iii) What are the phases in 60/40 brass at 200 °C? Find their compositions and proportions.

E.11    Use the phase diagram for the $SiO_2$-$Al_2O_3$ system in Figure GL2.17 to answer
the following:
(a) The intermediate compound mullite may be considered as having the
formula $SiO_2$ $(Al_2O_3)_x$. Find the approximate value of $x$. The atomic
masses of Si, Al and O are 28.1, 26.9 and 16.0, respectively.
(b) Use the lever rule to find the equilibrium constitution of a 50 wt% $Al_2O_3$
alloy at 1700 °C. Is it valid to measure directly from the diagram in this
case? Why?

## Answers to exercises, Part 4

E.9     See Figure GL2.20 (a): Al-Si: eutectic at 580 °C, composition 12 wt% Si.
Scc Figure GL2.20 (b): Cr-Ni: eutectic at 1350 °C, composition 47 at% Ni.
Conversion to wt%: wt% Ni $= (0.47 \times 58.71)/(0.53 \times 52.0 + 0.47 \times 58.71) =$
50.0 wt% Ni.

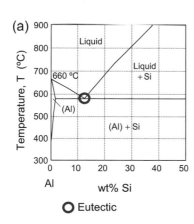

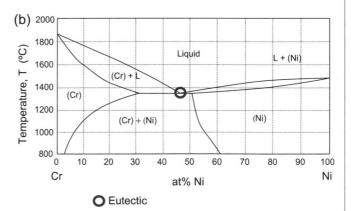

**Figure GL2.20**   (a) Aluminium-silicon Al-Si, and (b) chromium-nickel Cr-Ni phase diagrams.

E.10

(a) See Figure GL2.21: Eutectoid: $W_{Zn} = 73$ wt%, $T = 555$ °C

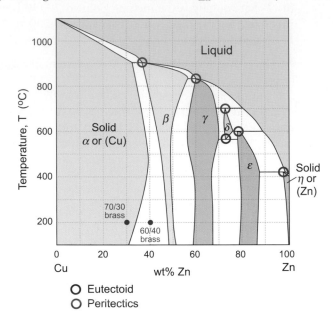

**Figure GL2.21** Cu-Zn phase diagram.

Peritectics:

$W_{Zn} = 37$ wt%, $T = 900$ °C.
$W_{Zn} = 60$ wt%, $T = 825$ °C.
$W_{Zn} = 73$ wt%, $T = 700$ °C.
$W_{Zn} = 79$ wt%, $T = 600$ °C.
$W_{Zn} = 99$ wt%, $T = 420$ °C.

(b) 70/30 brass is single phase, but 60/40 brass is two phase.
70/30 brass starts to melt at 920 °C and is completely liquid at 960 °C.
At 200 °C: $\alpha$ (copper-rich solid) and $\beta$ (roughly CuZn). $W_{Zn} \approx 33\%$, $W_{Zn} \approx 48\%$.

   Proportions roughly $50 - 50$; more precisely: proportion of $\alpha = (48-40)/(48 - 33) = 53\%$, and 47% $\beta$.

E.11

(a) See Figure GL2.22: Mullite contains approximately 73 wt% $Al_2O_3$ (27% $SiO_2$).
The molecular weight of $SiO_2$ is $28.1 + (16 \times 2) = 60.1$.
The molecular weight of $Al_2O_3$ is $(26.9 \times 2) + (16 \times 3) = 101.8$.
Molecular weight of mullite $= 1$ mol of $SiO_2 + x$ mols of $Al_2O_3 = 60.1 + 101.8x$.

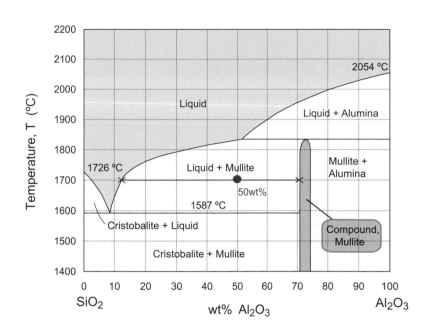

**Figure GL2.22**   $SiO_2$-$Al_2O_3$ phase diagram.

Alumina fraction by weight $\approx 0.73 \approx 101.8x/(60.1 + 101.8x)$.
Hence $x \approx 1.6$.
(b) At 1700 °C and 50 wt% $Al_2O_3$, the phases are liquid containing 13 wt% $Al_2O_3$ and mullite containing 71 wt% $Al_2O_3$.
Weight fraction of liquid = $(71 - 50)/(71 - 13) = 36\%$.
Weight fraction of mullite = $(50 - 13)/(71 - 13) = 64\%$.
It would be valid to measure the lengths of the tie-line directly in this case because the phase diagram has a *linear* weight % scale.

## PART 5: Phase transformations and microstructural evolution

After completing Parts 1–4 you should be able to:

- Recognise the important features of binary phase diagrams, and to identify whether fields are single phase or two phase.
- Find the proportions and compositions of the equilibrium phases, at a given temperature and alloy composition.

Parts 5–8 link this understanding of phase diagrams to microstructure evolution in important industrial processes. Almost all manufacturing processes involve some combination of heating and cooling. For example:

- Casting is filling a mould with molten metal, which solidifies and cools.
- Hot forming shapes metal billets, giving complex temperature histories due to heat conduction into the tooling and surrounding air.
- Welding often causes a thermal cycle of heating followed by cooling.
- A final stage of manufacture is often a separate heat treatment, of the whole component or just its surface.

When the temperature varies in a process, the equilibrium condition of the material keeps changing (e.g. as boundaries on the phase diagram are crossed)—hence *phase transformations* take place. These transformations determine which phases are present after processing, and how they are distributed among one another; that is, the final *microstructure*. This in turn controls the material properties. So, for a complete understanding of properties and processing, we need to know more about the microstructure than is given by a phase diagram. As we will see, controlling properties relies on managing not just which phases are present, but also their *morphology*. For instance, in a two-phase solid the *shape* of the phases and their *dispersion* within one another often have a strong influence on the properties.

*Observing phases and phase transformations*   Phase diagrams tell us the equilibrium phases, their compositions and proportions. But you might wonder how this information was produced in the first place. And how do we measure whether a real process is actually following equilibrium? Many techniques are available to quantify the phases present in a microstructure, with the appropriate method to use depending on what samples are available and the different length scales of the microstructural features (seen in overview in Chapter 19). Table GL2.1 summarises the various methods and their physical basis.

Phase diagrams have been built up experimentally using many of these techniques. Increasingly they are also determined by thermodynamic computation, from first principles. Many software packages are available to compute sub-domains in alloy phase diagrams containing many different elements. They are nonetheless dependent on good experimental data for calibration of the computations.

*Key concepts in phase transformations*   Phase diagrams give important information needed to predict the phase transformations and final microstructure that result from a given thermal history. The real microstructure may not be at equilibrium, but phase diagrams give a starting point from which other (non-equilibrium) microstructures can often be inferred.

Table GL2.1  Techniques for quantifying phases

| Technique | Physical basis |
| --- | --- |
| Dilatometry | Measurement of dimensional changes at high resolution to detect changes of density and symmetry caused by phase change. |
| Electrical resistivity | Measurement of electrical resistivity changes associated with difference in electron mean free path before and after phase change. |
| Calorimetry: differential thermal analysis | Sensitive differential measurement of release or take-up of latent heat associated with phase change. |
| Optical microscopy | Differential reflection of light, either in color or intensity, by phases or by a surface film created by chemical or electro-etching. |
| X-ray diffraction | Diffraction of X-radiation by the crystal lattice of each phase, giving diffraction patterns from which crystal structure and volume fraction of phases can be inferred. |
| Scanning electron microscopy (SEM) | Differential back-scattering of electrons by differing phases giving both an image and compositional information. |
| Transmission electron microscopy (TEM) | Diffraction of an electron beam by the crystal lattice of each phase, giving both an image of the structure and, from the diffraction pattern, its crystal structure. |

The fundamental thermodynamics and kinetics of phase transformations are described in Chapter 19. The key concepts in phase transformations are as follows:

- Phase transformations are driven by the resulting change in free energy (defined in Part 1), also known as the *driving force*.
- At the phase boundaries, the free energies of the states on either side of the boundary are equal—the driving force $\Delta G = 0$.
- Phase changes almost always involve diffusion as the kinetic mechanism by which atomic rearrangement occurs.
- Phase transformations occur via a two-stage process of *nucleation* and *growth*, in which nucleation may be spontaneous (homogeneous), or take place on some kind of interface (heterogeneous).
- *TTT diagrams* capture the extent of an isothermal transformation as a function of hold temperature and time, giving characteristic C-curves for diffusion-controlled phase transformations.
- In continuous cooling, there is a *critical cooling rate* that will just avoid the onset of the diffusional transformations.

To relate these concepts to phase diagrams and cooling in real industrial processes, we begin with examples of slow cooling, in which it is reasonable to assume that the material state can always evolve to maintain equilibrium: Part 6 considers solidification, and Part 7 covers solid-state phase changes. Finally Part 8 looks at examples of heat treatments in which we deliberately cool quickly to bypass equilibrium, manipulating the phases formed and their morphology.

## PART 6: Equilibrium solidification

*Latent heat release on solidification*   Figure GL2.23(a) shows the phases found in pure iron (i.e. the phase diagram becomes a single temperature axis). If we cool iron slowly from above its boiling point, the temperature as a function of time shows two shelves in the cooling curve, called *arrest points*, at the boiling and melting points (at atmospheric pressure, iron boils at 2860 °C and melts at 1536 °C). At each temperature there is a *phase change*: vapour-to-liquid at the boiling point, and liquid-to-solid at the melting point. The arrest point on cooling is due to the release of the *latent heat*. On heating the reverse occurs—an arrest in the temperature rise while heat is absorbed to melt or boil the material.

Pure iron solidifies initially to BCC δ-iron, but undergoes further solid-state phase transformations on cooling, first to FCC γ-iron at 1391 °C, and then back to BCC α-iron at 914 °C. These transformations *also* release latent heat, but the amount is much smaller, as indicated by the modest arrests in the cooling history in Figure GL2.23(b). The mechanisms of these solid-state changes are considered later.

Alloys frequently solidify over a *range* of temperature (between the liquidus and solidus lines). In this case the latent heat is released progressively as the temperature falls between the liquidus and solidus. The cooling curve therefore does not show a shelf at constant temperature, but the cooling rate is reduced by the progressive release of latent heat. This is illustrated in Figure GL2.24, for two compositions in the Cu-Ni system.

*Solidification of pure metals*   The mechanism of solidification is illustrated in Figure GL2.25. For *homogeneous nucleation*, solid colonies form spontaneously within the melt

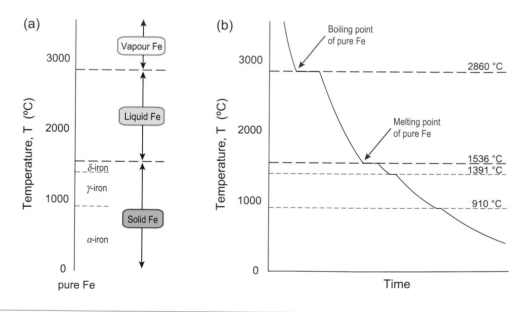

**Figure GL2.23**   (a) One-dimensional phase diagram for pure iron; (b) corresponding cooling curve for condensing then solidifying iron from vapour to liquid to solid.

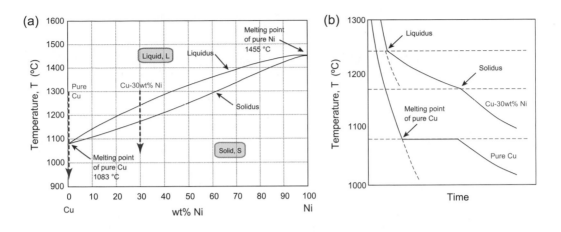

Figure GL2.24   (a) Cu-Ni phase diagram, and (b) corresponding cooling curves, for solidifying pure Cu and a Cu-30 wt% Ni alloy.

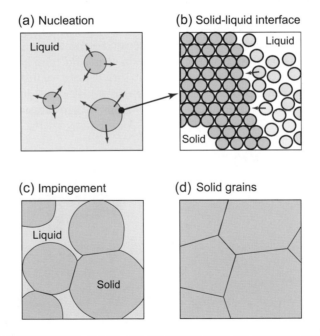

Figure GL2.25   Solidification mechanism: (a) homogeneous nucleation; (b) magnified view of atomic transfer at the solid-liquid interface; (c) growth of nuclei and the onset of impingement; (d) final solid grain structure.

(Figure GL2.25(a)). They grow stably provided they can reach a critical radius. This initial barrier reflects the surface energy of the solid-liquid interface, which uses up some of the free energy released by the transformation—the probability of forming stable nuclei increases rapidly as the liquid is undercooled below the transformation temperature. *Heterogeneous nucleation* facilitates the process, with solid nuclei forming more readily on a pre-existing solid in contact with the liquid (e.g. the walls of the mould, or high melting point particles of another solid mixed into the liquid).

Each region grows by atoms transferring at the solid-liquid interface (Figure GL2.25(b))—the interface advancing in the opposite direction to the atomic transfer. Growth continues until *impingement* of the solid regions occurs (Figure GL2.25(c)). Since each nucleus has its own independent crystal orientation, there is a misfit in atomic packing where they impinge. The individual crystallites remain identifiable once solidification is complete (Figure GL2.25(d)). We call these *grains*, and the surfaces where they meet are *grain boundaries*. Grains are typically on the length-scale of 1 μm−1mm, and are easily revealed by optical microscopy. Polished surfaces are chemically etched to 'round off' the boundaries where they meet the surface, and light is then scattered at the boundaries, which appear dark. Figure GL2.26 shows optical micrographs of pure metals after solidification, showing equiaxed grain structures. Note that some techniques also generate colour contrast between grains, by applying surface etches that change the polarity of the reflected light (see Figure GL2.26(b)). Here is an early warning of the need for caution in interpreting micrographs—the image of pure Al in the figure shows a single phase, so multiple colours do not necessarily indicate different phases!

So grain structure is our first example of a microstructure formed by a phase transformation. Let's emphasise a key idea that will recur from here onward. The phase diagram tells us that the liquid should transform to a single-phase solid—and that's all. It *doesn't* tell us the grain size—this depends on the density of nucleation sites, and the kinetics of the diffusive mechanism of forming solid at the interface. These in turn depend on just how slowly the liquid cools. So to describe the microstructure for *any* phase transformation, we need to bring in some additional knowledge. This example also highlights why processing details are so important in determining microstructure and properties. And it suggests the use of processing 'tricks' to

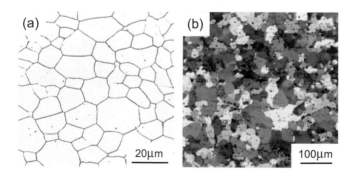

Figure GL2.26  Optical micrographs of (a) pure Fe and (b) pure Al. In (b) the etching technique produces colour contrast between different crystal orientations, but it is still all one phase. (Images courtesy: ASM Micrograph Center, ASM International, 2005).

modify what happens (e.g. adding lots of fine particles to the melt to promote copious heterogeneous nucleation; the result—a much finer grain size). Process-microstructure interactions such as these are discussed in Chapter 19.

*Solidification of simple binary alloys, and phase reactions*  Alloys show a freezing range, between the liquidus and solidus lines. Consider solidification of Cu-30 wt% Ni on the phase diagram of Figure GL2.24. Solidification starts at 1240 °C—nuclei form as in Figure GL2.25(a). In contrast to a pure metal, further cooling is needed for more solid to form, for example, at 1200 °C a tie-line indicates 50% solid and 50% liquid are stable, and the microstructure would appear as in Figure GL2.25(b). To reach 100% solid (Figure GL2.25(c)) we need to keep cooling to 1170 °C, and the end result looks much the same as in the pure case (with the difference that the phase forming these grains is a Ni-rich solid solution, not something we can detect optically). Note that the compositions of the solid and liquid both evolve, following the ends of the tie-line.

So when an alloy is cooled, the constitution point for the alloy drops vertically on the phase diagram. Phases transform when we cross boundaries on the diagram, but the Cu-Ni example shows that phases may also evolve in composition, *without* a change in the phases present. We call this a *phase reaction*.

**DEF.** When any phase compositions change with temperature, a *phase reaction* is taking place.

In a single-phase field, the composition of the phase is always that of the alloy; and no phase reaction takes place on cooling. In a two-phase region the compositions of the two phases are given by the ends of the tie-line through the constitution point. In general, as the constitution point falls vertically, the ends of the tie-line do *not*—instead they run along oblique phase boundaries. The compositions of the two phases change with temperature, and phase reactions occur. The exception would be where both boundaries at the end of the tie-line are vertical. This is not something associated with liquidus and solidus boundaries, but is seen in the solid state, particularly with compounds (recall the unusual Ag-Sr phase diagram in Figure GL2.9(a)).

*Solidification of dilute alloys*  The Cu-Ni system is unusual in that the solid that forms initially is unchanged on further cooling—no transformations or reactions take place. But most binary systems show eutectics, with two-phase fields over a wide range of composition at room temperature. First we will consider solidification of a relatively dilute alloy in the Pb-Sn system, Pb-10 wt% Sn (see Figure GL2.27, alloy A).

The transformations and reactions are as follows, leading to the microstructure evolution shown in Figure GL2.28.

1. *Above 305 °C*: Single-phase liquid of composition identical to that of the alloy; no phase reaction.
2. *From 305 °C to 270 °C*: The liquidus line is reached at 305 °C; the reaction liquid → solid (Pb-rich solid solution) starts. The solid contains less tin than the liquid (see first tie-line), so

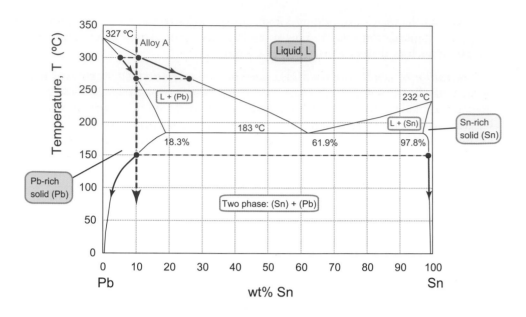

**Figure GL2.27** The Pb-Sn phase diagram: solidification of a dilute alloy A.

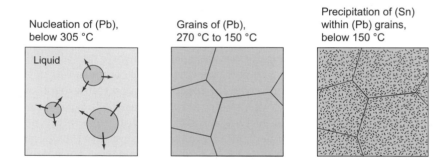

**Figure GL2.28** Schematic microstructure evolution in solidification of a dilute Pb-10 wt% Sn alloy.

the liquid becomes richer in tin and the composition of the liquid moves down the liquidus line, as shown by the arrow. The composition of the solid in equilibrium with this liquid changes too, becoming richer in tin also, as shown by the arrow on the solidus line: a *phase reaction* is taking place. The *proportion* of liquid changes from 100% (first tie-line) to 0% (second tie-line).

3. *From 270 °C to 150 °C:* Single-phase solid of composition identical to that of the alloy; no phase reaction.

4. *From 150 °C to room temperature.* The Pb-rich phase becomes unstable when the phase boundary at 150 °C is crossed. It breaks down into *two solid phases*, with compositions

given by the ends of the tie-line, and proportions given by the lever rule. On cooling the compositions of the two solid phases change as shown by the arrows: each dissolves less of the other, and a phase reaction takes place.

The formation of the Sn-rich phase takes place by *precipitation* from the solid solution. This mechanism too involves nucleation and growth, but on a much finer scale than the grain structure. Small clusters of Sn-rich solid nucleate spontaneously within the Pb-rich matrix. The fraction of this phase increases as the nuclei grow, by depleting the surrounding matrix of some of its Sn. The compositions of the phases (particles and matrix) continually adjust by inter-diffusion of Pb and Sn atoms. The practicalities of solid-state precipitation are discussed further in Part 7.

Note that the phase diagram tells us the proportions of the phases, in weight %. The density of the phases is not the same, so this does not convert directly into volume % (or area fraction in a metallographically prepared cross-section). Nonetheless, provided the phase densities are not too dissimilar, the phase proportions give some idea of the proportions we expect to see in the microstructure (good enough for sketching, as in Figure GL2.28(a)).

## Exercise

E.12   Find the proportions and compositions of the phases formed on solidification of the Pb-10 wt% Sn alloy at 250 °C and at room temperature (20 °C).

*Eutectic solidification*   Next consider the solidification of the eutectic composition itself in the Pb-Sn system (Figure GL2.29, alloy B). The eutectic composition is Pb-61.9 wt% Sn. When liquid of this composition reaches the eutectic temperature (183 °C), the liquid can transform to 100% solid without further cooling. This is a unique characteristic of eutectic alloys, and in this respect they resemble pure components. But the final microstructure will be very different, since two solid phases form simultaneously (with proportions and compositions governed by the tie-line through the eutectic point). This transformation is called the *eutectic reaction*.

**DEF.** A *eutectic reaction* is a three-phase reaction by which, on cooling, a liquid transforms into two solid phases at constant temperature: Liquid, L $\rightarrow$ Solid $\alpha$ + Solid $\beta$.

## Exercise

E.13   Find the proportions and compositions of the solid phases formed on solidification at the eutectic point in the Pb-Sn system.

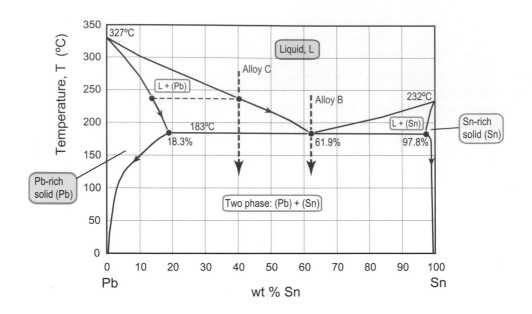

**Figure GL2.29** The Pb-Sn phase diagram: solidification of eutectic alloy (B) and an off-eutectic alloy (C).

So how does this transformation take place? The phase diagram provides clues—the Pb and Sn in the liquid are uniformly mixed, but after transformation we have Pb-rich solid and Sn-rich solid. Pb and Sn must therefore inter-diffuse to generate regions in which each dominates. This must happen at an interface between liquid and two-phase solid. It will therefore be easier if the diffusion distance is kept small; that is, the two phases separate out on a small scale, such that no single atom has to diffuse too far to be able to join a growing colony of (Pb) or (Sn). Eutectics usually therefore form as intimate mixtures of the two phases, on a length scale much smaller than a typical grain size. Figure GL2.30(a) shows a micrograph of eutectic Al-Si (the phase diagram for this system was shown in Exercise E.9).

The proportions of the phases in a eutectic can vary widely, depending on the position of the V along the eutectic tie-line. But in broad terms we can think of two characteristic dispersions of the phases. If the eutectic point is toward the middle of the tie-line, the proportions of the two phases are roughly equal, and neither can be thought of as a matrix containing the other phase. This is the case in Pb-Sn (the proportions were found in Exercise E.13). However, if the V is located toward one end of the tie-line, then one phase forms a matrix containing the second phase as isolated particles. This is the case in Al-Si (Figures GL2.18(a) and GL2.30(a)).

Note in Figure GL2.30(a) that the *shape* of the phases is elongated into plates or needles. There are two reasons why this often occurs. One is that growth of the interface is planar at the atomic scale, with plates extending in the growth direction to minimise diffusion distances. The other is that the boundary between the phases is also a crystallographic boundary, and some pairs of orientations have a lower surface energy that will grow more rapidly. Forming a fine mixture of phases does mean that there is a price to be paid in surface energy between

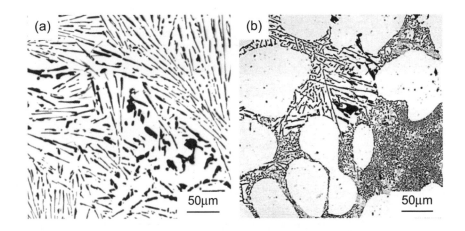

Figure GL2.30 Optical micrographs of (a) eutectic Al-Si alloy; (b) off-eutectic Al-Si alloy, showing grains of primary (Al) (white). (Images courtesy: ASM Micrograph Center, ASM International, 2005).

phases—the area of interface per unit volume is large. But this is a small energy penalty compared to the free energy release in the transformation as a whole. So again we find that the detail of the transformation mechanism has an important influence on the microstructure (and hence properties).

A further phase reaction occurs on cooling of the two-phase eutectic solid, for example, for Pb-Sn from 183 °C to room temperature. As the phase boundaries on the phase diagram are not vertical, the compositions and proportions of the two solid phases evolve on cooling. Diffusion distances are automatically small within the eutectic, so both phases can become purer and adjust their proportions, by inter diffusion.

*Solidification of off-eutectic compositions* Casting alloys are often off-eutectic, with a composition to one or other side of the 'V' on the phase diagram. The solidified microstructure in these cases can now be inferred—above the eutectic, partial solidification of a single phase solid solution occurs (as seen in Cu-Ni earlier), then at the eutectic temperature the remaining liquid undergoes the eutectic reaction (as seen in Pb-Sn earlier).

Consider solidification of Pb-40 wt% Sn (see Figure GL2.29, alloy C). The transformations and reactions are as follows, leading to the microstructure evolution illustrated in Figure GL2.31.

1. *Above 235 °C*: Single-phase liquid; no phase reactions.
2. *From 235 °C to 183 °C*: The liquidus is reached at 235 °C, and nuclei of Pb-rich solid solution appear first (Figure GL2.31(a)). The composition of the liquid moves along the liquidus line, that of the solid along the solidus line. This regime ends when the temperature reaches 183 °C. Note that the alloy composition (40 wt%) is roughly halfway between that of the new solid (18.3 wt%) and the residual liquid (61.9 wt%); so the alloy is about half liquid, half solid, by weight (and very roughly by volume, neglecting the difference in phase densities; Figure GL31(b)).

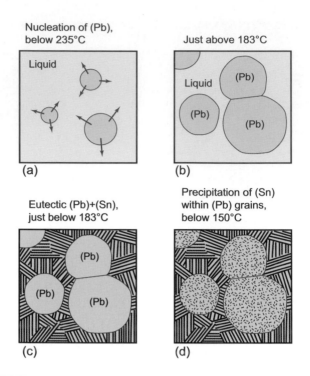

**Figure GL2.31**  Schematic microstructure evolution in solidification of a dilute Pb-40 wt% Sn alloy.

3. *At 183 °C:* The remaining liquid has reached the *eutectic point,* and this liquid undergoes solidification exactly as described before (Figure GL2.31(c)). Note that the proportions of primary (Pb) and eutectic microstructure are exactly the same as the solid-liquid proportions just above the eutectic temperature. Nucleation of the eutectic will be straightforward, since there are already solid grains present on which the eutectic can form.

4. *From 183 °C to room temperature:* The two types of microstructure each evolve:
   (a) The Pb-rich solid becomes unstable, and Sn-rich solid precipitates (exactly as before, in 10 wt% alloy).
   (b) The eutectic region evolves exactly as 100% eutectic did before: both phases change composition by inter-diffusion, becoming purer.

The final microstructure therefore combines all the features discussed so far (Figure GL2.31(d)). Note that in this final microstructure there are just two phases, with the Pb-rich and Sn-rich regions in the eutectic being exactly the same phases in the (Pb) grains containing (Sn) precipitates. The phase diagram only tells us the overall phase proportions (from the tie-line at room temperature). But their spatial dispersion (and hence properties) involves understanding of how the cooling history interacts with the phase diagram, and the mechanisms of phase transformations.

Figure GL2.30(b) shows a micrograph of an off-eutectic Al-Si alloy. The primary Al grains are surrounded by eutectic microstructure. The primary grains have a uniform colour, appearing therefore to be a single phase. This may be because the scale of the precipitates is too fine to resolve in this image—another word of warning about potential over-interpretation of micrographs. On the other hand, it is possible that the microstructure has not followed equilibrium in practice—the grains have remained as a solid solution due to some difficulty in nucleating the second solid phase. In this case the phase will be *supersaturated* in solute, and the phase is *metastable* (i.e. thermodynamically stable at room temperature at a higher free energy level than the equilibrium state).

## Exercises

E.14  Not all alloys in the lead-tin system show a eutectic reaction on slow cooling: pure lead, for example, does not. Examine the Pb-Sn phase diagram and identify the composition range for which you would expect a eutectic reaction to be possible.

E.15  (a) A eutectic reaction was defined in the text. Define what happens on heating a solid of eutectic composition. Over what temperature range does melting occur?
      (b) For a Pb-35 wt% Sn alloy, identify the temperatures at which melting starts and finishes on heating, and describe how the proportion of liquid evolves during melting.

E.16  Figure GL2.18(a) showed the phase diagram for the Al-Si system (the basis of aluminium casting alloys). Use this figure to answer the following:
      (a) Describe the solidification of an Al-20 wt% Si alloy, sketching the microstructure at key temperatures (e.g. onset of solidification, just above and just below the eutectic). Estimate the phase proportions in two-phase regions, and sketch the microstructures accordingly.
      (b) How would the final microstructure differ for Al-Si alloys of: (i) Al-1 wt% Si; (ii) eutectic composition?

*Segregation*   It has been noted that when solidification starts, either on the walls of a mould or spontaneously as nuclei throughout the melt, the first solid to form is purer in composition than the alloy itself (due to *partition* of the liquidus and solidus lines). For example, in Pb-10 wt% Sn (considered earlier), the first solid appearing at 305 °C has a composition of 5 wt% Sn. This means that tin is *rejected* at the surface of the growing crystals, and the liquid grows richer in tin: that is why the liquid composition moves along the liquidus line. From 305 to 270 °C the amount of primary (Pb) increases, and its equilibrium composition increases, following the solidus line. This means that lead must diffuse *out* of the solid (Pb), and tin must diffuse *in*. This diffusion takes time. If cooling is very slow, time is available and equilibrium may be maintained. But practical cooling rates are not usually slow enough to allow sufficient time for

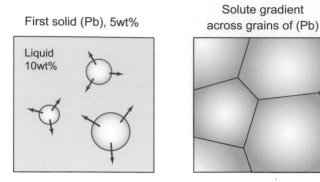

First solid (Pb), 5wt%

Solute gradient
across grains of (Pb)

Liquid
10wt%

Figure GL2.32    Segregation in solidification of binary alloys.

diffusion, so there is a *composition gradient* in each (Pb) grain, from the middle to the outside. This gradient is called *segregation*, and is found in almost all alloys (see Figure GL2.32).

Segregation can to some extent be 'smoothed out' after casting by *homogenisation*: holding a casting at temperature to allow some redistribution of solute by diffusion. It can never be eliminated completely.

One consequence of segregation is that the last liquid to solidify will be richer in solute than expected from the phase diagram (as the solid is purer than it should be). This may lead to the formation of eutectics at the grain boundaries (the last part to solidify), even in alloys that are not expected to show a eutectic (i.e. compositions beyond the range of the tie-line through the eutectic point). The temperature at which melting would start on heating is thus lower than expected from the phase diagram, that is, the eutectic temperature, not the solidus temperature.

Segregation is even more important with respect to *impurities*. No alloy is truly binary, ternary, or whatever, but will contain traces of other elements (e.g. from the ore used as the source of the metal). Impurities also dissolve more readily in the liquid than in the solid. Hence when alloys solidify, the impurities concentrate in the last bit to solidify. Even if the alloy contains only a fraction of 1% of a given impurity overall, if the solid will not dissolve it in solution the concentration can be locally much higher on the grain boundaries. This is damaging if the impurity forms brittle phases, which can be disastrous from the point of view of toughness. An example is sulfur in steel, which forms brittle iron sulfide. Rather than trying to remove the traces of impurity from the melt, the way to solve the problem is further alloying. Plain carbon steels all contain manganese, which forms MnS at higher temperatures during solidification and therefore ties up the sulfur before segregation takes place, rendering it harmless.

Another example of detrimental impurity segregation is in *fusion welding*. This is like casting on a small scale, with a melt pool under a moving heat source solidifying as it goes. Solidification finishes on the weld centre-line, and segregation concentrates impurities there. Contraction of the joint can generate high enough thermal stresses to crack the weld along the weakened grain boundaries down the weld centre: this is known as *solidification cracking*.

## Answers to Exercises, Part 6

| | |
|---|---|
| E.12 | At 250 °C, 100% Pb-rich solid (composition same as alloy, 10 wt% Sn). At 20 °C, two phases: Pb-rich solid (1 wt% Sn) and Sn-rich solid (99 wt% Sn). Proportion of Pb-rich solid = $(99 - 10)/(99 - 1) = 0.91$, or 91 wt% (hence 9 wt% Sn-rich solid). |
| E.13 | The phases are Pb-rich solid (18.3 wt% Sn) and Sn-rich solid (97.8 wt% Sn). The proportion of Pb-rich solid is $(97.8 - 61.9)/(97.8 - 18.3) = 0.45$, or 45 wt% (hence 55 wt% Sn-rich solid). |
| E.14 | From $W_{Sn} = 18.3\%$ to $W_{Sn} = 97.8\%$ (i.e. the ends of the tie-line through the eutectic point—alloys outside this range reach 100% solid above the eutectic temperature). |

E.15
(a) As this is an equilibrium diagram, on heating the reaction simply goes in reverse. The two solids 'react' to give a single liquid. In general, on heating: $\alpha + \beta \rightarrow$ Liquid. Eutectics melt at constant temperature.

(b) For Pb-35 wt% Sn, melting starts at the eutectic microstructure and finishes at 250 °C. At the eutectic temperature, all the solid eutectic microstructure melts, giving roughly 40% liquid. On heating to 250 °C, the remaining solid (the original Pb-rich grains) melts progressively.

E.16
(a) Cooling of Al-20 wt% Si casting alloy, illustrated by phase diagram and schematics in Figure GL2.33.

1. *Above 680 °C*: Single-phase liquid, with composition of the alloy; no phase reaction.

2. *From 680 to 577 °C*: The liquidus is reached at about 680 °C, when primary solid Si (pure, not a solution) starts to separate out. As the temperature falls the liquid composition moves along the liquidus line, and the amount of solid Si increases. At the eutectic temperature (577 °C), the proportion of solid is $(20 - 12)/(100 - 12) = 9\%$; see Figure GL2.33(b).

3. *At 577 °C*: The eutectic reaction takes place: the remaining liquid decomposes into solid (Al) mixed with solid Si, on a fine scale, Figure GL2.33(c). The proportion of Si in the eutectic is $(12 - 2)/(100 - 2) = 10\%$. So the Si forms as a dispersion within a matrix of (Al), usually in the characteristic needle-shape of a eutectic.

4. *From 577 °C to room temperature*: On further cooling the (Al) in the eutectic needs to grow purer and reject Si. So in the eutectic structure,

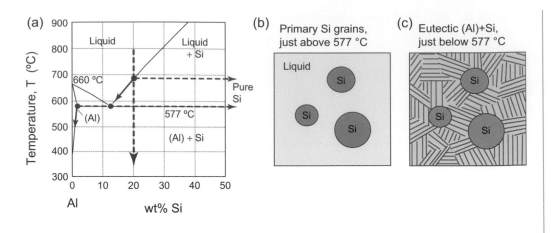

**Figure GL2.33**   (a) Al-Si phase diagram and (b) and (c) microstructure evolution on solidification of Al-20 wt% Si.

Si diffuses out of the (Al). The final appearance of the microstructure still looks like Figure GL2.33(c).

(b) (i)   An Al-1 wt% Si would solidify to 100% (Al) grains with no eutectic, and would be expected to precipitate Si within the grains on cooling to room temperature (assuming Si is able to nucleate and grow in the solid state—if not, the (Al) grains remain as metastable supersaturated solid solution).

   (ii)   A solidified alloy of eutectic composition would be 100% eutectic, with no primary Si.

## PART 7: Equilibrium solid-state phase changes

***Precipitation reactions***   Solid-state phase transformations were introduced in Part 6, following solidification of dilute binary alloys (Figures GL2.27 and GL2.28). The solid formed first from the liquid as grains of single-phase solid solution, but on further cooling the solvus boundary is crossed into a two-phase region. One of the new phases is present already, but the falling solubility means that the excess solute needs to be absorbed by the second phase. This type of transformation is a *precipitation* reaction: $\alpha \rightarrow \alpha + \beta$.

In many systems, the second phase in this two-phase region is a *compound*, with precipitates forming on slow cooling having a specific stoichiometry, rather than being another solid solution. This is important for making alloys with high strength—compounds are generally hard and resistant to dislocations. This offers the potential for *precipitation hardening* (Chapter 6)—the mechanism of formation of the second phase explaining the name.

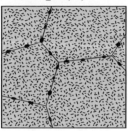

Precipitates of
$CuAl_2$ in (Al)

**Figure GL2.34** Schematic microstructure of a slow-cooled Al-4 wt% Cu alloy.

Refer to Figure GL2.9(b), the Al-rich end of the Al-Cu system. Exercise E.7(d) considered an Al-4 wt% Cu alloy at 550 °C and 250 °C. On cooling between these temperatures, at 490 °C the solvus boundary is crossed and the θ-phase ($CuAl_2$) starts to precipitate, nucleating homogeneously within all the grains and increasing in proportion as the temperature falls. Figure GL2.34 shows the final microstructure.

At modest magnification, the two-phase region may appear a uniform colour, and higher resolution is needed to distinguish the precipitates from the matrix. The scale of the microstructure is dictated by the kinetic mechanism. By forming very large numbers of small precipitates, the average diffusion distance for solute is short, enabling the transformation to occur in the solid state. The precipitate spacing is therefore much finer than in a eutectic. But in common with a eutectic microstructure, there is an energy penalty in forming a large area per unit volume of interface between the phases, but again the free energy release dominates over the surface energy.

From the point of view of precipitation hardening, it is the spacing of precipitates that matters. In practice the precipitates formed by slow cooling are still too far apart to be effective in obstructing dislocations. Useful strengthening comes from forming a much finer dispersion of precipitates—the means to achieve this in Al-Cu and other alloys is discussed in Part 8.

The precipitation illustrated here is expected on cooling due to the thermodynamics of the system. But nucleation of precipitates depends strongly on the kinetic mechanism. As noted earlier it can be difficult to form a new lattice within an existing solid crystal—the existence of a driving force may not be sufficient to cause the phase change, leaving a metastable supersaturated solid solution on cooling. Whether or not a given system will precipitate is an alloy-specific detail, not indicated by the phase diagram.

Figure GL2.34 illustrates another feature that may be observed in precipitation reactions, on the grain boundaries. Here the precipitates are larger than within the bulk of the grains. Two factors contribute to this effect, both due to the extra 'space' associated with the boundary: first, faster nucleation can occur, as atoms can rearrange more readily into the precipitate lattice (this is a form of heterogeneous nucleation); second, the boundaries provide faster diffusion paths for solute to be drawn in to build up the precipitates. This doesn't always

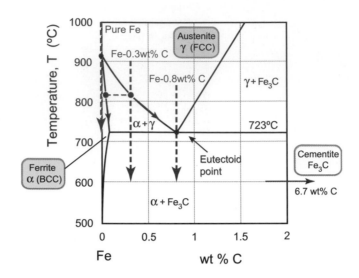

Figure GL2.35 The Fe-C phase diagram for compositions of carbon steels below 1000 °C, including the eutectoid point.

occur—again it depends on the alloy system—but it can have important practical consequences. Coarser precipitation on the grain boundaries may weaken the boundaries, with possible loss of toughness (intergranular fracture) or susceptibility to corrosive attack, forming grain boundary cracks.

***Phase transformations in carbon steels*** The iron-carbon system was introduced in Part 3, including the various single-phase forms of iron-carbon solid solutions, the compound iron carbide and the eutectoid point. Here we investigate the important phase transformations that occur in *carbon steels*; that is, Fe-C alloys with compositions between pure iron and the eutectoid composition. Figure GL2.35 shows the relevant section of the Fe-C phase diagram.

First consider cooling of pure iron, starting with FCC $\gamma$-iron at 1000 °C. At 910 °C, a solid-state change to BCC $\alpha$-iron occurs (at constant temperature). *Nucleation* starts at the grain boundaries, where there is more space for BCC nuclei to form. The $\alpha$ grains grow by atoms jumping across the boundary between the two phases, adopting the crystal structure of the growing phase as they do so (shown schematically in Figure GL2.36(b)). The interface migrates in the opposite direction to the atomic jumps. Figure GL2.36(a) and (c) show the solid-solid transformation under way and complete at 910 °C. When the growing $\alpha$ colonies impinge, new $\alpha$-$\alpha$ grain boundaries are formed. Since there is typically more than one $\alpha$ nucleus per $\gamma$ grain, the average grain size is reduced by this transformation.

***The eutectoid reaction*** The eutectoid point in the Fe-C system was introduced in Part 3. Cooling austenite of eutectoid composition (0.8 wt% C) leads to complete transformation to

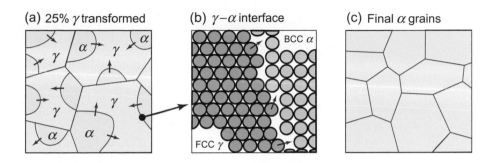

**Figure GL2.36** Schematic illustration of the transformation from FCC $\gamma$-iron to BCC $\alpha$-iron in pure iron: (a) $\gamma$-$\alpha$ grain structure after 20% transformation; (b) atomic transfer mechanism at the phase interface; (c) final $\alpha$-iron grain structure.

ferrite and cementite, at the eutectoid temperature, 723 °C (see Figure GL2.35). More generally we therefore define a eutectoid reaction as follows:

> **DEF.** A *eutectoid reaction* is a three-phase reaction by which, on cooling, a single-phase solid transforms into two different solid phases at constant temperature: Solid $\gamma \rightarrow$ Solid $\alpha +$ Solid $\beta$.

The mechanism of the eutectoid reaction must transform a single solid phase into two others, both with compositions that differ from the original (given by the ends of the tie-line through the eutectoid point). In the Fe-C eutectoid reaction, austenite containing 0.8 wt% C changes into ferrite (containing almost no carbon) and cementite (Fe$_3$C, containing 6.7 wt% C, or 25 *atomic* % carbon). Hence carbon atoms must diffuse to form regions of high and low concentration, at the same time as the FCC austenite lattice transforms into BCC ferrite and cementite lattices. Nuclei of small plates of ferrite and cementite form at the grain boundaries of the austenite, and carbon diffusion takes place on a very local scale just ahead of the interface. The plates of ferrite and cementite grow in tandem, consuming the austenite as they go (see Figure GL2.37(a)).

The resulting grains therefore consist of alternate plates of ferrite and cementite. This structure again has a very large area of phase boundary between ferrite and cementite, with associated surface energy penalty. However, kinetics dictate the mechanism of transformation, such that C atoms on average diffuse only one plate spacing, whereas the interface between the new phases and the austenite traverses whole grains.

The eutectoid ferrite-cementite microstructure has a special name: *pearlite* (because it has a pearly appearance). Note that pearlite is a two-phase microstructure, *not* a phase, in spite of the similarity in nomenclature. The micrograph in Figure GL2.37(b) shows pearlite, at high magnification. The length scale of the plates in pearlite is again much finer than in a eutectic, due to the solid-state diffusion. At lower magnification, pearlite may just appear as dark

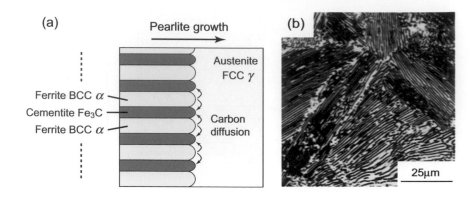

Figure GL2.37   (a) Schematic illustration of the eutectoid transformation from austenite to pearlite; (b) micrograph of pearlite. (Image courtesy: ASM Micrograph Center, ASM International, 2005).

etching grains. This is easily mistaken for a single phase—it is not, but this is another example of a dispersion of two phases at a scale below the resolution of the microscopy technique.

*Phase transformations in hypo-eutectoid steels*   Some commercial steels have a eutectoid composition—steel for railway track is an example of a 'pearlitic steel'. Most carbon steels are 'hypo-eutectoid', containing less than 0.8 wt%. Mild steels contain 0.1–0.2 wt% C, medium carbon steels around 0.4 wt%. Here we will look at how the equilibrium microstructures relate to the phase diagram.

Consider slow cooling of a medium carbon steel containing 0.3 wt% C (as indicated on Figure GL2.35). Starting with austenite at 900 °C, we have a solid solution of C in FCC austenite. At 820 °C, we enter the two-phase region: ferrite plus austenite. The formation of ferrite follows the same mechanism as in pure iron, nucleating on the austenite grain boundaries. The ferrite rejects carbon into the remaining austenite, the composition of which increases accordingly. Grains of ferrite grow until, just above 723 °C, the proportion of ferrite to austenite is roughly 2:1. The remaining austenite contains 0.8 wt% C—it is at the eutectoid point. This austenite then decomposes as before into pearlite, the two-phase mixture of ferrite and cementite.

The schematics in Figure GL2.38 show the reaction under way just above and just below the eutectoid temperature. Note that as before the final structure is still only two phase, but different grains have very different microstructures. The ferrite within the pearlite structure is the same stuff as the ferrite forming whole grains—it is all one phase.

Iron carbide is a hard phase, and the pearlite structure is effective in obstructing dislocation motion (due to the plate-like structure). But ferrite itself has a relatively high intrinsic strength, with very high toughness. Carbon steels that have been slow-cooled (or *normalised*) to a ferrite and pearlite microstructure provide an excellent combination of strength and toughness, widely exploited for structural and mechanical applications. As in pure iron the transformation from austenite leads to a reduction in grain size, as each old grain nucleates more than one new grain. This is used commercially for *grain refinement*, an important heat treatment as it simultaneously enhances both yield strength *and* toughness.

Just above 723 °C

Just below 723 °C:
ferrite + pearlite

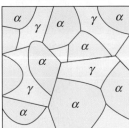 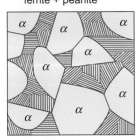

**Figure GL2.38** Schematic illustration of microstructure evolution in a hypo-eutectoid steel, from austenite to ferrite + pearlite.

Ferrite and pearlite microstructures are strong and tough, but we can do even better by *quenching* and *tempering* carbon steels. This heat treatment leads to the same phases, but in a different morphology, enhancing the strength without loss of toughness. We return to this in Part 8.

## Exercises

E.17    An Al-10 wt% Cu alloy was slow-cooled and the length scales measured by various techniques in the final microstructures, as follows: primary grain size = 100 μm; phase spacing in the eutectic = 1 μm; precipitates within the primary grains 0.05 μm. Identify the phases involved in each of these microstructural features, and explain the differences in length scale.

E.18    Pure iron cooled slowly contains 100% ferrite; the eutectoid composition contains 100% pearlite. Estimate the carbon content of the hypo-eutectoid steel shown in Figure GL2.39, and sketch the structure of a 0.2 wt% carbon steel after slow cooling to room temperature.

E.19    *Hyper-eutectoid* steels contain >0.8 wt% C. A high carbon steel containing 1.0 wt% C is cooled slowly from 1000 °C in the austenite field. Refer to the Fe-C phase diagram in Figure GL2.35 to answer the following:
   (i) At what temperature does a phase transformation begin, and what new phase then appears? Why do you think this phase tends to nucleate and grow along the austenite grain boundaries?
   (ii) What happens to the compositions of the phases on cooling to just above 723 °C?
   (iii) What phase transformation takes place at 723 °C?
   (iv) What are the final phases at room temperature, their compositions and proportions? Sketch the expected final microstructure.

**Figure GL2.39**   Optical micrograph of the ferrite-pearlite microstructure in a hypo-eutectoid carbon steel. (Image courtesy: ASM Micrograph Center, ASM International, 2005).

## Answers to Exercises, Part 7

E.17    Al-10 wt% Cu first forms primary grains of Al-rich solid, then a eutectic mixture of (Al) and $CuAl_2$. The (Al) grains then become unstable and $CuAl_2$ precipitates form. The scale of the grains is largest, being determined by the spontaneous nucleation of (Al) in the liquid. The eutectic microstructure is finer-scale, since the temperature is now lower, atomic mobility is reduced, and the microstructural scale is strongly influenced by the kinetics of inter-diffusion of Al and Cu, enabling transformation into two phases of widely different Cu contents. The precipitation of $CuAl_2$ is a solid-state process at lower temperature still, with the kinetics of nucleation dictating that large numbers of small precipitates form.

E.18    (a) Carbon content $\approx$ 0.4 wt%: the structure is about 50% pearlite, so the carbon content is about 50% of 0.8 wt%. To understand this, apply the lever rule *just above* the eutectoid temperature in the $\alpha + \gamma$ region: it is approximately 50% ferrite, 50% austenite. The pearlite all comes from the austenite at 0.8 wt%, so the ferrite:austenite ratio just above the eutectoid temperature is identical to the final ferrite:pearlite ratio.

   Similarly, 0.2 wt% C would give about 0.2/0.8 = 25% pearlite at equilibrium, as shown schematically in Figure GL2.40(a).

E.19    (i) Transformation begins at 765 °C, and the new phase to appear on the austenite grain boundaries is cementite, or iron carbide, $Fe_3C$. Cementite has a more complex crystal lattice, so it is easier to nucleate on the grain boundaries and to spread along them.

**(a)** 0.2wt% C:
ferrite + pearlite

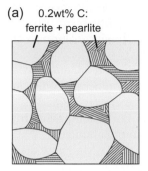

**(b)** 1.0wt% C:
cementite + pearlite

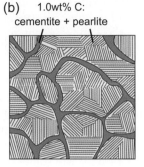

**Figure GL2.40** Schematic microstructures: (a) ferrite-pearlite in 0.2 wt% C steel; (b) cementite-pearlite in 1.0 wt% C steel.

  (ii) Carbon must diffuse out of the austenite to form the cementite, following the phase boundary down to the eutectoid composition. So just above the eutectoid temperature we have austenite containing 0.8 wt% C and $Fe_3C$ containing 6.7 wt% C.

  (iii) At 723 °C, the eutectoid reaction takes place—the austenite transforms to pearlite (two-phase mixture of ferrite and more cementite).

  (iv) Final phases at room temperature: ferrite, $\approx$ 0 wt% C; cementite, 6.7 wt% C.

Proportion of ferrite: 5.7/6.7 = 85 wt%; cementite: 1.0/6.7 = 15%.

The expected microstructure is shown in Figure GL2.40(b).

## PART 8: Non-equilibrium solid-state phase changes

The transformations considered thus far assume that the alloy is able to remain at equilibrium during cooling; that is, cooling is slow, and diffusion remains sufficiently rapid for phase fractions and compositions to adjust. In a couple of instances though, it has been noted that this may not be the case. The practical limitations of kinetic mechanisms operating at finite cooling rates intervene. For example, we noted in casting that solute gradients build up across grains (segregation). And precipitation reactions may not occur, due to barriers to nucleation of the second phase, leaving a metastable supersaturated solid solution.

In this section, a number of important heat treatments are explored. The common starting point in each case is to *solutionise* the alloy, dissolving the alloying additions in a high-temperature single phase. Then by deliberately applying non-equilibrium cooling, different microstructural outcomes with enhanced properties can be achieved. Further details of the resulting properties are discussed in Chapter 19.

***Non-heat-treatable aluminum alloys*** The *wrought* aluminum alloys are those that are shaped by deformation processing, as opposed to the *casting* alloys, which are cast directly to shape. Both are sub-divided into non-heat-treatable, and heat-treatable, reflecting the fact that only some respond to heat treatments to modify their properties (particularly strength). Wrought Al alloys are all relatively dilute in alloying additions, and below the upper limit of solid solubility in aluminum.

Virtually all wrought aluminum alloys contain magnesium—this is actually the single biggest application of Mg, greater than the manufacture of products in Mg alloys. The non-heat-treatable wrought Al-Mg alloys and the closely related Al-Mg-Mn alloys are familiar from the aluminum beverage can, but also find widespread structural use, including ship-building (partly attributable to good corrosion resistance).

Figure GL2.41(a) shows the Al-rich end of the Al-Mg phase diagram. It looks as though any alloy up to about 14 wt% Mg solutionised in the single-phase $\alpha$-field should form precipitates of the complex $Mg_5Al_8$ compound on cooling. But here is an example in which kinetics dominates over thermodynamics. Figure GL2.41(b) shows the corresponding TTT diagram for formation of this compound, for a typical alloy containing 6 wt% Mg solutionised at 400 °C. Long times are needed to start the phase transformation—the critical cooling rate is sufficiently low that a metastable Al-rich solid solution is easily retained at room temperature with a moderate quench. It is in this state that the alloys are used commercially, with cold deformation often used to strengthen the alloy further by work hardening.

***Heat-treatable aluminum alloys*** Commercial heat-treatable wrought aluminum alloys combine additions of magnesium with one or more of Si, Zn, Cu and other metallic elements. But the key behaviour exploited in heat treatment is found in some binary systems, such as Al-Cu. Comparing Figure GL2.41(a) for Al-Mg alloys with Figure GL2.9 for Al-Cu alloys, the diagrams appear superficially the same for dilute alloys. But in Al-Cu alloys, cooling at moderate rates follows equilibrium—the compound $CuAl_2$ precipitates out in a matrix of (Al) (recall Figure GL2.34). The difference is simply down to the greater ease of nucleation in this system—again, a technical detail beyond the scope of the phase diagram.

It was noted earlier that the $CuAl_2$ precipitates formed on slow cooling are actually rather far apart, from the perspective of interacting with dislocations, giving little strengthening. What we want is a much finer dispersion; that is, many more nuclei forming, with diffusion distances to the nuclei being much shorter. This can be achieved by a two-step heat treatment: first by *quenching* to room temperature, at a cooling rate above the critical cooling rate for precipitation to occur, then by *ageing* at an intermediate temperature. The quench 'traps' the high-temperature state as a metastable supersaturated solid solution at room temperature, from which we can then precipitate on a fine scale during ageing.

As so often in metallurgy, this precipitation during ageing is more complicated than expected. The quench achieves the desired result—a supersaturated solid solution of Cu in Al. But ageing at temperatures between room temperature and about 200 °C does *not* cause equilibrium $CuAl_2$ precipitates to form. Instead, metastable second phases form, and the TTT diagram for Al-Cu shows further C-curves for these phases at low temperature, as shown in Figure GL2.42(a).

This subtle difference in behaviour could not be anticipated from the phase diagram at all, but gives us the high strength precipitation hardened alloys on which aerospace depends, as increasingly do other transportation systems (rail and road). During the ageing treatment of Al-Cu after quenching, the yield strength evolves as shown in Figure GL2.42(b). For *artificial*

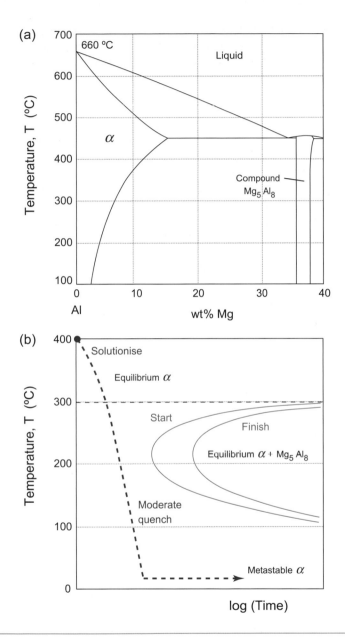

**Figure GL2.41**  (a) Al-rich end of the Al-Mg phase diagram; (b) schematic TTT diagram for precipitation of equilibrium $Mg_5Al_8$ in an Al-6 wt% Mg alloy.

*ageing* at temperatures above room temperature, the strength rises to a peak before falling to a value below the as-quenched value. The solid solution strength in the as-quenched state is replaced with precipitation hardening, and it is this that produces the peak strength. *Natural*

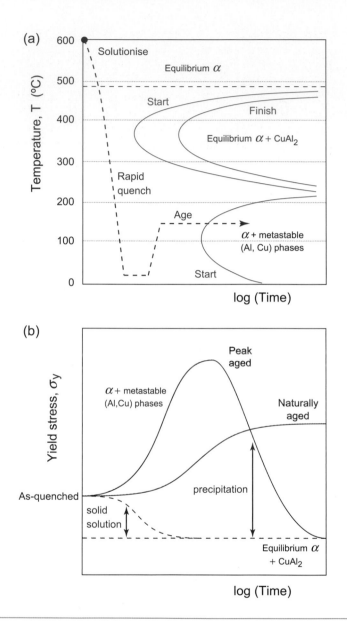

**Figure GL2.42**   (a) TTT diagram for precipitation of equilibrium and metastable phases in dilute Al-Cu alloys; (b) the 'ageing curve', showing how yield strength evolves with time.

*ageing* takes place at room temperature, with the strength rising to a plateau value below the peak of the curve for artificial ageing.

The full detail of the precipitate evolution during ageing need not concern us here (see Further reading for more information). In fact a whole sequence of metatstable second phases form, starting with clusters of pure Cu, and progressively re-nucleating and growing though various Al-Cu phases toward the equilibrium θ phase (CuAl$_2$). At the same time, the precipitates *coarsen* (i.e. their average size increases, while their number decreases), and the way they interact with dislocations changes. In combination this leads to the ageing curve shown in Figure GL2.42(b). Note that equilibrium is eventually achieved in the final 'over-aged' condition, with coarse precipitates of CuAl$_2$. The structure is then much as it would have been after slow-cooling from the solutionising temperature. It is clear from Figure GL2.42(b) how important it is to develop the right metatstable two-phase microstructure to achieve useful strength in these alloys.

## Exercises

E.20    Figure GL2.43 shows the aluminum-silver (Al-Ag) phase diagram.

(a) An entrepreneur sees an opportunity for casting decorative artefacts in 'budget silver', using an Al-Ag alloy. Suggest suitable compositions, explaining your reasoning.

(b) Al-Ag is known to exhibit age hardening, as illustrated earlier for Al-Cu. Identify suitable compositions for wrought heat-treatable Al-Ag alloys, and propose a possible heat treatment sequence to achieve high strength.

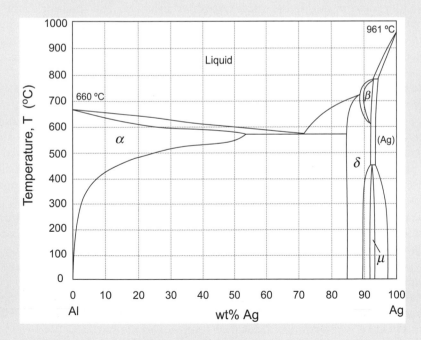

Figure GL2.43   Aluminum-silver Al-Ag phase diagram.

*Heat-treatment of carbon steels*   The Al-Cu example highlighted the key characteristics required for effective precipitation hardening: (a) high solubility of alloying elements at high temperature (to enable an initially uniform distribution of solute); and (b) low solubility at room temperature, with a two-phase region including a hard compound absorbing most of the alloying additions (to enable fine-scale hard precipitates to be formed). The Fe-C system shows exactly the same key characteristics. Figure GL2.35 shows the high solubility for carbon in iron in the austenite condition, with very low solubility in ferrite, and the prospect of forming precipitates of iron carbide.

A typical medium carbon steel can therefore be 'austenitised' (i.e. solutionised in the austenite field) at 850–900 °C. We have seen that slow-cooling in this case leads to the mixed ferrite-pearlite microstructure, in itself used extensively for its excellent strength and toughness. But we can do better still. Following the same reasoning as before, quenching at a rate above the critical cooling rate will trap the carbon in supersaturated solution. Reheating to temperatures around 500–600 °C for an hour or two enables fine-scale precipitates of iron carbide to be formed: in steels this stage is known as *tempering*.

In general terms therefore, quenching and tempering carbon steels parallels age hardening of aluminium alloys. But there are important differences. First, the phase that precipitates is in this case the equilibrium one—iron carbide, in a matrix of ferrite. But second, there is also the transformation from FCC iron to BCC iron when it is cooled from the solutionising temperature (which was not an issue in aluminum, which remains FCC throughout).

Phase transformations usually require diffusion to enable the atoms to re-arrange into the new phases. But here we have an unusual phase transformation, in which the FCC iron lattice is able to transform during the quench, *without* diffusion. The resulting single phase is called *martensite*. It is metastable, and is a super-saturated solid solution of carbon in iron. The iron lattice wishes to form BCC ferrite, but this has very low solubility for carbon, whereas martensite contains the whole carbon content of the alloy. As a result, martensite forms with a somewhat distorted body-centred lattice (called 'body-centred tetragonal').

The transformation has a conventional driving force, as there is a difference in free energy, $\Delta G$, between austenite and martensite. But the mechanism is quite distinct, as illustrated in Figure GL2.44(a). Small regions at austenite grain boundaries *shear* to form nuclei of the new lattice, and these propagate very rapidly across the grains forming thin lens-shaped plates of martensite (Figure GL2.44(b)). Figure GL2.44(c) shows how this is achieved at the atomic scale purely by straining the FCC lattice—there is a simple geometric relationship between the atom sites in FCC and a body-centred structure. This non-diffusive phase change is known as a *displacive transformation.*

The martensite transformation in a given steel has a characteristic temperature at which the transformation starts, and a second (lower) temperature at which it goes to completion. Since diffusion is not involved, the extent of transformation is independent of time—the start and finish of the transformation are shown on TTT diagrams by horizontal lines at the martensite start and finish temperatures; see Figure GL2.45(a).

Finally, what is the effect of quenching and tempering on properties? Since martensite contains far more carbon in solution than can normally be accommodated by BCC ferrite, the distorted lattice is resistant to dislocation motion and the hardness (and yield stress) is high. But as a result the microstructure is brittle, with a low fracture toughness similar to that of a ceramic. On tempering, fine precipitates of $Fe_3C$ nucleate and grow, removing the carbon from

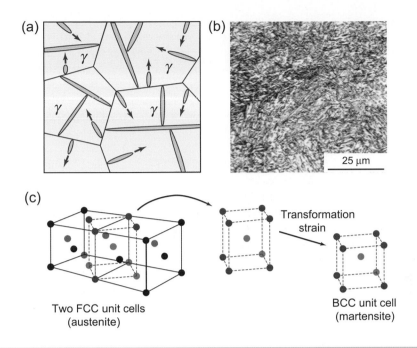

Figure GL2.44   The martensite transformation: (a) nucleation from austenite grain boundaries; (b) micrograph of martensite (image courtesy: ASM Micrograph Center, ASM International, 2005); (c) lattice relationship between FCC unit cells (blue) and the body-centred unit cell (red) which strains to form the distorted structure of martensite.

solution and giving precipitation strength instead. The resulting yield strength is below that of martensite—the yield stress evolves with tempering time as shown in Figure GL2.45(b). But tempering restores the fracture toughness to acceptable levels, comparable to that of the normalised ferrite-pearlite microstructure. The net effect is a yield stress of order two to three times greater in the tempered martensite.

This outcome in terms of properties explains the widespread use of the quench-and-temper treatment for tools and machinery. But this heat treatment has some practical consequences in manufacturing steel components. Quenching red-hot steel into cold water or oil induces thermal stresses, which may result in *quench cracking*, due to the inherent brittleness of the martensite. Furthermore, there is a limit to how fast the centre of a component can be made to cool, due to the timescale of heat conduction across the section—the centre cooling rate falls as the size of component increases. There will therefore be a limiting component size that can be quenched at a rate above the critical cooling rate for the steel. For plain carbon steels (Fe-C alloys), the critical cooling rate is high—only small components can form martensite throughout the whole section (and thereby be tempered afterwards). The solution to this problem turns out to be more alloying. Additions of a few wt% of elements such as Ni, Cr and Mo significantly retard the diffusional transformations from austenite to ferrite and pearlite, shifting the C-curves on the TTT diagram to the right. Martensite can then form at a greatly

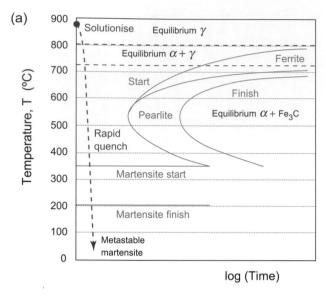

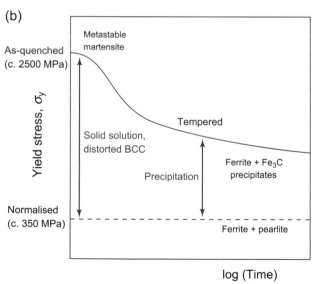

**Figure GL2.45**  (a) Appearance of martensite start and finish on a TTT diagram; (b) evolution of yield stress during tempering of carbon steels after quenching.

reduced cooling rate, so it can be formed throughout a larger cross-section prior to tempering (and with lower thermal stress). The ability of a steel to form martensite on quenching is known as its *hardenability*. This is one of the many reasons for producing *alloy steels*.

So the lesson of heat treatment of steels is again that the phase diagram points the way to a promising microstructure (due to high and low carbon solubility in iron at different temperatures) but the detailed behaviour is complicated. It also illustrates that phase transformations can be further manipulated by alloying—steels are prime examples of our mantra: composition + processing → microstructure + properties.

## Exercises

E.21   Three samples of high carbon steel containing 1.2 wt% carbon are to be heat treated. All three are first austenitised. One sample is cooled slowly to room temperature; another is quenched rapidly in cold water and the third is quenched and then tempered at 600 °C. The yield strength was measured in each condition, as follows: slow-cooled 800 MPa; as-quenched 3500 MPa and quenched and tempered 1100 MPa. Refer to the Fe-C phase diagram in Figure GL2.13 to answer the following:
(a) What is the minimum temperature required to fully austenitise the steel?
(b) What are the phases present, their proportions and compositions, in each condition after heat treatment?
(c) Describe the microstructure and hardening mechanisms responsible for the yield strengths in each condition.

## Answers to Exercises, Part 8

E.20   (a) Casting alloys are usually around the eutectic composition—Al-72 wt% Ag in this system. This is in order to exploit the minimum temperature for full melting at a eutectic, and the reasonable properties of a two-phase eutectic microstructure. As the slope of the liquidus (towards Al) is low, it would be possible to reduce the Ag content considerably (and thus the cost!) without a significant increase in casting temperature. For example Al-60 wt% Ag would have a mixed structure of primary Al and eutectic. Since Al-Ag is easily heat treated, it may be preferable to use less silver still and cast a dilute alloy (e.g. Al-10 wt% Ag), with a subsequent solutionising and ageing treatment; see (b).
(b) Aluminum dissolves up to 55.6 wt% of Ag. Any alloy up to this concentration could be solutionised at 560 °C. Lower concentrations would be a lot cheaper, and by analogy with Al-Cu it may be sufficient to use a

dilute composition such as 5 wt% Ag to achieve effective hardening. This alloy can also be solutionised at a lower temperature (400 °C), before quenching to room temperature, and ageing in the two-phase field (at 100–200 °C). Metastable precipitates are expected, as in Al-Cu, giving an ageing curve.

E.21

(a) From the phase diagram, the lower limit of the austenite field at 1.2 wt% carbon is 830 °C.

(b) After slow cooling, the steel is at equilibrium containing ferrite (pure iron) and iron carbide ($Fe_3C$, with 6.7 wt% carbon). From the lever rule, the proportion of iron carbide is 1.2/6.7 ≈ 18% (hence 82% ferrite). After quenching, the microstructure is metastable martensite containing 1.2 wt% C. After quenching and tempering, equilibrium is restored and the phases are exactly the same as after slow cooling.

(c) The slow cooled microstructure is similar to that shown in Figure GL2.40(b)—cementite on the prior austenite grain boundaries, with pearlite grain interiors. Ferrite is soft and iron carbide is hard, with the composite microstructure of pearlite giving moderate strength. After quenching, the martensite formed is very hard since it contains a high supersaturation of carbon, distorting the lattice severely. On tempering, a uniform dispersion of iron carbide precipitates forms in a matrix of ferrite. The resulting precipitation hardening is more effective than the same phases in the form of pearlite, giving higher yield strength.

## Further reading

Ashby, M. F., & Jones, D. R. H. (2005). *Engineering Materials II* (3rd ed.). ISBN 0-7506-6381-2. (Popular treatment of material classes, and how processing affects microstructure and properties).

*ASM Handbook Series* (1971–2004). Volume 4, *Heat Treatment*; Volume 5, *Surface Engineering*; Volume 6, *Welding, Brazing and Soldering*; Volume 7, *Powder Metal Technologies*; Volume 14, *Forming and Forging*; Volume 15, *Casting*; and Volume 16, *Machining*; ASM International, Metals Park, OH, USA. (A comprehensive set of handbooks on processing, occasionally updated, and now available online at <www.products.asminternational.org/hbk/index.jsp>.)

*ASM Micrograph Center* (2005). ASM International, Metals Park, OH, USA. (An extensive online library of micrographs documenting the composition and process history of the materials illustrated, available at <www.products.asminternational.org/mgo/index.jsp>.)

MATTER; <*www.matter.org.uk.*> <*www.matter.org.uk/steelmatter*> *and aluminum alloys* <*www.aluminium.org.uk*>. (Introduction to phase diagrams, and many other topics. Specific coverage of steels.)

## Further exercises

These further exercises are provided without solutions.

E.22    A special brazing alloy contains 63 wt% gold (Au) and 37 wt% nickel (Ni) (which is written Au-37 wt% Ni). The atomic weight of Au (197.0) is more than three times that of Ni (58.7). At a glance, which of the two compositions, in at %, is more likely to be correct?

(a) $X_{Au} = 0.34$, $X_{Ni} = 0.66$
(b) $X_{Au} = 0.66$, $X_{Ni} = 0.34$

E.23    The copper-tin system (which includes *bronzes*) is shown in Figure GL2.46.
(a) Shade the single phase regions.
(b) Highlight the four eutectoids in this system and write down their compositions and temperatures.
(c) Find the chemical formulae for the intermediate compounds $\varepsilon$ and $\eta$ (atomic weights of Cu and Sn are 63.54 and 118.69, respectively).

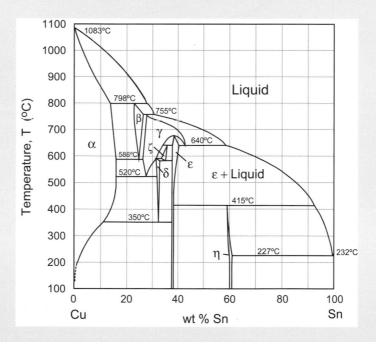

**Figure GL2.46**    Copper-tin Cu-Sn phase diagram.

E.24    (a) Figure GL2.47 shows the phase diagram for the Al-Zn system.
         (i) The phase diagram contains regions labelled A, B, C, D and E. Write
             down the phases in each of these regions.
         (ii) The phase diagram contains two reactions. For *each* reaction, state
              the type of reaction, temperature and composition.
         (iii) List the hardening mechanisms potentially available to Al-Zn alloys.
               Explain your reasoning.
        (b) An Al-Zn alloy is prepared by mixing 20g of Al and 30g of Zn. The alloy
            is melted and then allowed to cool slowly so that equilibrium is main-
            tained. At each of the following temperatures, write down the phases
            present, their composition and the proportions of each phase: (i) 500 °C;
            (ii) 300 °C; (iii) 250 °C.

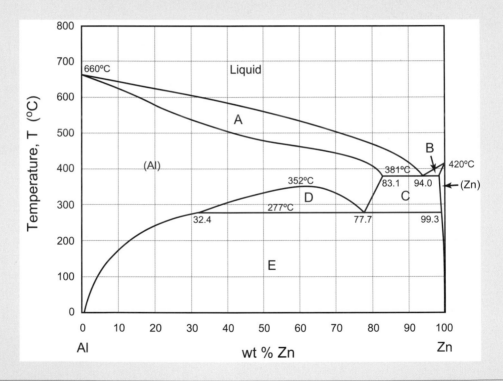

Figure GL2.47    Aluminium-zinc Al-Zn phase diagram.

E.25    By analogy with the definitions for eutectic and eutectoid reactions, and
        consideration of the phase diagrams in Part 2 of Guided Learning Unit 2, define:
        (a) A peritectic reaction.
        (b) A peritectoid reaction.

E.26 The Al–Ag phase diagram was shown in Figure GL2.43.

(a) The Al–Ag system has one eutectic reaction. Write down the temperature of the reaction, the composition and the phases involved.

(b) An Ag–20 wt% Al alloy is cooled slowly to room temperature from the melt. Describe the microstructural changes that occur, noting key temperatures and phase transformations.

E.27 Figure GL2.48 shows the Ti–Al phase diagram (important for the standard commercial alloy Ti-6% Al-4% V).

(a) Shade all single-phase fields and highlight three compounds with well-defined compositions. Find the formula for these three compounds (atomic weights of Ti and Al are 47.90 and 26.98, respectively). (Note that two other single phase fields are shown with a compound formula in brackets—i.e. these fields are solid solutions showing a spread of composition around that of each compound).

(b) Ring the five peritectic points, one peritectoid point and two eutectoid points.

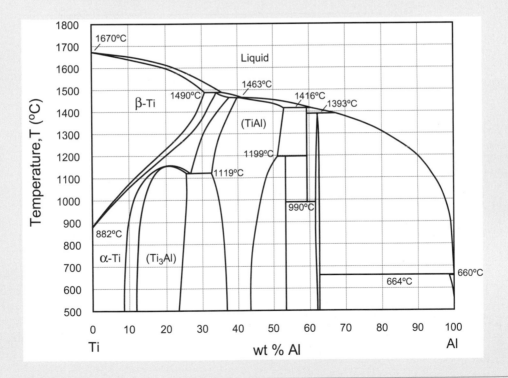

Figure GL2.48 Titanium-aluminium Ti-Al phase diagram.

(c) On heating, over what temperature range does a Ti-6 wt% Al alloy change from α-Ti (HCP) to β-Ti (BCC)? Over what temperature range does it melt?

E.28    Figure GL2.49 shows the copper-antimony Cu-Sb phase diagram.
(a) Find the chemical formula for the compound marked X (atomic weights of Cu and Sb are 63.54 and 121.75, respectively).
(b) The Cu-Sb system contains 2 eutectics, 1 eutectoid, 1 peritectic and 1 peritectoid. Mark them all on the figure, write down the temperature and composition of each point, and identify the phases involved in each reaction, on cooling.
(c) An alloy containing 95 wt% Sb is cooled slowly to room temperature from the melt. Describe the phase changes that occur during cooling, using schematic sketches of the microstructure at key temperatures to illustrate your answer.
(d) Sketch a temperature–time curve for the 95 wt% Sb alloy over the range 650 to 450 °C and account for the shape of the curve.

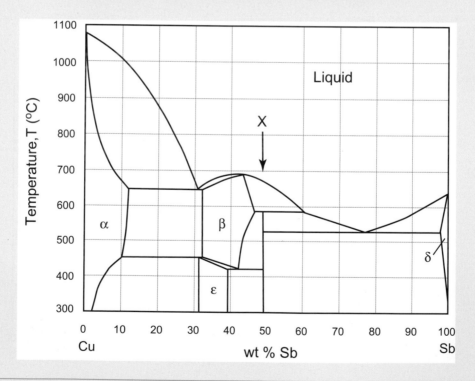

Figure GL2.49    Copper-antimony Cu-Sb phase diagram.

E.29    The phase diagram for the silver—copper Ag-Cu system is shown in Figure
GL2.50, with a linear scale of atomic % copper. The numbers in brackets
refer to weight % copper, which is shown in the upper scale. The liquid
single-phase region is indicated.

(a) On a rough sketch of the phase diagram indicate the phases that are
present in each region of the diagram.

(b) An alloy containing 95 wt% Cu is heated to 1100°C and then slowly
cooled to room temperature. Make labelled sketches at key temperatures
to show the evolution of the microstructure, and evaluate the final proportions and compositions of the phases present.

(c) Explain how segregation can occur during the non-equilibrium cooling
from the liquid state of the alloy described in (b).

(d) Describe, using the phase diagram, how you would choose an alloy
composition and a heat treatment schedule that might be suitable for
producing a precipitation-hardened silver-rich alloy.

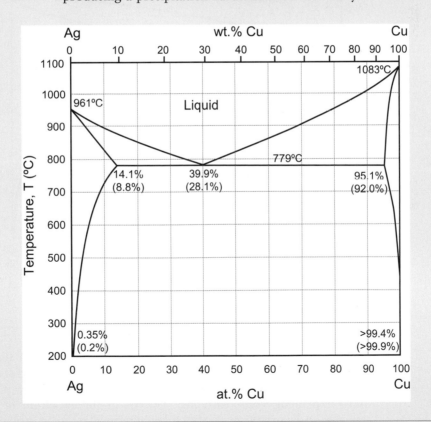

**Figure GL2.50**    Silver-copper Ag-Cu phase diagram.

E.30

(a) Figure GL2.51 shows the phase diagram of the Au-Ni binary system. This system shows complete solubility above 810 °C and immiscibility at lower temperatures. The phase diagram contains regions labelled A, B and C. Write down the phase(s) present in each of these regions. For an Au-Ni alloy with composition 70 wt% Au, write down the phase(s) present, their composition and proportions of each phase at 1200, 900 and 600 °C.

(b) An Au−10 wt% Ni alloy is cooled to room temperature from the liquid state. Describe the microstructural changes that occur during cooling if equilibrium is maintained, noting key temperatures and phase transformations. Provide details of the microstructure, compositions and relative proportions of the phases present at room temperature.

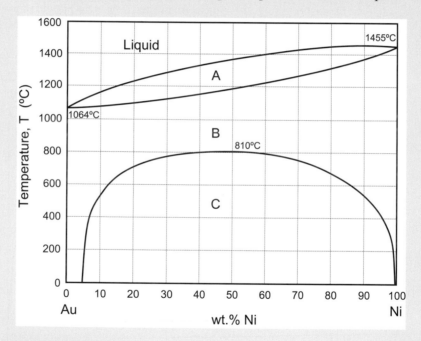

Figure GL2.51   Gold-nickel Au-Ni phase diagram.

E.31

Figure GL2.52 shows schematically the microstructure of a binary Pb-Sn alloy at room temperature, after slow cooling from the liquid state. Use the Pb-Sn phase diagram in Figure GL2.29 to estimate the composition of the alloy, explaining your reasoning in terms of the evolution of the microstructure in Figure GL2.52. Sketch key intermediate microstructures during the cooling history to illustrate your answer.

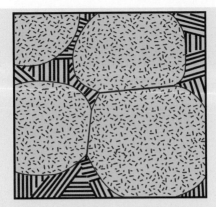

**Figure GL2.52**   Schematic room temperature microstructure of a Pb-Sn alloy after slow solidification and cooling.

# Appendix
# Data for engineering materials

**Contents**

This appendix lists the names and typical applications of common engineering materials, together with data for their properties.

A1   Names and applications: metals and alloys

| Materials | | Applications |
|---|---|---|
| Ferrous | Cast irons | Automotive parts, engine blocks, machine tool structural parts, lathe beds |
| | High carbon steels | Cutting tools, springs, bearings, cranks, shafts, railway track |
| | Medium carbon steels | General mechanical engineering (tools, bearings, gears, shafts, bearings) |
| | Low carbon steels | Steel structures ('mild steel')—bridges, oil rigs, ships; reinforcement for concrete; automotive parts, car body panels; galvanised sheet; packaging (cans, drums) |
| | Low alloy steels | Springs, tools, ball bearings, automotive parts (gears, connecting rods, etc.) |
| | Stainless steels | Transport, chemical and food processing plant, nuclear plant, domestic ware (cutlery, washing machines, stoves), surgical implements, pipes, pressure vessels, liquid gas containers |
| Non-ferrous | Aluminum alloys | |
| |   Casting alloys | Automotive parts (cylinder blocks), domestic appliances (irons) |
| |   Non-heat-treatable alloys | Electrical conductors, heat exchangers, foil, tubes, saucepans, beverage cans, lightweight ships, architectural panels |
| |   Heat-treatable alloys | Aerospace engineering, automotive bodies and panels, lightweight structures and ships |
| | Copper alloys | Electrical conductors and wire, electronic circuit boards, heat exchangers, boilers, cookware, coinage, sculptures |
| | Lead alloys | Roof and wall cladding, solder, X-ray shielding, battery electrodes |
| | Magnesium alloys | Automotive castings, wheels, general lightweight castings for transport, nuclear fuel containers; principal alloying addition to aluminum alloys |
| | Nickel alloys | Gas turbines and jet engines, thermocouples, coinage; alloying addition to austenitic stainless steels |
| | Titanium alloys | Aircraft turbine blades; general structural aerospace applications; biomedical implants |
| | Zinc alloys | Die castings (automotive, domestic appliances, toys, handles); coating on galvanised steel |

A2   Names and applications: polymers and foams

| Materials | | Abbreviation | Applications |
|---|---|---|---|
| Elastomer | Butyl rubber | | Tyres, seals, anti-vibration mountings, electrical insulation, tubing |
| | Ethylene-vinyl-acetate | EVA | Bags, films, packaging, gloves, insulation, running shoes |
| | Isoprene | IR | Tyres, inner tubes, insulation, tubing, shoes |
| | Natural rubber | NR | Gloves, tyres, electrical insulation, tubing |
| | Polychloroprene (neoprene) | CR | Wetsuits, O-rings and seals, footwear |
| | Polyurethane elastomers | el-PU | Packaging, hoses, adhesives, fabric coating |
| | Silicone elastomers | | Electrical insulation, electronic encapsulation, medical implants |
| Thermoplastic | Acrylonitrile butadiene styrene | ABS | Communication appliances, automotive interiors, luggage, toys, boats |
| | Cellulose polymers | CA | Tool and cutlery handles, decorative trim, pens |
| | Ionomer | I | Packaging, golf balls, blister packs, bottles |
| | Polyamides (nylons) | PA | Gears, bearings; plumbing, packaging, bottles, fabrics, textiles, ropes |
| | Polycarbonate | PC | Safety goggles, shields, helmets; light fittings, medical components |
| | Polyetheretherketone | PEEK | Electrical connectors, racing car parts, fibre composites |
| | Polyethylene | PE | Packaging, bags, squeeze tubes, toys, artificial joints |
| | Polyethylene terephthalate | PET | Blow moulded bottles, film, audio/video tape, sails |
| | Polymethyl methacrylate (acrylic) | PMMA | Aircraft windows, lenses, reflectors, lights, compact discs |

*Continued*

A2   Names and applications: polymers and foams *continued*

| Materials | | Abbreviation | Applications |
|---|---|---|---|
| | Polyoxymethylene (acetal) | POM | Zips, domestic and appliance parts, handles |
| | Polypropylene | PP | Ropes, garden furniture, pipes, kettles, electrical insulation, Astroturf |
| | Polystyrene | PS | Toys, packaging, cutlery, audio cassette/CD cases |
| | Polyurethane thermoplastics | tp-PU | Cushioning, seating, shoe soles, hoses, car bumpers, insulation |
| | Polyvinylchloride | PVC | Pipes, gutters, window frames, packaging |
| | Polytetrafluoroethylene (Teflon) | PTFE | Nonstick coatings, bearings, skis, electrical insulation, tape |
| Thermoset | Epoxies | EP | Adhesives, fibre composites, electronic encapsulation |
| | Phenolics | PHEN | Electrical plugs, sockets, cookware, handles, adhesives |
| | Polyester | PEST | Furniture, boats, sports goods |
| Polymer foams | Flexible polymer foam | | Packaging, buoyancy, cushioning, sponges, sleeping mats |
| | Rigid polymer foam | | Thermal insulation, sandwich panels, packaging, buoyancy |

A3   Names and applications: composites, glasses, ceramics, and natural materials

| Materials | | Applications |
|---|---|---|
| **Composites** | | |
| Metal | Aluminum/silicon carbide | Automotive parts, sports goods |
| Polymer | CFRP | Lightweight structural parts (aerospace, bike frames, sports goods, boat hulls and oars, springs) |
| | GFRP | Boat hulls, automotive parts, chemical plant |
| **Glasses** | Borosilicate glass | Ovenware, laboratory ware, headlights |
| | Glass ceramic | Cookware, lasers, telescope mirrors |
| | Silica glass | High-performance windows, crucibles, high-temperature applications |
| | Soda-lime glass | Windows, bottles, tubing, light bulbs, pottery glazes |
| **Ceramics** | | |
| Porous | Brick | Buildings |
| | Concrete | General civil engineering construction |
| | Stone | Buildings, architecture, sculpture |
| Technical | Alumina | Cutting tools, spark plugs, microcircuit substrates, valves |
| | Aluminum nitride | Microcircuit substrates and heat sinks |
| | Boron carbide | Lightweight armour, nozzles, dies, precision tool parts |
| | Silicon | Microcircuits, semiconductors, precision instruments, IR windows, MEMS |
| | Silicon carbide | High-temperature equipment, abrasive polishing grits, bearings, armour |
| | Silicon nitride | Bearings, cutting tools, dies, engine parts |
| | Tungsten carbide | Cutting tools, drills, abrasives |
| **Natural** | Bamboo | Building, scaffolding, paper, ropes, baskets, furniture |
| | Cork | Corks and bungs, seals, floats, packaging, flooring |
| | Leather | Shoes, clothing, bags, drive-belts |
| | Wood | Construction, flooring, doors, furniture, packaging, sports goods |

A4   Melting temperature, $T_m$, and glass temperature $T_g^1$

|  |  | $T_m$ or $T_g$ (°C) |
|---|---|---|
| **Metals** | | |
| Ferrous | Cast irons | 1130–1250 |
| | High carbon steels | 1289–1478 |
| | Medium carbon steels | 1380–1514 |
| | Low carbon steels | 1480–1526 |
| | Low alloy steels | 1382–1529 |
| | Stainless steels | 1375–1450 |
| Non-ferrous | Aluminum alloys | 475–677 |
| | Copper alloys | 982–1082 |
| | Lead alloys | 322–328 |
| | Magnesium alloys | 447–649 |
| | Nickel alloys | 1435–1466 |
| | Titanium alloys | 1477–1682 |
| | Zinc alloys | 375–492 |
| **Glasses** | Borosilicate glass (*) | 450–602 |
| | Glass ceramic (*) | 563–1647 |
| | Silica glass (*) | 957–1557 |
| | Soda-lime glass (*) | 442–592 |
| **Ceramics** | | |
| Porous | Brick | 927–1227 |
| | Concrete, typical | 927–1227 |
| | Stone | 1227–1427 |
| Technical | Alumina | 2004–2096 |
| | Aluminum nitride | 2397–2507 |
| | Boron carbide | 2372–2507 |
| | Silicon | 1407–1412 |
| | Silicon carbide | 2152–2500 |
| | Silicon nitride | 2388–2496 |
| | Tungsten carbide | 2827–2920 |
| **Composites** | | |
| Metal | Aluminum/silicon carbide | 525–627 |
| Polymer | CFRP | n/a |
| | GFRP | n/a |
| Natural | Bamboo (*) | 77–102 |
| | Cork (*) | 77–102 |
| | Leather (*) | 107–127 |
| | Wood, typical (longitudinal) (*) | 77–102 |
| | Wood, typical (transverse) (*) | 77–102 |

1. The table lists the melting point for crystalline solids and the glass temperature for polymeric and inorganic glasses.
(*) glass transition temperature.
n/a: not applicable (materials decompose, rather than melt).

*Continued*

A4 Melting temperature, $T_m$, and glass temperature $T_g^1$ *continued*

| | | $T_m$ or $T_g$ (°C) |
|---|---|---|
| **Polymers** | | |
| Elastomer | Butyl Rubber (*) | −73−−63 |
| | EVA (*) | −73−−23 |
| | Isoprene (IR) (*) | −83−−78 |
| | Natural rubber (NR) (*) | −78−−63 |
| | Neoprene (CR) (*) | −48−−43 |
| | Polyurethane elastomers (elPU) (*) | −73−−23 |
| | Silicone elastomers (*) | −123−−73 |
| Thermoplastic | ABS (*) | 88−128 |
| | Cellulose polymers (CA) (*) | −9−107 |
| | Ionomer (I) (*) | 27−77 |
| | Nylons (PA) (*) | 44−56 |
| | Polycarbonate (PC) (*) | 142−205 |
| | PEEK (*) | 143−199 |
| | Polyethylene (PE) (*) | −25−−15 |
| | PET (*) | 68−80 |
| | Acrylic (PMMA) (*) | 85−165 |
| | Acetal (POM) (*) | −18−−8 |
| | Polypropylene (PP) (*) | −25−−15 |
| | Polystyrene (PS) (*) | 74−110 |
| | Polyurethane thermoplastics (tpPU) (*) | 120−160 |
| | PVC | 75−105 |
| | Teflon (PTFE) | 107−123 |
| Thermoset | Epoxies | n/a |
| | Phenolics | n/a |
| | Polyester | n/a |
| **Polymer foams** | | |
| | Flexible polymer foam (VLD) (*) | 112−177 |
| | Flexible polymer foam (LD) (*) | 112−177 |
| | Flexible Polymer foam (MD) (*) | 112−177 |
| | Rigid polymer foam (LD) (*) | 67−171 |
| | Rigid polymer foam (MD) (*) | 67−157 |
| | Rigid polymer foam (HD) (*) | 67−171 |

1. The table lists the melting point for crystalline solids and the glass temperature for polymeric and inorganic glasses.
(*) glass transition temperature.
n/a: not applicable (materials decompose, rather than melt).

## A5 Density, $\rho$

| | | | $\rho$ (Mg/m$^3$) |
|---|---|---|---|
| **Metals** | | | |
| Ferrous | | Cast irons | 7.05–7.25 |
| | | High carbon steels | 7.8–7.9 |
| | | Medium carbon steels | 7.8–7.9 |
| | | Low carbon steels | 7.8–7.9 |
| | | Low alloy steels | 7.8–7.9 |
| | | Stainless steels | 7.6–8.1 |
| Non-ferrous | | Aluminum alloys | 2.5–2.9 |
| | | Copper alloys | 8.93–8.94 |
| | | Lead alloys | 10–11.4 |
| | | Magnesium alloys | 1.74–1.95 |
| | | Nickel alloys | 8.83–8.95 |
| | | Titanium alloys | 4.4–4.8 |
| | | Zinc alloys | 4.95–7 |
| **Glasses** | | Borosilicate glass | 2.2–2.3 |
| | | Glass ceramic | 2.2–2.8 |
| | | Silica glass | 2.17–2.22 |
| | | Soda-lime glass | 2.44–2.49 |
| **Ceramics** | | | |
| Porous | | Brick | 1.9–2.1 |
| | | Concrete, typical | 2.2–2.6 |
| | | Stone | 2.5–3 |
| Technical | | Alumina | 3.5–3.98 |
| | | Aluminum nitride | 3.26–3.33 |
| | | Boron carbide | 2.35–2.55 |
| | | Silicon | 2.3–2.35 |
| | | Silicon carbide | 3–3.21 |
| | | Silicon nitride | 3–3.29 |
| | | Tungsten Carbide | 15.3–15.9 |
| **Composites** | | | |
| Metal | | Aluminum/Silicon Carbide | 2.66–2.9 |
| Polymer | | CFRP | 1.5–1.6 |
| | | GFRP | 1.75–1.97 |
| Natural | | Bamboo | 0.6–0.8 |
| | | Cork | 0.12–0.24 |
| | | Leather | 0.81–1.05 |
| | | Wood, typical (longitudinal) | 0.6–0.8 |
| | | Wood, typical (transverse) | 0.6–0.8 |

*Continued*

A5    Density, $\rho$ *continued*

|  |  | $\rho$ (Mg/m$^3$) |
|---|---|---|
| **Polymers** | | |
| Elastomer | Butyl rubber | 0.9–0.92 |
| | EVA | 0.945–0.955 |
| | Isoprene (IR) | 0.93–0.94 |
| | Natural rubber (NR) | 0.92–0.93 |
| | Neoprene (CR) | 1.23–1.25 |
| | Polyurethane elastomers (elPU) | 1.02–1.25 |
| | Silicone elastomers | 1.3–1.8 |
| Thermoplastic | ABS | 1.01–1.21 |
| | Cellulose polymers (CA) | 0.98–1.3 |
| | Ionomer (I) | 0.93–0.96 |
| | Nylons (PA) | 1.12–1.14 |
| | Polycarbonate (PC) | 1.14–1.21 |
| | PEEK | 1.3–1.32 |
| | Polyethylene (PE) | 0.939–0.96 |
| | PET | 1.29–1.4 |
| | Acrylic (PMMA) | 1.16–1.22 |
| | Acetal (POM) | 1.39–1.43 |
| | Polypropylene (PP) | 0.89–0.91 |
| | Polystyrene (PS) | 1.04–1.05 |
| | Polyurethane thermoplastics (tpPU) | 1.12–1.24 |
| | PVC | 1.3–1.58 |
| | Teflon (PTFE) | 2.14–2.2 |
| Thermoset | Epoxies | 1.11–1.4 |
| | Phenolics | 1.24–1.32 |
| | Polyester | 1.04–1.4 |
| Polymer foams | | |
| | Flexible polymer foam (VLD) | 0.016–0.035 |
| | Flexible polymer foam (LD) | 0.038–0.07 |
| | Flexible polymer foam (MD) | 0.07–0.115 |
| | Rigid polymer foam (LD) | 0.036–0.07 |
| | Rigid polymer foam (MD) | 0.078–0.165 |
| | Rigid Polymer foam (HD) | 0.17–0.47 |

## A6  Young's modulus, *E*

| | | *E* (GPa) |
|---|---|---|
| **Metals** | | |
| Ferrous | Cast irons | 165—180 |
| | High carbon steels | 200—215 |
| | Medium carbon steels | 200—216 |
| | Low carbon steels | 200—215 |
| | Low alloy steels | 201—217 |
| | Stainless steels | 189—210 |
| Non-ferrous | Aluminum alloys | 68—82 |
| | Copper alloys | 112—148 |
| | Lead alloys | 12.5—15 |
| | Magnesium alloys | 42—47 |
| | Nickel alloys | 190—220 |
| | Titanium alloys | 90—120 |
| | Zinc alloys | 68—95 |
| **Glasses** | Borosilicate glass | 61—64 |
| | Glass ceramic | 64—110 |
| | Silica glass | 68—74 |
| | Soda-lime glass | 68—72 |
| **Ceramics** | | |
| Porous | Brick | 15—25 |
| | Concrete, typical | 25—38 |
| | Stone | 20—60 |
| Technical | Alumina | 215—413 |
| | Aluminum nitride | 302—348 |
| | Boron carbide | 400—472 |
| | Silicon | 140—155 |
| | Silicon carbide | 300—460 |
| | Silicon nitride | 280—310 |
| | Tungsten carbide | 600—720 |
| **Composites** | | |
| Metal | Aluminum/silicon carbide | 81—100 |
| Polymer | CFRP | 69—150 |
| | GFRP | 15—28 |
| Natural | Bamboo | 15—20 |
| | Cork | 0.013—0.05 |
| | Leather | 0.1—0.5 |
| | Wood, typical (longitudinal) | 6—20 |
| | Wood, typical (transverse) | 0.5—3 |

*Continued*

A6 Young's modulus, *E continued*

|  |  | *E* (GPa) |
|---|---|---|
| **Polymers** |  |  |
| Elastomer | Butyl rubber | 0.001–0.002 |
|  | EVA | 0.01–0.04 |
|  | Isoprene (IR) | 0.0014–0.004 |
|  | Natural rubber (NR) | 0.0015–0.0025 |
|  | Neoprene (CR) | 0.0007–0.002 |
|  | Polyurethane elastomers (elPU) | 0.002–0.003 |
|  | Silicone elastomers | 0.005–0.02 |
| Thermoplastic | ABS | 1.1–2.9 |
|  | Cellulose polymers (CA) | 1.6–2 |
|  | Ionomer (I) | 0.2–0.424 |
|  | Nylons (PA) | 2.62–3.2 |
|  | Polycarbonate (PC) | 2–2.44 |
|  | PEEK | 3.5–4.2 |
|  | Polyethylene (PE) | 0.621–0.896 |
|  | PET | 2.76–4.14 |
|  | Acrylic (PMMA) | 2.24–3.8 |
|  | Acetal (POM) | 2.5–5 |
|  | Polypropylene (PP) | 0.896–1.55 |
|  | Polystyrene (PS) | 2.28–3.34 |
|  | Polyurethane thermoplastics (tpPU) | 1.31–2.07 |
|  | PVC | 2.14–4.14 |
|  | Teflon (PTFE) | 0.4–0.552 |
| Thermoset | Epoxies | 2.35–3.075 |
|  | Phenolics | 2.76–4.83 |
|  | Polyester | 2.07–4.41 |
| Polymer foams |  |  |
|  | Flexible polymer foam (VLD) | 0.0003–0.001 |
|  | Flexible polymer foam (LD) | 0.001–0.003 |
|  | Flexible polymer foam (MD) | 0.004–0.012 |
|  | Rigid polymer foam (LD) | 0.023–0.08 |
|  | Rigid polymer foam (MD) | 0.08–0.2 |
|  | Rigid polymer foam (HD) | 0.2–0.48 |

A7   Yield strength, $\sigma_y$, and tensile strength, $\sigma_{ts}$

| | | $\sigma_y$ (MPa) | $\sigma_{ts}$ (MPa) |
|---|---|---|---|
| **Metals** | | | |
| Ferrous | Cast irons | 215—790 | 350—1000 |
| | High carbon steels | 400—1155 | 550—1640 |
| | Medium carbon steels | 305—900 | 410—1200 |
| | Low carbon steels | 250—395 | 345—580 |
| | Low alloy steels | 400—1100 | 460—1200 |
| | Stainless steels | 170—1000 | 480—2240 |
| Non-ferrous | Aluminum alloys | 30—500 | 58—550 |
| | Copper alloys | 30—500 | 100—550 |
| | Lead alloys | 8—14 | 12—20 |
| | Magnesium alloys | 70—400 | 185—475 |
| | Nickel alloys | 70—1100 | 345—1200 |
| | Titanium alloys | 250—1245 | 300—1625 |
| | Zinc Alloys | 80—450 | 135—520 |
| Glasses | Borosilicate glass (*) | 264—384 | 22—32 |
| | Glass ceramic (*) | 750—2129 | 62—177 |
| | Silica glass (*) | 1100—1600 | 45—155 |
| | Soda-lime glass (*) | 360—420 | 31—35 |
| **Ceramics** | | | |
| Porous | Brick (*) | 50—140 | 7—14 |
| | Concrete, typical (*) | 32—60 | 2—6 |
| | Stone (*) | 34—248 | 5—17 |
| Technical | Alumina (*) | 690—5500 | 350—665 |
| | Aluminum nitride (*) | 1970—2700 | 197—270 |
| | Boron carbide (*) | 2583—5687 | 350—560 |
| | Silicon (*) | 3200—3460 | 160—180 |
| | Silicon carbide (*) | 1000—5250 | 370—680 |
| | Silicon nitride (*) | 524—5500 | 690—800 |
| | Tungsten carbide (*) | 3347—6833 | 370—550 |
| **Composites** | | | |
| Metal | Aluminum/silicon carbide | 280—324 | 290—365 |
| Polymer | CFRP | 550—1050 | 550—1050 |
| | GFRP | 110—192 | 138—241 |
| Natural | Bamboo | 35—44 | 36—45 |
| | Cork | 0.3—1.5 | 0.5—2.5 |
| | Leather | 5—10 | 20—26 |
| | Wood, typical (longitudinal) | 30—70 | 60—100 |
| | Wood, typical (transverse) | 2—6 | 4—9 |

(*) NB: For ceramics, yield strength is replaced by *compressive strength*, which is more relevant in ceramic design. Note that ceramics are of the order of 10 times stronger in compression than in tension.

*Continued*

A7   Yield strength, $\sigma_y$, and tensile strength, $\sigma_{ts}$ *continued*

| | | $\sigma_y$ (MPa) | $\sigma_{ts}$ (MPa) |
|---|---|---|---|
| **Polymers** | | | |
| Elastomer | Butyl rubber | 2–3 | 5–10 |
| | EVA | 12–18 | 16–20 |
| | Isoprene (IR) | 20–25 | 20–25 |
| | Natural rubber (NR) | 20–30 | 22–32 |
| | Neoprene (CR) | 3.4–24 | 3.4–24 |
| | Polyurethane elastomers (elPU) | 25–51 | 25–51 |
| | Silicone elastomers | 2.4–5.5 | 2.4–5.5 |
| Thermoplastic | ABS | 18.5–51 | 27.6–55.2 |
| | Cellulose polymers (CA) | 25–45 | 25–50 |
| | Ionomer (I) | 8.3–15.9 | 17.2–37.2 |
| | Nylons (PA) | 50–94.8 | 90–165 |
| | Polycarbonate (PC) | 59–70 | 60–72.4 |
| | PEEK | 65–95 | 70–103 |
| | Polyethylene (PE) | 17.9–29 | 20.7–44.8 |
| | PET | 56.5–62.3 | 48.3–72.4 |
| | Acrylic (PMMA) | 53.8–72.4 | 48.3–79.6 |
| | Acetal (POM) | 48.6–72.4 | 60–89.6 |
| | Polypropylene (PP) | 20.7–37.2 | 27.6–41.4 |
| | Polystyrene (PS) | 28.7–56.2 | 35.9–56.5 |
| | Polyurethane thermoplastics (tpPU) | 40–53.8 | 31–62 |
| | PVC | 35.4–52.1 | 40.7–65.1 |
| | Teflon (PTFE) | 15–25 | 20–30 |
| Thermoset | Epoxies | 36–71.7 | 45–89.6 |
| | Phenolics | 27.6–49.7 | 34.5–62.1 |
| | Polyester | 33–40 | 41.4–89.6 |
| **Polymer foams** | | | |
| | Flexible polymer foam (VLD) | 0.01–0.12 | 0.24–0.85 |
| | Flexible polymer foam (LD) | 0.02–0.3 | 0.24–2.35 |
| | Flexible polymer foam (MD) | 0.05–0.7 | 0.43–2.95 |
| | Rigid polymer foam (LD) | 0.3–1.7 | 0.45–2.25 |
| | Rigid polymer foam (MD) | 0.4–3.5 | 0.65–5.1 |
| | Rigid polymer foam (HD) | 0.8–12 | 1.2–12.4 |

A8  Fracture toughness (plane-strain), $K_{IC}$

|  |  |  | $K_{IC}$ (MPa$\sqrt{m}$) |
|---|---|---|---|
| **Metals** |  |  |  |
| Ferrous |  | Cast irons | 22–54 |
|  |  | High carbon steels | 27–92 |
|  |  | Medium carbon steels | 12–92 |
|  |  | Low carbon steels | 41–82 |
|  |  | Low alloy steels | 14–200 |
|  |  | Stainless steels | 62–280 |
| Non-ferrous |  | Aluminum alloys | 22–35 |
|  |  | Copper alloys | 30–90 |
|  |  | Lead alloys | 5–15 |
|  |  | Magnesium alloys | 12–18 |
|  |  | Nickel alloys | 80–110 |
|  |  | Titanium alloys | 14–120 |
|  |  | Zinc alloys | 10–100 |
| **Glasses** |  | Borosilicate glass | 0.5–0.7 |
|  |  | Glass ceramic | 1.4–1.7 |
|  |  | Silica glass | 0.6–0.8 |
|  |  | Soda-lime glass | 0.55–0.7 |
| **Ceramics** |  |  |  |
| Porous |  | Brick | 1–2 |
|  |  | Concrete, typical | 0.35–0.45 |
|  |  | Stone | 0.7–1.5 |
| Technical |  | Alumina | 3.3–4.8 |
|  |  | Aluminum nitride | 2.5–3.4 |
|  |  | Boron carbide | 2.5–3.5 |
|  |  | Silicon | 0.83–0.94 |
|  |  | Silicon carbide | 2.5–5 |
|  |  | Silicon nitride | 4–6 |
|  |  | Tungsten carbide | 2–3.8 |
| **Composites** |  |  |  |
| Metal |  | Aluminum/silicon carbide | 15–24 |
| Polymer |  | CFRP | 6.1–88 |
|  |  | GFRP | 7–23 |
| Natural |  | Bamboo | 5–7 |
|  |  | Cork | 0.05–0.1 |
|  |  | Leather | 3–5 |
|  |  | Wood, typical (longitudinal) | 5–9 |
|  |  | Wood, typical (transverse) | 0.5–0.8 |

*Continued*

## A8 Fracture toughness (plane-strain), $K_{IC}$ *continued*

|  |  | $K_{IC}$ (MPa$\sqrt{m}$) |
|---|---|---|
| **Polymers** |  |  |
| Elastomer | Butyl rubber | 0.07—0.1 |
|  | EVA | 0.5—0.7 |
|  | Isoprene (IR) | 0.07—0.1 |
|  | Natural rubber (NR) | 0.15—0.25 |
|  | Neoprene (CR) | 0.1—0.3 |
|  | Polyurethane elastomers (elPU) | 0.2—0.4 |
|  | Silicone elastomers | 0.03—0.5 |
| Thermoplastic | ABS | 1.19—4.30 |
|  | Cellulose polymers (CA) | 1—2.5 |
|  | Ionomer (I) | 1.14—3.43 |
|  | Nylons (PA) | 2.22—5.62 |
|  | Polycarbonate (PC) | 2.1—4.60 |
|  | PEEK | 2.73—4.30 |
|  | Polyethylene (PE) | 1.44—1.72 |
|  | PET | 4.5—5.5 |
|  | Acrylic (PMMA) | 0.7—1.6 |
|  | Acetal (POM) | 1.71—4.2 |
|  | Polypropylene (PP) | 3—4.5 |
|  | Polystyrene (PS) | 0.7—1.1 |
|  | Polyurethane thermoplastics (tpPU) | 1.84—4.97 |
|  | PVC | 1.46—5.12 |
|  | Teflon (PTFE) | 1.32—1.8 |
| Thermoset | Epoxies | 0.4—2.22 |
|  | Phenolics | 0.79—1.21 |
|  | Polyester | 1.09—1.70 |
| Polymer foams |  |  |
|  | Flexible polymer foam (VLD) | 0.005—0.02 |
|  | Flexible polymer foam (LD) | 0.015—0.05 |
|  | Flexible polymer foam (MD) | 0.03—0.09 |
|  | Rigid polymer foam (LD) | 0.002—0.02 |
|  | Rigid polymer foam (MD) | 0.007—0.049 |
|  | Rigid polymer foam (HD) | 0.024—0.091 |

Note: $K_{IC}$ is only valid for conditions under which linear elastic fracture mechanics apply (see Chapter 8). The plane-strain toughness, $G_{IC}$, may be estimated from $K_{IC}2 = E\, G_{IC}/(1 - \nu^2) \approx E\, G_{IC}$ (as $\nu^2 \approx 0.1$).

A9    Thermal conductivity, $\lambda$, and thermal expansion, $\alpha$

| | | $\lambda$ (W/m.K) | $\alpha$ $(10^{-6}/C)$ |
|---|---|---|---|
| **Metals** | | | |
| Ferrous | Cast irons | 29—44 | 10—12.5 |
| | High carbon steels | 47—53 | 11—13.5 |
| | Medium carbon steels | 45—55 | 10—14 |
| | Low carbon steels | 49—54 | 11.5—13 |
| | Low alloy steels | 34—55 | 10.5—13.5 |
| | Stainless steels | 11—19 | 13—20 |
| Non-ferrous | Aluminum alloys | 76—235 | 21—24 |
| | Copper alloys | 160—390 | 16.9—18 |
| | Lead alloys | 22—36 | 18—32 |
| | Magnesium alloys | 50—156 | 24.6—28 |
| | Nickel alloys | 67—91 | 12—13.5 |
| | Titanium alloys | 5—12 | 7.9—11 |
| | Tungsten alloys | 100—142 | 4—5.6 |
| | Zinc alloys | 100—135 | 23—28 |
| **Glasses** | Borosilicate glass | 1—1.3 | 3.2—4.0 |
| | Glass ceramic | 1.3—2.5 | 1—5 |
| | Silica glass | 1.4—1.5 | 0.55—0.75 |
| | Soda-lime glass | 0.7—1.3 | 9.1—9.5 |
| **Ceramics** | | | |
| Porous | Brick | 0.46—0.73 | 5—8 |
| | Concrete, typical | 0.8—2.4 | 6—13 |
| | Stone | 5.4—6.0 | 3.7—6.3 |
| Technical | Alumina | 30—38.5 | 7—10.9 |
| | Aluminum nitride | 80—200 | 4.9—6.2 |
| | Boron carbide | 40—90 | 3.2—3.4 |
| | Silicon | 140—150 | 2.2—2.7 |
| | Silicon carbide | 115—200 | 4.0—5.1 |
| | Silicon nitride | 22—30 | 3.2—3.6 |
| | Tungsten carbide | 55—88 | 5.2—7.1 |
| **Composites** | | | |
| Metal | Aluminum/silicon carbide | 180—160 | 15—23 |
| Polymer | CFRP | 1.28—2.6 | 1—4 |
| | GFRP | 0.4—0.55 | 8.6—33 |
| Natural | Bamboo | 0.1—0.18 | 2.6—10 |
| | Cork | 0.035—0.048 | 130—230 |
| | Leather | 0.15—0.17 | 40—50 |
| | Wood, typical (longitudinal) | 0.31—0.38 | 2—11 |
| | Wood, typical (transverse) | 0.15—0.19 | 32—42 |

*Continued*

A9   Thermal conductivity, $\lambda$, and thermal expansion, $\alpha$ *continued*

| | | $\lambda$ (W/m.K) | $\alpha$ ($10^{-6}$/C) |
|---|---|---|---|
| **Polymers** | | | |
| Elastomer | Butyl rubber | 0.08–0.1 | 120–300 |
| | EVA | 0.3–0.4 | 160–190 |
| | Isoprene (IR) | 0.08–0.14 | 150–450 |
| | Natural rubber (NR) | 0.1–0.14 | 150–450 |
| | Neoprene (CR) | 0.08–0.14 | 575–610 |
| | Polyurethane elastomers | 0.28–0.3 | 150–165 |
| | Silicone elastomers | 0.3–1.0 | 250–300 |
| Thermoplastic | ABS | 0.19–0.34 | 84.6–234 |
| | Cellulose polymers (ca) | 0.13–0.3 | 150–300 |
| | Ionomer (I) | 0.24–0.28 | 180–306 |
| | Nylons (PA) | 0.23–0.25 | 144–150 |
| | Polycarbonate (PC) | 0.19–0.22 | 120–137 |
| | PEEK | 0.24–0.26 | 72–194 |
| | Polyethylene (PE) | 0.40–0.44 | 126–198 |
| | PET | 0.14–0.15 | 114–120 |
| | Acrylic (PMMA) | 0.08–0.25 | 72–162 |
| | Acetal (POM) | 0.22–0.35 | 76–201 |
| | Polypropylene (PP) | 0.11–0.17 | 122–180 |
| | Polystyrene (PS) | 0.12–0.12 | 90–153 |
| | Polyurethane thermoplastics | 0.23–0.24 | 90–144 |
| | PVC | 0.15–0.29 | 100–150 |
| | Teflon (PTFE) | 0.24–0.26 | 126–216 |
| Thermoset | Epoxies | 0.18–0.5 | 58–117 |
| | Phenolics | 0.14–0.15 | 120–125 |
| | Polyester | 0.28–0.3 | 99–180 |
| Polymer foams | | | |
| | Flexible polymer foam (VLD) | 0.036–0.048 | 120–220 |
| | Flexible polymer foam (LD) | 0.04–0.06 | 115–220 |
| | Flexible polymer foam (MD) | 0.04–0.08 | 115–220 |
| | Rigid polymer foam (LD) | 0.023–0.04 | 20–80 |
| | Rigid polymer foam (MD) | 0.027–0.038 | 20–75 |
| | Rigid polymer foam (HD) | 0.034–0.06 | 22–70 |

A10    Electrical resistivity and dielectric constant

| | | Resistivity ($\mu$ohm.cm) | Dielectric constant |
|---|---|---|---|
| **Metals** | | | |
| Ferrous | Cast irons | 49–56 | – |
| | High carbon steels | 17–20 | – |
| | Medium carbon steels | 15–22 | – |
| | Low carbon steels | 15–20 | – |
| | Low alloy steels | 15–35 | – |
| | Stainless steels | 64–107 | – |
| Non-ferrous | Aluminum alloys | 2.5–5.0 | – |
| | Copper alloys | 1.7–24 | – |
| | Lead alloys | 15–22 | – |
| | Magnesium alloys | 4.2–15 | – |
| | Nickel alloys | 6–114 | – |
| | Titanium alloys | 100–170 | – |
| | Tungsten alloys | 10.2–14 | – |
| | Zinc alloys | 5.4–7.2 | – |
| **Glasses** | Borosilicate glass | $3\times10^{21}$–$3\times10^{22}$ | 4.6–6.0 |
| | Glass ceramic | $2\times10^{19}$–$1\times10^{21}$ | 5.3–6.2 |
| | Silica glass | $1\times10^{23}$–$1\times10^{27}$ | 3.7–3.9 |
| | Soda-lime glass | $8\times10^{17}$–$8\times10^{18}$ | 7.0–7.6 |
| **Ceramics** | | | |
| Porous | Brick | $1\times10^{14}$–$3\times10^{16}$ | 7.0–10 |
| | Concrete, typical | $1.8\times10^{12}$–$1.8\times10^{13}$ | 8.0–12 |
| | Stone | $1\times10^{8}$–$1\times10^{14}$ | 6.0–18 |
| Technical | Alumina | $1\times10^{20}$–$1\times10^{22}$ | 6.5–6.8 |
| | Aluminum nitride | $1\times10^{19}$–$1\times10^{21}$ | 8.3–9.3 |
| | Boron carbide | $1\times10^{5}$–$1\times10^{7}$ | 4.8–8.0 |
| | Silicon | $1\times10^{6}$–$1\times10^{12}$ | 11–12 |
| | Silicon carbide | $1\times10^{9}$–$1\times10^{12}$ | 6.3–9.0 |
| | Silicon nitride | $1\times10^{20}$–$1\times10^{21}$ | 7.9–8.1 |
| | Tungsten carbide | 20–100 | – |
| **Composites** | | | |
| Metal | Aluminum/silicon carbide | 5–12 | – |
| Polymer | CFRP | $1.7\times10^{5}$–$1\times10^{6}$ | – |
| | GFRP | $1\times10^{16}$–$2\times10^{22}$ | 4.2–5.2 |
| Natural | Bamboo | $6\times10^{13}$–$7\times10^{14}$ | 5–7 |
| | Cork | $1\times10^{9}$–$1\times10^{11}$ | 6–8 |
| | Leather | $1\times10^{8}$–$1\times10^{10}$ | 5–10 |
| | Wood, typical (longitudinal) | $6\times10^{13}$–$2\times10^{14}$ | 5–6 |
| | Wood, typical (transverse) | $2\times10^{14}$–$7\times10^{14}$ | 5–6 |

*Continued*

A10   Electrical resistivity and dielectric constant *continued*

|  |  | Resistivity ($\mu$ohm.cm) | Dielectric constant |
|---|---|---|---|
| **Polymers** | | | |
| Elastomer | Butyl rubber | $1\times10^{15}-1\times10^{16}$ | 2.8–3.2 |
| | EVA | $3.2\times10^{21}-1\times10^{22}$ | 2.9–3.0 |
| | Isoprene (IR) | $1\times10^{15}-1\times10^{16}$ | 2.5–3.0 |
| | Natural rubber (NR) | $1\times10^{15}-1\times10^{16}$ | 3.0–4.5 |
| | Neoprene (CR) | $1\times10^{19}-1\times10^{23}$ | 6.7–8.0 |
| | Polyurethane elastomers | $1\times10^{18}-1\times10^{22}$ | 5.0–9.0 |
| | Silicone elastomers | $3\times10^{19}-1\times10^{22}$ | 2.9–4.0 |
| Thermoplastic | ABS | $3\times10^{21}-3\times10^{22}$ | 2.8–3.2 |
| | Cellulose polymers (ca) | $1\times10^{17}-4\times10^{20}$ | 3.0–5.0 |
| | Ionomer (I) | $3\times10^{21}-3\times10^{22}$ | 2.2–2.4 |
| | Nylons (PA) | $1.5\times10^{19}-1.1\times10^{20}$ | 3.7–3.9 |
| | Polycarbonate (PC) | $1\times10^{20}-1\times10^{21}$ | 3.1–3.3 |
| | PEEK | $3\times10^{21}-3\times10^{22}$ | 3.1–3.3 |
| | Polyethylene (PE) | $3\times10^{22}-3\times10^{24}$ | 2.2–2.4 |
| | PET | $3\times10^{20}-3\times10^{21}$ | 3.5–3.7 |
| | Acrylic (PMMA) | $3\times10^{23}-3\times10^{24}$ | 3.2–3.4 |
| | Acetal (POM) | $3\times10^{20}-3\times10^{21}$ | 3.6–4.0 |
| | Polypropylene (PP) | $3\times10^{22}-3\times10^{23}$ | 2.1–2.3 |
| | Polystyrene (PS) | $1\times10^{25}-1\times10^{27}$ | 3.0–3.3 |
| | Polyurethane thermoplastics | $3\times10^{18}-3\times10^{19}$ | 6.6–7.1 |
| | PVC | $1\times10^{20}-1\times10^{22}$ | 3.1–4.4 |
| | Teflon (PTFE) | $3\times10^{23}-3\times10^{24}$ | 2.1–2.2 |
| Thermoset | Epoxies | $1\times10^{20}-6\times10^{21}$ | 3.4–5.7 |
| | Phenolics | $3\times10^{18}-3\times10^{19}$ | 4.0–6.0 |
| | Polyester | $3\times10^{18}-3\times10^{19}$ | 2.8–3.3 |
| **Polymer foams** | | | |
| | Flexible polymer foam (VLD) | $1\times10^{20}-1\times10^{23}$ | 1.1–1.15 |
| | Flexible polymer foam (LD) | $1\times10^{20}-1\times10^{23}$ | 1.15–1.2 |
| | Flexible polymer foam (MD) | $1\times10^{20}-1\times10^{23}$ | 1.2–1.3 |
| | Rigid polymer foam (LD) | $1\times10^{17}-1\times10^{21}$ | 1.04–1.1 |
| | Rigid polymer foam (MD) | $3\times10^{16}-3\times10^{20}$ | 1.1–1.19 |
| | Rigid polymer foam (HD) | $1\times10^{16}-1\times10^{20}$ | 1.2–1.45 |

# Index